The Encyclopedia of World Geography

Grange
BOOKS

THE ENCYCLOPEDIA OF
WORLD GEOGRAPHY

Consultant Editor	Graham Bateman
Project Editor	Victoria Egan
Advisory Editor	Susan Kennedy
Text Editors	Lauren Bourque, Rita Demetriou, Peter Lewis
Art Editor	Steve McCurdy
Designers	Chris Munday Martin Anderson
Cartographic Manager	Richard Watts
Cartographic Editor	Tim Williams
Picture Research Manager	Claire Turner
Picture Research	Alison Floyd
Jacket Design	Ayala Kingsley
Typesetting	Brian Blackmore
Production	Clive Sparling Nicolette Colborne

Published 2007 by Grange Books
an imprint of Grange Books Ltd
The Grange
Kingsnorth Industrial Estate
Hoo, Near Rochester
Kent ME3 9ND

www.Grangebooks.co.uk

AN ANDROMEDA BOOK

The Brown Reference Group plc.
(incorporating Andromeda Oxford Limited)
8 Chapel Place
Rivington Street
London EC2A 3DQ

ISBN 978-1-84013-918-1

Printed in Thailand

Introductory photographs
Previous page: Fall scene at Half Dome,
Yosemite National Park, central California
(A. Gin/Powerstock Zefa)
This page: Mount Paektu, North Korea
(Magnum/H. Kubota)

CONTENTS

INTRODUCTION

THE WORLD IN WHICH WE LIVE IS CHANGING MORE RAPIDLY than at any time in human history. Politically, the collapse of communism in the former Soviet Union and Eastern Europe has ended the Cold War and altered the way that many countries are governed. Borders have been redrawn, new nations have emerged and old countries have been reborn. Notably there are now 15 republics in place of the Soviet Union, Yugoslavia has fragmented, Czechoslovakia has split in two and Germany has reunited. Although most of these changes have been comparatively peaceful, political realignments and economic reorganizations have created new tensions. In some instances, particularly in the Balkans and some of the former states of the Soviet Union, age-old hostilities between ethnic groups have re-emerged in the form of outright civil war. Environmentally, too, the world is at a crossroads; the speed and degree of our exploitation of global resources have increased dramatically in recent decades, and threaten the very future of the planet. In some cases these abuses have already caused irreversible damage, the ultimate consequences of which can only be surmised.

At the same time, the world is growing smaller; modern communication means that images of an event – whether an earthquake or a revolution – in one country are instantly flashed onto television screens on the other side of the world. The same communications network has inextricably linked the world's financial markets and created a truly global economy where no one currency or stock exchange is truly independent.

Despite the wealth of superficial information we are bombarded with on a daily basis, educators have become increasingly aware of the fact that most of us really know very little about the world beyond our immediate geographical boundaries. While better forms of communication allow us to see other cultures more easily, they do little to help us gain insight into and understanding of the different customs, religions, languages and the wealth and lifestyles of the many nations of the world. All these have been influenced by the physical environment, the actions of other countries and the deeds of particular individuals – soldiers and politicians, religious leaders, artists, scientists and industrialists – who have contributed to the development of each nation. The aim of the *Encyclopedia of World Geography* is to provide, in a single volume, a complete portrait of each of the countries in the world by looking in detail at the

The world's sovereign states

- sovereign by 1945
- sovereign since 1945
- sovereign since 1990
- dependency

The world political map Although all sovereign states are of equal status in the eyes of the United Nations, there are vast differences between them in terms of size and population, and in degrees of wealth and poverty.

climate and physical features of each, its vegetation and animal life, the major events of its history, the lifestyles of its people, and the components of each country's industry and economy. The result provides the reader with a better view of the world than can be gained from any atlas or almanac and contains more information on each country than any general encyclopedia.

STRUCTURE OF THE ENCYCLOPEDIA
The Encyclopedia of World Geography contains detailed individual country profiles for all 193 nation-states of the world, as well as many dependencies. At the primary level the book is divided into 22 regional sections. These regions have been defined either in geographic terms (South America, Northern Africa, Central Africa and Southern Africa, for example) or according to political criteria (United States, British Isles, Italy and Greece, France etc). Dependencies are given special coverage in one of two ways. If a dependency or island territory is located in the same region as the country to which it belongs, it is treated as a special feature within that country

GREENLAND
(Denmark)

SVALBARD
(Norway)

ICELAND

NORWAY SWEDEN FINLAND
ESTONIA
LATVIA
DEN LITH
UNITED
KINGDOM
IRELAND
NETH
BEL GER POLAND BELARUS
LUX CZ UKRAINE
FRANCE SL HUN MOLDOVA
SW AU ROMANIA
CR B YU BULGARIA
ITALY AL GEORGIA
PORTUGAL SPAIN GREECE TURKEY
MALTA AR AZ
TUNISIA CYP SYRIA
ISRAEL IRAQ
MOROCCO JORDAN
Canary Is
(Spain) KUWAIT
WESTERN
SAHARA
(Morocco)

RUSSIA

KAZAKHSTAN MONGOLIA

UZBEKISTAN KYRGYZSTAN
TURKMENISTAN TAJIKISTAN
AFGHANISTAN
IRAN PAKISTAN
NEPAL BHUTAN
BAHRAIN QATAR
SAUDI UAE
ARABIA OMAN INDIA
YEMEN

NORTH
KOREA
SOUTH
KOREA JAPAN

CHINA

TAIWAN

Tropic of Cancer

BANG
MYANMAR

NORTHERN
MARIANA
ISLANDS
(U.S.A.)

CAPE
VERDE MAURITANIA
SENEGAL MALI NIGER CHAD SUDAN ERITREA
GAMBIA
GUINEA- GUINEA BURKINA
BISSAU FASO BE
SIERRA LEONE CÔTE G NIGERIA
D'IVOIRE
LIBERIA CAMEROON CENTRAL
AFRICAN
REPUBLIC ETHIOPIA
SAO TOME & EQ
PRINCIPE REPUBLIC
GABON OF CONGO UGANDA
DEMOCRATIC RW KENYA
CABINDA REPUBLIC BU
(Angola) OF CONGO TANZANIA

LAOS
THAILAND VIETNAM
CAMBODIA

DJIBOUTI
Socotra
(Yemen)

SRI
LANKA

MALDIVES

SOMALIA

SEYCHELLES

GUAM
(U.S.A.)

PHILIPPINES

BRUNEI
MALAYSIA
SINGAPORE

MARSHALL
ISLANDS

FEDERATED STATES
OF MICRONESIA
PALAU

NAURU KIRIBATI Equator

INDONESIA
EAST
TIMOR

PAPUA
NEW GUINEA
SOLOMON
ISLANDS

TUVALU

ANGOLA MALAWI
ZAMBIA
MOZAMBIQUE MADAGASCAR
ZIMBABWE MAURITIUS
NAMIBIA REUNION
BOTSWANA (France)
SWAZILAND
SOUTH LESOTHO
AFRICA

COMOROS

VANUATU

SAMOA

FIJI TONGA
NEW
CALEDONIA
(France) Tropic of Capricorn

AUSTRALIA

NEW
ZEALAND

A ANDORRA
AL ALBANIA
AR ARMENIA
AU AUSTRIA
AZ AZERBAIJAN
B BOSNIA HERZEGOVINA
BANG BANGLADESH
BE BENIN
BEL BELGIUM
BU BURUNDI
CR CROATIA
CYP CYPRUS
CZ CZECH REPUBLIC
DEN DENMARK
DOM DOMINICAN REPUBLIC
EQ EQUATORIAL GUINEA
G GHANA
GER GERMANY
HUN HUNGARY
L LEBANON
LITH LITHUANIA
LUX LUXEMBOURG
M MACEDONIA
(Former Yugoslav Republic of)
NETH NETHERLANDS
R RUSSIA
RW RWANDA
S SLOVENIA
SL SLOVAKIA
SW SWITZERLAND
T TOGO
UAE UNITED ARAB EMIRATES
YU YUGOSLAVIA

Arctic Circle

entry. For example, Corsica is treated as a special feature within the France entry. However, if the dependency is located in a different region, it is featured at the end of its regional section. For example, French Guiana is featured at the end of the regional section on South America, not France.

Each region is introduced by a high-quality topographic map showing key physical features (rivers, mountains, lakes, etc), national boundaries and country names, plus major cities and towns. These maps give the fundamental physical background to the country profiles that follow.

COUNTRY PROFILES

The core of the Encyclopedia comprises individual entries on every country of the world, which vary in length from one page (or less for small island states) to 50 pages for the United States. The United States' section has been divided into two parts: an extensive national profile followed by a section covering each of the States separately. Whatever the length of the entry, the text has been prepared to present information in a consistent and logical fashion that enables the reader easily to make comparisons between countries. Each profile is divided into three major parts – GEOGRAPHY, SOCIETY and ECONOMY.

The GEOGRAPHY section describes the physical landscape including main areas of mountains, deserts and plains, and the main water features – rivers and lakes. The characteristic climatic conditions are outlined and native plants and animals described.

The second major section, SOCIETY, puts the people into the context of their environment. First the history of the country is told chronologically from earliest known times up to the present. This historical perspective is crucial to our understanding of the country today; subsequent subsections describe the present-day system of government, the ethnic groups that make up the population, their religious beliefs and other cultural distinctions.

The final section – ECONOMY – describes how the people have utilized the land and its resources to build an economy that funds (or fails to fund) their nation's requirements. It lists the key features of agriculture and industry, explains how local trade and commerce operate, and describes the transportation and communications systems. The last subsection looks at health, welfare and education.

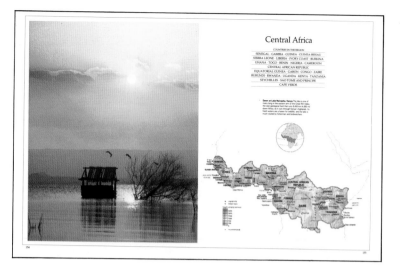

△ Opening pages to one of 22 regions with topographic map and locator within its continent.

▷ Typical two-page entry on a country.

1 The country's popular name followed by its official name in smaller type.

2 Map including major cities, rivers, lakes and seas, plus a locator map and national flag.

3 Authoritative text organized under three major headings: Geography, Society and Economy.

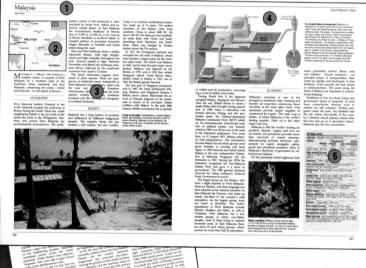

△ Typical pages from a longer country entry.

4 Diagram explaining the system of government of selected countries.

5 Fact panel with current statistics on population, geography, economy and government.

6 Special features focusing on key people, notable historical events and island territories.

FEATURE PANELS

Within many entries special features have been included to give a deeper appreciation of the country. These self-contained, illustrated articles focus on elements of history, culture or geography that are integral to that nation's "personality". They include discoveries such as Peary's race for the North Pole (Canada), monuments of culture such as Florence (Italy), lost civilizations, for instance, the Incas (Peru), technological achievements like the Space program (US) and biographies of the individuals who are forever linked to the history or heritage of their country – whether Lincoln (United States), Shakespeare

(United Kingdom) or Erik the Red (Canada). Features on geography include farflung territories such as Easter Island (Chile), unique agricultural regions, for instance, Madeira (Portugal) and pivotal sites like Egypt's Suez Canal.

FACTS AND STATISTICS

Reference data is incorporated throughout the Encyclopedia in text, maps and artwork. In addition, special fact panels have been compiled for every country. The structure of these panels broadly mirrors the sequence of information in each country profile. Summary details and the latest statistics are given for land area, climate, major physical features, population, form of government, composition of armed forces, largest cities (including the capital), official language, ethnic

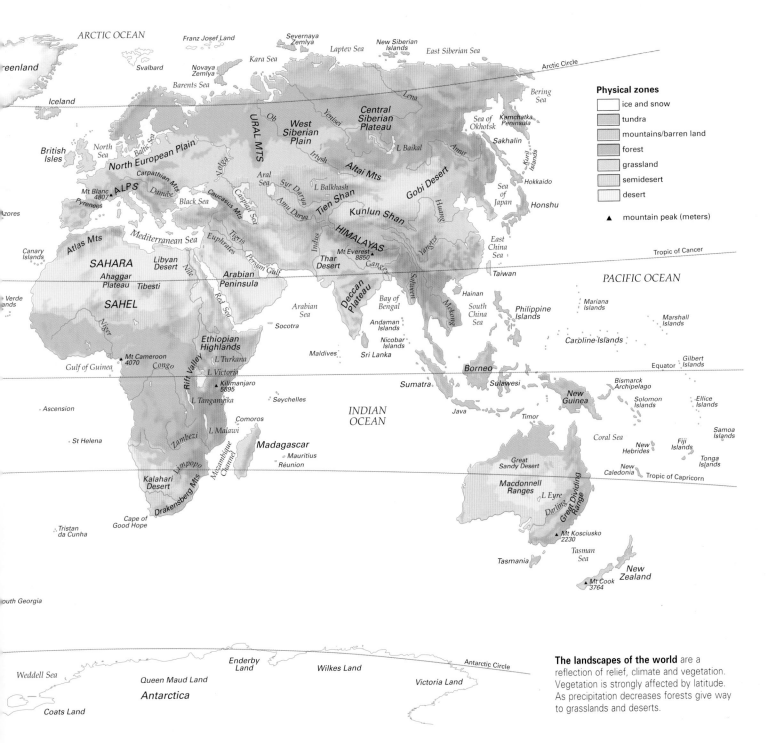

ARCTIC OCEAN

Franz Josef Land
Severnaya Zemlya
New Siberian Islands
East Siberian Sea
Laptev Sea
Kara Sea
Svalbard
Novaya Zemlya
Barents Sea
Arctic Circle
eenland
Iceland
Bering Sea
Kamchatka Peninsula
Sea of Okhotsk
Lena
Central Siberian Plateau
Ob
Yenisei
West Siberian Plain
URAL MTS
Irtysh
L Baikal
Altai Mts
Amur
Sakhalin
Kuril Islands
Hokkaido
British Isles
North Sea
Baltic Sea
North European Plain
Carpathian Mts
Volga
Aral Sea
Syr Darya
L Balkhash
Tien Shan
Gobi Desert
Sea of Japan
Honshu
Azores
Mt Blanc 4807
ALPS
Danube
Pyrenees
Black Sea
Caucasus Mts
Caspian Sea
Amu Darya
Kunlun Shan
Huang
Yangtze
East China Sea
Tropic of Cancer
Canary Islands
Atlas Mts
Mediterranean Sea
Tigris
Euphrates
Persian Gulf
HIMALAYAS
Mt Everest 8850
Ganges
Taiwan
PACIFIC OCEAN
SAHARA
Libyan Desert
Nile
Arabian Peninsula
Thar Desert
Verde ands
Ahaggar Plateau
Tibesti
SAHEL
Red Sea
Arabian Sea
Deccan Plateau
Bay of Bengal
Salween
Mekong
Hainan
South China Sea
Philippine Islands
Mariana Islands
Marshall Islands
Niger
Socotra
Andaman Islands
Nicobar Islands
Caroline Islands
Gilbert Islands
Mt Cameroon 4070
Ethiopian Highlands
L Turkana
Congo
Rift Valley
L Victoria
Maldives
Sri Lanka
Borneo
Equator
Ascension
Kilimanjaro 5895
L Tanganyika
Seychelles
Sumatra
Sulawesi
New Guinea
Bismarck Archipelago
Solomon Islands
Ellice Islands
St Helena
Zambezi
L Malawi
Comoros
INDIAN OCEAN
Java
Timor
Samoa Islands
Mozambique Channel
Madagascar
Mauritius
Réunion
Coral Sea
New Hebrides
Fiji Islands
Tonga Islands
Kalahari Desert
Drakensberg Mts
Limpopo
Great Sandy Desert
Macdonnell Ranges
L Eyre
Great Dividing Range
New Caledonia
Tropic of Capricorn
Tristan da Cunha
Cape of Good Hope
Darling
Mt Kosciusko 2230
Tasman Sea
Tasmania
New Zealand
outh Georgia
Mt Cook 3764

Physical zones
- ice and snow
- tundra
- mountains/barren land
- forest
- grassland
- semidesert
- desert
- ▲ mountain peak (meters)

Enderby Land
Wilkes Land
Antarctic Circle
Weddell Sea
Queen Maud Land
Victoria Land
Antarctica
Coats Land

The landscapes of the world are a reflection of relief, climate and vegetation. Vegetation is strongly affected by latitude. As precipitation decreases forests give way to grasslands and deserts.

composition and religious affiliations, currency, gross domestic product for the country and gross national product per person, average life expectancy of men and women and the major resources that contribute to the economy, natural and agricultural. Information for these panels has been researched from numerous sources providing a unique data resource for student and researcher alike.

MAPS AND OTHER ILLUSTRATIONS
No book on geography would be complete without maps to depict the location of features described. *The Encyclopedia of World Geography* is no exception. In the early pages there are maps showing all the nation-states of the world and the major global physical features. In addition to the topographic map for each regional section, each country entry

has a small map showing major cities (including the capital), rivers and lakes, and adjacent countries, oceans and seas. The position of the country within the continent appears on an inset locator map. The national flag of each country is also depicted. Other maps show major federal divisions of certain countries, where relevant.

While this Encyclopedia is bulging with reference information, it is also a visual celebration of the diversity of the lands and peoples that occupy this planet. Hundreds of color photographs have been specially selected to present a stunning panorama that displays scenic beauty and evokes local life. These are supplemented with detailed captions. Schematic diagrams showing how the systems of election and government work in selected countries are also included.

Canada and the Arctic

COUNTRIES IN THE REGION

CANADA

DEPENDENCIES IN THE REGION

GREENLAND · ST PIERRE AND MIQUELON

The maple leaf, Canada's national symbol, emblazoned on garage doors in Saskatchewan. The maple leaf's official status was only confirmed in 1965; it became the central motif of the new national flag introduced in that year. Syrup is made from the sweet sap of the sugar maples that are native to the deciduous forests of the eastern part of the country, and whose leaves in fall turn a brilliant red.

ARCTIC OCEAN

Ellesmere Island

GREENLAND
(Denmark)

Thule
(Qaanaaq)

North Magnetic Pole
1985

Queen Elizabeth Islands

Devon Island

Baffin Bay

Banks Island

Resolute

Somerset Island

Prince of Wales Island

Arctic Circle

Inuvik

Victoria Island

Boothia Peninsula

Baffin Island

Godthåb
(Nuuk)

Cape Farewell

Davis Strait

Great Bear Lake

Melville Peninsula

Mackenzie Mountains

Mt Logan
5951

Selwyn Mts

Mackenzie

Back

Iqaluit

Southampton Island

Hudson Strait

ATLANTIC OCEAN

Whitehorse

ROCKY MOUNTAINS

Great Slave Lake

Yellowknife

Ungava Peninsula

PACIFIC OCEAN

Cassiar Mts

Hay River

Lake Athabasca

Hudson Bay

Labrador

Goose Bay

Queen Charlotte Islands

Coast Mountains

Great Plain

Williston Lake

Peace

Athabasca

Reindeer Lake

Churchill

Nelson

Churchill

Newfoundland

St John's

Laurentian Highlands

Sept-Iles

Cape Race

Mt Robson
3954

Fraser

Edmonton

Saskatchewan

CANADA

Fort Albany

St-Pierre & Miquelon
(France)

Mt Waddington
3994

Selkirk Mts

Lake Winnipegosis

Lake Winnipeg

Albany

Appalachian Mts

Prince Edward I

Cape Breton Island

Vancouver Island

Saskatoon

Quebec

Vancouver

Calgary

Nova Scotia

Halifax

Regina

Lake Manitoba

Winnipeg

Thunder Bay

Sault Ste Marie

Montreal

Ottawa

Cape Sable

St Lawrence

Gulf of St Lawrence

Lake Superior

L Huron

Toronto

Lake Ontario

Niagara Falls

Lake Michigan

Lake Erie

Canadian Shield

capital city
major town

height of land (meters)

3000
2000
1000
500
200

permanent ice

mountain peak

Canada

ARCTIC OCEAN
Ellesmere I
Greenland (Denmark)
Beaufort Sea
USA
Baffin Bay
Baffin I
Victoria I
Davis Strait
Yukon
Mackenzie
Great Bear Lake
Labrador Sea
Great Slave Lake
Hudson Bay
CANADA
Newfoundland
Edmonton
Calgary
Vancouver
Winnipeg
L Winnipeg
L Superior
St Lawrence
Quebec
Ottawa
Montreal
L Michigan
Toronto
UNITED STATES OF AMERICA
ATLANTIC OCEAN

CANADA OCCUPIES MOST OF THE NORTHERN half of the North American continent. Its economy and culture show many similarities with the United States – so much so that some outsiders tend to think of it as no more than a satellite. However, nothing could be further from the truth: for Canada has its own distinct character and rich ethnic mix. Its unique bilingual and bicultural society can be traced back to the nation's French and British origins.

NATIONAL DATA - CANADA

Land area 9,093,507 sq km (3,511,023 sq mi)

Climate	Altitude m (ft)	Temperatures January °C(°F)	July °C(°F)	Annual precipitation mm (in)
Resolute	64 (200)	-32 (-26)	4 (40)	138 (5.4)
Vancouver	0 (0)	3 (38)	17 (63)	1,199 (47.2)
Winnipeg	248 (813)	-18 (0)	20 (67)	514 (20.2)
Montreal	30 (98)	-12 (10)	20 (67)	1,065 (41.9)
Halifax	30 (98)	-6 (21)	19 (65)	1,542 (57.1)

Major physical features highest point: Mount Logan 5,951 m (19,524 ft); longest river: Mackenzie 4,240 km (2,635 mi); largest lake: Lake Superior (part) 83,270 sq km (32,150 sq mi)

Population (2006 est.) 33,098,932

Form of government federal multiparty parliamentary monarchy with two legislative houses

Armed forces army 33,000; navy 12,000; air force 17,100

Largest cities Toronto (4,726,194); Montreal (3,290,804); Vancouver (1,840,441); Calgary (1,013,675); Ottawa (capital - 896,048); Edmonton (841,202); Hamilton (699,795); Quebec (650,026); Winnipeg (633,107); London (350,340)

Official languages English, French

Ethnic composition British 28%; French 23%; other European 15%; Amerindian 2%; other mainly Asian, African, Arab 6%; mixed background 26%

Religious affiliations Roman Catholic 42.6%; Protestant 23.3%; other Christian 4.4%; Muslim 1.9%; other and unspecified 11.8%; none 16%

Currency 1 Canadian dollar (CAD) = 100 cents

Gross domestic product (2006 est.) $1.165 trillion

Gross domestic product per capita (2006 est.) $35,200

Life expectancy at birth male 76.86 yr; female 83.74 yr

Major resources coal, natural gas, petroleum, hydropower, iron ore, nickel, zinc, copper, gold, lead, molybdenum, potash, diamonds, uranium, rock salt, fisheries, tourism, cereals, dairy and meat products, food stuffs, timber, tobacco

GEOGRAPHY

Canada covers a much greater expanse than the United States. This vast and often mountainous country is second only in land area to Russia, though its population is very much smaller.

The land
Canada can be divided into several geographical areas, of which the largest is the so-called Canadian Shield, occupying almost half the country. This bowl-shaped area centered on Hudson Bay consists of ancient rocks that were once raised into mountainous forms. It extends as far east as the Gulf of St Lawrence, and west as far as the lowlands around Great Bear Lake and Great Slave Lake in the Northwest Territories.

Southeast of the Shield lie the lowlands of the Great Lakes–St Lawrence river area, occupying southern Ontario and southwestern Quebec. This much smaller area is the most densely populated part of Canada. The gently rolling country of the Bruce Peninsula is bounded by Lake Huron and Georgian Bay, Lake Erie and Lake Ontario. It is traversed by the Niagara Escarpment over which the Niagara Falls cascade. The Ontario Lowlands to the northeast of the escarpment are an ancient glacial plain covered in

The Kicking Horse river (*above*) A canoe follows the path of early explorers through the Rocky Mountains in British Columbia. Today five of Canada's national parks lie within the spectacular scenery of the Rockies.

The Klondike highway (*left*) cuts through a forested landscape of snow-clad mountains that captures the romantic image of Canada's far north. In the two years between 1897 and 1899, some 30,000 people came this way in search of gold.

drumlins – low ridges of rocky debris left by the retreating glaciers. Farther northeast lie the lowlands of the Ottawa Valley and southern Quebec.

East of Quebec is the region known as Appalachian Canada, which includes the Atlantic provinces of New Brunswick, Prince Edward Island, Nova Scotia and the island of Newfoundland. Here ancient rocks have been folded and reshaped over time into craggy, mountainous country, now much eroded by weather and glaciers. Among the highest of the mountain chains are the Nôtre Dame Mountains that run northeast from Quebec, and the Long Range Mountains of Newfoundland.

Southwest of the Shield and Hudson Bay is the Great Plains area of Canada – a vast triangular expanse of more level country covering about one-fifth of the land area. The Manitoba Lowlands are seamed with a network of lakes, most of them remnants of one immense lake that covered the whole area during the last glacial retreat. The deep sediments left behind provide rich farming country.

Westward, the land rises over the dissected edge of the Manitoba Escarpment, beyond which are found the rolling plateaus of the Saskatchewan Plain. Here, torrential meltwaters from the ice-age glaciers have carved wide, steep-banked river channels through the countryside. Westward through Alberta the land con-

tinues to rise. It has been severely eroded since the glaciers last retreated to form a rough plateau area.

West of the Great Plains is the Canadian Cordillera – the mountainous area that borders Canada's Pacific coast. A great mountain belt some 800 km (500 mi) wide extends from Alaska through the Yukon Territory, western Alberta and British Columbia. The western spine of the Cordillera is formed by the Coast Mountains. The range then turns south, interrupted by plateaus, to form the Mackenzie Range and eventually the Rocky Mountains that form the eastern spine of the Cordillera. These mighty peaks of relatively young rock rise to heights of more than 3,000 m (10,000 ft).

Along the western seaboard lies a parallel and even higher chain. The St Elias Mountains to the north include Mount Logan, which at nearly 6,000 m (20,000 ft) is Canada's highest peak. The Coast Mountains run southward along the coast, cut by deep fjords or inlets of the sea. Offshore, Vancouver Island and the Queen Charlotte Islands are the peaks of yet another parallel mountain range, while north of the Shield lies the Arctic.

To the northwest the broad archipelago

of the Arctic islands border the permanent ice of the Arctic Ocean. Much of the ocean's surface is covered throughout the year by a layer of ice at least 3 m (10 ft) thick, yet less than half the surrounding land is ice-covered. The islands rise toward the northeast, becoming mountainous in Ellesmere Island and along the east coast of Baffin Island. Their barren, rocky terrain is largely covered with permanent or semipermanent fields of snow and ice, and around the mountain peaks there is extensive glaciation.

Climate

More than half of Canada has a subarctic climate, with cool summers and extremely cold winters – often colder than in the Arctic itself, where great extremes of temperature are moderated by the waters of the Arctic Ocean just below the ice. In the polar regions winter begins in August. Temperatures drop rapidly until December, then remain low until March. The more equable waters of the Atlantic Ocean also moderate the climate where they flow into the Arctic, producing warmer winters, colder summers and higher precipitation. Summer temperatures are usually more uniform throughout the Arctic, though some continental areas may enjoy short periods of almost Mediterranean warmth.

Farther south the climate is typically continental, with hot summers and cold winters. Southern Ontario, among the Great Lakes, has the warmest summers. The southeast coast and especially the west coast enjoy the mildest winters. Onshore winds bring heavy rain and snow to the Coast Mountains, but the air over the interior is dry, especially in winter. In the summer, the famous prairie thunderstorms help to redistribute moisture across the Great Plains. Farther east the Great Lakes provide some moisture, while moist Atlantic air supplies ample rainfall to the Atlantic seaboard, where the winter snowfall is particularly heavy.

Vegetation

The treeline in northern Canada arches northward from Labrador on Canada's east coast across the Ungava Peninsula, then down the east side of Hudson Bay. Across this bay it sweeps sinuously northwest to the lower Mackenzie Valley and Alaska. North of it there is little or no fertile soil, and most of the country, known as the Barren Grounds, is tundra.

Farthest from the pole the tundra vegetation includes low shrubs, grasses and sedges. In the most northerly lands, the cushion plants of the polar desert cover less than one-tenth of the ground surface.

South of the treeline, the boreal region forms one of the world's largest coniferous forests, stretching from Alaska to Newfoundland.

The eastern forests, from the Great Lakes to the coasts, are mostly mixed, with sugar maple, beech, birch, pine and hemlock. But the lowland plains of the far south are covered with broadleaf woodland: hickory, oak and elm flourish here, along with chestnut, maple and walnut. In the western mountains spruce, Douglas fir and lodgepole pine are the commonest, with aspen and yellow pine in

The icy grip of winter blankets Montreal in snow. Although lying in the same latitude as Florence in Italy, Montreal has bitter winters with temperatures falling below 0°C (32°F) for more than 150 days each year.

plateau areas. On the Pacific coast, with its high rainfall, are some of the world's most imposing forests – dense, towering ranks of Douglas fir, western red cedar and hemlock.

The prairie country is too dry for more than a scattering of tree cover and was originally wide rolling grassland. Today little of this remains, the rich black soils having been plowed up and turned into the famous wheat-growing country.

Wildlife

Canada's Arctic waters support whales, walruses and seals, along with the semi-aquatic Polar bear. On the tundra there are Musk oxen, caribou, wolves, Arctic foxes and lemmings; many migratory birds also spend the summer here, including auks, sea ducks, gulls, terns and

shorebirds. Caribou flourish in the northern forests, along with moose (known as elk in Europe), lynx, and Black and Brown bears. Beavers, martens, muskrats, mink and other fur animals were once, and to a degree remain, the basis of the fur trade.

Farther south the White-tailed deer is common, while the more settled country harbors mostly smaller mammals such as Gray and Red squirrels, chipmunks, weasels and otters. Among the varied and noisy bird life are cardinals, wood warblers, Baltimore orioles and catbirds.

The prairies support smaller creatures such as jackrabbits, gophers and Sharp-tailed grouse as well as some surviving bison and Pronghorn antelope. The western mountains support highly adapted species such as the Bighorn sheep and the Mountain goat.

Robert Peary and the race for the North Pole

As early as 1607 Henry Hudson (c.1565–1611) set out by ship in search of a route across the North Pole to China. He started by sailing northward near the east coast of Greenland, but turned east when he encountered the Arctic pack ice. For 300 years the North Pole remained unconquered. Many attempts were made to reach it but they all failed. In 1895–96 the Norwegian explorer Fridtjof Nansen (1861–1930) in a specially constructed

Robert Peary's 1905 expedition (*above*) This contemporary artist's impression shows the team that penetrated as far north as 87° 6' N. Another three years were to pass before he reached his goal.

Nansen's ship, the *Fram* (*left*), as shown on a Wills' cigarette card. The *Fram* was specially constructed to withstand the pressure of Arctic ice.

vessel, the *Fram*, proved for the first time that the polar seas were entirely covered in ice; but his attempt to reach the pole by sledge did not succeed.

By this time an American naval officer named Robert Edwin Peary (1856–1920) had begun his own preparations for a sledge journey to the pole. He used a "pyramid" approach whereby, beginning with a large group, he would set up a chain of depots, sending back some of the support sledges each time. After several close approaches, in 1908 he

finally achieved his goal. His expedition took him from Ellesmere Island's Cape Columbia all the way to the North Pole. There were five people on the final run: Peary himself, three Inuit, and his dog-driver Matthew Henson (1866–1955).

On his return in 1909, Peary learned that his former companion Frederick Albert Cook (1865–1940) was claiming to have reached the pole before him. Cook's proof was unsatisfactory, but the evidence is hard to judge, and the debate continues.

SOCIETY

Canada's democratic government derives from that of Britain, of which it was once a colony. It has also retained elements of its French heritage. It is a country that has come increasingly to accept its obligations toward its indigenous peoples. After centuries of immigration, its population has great diversity.

History
The first peoples to enter the Americas almost certainly came from Siberia via the

Bering Strait and Alaska during the last ice age between 30,000 and 10,000 years ago and gradually spread southward through the rich, uninhabited land over thousands of years into the heart of the continent. The ancestors of today's Inuit were the last people to move into North America from Asia about 4000 BC, reaching Greenland about 2500 BC. In the western Arctic the traditional Inuit culture, based on hunting for food in small family groups, still survives, but old patterns are changing: snowmobiles have replaced dog teams, while guns and whaling bombs have replaced harpoons, and today's Inuit boats are driven by outboard motors. Their summer tents and winter igloos have been replaced by modern buildings, and many have relocated to large settlements.

Europe's first definite contact with Canada came with the arrival of Viking adventurers such as the Norse explorer Leif Eriksson (originally from Greenland) and the Icelander Thorfinn Karlsefni in about AD 1000. The Vikings did not stay, however, perhaps because of the hostility of the local Native Indian peoples.

In 1497 an Italian navigator called John Cabot (1450–98) led an English expedition to discover unknown lands, and arrived at Newfoundland. In the 16th century explorers began to penetrate the interior. Among them were Henry Hudson (c. 1550–1611) from England, Jacques Cartier (1491–1557) who is supposed to have given Canada its name, and his fellow-Frenchman Samuel de Champlain (1567–1635), who is known as the Father of New France for his work in founding the colony that grew along the St Lawrence river from Quebec to Montreal.

French influence and colonization also invaded the Maritime Provinces, and from its St Lawrence core spread with the fur trade across the Great Lakes to the Mississippi and Saskatchewan river watersheds. However, French northern and western fur interests came increasingly to be rivaled by those of the British Hudson's Bay Company, founded in 1670. During the next century, growing friction between England and France culminated in the capture of Quebec and Montreal by the British, led by General James Wolfe (1727–59) in 1759. This was followed by the Treaty of Paris in 1763, by which New France was ceded to Britain as the colony of Quebec, with its Maritime territory.

During and after the American War of Independence (1775–83), when the North American colonies were in revolt against British rule, several thousand British loyalists fled to Canada, especially to Nova Scotia and Quebec. In 1791 the separate colonies of Upper Canada (Ontario) and Lower Canada (Quebec) were created by the Constitutional Act of 1791. In 1837 dissatisfaction in both Upper and Lower Canada with British economic policy, along with the continuing tight constraints on self-government, brought open revolt. The uprising failed, but it did bring about political reform and the reunification of the two Canadas. In 1849 the tiny colony of Victoria was created on Vancouver Island in the far southwest of the region. Following the subsequent

gold rush in the Fraser Valley in the southwest, the colony of British Columbia was established in 1858.

In 1867 the British North America Act created the Dominion of Canada, comprising the provinces of Quebec, Ontario, Nova Scotia and New Brunswick. The Dominion Government bought the Hudson's Bay Company lands, which became the Northwest Territories. In 1870 Manitoba joined the Dominion, followed in 1871 by British Columbia. Prince Edward Island followed in 1873, the Yukon Territory in 1898 and Saskatchewan and Alberta in 1905. Newfoundland remained nominally independent until 1949. Canada today is composed of 10 provinces and three national territories, including the new territory of Nunavut, which came into existence in 1999 as a response to Inuit land claims in the Northwest Territories.

Meanwhile, Canada's prosperity had risen, swelled by the enormous grain production across the recently settled prairies, by the exploitation of northern forests for lumber and pulp, and also by the development of other natural resources. The discovery of gold deposits along Klondike Creek in the Yukon in 1896 precipitated the famous Gold Rush.

All these developments disrupted the way of life of the indigenous population far more than the withdrawal of the fur trade or the destruction of the bison. There followed an armed uprising among the *Métis* – a rural people of mixed French and Native Indian descent – led by Louis Riel (1844–85). Riel was hanged in 1885.

After Canada's vigorous support of the Allies (Britain, France, Italy, Russia and the United States) during World War I, Britain gave the Dominion greater autonomy by the Statute of Westminster of 1931. Not until 1982 did Canada gain full control of its own constitution.

Since the 1960s French-Canadian nationalists in Quebec have campaigned vigorously for independence from Canada, and in 1976 the Parti Québecois won control of the provincial government. In 1980 Quebec voters decisively rejected independence in a referendum and in 1995 a similar referendum was defeated. Nationalists have continued to press for independence, but in 1998 the Supreme Court ruled that without a majority vote, Quebec had no right to unilaterally secede from Canada.

Government

Essentially Canada still remains the federation of states set up by the 1867 British North America Act. As a constitutional

monarchy, Canada is headed by a sovereign, now specifically called the sovereign of Canada and not the sovereign of the British Isles.

In practice the sovereign's power is exercised by a governor-general, usually appointed for a six-year term on the advice of the Canadian prime minister. The governor-general gives the royal assent to new legislation, without which it cannot become law. To receive assent, legislation must first pass both houses of the federal parliament: the senate (upper house) and the house of commons. Both houses can introduce legislation, but only the commons can introduce new taxation or expenditure. The federal parliament has authority over all international affairs and all major domestic ones.

Each of Canada's 10 provinces has a governmental structure mirroring that of the nation. A lieutenant-governor appointed by the governor-general summons a single-chamber elected legislative assembly, in which the majority leader becomes premier and appoints a cabinet. Provincial governments are responsible for local concerns such as civil law, local taxation, land management, local trade

The Canadian provinces (*below*) Quebec was first settled by the French, from 1534, but in 1763 the whole country was ceded to Britain. Names like Saskatchewan preserve Canada's Native Indian inheritance.

Provinces of Canada
- ■ national capital
- ● provincial capital

Francis McClintock's Arctic expedition (*below*) in 1859 cuts roads for sledges through heavy snow hummocks in Queen's Channel in Canada's far northwest. McClintock's expedition found the logbook of an earlier British explorer, John Franklin, detailing the illnesses that fatally overwhelmed his expedition in 1845.

and commerce, hospitals, health, welfare and education. The huge but very sparsely populated Yukon, Nunavut, and Northwest Territories are administered directly by the federal government.

People

Canadians of British descent form the largest ethnic group, accounting for somewhat less than half the total. French Canadians form about one-quarter of the

Canada's system of government has grafted the federalism of the neighboring United States onto a parliamentary system modeled on that of the British. The two-chamber federal parliament consists of the senate, whose 104 members are appointed for life, and the 282-member house of commons, elected for a maximum of five years. Legislation must be passed by both houses to become law.

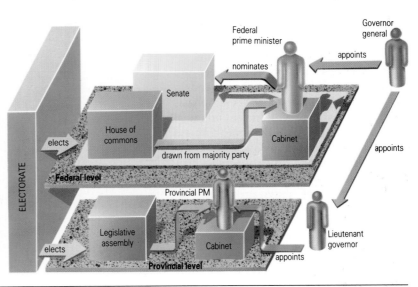

national population and some four-fifths of Quebec's. A further one-tenth of Canadians are of mixed origin, the largest group being the *Métis*. Most of the rest are descended from European immigrants, especially German, Italian, Ukrainian and Dutch. Native Indians and Inuit, the original inhabitants, now account for less than one-fiftieth of the population.

Most Canadians speak English as their first language. The exception is in Quebec, which has a French majority – rigorous laws here protect and encourage French language and culture. Roman Catholics slightly outnumber Protestants in Canada's current Christian population. However, heavy immigration from Asia in recent decades has greatly increased the variety of religions.

ECONOMY

Canada is rich in resources, with massive mineral deposits, enormous areas of commercial-grade timber and vast areas under cultivation. The resulting prosperity has created one of the highest standards of living in the world.

Agriculture and forestry

Agriculture is one of Canada's major industries, though today it employs only one-twentieth of the labor force on one-twentieth of the total land area. Most agricultural land lies in the south; in more northerly latitudes the growing season is too short.

British Columbia's extensively irrigated southern plateau and Fraser delta lands are almost entirely given over to fruit growing, while cattle and sheep ranching prevail farther north in parts of the interior. Almost four-fifths of Canada's farmland lies between the Rocky Mountains and the Great Lakes. In the prairie provinces hard wheat remains the dominant crop, though there is increased competition from other grains.

In Alberta, cattle, pigs and sheep are raised, principally in dryer areas and in the foothills. Similar livestock emphasis is characteristic of the more marginal parts of Manitoba. Southwestern Ontario is Canada's corn and soybean belt, but it also produces most of the country's tobacco and much of its fruit and vegetables. Apart from wheat, other agricultural export crops include feed grains, oilseeds, apples, potatoes and maple syrup.

Canada is the world's leading exporter of wood products. Sawn timber, plywood and pulp for newsprint account for more

than one-tenth of exports, and there is potential for even higher production levels. The most productive area is British Columbia, the majority of which is covered by forests that are largely exploited by major multinational companies. Quebec's massive forests are not yet fully exploited and most of the Atlantic provinces have a well-developed woodland industry.

Fishing and the fur trade

Historically, the Canadian fishing industry centered on Newfoundland. Today's catches are taken principally by large trawlers fishing the banks out to the extended 400 km (250 mi) limit set by the Canadian government in 1977. Ashore,

Downtown Toronto (*left*), and the CN tower. Modern buildings in the lower part of the city have replaced those that were destroyed by fire in 1904.

A baseball player slides into base (*right*) Baseball is one of the less controversial imports from the United States, whose economic domination is still sometimes seen as a threat in Canada.

French influence (*below*) in Montreal, the main commercial center in Quebec province, where French language and culture are protected and encouraged by law.

fish are frozen or canned for export. These limits and subsequent quota regulations have brought sharp disputes with France over fishing rights around the islands of St Pierre and Miquelon, but have helped to revive the ailing Pacific fisheries, famous for their salmon. Canada's extensive freshwater lakes are also exploited for both commercial and sport fishing.

Furs remain an important export, notably from Quebec, Saskatchewan, the Northwest Territories and the Yukon.

Industry and power

Canada's mineral wealth is very considerable and makes a major contribution to its exports. The western provinces, particularly Alberta, have the greatest part of

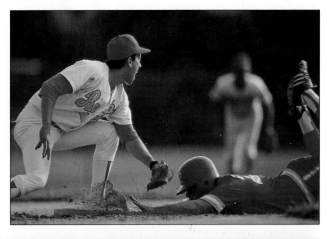

Canada's known fossil fuel deposits, including oil, gas and coal. Gas and oil have also attracted much recent attention in the Mackenzie Valley and under the Beaufort Sea. Northern Ontario and Quebec have rich iron mines exploited in the past, and they contain some of the largest untapped low-grade iron-ore deposits in the world. Other important minerals include zinc (notably in New Brunswick), nickel, copper, lead, uranium (in the north of Quebec, Ontario and Manitoba), potash (in Saskatchewan) and molybdenum (in British Columbia).

Manufacturing accounts for nearly one-fifth of Canada's gross national product and about one-sixth of the workforce. The most important industries are petroleum refining and automobile production. Apart from pulp and paper production, other significant activities include meat-packing and iron and steel production. Ontario is the country's most important industrial area. The prairie provinces produce food, beverages, petroleum and chemicals. British Columbia produces machinery and chemicals.

Hydroelectricity remains the most important source of power, but there are also thermal and nuclear power stations.

Trade and commerce
Canada's economy is heavily dependent on trade, and some three-quarters of all Canada's trade is with the United States. The economy achieved an average real growth rate of over 2 percent GDP in the late 1990s. Canada remains a rich country with prospects of solid economic growth.

Transportation and communications
Many northern communities are accessible only to all-terrain vehicles and aircraft where many so-called bush pilots operate. Air Canada and Canadian Airlines provide both domestic and international flights.

There are two transcontinental railroad systems: the privately owned Canadian Pacific Railway and the Canadian National Railway. The British Columbia Railway, which is controlled by the provincial government, provides a vital link to the province's central and northern areas. A similar provincial railway links Moosonee on James Bay at the southern tip of Hudson Bay to southern Ontario. The hub of the whole system is Montreal, which connects not only with the Canadian West, but also with the American Midwest, New York, the seaboard cities and the ports of the Atlantic provinces.

In the Atlantic provinces ferries still

The fishing village of Peggy's Cove on the coast of eastern Canada. It was to exploit the yearlong coastal fishing, notably around Newfoundland, that Europeans first settled permanently in Canada.

provide essential road, rail and passenger links. The ports of Halifax and St John in Nova Scotia and Newfoundland respectively carry much of Canada's external Atlantic trade, especially in winter, while Vancouver and Prince Rupert on the west coast are Canada's year-round Pacific outlets. The St Lawrence Seaway connects the Atlantic Ocean lanes directly with Great Lake ports such as Chicago and Thunder Bay in the heart of the continent.

In British Columbia one of the largest ferry fleets in the world plies between Vancouver Island and the mainland, where Vancouver itself serves the northwestern coastal ports with freight and passengers. In the Northwest Territories the Mackenzie water system connects the Arctic Ocean with the Mackenzie river, the Great Slave Lake and Lake Athabasca; the Great Slave Lake Railway and the Mackenzie Highway connect with the south.

About half Canada's 885,000 km (550,000 mi) of roads are in the prairie provinces. The Trans-Canada Highway links the capitals of the four Atlantic provinces to Toronto, Montreal, the prairies and Vancouver, while superhighways serve the Windsor–Quebec City corridor and major western centers.

The media
Canada has over 100 daily newspapers, mostly privately owned by a number of multimedia conglomerates. Broadcasting is controlled by the Canadian Radio-Television and Telecommunications Commission, an independent authority established by parliament in 1968. Two national TV networks are operated by the publicly owned Canadian Broadcasting Corporation, one broadcasts in English and one in French. There are three private television networks and over 2,000 cable TV systems. There are also hundreds of independent radio stations. Advances in telecommunications technology have been of benefit to isolated northern and Arctic communities.

Health and welfare
Federal government provides pensions, unemployment insurance and family allowances, while the provinces provide everything else. Most people belong to the federal medical insurance program, and are covered by provincial hospital insurance schemes.

Each province has its own education system, reflecting local needs. Children usually attend pre-school classes, then elementary and secondary education leading to college or one of 60 or so universities. Most education is free and compulsory from the age of six to secondary level.

An engineering triumph

The great Saint Lawrence river has been a trade route since the earliest days of settlement. Today ocean-going cargo vessels ply the waters where once the fur-trappers paddled their canoes. Flowing 1,200 km (750 mi) from Lake Ontario to the sea, the Saint Lawrence travels through the attractive Thousand Islands area before narrowing as it passes over a series of turbulent rapids above Montreal. It widens beyond Montreal to form Lake Saint Francis and Lake Saint Louis, and flows into Lake Saint Pierre. The final section follows a gently meandering course, eventually broadening out at the Gaspé Peninsula to enter the Atlantic via the Gulf of Saint Lawrence.

Until the 1950s the rapids denied passage from the Great Lakes to the sea to all but smaller river craft. In 1954 the United States and Canada began an ambitious joint project to open the river to deep-sea vessels, at a total cost of over $340 million to the Canadians and over $130 million to the Americans. The project involved building a total of 16 locks and a complex of canals and deepened waterways. Far to the northwest of the Saint Lawrence itself, Lake Huron and Lake Superior are linked by the Sault Sainte Marie Canal with its four locks. The Welland Ship Canal between Lake Ontario and Lake Erie has eight locks, and bypasses the Niagara Falls. The main seaway has seven locks between Lake Ontario and Montreal, and was opened in 1959.

Tolls are levied on vessels passing through the system. These are intended to cover the building costs (which must be met by the year 2009), interest on the funds raised for construction, and the normal operating and maintenance costs of the system. The seaway can accommodate most ocean-going vessels apart from the very largest. This makes the whole stretch from Montreal to the Great Lakes accessible to about four-fifths of the world's merchant shipping.

The seaway is closed by ice from mid-December to early April – about 115 days a year – but at other times it provides passage to the Atlantic from numerous inland cities in both Canada and the United States. Altogether 66 American and 12 Canadian ports have access to the seaway. The routes from these ports to western Europe are shorter than the transatlantic routes from the United States' eastern seaboard. About nine-tenths of the cargoes carried on the seaway are bulk commodities such as

Crashing through frozen water, an icebreaker in the St Lawrence Seaway clears a passage for a merchant ship in early spring. The seaway is closed by ice for about four months from mid-December.

grain, iron ore and coal, but other vessels also use it to transport a great variety of general cargoes.

Dependencies in the region

GREENLAND
DENMARK

Greenland, the world's largest island, lies largely within the Arctic Circle. Some 85 percent of the land area is covered by the northern hemisphere's biggest ice cap. Near its center it is over 3,000 m (10,000 ft) thick and its base is about 300 m (1,000 ft) below sea level. At its edges, along the coast, glaciers have created a complex landscape of deep fjords and offshore islands. The ice sheet is partly enclosed by mountains to the west and east, where they rise to Greenland's highest peak, Mount Gunnbjørn.

The climate is arctic, but precipitation is noticeably higher in the south, which is warmed by the North Atlantic Drift. The east coast is cooled by the Labrador Current, which flows from the Arctic Ocean to the Atlantic. Wildlife and vegetation in the ice-free areas are characteristic of the Arctic tundra. There are salmon and trout in rivers and streams, while cod, salmon, halibut, flounder and capelin are found offshore.

A series of Inuit migrations probably reached Greenland between 4000 BC and 1000 AD, and Inuit have lived here ever since. In the 10th century Viking settlers from Norway established two colonies, but contact with Scandinavia was lost some 500 years later. Colonization was only resumed in 1721, when Greenland became a Danish colony. Hans Egede, a missionary from Norway, founded a trading post – with Danish permission – near Nuuk (Godthåb) (now the capital). In 1776 the Danes assumed a trade monopoly (free trade was not reestablished until 1951). During World War II, while Denmark was under Nazi German occupation (1940–45), the United States protected Danish interests in Greenland. After the war Denmark met local demands for extensive reform, and in 1953 Greenland became part of the Kingdom of Denmark, no longer a colony.

Home rule was instituted in 1979. However, Denmark retains control of foreign affairs, and Greenlanders are Danish citizens. The 31-member Landsting (parliament) is elected by all adults aged 18 or over, and is chaired by the leader of the majority party. Greenland also elects two members of the Danish parliament. Today more than four-fifths of the population are native Greenlanders, mostly descended from Inuit with an admixture of early European settlers. The remainder are largely Danes. The official languages are Greenlandic (Eskimo) and Danish, and most people belong to the Danish Lutheran church.

Settlement is effectively limited to the coastal fringe. Sheep and reindeer are raised in the extreme south, where hay is grown for fodder. Vegetables are cultivated in the south and along the warmer west coast. Fishing has displaced seal-hunting as the leading industry. It is largely state-financed, as are the onshore canning and freezing plants. Hunting in the north produces pelts for export.

Greenland's mineral resources are large but difficult to exploit. Mining of the world's largest cryolite deposits was abandoned as uneconomic in 1963. Exploration has revealed valuable metal deposits, of which zinc and lead are now being mined. Electricity, from thermal

The colorful corrugated houses of Ilulissat (Jakobshavn) (*above*) on the west coast of Greenland. Fishing is the main occupation, and there is a large processing and canning plant beside the bay.

The frozen sea (*below*) at the Taasiilaq (Ammassalik) trading post on Greenland's eastern coast, just south of the Arctic Circle, prevents the movement of cargo.

has its own hospital. Nine years of elementary education are free and compulsory; four years of secondary education are optional. Both vocational and university education are available for Greenlanders in Denmark.

ST PIERRE AND MIQUELON
FRANCE

This tiny archipelago lies off the south coast of Newfoundland on Canada's eastern seaboard. The largest island, Miquelon, consists of two islands linked by a narrow isthmus. St Pierre to the southeast is smaller, but accommodates most of the islands' small population. The scenery is barren, with few trees and extensive peat bogs. The tall cliffs and rocky islets are home to many sea birds.

The islands were first settled by fishermen from western France in the 17th century, and the population has remained staunchly French ever since. The islands became a French overseas *département* in 1976, but in 1985 they were promoted to the status of a *collectivité territoriale*. The people vote both in local and in French national elections; the governor is appointed from France.

There is little agriculture apart from the growing of a few vegetables for local consumption, and the economy depends primarily on fishing from the rich waters nearby. Most food is imported from Canada and France, with which good transportation links are maintained. The islanders enjoy the benefits of French citizenship, including free education and good health and welfare provisions.

stations, relies on imported fuels. Other major imports include machinery and transportation equipment, manufactured goods and food.

Greenland has no railroads. The small road network is well used, but dog sleds remain the chief form of surface transportation. Most freight arrives by sea, but there is a well-developed internal and external air service. A sophisticated telecommunications network is available to most of the population. Radio and television broadcasts, like Godthåb's one newspaper, *Atuagagdliutit*, are in both Greenlandic and Danish.

A full range of welfare services is funded by the Danish government. Free health care is available to everyone, and each of the three administrative districts

Erik the Red – Viking explorer

In about 982 AD a man called Erik the Red set out from Iceland on a dangerous voyage into the Atlantic. According to a 13th-century saga Erik was a violent man who had been sentenced to three years' exile for "some killings" – but he was also a brave man. His target was a little-known island far to the west of Iceland – a place that had brought disaster and death to a group of Icelandic colonists only a few years before.

Sailing due west, Erik made landfall near a massive glacier. However, recalling an earlier ill-fated expedition by another explorer, he decided to sail southward. After rounding the cape, he discovered fertile valleys hidden among the glaciers on the western coast, and called the place Greenland – believing "people would be much more interested in going there if it had a pleasant name".

Some four years after his first voyage he set sail again from Iceland with 300 prospective colonists in 25 ships. Just 14 of these vessels reached the fjord that Erik had named for himself – present-day Tunugdliarfik. On a steep, grassy hillside he founded his Eastern Settlement, where the site of his house can still be seen. A breakaway group then founded a Western Settlement farther up the coast.

Greenland in the 10th century was appreciably warmer than it is today. The Vikings raised cattle and sheep in the south, but also ventured north to hunt caribou, whales, seals and bears; Polar bears were much prized as curiosities at European courts. For a while the colony flourished. About 280 farms are recorded in written sources, and archaeologists believe there were many more. Christianity reached Greenland in about 1000 AD, brought by the Norse explorer Leif Eriksson on his return from the recently Christianized Norway, and over the next 200 years nearly 20 churches were built, along with a convent and a monastery. The remains of a 12th-century cathedral at Gardar still survive.

In 1261 Greenland, like Iceland, came under Norwegian rule, but within a century the Western Settlement had been abandoned, and the Eastern Settlement went into a long decline. No one knows exactly why this happened, but it is known that in the late Middle Ages the climate changed for the worse, and this may have had a serious impact on the settlers. Inuit stories also suggest battles between their ancestors and Viking settlers. Animals and people would certainly have died in the bitter cold of the Arctic winter. The last recorded contact was in 1406. After that, the last Viking settlement on Greenland vanished into the cold and darkness of the Arctic night.

The United States

COUNTRIES IN THE REGION

THE UNITED STATES OF AMERICA

The torch of freedom "Give me your tired, your poor, your huddled masses yearning to breathe free" run the famous lines on the Statue of Liberty in New York Harbor. Erected in 1886, and partly paid for by the French people to symbolize their friendship with the United States, it was the first sight that greeted generations of European immigrants, signaling their safe arrival in the land of opportunity.

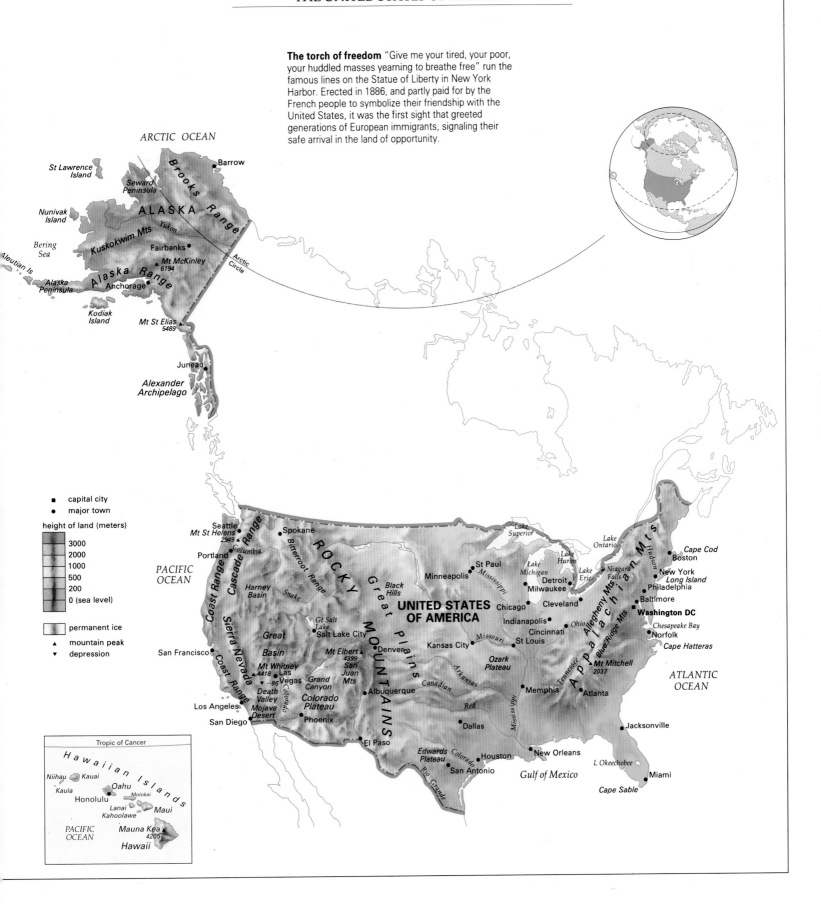

United States

THE UNITED STATES OF AMERICA

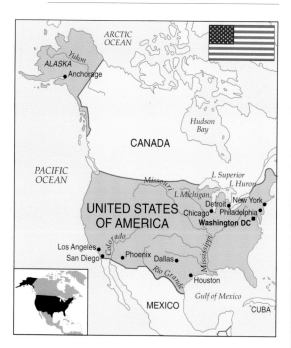

T HE UNITED STATES OF AMERICA OCCUPIES part of North America, bordered by Canada to the north and Mexico to the south, and washed by the Atlantic Ocean to the east, the Gulf of Mexico to the south and the Pacific Ocean to the west. This area contains 48 of the 50 American states, and is known as the coterminous United States. The other two states are Alaska at the northwestern tip of North America, and the island group of Hawaii in the central Pacific. Other outlying territories include Puerto Rico and the Virgin Islands of the United States in the Caribbean, and Guam, American Samoa and the Trust Territory of the Pacific Islands.

The United States has the third largest population after China and India; in geographical extent it is the fourth largest country in the world, after Russia, Canada and China. But in terms of economic, political and cultural influence, it is the leading nation of the world. It owes its success to its plentiful natural resources, a rich cultural mix, and a strong sense of national identity.

GEOGRAPHY

Great diversity is, not surprisingly, a dominant characteristic of the landscape of the United States. The towering peaks of the west give way suddenly to the grasslands of the central lowlands, which in turn merge gradually into the more varied landscapes of the east. The climate is similarly diverse, ranging from Arctic in northern Alaska to subtropical in the southeast; the warmest areas include both the arid heat of the Arizona deserts and the balmy everglades of Florida.

The land

There are five main topographic areas within the coterminous United States. To the east and southeast, along the Atlantic and Gulf coasts, a broad coastal plain rises inland to the Appalachian Mountains. West of the Appalachians is a vast lowland area that rises steeply in the west to the mountains of the cordillera – a continuation of Canada's Rocky Mountains. The eastern branch of the cordillera flanks the central lowlands; its western branch runs parallel to the Pacific coast, and between these two branches lies a variegated landscape of plateaus, basins and mountain ranges.

The eastern and southeastern coastal plain is broad and generally low-lying, with many areas of swamp. The northern portion is largely submerged beneath the Atlantic; almost all that remains is a series of islands, most notably Long Island in New York State, and the peninsula of Cape Cod in Massachusetts. South of Long Island the lowlands become wider, reaching their greatest extent in the southern coastal states. Offshore, a line of barrier islands and sandbars runs for

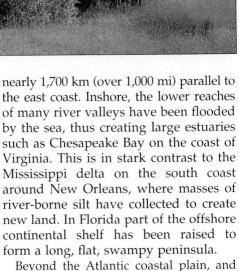

Big Sur, California (*above*), on the Monterey Peninsula. The Santa Lucia range is separated from the Coast Range inland by a narrow valley and the main railroad linking San Francisco and Los Angeles.

Vermont in the fall (*left*) Vast tracts of New England forest were cleared by early settlers, and by the 1850s some 50 percent of the area was farmland. Today only 12 percent is farmed.

nearly 1,700 km (over 1,000 mi) parallel to the east coast. Inshore, the lower reaches of many river valleys have been flooded by the sea, thus creating large estuaries such as Chesapeake Bay on the coast of Virginia. This is in stark contrast to the Mississippi delta on the south coast around New Orleans, where masses of river-borne silt have collected to create new land. In Florida part of the offshore continental shelf has been raised to form a long, flat, swampy peninsula.

Beyond the Atlantic coastal plain, and parallel to it, the Appalachians extend

southwestward from Canada to northeastern Alabama. They form a generally low but almost unbroken mountain chain largely covered with deciduous forests. Toward the southeast, the Blue Ridge Mountains rise in a steep escarpment above the rolling landscape of the Piedmont, once a highly productive agricultural area. The Blue Ridge Mountains reach their highest point at Mount Mitchell in North Carolina. Northwest of the geologically older Blue Ridge, and parallel to it, is a series of valleys separated by mountain ridges. The largest of the valleys is the Cumberland Valley, which is part of a long trench system that runs across the main Appalachian ridge to meet the Shenandoah Valley in Virginia, continuing through Pennsylvania and New York State to the Valley of Vermont in New England.

Westward, beyond the Appalachian chain, is a vast continental basin up to 2,500 km (1,500 mi) wide. It extends south from the Canadian border and the Great Lakes to meet the coastal plains that

NATIONAL DATA – UNITED STATES OF AMERICA

Land area   9,161,923 sq km (3,537,438sq mi)

Climate	Altitude m (ft)	Temperatures January °C(°F)	July °C(°F)	Annual precipitation mm (in)
Barrow	42 (13)	-27 (17)	5 (41)	107 (4.2)
San Francisco	5 (16)	9 (49)	17 (63)	500 (48.4)
New Orleans	9 (30)	11 (51)	28 (82)	1,572 (61.8)
Chicago	190 (623)	-6 (21)	23 (73)	910 (35.8)
Washington, DC	22 (72)	-1 (35)	27 (80)	981 (38.6)

Major physical features   highest point: Mount McKinley 6,194 m (20,320 ft); longest river: Mississippi-Missouri 6,020 km (3,740 mi); largest lake: Lake Superior (part) 83,270 sq km (32,150 sq mi)

Population   (2006 est.) 298,444,215

Form of government   federal multiparty republic with two legislative houses

Armed forces   army 595,946; navy 376,750; air force 347,400

Largest cities   New York (8,141,241); Los Angeles (3,945,097); Chicago (2,818,628); Houston (2,059,430); Phoenix (1,474,184); Philadelphia (1,426,259); San Diego (1,311,612); San Antonio (1,299,646); Dallas (1,222,053); San Jose (898,624); Detroit (858,803); Jacksonville (822,117); Indianapolis (770,634)

Official language   English

Ethnic composition   White 81.7%; Black 12.9%; Asian 4.2%; Amerindian and Alaska native 1%; Native Hawaiian and other Pacific islander 0.2%

Religious affiliations   Protestant 52%; Roman Catholic 24%; Mormon 2%; Jewish 1%; Muslim 1%; other 10%; none 10%

Currency   1 United States dollar (USD) = 100 cents

Gross domestic product   (2006 est.) U.S. $12.98 trillion

Gross domestic product per capita   (2006 est.) U.S. $43,500

Life expectancy at birth   male 75.02 yr; female 80.82 yr

Major resources   coal, copper, lead, molybdenum, phosphates, uranium, bauxite, gold, iron, mercury, nickel, potash, silver, tungsten, zinc, petroleum, natural gas, timber, fisheries, tourism, cereals, citrus fruits, cotton, fish, livestock, oilseeds, potatoes, soybeans, sugar beet, timber, tobacco, vegetables

border the Gulf of Mexico. The whole of this enormous area is drained by the great Mississippi–Missouri and Ohio river systems. In the north there are upland extensions of the Canadian Shield, such as the hills bordering Lake Superior. The eastern side of the basin consists of the Central Lowlands, whose southern limits are marked by the low plateaus of Kentucky and Tennessee, and farther west by the isolated uplands of the Ozark Plateau. The western side of the basin is formed by the semiarid Great Plains, which rise gradually from east to west to meet the western cordillera.

The western cordillera covers almost one-third of the North American continent. Within the coterminous United States it forms two main mountain systems that run south from the Canadian border. On the eastern side the Rocky Mountains face out across the Great Plains. In the north they are really an extension of the Canadian Rockies. Farther south, the Wyoming Basin provides the main corridor through the Rockies, used by pioneer trails and interstate highways alike. The nearby Grand Teton and Yellowstone National Parks include some spectacular scenery. South again, in Colorado and New Mexico, the Rockies rise steeply from the Great Plains, providing some breathtaking views.

West of the Rockies in the northwest lies the Columbia Basin, where the rocks are still being formed by a continuing upflow of lava that has buried many older landscape features. The rocks of the basin have been cut by deep river valleys and by meltwaters from ice-age glaciers. Farther south the landscape is dominated by low mountain ranges interspersed with desert basins floored with alluvium. One such basin has been filled by Utah's Great Salt Lake. East and south of this area the Colorado Plateau extends south into Arizona, with a remarkable landscape of mesas, buttes and canyons. The Grand Canyon of the Colorado river is just one of many national parks in this region.

The western branch of the western cordillera has two main strands. The Coast Ranges rise close to the shoreline, while the eastern line consists of the Cascade Range and the Sierra Nevada. Between them lies part of a discontinuous trench that appears all along the Pacific coast of North and South America. It includes Puget Sound in Washington, Oregon's Willamette Valley and the Central Valley of California.

The Cascade Range extends from Washington through Oregon to Lassen

The Bald eagle (*above*), America's national bird, is fully protected and has been reintroduced into New York State from captive-bred stock.

Snowcapped Mount Tom (*left*), in California's Sierra Nevada range, has the rough, jagged profile of relatively young mountains.

The stark beauty of Utah (*below*), where stratified rock formations surround Lake Powell. The buttes and mesas are familiar from many Western movies.

Peak in California, and includes a chain of high volcanoes. Among them is Mount Saint Helens, which erupted with devastating force in 1980. California's Sierra Nevada range includes Mount Whitney, the highest mountain in the coterminous United States, and is cut by spectacular glacial valleys. The eastern side forms a steep escarpment, whereas the western side slopes more gradually toward the Central Valley of California.

The northern outliers of the Coast Ranges are the glacier-clad Olympic Mountains of western Washington. A much lower section in western Oregon climbs to dramatic outcrops on the Californian state line. From here to Point Conception on the coast of southern California the ranges become low and linear, with major fault lines running parallel to them. An earthquake on the San Andreas Fault brought devastation to San Francisco in 1906, and a rather smaller earthquake in 1989 provided a grim reminder of this. In southern California the Coast Ranges become higher again but more fragmented.

Climate

The coterminous United States lies in a region of prevailing westerly winds, with the landmass of Canada to the north, a warm, shallow sea to the south, and broad oceans to both east and west. In general, however, the sheer size of the North American continent produces a continental climate throughout most of the country, marked by cold winters, warm or hot summers and a broad diurnal range of temperatures.

The west coast benefits from its maritime position, which brings much milder winters than elsewhere. Cool, moist winds from the Pacific shed much of their precipitation as they rise over the Coast Ranges, so that Oregon and Washington have the heaviest rainfall in the country, especially during the fall. Farther south, in California, the summers tend to be hot and very dry. Rainfall is consequently much lower, especially in the Central Valley; agriculture here is dependent on irrigation. In the far south there is little rain at all. Farther inland much of the remaining moisture falls as rain or snow over the Cascade Range and the Sierra Nevada. The westerlies then become warm and dry as they descend the eastern slopes, creating the deserts of Nevada and Arizona.

East of the Rockies the Great Plains are semiarid, but farther east the Central Lowlands receive more rain in summer from hot, humid air flowing north from the Gulf of Mexico. Less beneficial, though, are the spectacular weather conditions produced throughout the central United States when humid air from the south meets colder air from the western cordillera. The result may be anything from thunderstorms and tornadoes to hailstorms and blizzards. Not surprisingly, this American heartland suffers the most violent extremes of temperature, from season to season and sometimes even from one hour to the next. Spring and fall are pleasant but very short, and winters are longer toward the north.

Areas near to the Gulf of Mexico have much shorter winters, but are liable to hurricanes in the late summer and early fall. Equally vulnerable are the states in the southeast adjoining Florida, where warm, humid conditions prevail all year round. North of here, the Appalachians receive plentiful rain throughout the year. The northeast, however, has a largely continental climate, only slightly modified by the nearby ocean. This is because the prevailing westerly winds blow offshore. Low winter temperatures combined with unstable air conditions lead to some spectacular snowfalls, especially over the mountains and along the coast.

Vegetation

The first European settlers in the eastern United States found a land covered in rich deciduous forest, including hardwoods such as hickory, oak and walnut. Relatively few of these trees now remain; most of the forest cover in the east is secondary growth, though a few elms, maples and beeches can still be seen. Conifers grow in the colder northern areas and on high ground.

About halfway across the central basin the tree cover gradually gives way to

The gold rushes

A pioneering prospector from Minnesota, George Northrup, poses for a daguerreotype – an early type of photograph – in about 1860, with the tools and pistols of his trade, and its reward: a bag of gold dust worth the then amazing sum of $90,000.

It is largely thanks to the Hollywood film industry that the term gold rush evokes such vivid images – whiskery prospectors tramping through northern snows, panning the sand of icy Alaskan rivers, starving in lonely cabins, or "playing hard" in rough boom towns of hastily erected clapboard houses. But in fact gold rushes have happened all over the world, and the pattern is generally much the same. Their importance for the United States, both in its real history and in its mythology, lies in the fact that there were several gold rushes within a very short time, all of them along the pioneer routes of the West.

The first of them – the so-called Fortynine – occurred beyond the mountains of the Sierra Nevada in central California, where John Augustus Sutter (1803–80), a bankrupt Swiss emigrant, had founded the trading colony of Nueva Helvetia (New Switzerland; later Sacramento). In 1848, one of his workmen found particles of gold in the tailrace of his sawmill. Sutter tried to keep this discovery secret, but failed, and in the next year at least 80,000 prospectors swarmed over the area now known as Sutter's Mill in a frantic search for the "Mother Lode" – the rich gold seam from which other seams were thought to branch.

The land could not possibly support such hoards of settlers; many Fortyniners lived in hardship and died in squalor. Sutter's trade was shattered, his land devastated, his property and livestock stolen. He became bankrupt yet again, and returned east in 1851. When he died many years later, he was still fruitlessly seeking government compensation. He had at least a moral right to it, for the gold rush swelled settlement in the area, creating the prosperous city of Sacramento.

Other gold rushes followed in Colorado (1858) and South Dakota (1876). Hastily built townships were abandoned when local deposits ran out, leaving many ghost towns scattered across the countryside. These later became famous, and some have been restored as tourist attractions, such as Deadwood below the Black Hills of western South Dakota.

It was the Alaskan Rush, however, that really caught the world's imagination. It began in 1886, when gold was struck at Fortymile Creek on the Yukon river, and was fueled by further discoveries in 1896, both in the Yukon and in the far west of Alaska on the site of present-day Nome. Pioneering was by now in decline, with less land to be settled and fewer fortunes to be made. Alaska promised a last chance, both for the rootless men of the west and for discontented migrants from the east.

Few of them could afford to make

the hazardous sea passage to Alaska around South America, so most took the landward route. It was a path strewn with dangers that were hardly mentioned in the attractive though largely fictional manuals produced by opportunistic publishers in the east. Thousands set off, but many perished on the snow-covered mountain crossings, and those who arrived found conditions little better. Many of them starved, and some considered themselves lucky to be able to revert to their former occupations. One such person was bank clerk Robert William Service (1874–1958), whose once popular ballads captured much of the atmosphere created by the men who joined the Alaskan Rush.

By 1900 some 22 million dollars' worth of gold was being panned annually, but at a high cost. Few real fortunes were made, and as early as 1911 yields were declining. However, while the rush lasted, it spurred the settlement of inhospitable Alaska as almost nothing else could, making possible today's exploitation of Alaska's many other valuable resources. But for the gold rushes, the Last Frontier might have remained unconquered for a much longer period.

grassland, although much of the area concerned is now agricultural land. As rainfall decreases to the west, the long prairie grassland is replaced by shorter, thinner steppe grasses.

Alpine vegetation covers the western mountain ranges, with desert vegetation in the arid areas in between. Plant life here is surprisingly rich, with a wide variety of cacti and succulents. Westward again, on the coasts of Oregon and California, are Douglas firs and the last surviving stands of redwood, the world's tallest tree. Redwoods can reach as high as 90 m (300 ft), and some living specimens are thousands of years old.

Central and southern California are characterized by chaparral vegetation, with its dense thickets of shrubby plants that are resistant to the long summer droughts. Palm trees are native to southern California and Florida, though they are also grown elsewhere.

Wildlife

Animal life in the United States shows the dramatic effects of human settlement. The first European colonists brought horses, cattle, sheep and other European species such as sparrows and starlings. Settlers moving west in the 19th century all but eliminated the huge herds of buffalo and pronghorn that once roamed the Great Plains. Very few of these now protected species survive, and today even the Bald

Magnificent buffalo in Yellowstone National Park, Wyoming. These animals once dominated the landscape in herds that reached from horizon to horizon, but few survived. Today they live in small, isolated populations where inbreeding is a danger.

eagle, symbol of the United States, is in danger of extinction, as are the American bighorn sheep and the cougar.

Mammals with a wide distribution include White-tailed deer, American black bears, bobcats, raccoons, skunks, opossums, beavers and muskrats. Moose, Red foxes, otters and wolverines inhabit the northern coniferous forests. The forest rivers of the east and southeast support an abundance of fish species, and there are seven different species of salamander in the Appalachians. The deserts of southern California, Nevada and Arizona boast many reptile species, including the poisonous Gila monster. The Great Plains are the home of prairie dogs (gregarious rodents), while the Rockies harbor marmots, Mountain goats and pikas.

Among marine mammals, seals are found on both coasts, but sea lions only in the Pacific, and the Florida manatee only in the larger rivers of the southeast.

Conservation and management

The landscapes of the United States show startling contrasts between areas of man-made urban and rural wilderness and vast tracts of natural wilderness carefully conserved for future generations.

Armed for attack (*above*) Pronghorn bucks use their sharp, hooked antlers for sparring with rival bucks during the breeding season. Some 35 million once roamed the plains, but – like the buffalo – they were all but exterminated by European settlers.

A manatee (*above*) browses in a Florida stream. These huge, slow-moving creatures are in danger of extinction from sewage pollution of their freshwater habitats, and from injuries caused by speedboat propellers.

Zigzagging to the horizon (*right*) the Florida Keys are a chain of offshore islands linked by the Overseas Highway. Completed in 1938, this runs southwest from the mainland across Card Sound to the principal islands of Key Largo and Key West.

The first national parks, both of them in California, were established in 1890. The Sequoia National Park, which has over 300 lakes and some of the highest peaks of the Sierra Nevada, contains a fine stand of giant sequoias. The Yosemite National Park includes Half Dome, a peak that has been sliced in half by a glacier. The other national parks include Mount Rainier in Washington, Yellowstone in northwest Wyoming, and also the Great Smoky Mountains National Park, which straddles the Tennessee–North Carolina border in the southern Appalachians.

Many other areas enjoy similar levels of protection, but many are also suffering increasing pressure from millions of

visitors. This pressure, combined with atmospheric pollution and the effects of fertilizers and pesticides, means that even wilderness areas are now under threat. The population of redwoods, for instance, is declining. Pollution controls such as those initiated in Oregon may yet be able to turn the tide, but the fact remains that even a national park cannot be completely protected from the highly industrialized society that surrounds it.

Alaska

The state of Alaska is located in the far northwest of North America, with Canada to the east and the Bering Strait, which separates it from the Soviet Union, to the west.

The southernmost part of Alaska is called the Panhandle – a narrow strip of land that runs south along the northern Pacific coast bordering Canada inland. It is mountainous and flanked by offshore islands. The coast along the Gulf of Alaska is lined by an almost continuous mountain chain characterized by glaciers, volcanoes and severe earthquakes. It runs from the Boundary Range in the Panhandle via the Wrangell Mountains and the Alaska Range to the southwestward line of the Aleutian chain.

North of the Alaska Range is the broad central plateau, whose tablelands are greatly fragmented by the Yukon river and its tributaries. North of here, the Brooks Range runs clean across the state from the Canadian border to the Bering Strait. Northward again, there is a broad coastal plain, covered in tundra, bordering the Arctic Ocean.

The climate is generally cold, but varies greatly with altitude and latitude. The greatest extremes of temperature are to be found in the drier areas of the interior and north. Southern coastal areas are less cold in winter, but are generally subject to heavy rain and snow.

Hawaii

The Hawaiian Islands are the peaks of a submerged volcanic chain that stretches in a wide arc across the central Pacific Ocean. The largest of the eight main islands are Hawaii and Maui, and there are some 124 smaller islands. The climate is warm and humid, and rainfall varies from sparse in some areas to extremely heavy on slopes facing the northeasterly trade winds, where it supports a rich vegetation. There are no native reptiles or mammals, but the bird life is remarkable and varied; most resident native species are unique to these islands.

SOCIETY

In the course of the 20th century, the United States has achieved worldwide cultural dominance. Spread by the mass media, its popular culture enjoys worldwide appeal, helping to make English the most international of languages. Its standard of living is among the highest in the world, and its ethos of individual liberty is widely admired.

Nevertheless the country has considerable social and economic problems. Organized crime, political corruption and the wasteful use of resources characterize American society as much as the skill and enterprise that produced the Moon landings, medical research, the Philadelphia Orchestra, Hollywood or Mickey Mouse.

History

Archaeological evidence suggests that America's first settlers probably crossed from Siberia to Alaska via the Bering Strait during the ice ages between 30,000 and 10,000 years ago, and spread rapidly across the continent. Their descendants, the Native-Americans, developed a wide variety of societies and cultures. Their homes, for example, ranged from the cliff villages of the Pueblos in the southwest to the bark lodges of the Iroquois in the northeast and the huge decorated wooden halls of the sophisticated Northwest Coast Indians.

The question as to which European peoples reached the Americas first has caused much controversy. There is con-siderable evidence pointing to a Viking seafarer from Greenland, Leif Eriksson (c. 1000 AD), who with others settled briefly in a land they called Vinland; this has been variously located along the north-east coast from Labrador (Canada) to New Jersey. The Italian navigator and explorer Christopher Columbus (1451–1506), who arrived almost 500 years later in 1492, was attempting to find a westward route to India; he christened the Caribbean islands the West Indies and their inhabitants Indians. It soon became clear, however, that these lands were not Asia but a new continent; it received its name from Amerigo Vespucci (1454–1512), one of several navigators who followed Columbus west. Vespucci published a widely read account of his explorations.

The earliest settlements were Spanish and Portuguese. The first English settlement (following a number of abortive efforts such as that of Sir Walter Raleigh (1552–1618) at Roanoke, Virginia), was at Jamestown, Virginia, established by the Virginia Company in 1607. The first of the New England colonies, Plymouth, followed in 1620. Its founders, the so-called Pilgrim Fathers, were strict religious dissenters hoping to found an ideal community free of political harassment. The huge Massachusetts Bay colony, which eventually incorporated Plymouth, was founded by even more authoritarian Puritans. Maine and New Hampshire began as its offshoots, while the New Haven and Rhode Island settlements were established by opponents of their strict regime. Maryland, by contrast, was

Christopher Columbus (*above*) explorer and visionary. The wreck of his flagship, Santa Maria, which sank off the island of Hispaniola in 1492, was located in 1968.

A painting of William Penn (*left*), making a treaty with the Native Americans. Penn is renowned for his fair and friendly dealings with them.

founded in the 1630s by Roman Catholics.

Meanwhile, in 1624 the Dutch had established the prosperous colony of New Amsterdam, centered on Manhattan Island. In 1664 it was captured by the British and renamed New York; New Jersey then separated from it in 1665. Pennsylvania was in many ways the most successful of the early colonies. It was founded in 1682 by a moderate Quaker from England, William Penn (1644–1718). Pennsylvania was unusual in maintaining fairly good relations with the Native Americans, though this was eventually undone by Penn's son, who swindled the Delaware Indians out of a large area of their land in a deal known as the Walking Purchase. Part of Pennsylvania later became the colony of Delaware. North and South Carolina were established in 1663 by English investors, and Georgia in 1732 as a philanthropic alternative to the debtors' prison.

During the 18th century these colonies remained dependent on Britain in order to survive and expand. There was a mutual need for defense both against the French and the Native-Americans, upon whose territories the colonists were encroaching. By 1763, when the French had been finally defeated, the British colonies were close to being self-governing republics, and they began to resent paying British taxes. The British government's ban on expansion into Native-American lands beyond the Appalachians, together with economic decline, fueled the growing unrest, and criticism at home led the

British authorities to impose even sterner controls. This only lent support to the radical separatist minority. Incidents such as the so-called Boston Massacre of 1770, in which five people were killed by British troops, became propaganda showpieces. In 1776 the Second Continental Congress duly declared the 13 colonies independent, and proudly proclaimed their confederation as a new nation – the United States.

In the War of Independence that followed, the American Continental Army, backed by state militia, faced British garrisons, later reinforced by German mercenaries. The ill-equipped American forces were often rescued from defeat only by capable generalship, especially that of George Washington (1732–99). The turning point came when revolutionary France entered the war, sending military and financial aid to the rebels, harrying transatlantic supply lines and threatening the British mainland. The surrender of the forces under the command of the British military leader Lord Cornwallis (1738–1805) at Yorktown in 1781 marked the effective end of Britain's hold on the United States, although the war did not end until 1783. The American victory was marked by savage reprisals against the loyalists, many of whom retreated north to British Canada.

Even during the later stages of the war America's great expansion had begun. During the next few years, Vermont, Kentucky, Tennessee and Ohio joined the Union. In 1803 the Louisiana Purchase

George Washington, first president of the United States. His characteristically stern, set mouth in this and other portraits was due to his swollen gums, caused by primitive false teeth. His sets, carved in wood or ivory, have been preserved.

added extensive but sparsely settled French territories in the west and south, which came to form the heartland of modern America. A year later Thomas Jefferson (1743–1826), third president of the United States, despatched the American explorers Meriwether Lewis (1774–1809) and William Clark (1770–1838) on an expedition to explore the new territory. They followed the Missouri river to its source, crossed the Rocky Mountains, and were the first Europeans to make the overland journey to the Pacific coast. In 1806 they returned to Saint Louis on the Mississippi, and their account of the journey helped to establish the strong

Wharfside laborers, Virginia (*above*) At the end of the American Civil War in 1865, the 13th, 14th and 15th amendments to the US constitution abolished slavery altogether and guaranteed citizenship to former slaves.

The Battle of Kennesaw Mountain, Georgia (*left*), June 1864; a major Civil War confrontation that cost the Union some 3,000 men, and the Confederates 600.

Abraham Lincoln – patron saint of common sense

"In the present civil war it is quite possible that God's purpose is something different from the purpose of either party – and yet the human instrumentalities, working just as they do, are of the best adaptation to effect His purpose." So wrote Abraham Lincoln (1809–65) in 1862, during the second year of his presidency – and his words suggest the true character of the man known to history as the Great Emancipator.

Abraham Lincoln's life story is the stuff of legend, but the legend often obscures the known facts about his life. Born in a backwoods cabin in Kentucky on 12 February 1809, he was six when his family moved to Indiana; three years later his mother died and his father remarried. In 1830 the family moved to Illinois, where he worked in a variety of occupations. In 1832 Lincoln volunteered for the Black Hawk War, in which the Sauk and Fox Indians came into conflict with the Illinois militia over contested lands. In 1834, after the first of his four elections to the state assembly, he began to

Abraham Lincoln, seen here in 1860, belonged to the first generation of historical figures whose features were accurately recorded for posterity in photographs rather than paintings.

study law; he passed his bar examination two years later. Lincoln soon built a prosperous legal practice and a reputation for straightforwardness and common sense. He married in 1842, and had four children.

During his term in Congress (1847–49), Lincoln proposed a limited form of emancipation for slaves, but nothing came of his suggestions. Then, in 1854, Stephen Arnold Douglas (1813–61) steered the Kansas–Nebraska Act through Congress, which gave Kansas and Nebraska settlers the option to permit slavery in their territories. Lincoln and his fellow Republicans thought the territories should be free soil where anyone, black or white, could better themselves. In 1858 Lincoln contested Douglas' seat in the Senate in a series of sharp public debates. He failed to win the seat, but in May 1860 he won the presidential nomination, and on 6 November he was elected. Shortly after, South Carolina left the Union.

Lincoln held to his principles and to the constitution, resisting all efforts to compromise on slavery in the territories. Before his inauguration on 4 March 1861, six more states had seceded, forming, with South Carolina, the Confederate States of America. On 12 April Confederate guns fired on Fort Sumter in Charleston harbor on the coast of South Carolina. The American Civil War had begun.

Lincoln applied himself to the war with the same common sense that had served him as a lawyer, and his careful management of the high command brought him eventual victory. His chief object was always to save the Union, but in 1863 he declared freedom for all slaves in rebel territories. Lincoln always insisted that this Emancipation Declaration was a war measure, but it opened the way to the abolition of slavery. In 1864 Lincoln was reelected on the promise of a constitutional amendment that would outlaw slavery for ever. State after state ratified the amendment, but before it could be adopted Lincoln was assassinated. He died on 15 April 1865, and the Thirteenth Amendment was adopted later the same year.

Lincoln undoubtedly hated slavery – but only such an honest and clear-thinking statesman could have turned that hatred into such a lasting and positive achievement.

sense of "Manifest Destiny" that was to fuel westward settlement.

In 1812 anti-British resentment led the United States Congress to declare war again. Despite American military success, the war brought the country close to fragmentation in 1814; but Britain was unwilling to fight another long war, and agreed to equitable peace terms. The United States settled down to a period of consolidation in the east, with the growth of great cities such as New York, Boston (Massachusetts), Philadelphia (Pennsylvania) and Charleston (South Carolina). Meanwhile, exploration and expansion continued westward. Texas and other territory was acquired from Mexico in 1848, followed by further gains in the Gadsden Purchase of 1853; Alaska was bought from Russia in 1867.

In the meantime, however, conflict over state rights was increasing between the northern and southern states, with the south resisting the centralizing authority of the north. The plantation economy of the southern states depended on slave labor, while the religious idealists of the more industrialized northern states found slavery repugnant. Matters at last came to a head in 1860 under the presidency of Abraham Lincoln (1809–65). The southern states seceded and formed the Confederacy. The Civil War of 1861–65 devastated the south, which suffered further privations after Lincoln's assassination in 1865. Slavery had been abolished, but crushing poverty led to interracial resentment that endured for more than a century.

The expansion westward continued, helped by the spread of the great railroad networks. A flood of immigration from Europe was encouraged in order to open up the new lands to farming. The westward movement was hastened by the apparent freedom and opportunity of frontier life, and by the prospect of gold on the west coast. The reality was often cruelly disillusioning, but the potent legends of the Wild West were born. The treatment of the Native-American population at this time was insensitive and often brutal, leading to atrocities on both sides. Meanwhile, industrial expansion in the northeast produced the first great industrial tycoons – selfmade men such as John D. Rockefeller (1839–1937) and Andrew Carnegie (1835–1919). It also created serious social problems.

The United States established itself as the dominant power in the region, following the declaration of the Monroe Doctrine in 1823, which denied Europe

Driving the Golden Spike, a painting by H. Charles McBarron commemorating the ceremonial linking of the Union Pacific and Central Pacific railroads at Promontory, Utah, in May 1869.

any right to intervene in the New World states. The United States encouraged revolutions in Latin American colonies, including one in Cuba, which was supported by a US invasion force. This led to war with Spain in 1898, which America won. As a result they acquired other Spanish territories, notably Puerto Rico and the Philippines. The islands of Hawaii were also annexed during this period.

As World War I raged throughout Europe, President Woodrow Wilson (1856–1924) committed the United States to neutrality, but in 1917 the German U-boat campaign, which hit American shipping, forced him to declare war. A strong isolationist reaction followed once peace was won, causing the United States to reject membership of the League of Nations – an international organization created to preserve the territorial integrity of member countries.

At home, the passing of the Volstead Act in 1920 banned the sale of alcohol. This proved unenforceable, encouraging the unprecedented growth of organized crime, and creating another American legend – the Prohibition era. The apparent prosperity of these years ended abruptly with the stock-market crash of 1929. This precipitated the Great Depression – a spiral of decline that was worsened by the European financial crisis of 1931. When Franklin D. Roosevelt (1882–1945) was elected president in 1932, he introduced emergency measures that effectively halted the decline. However, his New Deal policies had limited success in restor-

ing growth until the forthcoming war effort regenerated production.

The political unrest in Europe strengthened isolationist sentiment. When World War II broke out, the United States did no more than supply arms to Britain and France. Controversy continued at home until the end of 1941, when an attack on Pearl Harbor, Hawaii, by Japan, Germany's ally, brought the United States into the war. American industrial might was instrumental in the final victory, culminating in the use of the first atomic bombs to bring about Japan's surrender.

In the postwar period the United States established itself as the leading nation of the West in opposition to the Soviet Union. Financial aid programs such as the Marshall Plan reconstructed Western Europe and Japan as buffers against communism. As the Cold War tensions between the Soviet Union and the United States increased, so the spread of communism became the dominant fear in American society. This led to the political purges of communists of the early 1950s associated with Senator Joseph McCarthy (1909–57). Under United Nations auspices, the United States took the lead in the Korean War, repelling communist forces invading South Korea.

A period of prosperity followed, paving the way for the liberal aspirations of President John F. Kennedy (1917–63). In

foreign policy, however, Kennedy approved an abortive invasion of Cuba and faced down a Soviet attempt to establish missile sites there. He also committed the United States to protecting South Vietnam against communist invasion from the north. After Kennedy's assassination in 1963, this escalated into the Vietnam War, which eventually ended in political defeat for the US and withdrawal in 1974.

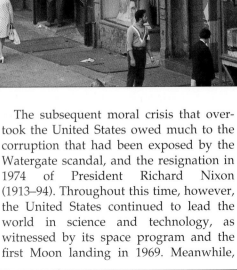

A victim of the Great Depression (*above*) with a handwritten placard proclaiming his need of a job. When this photograph was taken in 1932 the Depression was at its height, and one out of every four workers in the United States was unemployed.

Enforcing the law (*above*) – bootleg whiskey goes into the sewer. The blackmarket trade in illicit liquor was the inevitable result of Prohibition, and gave rise to a wave of organized crime and gang warfare.

City neighborhood (*left*) Iron fire escapes decorate the tenement buildings on Delancey Street on New York's Lower East Side, a dilapidated neighborhood of bars, hotels and small, oldfashioned businesses, reflecting the ethnic mix of urban society.

The subsequent moral crisis that overtook the United States owed much to the corruption that had been exposed by the Watergate scandal, and the resignation in 1974 of President Richard Nixon (1913–94). Throughout this time, however, the United States continued to lead the world in science and technology, as witnessed by its space program and the first Moon landing in 1969. Meanwhile,

civil rights leaders such as Martin Luther King Jr. (1929–68) campaigned to remove social and political inequalities from the United States' black population.

During the 1970s America's international standing suffered as it entered an economic recession, partly as a result of an oil crisis in the Middle East. In 1990 the United States became involved in military action in the Middle East after Iraq seized Kuwait. On

September 11 2001 a recession-struck US suffered great loss when suicide terrorists used commercial passenger jets to attack New York City and Washington, D.C., annihilating the World Trade Center, and partially destroying the Pentagon. The US declared total war on terrorism and on Afghanistan's Taliban government for failing to deliver the suspects, Osama bin Laden and his al-Qaeda terrorist network.

Government

The governmental system of the United States was based on that of 18th-century Britain, in the sense that its creators were attempting to reproduce its democratic strengths while correcting its flaws. It differs in that it is a federal system, in which certain powers are delegated to the national government and all other powers fall to the legislators of the individual states. The use of names such as Capitol and Senate for its institutions sprang from their intention to embody idealized notions of Roman and Greek democracy.

A written constitution was chosen in preference to the half-statutory, half-traditional British model as a better guarantor of rights. Included in the constitution are provisions that allow it to be amended. Amendments may be proposed by a two-thirds vote of both houses and ratified by three-quarters of the state legislatures – the method used for all 26 amendments adopted so far. A constitutional convention to consider amendments may be called if two-thirds of the states wish it.

The first 10 amendments are known collectively as the Bill of Rights and were adopted in 1791. They explicitly guarantee freedom of speech, religion, and the press and the right to a speedy and fair trial. The Tenth Amendment reserves all governmental powers not explicitly mentioned in the constitution to the states or to the people.

Three amendments – the Thirteenth, Fourteenth and Fifteenth – were adopted after the Civil War. These extended political rights to black Americans by abolishing slavery, establishing the citizenship rights of native-born and naturalized Americans and recognizing the rights of blacks to vote. The Fourteenth Amendment guarantees "equal protection under the laws" to all Americans.

Later amendments have continued to enlarge the political rights of ordinary Americans and to clarify archaic and confusing clauses in the original constitution. The Nineteenth Amendment (passed in 1919) extended the franchise to women in federal elections (some states had done so earlier). The Twentieth streamlined the period of transition between administrations after a presidential election, and the Twenty-fifth clarified the succession process following the death or resignation of a president in mid-term.

The drafters of the constitution tried to limit the power of the federal government and retain that of the states. Central to their effort was the concept of separation of powers, by which lawmaking, administration and enforcement are kept separate. Instead of a monarch as head of state, a president was to be elected. This was to be done via an electoral college, whose elected representatives would choose the president, for it was feared that a simple contest for public favor might easily lead to corruption and the election of demagogues. The electoral college soon became a formality, and it has been suggested that the worst fears of the constitution-makers have come true. The prolonged dispute in the 2000 election over the victory of George W. Bush – who lost the popular vote but won the electoral college vote – led to calls to abolish the electoral system.

The president must be at least 35 years old, and must be a native-born United States citizen who has been resident there for at least 14 years. In practice he must normally possess considerable wealth and political support. In theory he heads only the executive branch of government, but in practice his authority has increased to cover foreign policy and (as party leader) the legislative program in Congress. Because of the separation of powers, however, he often lacks direct authority to enforce his decisions, especially when his party is in the minority in Congress. In such cases, his limited powers of veto provide opportunities for mutual obstruction, leading hopefully to compromise but sometimes stalemate.

Congress alone can make laws, levy taxes, declare war and determine its own operations. In extreme cases it can impeach the president. It consists of two houses: the House of Representatives and the Senate. Representatives are elected for a two-year term from each state, the number being determined by each state's population, up to an overall total of 435. They must be at least 25 years old, state residents and United States citizens of seven years' standing. Senators must be at least 30, state residents and United States citizens of at least nine years' standing. Each state elects two senators to serve six-year terms, and one-third of the Senate is reelected every two years. The Senate possesses greater power and status than the lower house, and its 16 standing committees have now become powerful organs of government.

At the head of the judiciary is the United States Supreme Court. This is the final authority on the interpretation of laws and of the constitution, and the ultimate appeal court for decisions of the

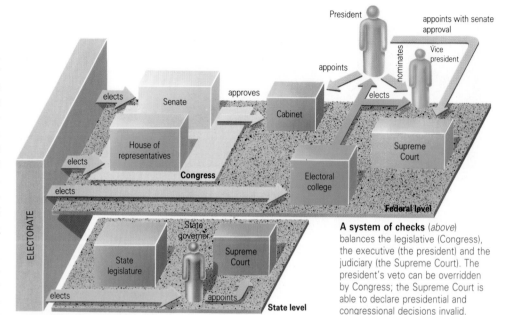

A system of checks (*above*) balances the legislative (Congress), the executive (the president) and the judiciary (the Supreme Court). The president's veto can be overridden by Congress; the Supreme Court is able to declare presidential and congressional decisions invalid.

Miami cop (*above*) Most law enforcement in the United States is entrusted to city, town or state police authorities, and is paid for by local taxes. Television movies give the job a glamorous image, and recruits are attracted from every social group.

Native American culture The Pueblos of New Mexico live in the same type of adobe-built dwellings that their pre-Columbian ancestors did. They preserve much of their traditional lifestyle, but have adopted some features of modern life, such as television.

Life in the sun (*above*) A volleyball tournament on the beach at Santa Barbara, California, attracts a large crowd of young people. The relaxed and affluent way of life that can be found on America's west coast continues to attract immigrants, most of them young, to the state.

Fast food (*right*) A New York street vendor caters for the lunchtime needs of city workers. Foods such as pretzels, hot dogs and knish were introduced by German or East European Jewish immigrants, but are now regarded as typically American.

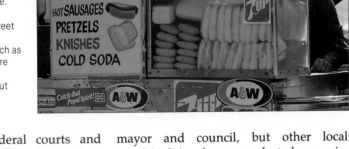

various specialized federal courts and state supreme courts. The nine justices are appointed for life by the president with the consent of the Senate. A majority decision is conclusive, but a tie will sustain the decision of a lower court. Political considerations almost always affect the appointment of judges.

Below the federal government are the individual state governments, whose structures generally mirror those of the federal government. Each has its own constitution, with a governor at the head and (apart from Nebraska) a two-chamber legislature. The governor is elected, as are most state officials, including judges. Municipal governments in the larger cities are most often headed by an elected

mayor and council, but other local municipalities favor an elected commission or a professional manager appointed by an elected body.

There are only two significant political parties, because the enormous electorate makes it hard for any third party to achieve a foothold nationally. Broadly speaking, Republicans are conservative and Democrats liberal. However, both parties reflect a wide spectrum of opinions, while local differences may modify or even reverse this balance.

People

The United States, with its millions of settlers from all over the world, is remarkable for its ethnic diversity – so much so that it has been characterized as a melting pot of peoples and cultures. But this is in some ways misleading. The years of heavy immigration were marked by a strong movement toward assimilation, which was seen as the best way to find employment and fulfill the "American dream". But as society has become more uniform, the process has reversed; ethnic pride is now a legitimate means of self-assertion, especially for marginalized minority groups.

There was considerable linguistic and cultural diversity among the original Native American inhabitants, but the impact of the European invasion was such that only a few have retained their distinctive cultural identity. The first Old World settlers came largely from northwestern Europe, especially Great Britain, Ireland, the Netherlands, Germany, France and Spain, and from Africa (chiefly the west coast regions) brought to work on plantations as slaves.

After the Civil War, however, a deliberate program of mass immigration brought in large numbers from Italy, Scandinavia and the troubled countries of Eastern Europe, including Russia; many of these were Jewish immigrants fleeing poverty and pogroms. Settlement patterns varied: many Italians and Jews settled in the big cities, founding their own distinctive neighborhoods such as Little Italy in New York; Scandinavians and some Balkan peoples took farmland in areas such as Minnesota. The degree of cultural acceptance and assimilation also varied very widely.

The economic collapse of the south following the Civil War swelled the class of so-called "poor whites", who lived in much the same conditions as the newly liberated black slaves, providing a potent breeding ground for racial hatred. Both in the agrarian south and in the more industrial north, a large, generally poor black community grew up, undereducated and exploited for cheap labor; in the south they were often denied basic civil rights. A gradual change in attitudes, aided by legislation, especially following World War II, has removed some of these distinctions; in many areas of society a degree of "positive discrimination" toward black (or African) Americans has been favored. But thousands of African-Americans are still effectively segregated in inner-city neighborhoods, or ghettos.

Later in the 19th century, an influx of indentured Chinese workers, especially to the west coast, created a strong Asian community that has spread far beyond areas such as Chinatown in San Francisco, California, and has been added to in recent years by many immigrants from Southeast Asia, especially the Philippines and Korea. Immigration from Central and South America has created a large Hispanic population, especially in the big cities of both coasts. Much of this immigration is illegal, creating serious social problems. Immigration is now limited by law, though the United States still has the highest level of immigration anywhere in the world. It welcomes refugees, and large influxes of people from countries such as Cuba and Vietnam.

All of these very diverse people are united by one major factor: the English language. It is the first language for most people, although the growth in immigration from Hispanic countries has made Spanish a second official language in some states. Nevertheless, a command of English opens the door not only to work and advancement, but to the enjoyment of popular culture – sports, music, movies and television.

Religious freedom is guaranteed under the Constitution, and there is remarkable religious diversity. The largest group are the Protestant Christians. Denominations such as Baptists, Methodists, Lutherans and Episcopalians are well represented, with a higher proportion of worshiping (as opposed to merely nominal) Christians than in much of Protestant Europe.

Roman Catholics form the next-largest group; they are largely of Irish, Italian or Hispanic descent. Jews account for only a small minority of the population as a whole, but form a very much higher proportion of the urban population of cities, particularly New York. All the other major world faiths are found here, and there are innumerable sects and cults, from Black Muslims to Scientologists and the American Indian Church. Some of these are identified with the particular areas where they have settled, especially the Mormons of Utah.

Tele-evangelist Jimmy Swaggart (above), at his Family Worship Center Church, is one of several popular preachers who have in recent times publicly confessed to misconduct in their private lives. But "born again" Christianity continues to attract huge numbers of followers.

Rice-Eccles Stadium, Salt Lake City, Utah (below) Fireworks herald the start of the 19th Winter Olympic Games in February 2002 and the end of three long years of preparations for the two-week event.

The United States possesses the largest economy in the world, and its people enjoy almost the highest standard of living, exceeded only by Japan. Nevertheless, not all Americans are in a position to benefit from its vast wealth.

The country's economy is dominated by the industrial and business sectors, and its agriculture, though vast, plays only a small part in wealth creation. Recent years have seen an increase in service industries at the expense of the traditional manufacturing sector. A consequent decline in the industrial northeastern states has been offset by the rise of new industries and technologies in the south and southwest.

The US has immense natural resources, yet even these cannot supply all the energy and raw materials demanded by the massive industrial output and adventurous foreign policy. As a result, recent years have been characterized by increasing trade and domestic deficits while in 2001 the economy spiralled into a recession.

Agriculture

About half the land area of the United States is used for agriculture. Production is well above domestic demand, and the country is a leading food exporter. The use of machinery and scientific farming methods means that only a small part of the labor force works on the land.

Arable farming accounts for less than half the agricultural area, but the vast tracts of land available have made the country a major world producer. Crop yields have risen, but so has the use of fertilizers, pesticides and herbicides. There are plans to reverse this trend by modifying existing government subsidies and incentives. However, measures to deal with the droughts of the late 1980s delayed the implementation of this policy. Soil erosion is a major problem in some areas of the country.

Crops vary greatly across the country on account of the vast differences in climate and topography. The main arable area is the great central basin between the Appalachians and the Rockies. This includes the Wheat Belt of the western prairies and the Corn Belt of the Central Lowlands, where maize – generally called corn in the United States – is still the principal crop. There is a wide variety of other crops, including soybeans, sunflowers and sugar beet. Cotton is grown in the Mississippi delta area, and farther

Vast orange groves California's Mediterranean climate makes it a center for truck farming: fruit, vegetables and salad crops, grown under irrigation almost all the year round, are transported by air and road to urban destinations all over the country.

west with the help of irrigation. Other crops in the south include groundnuts (peanuts), sugar cane and tobacco.

Fruit and vegetables are grown all over the country, but California is the leading producer, especially in the winter. The state's mild, dry climate is ideal for an expanding wine industry.

Livestock production varies similarly depending on the area. Dairying is the prime occupation in the north and in the mainly industrial northeast, though poultry is important here too. Farther south is a mixed farming area, but dairying is

again important in the west. Ranches and irrigated farms dominate an area extending from southern Nevada to the Rockies. In recent years the United States has found itself embroiled in trade wars with the European Union over agricultural produce. Disputes have ranged from tariff wars on bananas to European attempts to ban imports of US beef and genetically modified (GM) foods on health grounds.

Forestry and fishing

About one-third of the land area is forested, and of this area about two-thirds are commercially exploited or available for exploitation. Commercial forests are mainly in private ownership, but a significant proportion is under local, state or federal control. Between

Drilling for oil (*above*) The oil industry in the United States has led the world in developing new technology for discovering and exploiting petroleum and natural gas reserves.

The mighty Hoover Dam (*below*), also known as the Boulder Dam, is built across the Colorado river on the Arizona–Nevada border, and provides power for the three states of Arizona, Nevada and California

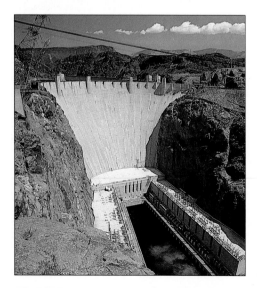

1990 and 1995 the United States planted more trees than any other country. It is the world's largest producer and exporter of roundwood (uncut timber).

The United States is among the five leading fishing nations of the world. Its catch is largely dominated by marine fish from fleets in the Pacific, Atlantic and Gulf of Mexico. Stocks are beginning to recover from the overfishing of the 1970s, but pollution remains a major threat. In 1989 a vast oil spill from the tanker *Exxon Valdez* devastated part of the shoreline of southern Alaska.

Industry and resources

The United States is rich in mineral deposits, and is a leading producer of valuable metals such as copper, lead, silver, gold and zinc. There are large accessible reserves of iron ore in the Great Lakes area, and especially around Lake Superior. There are also large copper-mining operations in several of the western mountain states. Lead mining is mainly

concentrated in southern Missouri, whereas zinc is extracted in a variety of scattered locations. Gold and silver production is limited, but many other minerals are extracted in significant quantities, notably uranium.

The United States has about one-third of the world's recoverable high-quality coal and one-eighth of its lignite. Coal deposits are largely concentrated in the Appalachian states of West Virginia, Kentucky and Pennsylvania, but there are significant fields in Illinois and Ohio. The country also possesses vast resources of natural gas and petroleum. The main petroleum and gas fields are in northern Alaska, Texas and Louisiana, but there are additional reserves off California and in some states bordering the Rockies.

Even these resources are not enough to meet demand within the United States, although the majority of the country's electricity comes from thermal power plants. The choice of alternatives is very limited. Virtually all hydroelectric

The space program

"I believe that this nation should commit itself to achieving the goal, before this decade is out, of landing a man on the moon and returning him safely to earth. No single space project in this period will be more exciting, or more impressive to mankind, or more important for the long-range exploration of space; and none will be so difficult or expensive to accomplish."

This dramatic announcement by President John Fitzgerald Kennedy (1917–63) in May 1961 marked the official beginning of the "space race" – but the starting pistol had already been fired in 1957, when the Soviet Union had launched Sputnik 1. The world's first artificial satellite had been little more than a spherical case broadcasting a simple electronically generated signal. Yet the United States Moon-landing program was partly driven by the fear that the Soviet Union might otherwise gain superiority in space. Big budgets, and the massive Saturn V launch vehicle, took the program rapidly through Earth-orbit and Moon-approach tests to the dramatic Apollo 11 mission of 1969. In July of that year, television viewers all over the world watched transfixed as Neil Alden Armstrong (b. 1930) put the first human footprints on the Moon.

Apollo 11 was the first of seven similar flights, but of these only the near-disastrous Apollo 13 mission captured the public imagination. Lunar landings were expensive, moreover, and their scientific and military value was limited. The Soviet Union, meanwhile, had made no apparent attempt to reach the Moon; instead, it was building and researching earth-orbit space stations. Skylab, the only American space station, was launched in May 1973, partly with equipment left over from canceled Moon-landing missions. It was never used as intensively as the Soviet stations, and in 1979 it fell back to Earth after five years of neglect.

The focus of the American manned space program had by now shifted to reusable vehicles. Budgets, however, were dwindling, and progress was slow. The first Space Shuttle was not launched until 12 April 1981. After four successful test flights the Shuttle came into its own as a satellite launch and repair platform, an experimental station and a military spacecraft. But on 28 January 1986 the orbital vehicle Challenger exploded just 75 seconds after the 25th Shuttle lift-off. There were no survivors, and further Shuttle flights did not resume until September 1988.

Ironically, some of the most impressive achievements of the American space program have been the least expensive ones. The unmanned Ranger Moon probes helped Apollo mission planners to select suitable landing sites. Today's communications and survey satellites support global television networks, and provide up-to-the-minute information on a range of subjects, from weather patterns to land and water resources, mineral deposits, agriculture, forestry and military maneuvers. Satellite data are also used in town planning, earthquake prediction and measurement, mapping and exploration, astronomical observation and many other widely varied applications.

Meanwhile, out in the wider solar system, Mariner spacecraft also made studies of the inner planets, and the Viking 1 and 2 missions put landing craft on the surface of Mars; there was no convincing sign of life, but the missions provided new and important information about the geology and geography of the planet. The Voyager missions discovered rings around Jupiter and Uranus, explored the rings of Saturn, and revealed a wealth of new information about the moons orbiting the outer planets.

In the 1990s highlights of the space program included the spacecraft Galileo reaching Jupiter in 1995 and Pathfinder landing on Mars in 1997. Construction began in 1998 on an international space station which will support long-term stays by US and Russian crews conducting research.

The dream comes alive The Space Shuttle, one of the most dramatic achievements yet in the United States' space program, rises in a plume of exhaust from the launch pad at Cape Canaveral, Florida.

potential has already been harnessed. Nuclear power has become increasingly unpopular since a serious accident in 1979 at Three Mile Island (Pennsylvania). Ten years later, California voters became the first to bring about the closure of a nuclear plant. The disposal of radioactive waste also presents significant problems. Aside from solar, tidal and wind power, geothermal energy is the only resource that is still largely untapped.

Manufacturing industry accounts for a further significant portion of gross domestic product. In recent decades the traditional heavy industries – aluminum, iron and steel products, machinery, cement and building materials – have been strongly challenged by the rise of many new electronics-based industries. With the so-called microchip revolution, electronics and telecommunications became the country's fastest-growing industry in the 1990s, making California's "Silicon Valley", south of San Francisco, world-famous as a high-tech industrial center. By 2000, three of the world's top five companies were American and their size related directly to the exponential growth of personal computers and the Internet. Between them, Cisco Systems, Intel and Microsoft have a market value of over $1.1 trillion.

Aerospace and associated industries were temporarily affected by a slowdown in the early 1990s. The backbone of manufacturing industry remains transportation equipment, especially automobiles, trucks and buses. Today most of this equipment – along with refined petroleum, paper, cigarettes and pharmaceuticals – comes from only a few major corporate producers. Other important manufactures include rubber and plastic products, textiles and clothing. Construction also remains a major industry, vital for providing and maintaining the vast interstate road network.

Trade and commerce

The United States handles a higher volume of trade than any other country in the world. Since the signing of the controversial North American Free Trade Agreement (NAFTA) in 1992, Canada has become the country's largest trading partner by far, overtaking the previous leader, Japan. Restrictions on United States imports by Japan, however, caused political friction between the two countries throughout the 1990s. Other major trading partners are Mexico (the third member of NAFTA), the European Union, and, increasingly, China.

The tangled skein of a freeway intersection (*above*) More than any other country in the world, the United States depends on the automobile – and is suffering for it in terms of congestion, pollution and cost.

Twin banners (*right*) Wall Street, heart of the United States' financial community, signifies its commercial might as clearly as the Stars and Stripes represents its national identity.

The United States is a major producer of fuel and raw materials, yet imports are substantially greater than exports. This is mainly due to unusually high industrial production and consumer demand. Machinery, transportation and telecommunications equipment dominate both exports and imports.

That the US was able to sustain growing current-account deficit through the 1990s indicates the pace of technology-driven expansion that the country experienced and the high level of foreign investment.

Business and finance provide the mainstay of the American economy. The country is responsible for around 15 per-

Desert storage After the terrorist attacks on the US in 2001, the downturn in the airline industry forced airlines to cut costs and store many older planes in airfields such as this one at Victorville in the Mojave Desert.

cent of world trade. In this area above all the United States exercises its influence over the rest of the world as the leading source of investment capital. Canada and Central and South America in particular are dependent on US investment.

Tourism brings with it significant revenue earnings. Visitors are attracted to coastal and mountain resorts, to major cities, and to national parks and monuments. Other major attractions include Disney-World, Disneyland and Hollywood. Service industries are growing in importance, and cities such as Washington DC and Denver, Colorado, earn most of their revenue in this way. Suburban superstores and shopping malls have changed the pattern of consumer spending, accelerating the flight from city centers to suburbs.

Transportation
The United States possesses the best-resourced transportation network in the world. The country depends for its prosperity on the efficient movement of passengers and freight across vast distances, and an emphasis on the development of its infrastructure has been a significant factor in its economic success.

The national road network has the longest mileage of any in the world. Most cities and towns are linked by a carefully planned network of interstate highways. The country is dominated by the automobile, and the majority of households own at least one vehicle. Most people travel to work in automobiles, cramming the commuter highways with rush-hour traffic from ever more distant suburbs. Pollution from exhaust fumes has forced many states to follow Oregon's lead and to introduce legal limits. Yet as automobile ownership increases, some public and city transit systems are declining.

The railroad network, too, is the largest in the world, but since the 1920s it has seen a marked decline both in mileage and in passenger traffic. Yet the railroads still carry a significant portion of the nation's freight traffic, including bulk freight, containers and "piggyback" loads (highway trailers carried on flatcars). Despite these general trends, railroads and subway systems still handle much commuter traffic in long-established cities such as Chicago, Illinois, and New York. As towns and cities face increasing congestion, some advances have been made in creating practical alternatives to the automobile. Examples include San Francisco's BART (Bay Area Rapid Transit) system and the "people movers" of cities such as Pittsburgh, Pennsylvania.

Long-distance passenger traffic within the United States is largely controlled by the airlines, though a comprehensive network of bus routes provides a cheap and easily accessible alternative. Railroads are scarcely contesting this market, as transcontinental passenger trains are a thing of the past. The twin hubs of the air transportation network are the airports at Atlanta, Georgia, and Chicago, but there are more than 800 other airports receiving scheduled flights. Thousands more serve local traffic, including private airlines and company-owned air fleets. Recent years have seen increasing financial pressure on the airlines, but this seems unlikely to diminish their importance.

The country's extensive network of rivers, canals and seaways serves more than 50 ports, and carries a significant portion of freight traffic. The busiest inland waterway is the Mississippi river, while New York and New Orleans, Louisiana, are the main ports. The Great Lakes in the north are connected by canal to New York, and are accessible to ocean-going vessels through Canada's Saint Lawrence Seaway.

Media and communications
All media in the United States are free of government control, and on controversial issues they have frequently challenged government policy. Media coverage of the Vietnam War helped bring American involvement to an end, and media investigation of the Watergate affair toppled the Nixon administration.

Television is the dominant medium, and almost every home has a television receiver. There are five major networks – NBC, ABC, CBS, Fox and PSBS – and over

1,500 commercial stations as well as 9,000 cable networks. Satellite broadcasting enables transmission to a worldwide audience. In 1997 there were 575 million radios, vastly more than in any other country, and around 12,000 commercial radio stations. Legislation of 1996 deregulated the media and telecommunications market, although there were restrictions on maximum network coverage.

There are more than 1,500 daily newspapers with a combined circulation of 56.7 million, the second highest in the world behind Japan. Newspapers have traditionally served regional or local readerships with only a few national dailies.

Prior to its dissolution in 1995, the American Telephone and Telegraph Company (AT&T) carried out most manufacturing, technology and research into communications. It has been replaced by regional private companies who operate telephone, telegraph and electronic transmissions. Some companies now provide both cable TV and long-distance telephone services, and the entire network has been expanded to include transmission by fiber-optic cable, microwave radio relay, coaxial cable and domestic satellites. There were also 86 million cellular phone subscribers in 1999, the world's highest national average.

By 2000, over 50 percent of American homes had a personal computer (PC) and 40 percent had access to the Internet. The United States led the world in the development of the technology which created the World Wide Web in the 1990s. By the end of the century the 300 million pages estimated to exist on the Web underlined the significance of the technology that had revolutionized commerce, education, and leisure equally. The possibility of interconnection between television, telephones and computers via the Internet is part of a dynamic, still-expanding field.

Health and welfare
Poverty and racial issues still affect many aspects of welfare provision, and around one-tenth of the population live below the poverty line. Welfare benefits cover unemployment and work injury, sickness and disability, old age, maternity and widowhood. They are funded both by the private sector and by government, and include cash payments and help to poor families. However, the amount of help available varies between states. Anyone who has been employed is entitled to a federal pension, but the scheme is proving difficult to fund and its future is in some doubt. Employers often provide extra

benefits, as well as health and life insurance, though employees normally contribute to such schemes.

Medical facilities are good in most areas, and people's health is generally improving. The federal government provides some subsidies for medical care, but for most people it is a commodity that has to be bought, and this produces inequalities. The mortality rate of black infants, for example, is twice as high as for white infants, and many people cannot afford health insurance of any kind. Medical facilities are good in most areas but are being strained by the growing problem of drug abuse and the incidence of AIDS/HIV infection, which is the highest in any country outside Africa. Health insurance is often denied to those infected with AIDS/HIV.

Education
Education is largely the responsibility of local and state authorities, though some schools are run by private or religious institutions. In most states it is free and compulsory from the ages of 6 to 16, and students generally stay on to complete their secondary education at 17 or 18. The United States has many world-renowned universities and institutes of higher education. These, too, are mainly state-supported, though the federal government underwrites student loans to cover course and other fees. It also provides funds for teaching materials, school meals, research grants, and education programs for Native-American children.

Federal laws have had a profound effect on many schools. For instance, the insistence on integrated schools in the 1964 Civil Rights Act led to the controversial practice of "bussing in" black children over long distances.

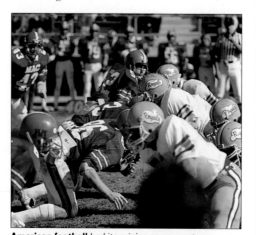

American football had its origins as an amateur intercollegiate sport. Now it is the United States' major professional sport. The campus football stars of today still rise to become the most valued players of the professional league tomorrow.

The worlds of Walt Disney

"I don't want the public to see the world they live in ... I want them to feel they're in another world."

The dawn of 17 July 1955 saw an 11-km (7-mi) traffic hold-up on the Santa Ana Freeway at Los Angeles in southern California, and there was similar congestion on other roads for miles around. This temporary chaos resulted from the vision and determination of one man – Walter Elias Disney (1901–66) – and its cause was the opening of Disney's newest venture, Disneyland.

In the early 1950s Disney was known as the man who had turned the animated cartoon into a highly profitable art form. For some people that would have been enough, but Disney wanted much more. He began to visit places such as amusement parks, zoos, circuses, museums and exhibitions, collecting ideas for what he called "a new concept in entertainment". Financial backers were hard to come by, but a Disney television series provided the necessary income.

When the first invited guests streamed into Disneyland in 1955, they found all that had been promised – and more. Adventureland offered a boat trip through a jungle landscape filled with mechanical animals. Frontierland took them back to the days of the Wild West, with scenes made famous by Disney's three films about the folk hero Davy Crockett (1786–1836), who died defending fort Alamo during the war for Texan independence. Dominating the whole scene was Sleeping Beauty's castle, the heart of Fantasyland, while Tomorrowland featured a rocket ride to the moon.

Despite a series of dispiriting problems on opening day, the world's first theme park was an unqualified success, and Disney's ambitious ideas for it continued to proliferate. The following year saw the arrival of Storybook Land and Tom Sawyer's Island. America's first daily monorail opened there in 1959, along with a scale replica of the Matterhorn in Switzerland and an imaginative undersea ride that owed something to the book *Twenty-thousand Leagues Under the Sea* (1864) by Jules Verne (1828–1905).

Walt Disney (*above*) is acclaimed by Mickey Mouse, his first and most famous creation, now known and loved by children the world over.

Fairytale magic (*right*) The floodlit towers of Cinderella's castle are a beacon for visitors escaping to the romantic fictions of Disneyland.

Disneyland, however, was only the beginning. When Walt Disney died in 1966, his plans for a new project in Florida were already well advanced. Walt Disney World finally opened near Orlando in central Florida in 1971, and the first phase was fully completed in 1975. It featured a host of new rides and attractions, including a fascinating audio-animatronic gallery of American presidents.

Then 1982 saw the opening of the Experimental Prototype Community of Tomorrow, better known as EPCOT, at nearby Lake Buena Vista. Disney had envisaged "a community of tomorrow that will never be completed ... a living blueprint of the future, where people actually live a life they can't find anywhere else today." What finally emerged was a marriage between entertainment, education and technology. Future World included a series of themed pavilions, including a time journey to the age of dinosaurs and a history of transportation. Meanwhile, the World Showcase featured pavilions in different national styles, and Spaceship Earth provided a ride through the history of communication.

Other highly successful theme parks have opened in Tokyo and near Paris, France, while Hong Kong Disneyland is expected to open in 2005. The company earned $25 billion in 2000 and is a major producer of home videos in addition to its successful films and TV shows. Future expansion is likely to include e-commerce and delivery of pay-per-view movies via the Internet.

The States of the Union

THE STATES OF THE UNITED STATES OF AMERICA

Pacific Coast
Alaska
Washington
Oregon
California
Hawaii

Mountain Region
Idaho
Nevada
Montana
Wyoming
Utah
Colorado

The Southwest
Arizona
New Mexico
Oklahoma
Texas

The Middle West
North Dakota
South Dakota
Nebraska
Kansas
Minnesota
Iowa
Missouri
Wisconsin
Illinois
Michigan
Indiana
Ohio

New England
Maine
New Hampshire
Vermont
Massachusetts
Connecticut
Rhode Island

Middle Atlantic Region
New York
Pennsylvania
New Jersey
Delaware
Maryland
District of Colombia

The South
Arkansas
Louisiana
Kentucky
Tennessee
Mississippi
Alabama
West Virginia
Virginia
North Carolina
South Carolina
Georgia
Florida

Sunset on the Grand Canyon, the spectacular gorge cut by the Colorado river through the mountains of northwestern Arizona. Its majestic beauty attracts millions of visitors each year.

States of the United States

- ■ national capital
- ● state capital

CONN	CONNECTICUT
DC	DISTRICT OF COLUMBIA
DEL	DELAWARE
MD	MARYLAND
MASS	MASSACHUSETTS
MISS	MISSISSIPPI
NJ	NEW JERSEY
RI	RHODE ISLAND
W VA	WEST VIRGINIA

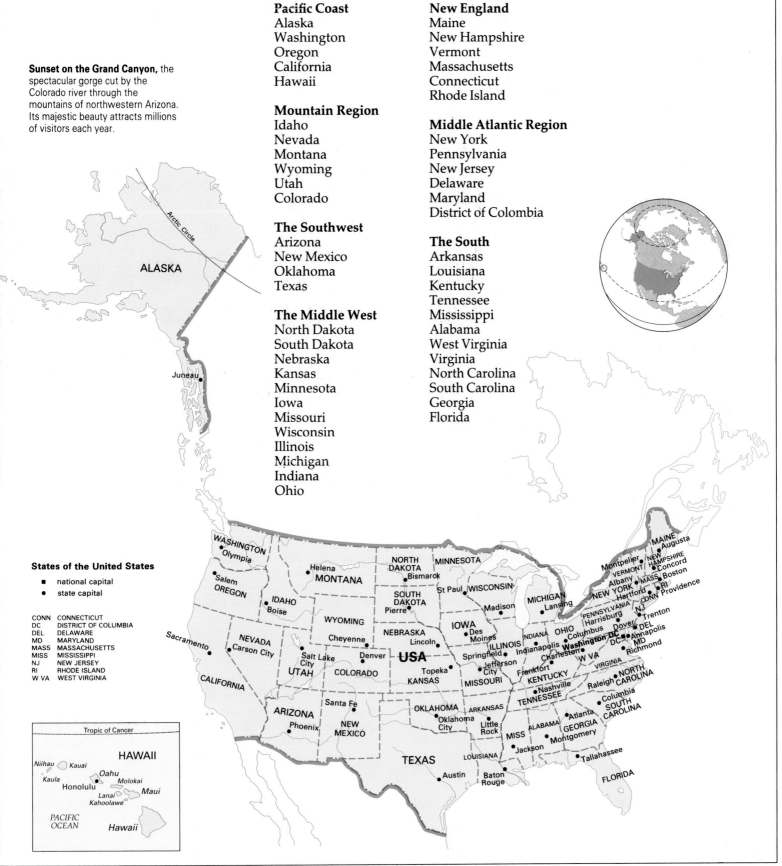

ALASKA

Nickname	The Last Frontier
Joined the Union	1959
Capital	Juneau
Area	1,530,693 sq km (591,004 sq mi)
Population	670,053
Outstanding features	Mount McKinley, Glacier Bay National Park
Principal industries	gas and petroleum mining, tourism, fishing

Alaska is America's largest state, and the newest apart from Hawaii. It is a wild and often mountainous country, lying at the northwestern tip of the continent, facing Russia across the Bering Strait.

It was across this Strait that the first peoples are thought to have entered the Americas some 30,000 years ago. Their descendants, the Inuit and Aleut peoples, still inhabit the remoter parts of Alaska. In the 18th century the first European settlers also arrived from Russia, and Alaska was controlled by the Russian-American Company until United States Secretary of State William Henry Seward (1801–72) negotiated its purchase from Russia in 1867.

Since then there has been a steady stream of immigrants, attracted by its vast mineral wealth and by huge projects such as the Trans-Alaska Oil Pipeline. Towns such as Anchorage in the south and Fairbanks in the center have expanded to become considerable conurbations. However, most of the state remains a vast and magnificent wilderness.

WASHINGTON

Nickname	Evergreen State
Joined the Union	1889
Capital	Olympia
Area	176,479 sq km (68,139 sq mi)
Population	6,395,798
Outstanding features	Mount Rainier (Cascade Range), Puget Sound
Principal industries	aircraft, wood and paper products, foodstuffs, information technology

During the 16th century European navigators were searching for a Northwest Passage to Asia. Their search was fruitless – but they did discover the existence of America's Pacific Northwest.

At first the area that is now Washington state was part of the Oregon Country – the whole of the Pacific Northwest – whose northern boundary was set in 1846 along the present Canadian frontier.

From the 1830s to the 1850s, white settlers moved into the area from the Midwest along the Oregon Trail. Washington finally became a separate territory in 1853, benefiting from the demand for construction materials created by the California gold rush. Railroad construction in the 1880s brought rapid growth, and during the 1890s Washington was once again a base for miners and prospectors, this time heading for Alaska and the Klondike. The 20th century saw the damming of the Columbia river to provide power, irrigation and improved navigation.

Washington's economy has been largely dependent on natural resources, such as forestry. However, in recent years, Seattle has become a major economic center, being the home of successful high-tech and software companies. The state is also a center for the aerospace industry, and there are numerous military establishments along the shores of Puget Sound.

OREGON

Nickname	Sunset State
Joined the Union	1859
Capital	Salem
Area	251,418 sq km (97,073 sq mi)
Population	3,700,758
Outstanding features	Mount Hood (Cascade Range), Crater Lake National Park, Columbia River Gorge
Principal industries	timber products, machinery, foodstuffs

Oregon has a dramatic landscape of high, densely forested mountain ranges, with plateaus and plains cut by deep valleys and gorges. The forests of the west and northeast provide much of its wealth, while in the east are arid deserts.

In the 16th century the area was claimed by both Britain and Spain. However, the large and diverse Native-American population that inhabited the area remained largely undisturbed until 1792,

Mighty mountains (*above*) The jagged flanks of the Cascade Range in the state of Washington were raised by movements of the Earth's crust, which also cause earthquakes or volcanoes in all the states along the Pacific coast.

Northern wilderness (*left*) Alaska's lakes and wetlands, havens for wildlife, are among Earth's last natural wildernesses. Seen here in the background, Mount McKinley, the tallest peak in North America, lies within Denali National Park, which has been declared a Biosphere Reserve.

CALIFORNIA

Nickname	Golden State
Joined the Union	1850
Capital	Sacramento
Area	411,047 sq km (158,706 sq mi)
Population	36,457,549
Outstanding features	Yosemite National Park (Sierra Nevada), Death Valley, Lake Tahoe
Principal industries	foodstuffs, aircraft, tourism, microelectronics, information technology

when Robert Gray (1755–1806) entered from the west via the Columbia river. Fur traders soon arrived from the east, and in 1811 John Jacob Astor (1763–1848) paved the way for white settlement when he established the trading post of Astoria. Subsequent decades saw a growing tide of settlement as pioneers arrived along the Oregon Trail; most settled in the Williamette Valley between the Cascade and Coast Ranges. In 1842 Britain abandoned its claim to Oregon, which later became the 33rd state of the Union.

Oregon depends largely on farmland and forestry for its prosperity, and was among the first states to introduce stringent control of vehicle exhaust emissions. In 1980 ash from the volcanic eruption of Mount Saint Helens in the neighboring state of Washington caused massive environmental damage. The state economy, however, has become more diversified in recent decades, and suffered less than might otherwise have been expected. Tourism is an increasingly important activity in the state.

Portland, Oregon, is the state's largest city, principal port and also its cultural and educational center. It grew up in the mid 19th century, boosted by its position as the main supply port to the west coast gold rushes.

California is the most populous of the United States, and its economy, though facing severe difficulties throughout 1994, can stand comparison with that of the world's largest nations. It has the world's heaviest concentration of motor vehicles, its most extensive freeway system and a consequent pollution problem – yet until the late 18th century it remained an unspoiled wilderness.

In 1542 the Spanish navigator Juan Rodríguez Cabrillo (d.1543) became the first European to see the coast of California. The first official Spanish expedition arrived in 1769. In the very same year Junípero Serra (1713–84), a Spanish priest, set up a Franciscan mission at Monterey on the rugged coast of central California.

In 1821 Mexico won its independence from Spain, and within 20 years Mexican settlers had taken over the mission ranches, expelling and exploiting the Native-American population. However, the area had also begun to attract white American settlers. In 1848, after a brief war with Mexico, the territory was ceded to the United States. Only a year later the discovery of gold to the east of Sacramento started a further massive influx of prospectors and settlers.

The state remained loyal to the Union during the Civil War. Waves of immigration from the east brought farmers to the inland valleys and industries to the growing cities. Hollywood, near Los Angeles, became the film and television capital of the world. California's population, which doubled on average once every two decades between 1860 and 1960, is ethnically very mixed, with large populations of Spanish-speakers, Chinese and Vietnamese, Native-Americans and African-Americans. More than 90 percent are urban dwellers.

Today California is not only the prime agricultural state; it also has enormous reserves of petroleum, a thriving industrial sector and a massive income from tourism. Silicon Valley, situated south of San Francisco Bay, has become famous in the field of information technology. The San Andreas Fault runs through to the north of San Francisco, making the area vulnerable to earthquakes. Movement along the fault caused the disastrous San Francisco earthquake of 1906 and that in southern California in January 1994.

HAWAII

Nickname Aloha State	
Joined the Union 1959	
Capital Honolulu	
Area 16,760 sq km (6,471 sq mi)	
Population 1,285,498	
Outstanding features Mauna Kea (the world's highest island peak), Mauna Loa	
Principal industries tourism, foodstuffs, defense industries	

Hawaii is America's most far-flung state, and perhaps its most unusual. It comprises a chain of steep volcanic islands in the Pacific some 3,800 km (2,400 mi) southwest of the Californian coast. The state is named after the largest island.

The islands were first settled about 1,500 years ago by Polynesian peoples. The first European visitor was the British explorer James Cook (1728–79) in 1778, who called them the Sandwich Islands. In about 1795 they were united for the first time under the rule of the formidable warrior monarch Kamehameha I (1758–1819). However, European settlement progressed, and in 1851 his successor Kamehameha III (1814–54) placed the islands under United States' protection. The native population declined considerably owing to introduced diseases among other causes; many new settlers

San Francisco's Golden Gate Bridge (*top*) frames the leisurely sailing craft in the bay, and the city's glittering skyline – a reminder of the fierce commercial drive behind California's easy-going lifestyle.

Icons of Hollywood culture (*above*) A movie star's hand and footprints left in the famous stretch of sidewalk outside Graumann's Chinese Theater, Los Angeles.

Hawaii's golden beaches (*right*) attract a multitude of vacationers to the island, to swim, surf or sunbathe on the beaches. Inland, other attractions include spectacular lava formations and tropical bird life.

arrived, especially from eastern Asia. Today there are more Hawaiians of Japanese than of Polynesian descent.

American influence grew until in 1900 the islands were finally annexed by the United States. A huge naval base was established at Pearl Harbor, and in 1941 this became the target of the Japanese attack that brought the United States into World War II. Hawaii was by now so firmly American in character that incorporation as the 50th state was inevitable.

Hawaii supports several industries, notably sugar refining; but its greatest asset is its natural beauty. The climate combines tropical sunlight with mild, temperate weather. Rainfall can be fierce, but it supports the lush forests that cover the slopes of the spectacular volcanic mountains, many of which are still active. Clear air and cloudless skies make their upper slopes an ideal site for astronomical observatories. Not surprisingly, the islands are a great tourist attraction, giving millions of mainland Americans each year a taste of Polynesia.

IDAHO

Nickname	Gem State
Joined the Union	1890
Capital	Boise
Area	216,430 sq km (83,564 sq mi)
Population	1,466,465
Outstanding features	Hell's Canyon, Craters of the Moon National Monument, Shoshone Falls
Principal industries	foodstuffs, timber processing, chemicals

Idaho lies in the northwest of the United States, its short northern border touching the Canadian border. Its center is dominated by several of the Rocky Mountain ranges, and the south by the strange lava landscapes of the Snake River Plain.

Native-American peoples such as the Shoshoni have lived here as hunter–gatherers for at least 10,000 years, and their descendants still inhabit the area. Idaho was originally part of Oregon Country, and was obtained by treaty from Great Britain in 1846. It was first settled by gold prospectors, and later by Confederate refugees after the Civil War. Many Mormons also settled here, fleeing persecution for their religion in the east; they did much to build up the state, and still form the majority in the southeast.

Idaho has a small but growing manufacturing base, but remains a strongly agrarian area, famous for its vast potato crop, its wheatfields and its lumber.

NEVADA

Nickname	Silver State
Joined the Union	1864
Capital	Carson City
Area	286,352 sq km (110,561 sq mi)
Population	2,495,529
Outstanding features	Boundary Peak, Carson Sink, Lake Mead (Boulder Canyon)
Principal industries	tourism, gold and mineral extraction, electronics, chemicals

Nevada lies east of the Sierra Nevada mountains (after which it is named). It is a state of surprising contrasts. The land is largely arid, unpromising and sparsely populated, and yet Nevada is one of the most prosperous states – for reasons that are often controversial.

Native-American peoples have lived here for more than 20,000 years. The first European settlers were Spanish missionaries and Canadian fur traders in the 18th and early 19th centuries. They were followed by American explorers such as John Charles Frémont (1813–56), who opened up the west to settlers and prospectors. In 1848 the area was ceded by Mexico to the United States. The Comstock Lode discovery in 1859 caused a silver rush, attracting settlers in great numbers. However, monetary reforms in the 1870s produced a dramatic slump in the silver industry, and the cattle ranching that replaced it declined shortly afterward, leaving large numbers of empty ghost towns. Mining and ranching revived early in the 20th century, only to be wiped out again during the Great Depression of 1929.

In the early 1930s, however, the state government passed the legislation that

A gambler's paradise Las Vegas began as a Mormon settlement in the 1850s, later becoming a small agricultural town in the desert. Now, as a result of the state's liberal gaming laws, it attracts millions of visitors each year to its casinos.

legalized gambling and prostitution, and minimized the statutory requirements for marriage and divorce. This led to an influx of investment from elsewhere in the United States; major resort centers grew up in the scenic areas around Las Vegas in the south and Reno in the west, offering luxurious hotels, night life and casinos. The resulting boom in tourism brought immediate prosperity to the area. Meanwhile, more orthodox industry was boosted by cheap hydroelectricity from the Hoover (Boulder) Dam on the Colorado river, and by the discovery of further mineral deposits. Most of the population remains crowded into the Las Vegas and Reno areas, leaving vast open spaces that have proved ideal for military and civil defense installations, including nuclear testing sites.

MONTANA

Nickname	Treasure State
Joined the Union	1889
Capital	Helena
Area	380,847 sq km (147,046 sq mi)
Population	944,632
Outstanding features	Granite Peak (Rocky Mountains), Glacier National Park
Principal industries	foodstuffs, copper and petroleum extraction, timber processing

In 1862 gold was discovered at Bannack in the northern Rockies, attracting a wave of prospectors and opportunists. Two years

Proud of the past (*above*) This reconstruction of a Wild West frontier town in Montana preserves a sense of the self-sufficiency of these isolated pioneer communities, where the bar and the bath-house stand side by side.

Reflected glory (*right*) Dark pines and golden aspens reach toward the snow-capped peaks of the majestic Maroon Bells, which tower above Maroon Lake near the famous mountain ski resort of Aspen, Colorado.

later the Montana Territory was created, and the first cattle ranchers moved in to graze their huge herds on the vast expanses of gently rolling land in the east of the territory.

The Native-Americans fought hard to defend their hunting grounds. Names such as Sioux and Cheyenne have become legendary, as have the battlefields where they won their last victories – Little Bighorn (1876) in southeastern Montana, and the Bighole Basin (1877) in the southwest. But shortly afterward the Native-Americans were defeated by United States troops, and this opened the way for white settlement.

The United States had acquired the territory as part of the Louisiana Purchase in 1803, but until the gold rush of 1862 little had been done to exploit it. With the Native-American population very greatly reduced, settlers turned their attention from gold to the massive local copper deposits. Offers of land brought a wave of immigrant homesteaders from northern Europe from 1909 onward, but the severe droughts of the 1920s ruined the livelihoods of many of the new farmers.

Agriculture and mining are among the mainstays of the economy. The chief crops are wheat, barley and sugar beet, but numerous minerals are extracted, apart from copper – notably lead, silver and phosphates. A growing number of tourists are attracted to the state's dramatic scenery and historic battle sites.

WYOMING

Nickname	Equality State
Joined the Union	1890
Capital	Cheyenne
Area	253,324 sq km (97,809 sq mi)
Population	515,004
Outstanding features	Yellowstone and Grand Teton National Parks
Principal industries	petroleum, gas and coal extraction, tourism, foodstuff

When the United States bought this area from France as part of the 1803 Louisiana Purchase, it was still a mountainous wilderness inhabited by the Arapaho and Shoshoni peoples. For many years the arid prairies and mountains discouraged settlement, and Wyoming was little more than a staging point for people heading farther westward along the Oregon and Overland Trails. The arrival of the telegraph and railroad gave rise to retaliatory raids by the Arapaho and Shoshoni, whose lands were taken and hunting grounds disturbed. By 1869, however, most of the Native-Americans had been confined to the Wind River Reservation on the eastern slopes of the Rockies.

Cattle ranchers were quick to move in once the railroads had arrived, and large areas of the state are still pastureland. However, mineral fuel production is the dominant industry, and easily available uranium and hydroelectricity have made Wyoming a leading producer of power. Wyoming was the first state to create national parks, and tourism has benefited correspondingly. It was also the first state to give women the vote – hence its nickname Equality State.

were resettled on reservations. By the 1890s the Mormons had established a solid, stable community, based largely on agriculture and small-scale cooperative businesses. A disapproving United States Congress refused successive statehood applications until 1896, when polygamy was officially renounced and the Mormon Church relinquished its political status. Its social dominance, however, remained, and its traditions of the work ethic and mutual assistance were a major factor that encouraged the state's industrial and commercial growth.

Today the Mormon influence endures, liberalized by pragmatic "revelations" such as on the admission of black people into the church. Utah's popular tourist centers, and the prosperity of its capital, Salt Lake City – a major financial and commercial center – are living testimonies to its worldly success.

COLORADO

Nickname	Centennial State
Joined the Union	1876
Capital	Denver
Area	269,594 sq km (104,091 sq mi)
Population	4,753,377
Outstanding features	Mount Elbert (Rocky Mountains), Black Canyon, Garden of the Gods
Principal industries	microelectronics, foodstuffs, machinery

Colorado, lying astride the Rockies, is the archetypal Western state. It was home to various Native-American peoples, ranging from the Mesa Verde cliff-dwellers to Plains peoples such as the Cheyenne and Arapaho. These were later overrun by the Spanish, who named the area Colorado ("red" or "ruddy") after its reddish rocks.

The eastern side, annexed by France, was sold in the Louisiana Purchase in 1803; the west passed from Mexico to the United States in 1848. The 1859 gold rush brought an immense influx of settlers, creating numerous frontier towns. Their need for food stimulated the development of irrigated agriculture, and cattle and sheep ranching replaced the exterminated buffalo. In the late 19th century discoveries of rich mineral reserves made the new state a major industrial center.

In recent years, Colorado's older industries have suffered setbacks, but have to a certain extent been replaced by modern computer and aerospace industries. The awesome scenic beauty of the Rockies, and the ideal skiing conditions at resorts such as Aspen, support a fast-expanding tourist industry.

UTAH

Nickname	Beehive State
Joined the Union	1896
Capital	Salt Lake City
Area	219,887 sq km (84,889 sq mi)
Population	2,550,063
Outstanding features	Great Salt Lake, Canyonlands National Park, Lake Powell
Principal industries	armaments, electronics, foodstuffs, tourism

The landscape of Utah is generally barren, arid and inhospitable, but it has nevertheless nurtured some hardy peoples and distinctive cultures. Among the earliest of these, about 400 AD, were the Anasazi or Pueblo peoples. Their society decayed long before the arrival of Spanish explorers in the 18th century, but their awesome cliff villages survive, along with pottery and rock paintings. Their place was taken by the Navajo and Ute peoples, many of whom still live in the state.

In 1848 the United States annexed the area from Mexico. At about this time, members of the Church of Jesus Christ of Latter-day Saints, or Mormons, were forced to flee from persecution, sparked off by their unconventional practices, in particular that of polygamy. In 1847 Brigham Young (1801–77), claiming a divine mission to transform the desert into a paradise, led wagon trains of Mormons into the Great Salt Lake Valley. With hard labor and missionary zeal, vast tracts of land were irrigated, and Young's vision was accomplished in less than 15 years.

The Native-Americans were seldom directly persecuted, but their food supplies suffered; clashes followed, and they

ARIZONA

Nickname	Grand Canyon State
Joined the Union	1912
Capital	Phoenix
Area	295,259 sq km (114,000 sq mi)
Population	6,166,318
Outstanding features	Grand Canyon, Painted Desert, Petrified Forest National Park
Principal industries	electronics, tourism, copper mining

Arizona is one of America's youngest states, but one of the oldest inhabited. Native-American peoples such as the Anasazi and Hohokam flourished in this arid but beautiful area for at least 25,000 years. They were succeeded by the Apache and Navajo not long before the Spanish came. The area remained Spanish until 1848, when it was ceded to the United States; further territory was added by the Gadsden Purchase of 1853. It is in Arizona that many of the legends of the Wild West are rooted; Tombstone, for example, was the site of the notorious gunfight at the OK Corral.

During the 1970s Arizona's population was nearly doubled by immigrants from other states, drawn by the sunny climate and employment opportunities in cities such as Tucson and Phoenix. Most areas, however, remain sparsely populated. Native-Americans are more numerous than in most other states, but are heavily outnumbered by Spanish-speakers.

Arizona's spectacular scenery, with its great mountains rising abruptly above the plains, is a major asset, and its many national parks form the basis for a thriving tourist industry.

NEW MEXICO

Nickname	Land of Enchantment
Joined the Union	1912
Capital	Santa Fe
Area	314,924 sq km (121,593 sq mi)
Population	1,954,599
Outstanding features	Wheeler Peak (Rocky Mountains), Carlsbad Caverns
Principal industries	mining, tourism, foodstuffs

New Mexico has a long history of habitation. Ruined cliff dwellings give evidence of the Pueblos' agricultural civilization; the nomadic Navaho and Apache peoples probably entered the area only a few centuries before the Spanish. In 1540 Francisco Vásquez de Coronado (1510–54) came here in search of the legendary "seven cities of gold", but left empty-handed. Early in the 17th century, Juan de Oñate (1550–1630) built the first European settlements, including Santa Fe in 1610. The colony was later destroyed by Native-Americans defending their way of life from the white settlers. In the 18th century Spain regained control, and in 1821 the area became part of Mexico.

The eastern section was annexed by the United States in 1845. During the subsequent Mexican War the whole area was occupied by the Army of the West, before Mexico formally ceded it to the United States in 1848. Conflict with the Navaho and Apache peoples fueled the legends of the Wild West. But the arrival of the railroad in 1880 brought a massive wave of immigration by white settlers.

During World War II the atomic bomb was developed at Los Alamos to the northeast of Santa Fe, and scientific research remains an important activity. Agriculture is limited by the low rainfall, but cattle ranching is still important. More than one-third of the population are of Spanish origin, about one-half of them Mexican-Americans, and Spanish is the state's second official language. The state attracts many tourists, and there is a thriving market for Native-American and Mexican craft souvenirs, such as pottery, blankets and silver jewelry.

OKLAHOMA

Nickname	Sooner State
Joined the Union	1907
Capital	Oklahoma City
Area	181,185 sq km (69,956 sq mi)
Population	3,579,212
Outstanding features	Black Mesa (Panhandle), Wichita Mountains
Principal industries	machinery, petroleum extraction, metals

The area of Oklahoma made up part of the Louisiana Purchase, sold by France to the United States in 1803. In 1828 Congress created an Indian Territory here, reserved

Monument Valley, Arizona (*left*), its buttes and mesas sculpted by nature from red sandstone. This stark but beautiful land is the ancestral home and present-day reservation of the Navajo people.

The Alamo fortress (*right*) at San Antonio, Texas, a small shrine to an immense legend. It was here, in 1836, that a tiny band of volunteers, notably Davy Crockett (1786–1836), sacrificed their lives to defend Texan independence.

A florid facade (*below left*) belonging to the missionary church of San Xavier del Bac at Tucson, Arizona. Established by Spanish Jesuits in 1700, this mission has served the Pagago Native-Americans down to the present day.

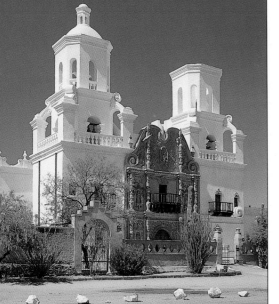

exclusively for Native-Americans and consisting of five republics or nations. It became known as Oklahoma, from the Choctaw words meaning "red people". In the Civil War the Native-Americans, unhappy with the federal government, sided with the Confederate army, and suffered accordingly. In 1889 Congress started to allow homesteaders to open up the Indian Territory land for farming, and Oklahoma was swamped by a series of "land runs". Today only a minority of the population are Native-Americans. In 1995 the terrorist bombing of a federal building in Oklahoma City took 168 lives; two local men were later convicted of the bombing.

A mining boom brought other settlers, but farming remained the most important occupation of the new state. However, many farmers lost their land and their livelihood through drought in the 1930s, but new prosperity came with the discovery of massive petroleum reserves. Modern Oklahoma remains primarily industrial, though less dependent purely on its extensive mineral resources.

TEXAS

Nickname	Lone Star State
Joined the Union	1845
Capital	Austin
Area	691,027 sq km (266,807 sq mi)
Population	23,507,783
Outstanding features	Mount Livermore, Guadalupe Peak, Big Bend National Park
Principal industries	petroleum, machinery, foodstuffs, clothing

Texas is the largest of the 48 coterminous states. Its sheer size and its distinctive character have done much to shape international perceptions of the United States as a whole. The wide open alluvial plains extend inland from the Gulf of Mexico, rising to meet the Great Plains to the north. The southwestern border with Mexico is formed by the waters of the Rio Grande, which flows down through the arid hill country of the west.

The first Spanish explorers found the area only sparsely inhabited by Native-American peoples. It was claimed by the United States following the Louisiana Purchase of 1803, but remained under Spanish rule. A coup in Mexico in 1833 drove the Texans, led by Anglo-American settlers, to stage a rebellion and declare independence. In 1836, following the famous siege of the Alamo at San Antonio in central-southern Texas, the Texans finally routed the Mexicans at San Jacinto in the east. Nine years later the Republic of Texas was annexed by the United States, and Mexican opposition was finally crushed in 1847.

Texas seceded from the Union in 1861, and came under martial law during the Reconstruction. In the 1870s and 1880s the Native-American population was forcibly expelled; railroad construction opened up the state to swift settlement, and a boom in cattle ranching created an image of Texas as the archetypal "Wild West" state.

The discovery of petroleum in 1901 began an even greater boom, in which Houston and Dallas in the east became two of the most affluent cities in the world. A wide range of other industries followed, making Texas a leading manufacturing center. Meanwhile, agriculture has retained its importance, especially in the areas of cotton and livestock production. The boom has slackened during the various periods of economic depression, but the economy has always revived. Recent years have seen a growth in the important electronics and consumer-product sectors.

NORTH DAKOTA

Nickname	Flickertail State
Joined the Union	1889
Capital	Bismarck
Area	183,117 sq km (70,702 sq mi)
Population	635,867
Outstanding features	North Dakota Badlands, Missouri River, Lake Sakakawea
Principal industries	foodstuffs, petroleum extraction, agricultural machinery

North Dakota is a leading agricultural state, the cool, subhumid climate of its northerly location creating conditions that are ideal for growing spring wheat and for cattle ranching.

The first European to arrive here was the French Canadian Pierre Gaultier de Varennes, sieur de La Vérendrye (1685–1749), in about 1740. Canadian traders were soon making regular visits. In 1803 the French sold their interest in the area to the United States as part of the Louisiana Purchase. White traders were quick to exploit the Native-Americans, who became dependent on the goods that the traders supplied, as well as highly vulnerable to diseases such as smallpox that they introduced. Gold prospectors traveled northwest along the Missouri river, their encroachment onto Native-American land causing serious conflict.

The arrival of the railroads in 1871 brought a massive wave of settlement and an agricultural boom. In 1889 the Dakota Territory was divided, and both North and South Dakota joined the Union. Farming remained the dominant interest, protected first by political means, and later through a cooperative movement.

Agriculture still dominates today's economy, and much of the industry is farming related. However, North Dakota is now a leading producer of petroleum and hydroelectricity.

SOUTH DAKOTA

Nickname	Coyote State
Joined the Union	1889
Capital	Pierre
Area	199,730 sq km (77,136 sq mi)
Population	781,919
Outstanding features	Mount Rushmore (Black Hills), Badlands National Park
Principal industries	foodstuffs, machinery, electronics

South Dakota was separated from North Dakota in 1889, 15 years after gold had been found in the west. The discovery had been made in the Black Hills – an area forbidden to Europeans by a treaty with the Teton Dakota peoples. But the Native-Americans were soon routed in battle, and in 1877 they were forced to cede the area. Before long, freight and stage lines into nearby Rapid City were bringing a steady flow of prospectors and settlers from the east. Between 1878 and 1887 the local population quadrupled; as a result, southern Dakota demanded, and ultimately achieved, separate statehood.

The Black Hills offered pastureland as well as gold, and more land became available farther west in 1889. Cattlemen came first, but homesteaders broke up the cattle range in the early 1900s, taking another huge parcel of Native-American land in the process. Since then, drought and depression have taken their toll, but farming remains the chief source of income. However, other sectors of the economy are now expanding, notably manufacturing and tourism.

KANSAS

Nickname	Sunflower State
Joined the Union	1861
Capital	Topeka
Area	213,096 sq km (82,277 sq mi)
Population	2,764,075
Outstanding features	Castle Rock, Horse Thief Canyon
Principal industries	foodstuffs, aircraft, agricultural machinery

Kansas lies at the geographical heart of the United States. At the time of the dinosaurs it was the bed of a shallow inland sea, and its landscape remains predominantly level today.

Until the arrival of Europeans, the Native-American peoples lived by hunting the vast buffalo herds that roamed the prairies. However, by the 1850s these had been largely exterminated, both to starve the Native-Americans and so that their hides could be exported to the east. The state of Kansas was part of the area that had been bought in the Louisiana Purchase of 1803, and later became a battleground over the slavery issue.

At first the state relied largely on cattle ranching, but hardy new wheat strains, introduced by Mennonite immigrants from Russia, opened up the vast wheatfields – many of which, however, became the infamous overfarmed "dustbowls" of the 1930s. Kansas nonetheless remains the leading wheat state and a major meat producer. Industry is important, but is largely agriculturally based.

The ages of the Earth (*above*) are revealed in the exposed rock layers of South Dakota's badlands. The fantastic rock formations and deep gullies of this extensive, elevated arid area were created by occasional heavy downpours – normal rainfall being insufficient to support a protective covering of grass.

Mount Rushmore (*right*) in the Black Hills of South Dakota preserves in monumental granite relief the faces of four American presidents: (from left to right) the first president, George Washington (1732–99); the third, Thomas Jefferson (1743–1826); the 26th, Theodore Roosevelt (1858–1919); and the 16th, Abraham Lincoln (1809–65). They each measure about 18 m (60 ft) in height.

MINNESOTA

Nickname	Gopher State
Joined the Union	1858
Capital	Saint Paul
Area	224,329 sq km (86,614 sq mi)
Population	5,167,101
Outstanding features	Lake of the Woods, Minnehaha Falls (Minneapolis)
Principal industries	foodstuffs, forestry, iron-ore extraction, machinery

Minnesota's landscape was once scoured by glaciers, which left it dotted with lakes, and covered its extensive, rolling prairies with fertile soils.

The area was originally inhabited by the Chippewa and Sioux peoples, and was acquired by the United States in the Louisiana Purchase of 1803. Settlers began to arrive in the 1820s, most of them New Englanders drawn by the great wealth of the immense hardwood forests. The two main settlements either side of the Mississippi river were to grow into the great Twin Cities of Minneapolis and Saint Paul.

Later immigrants, mostly from Scandinavia and Germany, began to farm the rich prairies, and iron-ore mines in the northeast formed the basis for industrial growth. In the 20th century, new large-scale farming methods opened up the vast grainfields on which Minnesota's prosperity is founded.

Trade and manufacturing are concentrated around the Twin Cities of Saint Paul and Minneapolis. Both are also major cultural and tourist centers, with their parks, colleges and museums.

IOWA

Nickname	Hawkeye State
Joined the Union	1846
Capital	Des Moines
Area	145,752 sq km (56,275 sq mi)
Population	2,982,085
Outstanding features	Okoboji Lakes, Effigy Mounds National Monument
Principal industries	foodstuffs, agricultural machinery, electronics

The Midwestern state of Iowa lies between the Mississippi and the Missouri rivers at the very heart of the North American continent. Ice-age glaciers once scoured its rolling landscape, but left rich soils behind them.

The original inhabitants of the area were Sioux and other Native-American peoples, whose Plains cultures reached their height after the horse had been introduced from Europe. The area was first claimed by the French, but was part of the vast Louisiana colony that was purchased by the United States in 1803. The eastern part of the state was bought from Native-Americans in the 1830s, opening the way for settlers from the east, and later from Central Europe; the German and Czech communities retain much of their traditional identity today.

Iowa is nowadays almost exclusively agricultural, producing a vast amount of livestock and feed crops annually. It is by no means a backwater, however, with modern cities such as Des Moines and Sioux City, and prestigious educational institutions such as the University of Iowa in Iowa city.

NEBRASKA

Nickname	Cornhusker State
Joined the Union	1867
Capital	Lincoln
Area	200,349 sq km (77,355 sq mi)
Population	1,768,331
Outstanding features	Platte and Missouri Rivers
Principal industries	foodstuffs, machinery, electronics

Nebraska, lying at the heart of the prairies, is not only a major agricultural state, but also forms an important transportation and communication link between east and west.

In 1803 the United States acquired the territory from France as part of the Louisiana Purchase. American explorers soon arrived along the Missouri river, opening a path for the fur trade. By the 1840s the Platte Valley had become a gateway to the west for thousands of pioneers. Railroad developers soon recognized Nebraska's agricultural potential, and 1869 saw the completion of the Union Pacific Railroad.

Nebraska's population spread rapidly after it achieved statehood, pushing back Native-American resistance as settlers migrated westward. Agriculture became depressed in the 1890s, then boomed in the early 20th century, only to decline once again in the Great Depression in 1929. However, it is still the dominant activity today, supplying raw materials for industries such as food processing and chemicals. The Platte Valley also retains its historic importance as a major artery for trade, transportation and various means of communication.

MISSOURI

Nickname	Show Me State
Joined the Union	1821
Capital	Jefferson City
Area	180,514 sq km (69,697 sq mi)
Population	5,842,713
Outstanding features	Ozark Plateau, Missouri River
Principal industries	foodstuffs, aircraft, motor vehicles, tourism

Throughout its history, Missouri has taken a middle line, straddling the divide between timberland and prairie, cornfields and cotton fields, slave-owners and abolitionists, east and west.

Named after a Native-American people, Missouri's first European settlement was Sainte Genevieve on the west bank of the Mississippi, established by French hunters and lead miners in about 1735. In 1803 the area was bought by the United States as part of the Louisiana Purchase. At first the population was largely French, but American settlers soon began to dominate the area. In 1821 Missouri joined the Union as a slave state, following the controversial Missouri Compromise produced by Congress to regulate the spread of slavery in the western territories. For many pioneers it became a gateway to the west. Officially Missouri stayed within the Union during the Civil War, but many Missourians fought on the Confederate side. Slavery was abolished in 1865, but racial discrimination was still very strong until the 1960s.

Since the late 19th century there has been a steady economic growth based on the great variety of local resources, including metal ores, forestry and agriculture. Recent decades have seen an increase in the financial and services sectors, and the establishment of a thriving aerospace industry.

WISCONSIN

Nickname	Badger State
Joined the Union	1848
Capital	Madison
Area	171,496 sq km (66,215 sq mi)
Population	5,556,506
Outstanding features	Lake Winnebago, Door Peninsula (Lake Michigan), Apostle Islands (Lake Superior)
Principal industries	foodstuffs, machinery, metals, wood and paper processing

Wisconsin was once an area of wild forests inhabited by Native-American peoples such as the Winnebago. Now, on the other hand, it has become a more urban and industrialized state.

In 1634 the French explorer Jean Nicolet (1698–1742) was the first European to visit the area, but no permanent settlements were established until 1717. In 1763 the whole area came under British control. The British in turn ceded the territory to the United States in 1783. Only four years later it became part of the Northwest Territory, and in 1800 it formed part of the Indiana Territory.

The early 19th century saw a steady flow of immigrant settlers, mostly from northwest and central Europe. The Black Hawk War of 1832 shattered local Native-American power, and in 1836 a separate Wisconsin Territory was created; statehood followed 12 years later.

Wisconsin is a leading supplier of dairy products, but manufacturing and processing are equally important. Service industries, especially tourism, are also a significant factor in the economy.

ILLINOIS

Nickname	Prairie State
Joined the Union	1818
Capital	Springfield
Area	149,885 sq km (57,871 sq mi)
Population	12,831,970
Outstanding features	Illinois and Mississippi Rivers
Principal industries	fabricated metals, electrical machinery, electronics, chemicals

Illinois is in many ways two states in one, urban Chicago on the shores of Lake Michigan contrasting sharply with the agricultural lands of the prairies to the south and west. This division is very much tied up with the history of Chicago.

When French explorers arrived in 1673, the only inhabitants were the Algonkins. The French soon established settlements along the Illinois river, which provided a base for southward exploration. In 1763 the area was ceded to Britain. The disruption that British rule caused to the Native-Americans' way of life led to conflict, and a lack of civil government resulted in a decline in European settlement. The area passed to America during the War of Independence, and Chicago was founded in about 1790 by the black pioneer merchant Jean-Baptiste-Point Du Sable (1750–1818). The Illinois Territory was created in 1809, when it was carved out of the Indiana Territory; a wave of settlement followed this.

Conflict with Native-Americans ended with the Black Hawk War of 1832, and slavery was abolished in the constitution of 1848. Despite Confederate sympathies in the south, the state remained in the Union during the Civil War; the then president, Abraham Lincoln (1809–65), had spent much of his life in Illinois.

Chicago prospered after the Civil War thanks to massive industrial expansion, and immigrants from Europe flocked to work there, settling in distinct ethnic neighborhoods. The city also attracted large numbers of freed slaves from the southern states. It remained a focus for militant labor movements well into the 20th century. During the Prohibition era of the 1920s and 1930s, rivalry between bootleg gangs made the city a byword for violence, overshadowing its considerable social and cultural achievements. Meanwhile, the agricultural south of Illinois shared in some of the antiblack violence of the Ku Klux Klan.

The distinction between north and south Illinois is still strong. Agriculture is important, but industry, trade, insurance and transportation are the major contributors to the state income; Chicago's O'Hare International Airport is one of the busiest in the world.

MICHIGAN

Nickname	Wolverine State
Joined the Union	1837
Capital	Lansing
Area	251,493 sq km (97,102 sq mi)
Population	10,095,643
Outstanding features	Lake Michigan sand dunes, Huron Mountains
Principal industries	motor vehicles, tourism, machinery, metals

Michigan occupies two great promontories of land largely surrounded by the waters of the Great Lakes. The state is usually associated with the United States automobile industry, but it also contains extensive wild forests in the north, and shares the rolling, richly fertile landscape of the Midwestern grain states that lie to the south.

The original inhabitants were Ojibwa, Ottawa and other Native-American peoples. French fur traders and missionaries began to arrive from the 17th century onward. The area passed to Britain in 1760, and became its power base until the end of the War of Independence. During the War of 1812 the British temporarily reoccupied Detroit, highlighting the need for ready links with the east. Road and rail systems were quickly established thereafter, and the completion of the Erie Canal in 1825 opened a shipping route via Lake Erie to New York. Detroit, in southeastern Michigan, became a major staging post for westbound migrants, many of whom settled in the rich farming land beyond.

Industry grew apace, encouraged by substantial ore reserves now accessible via the Great Lakes, and by the availability of labor. There was already a flourishing manufacturing base at Detroit when the Oldsmobile automotive plant was established in 1899, followed in 1903 by Ford and General Motors. The city's booming prosperity vanished in the Great Depression of 1929, only to recover again after World War II.

Michigan's industries are still suffering from subsequent economic recessions, aggravated by a decline in the United States automobile industry. Social tensions involving the large and expanding black population led to major riots in the 1940s and 1960s; meanwhile, Motown music, which originated as a distinct cultural form among the black communities, soared to nationwide and international popularity. The environmental problems of industrialization, notably the poisoning of the Great Lakes, are today proving so serious that they pose a major threat to the state's important lakeside and forest tourist industry.

INDIANA

Nickname	Hoosier State
Joined the Union	1816
Capital	Indianapolis
Area	94,309 sq km (36,413 sq mi)
Population	6,313,520
Outstanding feature	Indiana Dunes (Lake Michigan)
Principal industries	metals, transportation equipment, machinery, electronics

Indiana straddles the Midwest from Lake Michigan to the Ohio river, and shares many of the characteristics of both northern and southern states. The land consists mostly of rolling plains covered in fertile glacial soils.

The original inhabitants were the Algonkins who, to protect their land against incursions from the hostile Iroquois peoples, formed what was called the Miami Confederation. The first European settlers were French, but the area passed to Britain, and was ceded to the new United States in 1783. Homesteaders from the east then rapidly moved in to open up

The Windy City (*left*) Chicago is the third largest city in the United States after New York and Los Angeles.

Suburbia in downtown Detroit, Michigan (*below*) A busy industrial port and one of the automobile capitals of the world, Detroit has expanded in a rapid, orderly way.

its fertile land for farming, as improved transportation links made access possible. The Civil War brought industrialization to the north and an influx of refugees from the south, who strongly influenced the state's social and political character; they brought with them slavery and a strong distrust of federal government.

Modern Indiana is predominantly industrial. Its transportation network is highly developed including the Ohio river used to carry freight at low costs. Railroad and road density are higher than the national average. However, much of it is still relatively unspoilt, especially in the agricultural south.

OHIO

Nickname	Buckeye State
Joined the Union	1803
Capital	Columbus
Area	115,998 sq km (44,787 sq mi)
Population	11,478,006
Outstanding features	Sandusky Bay (Lake Erie), Miami River
Principal industries	transportation equipment, machinery

Ohio forms a bridge between the original states of the Union and the Midwest, and was the 17th state to be incorporated in the Union.

In prehistoric times Ohio was the home of the Hopewell peoples. Excavation of their large burial mounds and earthworks reveals them to have had a highly sophisticated culture that made finely crafted artifacts. In the 18th century possession of the area was contested by Britain and France. Eventual statehood was followed by swift settlement, notably by Scots–Irish and German-speaking immigrants. River, canal and railroad links made it readily accessible from the east, and though originally agricultural it quickly became industrialized, especially after the Civil War. This sped the growth of major cities such as Cleveland on the shores of Lake Erie and Cincinnati on the Ohio river in the south.

Modern Ohio still combines a rich, rolling agricultural landscape with a great weight of manufacturing industry. The town of Dayton was the home of the pioneer aviators Orville and Wilbur Wright (1871–1948 and 1867–1912) – hence the state's long association with the aviation industries. Traditional industries such as coal mining are in decline, but have to a certain extent been replaced by other heavy industries such as the manufacture of electrical machinery.

MAINE

Nickname	Pine Tree State
Joined the Union	1820
Capital	Augusta
Area	86,156 sq km (33,265 sq mi)
Population	1,321,574
Outstanding features	Mount Katahdin, Moosehead Lake
Principal industries	forestry, fisheries, paper

Maine lies on the Atlantic seaboard of New England next to the border with Canada. Its craggy coastline may have witnessed North America's first – but temporary – European visitors: 10th-century Vikings from Greenland.

From the 15th century onward, the area was disputed by both England and France, whose settlers met fierce resistance from the local Penobscot and Passamaquoddy peoples. The British at last

The Portland Head lighthouse (*above*) stands at the head of Casco Bay on the coast of Maine – the largest of the New England states. The numerous small islands and rocky inlets along the coast are particularly hazardous for Maine's fishermen.

prevailed in 1763, but the area belonged to Massachusetts until the War of 1812. It became a free state under the 1820 Missouri Compromise as a balance to Missouri's status as a slave state.

Maine's stony, heavily forested landscape restricted agricultural development; potatoes are still the main crop. However, as transportation improved during the 19th century, the rich resources of timber, stone and fish formed the basis for important industries; cheap water power attracted others. Maine today is still not very prosperous by New England standards, but its natural beauty remains relatively unspoilt, to the benefit of its flourishing tourist industry.

NEW HAMPSHIRE

Nickname	Granite State
Ratified the Constitution	1788
Capital	Concord
Area	24,032 sq km (9,279 sq mi)
Population	1,314,895
Outstanding features	Mount Washington, Lake Winnipesaukee
Principal industries	machinery, electronics, plastics, tourism

New Hampshire in northern New England was first settled by Europeans in the 17th century. The English Protestants who sailed here named their new home after the English county of Hampshire. Among the earliest settlements, founded in the 1620s, were Portsmouth (on the coast), Exeter and Dover – also named after English counterparts. From 1719 a second wave of Scots–Irish Presbyterian settlers began to arrive, founding the towns of Londonderry and Dublin.

New Hampshire became a separate crown colony in 1679, but boundary disagreements with neighboring colonies continued for another century. In 1774, New Hampshire citizens actively resisted the British at New Castle, where they seized Fort William and Mary. Having given strong support for the War of Independence, New Hampshire announced its declaration of independence several

Fall in Vermont (*below*) In Newfane, capital of Windham county, the trees take on their magnificent fall colors. The blazing change that spreads throughout Vermont's vast expanses of woodland attracts visitors from all over the United States.

weeks before the national declaration was made in 1776.

New Hampshire today adheres strongly to traditional democratic institutions, notably the town meetings that are held every year in each town. However, the state has also welcomed many new industries in recent decades, notably electrical and electronics manufacture, plastics and service industries. Meanwhile, tourism thrives on the old-world villages, the beautiful forest scenery and the winter skiing facilities.

VERMONT

Nickname	Green Mountain State
Joined the Union	1791
Capital	Montpelier
Area	24,900 sq km (9,614 sq mi)
Population	623,908
Outstanding features	Mount Mansfield, Lake Champlain
Principal industries	tourism, furniture, machine tools, microelectronics

Vermont is one of the most historic states, and also one of the most beautiful; the glorious colors of its woods in the fall are justly famous. The area was a hunting ground for various Native-American peoples when the first French explorers arrived – notably Samuel de Champlain (1567–1635). They named it Verts Monts, or Green Mountains, after the thick evergreen forest that covered many of its craggy slopes.

The French were the first to establish outposts in the area, but in the 18th century the Dutch and the British became permanent settlers. Following the French and Indian War (1754–63), settlers from Connecticut and Massachusetts moved into the area, possession of which was being contested by neighboring New York and New Hampshire. In 1770 Ethan Allen (1738–89) formed a band of armed irregulars called the Green Mountain Boys to drive out holders of New York land grants. They supported the Revolution at first, but resisted the new Union equally fiercely. In 1777 they declared Vermont an independent republic, and it remained so until 1791. A large influx of settlers followed, clearing extensive areas of the forests. The land, however, proved stony, making cultivation difficult, and by the 1830s most of the settlers had decided to move westward, allowing the forests to reestablish themselves. During the Civil War, Vermont saw the only action north of Pennsylvania – a Canada-based Confederate raid.

Vermont's industrial growth has been relatively limited apart from granite and marble quarrying. Industries are mostly light, including furniture, and more recently precision engineering and computers. Consequently, Vermont's cities have remained small and its character resolutely rural. Given its proximity to more populous areas, this has been greatly to its advantage, for Vermont has become a favorite holiday area, providing tranquillity in summer and good skiing in winter. It enjoys some of the richest cultural life of any state. Many artistic and literary communities have flourished, along with famous colleges such as Middlebury, Bennington, Marlboro and the University of Vermont at Burlington.

MASSACHUSETTS

Nickname	Bay State
Ratified the Constitution	1788
Capital	Boston
Area	21,455 sq km (8,284 sq mi)
Population	6,437,193
Outstanding features	Cape Cod, Martha's Vineyard (island)
Principal industries	electrical machinery, electronics, precision instruments

Massachusetts, lying on the Atlantic coast of New England, is a state with a long and illustrious history. The Viking seafarer Leif Eriksson is thought to have landed here in about 1003, several centuries before Europeans started to arrive in any numbers. Among the very earliest visitors were navigators such as Bartholomew Gosnold (1572–1605) in 1602 and Samuel de Champlain (1567–1635) in 1605. In 1620 a party of religious dissidents from the southwest of England anchored near Provincetown, later moving their ship the *Mayflower* to their new settlement, Plymouth. The settlers initially established friendly relations with the local Native-Americans, but mutual hostility developed, leading to the massacres of King Philip's War (1675–76), which takes its name from the son of one of the Native-American leaders.

From the early days, Massachusetts depended more on trade and industry than on agriculture, and local resentment of British trade restraints made the colony a leader in the independence movement, provoking incidents such as the Boston Massacre (1770) and the Boston Tea Party (1773). Many important battles in the ensuing War of Independence took place on Massachusetts soil, as did Shays' Rebellion in 1786 – part of the process leading to the eventual ratification of the Federal Constitution. During the 19th century the state became increasingly industrialized, and immigrants, especially from Ireland, swelled the labor force. The Abolitionist movement also became strong here, and the state sided firmly with the Union in the Civil War.

Massachusetts today maintains many legacies of its past, whether in the wealth and influence of older families (the "Proper Bostonians"), its Irish heritage, or the many historical sites and monuments. However, immigration from many parts of the world has created a highly cosmopolitan society. The traditional industrial base in textiles has declined in recent decades, but the gap has been filled by many new "high-tech" industries, often stimulated by the presence of internationally famous academic institutions such as Harvard University, Boston University and the Massachusetts Institute of Technology (MIT).

The Old State House, Boston, Massachusetts
(*below*) stands at the heart of this crowded city of skyscrapers – a reminder of its colonial past. Today the city is host to sizable Irish, Italian, African-American and Puerto Rican communities.

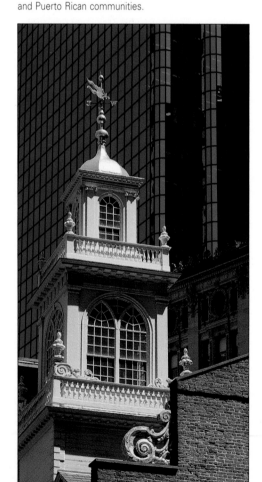

CONNECTICUT

Nickname	Constitution State
Ratified the Constitution	1788
Capital	Hartford
Area	12,997 sq km (5,018 sq mi)
Population	3,504,809
Outstanding features	Connecticut River, Long Island Sound
Principal industries	military equipment, insurance, electrical equipment

Connecticut is one of the United States' oldest and proudest communities, lying on the New England coast between the great cities of New York and Boston.

The colony was founded during the 1630s by Puritans – originally from Britain – who had left the Massachusetts Bay Colony. The colonists enjoyed extremely good relations with the local Algonkin peoples. They settled in two main areas: the Connecticut River Valley (1633–35), and the coastal strip between Saybrook and New Haven (1635–38). In 1662 a royal charter provided effective self-government, and Connecticut citizens continued to value their independence. In 1776 more than half of the army under the command of George Washington (1732–99), based at New York, was drawn from Connecticut; and later that year Connecticut citizens adopted the original 1662 charter as their

state constitution. In 1814–15 Connecticut joined four other New England states in the Hartford Convention – an attempt to revise the federal constitution. During the Civil War, however, the state strongly supported the Union.

The 19th century saw radical changes. A stream of Irish immigrants began in the 1840s, and this was swelled later in the century by an influx of French Canadians and immigrants from Europe. Textile factories were built, and industry soon displaced agriculture as the main source of local employment.

Today farming plays only a small and dwindling part in the state economy. Some of the original 19th-century industries still thrive, notably gun- and clock-making, silverware and the manufacture of sewing machines. However, textile industries have largely disappeared as the economy has become more service-based. The ancient forest that once covered most of Connecticut has long been felled, and today's mixed woodland is a later growth. The original towns and villages have gradually merged along the main traffic arteries, and most people are urban-dwelling. However, there are no large cities. Local autonomy is fiercely prized, and local history is vehemently preserved in the form of historic buildings, sites and monuments.

Unspoilt beauty (*above*) in upstate New York, seen at Lows Lake in the Adirondack Forest Preserve.

South Manhattan, New York City (*left*) Two days after the terrorist attacks of September 11 2001, smoke and dust still rise from the site of the World Trade Center. The cost of the damage was an estimated $105 billion.

RHODE ISLAND

Nickname	Little Rhody
Ratified the Constitution	1790
Capital	Providence
Area	3,139 sq km (1,212 sq mi)
Population	1,067,610
Outstanding feature	Narragansett Bay
Principal industries	jewelry, machinery, textiles, electronics

The smallest state in the Union is Rhode Island on New England's Atlantic seaboard; it takes its name from one of several islands situated in Narragansett Bay. The first European settlement was the mainland town of Providence, founded in 1836 by the Puritan minister Roger Williams (1603–83) after his expulsion from Massachusetts. Williams was a strict disciplinarian, but maintained a spirit of religious tolerance and dealt honestly with the local Native-Americans. This highly independent tradition made the colony swift to rebel but slow to join the new nation. Rhode Island was the last of the original 13 colonies to ratify the Constitution, and the last to introduce universal adult suffrage – the subject of Dorr's Rebellion in 1842.

Economically, Rhode Island has always been progressive. It was at the forefront in the Industrial Revolution, and today its economy relies mainly on sophisticated light industries and the service sector. The state retains strong historical roots, however, particularly in the old seaports of Providence and Newport, where whole districts are carefully preserved, including some of the oldest buildings in the United States.

NEW YORK

Nickname	Empire State
Ratified the Constitution	1788
Capital	Albany
Area	136,583 sq km (52,735 sq mi)
Population	19,306,183
Outstanding features	Mount Marcy (Adirondack Mountains), Long Island, Niagara Falls
Principal industries	publishing, clothing, pharmaceuticals, machinery, finance

New York was one of the 13 original American colonies, and the state, like the city of New York, reflects the diversity of its many immigrant settlers. The largest groups came originally from Italy, Germany, the Soviet Union, Poland, Ireland and Puerto Rico.

The first European to arrive was the Englishman Henry Hudson (1565–1611), who explored the river that was later named after him. The first settlers, however, were the Dutch, who founded the colony of New Amsterdam on Manhattan Island. In 1664 the British took the colony by force, renaming it New York. A troubled period followed, during which the conflicting claims of France and Britain were further complicated by rivalries between Native-American groups; the situation was largely resolved with the Treaty of Paris in 1763. During the War of Independence, much of the colony remained loyal to Britain. New York City remained in British hands until the war ended, but then became the capital of the new nation from 1789 to 1790.

The state's already cosmopolitan population was constantly swelled by an uninterrupted stream of immigrants. Immigration continued unabated after the Civil War, doubling the population of New York City between 1880 and 1890. The city today is among the world's most populous cities, and a national center for finance and the arts. It is also the nation's largest seaport, and there are three international airports located within the metropolitan area. So dominant is New York City that the rest of the state is often forgotten. Upstate New York contains a variety of landscapes, from the Great Lakes and the Niagara Falls in the west, via the mountains and lakes of the Appalachians, to the Hudson Valley in the east.

On September 11 2001 New York City was the target of a brutal terrorist attack which completely destroyed the twin towers of the World Trade Center – a potent symbol of the financial world and a great landmark recognized the world over. Over 3,000 people were killed.

PENNSYLVANIA

Nickname	Keystone State
Ratified the Constitution	1787
Capital	Harrisburg
Area	119,251 sq km (46,043 sq mi)
Population	12,440,621
Outstanding features	Mount Davis (Allegheny Mountains), Susquehanna River
Principal industries	steel and other metals, foodstuffs, machinery

Pennsylvania is one of the oldest and most populous states, extending westward from the Delaware river as far as the shores of Lake Erie. Its official name, the Commonwealth of Pennsylvania, reflects the idealism of the state's founder, William Penn (1644–1718).

In 1681 Penn was granted a large expanse of American land as payment of a debt owed to his father by the British monarch. The following year he founded a colony based on Quaker principles of tolerance and democracy on the site of present-day Philadelphia by the Delaware river. Penn was notably honest in his dealings with the Native-American inhabitants, and the Quaker policy of tolerance encouraged many religious minorities to settle in the colony. A significant proportion were Germans, known today as Pennsylvania Dutch, and included communities such as the Mennonites and the Amish. Pennsylvania played a central role in the independence struggle; the Declaration of Independence was signed at Philadelphia, and the city became the first capital and seat of government. The state was also a Civil War battleground, and saw a major Union victory at Gettysburg in 1863.

Pennsylvania's fortunes were founded on its rich agricultural land, but huge deposits of coal and other minerals helped to make it the United States' first really great industrial and manufacturing center, especially around Pittsburgh, where "steel kings" such as Andrew Carnegie (1835–1919) established gigantic works. These traditional industries have since declined, but Pennsylvania remains a primarily industrial state. Even today, though, much of the land is still cultivated, and there are beautiful stretches of farm and woodland. The city of Philadelphia, with its great art collections, orchestras and museums, is a world-famous cultural center, while Pittsburgh, its industrial environment dramatically transformed by a clean-up campaign, is rapidly achieving the same status.

NEW JERSEY

Nickname	Garden State
Ratified the Constitution	1787
Capital	Trenton
Area	20,168 sq km (7,787 sq mi)
Population	8,724,560
Outstanding features	Kittatinny Mountains, Pine Barrens (heath, dunes, and woodland)
Principal industries	chemicals, electrical machinery, scientific research

New Jersey lies on the Atlantic seaboard, sandwiched between New York and Pennsylvania. It calls itself the Garden State, but to travelers on its highways it may seem crowded and industrialized, plagued by pollution, organized crime and urban decay. In fact both aspects are true; as one of the smallest and most densely populated states, New Jersey harbors many contradictions.

European explorers found the area inhabited by the Lenni Lenape peoples. The Swedes and the Dutch established the first settlements, but the English took over in 1664. New Jersey, created in 1702 from East and West Jersey, became a separate colony in 1738. The War of Independence sharply divided the colony, and some of the fiercest battles were fought here, notably at Trenton (1776) and at nearby Princetown (1777).

New Jersey was in the forefront of the Industrial Revolution, site of the United States' first factory town (albeit unsuccessful) in 1791, and a center of canal,

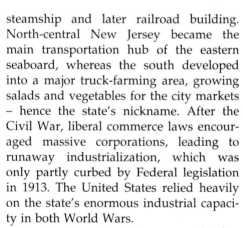

Elfreth's Alley (*above*), a historic enclave in downtown Philadelphia – very British in architectural style.

Rural tranquility in Maryland (*right*) The pace of life here is relatively slow, the climate mild, and on the undulating hills dairy and poultry farming are common.

steamship and later railroad building. North-central New Jersey became the main transportation hub of the eastern seaboard, whereas the south developed into a major truck-farming area, growing salads and vegetables for the city markets – hence the state's nickname. After the Civil War, liberal commerce laws encouraged massive corporations, leading to runaway industrialization, which was only partly curbed by Federal legislation in 1913. The United States relied heavily on the state's enormous industrial capacity in both World Wars.

Northeastern New Jersey rapidly became an overspill area for neighboring New York City, so that today a majority of the population lives within a 30-mile radius of that city, along highways that are among the world's busiest. Farmland is rapidly diminishing in this area, but elsewhere – both in the heavily cultivated south and in the lush pastures of the northwest – New Jersey can still lay claim to its nickname. The long stretches of good beaches along the Atlantic Ocean are a major attraction for the thousands of visitors to the area, and Atlantic City has become a world-famous resort.

Nickname Old Line State

Ratified the Constitution 1788

Capital Annapolis

Area 27,091 sq km (10,460 sq mi)

Population 5,615,727

Outstanding features Chesapeake Bay, Backbone Mountain

Principal industries electrical and electronic equipment, foodstuffs, chemicals

In 1632 a royal charter from Charles I of England (1600–49) granted land in the Chesapeake Bay area to Cecil Calvert, Lord Baltimore (1605–75). He named the new colony Maryland after Henrietta Maria (1609–69), the king's wife, who was a Roman Catholic and made it a haven for Roman Catholics suffering religious persecution in England. However, the Calverts imposed no corresponding restrictions on non-Catholics, welcoming all Christians who believed in the Trinity.

Calvert's brother Leonard (1606–47) founded Saint Mary's City, the first capital, in 1634. However, the rapid growth of tobacco plantations shifted the main centers of population northwestward, and in 1694 the capital was moved to Annapolis; 1729 saw the founding of the great city of Baltimore. Although many plantations used African slaves, the slave trade was penally taxed from 1783; even so, full abolition did not follow until 1864.

Disputes between the Calverts and the Penns were settled in the 1760s by the creation of the Mason and Dixon Line, marking the boundary between Maryland and Pennsylvania. Maryland was active in the War of Independence; the Treaty of Paris, which acknowledged the independence of the colonies, was ratified in Annapolis in 1783. In 1814 a British attack on Baltimore was decisively repulsed. The next decades saw the construction of the first major road and rail routes across the Appalachian Mountains – both of them on Maryland soil. The loyalties of Marylanders were divided in the Civil War, but the state remained in the Union. Maryland never disenfranchised its black population, and this encouraged an influx of black people from the less tolerant states of the South.

The port of Baltimore remains one of the nation's busiest, supported by a strong industrial base that includes many new industries. Service industries are increasing in importance, and the state's many historical sites form the basis for a thriving tourist industry.

DELAWARE

Nickname First State

Ratified the Constitution 1787

Capital Dover

Area 5,294 sq km (2,044 sq mi)

Population 853,476

Outstanding features Delaware River, Delaware Bay

Principal industries chemicals, textiles, clothing

Delaware today is heavily industrialized and densely populated. The north of the state forms part of the Atlantic megapolis – the long urban corridor that runs from Washington to Boston. Nothing could be more different from the primitive log cabins built by the Swedish settlers who established Fort Christina – modern Wilmington – in 1638.

An earlier Dutch colony had been destroyed by local Native-Americans, but in 1655 Dutch from New Amsterdam (later New York) reclaimed the area, only to lose it to the British in 1664. In 1682 the area was ceded to William Penn (1644–1718), providing his inland colony of Pennsylvania with a corridor to the coast. Delaware gained its own assembly in 1704, and its present name in 1776. The origins of that name went back to early colonial days. Delaware Bay had been so named in 1610 after Thomas West (1577–1617), 12th Baron De la Warr, upon his appointment as governor of Virginia.

During the War of Independence, Delaware was invaded by a British army, and its coastline was constantly under threat from the British navy. Following independence, Delaware's prompt acceptance of the Constitution on 7 December 1787 earned it the title of First State. Although Delaware was strictly speaking a slave state, many slaves had already been granted their freedom before the outbreak of the Civil War, and the state therefore remained a slightly reluctant part of the Union.

The development of industry has had a marked effect on the demographic balance of the state. The older rural communities around Dover and Georgetown in the mainly agricultural south have now declined, and extensive suburban areas have grown up in the industrial north, especially around Wilmington, which is now a major manufacturing center for artificial fibers.

The fountainhead of American power – the elegant balconies of the White House in Washington, D.C., residence of the United States president. The original building was first painted white to hide the scars of its burning by the British in 1814.

DISTRICT OF COLUMBIA

Federal territory authorized	1790
Coterminous with the United States capital	Washington, DC
Area	179 sq km (69 sq mi)
Population	581,530
Outstanding feature	Potomac River
Principal industries	federal administration, tourism

Columbia is the district that was chosen by the first president, George Washington (1732–99), to house a purpose-built capital city – the first in any Western nation.

The French military engineer Pierre-Charles L'Enfant (1754–1825) laid out the city's distinctive plan, based on French principles of formal landscaping, with avenues radiating out from the Capitol and the presidential mansion (the latter became known as the White House following its renovation after the War of 1812). Washington, D.C. also houses the headquarters of America's military, the Pentagon. It was this building which was partially destroyed in the terrorist attacks of September 11 2001, targeting both the capital and New York City. The attack on the Pentagon killed about 200 people.

The District now enjoys most rights of statehood, but is only part of a larger metropolitan area, with all the associated problems. The imposing embassies and government officials' homes almost rub shoulders with slums, which saw serious race riots in the 1960s; a few areas have been cleared in urban renewal programs.

ARKANSAS

Nickname	Land of Opportunity
Joined the Union	1836
Capital	Little Rock
Area	137,754 sq km (53,187 sq mi)
Population	2,810,872
Outstanding features	Boston Mountains, Hot Springs National Park
Principal industries	foodstuffs, chemicals, electronic equipment
Principal industries	electronics, tourism, copper mining

Arkansas lies on the west bank of the great Mississippi river, extending westward to the forested uplands of the Ozark Plateau and Ouachita Mountains.

There were well-established Native-American cultures in the area as early as 500 AD. The first European settlers to arrive were the French, in the 17th century. They then sold the territory to the United States in the Louisiana Purchase of 1803. The lower Mississippi plains became the site of great cotton plantations worked by black slaves transported from Africa. From 1820 until 1836, when statehood was granted, Arkansaw Territory was the area south of the Missouri Compromise line, which separated slave and free states.

Arkansas did not secede from the Union until 1861, when it joined the Confederacy. The Civil War left it impoverished and overly dependent on cotton, and with racial-segregation policies that were maintained until the 1960s. The state today is much more prosperous, but maintains its largely rural character. More varied agriculture and new industries have replaced cotton, and tourism is growing fast in the uplands, which are becoming a popular retirement area.

LOUISIANA

Nickname	Pelican State
Joined the Union	1812
Capital	Baton Rouge
Area	123,677 sq km (47,752 sq mi)
Population	4,287,768
Outstanding features	Mississippi River and Delta
Principal industries	petroleum extraction, foodstuffs, chemicals

When, in 1682, René-Robert Cavelier, sieur de La Salle (1643–87), sailed down the Mississippi and claimed the entire basin of one of the world's largest rivers for King Louis XIV of France (1638–1715), Native-American peoples had been living in the area for some 16,000 years.

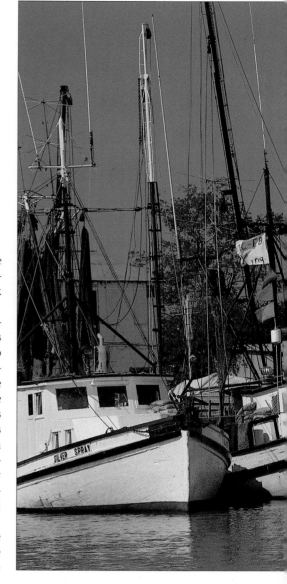

In 1718 Jean-Baptiste Le Moyne, sieur de Bienville (1680–1768), founded New Orleans near the Mississippi Delta, and in 1731 the crown colony of Louisiana was born. Many new settlers arrived to swell the colony. They included the Cajuns – French-speakers expelled from Nova Scotia by the British. In 1762 Louisiana fell into Spanish hands and remained Spanish until 1800, when Napoléon Bonaparte (1769–1821) bought it back for France. The effect of this French and Spanish heritage can be seen in the state's unique legal code, and in the architecture and culture of New Orleans.

The second French tenure was shorter even than the first. In 1803 the United States acquired Louisiana as part of the Louisiana Purchase. This very remarkable real-estate deal added more than 2 million sq km (over 800,000 sq mi) of land to United States territory for less than 3 cents per acre.

In Louisiana, as in other southern states, the cotton and sugar cane planters depended for their wealth on the use of large numbers of black slaves. Conflicts of interest soon occurred between wealthy

Mardi Gras (*above*) A reveler in New Orleans' famous parade peers from beneath his gaudy tiger mask. Like jazz, this Caribbean-style carnival is a highlight of Louisiana's African-American culture.

Shrimp boats (*left*) jostle along the wharves at the small fishing port of Delcambre, Louisiana. The shape of the coastline is constantly being changed by deposits of silt, brought down by the Mississippi river.

Competing in the Kentucky Derby (*below*) This world-famous race takes place every May at the Churchill Downs course in Louisville. Together with the Preakness and Belmont stakes it forms the coveted Triple Crown.

KENTUCKY

Nickname	Bluegrass State
Joined the Union	1792
Capital	Frankfort
Area	104,659 sq km (40,409 sq mi)
Population	4,206,074
Outstanding features	Black Mountain, Land of 10,000 Sinks
Principal industries	coal mining, machinery, foodstuffs

The Commonwealth of Kentucky unites the traditions of the frontier and the old South with a modern industrial economy.

The area was once the hunting ground of Native-American peoples such as the Iroquois and the Cherokee. After 1767, it was explored by the frontiersman Daniel Boone (1734–1820), who paved the way for the westward migration of settlers. A substantial black slave population was imported, and the state government opposed abolition. However, Kentucky did not join the Confederacy, and most of its soldiers fought for the Union in the Civil War. In the upland "hillbilly" country – poor and fiercely clannish – Civil War feuds continued into the 20th century.

Modern Kentucky has remained agricultural, famous for its tobacco, horse breeding and bourbon whiskey. However, since the late 19th century massive coal reserves have been exploited, and large areas laid waste by strip mining – controlled since 1966. Kentucky is also well known for its cultural heritage, ranging from the famous Kentucky Derby to folk music, including the celebrated Bluegrass style.

TENNESSEE

Nickname	Volunteer State
Joined the Union	1796
Capital	Nashville
Area	109,152 sq km (42,144 sq mi)
Population	6,038,803
Outstanding features	Clingmans Dome (Smoky Mountains), Great Appalachian Valley
Principal industries	chemicals, foodstuffs, machinery

In 1784 North Carolina offered to cede its western territories to Congress. Angry settlers reacted by creating their own state of Franklin, but its life was short. In 1789 the territory was formally separated from North Carolina, and seven years later it became the state of Tennessee.

Many Tennesseans achieved fame in the War of 1812, including future president Andrew Jackson (1767–1845). Before

planters and farmers on issues of land use and political suffrage. Louisiana seceded from the Union in 1861, but there was considerable opposition to its secession, and less than a year passed before Union troops reoccupied New Orleans. In 1868 Louisiana was readmitted to the Union, but a legacy of bitterness remained. The constitution of 1898 denied voting rights to most blacks – a situation that was to remain unchanged until the civil rights campaigns of the 1960s.

Modern Louisiana remains primarily agricultural, but cotton is no longer the main crop, having given way to soybeans, rice and other food crops. Beef cattle are also important. Along the coast the distinctive bayous, or creeks, of the Mississippi delta experience a subtropical climate. New Orleans and Baton Rouge are both major ports, and tourists flock to the French Quarter of New Orleans, with its fascinating colonial architecture and long cultural traditions; it is renowned as the birthplace of jazz, and is equally famous for its extremely colorful annual carnival parade or festival of Mardi Gras (Shrove Tuesday).

Country and Western music in full swing at the Opryland Park Country Music Show in Nashville, Tennessee, where the style first began. The Grand Ole Opry was the music show of the 1920s and 1930s where many legendary singers first appeared.

the Civil War there was strong loyalty to the Union, but deep divisions appeared on the eve of the conflict, and Tennessee became a battleground second only to Virginia. The state escaped the worst excesses of reconstruction. Even so, the white supremacist Ku Klux Klan was born at Pulaski (south of Nashville) in 1865, and blacks were effectively segregated from the 1870s until the 1960s.

From the 1940s onward, power from the massive Tennessee Valley hydroelectric scheme fueled rapid industrial development, ending the state's reliance on agriculture. Nowadays, however, there is a move toward the trade and services sector, and tourism has also become important. Many come to visit Nashville, which is world famous for being the heart of Country and Western music.

MISSISSIPPI

Nickname	Magnolia State
Joined the Union	1817
Capital	Jackson
Area	123,514 sq km (47,689 sq mi)
Population	2,910,540
Outstanding feature	Mississippi River
Principal industries	clothing, foodstuffs, timber processing

Mississippi lies at the very heart of the American cotton belt, a fact that is reflected in its history. The first colonists were French Canadians, who built Fort Maurepas on the coast in 1699. Other French settlements followed, but in 1763 the British took control. Spain occupied the south during the War of Independence, and the area remained in dispute until 1795, when it came to the United States in the Treaty of San Lorenzo.

The Mississippi Territory extended from Tennessee in the north to the Gulf coast in the south, and in 1817 the western part became the state of Mississippi. The Native-Americans were forcibly removed, and Mississippi became a prosperous cotton state. Entrenched attitudes to slavery led to its secession from the Union in 1861. It was devastated during the Civil War, and in 1890 its black population was virtually disenfranchised.

Agriculture suffered in the Depression of 1929, but revived strongly after World War II, and a long, slow process of racial integration and acceptance began. However, the state remains relatively backward economically, and has benefited less from modern industrial development than some neighboring states.

ALABAMA

Nickname	Cotton State
Joined the Union	1819
Capital	Montgomery
Area	133,915 sq km (51,705 sq mi)
Population	4,599,030
Outstanding features	Cheaha Mountain, Mobile Bay
Principal industries	pulp and paper, chemicals, electronics

Alabama is a state with a troubled history, torn by settlement disputes between European colonists and the Native-Americans, by issues of social reform in the 19th century and by civil rights protests in the 1960s.

The first European exploration was made by the Spanish, led by Hernando de Soto (1500–42), who came in search of precious metals. Although they failed to find any, they succeeded in opening up the whole area to European settlement, killing thousands of Native-Americans in the process. For the next three centuries the area was contested by Spain, France and Britain. A French expedition to the Mississippi area established the first permanent settlement on Dauphin Island at the mouth of the Alabama river in 1702. Seventeen years later the first shipload of black slaves from Africa disembarked at Port Dauphin.

In 1763 the British gained control over most of the area, and the Spanish foothold on the coast became increasingly tenuous. The Alabama Territory was created in 1817, achieving statehood two years later. Meanwhile the remaining Native-American populations had been subdued and transported west to Oklahoma, opening the way for another wave of European settlement.

The result was a wealthy cotton-growing state dependent on black slave labor. By 1860 half the population of

Alabama were black slaves, and secession from the Union in 1861 was all but inevitable; Alabama joined the Confederate States of America. During the Civil War the state suffered appalling casualties, and from 1867 to 1868, following the collapse of the Confederacy, it was placed under military rule. The period of reconstruction (1868–74) merely deepened the gulf between blacks and whites, with increasing support for the white supremacist creed of the Ku Klux Klan. Here, as in other Southern states, blacks had to wait until the 1960s to achieve basic civil rights, and even then progress was slow in the face of discrimination against black voters. Alabama was the scene of many civil rights protests in the 1950s and 1960s, including the march of Montgomery, led by Martin Luther King (1929–68) in 1965.

Agriculture remains important today, with cotton and groundnuts (peanuts) among the chief crops. Manufacturing has diversified from iron and steel and agriculture-based industries to include textiles and modern electronics.

The mighty Mississippi river seen from the air, snaking through dense forest along the western border of the state that bears its name. Its sluggish meanders become ever more tortuous until they meet, creating islands and oxbow lakes.

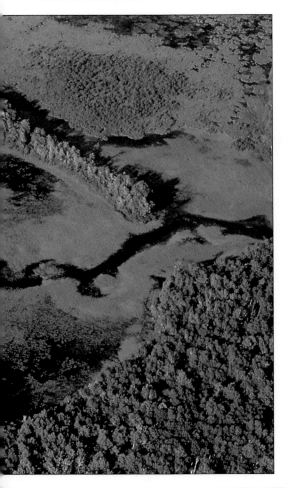

WEST VIRGINIA

Nickname Mountain State

Joined the Union 1863

Capital Charleston

Area 62,758 sq km (24,231 sq mi)

Population 1,804,200

Outstanding features Spruce Knob (Allegheny Mountains), Ohio river

Principal industries mining, chemicals, machinery, quarrying

West Virginia lies between the eastern and the midwestern states, across the Appalachian Mountains to the north and west of the state of Virginia. Whereas Virginia belongs firmly to the South, West Virginia has always had more in common with the northern states.

Most of West Virginia is mountainous, carved by swift rivers and covered with dense forest. This wild land was home to various Native-American peoples, notably the Hopewell, whose prehistoric earthworks can still be found in the Charleston area. Their successors – the peoples of the Iroquois League – put up strong resistance to the first European settlers when they arrived from Virginia across the Appalachian Mountains in the early part of the 18th century.

These new colonists established a very different order from the slave-worked plantations of Virginia nearer the coast. Cultivation in the mountains was largely limited to small farms, and slavery was virtually nonexistent. The mountain-dwellers soon became dissatisfied with the ruling authorities in the east, and their antipathy to slavery led in 1859 to the seizure of a Federal arsenal at Harper's Ferry (now the easternmost point of West Virginia). This was one of the incidents that precipitated the Civil War. When Virginia seceded from the Union in 1861, the people of the northwest refused to follow them and join the Confederacy; thus the state of West Virginia was born.

After the Civil War, the expansion of the railroad network brought industrial development to the mountains. The rich local resources of salt, limestone and especially coal have been exploited ever since, providing plentiful energy and raw materials for neighboring states – and bringing accompanying social and environmental problems. Agriculture and mining have both declined now, but petroleum and natural gas have since been added to the list of resources; tourism is also expanding.

VIRGINIA

Nickname The Old Dominion

Ratified the Constitution 1788

Capital Richmond

Area 105,586 sq km (40,767 sq mi)

Population 7,642,884

Outstanding features Mount Rogers (Blue Ridge Mountains), Chesapeake Bay

Principal industries textiles, transportation equipment, electronics

Virginia, lying on the Atlantic seaboard between Delaware and North Carolina, has one of the longest histories of any state. With strong links to both south and north, it has often played a pivotal role in the history of the United States.

Virginia, named after Queen Elizabeth I (1533–1603) of England (the "virgin queen"), was the first British colony in America, established in 1607 on the site of present-day Jamestown (southeast of Richmond). The settlers barely survived, plagued by hunger, disease and frequent raids by the Native-Americans who were defending their land from encroachment. However, the colonists' tobacco plantations soon brought increasing prosperity, enabling them to establish a tradition of culture and education, with prestigious institutions such as William and Mary College in Williamsburg (across the river from Jamestown).

The colony was chafing at British rule long before the War of Independence, in which several Virginians played a leading part, notably George Washington (1732–99) and Thomas Jefferson (1743–1826).

Colonial Williamsburg, Virginia, which was founded in 1632, has been the subject of a huge restoration project. More than 400 buildings have been restored, and 18th-century crafts such as spinning yarn have been revived and are demonstrated to the public.

Virginia hosted the first Continental Congress in 1774, and witnessed the decisive British surrender at Yorktown (east of Williamsburg) in 1781.

The importation of slaves from Africa was abolished in 1778, but slavery itself was maintained. In 1861 Virginia led the Confederacy, whose capital was at Richmond. As a result, the northwestern part of the state split away, to become West Virginia. The Civil War ended in 1865, soon after Confederate troops had surrendered at Appomattox (west of Richmond). Many years passed before the state's economy recovered. However, by World War I the old reliance on agriculture, especially tobacco, had been superseded by the growth of various manufacturing industries.

World War II brought very considerable prosperity, with the expansion of military bases and shipbuilding concerns along the coast; the federal government is still a major employer. But Virginia's proudest possession is its history – beautifully preserved at sites such as Williamsburg and Mount Vernon (to the south of Washington, D.C.), which attract a thriving tourist trade.

Captain Sam (*above*) in front of the oldstyle sternwheel riverboat in which he runs harbor cruises at Savannah, Georgia.

Classical elegance (*right*) in Charleston, South Carolina, where many plantation houses still stand. It was here that the popular back-kicking dance of the 1920s originated, and became a national, and international, craze.

NORTH CAROLINA

Nickname Tar Heel State

Ratified the Constitution 1789

Capital Raleigh

Area 136,412 sq km (52,669 sq mi)

Population 8,856,505

Outstanding features Mount Mitchell (Blue Ridge Mountains), Cape Hatteras, Cape Lookout

Principal industries textiles, tobacco, electronics

In 1663 King Charles II of England (1630–85) granted the area between the 31st and 36th parallels to eight "Lords Proprietors" (an earlier colony on Roanoke Island had vanished without trace). The new colony was named Carolina in the king's honor, and was extended northward in 1665 to include areas that had belonged to Virginia. In 1712 North Carolina became a separate colony with its own governor. In 1729 it returned to royal control – which ended in 1781 at the Battle of Guilford Court House.

During the 1830s most of the native Cherokee population were forcibly removed to the west of the Mississippi river. North Carolina seceded from the Union in 1861, and paid dearly for doing so. The constitution of 1868 was more liberal than some, but black people were effectively denied civil rights from 1900

until the 1960s. Today manufacturing has largely replaced agriculture as the chief source of the state's income, but North Carolina remains the center of the tobacco industry in the United States.

SOUTH CAROLINA

Nickname Palmetto State

Ratified the Constitution 1788

Capital Columbia

Area 80,582 sq km (31,113 sq mi)

Population 4,321,249

Outstanding features Sassafras Mountain (Blue Ridge Mountains), Sea Islands

Principal industries tourism, textiles, chemicals, machinery

In 1670 British colonists built Charles Town near the mouth of the Ashley river. Named after King Charles II of England (1630–85), it was the first lasting European settlement of South Carolina. Ten years later, Charleston was rebuilt across the river, and grew wealthy from its trade in rice, pelts and indigo. Crown control of the colony's finances caused deep resentment, and South Carolina was firm in its support for the American Revolution. In 1780 a British fleet captured Charleston, and many engagements were fought on South Carolina soil.

Cotton plantations brought new wealth inland, but dependence on slavery led to secession from the Union in 1861, when an engagement with Federal troops at Fort Sumter in Charleston Harbor triggered the Civil War. The reconstruction period created lasting racial tensions that were reflected in the 1895 constitution, and black civil rights were not restored until the 1960s.

Modern South Carolina is industrial rather than agricultural, though it is less dominated by textiles than it once was. The warm climate, and the old-world atmosphere of its coastal towns, encourage a thriving tourist industry.

GEORGIA

Nickname Empire State of the South

Ratified the Constitution 1788

Capital Atlanta

Area 152,576 sq km (58,910 sq mi)

Population 9,363,941

Outstanding features Brasstown Bald (Blue Ridge Mountains), Okefenokee Swamp

Principal industries foodstuffs, textiles, clothing, electrical equipment

In 1732 James Edward Oglethorpe (1696–1785) received a charter from King George II of England (1683–1760) to settle a new colony in North America. He named it Georgia in honor of the king. This, the last of the 13 original colonies, was founded to give Britain's poor greater opportunities in a new land. However, its attempts to produce silk, wine and spices had only limited success; in 1752 the colony reverted to crown control.

Other settlers began to make their homes along the banks of the Savannah river (the eastern border) before the War of Independence. Later on they came in such great numbers that the original inhabitants, the various Native-American peoples, were forcibly displaced beyond the Mississippi river. Black slaves were later brought in to work the state's rapidly developing cotton plantations.

The issue of slavery caused much heart-searching in Georgia, but in January 1861 the state convention voted overwhelmingly for secession from the Union. In 1864, after a series of blockades and invasions, the Union army under General William Tecumseh Sherman (1820–91) destroyed Atlanta and advanced to the coast, leaving a trail of devastation. In 1865 the president of the Confederacy, Jefferson Davis (1808–89), was captured on Georgian soil.

The war was over, but even during the reconstruction period that followed, Georgia's politicians resisted black enfranchisement – an attitude that was to remain entrenched for another century. Integration of black and white pupils in schools was a hotly debated issue as late as 1967. Since then, however, African-Americans have taken a far more active role in the community: the state capital Atlanta was the first major city in the South to elect a black mayor, in 1973.

Miami, Florida, is host to a diverse cultural mix: vacationers from all over the world come to enjoy the beach life; elderly people come here to retire; and thousands of immigrants from Cuba, Haiti, Mexico and South America come to begin a new life.

Cotton is no longer the main crop, having been overtaken by food crops such as groundnuts (peanuts) and maize. Textiles, however, remain important, together with many new industries such as electronics. Federal military installations supplement income from industry. Atlanta is the national headquarters of Coca-Cola, and its rapid growth in recent years makes it not only the commercial and financial center of the southeast, but also a leading national city. Atlanta hosted the 1996 Olympic Games.

FLORIDA

Nickname	Sunshine State
Joined the Union	1845
Capital	Tallahassee
Area	151,939 sq km (58,664 sq mi)
Population	18,089,888
Outstanding features	the Everglades, Florida Keys
Principal industries	tourism, electronics, electrical machinery, foodstuffs

Today tourists flock to the far southeastern state of Florida in search of sunshine, relaxation and entertainment, and large numbers of elderly people retire there from northern cities. In Easter 1513 the Spanish soldier and explorer Juan Ponce de León (1460–1521) came in search of the Fountain of Youth, but instead he began a 250-year cycle of conflict and exploitation that almost wiped out the original Native-American population.

In 1565 another Spanish expedition under the leadership of Pedro Menéndez de Avilés (1519–74), utterly destroyed a Huguenot colony on the Saint Johns river, and Menéndez de Avilés founded his own colony of Saint Augustine nearby. For 200 years, control of Florida was disputed between Great Britain, France and Spain; however, the British finally took control in 1763, at the end of the French and Indian Wars. During the War of Independence, Florida remained loyal to Britain, but in 1781 it was seized by the Spanish.

During the war of 1812 an American force seized Fernandina, hoping to prevent the British from using Florida as a military base. The force withdrew in the face of Spanish protests – and the British duly moved in, directing a guerrilla war against American settlements. In 1818 Andrew Jackson (1767–1845) captured Pensacola for the United States, and the following year Spain ceded the whole of Florida. Relations with the remaining Native-Americans deteriorated rapidly, and most were expelled in 1842. After the Civil War new railroads brought settlers and tourists alike, beginning a period of rapid expansion in population that has continued throughout the 20th century. After the Cuban revolution in 1959 many Cuban refugees settled in the state, especially in the southern city of Miami.

Much of the land is covered by forests, lakes and military reserves; there are large areas of farmland, including significant citrus fruit plantations, but most of the population is urban. Tourism is the leading industry, with many visitors to the Everglades National Park in the south and to Disney World near Orlando in central Florida. The state also hosts many major sporting events, including football at the Miami Orange Bowl Stadium, golf tournaments, and auto racing at Daytona Beach. The Kennedy Space Center at Cape Canaveral also makes a significant contribution to the economy, as does the production of electrical machinery.

Florida became the focus of world attention during the presidential elections of 2000, when a dispute over vote counting became crucial to the outcome.

Mexico, Central America and the Caribbean

COUNTRIES IN THE REGION

MEXICO · GUATEMALA · BELIZE · EL SALVADOR

HONDURAS · NICARAGUA · COSTA RICA · PANAMA

CUBA · BAHAMAS · HAITI · JAMAICA

DOMINICAN REPUBLIC · SAINT KITTS-NEVIS

ANTIGUA AND BARBUDA · DOMINICA · SAINT LUCIA

SAINT VINCENT AND THE GRENADINES · BARBADOS

GRENADA · TRINIDAD AND TOBAGO

DEPENDENCIES IN THE REGION

TURKS AND CAICOS ISLANDS · BERMUDA

CAYMAN ISLANDS · PUERTO RICO

UNITED STATES VIRGIN ISLANDS

BRITISH VIRGIN ISLANDS · ANGUILLA

MONTSERRAT · ARUBA · THE NETHERLANDS

ANTILLES · GUADELOUPE · MARTINIQUE

Casting his net A fisherman in Haiti. This impoverished state occupies the western third of Hispaniola, the Caribbean island that was discovered by Christopher Columbus in 1492 and was the site of the first European landing in the Americas.

capital city
major town

height of land (meters)

3000
2000
1000
500
200
0

▲ mountain peak

Great Abaco

Grand
Bahama

Eleuthera

Gulf of Mexico

Andros

Nassau

BAHAMAS

Long Island

Turks and
Caicos
Islands
(U.K.)

*ATLANTIC
OCEAN*

Acklins Island

Great Inagua

*Virgin
Islands
(U.S.A.)*

*Virgin Islands
(U.K.)*

**ANTIGUA AND
BARBUDA**

Tampico

Havana

CUBA

Camagüey

Hispaniola

San Juan

Basseterre

St John's

**DOMINICAN
REPUBLIC**

*Guadeloupe
(France)*

Mérida

*Isla de la
Juventud*

Santiago de Cuba

Guantánamo

HAITI

Santo
Domingo

*Puerto Rico
(U.S.A.)*

**ST KITTS-
NEVIS**

*Montserrat
(U.K.)*

DOMINICA

Bay of Campeche

Yucatán

*Cayman
Islands
(U.K.)*

Port-au-Prince

Jacmel

G r e a t e r   A n t i l l e s

Roseau

*Martinique
(France)*

Veracruz

Montego Bay

Kingston

ST LUCIA

Castries

*Citlaltépetl
5699*

JAMAICA

Lesser Antilles

BARBADOS

Bridgetown

*Isthmus of
Tehuantepec*

Usumacinta

Belmopan

**ST VINCENT AND
THE GRENADINES**

Kingstown

del Sur

Sierra Madre

BELIZE

Caribbean Sea

GRENADA

St George's

GUATEMALA

San Pedro Sula

Cape Gracias á Dios

*Netherlands
Antilles
(Neth.)*

Tobago

*Gulf of
Tehuantepec*

HONDURAS

Coco

*Aruba
(Neth.)*

Port-of-Spain

**TRINIDAD
AND TOBAGO**

Guatemala City

Tegucigalpa

Bonaire

Trinidad

San Salvador

Curaçao

EL SALVADOR

NICARAGUA

Lake Managua

Managua

Lake Nicaragua

**COSTA
RICA**

San José

*Panama
Canal*

**Panama
City**

▲*Chirripo
3820*

PANAMA

Mexico

UNITED MEXICAN STATES

MEXICO IS ONE OF THE LARGEST OF THE Latin American states, possessing a colorful diversity of cultures. However, rich natural resources and swift economic growth have been largely offset by the rapid rise in population.

GEOGRAPHY

Mexico forms a large tongue of mountainous land between the Pacific Ocean and the Gulf of Mexico. The southern half of the country, which lies within the tropics, has a wide variety of landscapes.

The land

Mexico lies in a geologically unstable part of the world. Many areas are subjected to frequent earthquakes and volcanic eruptions; a major earthquake in 1985 devastated Mexico City.

Much of the northern border with the United States is formed by the Rio Bravo del Norte (Rio Grande). From here the land arcs southeast toward the borders with Guatemala and Belize. The low-lying Yucatán Peninsula extends northward from the Belize border. The long and mountainous peninsula of Baja California runs parallel to the west coast of the Mexican mainland. Some 500 km (300 mi) out to sea is a group of volcanic islands called the Islas de Revillagigedo.

The broad coastal plain along the Gulf of Mexico is fringed by swamps, lagoons and sandbars, while the Pacific coastal plain is narrower and more interrupted. The area in between is occupied by the massive central plateau. This rises in the south to the Mesa Central, which is crossed by a series of mighty volcanoes, including Citlaltépetl and Popocatépetl. The plateau is bounded to the east, west and south by the mountain ranges of the Sierra Madre. In the south, beyond the Sierra Madre del Sur, the Isthmus of Tehuantepec links these highlands to the mountains on the Guatemalan border.

Climate

The climate varies both with latitude and altitude. In the far north low rainfall and extremes of temperature create desert landscapes. In the south most rain falls between May and August. Rainfall increases southeastward, with the heaviest falls along the Gulf coast and in the south.

The coastal plains and the foothills of the Sierra Madre are the *tierra caliente* ("hot land"), where temperatures are mostly uniformly high. Between August and October these areas are vulnerable to tropical hurricanes from the Gulf of Mexico and the Pacific. Above 1,000 m (3,000 ft) is the *tierra templada* ("temperate land"), which extends over much of the central plateau. The Mesa Central often reaches into the *tierra fria* ("cold land") above 2,000 m (6,000 ft). Above about 4,000 m (13,000 ft) is the *tierra helada* ("frozen land"), where temperatures are always below 10°C.

Plants and animals

Desert scrub predominates in the deserts of the north, with grasses, shrubs and succulents on higher ground. Farther south settlers have destroyed most of the ancient forests, but stands of conifers still remain on the higher slopes of the western mountains. Much of the central plateau is covered with grassland. Higher rainfall in the east allows tropical rainforest to flourish there, with temperate rainforest growing at higher altitudes. In southwestern areas , deciduous and semi-deciduous woodland grows along the Pacific coastline.

The rainforests of the south support a great diversity of animals, such as monkeys, parrots, jaguars, tapirs and anteaters. In the north, and in settled areas of the central plateau, domestic livestock have been introduced, taking the place of the native species. However, in the deserts and steppes native animals such

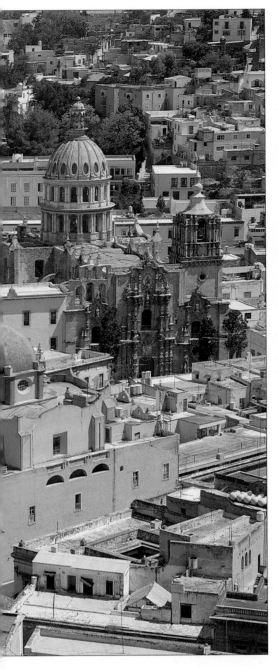

One for the road (*above*) Farm workers wearing traditional ponchos refresh themselves with a cup of pulque – a cloudy white beer made from the agave, a succulent plant.

Colonial riches (*left*) For centuries, gold and silver mines made the beautiful city of Guanajuato, with its fine baroque churches and colonial buildings, the wealthiest in Mexico. Today it is an important university town and cultural center.

as armadillos, deer, pumas and coyotes still thrive. The coastal seas abound with numerous varieties of fish.

SOCIETY

The cultural diversity and racial mix of modern Mexico owes much to its complex, colorful and often violent history.

History

Mexico had a succession of highly sophisticated early civilizations. The Olmecs reached their peak from 900–400 BC. They built impressive ceremonial centers elaborately decorated with mosaics and stone carvings. The Mayas rapidly expanded into the forests of southeastern Mexico from Guatemala, where they originated, in about 300 AD. They were a highly sophisticated, wealthy people; archaeological remains include not only majestic stone buildings and stepped pyramids, but also intricate jewelry, carved figures and hieroglyphic manuscripts that are not yet fully understood.

The Aztecs, originally a nomadic people from the arid northern highlands, migrated to Mexico's great Central Valley in about 1200. Here they became the dominant power and built their capital, Tenochtitlán, on the site of modern Mexico City, from which they established a vast empire. Their architecture, jewelry and textiles were outstanding; they used a complex calendar for time-keeping; and they practiced ritual human slaughter on a grand scale to propitiate their gods. The Aztec empire was largely destroyed by the 16th-century Spanish adventurer Hernán Cortés (1485–1547). Tenochtitlán fell in 1521, and in 1535 the conquered territory became New Spain.

Spanish rule was often harsh, especially in the 18th century, and independence was not achieved until 1821. Mexico then became a republic, adopting a federal constitution in 1824.

In 1845 Mexico tried and failed to resist the United States' annexation of Texas. A bitter war was to follow, and further territories were ceded by the Treaty of Guadalupe Hidalgo in 1848. In 1857 the introduction of a reformed constitution brought civil war. The reformers, led by Benito Juárez (1806–72), were victorious.

In 1910, following the overthrow of the dictatorial president Porfirio Díaz (1830–1915), there began a long power struggle by various revolutionary groups, who eventually formed a single National Revolutionary Party in 1929. In 1946 it

---

NATIONAL DATA - MEXICO

Land area 1,923,040 sq km (742,490 sq mi)

Climate	Altitude m (ft)	Temperatures January °C(°F)	July °C(°F)	Annual precipitation mm (in)
Mexico City	2,306 (7,564)	14 (56)	17 (63	816 (32.1)

Major physical features highest point: Ciltlaltépetl 5,610 m (18,406 ft); longest river: Rio Grande (part) 3,060 km (1,900 mi)

Population (2006 est.) 107,449,525

Form of government federal multiparty republic with two legislative houses

Armed forces army 144,000; navy 37,000; air force 11,770

Largest cities Mexico City (capital – 8,658,576); Ecatepec* (1,884,585); Guadalajara (1,632,171); Ciudad Juarez (1,496,551); Tijuana (1,482,294); Puebla (1,441,685); Nezahualcóyotl (1,229,673); León (1,156,053); Monterrey (1,123,799); Zapopan† (1,022,564)   * part of Mexico City Metropolitan area   † part of Guadalajara Metropolitan area

Official language Spanish

Ethnic composition Mestizo (Amerindian-Spanish) 60%; Amerindian or predominantly Amerindian 30%; White 9%; other 1%

Religious affiliations Roman Catholic 89%; Protestant 6%; other 5%

Currency 1 New Mexican peso (MXN) = 100 centavos

Gross domestic product (2006 est.) U.S. $1.134 trillion

Gross domestic product per capita (2006 est.) U.S. $10,600

Life expectancy at birth male 72.63 yr; female 78.33 yr

Major resources petroleum, silver, copper, gold, lead, zinc, natural gas, hydropower, timber; textiles; iron and steel; petrochemicals; beverages, fisheries, tourism, coffee, cotton, fruit and vegetables, aluminium, sorghum, sugarcane

was renamed the Party of the Institutionalized Revolution (PRI). This party carried out many important reforms, and remained the dominant political force during subsequent decades.

Government
According to the 1917 constitution, voting is mandatory for all adults over 18 years of age. The president is elected for one six-year term only. There are two legislative houses: a 500-member chamber of deputies and a 64-member senate. The deputies are elected for a three-year term – three-quarters of them directly and the rest according to the votes cast for individual parties. Senators are elected for a six-year term – two for each of the 31 states and two for the Federal District. Each state elects its own governor and legislature, but the appointment of the governor of the Federal District is the prerogative of the president.

People
About one-third of Mexico's population are Amerindians, while over half are mestizos of mixed European and native descent; there are several other minority groups. The official language is Spanish, but some 50 Amerindian languages are also spoken. Most of the population is Roman Catholic.

More than two-thirds of Mexicans live in or around the three major cities of Mexico City, Guadalajara and Monterrey. Mexico City, the capital, is probably the world's most populous city. Large-scale emigration across the northern border has caused difficulties with the United States, Mexico's chief trading partner.

ECONOMY

Mexico has little good agricultural land, and much food has to be imported. Its struggle for economic growth has been hampered by falling oil prices, the 1985 Mexico City earthquake and massive foreign debt. Inflation remains high, and economic controls have been only moderately successful. Emergency austerity measures were taken in 1995.

Agriculture
Most agricultural land is on the central plateau, though irrigation projects in the north have created new farmlands there. Since 1910 much of this land has been occupied by small subsistence farms. Export crops include coffee, sugar cane and cotton, while staples include maize, squash and kidney beans. Cattle ranching

is largely concentrated in the north, and some meat is exported. Fishing is largely for export, the most important being tuna, anchovies, sardines and shrimps.

About one-fifth of Mexico is forested, mostly in the east and south. These forests are exploited for hard and fragrant woods, and for chicle, the base material

for chewing gum. Many forested areas, however, are threatened by large-scale clearance for agriculture.

Industry
Mexico is the world's leading source of silver; other commercial minerals include zinc, copper and lead. Ore from Durango

Cultural revival (*above*) A member of an Aztec dance troupe helps to keep ancient traditions alive. Mexico's rich Amerindian culture inspires many modern artists, sculptors and architects.

Monte Albán (*left*), the center of ancient Zapotec culture, was built on a flattened hilltop rising 300 m (1000 ft) from the valleys below. This impressive site is a lasting tribute to the sophistication and skill of Mexico's indigenous civilizations.

in and around Mexico City. Products include industrial chemicals, machinery and transportation equipment.

Transportation and communications
Tourism is an important foreign-currency earner, and has stimulated the development of both road and air networks. The extensive railroad system is slow and unreliable, but the road network has almost doubled in extent since the 1940s and is continuing to expand. There are two state-owned airlines, Aeroméxico and Mexicana, with more than 75 airports serving major cities, resorts and international destinations.

Mexico has a free press, with several hundred daily newspapers.

Health and education
Subsidized health care is available to many, but poor facilities in rural areas are reflected in higher rates of death and disease, particularly malaria. Adequate housing is a major problem, particularly in the cities. There are minimum-wage laws, but rural poverty is widespread.

Education is free, and literacy levels are high as a result of an extensive campaign. However, rural areas lack facilities, and there are few secondary schools outside the cities.

in the west of the country is used for iron and steel production at Monterrey in the northeast, fueled by coking coal from the nearby Sabinas fields. Mexico's massive hydrocarbon resources now provide the majority of foreign-exchange earnings. The state-owned company Pemex, Mexico's largest employer, controls the exploration, production and marketing of natural gas and petrochemical products.

Petrochemicals are the chief energy source. Hydroelectricity is also significant, however, and geothermal energy is being increasingly exploited. There is a nuclear power plant at Laguna Verde. Manufacturing industry is concentrated

Guatemala

REPUBLIC OF GUATEMALA

T HE REPUBLIC OF GUATEMALA LIES ON THE southwestern border of Mexico, with coastlines on both the Atlantic and the Pacific. It also shares frontiers with Belize to the northeast, and Honduras and El Salvador to the southeast.

GEOGRAPHY

The heart of the country is crossed by two main mountain ranges. The Altos Cuchumatanes to the north are older and more eroded. The younger Sierra Madre range in the south includes 33 volcanoes, three of them active. The soils here are enriched by volcanic ash, making it the most fertile part of the country. In the north the highlands fall away to a large, flat, forested area called the Petén, and to the plains bordering the Gulf of Honduras.

The climate is mostly tropical with seasonal rains, though the mountains are both cooler and wetter. The lowland rainforests are rich in tropical hardwoods and rubber trees, with oaks and conifers higher up, and some mangroves along the coasts. Typical wildlife includes jaguars, peccaries and monkeys, while the rivers support crocodiles and manatees.

SOCIETY

For many centuries the area that is now Guatemala formed the heart of the ancient Maya civilization, which reached its peak from 300 to 900 AD. The empire then fell into decline, and the ruins of the Mayas' great stone cities can still be found in the remote northern lowlands. In 1523 the country was overrun by Spanish conquistadors under Pedro de Alvarado (c. 1485–1541). They set up large agricultural estates that were worked by Amerindian laborers. Between 1821 and 1847 the ruling landowners established Guatemala's independence. Their leader, Rafael Carrera (1841–65), became the first of a series of powerful dictator-presidents. A few were relatively liberal reformers whose aims were to modernize the country; Jorge Ubico (1878–1946), who ruled in the 1930s, was a protector of the Amerindians, but there were often periods of political repression.

The effects of World War II caused an economic decline, and subsequent decades were characterized by political extremism and constant divisive military coups. In 1986 democratic civilian rule was restored. However, several abortive coups and continued assassinations have shown that democracy is fragile. A new constitution was ratified in 1994 and in 1996 an accord was signed recognizing the rights of the indigenous population.

Over 50 percent of the population are of mixed Amerindian and Hispanic descent, and Roman Catholicism is the dominant religion in Guatemala.

ECONOMY

The economy is still largely agricultural. Coffee is the main crop and principal export. Bananas, cotton, sugar cane and maize are also important. Large numbers of cattle are raised on the Pacific coastal pastures. The forests provide timber and chicle, used in chewing gum. Guatemala has few mineral or energy reserves apart from small quantities of petroleum, copper and other metals. Industry is mostly confined to the processing of timber, sugar, tobacco and other agricultural products. Education is free, but resources are insufficient to provide adequate schooling or health care in rural areas.

NATIONAL DATA - GUATEMALA

Land area	108,430 sq km (41,865 sq mi)			

Climate	Altitude m (ft)	Temperatures January °C(°F)	July °C(°F)	Annual precipitation mm (in)
Guatemala	1,480 (4,856)	18 (64)	20 (68)	1,186 (46.7)

Major physical features	highest point: Tajumulco ken 4,220 m (13,845 ft)

Population	(2006 est.) 12,293,545

Form of government	multiparty republic with one legislative house

Armed forces	army 27,000; navy 1,500; air force 700

Largest cities	Guatemala City (capital - 1,024,101); Mixco* (540,348); Villa Nueva* (499,133); Quezaltenango (140,405) *Port of Guatemala City Metropolitan area

Official language	Spanish

Ethnic composition	Mestizo (Ladino) and European 59.4%; K'iche 9.1%; Kaqchikel 8.4%; Mam 7.9%; Q'eqchi 6.3%; other Mayan 8.6%; indigenous non-Mayan 0.2%; other 0.1%

Religious affiliations	Roman Catholic 75%; Protestant 25%

Currency	1 quetzal (Q) = 100 centavos

Gross domestic product	(2006 est.) U.S. $60.57 billion

Gross domestic product per capita	(2006 est.) U.S. $4,900

Life expectancy at birth	male 67.65 yr; female 71.18 yr

Major resources	petroleum, nickel, fisheries, chicle, hydropower, bananas, cardamom, cattle, coffee, cotton, lead, maize/corn, rice, sugar beet, timber, tobacco

Thriving Maya culture (*above*) Guatemala is famous for the colorful dress, folk art, village markets and religious festivals – which mingle traditional and Christian practices – of its large Maya population.

Lake Atitlán (*right*), fringed by three volcanoes, is one of Guatemala's most beautiful sights. On market days, people from miles around bring their goods in brightly colored bundles to sell in the villages along the lakeside.

Belize

BELIZE, FORMERLY BRITISH HONDURAS, IS A small country on the Caribbean coast of Central America, bordered by Mexico and Guatemala.

GEOGRAPHY

Northern Belize is mostly a low-lying and often swampy plain drained by the Belize and Honda rivers; the latter forms the Mexican border. Southern Belize rises sharply from a narrow coastal strip to the Maya Mountains – a fragmented plateau of serrated ranges; the highest point is Victoria Peak. Along the coast a chain of small coral islands forms a long barrier reef, second only to Australia's.

The climate is subtropical with marked seasonal rains, usually between June and November. The coastal area is particularly vulnerable to hurricanes; in 1961 Belize City was destroyed, and the capital was moved inland to Belmopan. Dense tropical rainforests cover nearly half the country, while drier areas are characterized by open forest and savanna. The abundant wildlife includes pumas, jaguars, crocodiles and manatees.

SOCIETY

Belize was once part of the ancient Maya empire, whose remains are still to be found there. Spanish settlement did not extend into this area, and the first recorded European colonists were British logwood cutters in the 17th century. They gradually migrated into the interior, clashing with the indigenous Amerindian peoples and repelling Spanish incursions from Guatemala. Sugar plantations were later established, worked by imported African slaves. British Honduras was declared a crown colony in 1862, but newly independent Guatemala continued to claim the country. The colony became

City on water Many of the buildings in Belize are wooden and raised on stilts above the mangrove swamps that surround the town. Once the capital, Belize is still the major port and commercial center.

independent as Belize in 1981; Britain relinquished responsibility for external defence in 1993. The country remains within the Commonwealth, and has parliamentary government on British lines.

The official language is English, but the majority of people speak a Creole dialect that reflects the complex racial mix. The largest element are descendants of black Africans. Then there are Maya Indians, Black Caribs of mixed black and Amerindian descent, and a variety of Europeans and North Americans, including farmers of the Mennonite religious sect.

ECONOMY

Agriculture is the mainstay of the economy, employing more than one-quarter of the labor force, and is being actively developed. Sugar cane is the main crop, while citrus fruits and bananas are also important. Maize, rice, kidney beans and sweet potatoes are the main domestic staples. Livestock is less important, except for beef and dairy cattle, especially on Mennonite farms.

Forestry, once the principal activity, is today less important but still yields valuable rosewood, mahogany and chicle, used in chewing gum. A great deal of timber is produced not only for export but also as a chief source of fuel, and reforestation schemes are encouraged. Cooperative fisheries provide some items such as lobsters and shrimps for export.

Mineral and fuel reserves (with the exception of timber) are scanty, so both have to be imported in quantity. Manufacturing industry is small but growing, producing food products, furniture and some clothing for export. There are virtually no railroads, but there is a good road network in the north. Education, though government-funded, is mostly run by the Roman Catholic church; literacy levels are relatively high. Medical care is free, and extensive social-security schemes are in operation.

NATIONAL DATA - BELIZE

Land area	22,806 sq km (8,805 sq mi)			
Climate	Altitude m (ft)	Temperatures January °C(°F)	July °C(°F)	Annual precipitation mm (in)
Belmopan	41 (135)	23 (73)	27 (81)	1,890 (74.4)

Major physical features	highest point: Victoria Peak 1,122 m (3,681 ft)
Population	(2006 est.) 287,730
Form of government	multiparty constitutional monarchy with two legislative houses
Armed forces	army 1,050
Largest cities	Belize City (66,140); San Ignacio (18,334); Belmopan (capital – 15,940); Orange Walk (15,927); Dangriga (11,512)
Official language	English
Ethnic composition	Mestizo 48.7%; Creole 24.9%; Maya 10.6%; Garifuna 6.1%; other 9.7%
Religious affiliations	Roman Catholic 49.6%; Protestant 27% (Pentecostal 7.4%; Anglican 5.3%; Seventh-Day Adventist 5.2%; Mennonite 4.1%; Methodist 3.5%; Jehovah's Witnesses 1.5%); other 14%; none 9.4%
Currency	1 Belize dollar (BZD) = 100 cents
Gross domestic product	(2006 est.) U.S. $2.307 billion
Gross domestic product per capita	(2006 est.) U.S. $8,400
Life expectancy at birth	male 66.43 yr; female 70.26 yr
Major resources	sugarcane,, tourism, timber, fisheries, hydropower, bananas, citrus fruits, coconuts, fish, maize/corn, rice

El Salvador

REPUBLIC OF EL SALVADOR

THE SMALLEST AND MOST DENSELY POPU-
lated country in Central America, El
Salvador lies between Guatemala to the
west and Honduras to the northeast.

GEOGRAPHY

In the south a narrow Pacific coastal plain
rises to a mountain range including more
than 20 volcanoes. Beyond a central plain,
cut by the river valleys of the Lempa, are
the Metapán and Chalatenango mountain
chains near the Honduran border. The climate is mainly hot, with heavy rains
between May and October. Temperatures
are lower in the mountains, and rainfall
there is generally much higher.

The coastal plains and foothills are
covered in savanna grasslands or deciduous forest. The trees become sparser in
the valleys and the central plains. Higher

The dangers of earthquakes Heavy debris from a
collapsing building landed on a car during this quake in
January 2001 in El Salvador. Like the rest of Central
America, the country is prone to earthquakes.

up the mountains there are temperate
grasslands, with scattered areas of pine
and deciduous forest. Large areas of the
land are cultivated, and mammals are
therefore less prolific than elsewhere in
Central America.

SOCIETY

Between 1524 and 1539 Spanish invaders
overran the Pipil Indian kingdom of
Cuzcatlán, creating the territories of San
Salvador and Sonsonate. In 1821 the area
became independent as part of the
Mexican empire. On the empire's collapse
in 1823, San Salvador and Sonsonate were
united to form the state of El Salvador.
However, even after independence, internal strife remained, and in 1931, when
there was a brief move toward democracy, there followed a succession of military dictatorships.

From 1979 to 1992 the left-wing Fara-
bundo Martí Front for National Liberation (FLMN) fought a guerrilla war
against successive governments. The conflict led to the deaths of tens of thousands
of people. A new constitution in 1983
brought in direct elections to a 60-
member legislative assembly but failed to
end the civil war. The FLMN boycotted
the 1989 elections, which were won by
the right-wing Arena party. In January
1992 the United Nations sponsored a
successful peace conference in Mexico,
and in the following January the last
FLMN troops were demobilized.

Most of the population are mestizos of
mixed Amerindian and European ancestry, and the majority are Roman Catholic.
The church played a leading role in
antigovernment movements in the early
1980s. The official language is Spanish.

ECONOMY

The economy is predominantly agricultural; the land has, in the main, been
owned by a small elite of white people,
but is now being slowly distributed by the
government. Coffee is a vital export crop,
along with cotton, maize and sugar cane.

There are few important mineral resources. Most power comes from a hydroelectric project on the Lempa river, east
of the capital, San Salvador. Massive
investment in the 1970s brought a host of
new manufacturing industries, but civil
war caused much damage to the infrastructure and the economy. In the early
1990s, fuel oil, petroleum and food processing were the major industries.

The El Salvador National Railroad operates a narrow-gauge railroad system.
There are three Pacific ports, and a road
and rail link to Puerto Barrios in
Guatemala.

The welfare system protects those in
paid employment, but unemployment is
high. Health and sanitation are often
poor, and disease is common in many
rural areas, where there are few trained
medical personnel.

Primary education is free and compulsory but the war reduced enrollment to
nearly half. There are three state universities and adult literacy had risen to 71.5
percent in 1995.

NATIONAL DATA - EL SALVADOR

Land area	20,720 sq km (8,000 sq mi)			
Climate	Altitude m (ft)	Temperatures January °C(°F)	July °C(°F)	Annual precipitation mm (in)
San Salvador	682 (2,238)	23 (73)	25 (76)	1,734 (68.2)

Major physical features	highest point: Izalco 2,386 m (7,828 ft); longest river: Lempa 320 km (200 mi)
Population	(2006 est.) 6,822,378
Form of government	multiparty republic with one legislative house
Armed forces	army 13,850; navy 700; air force 950
Largest cities	San Salvador (capital - 542,799); Soyapango* (340,110); Mejicanos* (164,407) * suburbs of San Salvador
Official language	Spanish
Ethnic composition	Mestizo 90%; White 9%; Amerindian 1%
Religious affiliations	Roman Catholic 83%; other 17%
Currency	1 United States Dollar (USD) = 100 centavos
Gross domestic product	(2006 est.) U.S. $33.2 billion
Gross domestic product per capita	(2006 est.) U.S. $4,900
Life expectancy at birth	male 67.88 yr; female 75.28 yr
Major resources	hydropower, geothermal power, petroleum, coffee, cotton, maize/corn, sugarcane, beans, cattle, poultry, rice, silver, sorghum, timber

Honduras

REPUBLIC OF HONDURAS

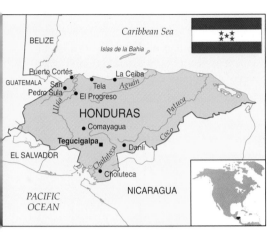

MOUNTAINOUS HONDURAS LIES BETWEEN Guatemala and El Salvador to the west, and Nicaragua to the southeast.

GEOGRAPHY

From the Caribbean coastal plain in the north, several deep river valleys cut southwestward into the central uplands; the coastal plains become broader toward the east. South of the mountains, a narrow lowland strip borders the very short Pacific coastline.

The hot, humid climate is moderated by altitude inland. The mainly seasonal rainfall is highest in the north and east, where destructive summer hurricanes may also strike, and is lowest in sheltered inland valleys and basins.

In the east, coastal mangroves and palms give way to sandy plains, pine-savanna and dense rainforest near the Nicaraguan border. The highland areas support grassland and broadleaf forests, with further savanna forest and mangroves on the Pacific coast. Although many forest areas have been cleared for agriculture, there is still a profusion of wildlife: insects, birds and reptiles as well as peccaries, pumas, jaguars and tapirs.

SOCIETY

Honduras was once an important center of the Maya civilization. It was first settled by Europeans in 1522. After a savage war with the indigenous Amerindians, the Spanish made Honduras part of Guatemala. Independence as part of Mexico came in 1821, and full independence in 1838. Government became increasingly unstable in the early 20th century, and the United States sent in marines to protect their investments, particularly in the banana trade. The 1932 election produced a dictator, General Tiburcio Carlas Andino (1876–1969), who ruled until 1949 despite the frequent uprisings. Later governments have been generally short-lived, and the military has often kept effective control. Until 1989 Honduras sheltered the Contra rebels who, with American backing, were fighting the Nicaraguan government.

The Honduran constitution has been rewritten at least 16 times since 1821. That of 1982 provides for a 128-member legislative assembly and a president, both directly elected. In practice ultimate power rests with the army. The population is still mostly rural, consisting of mestizos of mixed Amerindian and European ancestry. The official language is Spanish, and the majority are Roman Catholic.

ECONOMY

Just over one-third of the work force are farmers. Bananas and coffee are the major export crops, but Hurricane Mitch caused extensive damage in Honduras in 1998, affecting economic recovery and future growth badly. Maize, beans and rice are the most important staples, but the majority of the population is still affected by malnutrition. Mineral resources are quite considerable, but have so far remained largely unexplored. Energy comes mainly from hydroelectricity; a small amount comes from imported fuels. Small-scale manufacturing produces a wide variety of goods such as furniture, textiles, footwear, chemicals and cement.

There are few railroads, and freight and passengers rely mostly on the extensive road network, reducing the importance of domestic air services. Welfare benefits are mostly limited to urban workers. Health and sanitation are poor, and tropical diseases are common, with few doctors on hand to treat them. Education is free and compulsory, but often inaccessible to rural communities; illiteracy is relatively high.

Legacy of the Maya (*above*) A carved head at the ruins of Copán. This great city, first inhabited in 2000 BC, was the heart of Honduras' Maya civilization.

"The last paradise" (*left*) So-named by enchanted tourists, the Islas de la Bahía, 50 km (30 mi) off the coast of Honduras, were once a base from which pirates launched their attacks on Spanish ships.

NATIONAL DATA – HONDURAS				
Land area 111,890 sq km (43,201 sq mi)				
Climate		Temperatures		Annual
	Altitude m (ft)	January °C(°F)	July °C(°F)	precipitation mm (in)
Tegucigalpa	1,004 (3,294	20 (68)	23 (73)	917 (36.1)
Major physical features highest point: Cerro Las Minas 2,827 m (9,275 ft); longest river: Coco (part) 685 km (425 mi)				
Population (2006 est.) 7,326,496				
Form of government multiparty republic with one legislative house				
Armed forces army 8,300; navy 1,400; air force 2,300				
Largest cities Tegucigalpa (joint capital with Comayaguela - 893,589); San Pedro Sula (516,608); Comayagua (61,576)				
Official language Spanish				
Ethnic composition Mestizo 90%; Amerindian 7%; Black 2%; White 1%				
Religious affiliations Roman Catholic 97%; Protestant 3%				
Currency 1 Honduran lempira (HNL) = 100 centavos				
Gross domestic product (2006 est.) U.S. $22.13 billion				
Gross domestic product per capita (2006 est.) U.S. $3,000				
Life expectancy at birth male 67.75 yr; female 70.98 yr				
Major resources timber, gold, silver, copper, lead, zinc, iron ore, antimony, coal, fisheries, hydropower, bananas, beans, cattle, coffee, fruits, maize/corn, rice, shellfish, sugarcane, timber, tin, tobacco				

Nicaragua

REPUBLIC OF NICARAGUA

THE CENTRAL AMERICAN STATE OF NICAragua lies between Honduras to the northwest and Costa Rica to the south.

GEOGRAPHY

Its eastern plains are bordered with lagoons and swamps along the Caribbean or Mosquito Coast. The land rises westward toward a roughly triangular highland area bordering Honduras. Westward again are Lake Nicaragua and Lake Managua, and beyond them is a second, smaller range of about 40 volcanoes, flanked to the west by a narrow lowland belt. Some volcanoes are still active, and the area has severe earthquakes, as in 1992 when 16,000 people were made homeless.

Nicaragua's tropical climate is hotter and drier on the Pacific side, where the rainy season lasts from May to November. On the Caribbean side it lasts for at least nine months, and there is a greater risk of hurricanes. Temperatures are lower on the northern mountains. Most of the western half of the country is covered in savanna, and the eastern half in tropical rainforest.

SOCIETY

Nicaragua, like its neighbors, was conquered by Spain in the 16th century, and gained full independence in 1838. Later politics were dominated by Conservatives (the coffee and sugar plantation owners) and Liberals (artisans and small landowners). Great Britain and the United States both had strategic interests in Nicaragua, and the United States sent in military forces on two occasions to support Conservative regimes.

From 1934 until 1979 Nicaragua was ruled by a repressive military regime dominated by the Somoza family. However, in 1979 rebel forces seized power in a country devastated by civil war, and formed the left-wing Sandinista government. Until 1990 the United States supported the Contras – guerrilla opponents of the new Sandinista regime, trained in neighboring Honduras by United States advisers. Elections followed their withdrawal in 1990, in which the opposition coalition, the National Opposition Union (UNO), came to power. The president is elected by a majority, and the 92-member national assembly on a proportional basis. Constitutional deadlock ensued after the president blocked reforms passed by the national assembly in 1994.

Most of the population are mestizos of mixed Amerindian and European descent. Spanish is the official language, and most people are Roman Catholic.

ECONOMY

At the end of the 1990s, Nicaragua was showing signs of economic recovery from the effects of the civil war, but was devastated by Hurricane Mitch in 1998, estimated to cause US $1 billion damage. Despite international aid, growth will take some time to resume. The chief crops are cotton, coffee and bananas. Mineral resources are plentiful, but only gold and silver are intensively mined. Food processing, chemicals and textiles are the chief industries. Industrial development is limited.

Communications are poor except in the west, and even here only half the roads are passable when it rains. Railroads are also limited to the west, leaving eastern areas dependent on water and air transportation. Housing, sanitation and water supplies are inadequate, and tropical diseases are common. Health care is difficult to obtain despite reforms by the Sandinistas, and there are few trained doctors. There is limited social security for workers in Managua. Primary education is free, and enrollment is high. Few receive secondary education, but literacy levels are on the increase.

A fertile land Crops flourish on the rich volcanic soils of Nicaragua. Some volcanoes are active, but earthquakes are a worse threat; the capital Managua was almost totally destroyed in 1972.

NATIONAL DATA – NICARAGUA				
Land area	120,254 sq km (46,430 sq mi)			
Climate		Temperatures		Annual
	Altitude m (ft)	January °C(°F)	July °C(°F)	precipitation mm (in)
Managua	55 (180)	26 (79)	27 (81)	1,141 (44.9)
Major physical features	highest point: Cerro Mogotón 2,103 m (6,900 ft)			
Population	(2006 est.) 5,570,129			
Form of government	multiparty republic with one legislative house			
Armed forces	army 12,000; navy 800; air force 1,200			
Capital city	Managua (1,007,811)			
Official language	Spanish			
Ethnic composition	Mestizo 69%; White 17%; Black 9%; Amerindian 5%			
Religious affiliations	Roman Catholic 58.5%; Evangelical 21.5%; Moravian 1.5%; Jehovah's Witnesses 0.9%; other 1.6%; none 16%			
Currency	1 córdoba (C) = 100 centavos			
Gross domestic product	(2006 est.) U.S. $16.83 billion			
Gross domestic product per capita	(2006 est.) U.S. $3,000			
Life expectancy at birth	male 68.55 yr; female 72.81 yr			
Major resources	gold, silver, copper, tungsten, lead, zinc, timber, fisheries, bananas, cattle, coffee, cotton, maize/corn, rice, sugarcane			

Costa Rica

REPUBLIC OF COSTA RICA

COSTA RICA LIES ON THE CENTRAL AMERICAN isthmus between Nicaragua to the north and Panama to the east.

GEOGRAPHY

From the Caribbean lowlands in the northeast the land rises gently to a mountainous backbone. The Cordillera de Guanacaste, including four major volcanoes, runs southeastward from the Nicaraguan border to meet the much higher Cordillera Central. Between this and the volcanic Cordillera de Talamanca in the southeast lies the Meseta or Valle Central, where the capital, San José, lies in an area also subject to both volcanic eruptions and earthquakes. The narrow Pacific coastal belt and the Caribbean coast are lined with mangrove swamps and white, sandy beaches.

The climate is generally hot and rainy, but there is a marked dry season on the Pacific coast, whereas the Caribbean lowlands have rain throughout the year. Both areas are heavily forested, with an enormous variety of birds, insects and reptiles, and a mixture of South and North American mammals. The higher Central and Talamanca ranges are warm and temperate, with dry seasons on the more sheltered southern slopes.

SOCIETY

Costa Rica was not fully controlled by Spain until 1570. Its later history was broadly similar to that of the rest of Central America until independence in 1838. Coffee exports put the country on a sound financial footing, and its isolation largely protected it from the more dramatic events in neighboring lands. In 1890 the free election of President José Joaquin Rodríguez set a pattern of democracy that survived revolutions in 1917 and 1948. The new constitution of 1949 disbanded the army to prevent any future military coups. The president and the 57-member legislative assembly are each directly elected for a four-year term. More recently Costa Rica has suffered the effects of eathquake damage, falling coffee prices, foreign debt and an influx of refugees from its troubled neighbors.

The population profile is unusual for Central America in that most people are of European descent, with only a minority of mestizos of mixed Amerindian, European and African extraction. Most are Roman Catholic, and Spanish is the official language.

ECONOMY

The main cash crops are coffee and bananas. Maize, beans and rice are the chief staples, but food still has to be imported. Timber resources have enormous potential, and there are significant bauxite deposits. Hydroelectricity is being increasingly developed, but much electrical power still comes from diesel plants. Manufactures for export include medicines, paper products, sheet metal and electrical machinery.

Apart from the Pan-American Highway, the limited road network centers on the Valle Central, which is also linked by railroad to ports on the Pacific and Caribbean coasts. There is an international airport at Juan Santa Maria, near the capital, and a few other airports serve domestic flights. There are five daily

A craftsman at work in Sarchí, Costa Rica's handicrafts center. This prosperous and peaceful country – it has no army – has largely managed to avoid the political conflict that has troubled so much else of Central America.

newspapers and a single television network; there is no government censorship.

Welfare benefits are provided for those in employment, and a preventive health program is in place. Rural areas, however, are less well served and malnutrition and disease are a problem. Education is free and compulsory, enrollment is high, and so are literacy levels.

NATIONAL DATA – COSTA RICA

Land area	50,660 sq km (19,600 sq mi)			
Climate		Temperatures		Annual
	Altitude m (ft)	January °C(°F)	July °C(°F)	precipitation mm (in)
San José	1,146 (3,760)	19 (67)	21 (69)	1,866 (73.4)

Major physical features	highest point: Mount Chirripó 3,820 m (12,533 ft)

Population	(2006 est.) 4,075,261

Form of government	multiparty republic with one legislative house

Armed forces	paramilitary 8,400

Capital city	San José (342,977)

Official language	Spanish

Ethnic composition	White (including Mestizo) 94%; Black 3%; Amerindian 1%; Chinese 1%; other 1%

Religious affiliations	Roman Catholic 76.3%; Evangelical 13.7%; Jehovah's Witnesses 1.3%; other Protestant 0.7%; other 4.8%; none 3.2%

Currency	1 Costa Rican colón (CRC) = 100 céntimos

Gross domestic product	(2006 est.) U.S. $48.77 billion

Gross domestic product per capita	(2006 est.) U.S. $12,000

Life expectancy at birth	male 74.43 yr; female 79.74 yr

Major resources	hydropower, coffee, bananas, timber, bauxite, cattle, cocoa, gold, iron ore, maize/corn, oranges, rice, silver, sugarcane, sulfur

Panama

REPUBLIC OF PANAMA

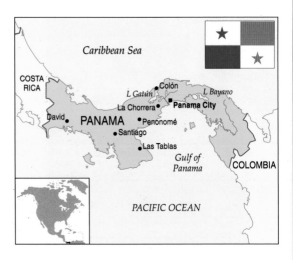

$\mathbf{P}$ANAMA IS A STRATEGICALLY POSITIONED country forming an S-shaped neck of land that links together Central and South America.

GEOGRAPHY

At its narrowest point, in the center of the country, the Panama Canal links the Caribbean Sea and the Pacific Ocean. West of the canal the mountains of the Serranía de Tabasara are flanked by narrow coastal plains on either side. In the southeast of this area the land rises once again to the mountains of the Azuero Peninsula. East of the canal, two mountain arcs run parallel to the Pacific and Caribbean coastlines.

The climate is generally hot and wet

The Panama Canal (*above*) links the Pacific and the Atlantic Oceans. The trade generated by this great feat of engineering – a tourist attraction in itself – is the mainstay of the Panamanian economy.

throughout the year, but the mountains are cooler, and there is a marked dry season along the Pacific coast. Most of the country is forested, with tropical rainforest predominating. Plant and animal life are rich and abundant except in the mainly deforested Canal Zone.

SOCIETY

After 300 years of Spanish rule, Panama achieved freedom from Spain in 1821, though it remained a part of neighboring Colombia. The California Gold Rush of 1849 provided the incentive for building a canal across the Central American isthmus that would make the journey from the eastern United States to the booming western states easier and faster than overland travel. The project finally got under way in 1879, but foundered ten years later. In 1903 the United States helped Panama to secure independence from Colombia in return for perpetual control of the 16 km (10 mi) wide Canal Zone. The canal was finally opened to traffic in 1914.

In 1977, after many years of unrest, a process began whereby formal control of the Canal Zone was gradually to be returned to Panama. Ownership of the canal itself was transferred in 2000. The 1983 constitution provided for the election of a president, two vice-presidents and a 72-member legislative assembly, all of these for a five-year term. However, political unrest continued, and in 1989 General Manuel Noriega (b. 1940), who had assumed effective power, declared a "state of war" against the United States,

who had accused him of drug-running and election rigging. An American invasion in 1989 ousted Noriega, who in January 1990 was flown to Miami in the United States to stand trial on drug-trafficking charges. In 1992 he was found guilty and jailed.

Most Panamanians are mestizos or mulattos of mixed Amerindian, European and African descent; the official language is Spanish. The birth rate is the lowest in Central America, but economic growth is slower than population growth.

ECONOMY

The chief export crops are bananas, coffee, and sugar. Rice, beans and corn are the main staples. Livestock is also important, and Panama is largely self-sufficient in food. Fishing is encouraged, and shrimps are now a major export. Mineral resources are scanty except for copper ore, and power comes largely from hydroelectricity. The industrial base around Panama City manufactures consumer goods, building materials and also petroleum products, as a result of the construction of a trans-Panamanian oil pipeline.

Most of the country's revenue comes from the service industries, including finance, trade and tourism. The main earners are the Panama Canal and the Colón Free Zone, which is one of the largest trading centers in the world. There is an extensive road system, and a railroad runs parallel to the canal. There are good port facilities on both coasts.

Welfare services are well developed, and medical facilities are adequate in most parts of the country, though tuberculosis is common in urban slum areas. Education is free and compulsory, and higher education is available.

NATIONAL DATA - PANAMA

Land area	75,990 sq km (29,340 sq mi)			
Climate		Temperatures		Annual
	Altitude m (ft)	January °C(°F)	July °C(°F)	precipitation mm (in)
Panama City	36 (118)	26 (79)	27 (81)	2,907 (114.4)

Major physical features highest point: Volcán Barú 3,475 m (11,401 ft); largest lake: Gatún 430 sq km (166 sq mi)

Population (2006 est.) 3,191,319

Form of government multiparty republic with one legislative house

Armed forces paramilitary 11,800

Capital city Panama City (403,808)

Official language Spanish

Ethnic composition Mestizo 70%; Amerindian and mixed (West Indian) 14%; White 10%; Amerindian 6%

Religious affiliations Roman Catholic 85%; Protestant 15%

Currency 1 Panamanian balboa (PAB) = 100 centesimos

Gross domestic product (2006 est.) U.S. $25.29 billion

Gross domestic product per capita (2006 est.) U.S. $7,900

Life expectancy at birth male 72.68 yr; female 77.87 yr

Major resources copper, shrimps, hydropower, tourism, coffee, bananas, cattle, maize/corn, rice, sugarcane, timber

Cuba

REPUBLIC OF CUBA

CUBA IS AN ISLAND STATE IN THE CARIBBEAN lying directly south of Florida. It consists of a single long island, flanked by the much smaller Isla de la Juventud (Isle of Youth) to the south.

GEOGRAPHY

The main island of Cuba is characterized by wide stretches of lowland divided by three main mountain areas. The largest of these, in the east, comprises the Sierra del Cristal and the Sierra Maestra along the southeast coast, the highest point being Pico Turquino. The lower central chain crosses the island to the east of Cienfuegos, while the Sierra de Los Organos runs along the northwestern tip beyond the capital, Havana. The coastline is irregular, and is lined with numerous mangroves, beaches and coral reefs.

The climate is semitropical, with heavy seasonal rainfall and a tendency to hurricanes. Rivers also vary with the seasons, and only a few are navigable. Although there has been considerable deforestation, the island's forests are still extensive, including tropical near-jungle.

SOCIETY

Cuba was among the earliest Spanish settlements after Christopher Columbus (1451–1506) landed there in 1492. From the early 16th century it was divided into large estates, worked at first by enslaved Amerindians, then by African slaves until slavery was finally abolished in 1886. Independence from Spain was achieved in 1899 after two uprisings (1868–78, 1895–99), the latter led by Jose Marlí (1853–95) and in which US troops became involved – they finally left in 1902. In 1959 the decaying regime of Fulgencio Batista (1901–73) was ousted by the communist revolutionary movement led by Fidel Castro (b.1927). Castro remodeled the Cuban state as a personal dictatorship on Soviet lines, but significantly improved the living standards of the poor. The United States imposed a trade embargo, and supported an unsuccessful countercoup in the Bay of Pigs on Cuba's south coast in 1961. From then onward, Cuba became increasingly dependent on the former Soviet Union. First direct elections were held in 1994, but only approved Communist Party candidates stood. Further relaxation of hard-line policies were seen in January 1998 when the Pope visited Cuba .

ECONOMY

The suspension of Soviet financial aid and technical advice in the early 1990s severely affected the Cuban economy. Gross domestic product declined by 35 percent during the period 1989–93 and growth was only slowly resumed in 1994. However, severe hurricanes in 1993, 1994 and 1996 devastated the sugar cane harvest and the economy slumped badly. The main cash crops are sugar cane, tobacco, citrus fruits and coffee. There has been some relaxation of state control of agriculture and limited free trading is allowed. Food has to be imported to meet domestic needs. Fishing, chiefly for tuna, has been developed with government support.

Large nickel deposits supply Cuba's other main export. Other reserves include copper, chromite and iron ore. Some petroleum has been found, but most fuel has to be imported. The traditional industries are food and tobacco processing, but this sector has been greatly diversified to include such items as steel, fertilizers, machinery and electrical goods.

The transportation network is well developed, with an emphasis on public service. Social benefits and health care are available, and literacy is high.

Keeping the revolution alive (*above*) Fidel Castro – Central America's most charismatic and durable leader – holds the local crowd spellbound with his oratory.

Money to spend (*left*) Automobiles parked proudly outside their owners' homes attest to good wages; yet despite many advances in living standards, Cuba has failed to industrialize and diversify its economy.

NATIONAL DATA – CUBA

Land area	110,860 sq km (42,803 sq mi)			
Climate		**Temperatures**		**Annual**
	Altitude m (ft)	January °C(°F)	July °C(°F)	precipitation mm (in)
Havana	49 (161)	22 (72)	28 (82)	1,189 (46.8)

Major physical features highest point: Turquino 2,000 m (6,560 ft); longest river: Cauto 249 km (155 mi)

Population (2006 est.) 11,382,820

Form of government multiparty republic with one legislative house

Armed forces army 38,000; navy 3,000; air force 8,000

Largest cities Havana (capital - 2,164,362); Santiago de Cuba (570,503); Camagüey (355,027); Holguín (330,116); Guantánamo (280,024); Santa Clara (256,184);

Official language Spanish

Ethnic composition Mulatto 51%; White 37%; Black 11%; Chinese 1%

Religious affiliations nonreligious 55.2%; Roman Catholic 39.7%; Afro-Cuban 1.7%; Protestant 3.4%

Currency 1 Cuban peso (CUP) or Convertible peso (CUC) = 100 centavos

Gross domestic product (2006 est.) U.S. $44.54 billion

Gross domestic product per capita (2006 est.) U.S. $3,900

Life expectancy at birth male 75.11 yr; female 79.85 yr

Major resources cobalt, nickel, iron ore, chromium, copper, salt, timber, silica, petroleum, tourism, beans, cassava, coffee, livestock, maize, oranges, rice, sugarcane, sweet potatoes, tobacco, tropical fruits

Bahamas

THE BAHAMAS ARE A LONG ARCHIPELAGO OF some 700 islands and numerous cays (reefs) lying in the Caribbean Sea southeast of Florida, and north of Cuba.

GEOGRAPHY

The islands are very much alike; they are all exposed segments of a single coralline limestone shelf. They are low-lying, apart from a few hills formed by windblown sand, and their coastlines are fringed with mangroves, lagoons and coral reefs. The climate is mild and humid, but the porous limestone of the landscape means that rainfall is easily lost. Caribbean pine forest covers much of the larger islands, such as Andros, New Providence, Grand Bahama and Great Abaco. The teeming birds and colorful sea life along the reefs attract numerous tourists, who create considerable environmental problems.

SOCIETY

The native Arawak people were forcibly removed by Spanish slave raiders, who put them to work on Hispaniola. In the 17th century British settlers arrived from Bermuda, but the colony degenerated into a pirate haven. Numbers were later swelled by Loyalist refugees from the American War of Independence (1775–83). The islands thrived briefly on illicit trade with the United States during the Civil War (1861–67) and later during Prohibition in the 1920s, but they gained more lasting prosperity from tourism following World War II. Internal self-government was granted by Britain in 1967, and full independence in 1973. Government is along British lines, with a governor-general representing the sovereign as head of state.

ECONOMY

Since only a tiny percentage of the land is arable, there is little agriculture beyond fruit and vegetable growing. Salt is the main mineral produced, although refining of imported petroleum for export is a major industry; others include pharmaceuticals, rum distilling and fruit canning. International banking is important.

Tourism is the country's economic mainstay. Poor road provision is offset by good air and sea communications. Education, social benefits and health care are widely available.

NATIONAL DATA - BAHAMAS				
Land area 10,070 sq km (3,888 sq mi)				
Climate		Temperatures	Annual	
	Altitude m (ft)	January °C(°F)	July °C(°F)	precipitation mm (in)
Nassau	4 (13)	21 (70)	28 (82)	1,389 (54.6)
Major physical features highest point: Cat Island 63 m (206 ft); largest island: Andros 5,957 sq km (2,300 sq mi)				
Population (2006 est.) 303,770				
Form of government multiparty constitutional monarchy with two legislative houses				
Armed forces Royal Bahamian Defense Force 860				
Capital city Nassau (235,105)				
Official language English				
Ethnic composition Black 85%; White 12%; Asian and Hispanic 3%				
Religious affiliations Baptist 35.4%; Anglican 15.1%; Roman Catholic 13.5%; Pentecostal 8.1%; Church of God 4.8%; Methodist 4.2%; other Christian 15.2%; none or unspecified 2.9%; other 0.8%				
Currency 1 Bahamian dollar (BSD) = 100 cents				
Gross domestic product (2006 est.) $6.476 billion				
Gross domestic product per capita (2006 est.) $21,300				
Life expectancy at birth male 62.24 yr; female 69.03 yr				
Major resources salt, calcium carbonate, timber, tourism, fruit and vegetables, limestone, okra, shellfish				

Haiti

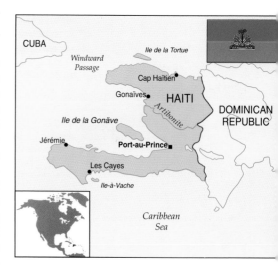

HAITI, ONE OF THE POOREST COUNTRIES IN the world, occupies the western third of the island of Hispaniola in the central Caribbean east of Cuba.

GEOGRAPHY

Two mountainous peninsulas enclose the central plain of the Artibonite river. The northern range, called the Massif du Nord, extends into the heart of the Dominican Republic, where it becomes the Cordillera Central. The climate is generally hot with seasonal rains, but temperature and rainfall patterns vary. Some areas are practically desert.

Echoes of Africa Over 90 percent of Haiti's people are descendants of African slaves. Haiti became the world's first black republic in 1804. Today, social services and health care are poor, and education, though free and compulsory, is poor outside the capital.

Jamaica

SOCIETY

Hispaniola's indigenous Arawak peoples were wiped out by the first Spanish conquistadors, through enslavement and disease. In the 17th century the west of the island became the French colony of Saint Domingue. Hundreds of thousands of slaves were used to work the plantations. The majority of today's population are Creole-speaking descendants of these Africans. Their animist religions mixed with the Roman Catholic faith, giving rise to the religious cult known as Voodoo.

The French were driven out in 1804, but in the next century Haiti experienced numerous brutal dictatorships and revolutions until the United States assumed control from 1915 to 1934. More brief dictatorships followed, until in 1957 François "Papa Doc" Duvalier (1907–71) was elected president and established a police state. A military coup in 1986 deposed his son. In 1991 Jean Bertrand Aristide was elected president, but was overthrown by a military coup. He was restored to power in 1994 with US support. In 1995 a UN peacekeeping force took over and Renee Préval succeeded Aristide.

ECONOMY

Haiti's impoverished population depends largely on subsistence agriculture. Some coffee and sugar cane are grown for export. Small amounts of bauxite, copper and other metals are mined.

NATIONAL DATA – HAITI

Land area	27,560 sq km (10,641 sq mi)			
Climate		Temperatures		Annual precipitation
	Altitude m (ft)	January °C(°F)	July °C(°F)	mm (in)
Port-au-Prince	37 (121)	26 (79)	29 (84)	1,353 (53.2)
Population	(2006 est.) 8,308,504			
Form of government	multiparty republic with two legislative houses			
Armed forces	no armed forces			
Capital city	Port-au-Prince (1,321,522)			
Official language	English			
Ethnic composition	Black 95%; Mulatto and White 5%			
Religious affiliations	Roman Catholic 80%; Protestant 16% (Baptist 10%; Pentecostal 4%; Adventist 1%; other 1%); none 1%; other 3% (approximately half the population practices Voodoo)			
Currency	1 gourde (HTG) = 100 centimes			
Gross domestic product	(2006 est.) U.S. $14.56 billion			
Gross domestic product per capita	(2006 est.) U.S. $1,800			
Life expectancy at birth	male 51.89 yr; female 54.6 yr			
Major resources	bauxite, copper, calcium carbonate, gold, marble, hydropower, coffee, sugarcane, bananas, cassava, maize/corn, rice, sisal, sorghum, timber			

THE LUSH, TROPICAL ISLAND STATE OF Jamaica lies in the Caribbean south of Cuba and west of Haiti.

GEOGRAPHY

Apart from the irregular coastal plains and several fertile river basins, mostly in the south and west, the land rises sharply to a limestone plateau that is heavily eroded. This is especially true of the Cockpit Country toward the northwest, with its typical karst scenery of bare limestone pavements and underground cave systems. Several mountain ranges rise above the plateau, the highest being the Blue Mountains in the east.

The climate is mainly tropical, with even temperatures all the year round. The mostly seasonal rains fall heavily on the northeastern slopes. These are still clad in the tropical rainforest that once covered most of the island, but the southern lowlands in the rainshadow of the mountains, support only savanna scrub.

SOCIETY

Jamaica's indigenous Arawak peoples were wiped out by the original Spanish settlers, and African slaves were imported to work the land. The island was captured by Britain in 1655; it soon became an important base for buccaneers, who mounted attacks on Spanish ships. The Spanish, in the meantime, gave assistance to the Maroons – escaped slaves who had taken refuge in the Cockpit Country, and who harried the British for nearly two centuries. Jamaica became an immensely profitable colony through its slave-worked sugar cane plantations, but this prosperity ended with the abolition of slavery in 1833. In 1866 Jamaica became a crown colony, and banana-growing was introduced.

Local self-government was introduced after World War II, leading to full independence in 1962. Jamaica is a constitutional monarchy within the Commonwealth. The sovereign is represented by the governor-general, and a two-chamber parliament on the British model is headed by a prime minister and cabinet.

The population is mostly English-speaking and of African descent, but there are several ethnic minorities, including Indians and Chinese. Life for many of the inhabitants is harsh, and poverty-stricken slums sit side by side with the wealthy tourist areas. Rastafarianism, the religious-political movement that regards Ethiopia, in Africa, as

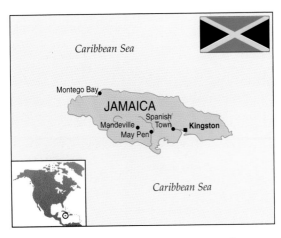

the spiritual homeland of black people, originated in the island.

ECONOMY

Agriculture has declined but remains an important sector. Sugar cane and bananas are still the main cash crops, along with coffee, cocoa and fruit; peppers and ginger are also grown. Staples are grown mostly at subsistence levels and for local sale. Fisheries are significant. The main mineral resource is bauxite, and alumina processing is one of the main industries. Other varied export activities include printing, textile manufacture and food processing, including rum distilling and sugar production. Tourism provides the largest source of foreign income.

The transportation network is generally good, and health, welfare and housing are government priorities. Education is free and widely available at all levels.

NATIONAL DATA – JAMAICA

Land area	10,831 sq km (4,182 sq mi)			
Climate		Temperatures		Annual precipitation
	Altitude m (ft)	January °C(°F)	July °C(°F)	mm (in)
Kingston	34 (112)	26 (79)	29 (84)	813 (32)
Major physical features	highest point: Blue Mountain Peak 2,256 m (7,402 ft)			
Population	(2006 est.) 2,758,124			
Form of government	constitutional monarchy with two legislative houses			
Armed forces	army 2,500; navy 190; air force 140			
Capital city	Kingston (586,630)			
Official language	English			
Ethnic composition	Black 90.9%; East Indian 1.3%; White 0.2%; Chinese 0.2%; Mixed 7.3%; other 0.1%			
Religious affiliations	Protestant 61.3% (Church of God 21.2%; Seventh-Day Adventist 9%; Baptist 8.8%; Pentecostal 7.6%; Anglican 5.5%; others 9.2%); Roman Catholic 4%; other including some spiritual cults 34.7%			
Currency	1 Jamaican dollar (JMD) = 100 cents			
Gross domestic product	(2006 est.) U.S. $12.71 billion			
Gross domestic product per capita	(2006 est.) U.S. $4,600			
Life expectancy at birth	male 71.54 yr; female 75.03 yr			
Major resources	bauxite, gypsum, limestone, rum, coffee, bananas, cocoa, citrus fruits, coconuts, ginger, molasses, pimiento, sugarcane, tourism			

Dominican Republic

THE DOMINICAN REPUBLIC OCCUPIES THE eastern part of the Caribbean island of Hispaniola. It is twice as large as its western neighbor Haiti.

GEOGRAPHY

The landscape is mountainous but varied. At its heart lies the mighty Cordillera Central, which extends into Haiti as the Massif du Nord. The highest peak, Pico Duarte, is also the highest point in the Caribbean islands. There are several lesser ranges along the north coast, in the east and in the southwest bordering Haiti. These are interspersed by extensive lowlands, including the long Cibao Valley in the north, and much of the south coast to the east of the capital, Santo Domingo.

The climate is mainly tropical with seasonal rains, but the mountainous terrain produces considerable variations in temperature and rainfall. The north and east have high rainfall, fertile soils and lush evergreen forests. Low-lying areas in the south are dry to the point of desert, supporting only savanna scrub.

SOCIETY

Hispaniola was first colonized by Spain, but in the 17th century the western third was ceded to France. The rest remained Spanish as Santo Domingo, but it later came under French control too. The French imported African slaves in great numbers, and established a profitable plantation economy. In 1821 Santo Domingo declared its independence as the Dominican Republic. However, it was immediately occupied by Haiti until Juan Pablo Duarte (1813–76) secured independence for the Dominican Republic in 1844. In the 1860s the country became Spanish again for a short time.

The republic has largely been ruled by dictators, with only brief periods of democratic rule. From 1916 to 1924 it was occupied by the United States, who did much to modernize the country. In 1930 Rafael Trujillo Molina (1891–1961) seized power, but his relentless suppression of his opponents led to his assassination in 1961. In 1978, for the first time ever, the government changed hands in a democratic election. Democracy has since been maintained, but with much civil unrest.

The majority of the rapidly growing population are mestizos of mixed European and African descent, but there are several other ethnic groups.

ECONOMY

The economy is still largely agricultural. Exportable cash crops are grown on larger plantations; these include cocoa, coffee, tobacco and especially sugar cane. Staples such as rice and corn are produced mostly by low-output subsistence farming.

Mineral reserves are varied, but only rock salt, bauxite, nickel and some other metals are exploited in quantity. Industries are mostly light, ranging from sugar refining to textiles and fertilizers. Tourism and telecommunications led to vigorous economic growth in the late 1990s. A social-security system offers reasonable benefits, while health care is limited. Education, though free, is not available in isolated rural districts.

NATIONAL DATA - DOMINICAN REPUBLIC

Land area 48,380 sq km (18,680 sq mi)

Climate	Altitude m (ft)	Temperatures January °C(°F)	July °C(°F)	Annual precipitation mm (in)
Santo Domingo	17 (56)	24 (76)	27 (81)	1,447 (56.9)

Major physical features highest point: Pico Duarte 3,175 m (10,417 ft)

Population (2006 est.) 9,183,984

Form of government multiparty republic with two legislative houses

Armed forces army 15,000; navy 4,000; air force 5,500

Largest cities Santo Domingo (capital - 2,306,138); Santiago (589,358); San Pedro de Macorís (235,467)

Official language Spanish

Ethnic composition Mixed 73%; White 16%; Black 11%

Religious affiliations Roman Catholic 95%; other 5%

Currency 1 Dominican peso (DOP) = 100 centavos

Gross domestic product (2006 est.) U.S. $73.74 billion

Gross domestic product per capita (2006 est.) U.S. $8,000

Life expectancy at birth male 70.21 yr; female 73.33 yr

Major resources nickel, bauxite, gold, silver, tourism, ferronickel, cocoa, coffee, cotton, flowers, mangoes, maize/corn, oranges, platinum, rice, salt, sugar beet, tobacco, tomatoes

Saint Kitts-Nevis

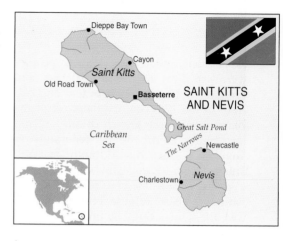

THE ISLANDS OF SAINT KITTS AND NEVIS FORM part of the western chain of the Leeward Islands in the eastern Caribbean.

GEOGRAPHY

The capital, Basseterre, is on Saint Kitts (Saint Christopher) – a mountainous volcanic island rising to Mount Misery. Nevis, to the south of a channel called The Narrows, is similarly volcanic and surrounded by coral reefs. The warm, humid climate is cooled by northeasterly trade winds. The interior has lush vegetation, including tropical rainforest on some mountain slopes. The lower slopes have mostly been cleared for cultivation.

The volcanic island (below) of Saint Kitts attracted early sugar cane planters with its verdant vegetation. Tourism now supplements the island's sugar economy.

Antigua and Barbuda

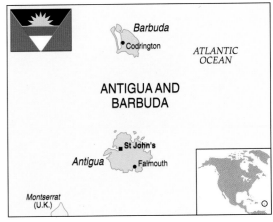

SOCIETY

The British colony of Saint Kitts was founded by Sir Thomas Warner in 1623, but ownership was disputed by the French until 1783. In 1882 Saint Kitts and the neighboring islands of Nevis and Anguilla were united in a federal state. Anguilla became unhappy with the Saint Kitts administration, and declared unilateral independence in 1967. Saint Kitts-Nevis became fully independent in 1983. The British sovereign, represented by a governor-general, is head of state. The prime minister heads a 15-member parliament; most are elected, but a few are appointed. Nevis enjoys a degree of autonomy. Most of the population are black or are mixed-race descendants of former slaves. The official language is English, and the main religious groups are Anglican and Methodist Christians.

ECONOMY

The main cash crops are sugar cane on Saint Kitts, cotton and coconuts on Nevis. Other major exports include footwear, fabrics and electronic goods. Tourism is a growing source of revenue, thanks to the fine beaches. There are good roads and ferry connections, a deep-water harbor at Basseterre and an airport on each island. Education is free and compulsory, and literacy levels are high.

Whiling the time away (*above*) Now at leisure after a lifetime working in the canefields, an old man surveys the scene. The sugar cane industry, once the sole basis of Antigua's economy, has been closed down and is now supplanted by tourism.

T HE ISLAND STATE OF ANTIGUA AND BARBUDA is part of the Leeward Islands group in the eastern Caribbean, and is of considerable strategic importance for the United States, which operates two military bases from the islands.

GEOGRAPHY

The low-lying island of Antigua has been mostly deforested, and includes an area of volcanic rock around Boggy Peak in the southwest. The coral island of Barbuda to the north is mostly flat and well wooded. The tiny islet of Redonda to the southwest is steep and totally uninhabited. The climate is warm and humid, but there is little fresh water, and the islands are subject to seasonal hurricanes.

SOCIETY

Barbuda was colonized by the British in 1628, and Antigua four years later. In 1685 the Codrington family leased Barbuda from the British, at a rent of one fat pig a year, but it reverted to the crown in the 19th century after the ending of slavery. In 1967 Antigua achieved internal self-government, but in 1979 demanded full independence. At the same time Barbuda sought secession from Antigua. The islands achieved their full independence together in 1981. The British sovereign, represented by a governor-general, is head of state. The parliament has two chambers and consists of an appointed senate and an elected lower house. Most of the population live on the island of Antigua. The official language is English, and the majority of the people are Anglican Christians.

ECONOMY

Tourism accounts for the majority of the islands' revenue; there are numerous white, sandy, palm-fringed beaches, a number of casinos, and some fine 18th-century architecture in the capital, St John's, which also has a splendid cathedral in the gothic style and a deep-water harbor. The roads are surfaced, and there are two airports. Good health-care and social-security schemes have been established. Basic education is free and compulsory, literacy levels are fairly high, and higher education is also available.

NATIONAL DATA – SAINT KITTS AND NEVIS

Land area 261 sq km (101 sq mi)

Climate	Altitude m (ft)	Temperatures January °C(°F)	July °C(°F)	Annual precipitation mm (in)
Basseterre	15 (49)	24 (75)	27 (81)	1,375 (54.2)

Major physical features highest point: Mount Misery 1,156 m (3,792 ft)

Population (2006 est.) 39,129

Form of government multiparty constitutional monarchy with one legislative house

Armed forces no armed forces

Capital city Basseterre (12,849)

Official language English

Ethnic composition Black 90.5%; Mixed 5.0%; Indian 3.0%; White 1.5%

Religious affiliations Protestant 76.4%; Roman Catholic 10.7%; other 12.9%

Currency 1 East Caribbean dollar (ECD) = 100 cents

Gross domestic product (2002 est.) U.S. $339 million

Gross domestic product per capita (2005 est.) U.S. $8,200

Life expectancy at birth male 69.56 yr; female 75.42 yr

Major resources tourism, coconuts, livestock, sugarcane, vegetables, cotton

NATIONAL DATA – ANTIGUA AND BARBUDA

Land area 443 sq km (171 sq mi)

Climate	Altitude m (ft)	Temperatures January °C(°F)	July °C(°F)	Annual precipitation mm (in)
St. John's	10 (33)	25 (77)	28 (83)	1,052 (41.4)

Major physical features highest point: Boggy Peak 405 m (1,330 ft); largest island: Antigua 280 sq km (108 sq mi)

Population (2006 est.) 69,108

Form of government multiparty constitutional monarchy with two legislative houses

Armed forces army 125; navy 45

Capital city St. John's (25,498)

Official language English

Ethnic composition Black 94.4%; Mulatto 3.5%; White 1.3%; others 0.8%

Religious affiliations Anglican 33%; other Protestants 31%; Roman Catholic 12%; other 23%; none 1%

Currency 1 East Caribbean dollar (ECD) = 100 cents

Gross domestic product (2002 est.) U.S. $750 million

Gross domestic product per capita (2005 est.) U.S. $10,900

Life expectancy at birth male 69.78 yr; female 74.66 yr

Major resources tourism, cotton, rum, shellfish, sugarcane, vegetables, fruit, chickens

Dominica

COMMONWEALTH OF DOMINICA

T HE ISLAND OF DOMINICA IS AN INDEPENDENT republic in the eastern Caribbean, situated in the Leeward Island chain between Guadeloupe and Martinique.

GEOGRAPHY

Dominica has a backbone of steep mountains, rising to Morne Diablotin. Vents and hot springs are a reminder of their volcanic origins. The climate is warm and humid, with a rainy season from June to October, when there is a risk of hurricanes. Rainfall is much higher in the mountains. The island is covered with lush tropical vegetation that supports abundant wildlife. The Morne Trois Pitons National Park is a wildlife preserve and an important tourist attraction.

NATIONAL DATA - DOMINICA

Land area	754 sq km (291 sq mi)			
Climate		Temperatures		Annual
	Altitude m (ft)	January °C(°F)	July °C(°F)	precipitation mm (in)
Roseau	16 (60)	24 (75)	27 (81)	1,981 (77.9)
Major physical features	highest point: Morne Diablotin 1,447 m (4,747 ft)			
Population	(2006 est.) 68,910			
Form of government	multiparty republic with one legislative house			
Armed forces	no armed forces			
Capital city	Roseau (16,582)			
Official language	English			
Ethnic composition	Black 91.2%; Mixed 6.6%; Amerindian 1.5%; White 0.5%; others 0.2%			
Religious affiliations	Roman Catholic 80%; Protestant 16%; other 4%			
Currency	1 East Caribbean dollar (ECD) = 100 cents			
Gross domestic product	(2003 est.) U.S. $384 million			
Gross domestic product per capita	(2005 est.) U.S. $3,800			
Life expectancy at birth	male 71.95 yr; female 77.93 yr			
Major resources	timber, hydropower, bananas, cattle, cocoa, coconuts, citrus fruits, pigs, pumice, tourism, vanilla			

Untamed beauty (*above*) Dominica's mountainous interior is covered with dense rainforest; the clouds that shroud its peaks send cascades of water into the valleys below. Some parts are still unexplored.

SOCIETY

For many years the British and the French fought for possession of Dominica, but the British eventually took over in 1783. The island was made a separate colony in 1771, but did not become self-governing until 1967; full independence followed in 1978. Soon afterward the island was devastated by a series of hurricanes. Despite two failed coup attempts, the 1980s were mainly a period of rebuilding. The president is head of state, and appoints a prime minister from the majority party. The president is appointed by parliament, which has a majority of elected members. The people are mostly the descendants of former African slaves, though several hundred Caribs (descendants of the original inhabitants) also live on the island. The official language is English, but a French patois is spoken, and there is a Roman Catholic majority with Protestants next.

ECONOMY

The chief cash crops are bananas, citrus fruits and coconuts, while cocoa, bay leaves and vanilla also produce revenue. Other major exports include galvanized sheets and coconut products. Tourism is an ever-increasing source of further revenue. Food, fuels and most manufactured goods have to be imported. The roads are good, and the capital, Roseau, has a port and two airfields nearby. Medical services are adequate, but disease continues to be a problem. Education is free, and literacy levels are high.

Saint Lucia

SAINT LUCIA

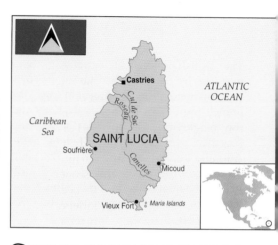

S AINT LUCIA IS A SMALL ISLAND STATE IN THE eastern Caribbean – part of the Lesser Antilles chain that stretches from Puerto Rico to Venezuela.

GEOGRAPHY

The island's rugged landscape was created by volcanic activity. The mountains at its heart rise to Mount Gimie, its highest point. Near the west coast is the volcano Qualibou, which was last active in 1766. It still vents gases, and is responsible for the boiling sulfurous springs that gave the neighboring town of Soufrière its name. The climate is hot and humid, and the volcanic soils are fertile. The lush rainforest that formerly covered the island has been cleared.

NATIONAL DATA - SAINT LUCIA

Land area	606 sq km (234 sq mi)			
Climate		Temperatures		Annual
	Altitude m (ft)	January °C(°F)	July °C(°F)	precipitation mm (in)
Castries	3 (10	25 (77)	27 (81)	2,199 (86.5)
Major physical features	highest point: Mount Gimie 950 m (3,117 ft)			
Population	(2006 est.) 168,458			
Form of government	multiparty constitutional monarchy with two legislative houses			
Armed forces	no armed forces			
Capital city	Castries (12,819)			
Official language	English			
Ethnic composition	Black 90%; Mixed 6%; East Indian 3%; White 1%			
Religious affiliations	Roman Catholic 75%; Protestant 13%; Hindu 1%; other 11%			
Currency	1 East Caribbean dollar (ECD) = 100 cents			
Gross domestic product	(2002 est.) U.S. $866 million			
Gross domestic product per capita	(2005 est.) U.S. $4,800			
Life expectancy at birth	male 70.29 yr; female 77.65 yr			
Major resources	timber, minerals (pumice), geothermal potential, bananas, cocoa, coconuts, fruit, spices, vegetables, tourism			

Saint Vincent and the Grenadines

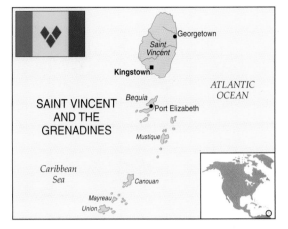

SOCIETY

The indigenous Arawak peoples were driven from Saint Lucia by the Caribs, who were themselves killed off by the diseases introduced by 17th-century French settlers. The island changed hands frequently between Britain and France, but was finally ceded to Britain in 1814. The population are mainly descended from African slaves. Saint Lucia became self-governing in 1967 and independent in 1979. The British sovereign remains head of state, but power lies with the prime minister and cabinet, responsible to an elected two-chamber parliament.

ECONOMY

Saint Lucia's economy depends chiefly on tourism and agriculture. Sugar cane has been replaced as the main cash crop by bananas, which provide some two-thirds of exports. Food crops such as cassava and sweet potatoes are produced by subsistence farmers, and much food has to be imported. Industry is developing, aided by the use of geothermal power. Primary education is free and compulsory.

The Pitons (below), twin peaks rising dramatically from the sea to a height of over 750 m (2,560 ft) on the southwest coast of Saint Lucia. Exposed by erosion, they are formed from the lava plugs of volcanoes.

THE ISLAND STATE OF SAINT VINCENT AND the Grenadines forms part of the Windward Islands in the Lesser Antilles and comprises Saint Vincent and all but a few of the Grenadines.

GEOGRAPHY

Saint Vincent, the main island, is a thickly forested cluster of volcanic mountains. Soufrière, its highest peak, is still very much active. It caused severe damage to the island in 1891 and 1902, and resumed activity in 1979.

The Grenadines are a string of low-lying coral islands, which extend south of Saint Vincent toward Grenada. The climate is hot with seasonal rainfall and occasional destructive hurricanes.

SOCIETY

When Christopher Columbus (1451–1506) visited the islands on Saint Vincent's Day in 1498, he found them inhabited by the fierce Caribs, who successfully resisted European settlement until the 18th century. After a brief period in French hands, Saint Vincent came under British rule from 1783. African slaves were brought in to work the fertile plantations. After the abolition of slavery, Indonesian and Portuguese laborers were also settled.

In 1958 Saint Vincent joined the Federation of the West Indies. In 1979 it became an independent constitutional monarchy, headed by a governor-general representing the British sovereign. The prime minister is the leader of the majority party.

ECONOMY

The economy of the islands rests largely on agriculture, mostly small-scale or subsistence farming on the lower slopes or terraces. Bananas are the chief cash crop and export, followed by arrowroot – used in the making of medicines, fine flour and computer paper. The 1979 volcanic eruption, followed by a hurricane in 1980, devastated agriculture and discouraged tourism, but both have since recovered. Industry is confined to food processing.

NATIONAL DATA – SAINT VINCENT AND THE GRENADINES			
Land area 389 sq km (150 sq mi)			

Climate		Temperatures		Annual
	Altitude m (ft)	January °C(°F)	July °C(°F)	precipitation mm (in)
Kingstown	1 (3)	26 (45)	28 (77)	1,524 (60)

Major physical features highest point: Soufrière 1,234 m (4,048 ft); largest island: Saint Vincent 344 sq km (133 sq mi)

Population   (2006 est.) 117,848

Form of government   multiparty constitutional monarchy with one legislative house

Armed forces   no armed forces

Capital city   Kingstown (18,323)

Official language   English

Ethnic composition   Black 66%; Mixed 19%; East Indian 6%; Carib Amerindian 2%; other 7%

Religious affiliations   Anglican 47%; Methodist 28%; Roman Catholic 13%; Hindu, Seventh-Day Adventist; other Protestant 12%

Currency   1 East Caribbean dollar (ECD) = 100 cents

Gross domestic product   (2002 est.) U.S. $342 million

Gross domestic product per capita   (2005 est.) U.S. $3,600

Life expectancy at birth   male 71.99 yr; female 75.77 yr

Major resources   hydropower, arrowroot, starch, bananas, coconuts, spices, taro, timber, tourism

Barbados

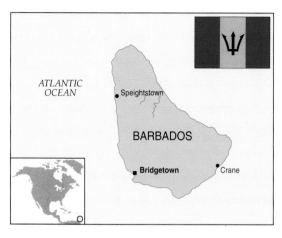

ATLANTIC OCEAN

Speightstown

BARBADOS

Bridgetown Crane

BARBADOS IS AN ISLAND STATE IN THE Caribbean, lying outside the Lesser Antilles some distance from Venezuela.

GEOGRAPHY

The island is founded on coral deposits around a rocky core. The rolling terrain rises to a hilly region in the north and center. The west and south coasts have extensive sandy beaches. The climate is hot with seasonal rains and occasional hurricanes. Trees are plentiful, chiefly mahogany, palm and other tropical varieties, but there are no large forests, and not much wildlife apart from birds.

SOCIETY

Arawak peoples may once have lived on Barbados, but modern settlement was begun by British sugar planters in the 17th century. Present-day Barbadians are

Colonial and modern (*above*) British colonial architecture rubs shoulders with modern office blocks and hotels in Bridgetown, the capital, main port and tourist center of Barbados.

descended from their slaves, liberated in the 1830s. Barbados is culturally the most British of the Caribbean states. From the 1930s onward, the island was given increasing self-government. It headed the short-lived West Indies Federation before becoming fully independent in 1966. The British sovereign, who is represented by a governor-general, remains the head of state, but power resides with the two-chamber parliament, which is constituted on British lines; it is headed by a prime minister and cabinet.

ECONOMY

Agriculture is dominated by sugar cane production, to which most of the flat land is devoted; despite a recent decline it remains by far the largest export. Sugar cane and its by-products, molasses and rum, are processed on the island. Diversification into other cash crops such as citrus fruits and cotton is officially encouraged in order to end the reliance on a single crop, as is fishing. Small petroleum and gas reserves are exploited. Light industries produce electrical components, clothing and many other items for export and for domestic use. Tourism, however, is now the single largest earner, visitors being attracted to the glittering white beaches and the coral reefs. Education is free and compulsory up to the age of 14, and available up to degree level. Social-security and health-care systems have also been introduced.

NATIONAL DATA – BARBADOS

Land area	431 sq km (166 sq mi)			
Climate		Temperatures		Annual
	Altitude m (ft)	January °C(°F)	July °C(°F)	precipitation mm (in)
Bridgetown	55 (180)	25 (77)	27 (81)	1,278 (50)

Major physical features	highest point: Mount Hillaby 337 m (1,104 ft)

Population	(2006 est.) 279,912

Form of government multiparty constitutional monarchy with two legislative houses

Armed forces army 500; navy 110

Capital city Bridgetown (98,947)

Official language English

Ethnic composition Black 90%; White 4%; Asian and mixed 6%

Religious affiliations Protestant 31%; other Anglican 29%; other Christian 24%; Roman Catholic 4%; other 12%

Currency 1 Barbadian dollar (BDS) = 100 cents

Gross domestic product (2006 est.) $5.108 billion

Gross domestic product per capita (2006 est.) $18,200

Life expectancy at birth male 70.79 yr; female 74.82 yr

Major resources bananas, fish, goats, limestone, natural gas, petroleum, onions, pigs, poultry, shellfish, sheep, sugarcane, vegetables, tourism, cotton

Grenada

GRENADA

G RENADA IS AN ISLAND STATE IN THE LESSER Antilles, centered on the most southerly of the Windward Islands off the coast of Venezuela. The territory also includes the southernmost islands of the neighboring Grenadines group.

GEOGRAPHY

Grenada itself consists largely of a ridge of volcanic mountain peaks. Their slopes are thickly forested, thanks to the hot, humid climate and the fertile volcanic soils. The southern coasts have many bays and harbors, including that of the capital, Saint George's.

SOCIETY

Grenada's first inhabitants were the Arawak peoples, exterminated by the fierce Caribs whom Christopher Columbus (1451–1506) encountered in 1498. The island was settled by the French in 1650, but was seized by Britain in the 18th century. The people are mostly descended from former African slaves, and some still speak a French patois. Grenada became fully independent in 1974 with a parliamentary democracy based on the British system. In 1979 the New Jewel Movement, led by Maurice Bishop (1944–83), seized power in a left-wing coup. Bishop made some economic progress, as well as establishing links with Cuba, Libya and Algeria, but in 1983 he was deposed and murdered by rival Marxists. The United States, with some neighboring Caribbean states, launched a successful invasion. Constitutional government was restored, and the governor-general instituted free elections to the lower house of the two-chamber parliament. US troops finally withdrew in 1995.

ECONOMY

Grenada's economy depends almost entirely on agriculture and tourism. Agriculture is mostly small scale except for a few cooperatives. The main cash crops and exports are bananas, cocoa, and the spices such as nutmeg and mace that once earned Grenada the name Isle of Spice. There is little industry apart from various food-processing plants. Tourism, shattered by the 1979 coup, is gradually recovering. There is an effective healthcare program and free education facilities.

St George's, Grenada's capital, is attractively set on wooded hillsides surrounding a horseshoe bay. Traditional exports of spices, bananas and cocoa still leave the island from St George's port.

NATIONAL DATA – GRENADA

Land area	344 sq km (133 sq mi)			
Climate		Temperatures		Annual
	Altitude m (ft)	January °C(°F)	July °C(°F)	precipitation mm (in)
St. George's	1 (3)	25 (77)	27 (81)	1,560 (61.4))

Major physical features	highest point: Mount St. Catherine 840 m (2,757 ft)
Population	(2006 est.) 89,703
Form of government	multiparty constitutional monarchy with two legislative houses
Armed forces	no armed forces
Capital city	St. George's (4,298)
Official language	English
Ethnic composition	Black 82%; Mixed Black and European 13%; European and East Indian 5%
Religious affiliations	Roman Catholic 53%; Anglican 13.8%; other Protestant 33.2%
Currency	1 East Caribbean dollar (ECD) = 100 cents
Gross domestic product	(2002 est.) U.S. $440 million
Gross domestic product per capita	(2005 est.) U.S. $3,900
Life expectancy at birth	male 63.06 yr; female 66.68 yr
Major resources	timber, tropical fruits, cocoa, vegetables, mace, nutmeg, tourism

Trinidad and Tobago

REPUBLIC OF TRINIDAD AND TOBAGO

T RINIDAD AND TOBAGO IS MADE UP OF TWO islands lying between the Caribbean Sea and the Atlantic Ocean just off the coast of Venezuela.

GEOGRAPHY

Venezuela's coastal mountains effectively continue into Trinidad's Northern Range and Tobago's Main Ridge. The Northern Range rises steeply to the Cerro del Aripo, and its rivers create some spectacular waterfalls. Otherwise Trinidad is mostly rolling or flat, with many swampy areas along the coast. South of Tobago's Main Ridge is a coral plain that extends out to sea, creating some magnificent reefs. The climate is hot and humid, and much of Trinidad is still covered with thick rainforest. Both islands are rich in bird life, especially flamingoes and the scarlet ibis.

NATIONAL DATA - TRINIDAD AND TOBAGO

Land area	5,128 sq km (1,980 sq mi)			
Climate		Temperatures		Annual
	Altitude m (ft)	January °C(°F)	July °C(°F)	precipitation mm (in)
Port-of-Spain	20 (66)	26 (77)	27 (81)	1,408 (55.4)

Major physical features highest point: Mount Aripo 940 m (3,084 ft)

Population (2006 est.) 1,065,842

Form of government multiparty republic with two legislative houses

Armed forces army 2,000 navy 700

Capital city Port-of-Spain (49,865)

Official language English

Ethnic composition Indian (South Asian) 40%; African 37.5%; Mixed 20.5%; other 1.2%; unspecified 0.8%

Religious affiliations Roman Catholic 26%; Hindu 22.5%; Anglican 7.8%; Baptist 7.2%; Pentecostal 6.8%; other Christian 5.8%; Muslim 5.8%; Seventh-Day Adventist 4%; other 10.8%; unspecified 1.4%; none 1.9%

Currency 1 Trinidad and Tobago dollar (TTD) = 100 cents

Gross domestic product (2006 est.) U.S. $20.99 billion

Gross domestic product per capita (2006 est.) U.S. $19,700

Life expectancy at birth male 65.71 yr; female 67.86 yr

Major resources asphalt, bananas, citrus fruits, cocoa, coffee, oil and natural gas, rice, sugarcane, tourism

SOCIETY

The original inhabitants were the Arawak peoples, exterminated on Tobago by the fierce Caribs. After colonization by Spain and France, both peoples were wiped out by disease and enslavement on tobacco and sugar plantations; they were replaced by African slaves. In the early 19th century both islands were ceded to Britain. With the ending of slavery laborers were brought in from Asia to work the plantations. Full independence was granted in 1962. In 1976 the islands became a Commonwealth republic, which survived an abortive coup led by Asian Trinidadians in 1980. The islands' ethnic mix of peoples has created a colorful local culture including carnival and calypso music. There is a large Hindu minority.

ECONOMY

The islands are among the most prosperous in the Caribbean. The economy depends heavily on the refining and export of the extensive petroleum reserves, but a fall in petroleum revenues has led to increased diversification in the industrial sector. Agriculture concentrates on cash crops for export – principally sugar cane, cocoa and coffee – with less emphasis on staple crops such as rice and vegetables. Tourism is an important source of revenue, and there are good education and health care facilities.

Tropical mix Trinidad is full of variety. Industrial in parts, it also has fertile agricultural land. In the towns, mosques and temples, English churches, the sounds of calypso and steel bands, all combine to form a vibrant cosmopolitan society.

Dependencies in the region

TURKS AND CAICOS ISLANDS
UNITED KINGDOM

The Turks and Caicos Islands are a British colony consisting of a string of about 30 small islands at the end of the Bahamas group, which they resemble in landscape and climate. Rainfall, however, is scantier except during hurricanes. The terrain is dry, and the vegetation is scrubby except in the coastal saltmarshes.

The inhabitants are descended from African slaves imported by the British in the 17th century to work saltpans. The islands have on several occasions been annexed both to the Bahamas and to Jamaica. In 1976 they were given self-government under the British crown, but their declared goal of full independence has so far been delayed indefinitely.

Salt production has become uneconomic, and agriculture is hindered by lack of water. Thus the island's economy rests largely on fishing, chiefly crayfish and conch, and on the development of tourism. Many offshore financial companies are also based here, taking advantage of the islands' status as a tax haven. Education and health care are government priorities, and are generally good.

BERMUDA
UNITED KINGDOM

The British colony of Bermuda occupies a string of coral islands in the North Atlantic. About 20 are inhabited, the largest being Great Bermuda. The climate is warm, mild and humid, with moderate rainfall. The vegetation is luxuriant, with whispering pines and many flowering plants. Lizards are the only reptiles, and there are no native mammals.

In 1609 the crew of a shipwrecked English vessel landed in the islands, and in 1612 Bermuda became a colony of the British Virginia Company. It passed to the crown in 1684, but remained self-governing. In 1941 Bermuda leased air and naval stations to the United States. The 1968 constitution allowed greater self-government. The 40-member House of Assembly is directly elected, but the 11-member senate is chosen by the governor, who is appointed by the British crown. The majority of the people are descended from former African slaves, others from British or Portuguese settlers. The official language is English, but some Portuguese is spoken. Christianity is the dominant religion; the majority of people belong to one of the denominations of the Protestant churches.

A thriving British colony, Bermuda's picturesque towns are a magnet for tourists; lenient tax laws bolster the economy by attracting foreign investors.

There is little scope for much agriculture or for large-scale industry, and food and fuel must be imported. The colony relies almost entirely on the income from tourism, and from military and financial interests. The standard of living is high. There are no railroads, but there is a good road network, which links the islands by bridges and ferries. The main port at Hamilton is well served by international shipping lines, and the American airport base at Kindley Field serves both military and international traffic. Health care is readily available, and there is social security for all citizens. Education is free and compulsory, and literacy levels are high. Overseas scholarships are available for higher education.

CAYMAN ISLANDS
UNITED KINGDOM

The Cayman Islands are a British colony in the Caribbean to the northwest of Jamaica. Grand Cayman, Cayman Brac and Little Cayman are low-lying coral islands on a rocky base. The coasts are lined with magnificent beaches and reefs, and the tropical climate is pleasantly cooled by the sea.

Though visited by Christopher Columbus (1451–1506), the islands remained no more than a pirate lair until ceded to Britain with Jamaica in 1670. Later settlers from Jamaica lived mostly from turtle fishing. The Caymans became a separate colony after Jamaica's independence in 1962. They enjoy some local autonomy under a governor, but have yet to become fully independent.

There is little agriculture or industry, although some turtle farming is being developed. Tourism is one major earner, and offshore finance another, encouraged by liberal tax and confidentiality laws. The Caymans now house hundreds of financial companies. These help to create high living standards, and good health and education services.

PUERTO RICO
UNITED STATES

Puerto Rico, an island state in the West Indies, is a self-governing commonwealth in association with the United States.

Most of the main island is upland, rising from the narrow southern coastal plains to the Cordillera Central, a mountain range that runs along the island like a backbone; Cerro del Punta is the highest peak. To the north it flattens out into hill country, through which the island's main rivers flow down to the much wider northern coastal plain. The two adjoining islands to the east are similarly rugged.

The climate is mainly warm and humid, but rainfall is unevenly distributed. The north receives most of the island's rainfall, and is dotted with marshlands, small lakes and mangrove swamps. By contrast the south, in the rainshadow of the mountains, is too dry for farming without irrigation, and in the southwest there are several saltpans. The island was once covered in tropical

hardwood rainforest, but so much of this has been cleared that it now survives only in the Caribbean National Forest enclave. Wildlife consists mostly of small reptiles and birds.

When Christopher Columbus (1451–1506) reached Puerto Rico, the islands were already inhabited by the Arawaks and Caribs. Spanish settlers wiped out both these peoples, by disease and enslavement in gold mines and plantations. Puerto Rico, unlike nearby Hispaniola, remained Spanish throughout the 19th century. African slaves were introduced to work the plantations until slavery was abolished in the 19th century.

The island was granted self-government in 1897, only to be occupied and ceded to the United States in the Spanish–American war (1898); its people were given United States citizenship in 1919. Various independence movements developed, but disagreed on whether to become wholly independent or seek statehood within the United States. In spite of strong American influences, the culture has remained primarily Hispanic.

After World War II the office of governor was made elective, and in 1952 a commonwealth was established. However, a fierce debate still continues regarding possible independence or statehood. The island returns a nonvoting member to the United States House of Representatives, but elects its own governor, who appoints an executive cabinet.

El Morro (*above*) One of the two 16th-century forts on the clifftops of San Juan, Puerto Rico, this is an impressive reminder of the days of Spanish control.

Holiday bliss (*right*) The clear blue waters and sandy beaches of the United States Virgin Islands have brought a thriving tourist trade. The fine harbor at St Thomas, once frequented by pirates, is now the haunt of luxury cruisers.

Puerto Rico's close links with the United States have made it one of the most prosperous Caribbean countries. Since World War II investment policies have made industry more important than agriculture. The main cash crops are sugar cane, coffee and tobacco, along with dairy products and livestock, but large amounts of food have to be imported. There are few mineral resources, although copper and nickel deposits have been found. Industry is based on a wide range of products, from petrochemicals, plastics and pharmaceuticals to textiles, machinery and electronics. Considerable amounts are exported, chiefly to the United States, but there is also a high volume of imports. Tourism, however, has become a major earner.

Social-security and welfare benefits operate to United States standards, and health care is improving rapidly toward that goal. Education is a priority, freely available at all levels, up to and including university.

VIRGIN ISLANDS OF THE UNITED STATES
UNITED STATES

The United States Virgin Islands are an unincorporated territory of the United States, comprising Saint Croix, Saint John and Saint Thomas and some 50 smaller islets of the Virgin Islands group to the east of Puerto Rico. Their rugged landscapes are formed by the peaks of an undersea mountain range.

Christopher Columbus (1451–1506) named the islands after Saint Ursula's martyred virgins. They were settled by the Spanish and French before being acquired by Denmark in the 17th and 18th centuries. African slaves were introduced to work the sugar cane plantations. In 1917 the islands were bought by the United States. In 1981 a constitution, adopted by referendum, introduced self-government to the islands by an elected governor and senate.

Agriculture has shifted from sugar cane to food crops and livestock, mostly for domestic use. Manufacturing ranges from rum distilling to petroleum refining. Tourism has been developed since 1945, and now dominates the economy, with game fishing being one of the major attractions. Education and health care are widely available, reflecting the islands' relative prosperity.

BRITISH VIRGIN ISLANDS
UNITED KINGDOM

The British Virgin Islands are a British colony comprising Tortola, Anegada, Virgin Gorda, Jost Van Dyke and more than 30 islets of the Virgin Islands group in the Caribbean. Most are the peaks of a submerged mountain chain, and share a subtropical climate. The landscape is mainly hilly, with sandy beaches and coral reefs around the coasts. Anegada, however, is very flat.

The original inhabitants, the Arawaks, were driven out by the Caribs, who were in turn wiped out by the Spanish. The islands were used as havens for piracy until they were annexed by Britain in the 17th century. They now have effective self-government through an elected council headed by an appointed governor.

Tourism is expanding to dominate the economy; the sheltered waters around the islands attract a great number of visitors who stay on yachts. Agriculture produces livestock for local use, and cash and food crops such as bananas and sugar cane. Fishing is also important, but manufacturing industry is very small. Despite water shortages, health care is excellent, as is primary education.

ANGUILLA
UNITED KINGDOM

Anguilla is the northernmost of the Leeward Islands in the eastern Caribbean. It consists of a long, thin coral island with a bare landscape of scrubland and occasional fruit plantations. The climate is warm and humid, and the island is subject to hurricanes.

Anguilla has been a British colony since 1650. For several years it was part of the federal state of St Kitts–Nevis–Anguilla, but since 1980 it has been a dependent territory of the United Kingdom. The governor is appointed by the crown, but internal affairs are managed by an elected assembly. Most Anguillans are descended from former African slaves. English is the chief language, and the main religions are Anglican and Methodist Christianity.

The chief exports are fish, salt and livestock. There are significant saltpeter deposits, and tourism is being developed, along with transportation and communications. Housing, health care and educational facilities are good.

MONTSERRAT
UNITED KINGDOM

The British colony of Montserrat is one of the Leeward Islands in the eastern Caribbean. Volcanic peaks, several of them active, dominate a mountainous landscape of pastureland and lush vegetation. The climate is warm and humid, with seasonal hurricanes.

Montserrat was colonized in the 17th century by Irish settlers from nearby Saint Kitts. It belonged to the Federation of the West Indies between 1958 and 1962, and hopes to gain full independence from the United Kingdom. The governor, who is appointed by the crown, is head of an executive council and a partly elected legislative council. Most people are descendants of former African slaves. English is the chief language, and Christianity, with its various denominations, is the predominant religion.

The chief exports are vegetables and cotton, but tourism is a growing source of revenue. There are surfaced roads, a seaport at the capital, Plymouth, and an air link with neighboring Antigua. Sanitation is good and health care is improving. The island was devastated in June 1997 when the Soufriere Hills volcano erupted.

ARUBA
THE NETHERLANDS

The island of Aruba lies at the southwestern end of the Lesser Antilles off the Venezuelan coast. The terrain is mostly flat, but is generously scattered with volcanic boulders. The climate is warm and humid, but rainfall is generally poor, and vegetation is sparse apart from cacti and succulents.

Aruba became part of the Netherlands Antilles in 1634, but seceded in 1986, becoming an internally self-governing part of the Netherlands. The monarch is represented by a governor, but the 21 members of the States of Aruba are directly elected. Arubans are mainly of African, European and Asian descent, and most are Roman Catholic. Dutch is the official language, but Papiamento, a Spanish creole, is widely spoken.

Tourism is almost the sole source of revenue. Aruba has surfaced roads, deep-water harbors and an international airport. Most drinking water is distilled from seawater. Education is not compulsory, but literacy levels are high.

THE NETHERLANDS ANTILLES
THE NETHERLANDS

The Netherlands Antilles are made up of two island groups within the Lesser Antilles. The southern group includes Curaçao and Bonaire – and formerly also Aruba – off the Venezuelan coast. The northern group is part of the Leeward Islands, and comprises Sint Eustatius, Saba and the southern part of Sint Maarten, the rest of which is French. The southern islands are generally flat, but those in the north are mountainous and volcanic. The climate is warm and humid; Curaçao and Bonaire have poor rainfall, while the northern islands are subject to seasonal hurricanes. The vegetation is sparse apart from cacti and succulents.

In the 16th century Spanish settlers killed most of the original Amerindian inhabitants. The Dutch claimed these islands in the early 17th century. Curaçao and Bonaire were plantation colonies of the Dutch West India Company, and Sint Eustatius was the center of the local slave trade until the late 18th century. Bonaire was given a separate political status in 1868. In 1954 the islands became self-governing as an integral part of the Netherlands. The governor is appointed by the Dutch crown, but the 22 members of the States of the Netherlands Antilles are directly elected.

The people are mainly of African and European descent, and most are Roman Catholic. Dutch is the official language, but English is spoken in the north, and Papiamento – a Spanish-based creole language – on Bonaire and Curaçao.

Since the discovery of oil on Curaçao, petroleum and its derivatives have accounted for the vast majority of exports. Tourism and banking have also been successfully developed and are now an important source of revenue. Some fruit and vegetables are grown, including oranges for the famous Curaçao liqueur, but most food is imported.

Curaçao has an excellent road system, and there are international airports on Curaçao and on Sint Maarten. Most drinking water is distilled from seawater. Health care is good, and although education is not compulsory, literacy levels are relatively high.

Simple luxury (*above*) A modest beach-house in Guadeloupe is beautified by its idyllic location.

Mount Pelée (*right*), an active volcano that last erupted in 1929, dominates the lush tropical landscape of Martinique.

Running repairs (*below*) A fisherman checks his nets for damage. Martinique's fishing, farming and tourist economy is heavily dependent on subsidies from mainland France.

GUADELOUPE
FRANCE

The French overseas *département* of Guadeloupe consists of two groups of islands within the Lesser Antilles in the eastern Caribbean. The main islands are Basse-Terre and Grand-Terre, separated by a narrow channel. Nearby are Marie Galante, La Désirade and the Iles des Saintes. A separate northwestern group comprises Saint Barthélemy and Saint Martin, whose southern part belongs to the Netherlands.

Basse-Terre is a mountainous island, culminating in the still-active volcano of Soufrière. Grande-Terre is generally low-lying, but Saint Barthélemy and Saint Martin are both rugged. The climate is warm and humid, with seasonal rains and occasional hurricanes. The vegetation

but Creole is widely spoken on the island.

Income from agriculture, industry and tourism leaves a large deficit that is mostly made up by grants from France. The main islands have excellent roads, and ferries connect all the islands. There are ports at Basse-Terre, at Pointe-à-Pitre (near an international airport) and at Marigot on Saint Martin. Education, welfare and health care are the same as in mainland France.

MARTINIQUE
FRANCE

Martinique is a Caribbean island in the Lesser Antilles chain, situated between Dominica and Saint Lucia. Formerly a French colony, it is now an overseas *département* within France.

The island is mountainous and volcanic in origin, and is dominated by the immense and active Mount Pelée near its northernmost cape. The basin of the river Lézarde is the only significant lowland area. The hot, humid climate and the fertile volcanic soils have created lush vegetation, with tropical rainforests and mangrove swamps on the coasts.

When Christopher Columbus (1451–1506) discovered Martinique in 1493, he found it occupied by the Caribs who had exterminated the indigenous inhabitants, the Arawaks. In the 17th century the island was settled by French sugar cane growers, who imported large numbers of African slaves. These were not freed until 1848. The Creole dialect spoken by the islanders reflects their mixed African and European origins. Martinique became a French *département* after World War II. Since then, beset by economic problems, it has given birth to a vociferous independence movement, which has launched a number of terrorist campaigns in mainland France. However, the island's economic reliance on France makes genuine independence difficult to achieve.

Agriculture and tourism are the island's mainstays. The main cash crops include sugar cane, bananas and exotic flowers. Sweet potatoes and cassava, the chief food crops, are mostly grown at subsistence level. Forestry and fisheries represent largely undeveloped potential. Industry, mostly food processing, has been developed with French aid. Tourism from the United States and France is far more important, and the capital, Fort-de-France, has good air links with both countries. Education is widely available and standards are high. Welfare benefits operate on the same basis as in France.

is often lush, ranging from mangrove swamps near Pointe-à-Pitre on Grande-Terre to dense forests on Basse-Terre. On the smaller islands, however, cacti and dry forest are more usual.

In 1626 the French displaced small groups of Spanish settlers, and in 1674 the colony passed to the French crown. It fell into British hands several times, but was finally returned to France in 1816. In 1946 Guadeloupe became an administrative district of France. Despite increasing degrees of self-government, separatist violence has grown since the 1970s. The prefect is appointed by the French government, but the 36-member legislative council is elected, and Guadeloupe is represented in both chambers of the French parliament. Most inhabitants are the black or mixed-race descendants of slaves, and the majority are Roman Catholic. French is the official language,

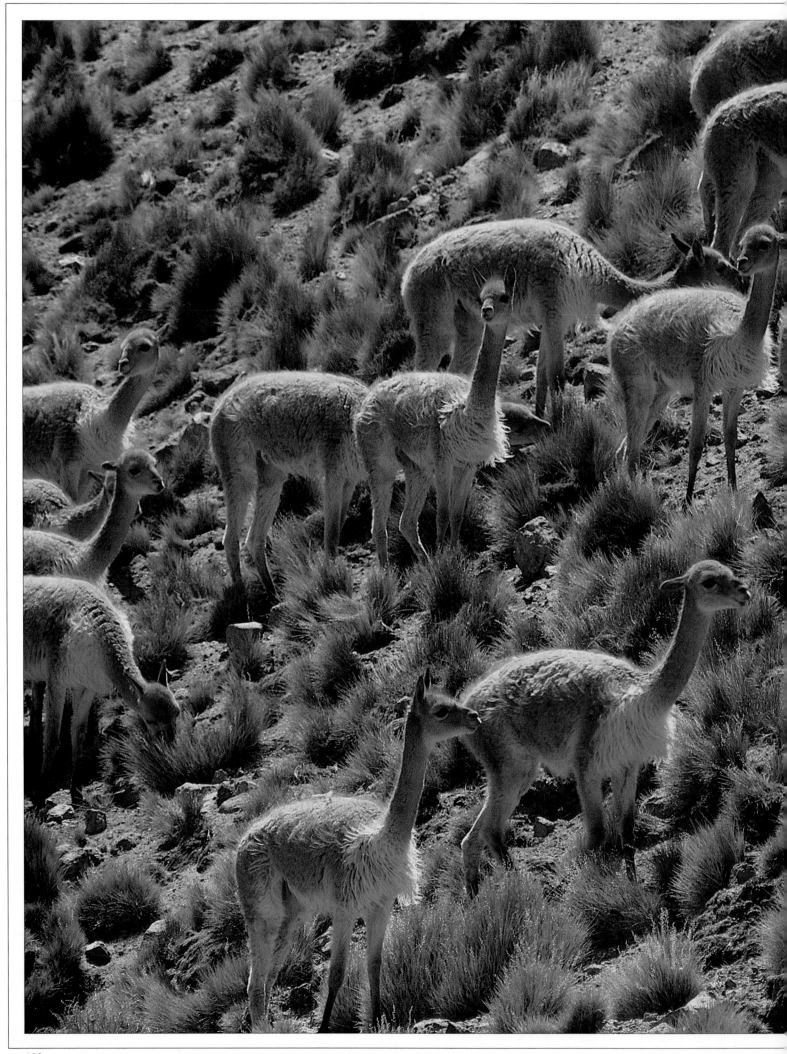

South America

COUNTRIES IN THE REGION

COLOMBIA · VENEZUELA · GUYANA · SURINAME
BOLIVIA · ECUADOR · PERU · BRAZIL · PARAGUAY
CHILE · ARGENTINA · URUGUAY

DEPENDENCIES IN THE REGION

FALKLAND ISLANDS · FRENCH GUIANA

Caribbean Sea

Barranquilla
Maracaibo
Caracas
Lake Maracaibo
Orinoco
Ciudad Bolívar
Georgetown
Paramaribo
Cayenne
VENEZUELA
GUYANA
Roraima 2810
SURINAME
FRENCH GUIANA (France)
Medellín
Bogotá
Guiana Highlands
COLOMBIA
Cali
Néblina Peak 3014
Equator
Quito
ECUADOR
Chimborazo 6310
Putumayo
Japurá
Negro
Amazon
Belém
Guayaquil
Iquitos
Manaus
Marañón
Juruá
Madeira
Tapajós
Xingu
Fortaleza
Cape of São Roque
PERU
Chiclayo
Ucayali
Selvas
Plateau of Borborema
Recife
Huascarán 6768
Araguaia
Tocantins
São Francisco
BRAZIL
Lima
Cuzco
Plateau of Mato Grosso
Plateau of Brazil
Salvador
Arequipa
Lake Titicaca
BOLIVIA
La Paz
Cuiabá
Brasília
Belo Horizonte
Sajama 6542
Santa Cruz
Serra da Mantiqueira
Arica
Sucre
Paraguay
PACIFIC OCEAN
Atacama Desert
Gran Chaco
Paraná
Serra do Mar
Rio de Janeiro
PARAGUAY
Tropic of Capricorn
Antofagasta
Llullaillaco 6723
Asunción
São Paulo
Iguaçu Falls
San Miguel de Tucumán
Salado
Paraná
Uruguay
CHILE
Pôrto Alegre
ATLANTIC OCEAN
ARGENTINA
Córdoba
Aconcagua 6960
Rosario
URUGUAY
Santiago
Colorado
Buenos Aires
Montevideo
Plate
ANDES
Concepción
Pampas
Bahía Blanca
Puerto Montt
Chiloé Island
Patagonia
Chronos Archipelago
Falkland Islands (U.K.)
Wellington Island
Stanley
Strait of Magellan
Punta Arenas
Tierra del Fuego
Cape Horn

■ capital city
● major town

height of land (meters)

5000
3000
2000
1000
500
200
0

▲ mountain peak

South American specialities Vicuna in Lanca National Park, Chile. Vicuna are the smallest of the South American members of the camel family, which includes llamas, guanacos and alpacas. South America contains an enormous variety of animals and plants that are found nowhere else in the world.

Colombia

REPUBLIC OF COLOMBIA

COLOMBIA IN THE NORTHWEST OF SOUTH America takes its name from the explorer Christopher Columbus (1451–1506). This beautiful country has a rich colonial heritage, but is plagued by internal strife.

GEOGRAPHY

Colombia's short frontier with Panama to the northwest divides the western (Pacific) coast from the northern (Caribbean) shoreline. Snowcapped Pico Cristóbal Colón forms part of an isolated massif overlooking the Caribbean to the north and a swampy basin to the west.

The two main rivers that flow into the basin from the south, through deep trenches separating the three main ranges of the mighty Colombian Andes, are the Magdalena and the Cauca. The ranges, which run roughly northeastward from the southeastern border with Ecuador, are the Cordillera Occidental, overlooking the narrow Pacific coastal plain, the volcanic Cordillera Central east of the Cauca river, and farther east still, beyond the broader Magdalena valley, the Cordillera Oriental. This divides at the Venezuelan border. Recent earthquakes and a devastating eruption in 1985 show that the area is geologically active.

More than half the country consists of a sparsely inhabited plain that drops gently southeast from the snowcapped Andean ranges toward the Peruvian and Brazilian borders. The northern lowlands – the *llanos* – are drained by the Orinoco river, which runs along the western border with Venezuela. The southern lowlands form part of the Amazon river basin.

Colombia's climate is very varied. The Caribbean lowlands are relatively dry with two short rainy seasons, though they are vulnerable to hurricanes. Rainfall is highest on the Pacific coast, in the central valleys and in the Amazonian lowlands, where tropical rainforest is characteristic. This gives way to savanna on the *llanos* and in the northern valleys. The vegetation changes with altitude, giving way on higher mountain slopes to cloud forests and tundra.

Human settlement and agriculture have stripped away much of the original forest cover, but the moist mountain forests contain many orchids and epiphytes, as well as balatas, from which nonelastic rubber is obtained, and ivory nuts, used for making ornaments. Animal life is extremely varied. Typical examples include tapirs, sloths, anteaters, jaguars and howler monkeys. Hummingbirds tap the nectar of innumerable flowers, while the rivers support caimans and capybaras, large, semiaquatic rodents.

SOCIETY

In the early 16th century Spanish invaders overcame the sophisticated farming and trading culture of the indigenous Chibcha peoples, craftsmen highly skilled at working gold and embellishing their artifacts with jewels. The colonists founded the city of Bogotá in the Cordillera Oriental, and Colombia became the heart of the Viceroyalty of New Granada. Its independence was gained after the Spanish defeat at Boyacá in 1819 and until 1830 it was part of the Republic of Gran Colombia, established by Simón Bolívar (1783–1830), which included modern Ecuador, Venezuela and Panama.

The modern history of Colombia has been one of continual unrest, beginning with a revolt in 1840 and the emergence of bitterly opposed liberal and conservative

The beauty of the high Andes (*left*) The river Cauca has carved a deep valley through the mountains on its journey north to enter the Caribbean Sea.

A pre-Columbian statue (*below*) from high in the Andes. Its fangs, and the tiny figure clasped in its hands, are indications that it is a sacred figure.

ural gas reserves and produces 60 percent of the world's emeralds. In the late 1990s, the government began to privatize coal, oil and electricity production. Production and investment by foreign companies are seriously affected by guerrilla activity. Manufacturing declined in the late 1990s after import subsidies were largely removed. Coffee and food processing are important, followed by chemicals and motor vehicles. Colombia has a well-developed telecommunications sector and this was expected to expand strongly into the 2000s.

Overland transportation is hampered by the difficult terrain, and most roads are unsurfaced, but buses and trucks carry more passengers and freight than the limited railroad system. The thriving airline network is heavily used, with six international airports.

Newspapers are the dominant news media. They are generally free of government censorship, and tend to reflect party loyalties. Radio and television broadcasting are subject to stricter controls. Welfare provision includes a range of unemployment, maternity and disability benefits, and retirement pensions. Health facilities are limited, but there is an emphasis on health education and improving sanitation. Among the poor, however, malnutrition and inadequate housing threaten general health. Primary education is free and compulsory for five years, and literacy is relatively high.

Colombia's colonial past is reflected in today's population, the majority of whom are of mixed ancestry. Most of these are mestizos of European and Amerindian descent, but mulattoes (of mixed African and European origin) form the majority along the coast. There is a large white minority, and small numbers of blacks and Amerindians, the latter divided into as many as 400 different groups. Though Spanish is the official language, many Amerindian languages are also spoken. The vast majority of the population are Roman Catholic.

political factions: a liberal uprising in 1899 triggered the murderous War of the Thousand Days, which lasted until 1903. In 1948 La Violencia broke out – a long struggle between the factions that led in 1958 to the creation of the National Front. In this uneasy coalition, presidents from each faction were elected in rotation. Multiparty elections since 1974 have brought some prosperity but little peace. Guerrilla groups and paramilitaries have grown in strength through the 1990s and have been responsible for large-scale violence and thousands of civilian deaths. Much of their income comes from drug trafficking and extortion.

Legislative power rests with a congress consisting of a senate and a house of representatives, each of which is directly elected every four years. The head of state is the president, elected for a single four-year term; he appoints the cabinet.

ECONOMY

The Colombian economy is founded largely on private enterprise, and though mainly agricultural it is becoming increasingly industrialized.

Only a small portion of the land is cultivated, but the crops produced are vital to the economy. The wide range of climate and terrain gives rise to a great range of food crops, from bananas and cassava in the lowlands to maize and potatoes in the mountains. Coffee accounts for one-third of all official export revenue, but cannabis and coca, from which cocaine is derived, provide the drug cartels with twice this amount.

Nearly half the land area is forested, and the lumber industry is growing fast. Much of the remaining available land is occupied by pasture, mostly for cattle. River fishing, once plentiful, has suffered from silting and pollution, and the potential for offshore fisheries has not yet been fully developed. Colombia has the largest coal reserves in Latin America, huge nat-

NATIONAL DATA – COLOMBIA

Land area	1,038,700 sq km (401,044 sq mi)			
Climate		**Temperatures**		**Annual**
	Altitude m (ft)	January °C(°F)	July °C(°F)	precipitation mm (in)
Bogotá	2,645 (8,678)	11 (55)	11 (55)	799 (31.4)

Major physical features	highest point: Pico Cristóbal Colón 5,800 m (19,029 ft)

Population	(2006 est.) 43,593,035

Form of government	multiparty republic with two legislative houses

Armed forces	army 178,000; navy 22,000; air force 7,000

Largest cities	Bogotá (capital - 7,363, 492); Cali (2,498,074); Medellin (2,042,093); Barranquilla (1,429,031)

Official language	Spanish

Ethnic composition	Mestizo 58%; White 20%; Mulatto 14%; Black 4%; Mixed Black-Amerindian 3%; Amerindian 1%

Religious affiliations	Roman Catholic 90%; other 10%

Currency	1 Colombian peso (COP) = 100 centavos

Gross domestic product	(2006) U.S. $366.7 billion

Gross domestic product per capita	(2006) U.S. $8,400

Life expectancy at birth	male 68.15 yr; female 75.96 yr

Major resources	petroleum, natural gas, coal, iron ore, nickel, gold, copper, , hydropower, bananas, cassava, cattle, coffee, cut flowers, emeralds, maize/corn, platinum, potatoes, rice, silver, sugarcane, sorghum, soybeans, timber

Venezuela

REPUBLIC OF VENEZUELA

V ENEZUELA IS A FEDERAL REPUBLIC ON THE northern (Caribbean) coast of South America, bordering Colombia to the west, Brazil to the south and Guyana to the east. Once a poor country, it has been radically transformed by the harnessing of its vast petroleum reserves – though not all sections of the population have shared in this newfound wealth.

GEOGRAPHY

The landscape of northwestern Venezuela is shaped by two northern branches of the Andes. The western branch, the Sierra de Perijá, defines part of the Colombian frontier. The eastern branch, the Cordillera de Mérida, enters Venezuela from the south and swings northeast toward the island-fringed Caribbean coast; it includes Venezuela's highest peak, Pico Bolívar. Between the two Andean ranges lie the swampy Maracaibo Lowlands, surrounding the huge, shallow Lake Maracaibo – an irregular extension of the Gulf of Venezuela.

Further mountain ranges run eastward along the coast from the Cordillera de Mérida. The basins and valleys between them shelter several important cities, including the capital, Caracas. South of the mountains are the extensive lowland plains, or llanos, of the Orinoco basin, which cross the country from Colombia to the Orinoco Delta area near the Guyanan border. Farther south is the vast, irregular granite plateau of the Guiana Highlands, where the Orinoco river rises.

The climate is generally tropical with a marked rainy season from April to October. However, the mountain areas are cooler, and rainfall varies enormously. The driest areas are in the lee of mountains, especially on the coast; the Orinoco Delta receives the highest rainfall. The llanos are flooded during the rainy season and arid for much of the rest of the year.

Forests still cover much of the country, ranging from semitropical evergreen in the northern mountains to tropical rainforest around the Orinoco Delta and in the far south. Wild grassland covers most of the rest of the land, including the extensive llanos areas, the Guiana Highlands and the high alpine meadows of the Andean mountain ranges.

The range of altitudes has created an enormous diversity in both the plant and the animal life. Human settlement has confined many animals to more sparsely populated areas, but they remain varied nonetheless, and include bears, caimans, ocelots, opossums and peccaries. Among the more remarkable bird species is the strange cave-dwelling oilbird. Manatees and dolphins are found in coastal waters. The threat to the many rare species has long been recognized, and hunting is forbidden to all except Amerindians, who depend on it for their way of life.

SOCIETY

When European explorers first arrived in the 15th century, they found a number of different Amerindian peoples in the area, including a sophisticated farming culture in the Andean highlands. In 1523 the Spaniards founded the first European settlement, situated at Cumaná, and a colonial economy was established using local slave labor.

Until the beginning of the 19th century, Venezuela was effectively in the hands of Spanish colonial administrators and priests. It was in Venezuela that Francisco Miranda (1750–1816) and his successor Simón Bolívar (1783–1830) began South America's struggle for independence. After the final defeat of the Spanish at Carabobo in 1821, Venezuela, Colombia and Ecuador were briefly united in the Republic of Gran Colombia. Venezuela left the union in 1829, and for more than a century was ruled by military dictators.

The establishment of a major petroleum industry in the 1920s initiated an era of greater prosperity and growth. Between 1945 and 1948, Rómulo Betancourt (1908–81) paved the way for a short-lived democratic radical government. In 1959, after a further period of dictatorship, he established a second, more moderate administration, which laid the foundations for lasting democracy. In 1990 the government agreed to United Nations mediation regarding Venezuela's claim to parts of neighboring Guyana. An austerity program was announced in 1994.

The Guiana Highlands, Venezuela (above) This vast area of rolling savanna and sandstone outcrops – occupying nearly half the country – is largely uninhabited and unexplored. Eighty-five percent of Venezuelans live in urban areas.

The petroleum industry (below) is the mainstay of the Venezuelan economy, producing about three-quarters of all government revenue. The country's largest oilfields are located in Lake Maracaibo, a shallow inland sea on the northwest coast.

the *llanos* provide good grazing.

Fishing potential is considerable but largely unexploited. However, anchovies, sardines and shellfish are caught off the coast by local fishing fleets. Most forestry land is in government ownership.

The road network is well developed in the north, with three major highways. Railroads make only a small contribution, chiefly to the iron and steel industries. Shipping carries most overseas trade, and there are dredged channels into Lake Maracaibo and the Orinoco river. There are three national airlines and some 30 airports operating scheduled flights, including internal links to areas that would otherwise be inaccessible. The press, radio and television are privately owned and operated, and are not subject to government censorship.

The welfare system is well established, with social security for the old, disabled and unemployed, and increasing public health provision. The quality of health care is good, but the rising population and a general move to urban centers have created shanty towns with inadequate housing and sanitation.

Primary education is free and compulsory, but secondary education is less well developed; however, adult literacy is very high. Higher education is free to students who qualify to enter a university, teacher-training college or polytechnic institute.

Venezuela is a federal republic made up of 23 states, a federal district (around Caracas), two federal territories and 34 dependencies. Legislative power rests with the two-chamber parliament, whose members belong to several political parties. The president, the 49 senators and the 201 deputies are each directly elected every five years.

The majority of the population are mestizos of mixed European and Amerindian origin, and there are substantial white and black minorities. Few pure Amerindians remain, but they have preserved as many as 25 native languages. Although Spanish is the official language, English and Italian are also spoken. The Venezuelan constitution guarantees freedom of religion to its citizens, but the great majority of the population belong to the Roman Catholic church.

ECONOMY

Venezuela's economy is heavily dependent on its petroleum and natural resources, which account for the majority of its export income. Sharply fluctuating international oil prices in the late 1990s had severe economic effects, forcing the government to impose austerity measures and hampering efforts to modernize the economy through diversification and free market policies. Petroleum, natural gas and hydroelectricity enable Venezuela to be self-sufficient in meeting industrial and domestic power needs. The petrochemical industry is based mainly around Morón, to the west of Caracas, and near the Gulf of Venezuela and Lake Maracaibo. The city of Maracaibo on its western shore produces food, pharmaceuticals, machinery and electrical equipment. Ciudad Guayana in the east is a major center for the processing of local iron ores. Significant amounts of diamonds and gold are also extracted.

Only a small proportion of the land is used for agriculture. Despite various agrarian reforms, modernization programs and extensive irrigation in the *llanos*, some food still has to be imported. The principal food crops include bananas, sorghum and maize, while cash crops include coffee, sugar cane and tobacco. Cattle are the most important livestock;

NATIONAL DATA – VENEZUELA

Land area	882,050 sq km (340,561 sq mi)			
Climate		Temperatures		Annual
	Altitude m (ft)	January °C(°F)	July °C(°F)	precipitation mm (in)
Caracas	1,042 (3,419)	26 (79)	24 (75)	914 (35.9)

Major physical features highest point: Pico Bolívar 5,007 m (16,427 ft); longest river: Orinoco 2,061 km (1,281 mi)

Population (2006 est.) 25,730,435

Form of government federal multiparty republic with two legislative houses

Armed forces army 34,000; navy 18,300; air force 7,000

Largest cities Caracas (capital – 1,801,562); Maracaibo (2,054,039); Valencia (1,457,912); Barquisimeto (833,338); Ciudad Guayana (792,508); Maracay (399,446); Petare (386,712); Barcelona (459,148); Maturin (445,451); Turmero (373,871)

Official language Spanish

Ethnic composition Mestizo 69%; White 20%; Black 9%; Amerindian 2%

Religious affiliations Roman Catholic 96%; Protestant 2%; other 2%

Currency 1 bolivar (VEB) = 100 centimos

Gross domestic product (2006) U.S. $176.4 billion

Gross domestic product per capita (2006) U.S. $6,900

Life expectancy at birth male 71.49 yr; female 77.81 yr

Major resources petroleum, natural gas, iron ore, gold, bauxite, manganese, other minerals, hydropower, diamonds, bananas, coal, cocoa, coffee, livestock, maize/corn, rice, sorghum, sugarcane, tobacco, vegetables

Guyana

COOPERATIVE REPUBLIC OF GUYANA

G UYANA, FORMERLY BRITISH GUIANA, IS A small state on the north coast of South America, lying between Venezuela to the northwest, Brazil to the west and south, and Suriname to the east across the Courantyne river. The country's name is derived from an Amerindian word meaning "land of waters".

GEOGRAPHY

Apart from the narrow coastal plain, which is mostly reclaimed land protected by dikes and crisscrossed by canals, most of the country consists of an irregular, thickly forested plateau. Straddling the western border are the ancient Pakaraima Mountains, including Mount Roraima –

inspiration for the famous novel *The Lost World* by the British writer Sir Arthur Conan Doyle (1859–1930). The Rupununi area in the southwest of the country supports savanna grassland interrupted by low mountain ranges.

The climate is hot and often humid. Heavy seasonal rains bring flooding to the poorly drained soils. The forests are mostly tropical hardwoods. Among the varied wildlife, manatees and capybaras are found in and beside the rivers, sloths and jaguars in the forests, and giant anteaters and armadillos on the savanna. The rich bird life includes toucans and brilliantly plumaged hummingbirds.

SOCIETY

The area was originally inhabited by Arawaks and Caribs. The first European settlers were the Dutch, who brought in thousands of African slaves to work the sugar plantations. The British occupied the area around the Demerara river during the Napoleonic Wars (1796–1815), and secured most of present-day Guyana in 1814. With the abolition of slavery, immigrant laborers were imported from Asia. A border dispute with Venezuela was temporarily settled in 1899, and in 1923 British Guiana became a crown colony. In 1953 it was granted home rule along British parliamentary lines, and in 1966 was granted full independence as Guyana. In 1970 it became a republic under Prime Minister Forbes Burnham (1923–85), who was made the country's executive president in 1980.

Guyanans are remarkable for their racial diversity. The great majority are descendants of Asian Indians or Africans. Amerindians are a minority, but are the dominant group in the villages of the sparsely populated interior. There are also small European and Chinese communities. Religion and language reflect the cultural mix: English is the main language, but creole and Hindi are also spoken and there are many Hindus and Muslims as well as Christians.

ECONOMY

Guyana's mixed economy experienced steady growth in the late 1990s following expansion in agriculture and mining. Agriculture is important, but is mostly confined to coastal areas. Plantation sugar cane is the principal crop, followed by rice and coconuts. There is some livestock ranching, chiefly of cattle, in the Rupununi valley in the southwest.

Guyana's industry is dominated by its world-class deposits of bauxite, which provide the chief export item. Other activities include sugar processing, gold mining and river dredging for diamonds. The main imports are fuel and manufactured goods. There is great hydroelectric potential, but its development has been hampered by poor road and railroad links in all but coastal areas. Light aircraft is the usual mode of transportation inland.

Health conditions are generally good, and endemic diseases are under control. Social welfare schemes are comprehensive, but inadequate housing remains a problem. Good education is generally available, and literacy levels are high.

The Kaieteur Falls, formed where the Potaro river plunges over an escarpment in the highlands of west-central Guyana. Some 230 m (750 ft) high and 110 m (350 ft) wide, the falls lie at the center of the Kaieteur National Park.

NATIONAL DATA – GUYANA

Land area	196,850 sq km (76,004 sq mi)			
Climate		Temperatures		Annual
	Altitude m (ft)	January °C(°F)	July °C(°F)	precipitation mm (in)
Georgetown	2 (7)	26 (79)	27 (80)	2,160 (85)

Major physical features	highest point: Roraima 2,810m (9,219 ft); longest river: Essequibo 1,014 km (630 mi)

Population	(2006 est.) 767,245

Form of government	multiparty republic with one legislative house

Armed forces	army 900; navy 100; air force 100

Capital city	Georgetown (238,747)

Official language	English

Ethnic composition	East Indian 50%; Black 36%; Amerindian 7%; White, Chinese, and mixed 7%

Religious affiliations	Christian 50%; Hindu 35%; Muslim 10%; other 5%

Currency	1 Guyanese dollar (GYD) = 100 cents

Gross domestic product	(2006) U.S. $3.62 billion

Gross domestic product per capita	(2006) U.S. $4,700

Life expectancy at birth	male 63.21 yr; female 68.65 yr

Major resources	bauxite, gold, diamonds, hardwood timber, shrimp, fish, coconuts, oranges, rice, semiprecious gems, sugarcane

Suriname

REPUBLIC OF SURINAME

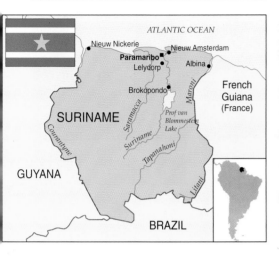

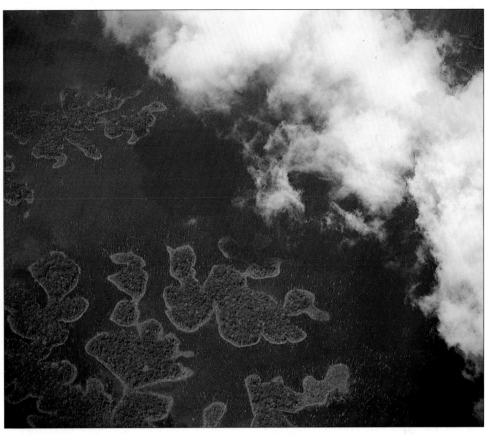

SURINAME, FORMERLY THE COLONY OF DUTCH Guiana, is a small state on the northeastern coast of South America, lying between Guyana to the west, French Guiana to the east and Brazil to the south.

GEOGRAPHY

Suriname's landscape falls into three main areas. The narrow coastal plain is low and swampy, consisting largely of artificially drained or reclaimed land. The coastal soils are rich and fertile, and most of the population live here. Farther inland the land rises to a low savanna plateau. In the south the plateau becomes higher and more rugged, with low mountains and thick forests. Several rivers rise near the Brazilian border and flow northward, the Corantijn forming the border with Guyana, while the Marowijne and Litani straddle that with French Guiana.

The climate is hot and very humid, tempered only by light coastal breezes, and there is heavy seasonal rainfall. The forests are chiefly of tropical hardwoods inland, with some mangroves along the coast. Wildlife is plentiful, and includes tapirs, sloths and ocelots, together with many unusual bats, snakes and birds.

SOCIETY

The Surinen peoples, after whom the country is named, had mostly been ousted by other indigenous groups before the arrival of European colonists. In 1667 the Dutch secured the colony from Britain in exchange for New Amsterdam (later New York). From 1682 the Dutch West India company brought in African slaves to work its tea and coffee plantations. Many escaped to the interior, where their descendants are known as bush blacks. When slavery was abolished in 1863, laborers from India, China and Java were brought over to work the land. In 1954

Suriname's largest lake – Professor van Blommestein Lake – was created by the damming of the river Suriname at Afobaka. The many small islands in the lake are covered in the tropical hardwood forest typical of northern Suriname.

Suriname was granted internal self-government; independence followed in 1975. Racial conflicts led to a military coup in 1980 under Lieutenant-Colonel Dési Bouterse. Foreign aid was withdrawn, the economy collapsed, and the Jungle Commandos, a bush blacks organization led by Ronnie Brunswijk (b.1961), began a guerrilla campaign. Bouterse contested elections in 1988 and lost. The new civilian government concluded a tentative peace with the Jungle Commandos, but Bouterse staged another coup in 1990. A peace accord was signed in 1992 and elections were held in 1996. The country's many political parties reflect its strong racial divisions. Bush blacks form the majority in the interior, with a small Amerindian minority. The two main groups, however, are Asian Indians and mestizos (of mixed descent). There are also significant Javanese and Chinese minorities. The official language is Dutch, but many others are spoken, including a lingua franca called Sranan. The main religions are Christianity, Hinduism and Islam.

ECONOMY

Suriname's economic potential is high. Agriculture is mainly concentrated on the coastal plain. The main crop is rice, some of which is exported; other crops include sugar cane and bananas. Livestock raising is small-scale, but forestry and fishing fulfill both domestic and export needs. Most exports, however, are provided by the country's rich bauxite deposits, which are smelted to aluminum using local hydroelectric power. The road network is mainly limited to coastal areas, and transportation elsewhere is mostly by water or air. Welfare and health-care standards are generally high, and free education is available to university level.

NATIONAL DATA - SURINAME

Land area	161,470 sq km (62,344 sq mi)			
Climate		**Temperatures**		**Annual**
	Altitude m (ft)	January °C(°F)	July °C(°F)	precipitation mm (in)
Paramaribo	4 (13)	26 (79)	27 (80)	2,312 (91)

Major physical features highest point: Juliana Top 1,230 m (4,035 ft)

Population (2006 est.) 439,117

Form of government multiparty republic with one legislative house

Armed forces army 1,400; navy 240; air force 200

Capital city Paramaribo (226,124)

Official language Dutch

Ethnic composition Hindustani 37%; Creole 31%; Javanese 15%; "Maroons" 10%; Amerindian 2%; Chinese 2%; white 1%; other 2%

Religious affiliations Hindu 27.4%; Protestant 25.2% (mainly Moravian); Roman Catholic 22.8%; Muslim 19.6%; traditional beliefs 5%

Currency 1 Surinam dollar (SRD) = 100 cents

Gross domestic product (2006) U.S. $3.098 billion

Gross domestic product per capita (2006) U.S. $7,100

Life expectancy at birth male 66.66 yr; female 71.47 yr

Major resources timber, hydropower, fish, kaolin, shrimp, bauxite, gold, nickel, copper, platinum, iron ore, bananas, cattle, citrus fruits, cocoa, coconuts, coffee, rice, sugarcane

Bolivia

REPUBLIC OF BOLIVIA

A devil-mask (*above*) shows the mix of Roman Catholic and pre-Columbian traditions in Bolivia.

A horse and cart (*left*) weaves a way through La Paz – at 3,627 m (11,900 ft) the world's highest capital.

BOLIVIA IS A LANDLOCKED STATE IN CENTRAL South America. It is a country with a rich culture and plentiful natural resources, but geographical isolation has hampered its development.

GEOGRAPHY

The population of Bolivia is mostly concentrated in the west on the Altiplano – a high plateau sandwiched between the two lofty ranges of the Andes. The Cordillera Occidental on the Chilean frontier includes several active volcanoes, while the Cordillera Oriental to the east is thickly forested. At the northern end of the Altiplano is Lake Titicaca. The southern Altiplano is an arid tableland scattered with salt flats. The Cordillera Oriental descends sharply to the northeast through the cliffs and cloud forests of the Yungas to the vast lowlands of the Oriente, which occupy the north and east.

Bolivia is a tropical country, but local climate varies enormously. Seasonal rainfall is high in the rainforests of the northern Oriente, moderate near Lake Titicaca and very sparse in the south.

SOCIETY

In the 7th century AD Tiahuanaco to the south of Lake Titicaca became the center of the first great Andean empire – that of the Aymará people. The Aymará culture survived despite the later arrival of the Quechua-speaking Incas in the 13th century and of the Spanish in the 16th. In 1545 the world's largest silver deposits were found at Potosí in the southwest.

In 1809 uprisings in Chuquisaca (Sucre) and La Paz signaled the start of South America's struggle for independence. In 1824 the rebel leader Antonio José de Sucre (1795–1830), a lieutenant of the revolutionary commander Simón Bolívar (1783–1830), routed the Spanish at Ayacucho in Peru, bringing independence the following year. The new country was named Bolivia after Bolívar, and Sucre became the first president.

Between 1836–39 Bolivia formed a confederation with Peru, but this was brought to an end after Chile declared war on it. Further disputes with Chile over nitrate extraction led in 1879 to the War of the Pacific; Bolivia and Peru were again defeated, and Bolivia lost the province of Atacama – its only outlet to the sea. By 1900 tin had replaced silver as the chief export, bringing substantial international investment. A disastrous war with Paraguay in the 1930s brought a further loss of territory in the southwest.

A series of bloody coups and counter-coups began in 1943, but in 1952 the military were ousted by civilian forces promising a new social order. For the first time Amerindians were given the vote. A succession of military regimes followed until civilian rule was restored in 1982. The constitution allows for a two-chamber multiparty assembly. The president is directly elected for a five-year term, as are the 27 senators and 65 of the 130 deputies, but in 1993 no majority candidate emerged and the president was elected by Congress.

Bolivia's ethnic and cultural mix is complex. The largest group are mestizos of mixed Spanish and Amerindian ancestry, but they are outnumbered by the Amerindian Quechua and Aymará together.

ECONOMY

Almost half the labor force work on the land; many are subsistence farmers. Coffee and sugar are the chief export crops. Coca, chiefly from the Yungas, is easy to grow, accounting for a large proportion of the world's cocaine supply. Though illegal, this trade has given a massive boost to the ailing economy. Bolivia has considerable mineral resources – notably tin, tungsten, zinc and antimony – but natural gas is the main legal export. Petroleum production meets local needs. Most electricity comes from hydroelectric plants.

Roads are largely confined to the Altiplano, and few are surfaced. Railroad links with Peru and Chile carry Andean minerals to Pacific ports. La Paz and Santa Cruz have international airports.

Health care is adequate in the cities but under-resourced elsewhere. Malaria and Chagas' disease are common in the Oriente. Primary education is free and compulsory between the ages of 6 and 14.

NATIONAL DATA - BOLIVIA

Land area	1,084,390 sq km (418,685 sq mi)			

Climate		Temperatures		Annual
	Altitude m (ft)	January °C(°F)	July °C(°F)	precipitation mm (in)
La Paz	4,103 (13,461)	12 (72)	9 (82)	571 (22.4)

Major physical features highest point: Sajama 6,520 m (21,391 ft); largest lake: Lake Titicaca (part) 3,810 sq km (3,200 sq mi)

Population (2006 est.) 8,989,046

Form of government multiparty republic with two legislative houses

Armed forces army 25,000; navy 5,000; air force 3,000

Largest cities Santa Cruz (1,468,658); Cochabamba (966,049); El Alto (911,525); La Paz (administrative capital - 820, 888); Sucre (judicial capital - 240,604); Oruro (211,217)

Official languages Spanish, Quechua, Aymará

Ethnic composition Quechua 30%; Mestizo (mixed White and Amerindian) 30%; Aymará 25%; White 15%

Religious affiliations Roman Catholic 95%; Protestant (Evangelical Methodist) 5%

Currency 1 boliviano (BOB) = 100 centavos

Gross domestic product (2006) U.S. $27.21 billion

Gross domestic product per capita (2006) U.S. $3,000

Life expectancy at birth male 63.21 yr; female 68.61 yr

Major resources tin, natural gas, petroleum, zinc, tungsten, antimony, silver, iron, lead, gold, timber, hydropower, alpacas, barley, bismuth, cassava, cocoa, coffee, copper, cotton, fruits, llamas, maize/corn, potatoes, quinol, rice, sheep, sugar beet, sugarcane, sulfur, tobacco

Ecuador

REPUBLIC OF ECUADOR

ECUADOR AS ITS NAME SUGGESTS, LIES astride the Equator on the west coast of South America between Colombia to the north and Peru to the south and east. It also includes the Galapagos Islands some 1,000 km (600 mi) out in the Pacific.

GEOGRAPHY

Mainland Ecuador is a country of high mountains that is prone to earthquakes. The coastal lowlands consist largely of fertile alluvial plains interspersed with volcanic hills. Inland the massive wall of the Andes crosses the country from north to south. It is made up of two main parallel chains divided by high valleys, one of which is the site of the capital, Quito. Many of the highest peaks are volcanic, notably Cotopaxi, the world's tallest active volcano. The eastern slopes of the Andes drop into the impenetrable rainforests of the Amazon basin. In this wild, little-known area the exact borders with Peru are disputed.

The climate is generally hot and humid, especially in the east, but the mountainous areas are much colder. The coastal lowlands are cooled by the Humboldt Current, which also reduces rainfall along parts of the coast. The forest here is sparser than on the mountain slopes and in the east, where wildlife is abundant.

The Galapagos are a complex archipelago formed by the cones of young volcanoes. Their shores consist of arid lava rocks, and only the peaks are forested. Kept cool and dry by the Humboldt Current around their shores, the islands have a unique wildlife, including marine iguanas, giant tortoises, flightless cormorants and 13 species of Darwin's finch, whose variations inspired the evolutionary theories of Charles Darwin (1809–82). The islands are protected for research, but have a limited tourist trade.

SOCIETY

In the 15th century a short-lived Amerindian kingdom on the mainland was conquered by the Incas, the ruling class from Peru. Civil war later weakened the Inca empire, which was quickly overrun by the Spanish conquistadors in the 1530s. Spanish settlers divided the country into large estates worked by landless Amerindian laborers, or peons. The situation remained unchanged up until the 19th century, when independence movements grew up among the emergent middle classes. Ecuador won its independence in 1830 after some two decades of war.

A divide remained between the conservative landowners of Quito and the more liberal commercial classes of Guayaquil, the chief port. The resulting instability brought about a series of ruthless dictatorships that lasted until World War II, when Ecuador lost some territory to Peru. The postwar period was dominated by the charismatic president, José María Velasco Ibarra (1893–1979), whose inconsistent policies sometimes hampered the country's economic development. Military rule in the 1970s was followed by a series of more liberal civilian presidents. An inconclusive four-week border war was fought with Peru in 1995.

The 1976 constitution provides for an executive president and vice president for a four-year term. There is a single legislative National Congress which is also elected for four years.

Spanish is the main language, although a large portion of the population are Quechua-speaking Amerindians. Most of the rest are mestizos (of mixed descent), with small African and European minorities. The majority are at least nominally Roman Catholic, though much Amerindian culture survives.

ECONOMY

Economically, Ecuador remains largely agricultural. Bananas are the principal cash crop, along with sugar cane, coffee and cacao. Rice and the indigenous potato are among the main staples. Despite land reforms, much food is still grown on massive hacienda estates. Livestock is raised both along the coast and in the mountain valleys. The great hardwood forests remain relatively unexploited, while the fishing industry is one of South America's largest. The chief mineral resources are petroleum and natural gas, which have overtaken bananas and shrimps as the main export.

Transportation is hindered by the difficult terrain. Road and rail networks are limited, and the mule is often the most effective means of land transportation. Social welfare schemes are in place, but there is much rural poverty, and health care and living conditions are often poor. Education is available to university level.

Women in Quito (*above*) carry their burdens in bundles on their backs. The Ecuadorean capital – which retains the air of a Spanish colonial town – lies almost on the Equator, situated in a high valley about 3,000 m (9,000 ft) above sea level.

NATIONAL DATA - ECUADOR

Land area	276,840 sq km (106,889 sq mi)			
Climate	Altitude m (ft)	Temperatures January °C(°F)	July °C(°F)	Annual precipitation mm (in)
Quito	2,879 (9,446)	15 (59)	15 (59)	1,116 (43.9)

Major physical features highest point: Chimborazo 6,267 m (20,561 ft); longest river: Napo (part) 1,100 km (700 mi)

Population (2006 est.) 13,547,510

Form of government multiparty republic with one legislative house

Armed forces army 37,500; navy 5,500; air force 4,000

Largest cities Guayaquil (1,952,029); Quito (capital - 1,399,814); Cuenca (276,974); Machala (198,123); Santo Domingo de los Colorados (200,421)

Official language Spanish

Ethnic composition Mestizo (mixed Amerindian and White) 65%; Amerindian 25%; Spanish and others 7%; Black 3%

Religious affiliations Roman Catholic 95%; other 5%

Currency 1 U.S. dollar (USD) = 100 cents

Gross domestic product (2006) U.S. $60.48 billion

Gross domestic product per capita (2006) U.S. $4,500

Life expectancy at birth male 73.55 yr; female 79.43 yr

Major resources petroleum, fish, timber, hydropower, bananas, cassava, cacao, coffee, natural gas, gold, limestone, livestock, maize/corn, oranges, potatoes, rice, sugarcane

Peru

REPUBLIC OF PERU

Peru, which lies astride the Andes on the western coast of South America, derives its name from a Quechua word implying abundance, recalling the former opulence of the Inca and Spanish empires. However, its modern development has been hindered by the hostile, though often spectacular terrain.

GEOGRAPHY

Peru is divided into three distinct geographical areas. The narrow coastal plain, or Costa, runs along the whole length of the Pacific coast from Ecuador in the northwest to the Chilean border in the south. Much is arid desert, but the Andean rivers have laid down patches of fertile alluvial soil. Most of the population lives in this part of the country.

The lofty Andes, or Sierra, form a continuous backbone that enters from Ecuador and broadens in the southeast on the border with Chile and Bolivia. The mountains are most dramatic in central Peru, where peaks such as Huascarán, the highest, tower above deep mountain trenches. In the south the two main ranges run either side of an extensive high plateau called the Altiplano, at the center of which is the great Lake Titicaca straddling the Bolivian border. The whole country is liable to earthquakes, and several volcanoes in the south are dormant rather than extinct.

The Andes fall away northeast to the Montaña – the deep, forest-clad valleys that lead out into the dense rainforests of the Amazon basin. This wild, undulating country bordering Colombia and Brazil is almost uninhabited.

The climate of the Amazon basin is hot and humid, typical of its latitude, but the

mountains are much colder with seasonal rain and snow. The coastal areas are kept dry by cold offshore currents; cloud and fog are common, but little rainfall normally occurs. However, an occasional switch from cold to warm sea currents creates a climatic phenomenon known as El Niño, causing severe flooding on land and a reduction in offshore marine plankton,

leading to loss of coastal fish and birds.

Plant and animal life vary tremendously. Coastal vegetation ranges from cacti to hardy spiny shrubs; the highlands support grasses and puna scrub; and the east has typical hardwoods such as mahogany and wild rubber. Wildlife includes tapirs and sloths in the forests, and chinchillas, vicunas and condors in the mountains.

Mestizo-style architecture (*above*) at Juli on the western shore of Lake Titicaca, Peru. The elaborate carving is Amerindian in style, but the subjects – the coat of arms above the door on the right and the floral motifs – are colonial.

A Peruvian river valley (*left*) shows the legacies of the Inca empire. Terraced fields make subsistence farming possible on the steep hillsides, while the road running parallel to the river was once part of a sophisticated communications network.

The cold offshore seas support rich fish populations and the many birds that feed on them such as pelicans and boobies.

SOCIETY

Various cultures once flourished among Peru's original Amerindian inhabitants, including most notably the Chavín, the valley-dwelling Nasca, the Chimú and the Huari. In the 15th century, however, the whole country was taken over by the Quechua-speaking Incas. They built massive stone cities, the most famous of which are at Cuzco – their capital and the center of their religion, which was based on sun-worship – and mountain-top Machu Picchu. Weakened by civil war, however, their wealthy empire was overthrown by the Spanish adventurer Francisco Pizarro (c. 1475–1541) in 1531. This marked the beginning of the Amerindian peoples' long history of exploitation.

Peru, with its immense natural wealth, remained Spain's treasurehouse for some three centuries, until its liberation by the Argentine general José de San Martín (1778–1850) in 1821. A brief federation with Bolivia collapsed in 1839. After the liberal presidency of General Ramón Castilla (1797–1867) military juntas alternated with civilian politicians, who failed in their attempts to stabilize the economy. Such problems were exacerbated by the War of the Pacific (1879–84), in which valuable territory was lost to Chile.

In 1963 Fernando Belaúnde Terry (b.1912) was elected president, and launched a program of reforms. In 1968 he was ousted by an army junta, which in turn was toppled by another junta in 1975, but in 1980 Belaúnde was returned to power in new elections. His attempts at reform were frustrated by economic difficulties, a devastating El Niño and ever-increasing

---

NATIONAL DATA – PERU

Land area	1,280,000 sq km (494,210 sq mi)			

Climate		Temperatures		Annual
	Altitude m (ft)	January °C(°F)	July °C(°F)	precipitation mm (in)
Lima	120 (394)	22 (77)	17 (81)	13 (0.5)

Major physical features	highest point: Huascarán 6,768 m (22,205 ft)

Population	(2006 est.) 28,302,603

Form of government	multiparty republic

Armed forces	army 40,000; navy 25,000; air force 15,000

Largest cities	Lima (capital – 7,979,965); Arequipa (878,129); Trujillo (788,911); Chiclayo (607,121); Iquitos (468,687)

Official languages	Spanish, Quechua

Ethnic composition	Amerindian 45%; Mestizo (mixed Amerindian and White) 37%; White 15%; Black, Japanese, Chinese, and other 3%

Religious affiliations	Roman Catholic 81%; Seventh Day Adventist 1.4%; other Christian 0.7%; other 0.6%; unspecified or none 16.3%

Currency	1 nuevo sol (PEN) = 100 centimos

Gross domestic product	(2006) U.S. $181.8 billion

Gross domestic product per capita	(2006) U.S. $6,400

Life expectancy at birth	male 68.05 yr; female 71.71 yr

Major resources	copper, silver, gold, petroleum, timber, fish, iron ore, coal, phosphate, potash, hydropower, natural gas, cattle, cotton, fruit, lead, livestock, potatoes, rice, sugarcane, tungsten, uranium, vegetables, wheat, zinc

Reed-gathering (*above*) on the shores of Lake Titicaca. The Uru – an ancient Amerindian people – live in floating villages made from huge rafts of totara, a reedlike papyrus. Their crescent-shaped boats are also fashioned from the reeds.

guerrilla activity. The two main guerrilla groups involved were the Tupac Amarú Revolutionary Movement (MRTA) and the neo-Maoist group known as Sendero Luminoso (Shining Path). The MRTA came to world attention in 1997 after besieging the Japanese embassy for 126 days.

President Alan García, elected in 1985, fared no better with the economy, the guerrillas or drug trafficking and lost the 1990 election to Alberto Fujimori. In April 1992 Fujimori appropriated dictatorial powers to enforce economic reforms and defeat the guerrillas, but their activity continues despite the capture of Shining Path leader Abimael Guzmán, with Shining Path joining forces with drug organizations in 1995 to provide funding. Fujimori had been reelected in November 1992, ending his dictatorship, and again in 1995, but was dismissed by Congress in November 2000 as being morally unfit to govern after he was implicated in a serious corruption scandal. He fled to Japan, where he remained as Peru prepared criminal charges. New elections in April 2001 led to a run-off between former president Garcia and Alejandro Toledo.

Peru's instability is partly due to ethnic and social divisions between the Spanish-speaking mestizos of the Costa and the Quechua-speaking Amerindians of the Sierra. Other minority groups include some whites, Aymará and other Amerindian peoples. Most of the population is Roman Catholic, but the influence of Amerindian traditions remains strong.

ECONOMY

Peru's economy is heavily dependent on the export of its rich mineral and other resources. Agriculture is limited by a lack of arable land, most of which is found in coastal areas, where irrigation is needed to grow crops such as rice and sugar cane. Highland farming is mainly at subsistence level. Crops here include wheat, potatoes and coca, which supplies an illicit cocaine trade. Sheep, llamas and alpacas are the main livestock animals, bred chiefly for their wool. There is some forestry in parts of the east, but the area is inaccessibe and ecologically fragile. The vast fishing industry periodically suffers drastic reductions in catch size due to the effects of El Niño.

Industry is powered mainly by hydro-electricity. Copper ore is the major mineral, but petroleum is becoming almost as important. These are the two main exports, followed by uranium and other metals such as iron and silver. Manufacturing industries are concentrated around the capital, Lima, and nearby Callao, the main port; many manufactured goods still have to be imported.

Transportation is limited by the rugged terrain and the climate, and there are few railroads. Air travel is hampered by the difficulty of constructing airstrips in swampy forest or steep mountain terrain.

Welfare provision and health care is available in the cities. However, housing and health conditions remain inadequate from impoverished mountain areas to the outskirts of cities such as Lima. Where education can be obtained it is free and compulsory, but it is often unavailable to poor or isolated rural communities.

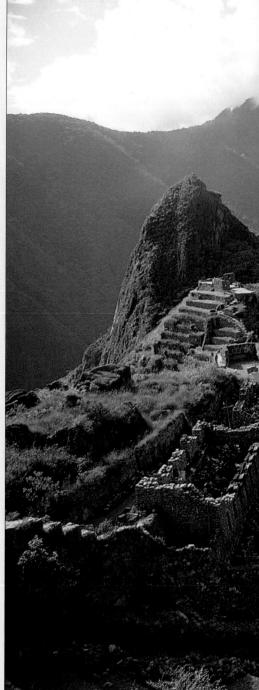

The Incas

The popular image of the Inca empire, the last of the great South American indigenous civilizations, is of a society as durable and mysterious as the mighty Inca fortifications that survive today. Yet the Inca empire was relatively short-lived. It was established only a few years before the birth of Christopher Columbus (1451–1506), and lasted little more than a century. The real mystery is how the Incas achieved so much in that short time.

They began as a band of Andean mountain dwellers who left their little

and colonization began; Cuzco became the capital of the growing Inca territory. His son Yupanqui (1438–71), who assumed the name Pachacuti, continued the expansion. To govern his large territories he created an entire administrative class of honorary Incas and undertook what may have been the Incas' most amazing achievement – an elaborate road system, thousands of kilometers long, that included tunnels and suspension bridges. Along it at regular intervals ran a system of relay stations that could send government communications as far as 250 km (150 mi) a day. The Incas achieved this without the use of either the wheel or the horse; messengers ran between the stations on foot. They had no writing system either, so messengers had to relay their messages from memory, with only knotted-string patterns called *quipu* to aid them.

Pachacuti and his successor Topa (1471–93) conquered an enormous area of western South America from present-day Ecuador in the north to Chile and Argentina in the south. Everywhere they passed, they built not only roads but towns, canals, roadside forts, resthouses and irrigation schemes in order to increase food production. Topa's successor, Huayna Cápac (1493–1525), continued the process, building a newer and more splendid capital on the site of the present-day Ecuadorean capital of Quito.

What made all this possible was sheer manpower; Huayna ruled some 12 million people, and every province owed the Inca not only tribute but levies of soldiers and workers. Service was a religious as well as a social duty, and may well have been given gladly. The results were impressive. The mountain-top city of Machu Picchu was surrounded by wide terraced fields, all the soil for which must have been carried up the mountainside on men's backs.

Every man was bound to his father's occupation, but no one starved; relief was generous, and storehouses provided support during famines. Although the result appeared to be a stable if authoritarian society, Huayna's sudden death precipitated a war between his two sons. No sooner had a winner emerged than he was seized and murdered by the Spanish conquistadors. Without a ruler at the top, the entire Inca edifice crumbled almost at once.

Machu Picchu (*above*), perhaps the best-known of Peru's Inca cities, was probably built no earlier than the 15th century, and is perched high above the Urubamba valley on the eastern flanks of the Andes.

Inca artifacts (*left*) An Inca stirrup-cup in the form of a jaguar savaging its victim, and a rare ceremonial gold figurine, one of the few pieces of gold to have survived the marauding Spanish conquistadors.

village to found a city at Cuzco in what is now south-central Peru. They grew into a dominant city-state that raided and feuded with its neighbors. Their chieftains – the Incas, for whom the empire was later named – became the prime focus of a divine cult.

Under the eighth Inca, Viracocha (reigned 1410–38) a process of conquest

Brazil

FEDERAL REPUBLIC OF BRAZIL

BRAZIL IS THE LARGEST COUNTRY IN SOUTH America, the fifth largest in the world and the sixth most populous To the northeast and southeast it faces the Atlantic along a 7,400-km (4,600-mi) coastline, while to the west it shares borders with most other South American countries. A land of wealth and opportunity, it has attracted settlers from all over the world. However, its development as a modern industrial state has brought with it problems of inadequate finance, overpopulation and environmental damage.

GEOGRAPHY

Brazil divides into two main areas. The vast Amazon basin to the north contains the largest river system in the world, while the highland areas of the south and east are the most populous.

The land
Brazil's northern frontier runs through an area of mountains and plateaus known as the Guiana Highlands, which extend into Venezuela, Guyana, Suriname and French Guiana. They include Brazil's highest mountain, Pico da Neblina, discovered as recently as 1962.

Immediately to the south is the great Amazon basin, which occupies most of the north of the country. Its landscape is generally low-lying, apart from the hillier fringes, and some flatter areas are flooded every year. The main river is some 6,570 km (4,080 mi) long – second in the world only to the Nile – and has over 1,000 known tributaries; most of the larger ones rise farther west in neighboring Colombia, Peru and Bolivia. To the east, the Amazon enters the Atlantic via a broad delta, and the lowlands extend eastward along the northern coast.

The Brazilian Highlands cover most of the rest of the country, rising gently from the Amazon basin along a line roughly from Bolivia to Cape São Roque, Brazil's northeasternmost point. The highlands consist mostly of broad tablelands cut by deep river valleys. Along the eastern seaboard they rise more sharply to form a steep escarpment overlooking a narrow coastal plain. Among the highest peaks is Pico das Agulhas Negras near Rio de Janeiro and São Paulo, the two largest cities. Most of the rivers drain southwest into neighboring Paraguay, Argentina and Uruguay. The Paraná and Uruguay rivers run along sections of the border, creating spectacular waterfalls as they leave the plateau.

The only other extensive lowland area is in the west bordering Bolivia and northern Paraguay, where the swampy plains of the Pantanal form the northernmost extension of the Gran Chaco.

Climate
The climate is generally warm and humid with seasonal rains, but varies somewhat across the country. Temperatures are higher around the Amazon basin, where they remain largely stable throughout the year. Conditions are similar along the east coast, while the Brazilian Highlands and the south are both cooler, with occasional winter frosts.

In the northeast the São Francisco basin has a unique microclimate of its own. This notorious drought pocket, known as the drought polygon, has virtually no rain in the dry season, and far less than the surrounding areas at other times. Elsewhere, rainfall is heaviest in parts of the Amazon basin and along the flanks of the eastern mountains.

Plants
Much of Brazil is forested; the tropical evergreen rainforests of the Amazon basin have a greater variety of plant species than any other habitat on earth. There are similar forests along the southeastern coast, and semideciduous forests on the heights of the escarpment. Inland, the highlands show a mixture of savanna and deciduous woodland, and in the far south are some of the original grass prairies. The dry northeastern areas are characterized by *caatinga* – a type of thorny woodland.

There has been international concern over the mass destruction of the Amazon rainforests for cattle ranching and mining. Some effort has been made to slow down the process of uncontrolled ex-

The many waters of the Amazon (*above*) The lushness of Amazonian vegetation reflects the vast amounts of fertile silt carried down through the world's largest river system, which receives up to one-quarter of the earth's run-off water every year.

The famous statue of Christ the Redeemer (*below*) with arms outstretched in blessing above the city of Rio de Janeiro. It stands on Corcovado peak, 787 m (2,310 ft) high, and looks across to the unmistakable outline of the Pão de Açúcar, or Sugar Loaf mountain.

Dutch briefly captured Bahia in 1624, and settled at Pernambuco (modern Recife) in 1630. They were driven out by Brazilian forces in 1654, 14 years after the Portuguese had once again wrested control of their homeland from Spain.

During the 17th and 18th centuries, numerous expeditions set out from São Paulo to capture Amerindians for slaves and to exploit the resources of the vast interior. The first rebellion against Portuguese rule came as early as 1789, but independence was to take another three decades. In 1808 the French invaded Portugal, and the prince regent, Dom John (1767–1826), took refuge in Brazil. He took a strong interest in his new home, and in 1815 he proclaimed the United Kingdom of Portugal, Brazil and the Algarve. A year later he became King John VI. In 1821 he returned to Portugal, leaving behind his son, Dom Pedro (1798–1834), as regent. The Portuguese court sought to return Brazil to its former colonial status, and Dom Pedro was faced with an angry population. In 1822 he declared independence, and became Brazil's first emperor.

His reign was troubled: war with Argentina (1825–28) led to the loss of Uruguay, and in 1831 he abdicated in

ploitation, but more effective conservation measures are essential if this fragile ecosystem is to be preserved.

Animals

Animal life in the Amazon Basin is as colorful and variegated as the plant life. Numerous species of monkeys, snakes and tree frogs share the trees with hundreds of different bird species. The rivers teem with caimans, capybaras, river dolphins and as many as 1,500 known species of fish.

Outside the Amazon basin, much of Brazil's once-abundant wildlife has all but vanished. There are still wildfowl in the marshlands of the Pantanal, and rheas (flightless birds) roam the tablelands, as do bustard-like seriemas. Snakes, lizards and armadillos abound, while Giant anteaters feed on termites in their pillar-like nests. Ocelots and jaguars, the predators of the Brazilian savanna, have been hunted to virtual extinction.

SOCIETY

Brazil has produced a remarkably integrated society since the early days of Portuguese colonial settlement. The country was named after its first export, *brasil*,

a red dye made from wood. As the interior was opened up, other valuable items were exploited, including gold, diamonds and coffee. After World War II, Brazil's predominantly rural society was dramatically changed into an urban, industrial society beset with the problems common to many developing countries.

History

Pedro Álvares Cabral (1467/8–1520), a Portuguese naval commander, was probably the first European to set eyes on Brazil. He recorded his discovery in 1500, but it was not until 1532 that Martim Afonso de Sousa (c.1500–64) founded the Portuguese colony of São Vicente – the first of 15 hereditary captaincies. The colonists created large estates, worked by Amerindians and later by African slaves.

In 1549 King John III of Portugal (1502–57) appointed a governor-general, who controlled the captaincies from his base at Bahia (later Salvador). Meanwhile, other colonial countries began to take an interest in Brazil. The French established a settlement at Rio de Janeiro in 1555, and it took the Portuguese five years to evict them. By 1580 the Spanish had gained control of Portugal, and colonial administration in Brazil was being neglected. The

Land area	8,456,510 sq km (3,265,077 sq mi)			
Climate		**Temperatures**		**Annual**
	Altitude m (ft)	January °C(°F)	July °C(°F)	precipitation mm (in)
Manaus	83 (272)	27 (81)	27 (81)	2,386 (93.9)
Brasilia	1,006 (3,301)	22 (72)	19 (66)	1,632 (64.2)
Rio de Janeiro	61 (200)	27 (81)	22 (72)	1,173 (46.1)

Major physical features highest point: Pico de Neblina 3,014 m (9,889 ft); longest river: Amazon (part) 6,436 km (4,000 mi)

Population (2006 est.) 188,078,227

Form of government federal multiparty republic with two legislative houses

Armed forces army 189,000; navy 32,850; air force 65,309

Largest cities Sao Paulo (10,095,244); Rio de Janeiro (6,086,917); Salvador (2,810,334); Belo Horizonte (2,424,579); Fortaleza (2,390,058); Brasilia (capital - 2,314,172); Curitiba (1,776,074); Manaus (1,690,331); Recife (1,500,293); Belem (1,465,670)

Official language Portuguese

Ethnic composition White 53.7%; Mulatto (mixed White and Black) 38.5%; Black 6.2%; other (includes Japanese, Arab, Amerindian) 0.9%; unspecified 0.7%

Religious affiliations Roman Catholic 73.6%; Protestant 15.4%; Spiritualist 1.3%; Bantu/voodoo 0.3%; other 1.8%; unspecified 0.2%; none 7.4%

Currency 1 real (BRL) = 100 centavos

Gross domestic product (2006) U.S. $1.616 trillion

Gross domestic product per capita (2006) U.S. $8,600

Life expectancy at birth male 68.02 yr; female 76.12 yr

Major resources bauxite, gold, iron ore, manganese, nickel, phosphates, platinum, tin, uranium, petroleum, hydropower, timber, asbestos, bananas, beryllium, cassava, cereals, chromium, coffee, coal, cocoa, cotton, crude oil, diamonds, fish, graphite, livestock, natural gas, oranges, quartz crystal, rice, rubber, salt, silver, soybeans, sugarcane, titanium, tobacco, tourism, tungsten, zinc

favor of his five-year-old son. After a difficult period of regency, Pedro II (1825–91) came to the throne in 1840, ruling for almost 50 years. His continuing involvement in Uruguayan affairs drew Brazil into the long and bloody War of the Triple Alliance (1865–70). A military coup in 1889 forced the emperor's abdication, and a republic was proclaimed; in 1891 this became the United States of Brazil.

Between 1895 and 1909 a series of boundary disputes were settled peacefully in Brazil's favor, while coffee and rubber brought new levels of prosperity. World War I was followed by a period of growing unrest, which ended, in 1930, in a successful rebellion. The de facto president, Getúlio Vargas (1883–1954), revised the constitution twice, giving himself dictatorial powers. In 1942 he entered World War II, siding with the Allies. After the war, Vargas lost power, only to return in 1950. He battled, with limited success, against spiraling inflation, until revelations of murder and corruption led him to commit suicide in 1954.

In 1960 the newly built city of Brasília in central Brazil became the country's capital in place of Rio de Janeiro – the climax of a campaign to develop the resources of the interior. Yet in spite of a rapid rise in productivity and material progress, inflation and foreign borrowing reached record levels. Instability led to a military coup in 1964, and civilian government did not return until 1985. In 1992, President Fernando Collor (b. 1949) resigned from office in order not to impeached for corruption.

Government

Brazil is a federal republic. The 1988 constitution – the eighth in the country's history – endorsed a two-chamber multiparty federal assembly. The senators are elected for eight years, the deputies for four years, and the president for four.

The country is divided into 26 states and a federal district (Brasília). Each state elects at least three senators, as does the federal district, but the number of deputies from each depends on the population.

People

Brazil's population shows the results of large-scale immigration and integration. Just over one-half the population are of mainly European descent, but most of the rest are mixed European, African and Amerindian. More recent immigrants include the Japanese. Only a few pure-blooded Amerindians remain, chiefly in

the remoter parts of the Amazon basin.

An important unifying factor is the Portuguese language, which was adopted by virtually all immigrants. However, Brazilian Portuguese has become very distinct from its European counterpart, strongly influenced by Amerindian as well as other immigrant languages.

The prevalence of Roman Catholicism is another unifying factor, although there is a significant Protestant minority. A number of African religions are held more or less in parallel with Catholicism; the *candomblé* cult in particular has a large following among many social groups.

ECONOMY

Brazil has enormous natural resources, but lacks enough capital to exploit them effectively. Successive governments have struggled with the double burden of foreign debt and massive inflation. The situation became critical in the late 1990s; Brazil had the highest foreign debt in the world and imposed stringent fiscal control

and structural reforms as part of an IMF-led support program. Such severe measures allowed little possibility of alleviating Brazil's massive poverty and income inequality in the near future. Following an economic crisis in early 1999 the Central Bank was forced to devalue the Brazilian currency.

Agriculture and fisheries

Brazil grows enough food to meet domestic needs, and agriculture provides raw materials for a significant proportion of exports.

Most of the arable land is along the coast and in the south. The main export crops are soybeans, sugar cane and especially coffee, of which Brazil is the world's leading producer. Sugar cane is used to produce gasohol, an automobile fuel consisting of ethanol and gasoline – a process that Brazil has pioneered. Maize, rice and wheat are the chief food crops.

Pastureland is mainly concentrated in the south, which is the center of the meat-packing industry. Brazil's livestock

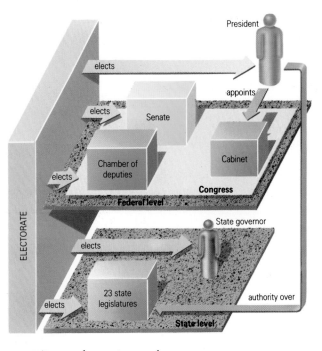

Virgin rainforest (*left*) Beneath the tangled tree canopy lies an unimaginable diversity of plants and animals potentially valuable to humans. Yet the rainforest is rapidly being destroyed: this area in Rondonia, western Brazil, is to be flooded by one of 60 planned dams.

A federal constitution (*right*) Alone of the states of South America, Brazil is a federal republic, with 26 constituent states and a federal district (Brasília). The two-chamber national assembly consists of a 81-member senate elected for eight years and a 513-member chamber of deputies elected for four years. The cabinet is chosen by the president.

population – mainly pigs and cattle – is among the world's largest. The creation of cattle ranches in the Amazon basin has led to widespread destruction of the forests, and there has been violent conflict with rubber tappers. Most trees are cleared by burning, leaving unstable soil that quickly becomes exhausted.

Almost two-thirds of the country is forested. Lumber from the northern forests generally goes for fuel. Most commercial timber, used for cellulose and paper, comes from eucalyptus and pine plantations in the south and southeast.

Fish resources are large but underexploited. There is good freshwater fishing in the Amazon basin, and in government-stocked reservoirs in the northeast. Most saltwater fishing ports are in the south, but a smaller fleet in the northeast catches lobsters and shrimps for export.

Industry and resources
Brazil has enormous mineral resources, with massive reserves of aluminum, manganese and especially iron ores; the extracted metals form a major export item. Other minerals are used by domestic industries. Coal resources in the far south are mostly low grade. There are productive petroleum and gas wells, both inland and offshore, all along the southeastern coast. Drilling in the Campos basin, off Rio de Janeiro, accounts for about half the country's petroleum production.

Exploration for new mineral resources sometimes has disastrous effects: multinational corporations have bulldozed the Amazon rainforest to create open-cast mines, and a gold rush in the late 1980s brought mercury pollution to the rivers, poisoning the local Amerindians.

Most of Brazil's electricity comes from hydroelectric stations, mainly in the densely populated south and southeast. A nuclear power station near Rio de Janeiro has been in operation since 1982.

The manufacturing sector is varied and expanding. Many industries are based on local raw materials – steel and petroleum refining, food processing, paper, textiles and footwear manufacture. Other major products include fertilizers, transportation equipment, armaments, construction materials and microelectronics.

Service industries are expanding, and there is an increasing influx of tourists lured by the warm waters, attractive scenery and glamorous city nightlife of the eastern seaboard. Copacabana Beach, in Rio de Janeiro, is the country's most famous resort.

Transportation and communications
Brazil's road network has been greatly improved in recent years, with surfaced roads linking all the major population centers; many of the smaller roads are unsurfaced. Railroads are few outside the Rio de Janeiro–São Paulo region, apart from freight lines for ore transportation. Many northern areas can only be reached by water or air.

The Amazon river is navigable for most of its length, and coastal traffic is served by a sizable merchant fleet. Air transportation provides a vital link between Brazil's widely scattered and often isolated communities. Most cities have an airport, and there is usually a dirt landing strip in even the smallest villages. Most international flights run from São Paulo and Rio de Janeiro.

There are over 300 daily newspapers with a circulation of 6.5 million. The main newspapers are published in São Paulo and Rio de Janeiro; censorship has been relaxed since the return to civilian government. There are several thousand radio stations and over one hundred television stations.

Welfare and education
Health care is well resourced in major cities, but underfunded and understaffed in rural areas, where malaria, yellow fever and other diseases are rife. Malnutrition is common, especially among children. Urban housing shortages mean that many of the people moving to the cities in search of work are compelled to take up residence in vast shanty towns, where poor sanitation aids the spread of disease.

All employers contribute to the National Social Security Institute, which pays benefits for illness, unemployment, and retirement. For many, however, the system is costly yet ineffective.

Education is free and compulsory from the ages of 7 to 14, and free secondary education is also available, but it is only a small proportion of Brazilian students – usually those from wealthier families – who are able to take full advantage of it; very few enrol for university or college education. Literacy levels are correspondingly low throughout the whole of the country.

Paraguay

REPUBLIC OF PARAGUAY

PARAGUAY IS A SMALL LANDLOCKED STATE IN South America, lying between Argentina to the south, Bolivia to the northwest and Brazil to the northeast. The Paraguay river flows north–south through the country, dividing it into two very distinct areas. The river Paraná forms part of the border with Argentina.

GEOGRAPHY

The Región Occidental, or Western Region, forms part of the flat plains of the Gran Chaco, which extend into Bolivia and Argentina. The Región Oriental, or Eastern Region, rises gently from the floodplains of the Paraguay and Paraná rivers to a series of low mountain chains that form part of the Brazilian plateau. At the eastern border the Paraná river descends from the plateau in a series of spectacular waterfalls.

The climate is subtropical with seasonal rainfall, heavy in the east, which is prone to flooding, but sparse in the west, where droughts are frequent. The country is well forested, with tropical hardwood species in the eastern hills, and open savanna woodland on parts of the Chaco. The wildlife includes peccaries, jaguars, armadillos and anteaters.

SOCIETY

The first European explorers to Paraguay found the country inhabited by the friendly Guaraní peoples. Spanish settlers lived peaceably among them, as did the Jesuit missionaries. Later slave-traders were much harsher toward the indigenous peoples, though they were generally treated better than elsewhere. In 1776 Paraguay effectively became part of the Viceroyalty of Río de la Plata, centered on Buenos Aires. Resentment at this eventually led to independence in 1811, followed by the establishment of an isolationist dictatorship under José Gaspar Rodríguez de Francia (1766–1840), known as El Supremo.

For many years Paraguay was ruled by sometimes brutal dictators. Carlos Antonio Lopez (1790–1862) and his son Francisco Solano Lopez (1827–70) ended isolationism and encouraged Paraguay's development, but embroiled the country in conflicts with its neighbors. The War of the Triple Alliance (1865–70) ended in total devastation and occupation by a Brazilian army. The subsequent painful period of reconstruction saw the introduc-

Asunción, the capital of Paraguay, was founded as a fort in 1537 by Spanish colonists, since when the church and the military have dominated the country. Here the military police stand guard outside the offices of the archdiocese of Asunción.

tion of greater democracy. Paraguay fought the Chaco War with Bolivia from 1932 to 1935. In 1954 General Alfredo Stroessner (b.1912) seized power. He ruled as an autocrat until overthrown in 1989. A new democratic constitution was adopted in 1992 which endorsed a two-chamber multiparty assembly. The president is elected for five years; parliament consists of 80 departmental deputies and 45 national senators.

The Paraguayan people are proud of their mainly Guaraní descent, though there is also a strong Spanish element. There are few pure-blooded Amerindians, but Guaraní is the main spoken language alongside the official language, Spanish. Most Paraguayans are nominally Roman Catholic.

ECONOMY

Paraguay has a market economy featuring a service-based formal sector and a large informal sector. Agriculture sustains a quarter of the workforce. Cattle ranching is the main pursuit, but crops such as cotton, sugar cane, soybeans and timber are also important. There are no significant mineral or energy reserves, but the hydroelectric potential is enormous, and Paraguay is already a leading power exporter. Water and road transportation are both well developed. Health care and sanitation are adequate in the capital, Asunción, but poorly developed elsewhere. Education is generally good, and literacy levels are relatively high.

NATIONAL DATA – PARAGUAY				
Land area 397,300 sq km (153,398 sq mi)				
Climate	**Altitude** m (ft)	**Temperatures** January °C(°F)	July °C(°F)	**Annual precipitation** mm (in)
Asunción	139 (456)	23 (73)	18 (64)	1,404 (55.2)
Major physical features highest point: Mount San Rafael 850 m (2,789 ft)				
Population (2006 est.) 6,506,464				
Form of government multiparty republic with two legislative houses				
Armed forces army 7,600; navy 1,600; air force 1,100				
Largest cities Asunción (capital – 506,094); Ciudad del Este (287,311); San Lorenzo* (245,468); Luque* (239,842); Capiatá* (232,831); Lambaré* (130,168); Fernando de la Mora* (123,620) (* adjacent to Asunción)				
Official languages Spanish, Guarani				
Ethnic composition Mestizo (mixed Spanish and Amerindian) 95%; other 5%				
Religious affiliations Roman Catholic 90%; Mennonite and other Protestant 10%				
Currency 1 guarani (PYG) = 100 centimos				
Gross domestic product (2006) $30.64 billion				
Gross domestic product per capita (2006) U.S. $4,700				
Life expectancy at birth male 72.56 yr; female 77.78 yr				
Major resources hydropower, timber, iron ore, manganese, limestone, cassava, cattle, coffee, cotton, fruit, maize/corn, rice, soybeans, sugarcane, tobacco				

Chile

REPUBLIC OF CHILE

C HILE OCCUPIES A NARROW AND MOUNTAIN-ous strip of land along the south-western seaboard of South America, bordering Peru to the north, Bolivia to the northeast and Argentina to the east. It was one of the first South American countries to develop democratic political and economic structures, but in recent times went through a period of military rule, when public and private freedoms were greatly curtailed.

GEOGRAPHY

Chile's shape is unusual: it measures on average little more than 160 km (100 mi) from east to west, but extends south from the tropics to within about 1,000 km (600 mi) of Antarctica, with the result that there are marked climatic differences.

The land

Perhaps surprisingly for such a long thin country, the landscape divides lengthways; the three main divisions are the coastal mountains, the Andean ranges and the long trench between them. The entire country is subject to severe earthquakes; the city of Concepción on the central west coast has been rebuilt four times following earthquake damage.

The coastal ranges are older and less steep than the Andes. They follow the Pacific coastline from near the Peruvian frontier as far south as the Gulf of Ancud in central-southern Chile. From here onward they are partly submerged, their peaks forming a long chain of islands.

The central depression extends from the salt flats of the Atacama Desert in the north to the Gulf of Ancud. The central-

A herd of llamas forages for sparse grazing in the harsh mountain landscape of far northern Chile, beneath the snowcapped Parinacota volcano. Their thick woolly coats enable these New World relatives of the camel to survive in the chill Andean highlands.

southern section between the Aconcagua and Bío-Bío rivers is known as the Central Valley. This is the most populous part of Chile, where the capital, Santiago, lies.

The mighty Andean cordillera includes many volcanoes – some active – both in the far north and in the central zone, where the highest peaks are to be found. The southern Andes have a glaciated landscape of lakes and fjords, and the far south is fragmented into innumerable islands, including Tierra del Fuego (the western part of which is Chilean) and Cape Horn, the southernmost point.

Climate

The climate of Chile is predominantly temperate, moderated by the proximity of the sea, but with extremes ranging from tropical to antarctic. Temperatures are lower in the high Andes, and rainfall increases southward. There is virtually no rain in the Atacama Desert, and in central Chile most of it falls in winter. In the south strong westerly winds blow all year round, and rainfall levels here are among the highest outside the tropics.

Plants and animals

The plant life is varied, with a diversity of species having adapted to the different types of climate and topography. In the north, tamarugo trees are able to survive the harsh desert conditions, while the high Andes support mountain grasses;

NATIONAL DATA - CHILE				
Land area 748,800 sq km (289,113 sq mi)				
Climate		Temperatures		Annual
	Altitude m (ft)	January °C(°F)	July °C(°F)	precipitation mm (in)
Santiago	520 (1,706)	22 (72)	9 (48)	313 (12.3)
Major physical features highest point: Llullaillaco 6,723 m (22,057 ft)				
Population (2006 est.) 16,134,219				
Form of government multiparty republic with two legislative houses				
Armed forces army 47,700; navy 19,398; air force 11,000				
Largest cities Santiago (capital - 4,949,540); Valparaíso (287,194); Concepción (666,381); La Serena (159,081); Antofagasta (319,422)				
Official language Spanish				
Ethnic composition White and White-Amerindian 95%; Amerindian 3%; other 2%				
Religious affiliations Roman Catholic 89%; Protestant 11%; some Jewish				
Currency 1 Chilean peso (CLP) = 100 centavos				
Gross domestic product (2006) U.S. $203 billion				
Gross domestic product per capita (2006) U.S. $12,600				
Life expectancy at birth male 73.49 yr; female 80.21 yr				
Major resources copper, timber, iron ore, nitrates, precious metals, molybdenum, hydropower, natural gas, petroleum, beans, coal, fish, fruit, grapes, lead, livestock, maize/corn, manganese, oil, onions, potatoes, rice, sugar beet, wheat, wine, zinc				

frequent fog along the coast provides moisture for cacti and shrubs. Farther south, the vegetation becomes gradually more prolific, ranging from hardy algarobas to the hardwoods and shrubs of the Central Valley, with stands of monkey puzzles on the neighboring Andean slopes. The slopes surrounding the lakes farther south are cloaked in forests, as is much of the mountainous south. The harsher climate of the far south and Tierra del Fuego supports only dwarf trees and hardy grasses.

Animals are less abundant than elsewhere in South America, but they include guanacos, vicunas, coypus, pumas and Andean cats. Bird life is diverse, thanks to the many migratory species that pass through the country. The Andean condor – a permanent resident – lives high in the mountains. There are no spiders or poisonous reptiles.

SOCIETY

The Chilean people have a strong sense of national identity, and can boast a history of relative freedom and democracy that was interrupted rather than extinguished by military rule in the 1970s and 1980s.

History

Little is known of Chile's precolonial history except in general terms. In the 15th century Inca invaders established a foothold in the north, but were halted by the fierce resistance of the Araucanian peoples farther south.

The Spanish arrived from Peru in 1536 under the command of Diego de Almagro (1475–1538), but returned northward the following year, having found none of the wealth that had provided such rich plunder in Peru. In 1541 Pedro de Valdivia (c. 1498–1554) established a colony at Santiago. The settlement grew, though efforts to push southward were hampered by Amerindian resistance, which continued well into the 19th century, by which time their population had been drastically reduced. For most of the colonial period, Chile remained isolated and dependent on Spain. In an economy relying on agriculture, most available land was divided into large estates held by a few wealthy families.

The struggle for independence began in 1810 under the influence of neighboring Spanish colonies. It started relatively peacefully, but the Battle of Rancagua (1814) inspired spirited resistance. In 1817 an army led by José de San Martín (1778–1850) crossed the Andes from

The birth of a new peak in the Chilean Andes is heralded by a plume of steam rising from a newly erupting cone. Such outbreaks of volcanic activity are common throughout this young range of mountains, and snow-covered peaks stud the horizon.

Argentina and defeated the Spanish at Chacabuco. Bernardo O'Higgins (1776/8-1842), a Chilean patriot who had fought with San Martín, was elected to the post of supreme director, and in 1818 he proclaimed Chile's independence.

Conflict between the army and the landowning families was resolved in 1830, initiating a period of democratic rule that was to continue until 1973. In the War of the Pacific (1879–84), Chile gained large areas of land from its northern rivals, Peru and Bolivia, securing the rich nitrate deposits of the Atacama Desert.

After World War I, the demand for naturally occurring nitrates (used in fertilizers) fell as the means of forming them chemically were developed, and the economy declined. Political instability increased despite land reforms in 1964. The election of a Marxist president, Sal-

Easter Island

A tiny, remote volcanic island, 3,780 km (2,350 mi) to the west of the South American coast, Easter Island was annexed to Chile in 1888.

The island was named by the Dutch admiral Jacob Roggeveen (1659–1729), who landed there on Easter Sunday 1722. Long before it came to the attention of European explorers, Easter Island had a thriving Polynesian community, now sometimes referred to as the "People of a Lost Culture". They left behind some quite spectacular ancient monuments. Most are monolithic statues and huge faces carved from tuff, a soft volcanic stone. Some stand 20 m (65 ft) tall and weigh as much as 50 tons. The oldest have been there for 1,000 years and the most recent date back to the 17th century.

vador Allende Gossens (1908–73), in 1970 led to a United States boycott, and in 1973 the government was overthrown in a military coup led by General Augusto Pinochet Ugarte (b. 1915). Chile remained under repressive military rule until the election of President Patricio Aylwin Azócar (b. 1918) in 1990, followed by Eduardo Frei in 1993.

Government

The 1981 constitution provided for the reestablishment in 1989 of a multiparty two-chamber parliament. The president, most senators and all 120 deputies are elected for six-year terms. However, conservative appointments to the senate have denied the government a majority in that chamber. This showed the continuing influence of the military.

People

The culture of Chile's largely mestizo population combines their Spanish heritage with Amerindian elements. Spanish is the dominant language even among the Amerindian minority and Roman Catholicism is the majority religion.

ECONOMY

Chile is a country with plentiful natural resources, but fluctuating world prices have made it economically vulnerable; it relies increasingly on its developing manufacturing industry.

Few of the stone carvers' descendants survive on the modern island, Peruvian slave raids in 1862–63, combined with major epidemics, virtually wiped out the Polynesian population. Christian missionaries in the 19th century contributed further to the decline of the indigenous culture, and with its annexation to Chile 26 years later, the island was repopulated with Spanish-speakers from the mainland.

Most of the 1,400 inhabitants are concentrated in the village of Hanga Roa on the sheltered western coast. Traditionally they earn their living by farming, particularly sheep ranching, which generates some trade in wool, but since the mid 1980s the mainstay of the economy has been tourism.

Man of mystery One of the strange stone faces carved by the original Polynesian inhabitants of Easter Island over 1,000 years ago.

Agriculture and industry

Agriculture is a flourishing sector, but much food is imported. Arable farming is concentrated in the fertile Central Valley. The variety of crops grown includes wheat, potatoes and rice. Fruit (including apples), vegetables and wine are exported. The most common livestock animals are sheep and beef cattle.

Fish stocks off the coast are enormous, and the market has expanded greatly. Sardines and anchovies are among the chief catches, and fish products such as oil and meal are a valuable export. Forestry, which is concentrated mainly in the south, is a source both of timber and paper for export.

Chile is the world's leading supplier of copper; other minerals include iron and molybdenum. The main energy source is hydroelectric but some coal provides a supplementary source, together with petroleum and natural gas from Tierra del Fuego. Manufacturing industry is largely concentrated in central Chile. Varied activities include iron, steel and petroleum refining, fish processing and petrochemicals. Despite growth in this sector, manufactured goods are among the chief imports of the country.

Transportation and communications

Historically, Chile has depended on the sea for its communications; the southernmost provinces are accessible only by sea or air. The Pan-American Highway runs south the Gulf of Ancud, but there are few other surfaced roads outside the Central Valley. Railroad services are good here but limited elsewhere. Two main airlines provide internal services, with international flights from Santiago.

Chile has 131 television stations and 168 radio stations. There are more than 30 national daily newspapers with a combined circulation of only 1.4 million.

Welfare and education

Chile has a well-established health service based on decentralized local units, and health conditions are generally good; a parallel private system was introduced in 1980. The welfare system is equally well developed; since 1973 the comprehensive social insurance system has been gradually converted into a savings plan operated by private companies.

Education is free and compulsory from the ages of 6 to 13, followed by four years of optional secondary education. Literacy levels are high, and there is a well-established tradition of education at university and college level.

Argentina

ARGENTINE REPUBLIC

ARGENTINA'S NAME DERIVES FROM THE Latin meaning "land of silver", and from the earliest colonial days the country has attracted settlers in search of wealth. Although its natural and human resources have transformed it into a modern, developed country, its social and economic development has been hampered by political instability.

GEOGRAPHY

Argentina is shaped like a long, narrow triangle occupying much of the southern end of South America. The scenery not only differs widely from north to south, but shows a marked contrast between the high mountains of the west and the broad plains of the east.

The land

The impenetrable barrier of the Andean mountain range runs the whole length of the western border with Chile, and carries on into Bolivia to the northwest. Some of the mountains are volcanic, including South America's highest peak, Aconcagua. The area is also subject to frequent earthquakes.

In the north, bordering Paraguay, are the plains of the Gran Chaco, which are poorly drained and covered in thorny subtropical scrubland. Southeast of here is a low-lying expanse of rainforests, known as Mesopotamia – from the Greek meaning "land between the rivers" – because it is sandwiched between the river Paraná to the west and the river Uruguay to the east, which forms the border with Brazil and Uruguay.

Southward again are the flat, fertile

grasslands of the Pampas, which extend southeast to meet the Atlantic coast south of the river Plate. This is by far the most densely populated area in the whole of the republic, especially around the capital, Buenos Aires.

The southernmost part of the country is Patagonia – a barren, arid plateau that is rich in mineral deposits. The Patagonian coastline has cliffs along its whole length. The scenery becomes generally more rugged toward the south, especially on the island of Tierra del Fuego, the eastern half of which belongs to Argentina.

Climate

Almost the whole of the country lies within the southern temperate zone. However, the northeastern plains are subtropical and humid, while the extreme south is subpolar.

Rainfall is light, and tends to be concentrated in Mesopotamia, the eastern Chaco and the Pampas, where winters can be damp and cold. Thunderstorms are common here, and snow falls occasionally. Patagonia, which is colder and more arid, is subject to strong winds from the Andes carrying dust and sand.

Plants and animals

Vegetation varies greatly throughout the country. The scrublands produce mainly dwarf shrubs, while the rainforests support giant cedar and laurel trees. Quebracho trees, from which tannin is extracted, are native to Mesopotamia.

The wildlife is typically South American, with llamas, monkeys, jaguars, armadillos and many snakes. Bird life includes toucans, hummingbirds, parrots and the ostrich-like rheas. Fish, including piranhas, abound in the many streams, and there are sea lions off the coast. Pigs, introduced by European settlers, are to be found wild in Patagonia.

SOCIETY

From the 16th century down to the present day, Argentinian society has been dominated by the European culture introduced by the colonists.

History

The indigenous Amerindian population of Argentina were mainly nomadic hunters, although in the northwest they had established settled agrarian societies.

Spanish settlers, mainly from Peru, Chile and Paraguay, began to colonize northern Argentina from the beginning of the 16th century. Asunción in neighbor-

ing Paraguay became the first colonial center. Buenos Aires, founded in 1536 and subsequently abandoned, was refounded in 1580. Other cities were established too, but for two centuries the colony remained effectively part of Peru. In 1776 the Viceroyalty of the Río de la Plata was created, with Buenos Aires as its capital. Forty years later the viceroy was deposed, and Argentina declared its independence from Spain. The 19th century was a period of great economic

The artist's quarter of Buenos Aires (*above*) Brightly painted houses and Italian restaurants crowd the area – called La Boca – where the city's first settlers landed.

The masts of an early steamship (*left*) in Buenos Aires. Spanish sailors named the port after Santa María del Buen Aire, meaning "St Mary of the good wind".

Guarding the Casa Rosada (*below*), the presidential palace in Buenos Aires. The uniform is little changed since the days of prosperous military rule in the 19th century.

(b. 1926). Elections were held in 1983, and the Radical Party's Raúl Alfonsín (b. 1926) became president, introducing many social and economic reforms.

In December 2001 Argentina's economic crisis peaked when popular protests against austerity measures forced President Fernando de la Rua to resign. A political crisis ensued with five presidents voted in two weeks!

Government

Argentina is a federal republic. The head of state is the president, who is directly elected for a four-year term, but can stand for two terms. The legislature consists of a senate, which includes three representatives from each province and territory,

growth, but the country was led by military rulers, who were backed by the most prosperous ranchers.

From 1916 to 1930 government was democratic and moderate. This situation was ended by a military coup. During the troubled years that followed, Juan Domingo Perón (1895–1974) emerged as a popular leader, becoming president in 1946. He established social reform and economic stability, but was deposed in 1955. After a period of great instability,

Perón was recalled from exile in 1973, but died a year later, to be succeeded by his widow, Isabel (b. 1931). The Perónist regime was overthrown by a coup in 1976, and there followed seven years of oppressive rule by military junta.

In 1982 Argentina invaded the British colony of the Falkland Islands (or Islas Malvinas), which it had claimed since 1820. Argentina's defeat in the ensuing conflict brought about the resignation of the president, General Leopoldo Galtieri

NATIONAL DATA – ARGENTINA

Land area 2,736,690 sq km (1,056,642 sq mi)

Climate	Altitude m (ft)	Temperatures		Annual precipitation mm (in)
		January °C(°F)	July °C(°F)	
Buenos Aires	25 (82)	25 (74)	12 (90)	1,205 (47.4)

Major physical features highest point: Aconcagua 6,960 m (22,834 ft); longest river: Paraná (part) 4,500 km (2,800 mi)

Population (2006 est.) 39,921,833

Form of government federal multiparty republic with two legislative houses

Armed forces army 41,400; navy 17,500; air force 12,500

Largest cities Buenos Aires (capital – 11,612,214); Córdoba (1,475,447); Rosario (1,178,407); Mendoza (890,521)

Official language Spanish

Ethnic composition White (mostly Spanish and Italian) 97%; Mestizo (mixed White and Amerindian ancestry); Amerindian, or other non-white groups 3%

Religious affiliations Roman Catholic 92% (less than 20% practising); Protestant 2%; Jewish 2%; other 4%

Currency 1 Argentine peso (ARS) = 100 centavos

Gross domestic product (2006) U.S. $599.1 billion

Gross domestic product per capita (2006) U.S. $15,000

Life expectancy at birth male 72.38 yr; female 80.05 yr

Major resources aluminum, cement, lead, zinc, tin, copper, hydropower, iron ore, manganese, petroleum, uranium, apples, cattle, citrus fruits, coal, cotton, fruit, fish, grapes, gold, limestone, maize/corn, mica, natural gas, olives, potatoes, peanuts, rice, sheep, silver, sorghum, soybeans, sugarcane, sunflower seeds, timber, tungsten, tobacco, wheat, wool

The high life (*above*) Buenos Aires – one of the most populous cities in the world – offers elegance and wealth to a lucky few.

Soccer fans (*left*), in Argentina's far northwestern Jujuy province, selling flags for their national team. Soccer in South America is treated almost as a religion.

and a chamber of deputies. The 72 senators are directly elected for nine-year terms, the 257 deputies for four-years.

There are 23 provinces and the federal capital (Buenos Aires). The local government for each province is based on that of the federal government, with legislative, judicial and executive branches.

People

The population is predominantly of European extraction. Most Argentinians are descendants of the Spanish and Italian immigrants who entered the country from about 1870 onward but there are also substantial numbers of people of German, Russian, French and British descent. The official language is Spanish, but a few people speak other languages, including some Welsh around Trelew, Patagonia, where there is a large Welsh settlement dating from the 19th century.

The customs and culture are heavily dominated by European influences. This is reflected in the country's art, music and literature, and also in the popularity of soccer and polo. The great majority of the people are Roman Catholics, with small Protestant and Jewish minorities.

ECONOMY

Argentina has a free market economy and plentiful resources, but the instability associated with high inflation and foreign debt has led to bankruptcy.

Agriculture and fisheries

Farmland occupies more than half the total land area, and much of this is pasture. Beef production, the traditional mainstay, is concentrated mainly in the large ranches of the Pampas, and beef is still one of Argentina's major commodities. Sheep, pigs, and goats are also raised, and wool provides yet another important export.

The principal crops include wheat, maize, soybeans and sugar cane. Argentina is a major exporter of vegetable oils, soybeans and animal feedstuffs. Grapes are also grown to make wine mainly for the domestic market.

Although more than one-fifth of the total land area is forested, the forestry industry is not sufficient to supply the country's needs. Most wood is used for timber, though quebracho trees supply tannin for the leather industry. Although the fishing catch is quite small, both fish and oil are exported.

Industry and resources

Argentina is so rich in energy resources that it is now virtually self-sufficient in power. The main sources are hydroelectricity, natural gas and petroleum, the latter occurring mainly in Patagonia and the Mendoza area bordering the central Andes. There are several nuclear power plants. Over 200,000 tonnes of coal was produced in 1996. Reform of the mining laws has encouraged investment, and in 1996 the US $1 billion Alumbrera copper and gold mine started production.

About one-fifth of the population is involved in the manufacturing industry, of which food processing is still a major sector. Beef in particular is processed for both the domestic and export markets; products include canned meat, meat extracts, tallow and leather. Petroleum-refining and petrochemical industries have been established around Mendoza and Buenos Aires. There is a growing iron and steel industry and paper, aluminum, cement and polyethylene are produced in significant quantities.

Transportation and communications

Argentina has the best transportation system in the whole of South America. The road network is extensive, as is the railroad system, especially in central Argentina. Most freight is carried by road, though some is taken by rail or in the small ships that ply between the coastal ports and along the main river arteries. Air transportation, both for internal and international travel, is well developed, with airports serving all the major cities.

Most radio and television stations are under private ownership, although at various times they have been subject to strong state influence. There are over 200 daily newspapers.

Welfare and education

Social welfare systems were developed during the first Perón presidency. However, many social problems remain. Housing in particular is often stretched to maximum capacity in the cities, where workers from rural areas have moved in search of work. Health provision includes both public and private services, but standards vary across the country.

Education is compulsory at primary level (up to the age of 13), and literacy levels are relatively high. Primary, secondary and higher education are all free and there are over 50 universities, the largest being at Buenos Aires.

Uruguay

EASTERN REPUBLIC OF URUGUAY

U RUGUAY IS A SMALL COUNTRY ON THE EAST coast of South America, bordered by Brazil to the north and by Argentina to the west across the river Uruguay. The country is still often known locally as the Banda Oriental, or "East Bank" (of the Uruguay river).

GEOGRAPHY

The country is mostly low-lying apart from two ranges of hills that extend south from the Brazilian border. The climate is mild and temperate, with plentiful rain throughout the year. Most of Uruguay is

Montevideo, a city European in its architecture and population, lies on the northern bank of the Plate estuary. It originally made its wealth from the meat industry, but has since grown as a financial center. Nearly all of Uruguay's trade passes through the port.

covered with lush grassland, and there are few trees apart from those along the river banks. The natural vegetation is tall pampas grass with a few flowers and shrubs. Wildlife is varied but no longer abundant, and includes armadillos and capybaras. Birds are still common, however, including the burrowing owls of the pampas and many river fowl.

SOCIETY

The original inhabitants were the fierce Charrúa Amerindians, who drove away the first Spanish explorers in the 16th century. The first settlers were Portuguese from the north in about 1680. In 1726 the Spanish founded Montevideo, and went on to drive out the Portuguese, effectively exterminating the Amerindians in the process. In 1776 they incorporated the country into the Viceroyalty of Río de la Plata. After 1811 a growing independence movement staged a revolt, but it was soon followed by a period of Brazilian–Portuguese occupation. This was ousted in 1828 with help from Argentina and Britain, and Uruguay became independent.

For many years the country was torn by war and civil unrest, until it became a military dictatorship. A short period of civilian rule (1890–1903) again degenerated into civil war. In 1903 the country was reunited under the social reformer José Batlle y Ordóñez (1856–1929). A period of prosperity was followed by economic

collapse in the 1930s, leading to a brief dictatorship. In World War II Uruguay remained neutral under a liberal president, heralding a further economic boom. Then growing economic problems fueled a campaign by the Tupumaro urban guerrillas, leading to a gradual military takeover in the 1970s. Civilian rule was restored in the 1980s.

The government is made up of a two-chamber elected assembly and an elected president, who appoints his own ministers. The people are mostly descendants of Spanish and Italian immigrants. The main language is Spanish, with a Portuguese element near the Brazilian border.

ECONOMY

Traditionally agriculture dominated Uruguay's economy especially the livestock industry. Crop farming is less important. By 1992 agriculture's share of export earnings had fallen to 23 percent and hydroelectric power and tourism had become the major earners. Manufacturing is expanding with imports of raw material and fuels. There is a growing iron and steel industry. Paper, aluminum, cement and polythylene are also produced in significant quantities. Brazil and Argentina are Uruguay's main trading partners. Mineral resources are poor.

Uruguay can boast South America's oldest welfare state, and has led its neighbors in education, welfare and health care.

NATIONAL DATA – URUGUAY

Land area	173,620 sq km (67,035 sq mi)			
Climate		**Temperatures**		**Annual**
	Altitude m (ft)	January °C(°F)	July °C(°F)	precipitation mm (in)
Montevideo	22 (72)	23 (73)	11 (52)	1101 (43.3)

Major physical features	highest point: Mount Catedral 514 m (1,685 ft)
Population	(2006 est.) 3,431,932
Form of government	multiparty republic with two legislative houses
Armed forces	army 15,200; navy 5,700; air force 3,100
Largest cities	Montevideo (capital - 1,272,411); Salto (101,318); Paysandú (73,135); Las Piedras (70,571); Rivera (65,010); Maldonado (57,205); Tacuaraembó (53,147); Melo (51,925)
Official language	Spanish
Ethnic composition	White 88%; Mestizo 8%; Black 4%
Religious affiliations	Roman Catholic 66%; Protestant 2%; Jewish 1%; nonprofessing or other 31%
Currency	1 Uruguayan peso (UYU) = 100 centesimos
Gross domestic product	(2006) U.S. $36.56 billion
Gross domestic product per capita	(2006) U.S. $10,700
Life expectancy at birth	male 73.12 yr; female 79.65 yr
Major resources	hydropower, fisheries, amethysts, barley, cattle, maize/corn, marble, rice, sheep, topaz, wheat

Dependencies in the region

FALKLAND ISLANDS
UNITED KINGDOM

The Falkland Islands, or Islas Malvinas, are a self-governing British colony in the South Atlantic consisting of an archipelago off the southeastern coast of South America. The colony also administers a series of island dependencies. These include the volcanic South Sandwich Islands and South Georgia – the largest of the dependencies – a whaling settlement some 1,300 km (800 mi) to the southeast.

The islands of East and West Falkland comprise most of the land area. Both are heavily indented, hilly and often swept by cold westerly winds. The landscape is mostly peat-covered moorland, with low grasses, scrubby vegetation and no native trees. Animal life includes elephant seals, sea lions and penguins.

These previously uninhabited islands were first sighted by British navigators in 1592. It was not until 1765, however, that they were settled by a British garrison. This was removed in 1774, but Britain maintained its claim to the islands. Spain kept a settlement there until 1811, and in 1820 the newly independent Argentina claimed the islands on the basis of previous Spanish claims – marking the start of a long-running conflict between the two nations over the islands. An Argentinian attempt to settle the islands in 1832 soon collapsed in mutiny, and in 1833 a British garrison was reestablished. An almost entirely British population became established with a limited economy based, as it is now, on fishing and sheep-rearing. In 1892 colonial government was introduced, with a governor appointed by the British monarch.

Argentina, however, continued to uphold its claim to the islands. Intervention by the United Nations in 1964 led to talks, in which Britain sought to settle the sovereignty issue on the basis of the islanders' wishes to remain British. Nevertheless, in 1982 Argentina invaded the islands. British forces were brought in, and after six weeks of fierce fighting they expelled the occupying forces. Hostilities were formally ended in 1989, but Argentina still claims the colony.

The local economy and way of life have been greatly affected by the war and its aftermath. An exclusion zone was enforced around the islands to prevent overfishing, and fishing licenses have now become the Falkland Islands' largest source of income. A 320 km (200 mi) oil exploration zone was imposed in 1993.

Port Stanley (*above*) on East Falkland, the largest of the Falkland Islands. In June 1982 fighting between British and Argentinian forces raged around this little town – the colony's principal settlement and main harbor – during the last days of a short war that was precipitated, in part, by the prospect of rich offshore oil fields.

Parasols in the jungle (*right*) These children belong to one of several Asian groups that have settled in French Guiana, notably the Hmong people from Laos, formerly a French protectorate in Southeast Asia. The Hmong, refugees from war-torn Indochina, were settled here in the late 1970s.

A tropical idyll (*far right*) French Guiana was once the home of the infamous penal colony of Devil's Island, to which France until 1945 exiled its worst offenders for long sentences of hard labor. Today this small tropical island is becoming a tourist resort.

FRENCH GUIANA
FRANCE

French Guiana is a French overseas *département* lying just north of the Equator on the northeast coast of South America. It is bordered by Suriname to the west, and Brazil to the south and east.

There are two main geographical areas: the low-lying coastal plains and the forested plateau of the interior, which rises to form mountains along the southern border. The climate is hot and humid, with heavy seasonal rains. Most people live along the coast, which is fertile but often swampy, with savanna in the west. Much of the country is covered by impenetrable hardwood forests that support a wide variety of wildlife.

French merchants first founded Cayenne, now the capital, in 1643, but their

initial attempts at colonization were frustrated by other European colonists. However, by the late 17th century the French were fully established, importing African slaves to work their plantations. After the French Revolution, political prisoners were sent here to die in appalling conditions. In the 1850s the abolition of slavery caused economic collapse in spite of the introduction of indentured laborers from India. French missionaries,

however, set up educational institutions for freed slaves. Penal settlements were again established, including the notorious Devil's Island; they were not abolished until after World War II. French Guianans have had French citizenship since 1848 and parliamentary representation since 1870; moves toward independence have had little success. Most people are of mixed descent, speaking French and Creole; most are Roman Catholic.

French Guiana remains economically dependent on France. Agriculture is concentrated in coastal areas, producing both staples and cash crops such as bananas and sugar cane. Fish are the other main export apart from minerals, particularly gold and bauxite. The European Space Agency has a rocket-launching base at Kourou. The roads are few, and most internal transportation is by air. Welfare and healthcare are to French standards.

Northern Africa

Desert oasis Surrounded by the encroaching sands of the Sahara, a town of baked clay houses has grown up beside an oasis in southern Algeria, where underground springs give life to date palms and other fruit trees.

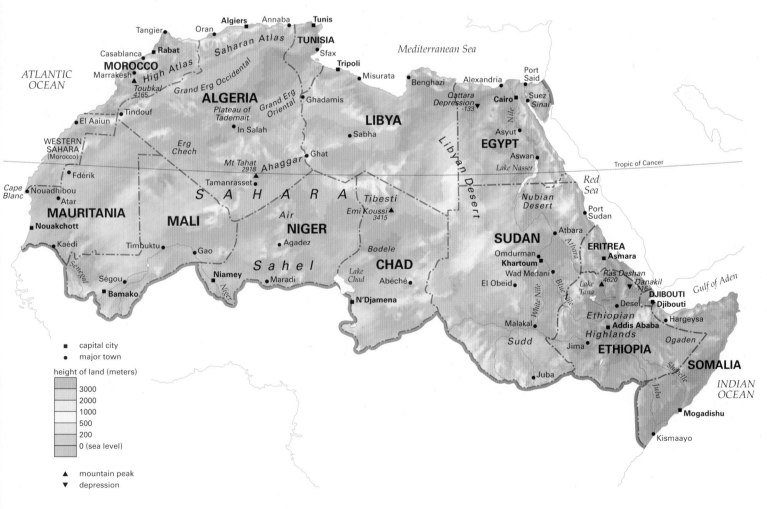

ATLANTIC OCEAN

Mediterranean Sea

Tangier · Oran · **Algiers** · Annaba · **Tunis**

Saharan Atlas · **TUNISIA**

Casablanca · **Rabat** · Sfax

MOROCCO · High Atlas · **Tripoli**

Marrakesh · Grand Erg Occidental · Misurata

Toubkal 4165 · **ALGERIA** · Ghadamis · Benghazi · Alexandria · Port Said

Plateau of Tademait · Grand Erg Oriental · **LIBYA** · Qattara Depression -133 · **Cairo** · Suez · Sinai

Tindouf · In Salah · Sabha · Asyut · **EGYPT** · Aswan · Lake Nasser

El Aaiun

WESTERN SAHARA (Morocco) · Erg Chech · Ghat · Libyan Desert · Tropic of Cancer

Mt Tahat 2918 · Ahaggar · Red Sea

Cape Blanc · Fdérik · S A H A R A · Nubian Desert · Port Sudan

Nouadhibou · Atar · Tamanrasset · Tibesti · Emi Koussi 3415

MAURITANIA · **MALI** · Air · **SUDAN** · Atbara · **ERITREA**

Nouakchott · **NIGER** · Bodele · Omdurman · **Asmara**

Kaédi · Timbuktu · Gao · Agadez · Lake Chad · **CHAD** · **Khartoum** · Wad Medani · Ras Dashan 4620 · Danakil -116

Ségou · Maradi · Abéché · El Obeid · Lake Tana · Desel · **DJIBOUTI**

Bamako · **Niamey** · **N'Djamena** · White Nile · Blue Nile · **Addis Ababa** · **Djibouti** · Hargeysa

Senegal · Sahel · Malakal · Ethiopian Highlands · Ogaden

Niger · Sudd · Jima · **ETHIOPIA** · **SOMALIA**

Juba · INDIAN OCEAN

Mogadishu

Kismaayo

Gulf of Aden

- ■ capital city
- ● major town

height of land (meters)

3000
2000
1000
500
200
0 (sea level)

▲ mountain peak
▼ depression

131

Morocco

KINGDOM OF MOROCCO

Morocco, in northwestern Africa, is one of the few remaining monarchies in the Arab world. Western Sahara (formerly Spanish Sahara) is today under Moroccan control, though its status continues to be a matter for dispute.

GEOGRAPHY

Most of Morocco is mountain or plateau. In the northeast a crescent-shaped limestone range called Er Rif runs along the Mediterranean coast. West of Er Rif lies the Rharb, a wide alluvial lowland.

The Atlas Mountains extend from the southwest to the northeast across the country like a backbone, spreading out into the high plateaus toward the

Algerian border. In the northwest they slope away via broad foothills to the Atlantic coastal plains – the most fertile area and the site of the principal towns. The Atlas Mountains form three distinct ranges: the High Atlas in the center, the Middle Atlas to the northeast and the much lower Anti-Atlas range in the southwest. The mountains slope away southward into the wide desert plateaus of Algeria and the Western Sahara.

The main rivers are all torrential. A few, such as the Loukos, rise in Er Rif, but most rise in the Atlas Mountains. Some flow north or west into the Atlantic (though the Moulouya runs northeast into the Mediterranean). Others, such as the Ziz and the Rheris, run southward through deep gorges toward the Sahara, where after feeding oases they gradually disappear.

Much of Morocco enjoys a temperate Mediterranean climate, with hot summers and mild, wet winters. However, toward the south and the interior the climate becomes increasingly arid and extreme. The rainfall is often uneven, and drought and floods are common. Some rivers in the south, such as the Drâa, only flow seasonally.

The lowland areas are mostly rough bush and esparto grasslands, turning to thorn scrub in the more arid south. In desert oases date palms flourish. On northern mountain slopes there are extensive forests of evergreen oak and cedar, and lower down grow thuja, juniper and gorse. In the Rharb and Er Rif extensive areas are covered in cork oak. Wildlife is largely Mediterranean, though populations of Cuvier's gazelle and the Barbary macaque also thrive here, with desert animals such as the Fennec fox in the south. Mouflon (wild sheep) live on the slopes.

SOCIETY

Morocco, like the rest of North Africa's Mediterranean countries, was a former Carthaginian and later a Roman province that was overrun in the 7th century AD by Muslim Arabs from the east. However, in the 11th century, the indigenous people – the Berbers – established their own Almoravid and Almohad empires, which at their height ruled much of Spain as well. Later sultans were much weaker, although in the 17th century Mawlay Ismail (1645–1727) did much to reunite his country, expelling British and Spanish settlers in favor of his French allies.

The 19th century was a time of weak rule and war between the indigenous

peoples, and France, having seized Algeria, extended its influence to Morocco. During 1896 Spain took over much of Morocco's Saharan territory, and in 1912 Morocco was made a French protectorate, though Spain kept control of areas in the north and south. Fierce resistance, especially by the Berbers of Er Rif, continued into the 1930s. The nationalist movement gained new impetus during the turmoil of World War II, in which Sultan Mohammed V (1909–61) sided with the Allies against Germany. When Morocco gained independence from France in 1956, the Sultan became king.

He was succeeded in 1961 by his son, Hassan II (1929–99), whose rule was marked by a moderate approach to international relations. Despite several coup attempts during his reign, Hassan generally enjoyed great popularity. He led a campaign to regain the Spanish Sahara, valued for its mineral resources, and in 1975 marched 350,000 unarmed volunteers into the territory. Spain thereupon

The Souk of the Dyers (*above*) at Marrakesh in western Morocco. Formerly one of the great Islamic cities its palace and mosques make it a popular tourist center.

The Casbah of Tifoultoute (*left*) in the town of Ouarzazate at the heart of Morocco. The word *casbah* is a French corruption of the Arabic word for citadel.

handed over the area, now known as the Western Sahara, to divided Moroccan and Mauritanian administration. A local nationalist force, Polisario, aided by Libya and Algeria, waged a guerilla war against this occupation and in 1979 Mauritania withdrew. Morocco immediately annexed the rest of the Western Sahara and built an 'electronic' wall protected by land-mines to defend key towns and the phosphate mines. A UN-sponsored peace plan led to the cessation of hostilities in 1991, but a planned referendum on the future of the province, postponed many times, is not now expected to take place until 2003.

The present king, Mohammed VI, succeeded on his father's death in 1999. Morocco is a constitutional monarchy in which the king wields considerable power. He appoints the prime minister and the supreme court. There are two elected chambers, the upper house, the Chamber of Counselors, and the lower house, the Chamber of Representatives, which is elected every 5 years by direct vote. The next election is due to be held in 2002.

Moroccans today are largely of Berber descent, but with a strong Arab overlay, both ethnically and culturally. The Berber element has remained very much stronger among mountain peoples such as those of Er Rif. About one-third of the population is still Berber-speaking, though most Moroccans speak Arabic and many speak French. Islam is the state religion, but there is also a historically important Jewish population.

ECONOMY

Agriculture produces about one-third of Morocco's exports and employs about two-fifths of the work force. Following the redistribution of many colonial farms, most are small-scale family concerns; others are worked as government-subsidized cooperatives. Irrigation, usually local and small-scale, is essential over much of the country. The chief crops are barley and wheat, along with citrus fruit, potatoes and other vegetables. Cedar and other wood products, especially cork, are also important. Dates are grown in desert oases, and some nomads still herd camels. Livestock production – chiefly goats, sheep and some cattle – is limited by poor pasture. Western Sahara has little agriculture of any kind.

Morocco's special wealth lies in its reserves of phosphates, of which it is one of the world's largest producers. There are vast deposits in Western Sahara. Coal, iron ore, zinc and other metals are also mined. Some natural gas has been found but little petroleum. The chief manufacturing industries are the processing of phosphates and agricultural produce for export. Otherwise industries are small scale, producing cement, motor vehicles, tires, textiles and clothing – a significant export.

Morocco's road and rail networks are well developed, as are port and airport facilities at Casablanca and the capital, Rabat, both on the northwest coast. The Tangier–Gibraltar ferry may one day be replaced by a bridge. Education is now compulsory from the ages of 7 to 13, and is funded by one-fourth of the national budget. As a result the literacy rate is improving. Health care, however, is largely restricted to urban areas, as is a limited social–welfare program. Housing shortages have led to the growth of shanty towns in some cities.

Algeria

DEMOCRATIC AND POPULAR REPUBLIC OF ALGERIA

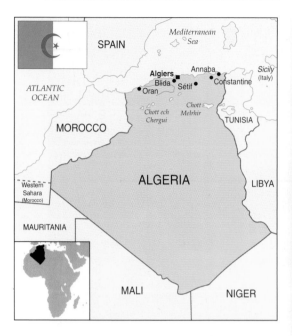

ALGERIA IS THE SECOND-LARGEST COUNTRY in all Africa, extending from the Mediterranean coast deep into the Sahara.

GEOGRAPHY

Algeria is divided into two contrasting areas by the Atlas Mountains. These are part of the great east–west chain that crosses North Africa – a relatively young and craggy range, upthrust by the collision of the African and European tectonic plates.

The coastal land to the north of the Atlas Mountains is called the Tell (meaning "hill"). The western half of this is a relatively dry but fertile area of valleys and plains. East of Algiers, however, toward the cities of Constantine and Annaba, the land becomes more hilly, rising to a series of mountain ranges. Their river systems help make this the most fertile part of the country. The climate in the north is temperate Mediterranean, but the winter rains are only reliable on the coast itself, and can vary widely from year to year.

South of the Atlas Mountains is the much hotter Sahara, which occupies more than four-fifths of Algeria's land area. This is a landscape of flat and ancient rocks supporting pure desert of stones and sand dunes. Apart from rivers at its northern fringes and a few widely separated oases farther south, this vast area has little natural water, and parts of it see no rain at all for years.

In all, only about one-thirtieth of the country is well enough watered to be arable, and almost all of that is exploited. Forests are mostly confined to higher slopes and consist of oak and evergreens.

Many woodlands, especially of the commercially valuable cork oak, have been seriously overexploited, but reforestation programs have had some success.

Open land is mostly covered with scrub (frequently overgrazed), whereas on the drier high plateaus toward the mountains is typical steppe vegetation such as esparto grass and wormwood. Wildlife here includes Wild boar and gazelles, together with the leopard and the cheetah (both endangered species). Desert plant life is highly specialized, flourishing briefly

Two faces of Algeria (*above*) The mosques and minarets of a typical Islamic city look out on the drifting sands of the Sahara, where the nomadic Tuareg wander the desert with their herds.

Prehistoric rock paintings (*right*) of the Tassili N'Agger in southern Algeria, northeast of the Ahaggar mountains. The paintings were discovered in the 19th century, but archaeologists have not yet been able to date them precisely.

after rare rain showers. Small reptiles and insects thrive here, along with a number of desert mammals such as jerboas and the Saharan hare.

SOCIETY

The ancient Roman colonies along the North African coast were overrun by the Muslim Ummayad Arabs in the 7th century AD, and later became part of the Moorish Almoravid empire. In the 16th century the area was taken over by the Ottoman empire. Algiers was a semi-independent Ottoman city-state, and home of the notorious Barbary Pirates, who preyed on the merchant ships that plied the Mediterranean Sea.

In 1830 France, which was in dispute with the Ottomans, launched an invasion and eventually deposed them. Army involvement and political pressure swiftly extended this into a colonizing venture, and European settlement began in the area that was to become known as Algeria. By the 20th century the dominant European population numbered over a million, maintaining strong and influential links with France.

During World War II the occupation of Algeria by British and American soldiers gave new impetus to Muslim nationalist movements. In 1954 a series of conflicts led to the outbreak of the eight-year Algerian War – a cycle of rebellions initiated first by Muslims and later by European settlers. In 1962 a huge majority voted for independence in a plebiscite, and a massive exodus of Europeans heralded the establishment of a new government under Ahmed Ben Bella (b. 1916). In 1965 he was deposed and imprisoned by his minister of defence General Houari Boumedienne (1925–78). Economic difficulties have contributed to social unrest and the rise of militant Muslim groups. In June 1992 the outlawed Islamic Salvation Front (FIS) claimed responsibility for the assassination of President Muhammad Boudiaf. In 1994 Liamine Zeroual (b. 1941) was appointed President and in 1995 tried unsuccessfully to reach political agreement with the FIS.

An amended constitution was approved by referendum in 1996. This defined the nation as made up of Islamic, Arabic and Berber cultures and banned political parties based on race or religion. The new constitution provoked fresh outbreaks of violence in which an estimated 10,000 people were killed. The FIS waged a constant campaign of terrorism lasting until 2000, when the armed wing of the FIS dissolved itself and many insurgents surrendered under an amnesty program. Nevertheless, sporadic unrest continued.

Most Algerians today are Arab, but a significant rural minority are descendants of earlier Berber settlers. Some country-dwellers are still nomads, as are the Tuareg of the Sahara. The European population dwindled when the country gained independence, but still includes French, Spanish and Italians. Islam is the predominant religion.

ECONOMY

Agriculture employs one-fifth of the work force, but produces less than one-third of Algeria's food requirements; the rest must be imported. Important crops include wheat and barley, potatoes and fruit. Drought and locust swarms have posed serious problems in recent years. Algeria is one of the world's largest cork producers, from its cork oaks. Livestock, often raised in seasonal pastures, is less important, but includes sheep, goats and some cattle, as well as draft animals such as donkeys and camels. There are plans to modernize fisheries along the coast.

Mineral resources, especially petroleum and natural gas, have become the economic mainstay. Petroleum and gas production has recently been cut back to extend the life of the reserves, but new reserves are being sought. Other important minerals include iron ore, phosphates, zinc, lead and coal. Uranium deposits have also been found in the Ahaggar mountains in the southwest of Algeria. Steel- and petroleum-related industries dominate the industrial sector.

There are plans to update and extend transportation, especially the rail system. The present railroad network extends from the Tunisian to the Moroccan border, with lines serving the main ports. The mainly French-built road system is good in the north, but there are few routes south into the mountains and desert. Air Algérie provides both international and domestic air services.

The government has established a welfare state including a national health

A market festival in Ghardaia, chief town of the Mzab oasis in the northern Sahara. This town was founded in the 11th century near the cave (*ghar*) of the female saint Daïa. Arcaded clay houses surround the busy central square.

service that is wholly or partly free depending on a means test. It is based on state hospitals, pharmacies and other medical centers, where all medical staff are required to serve for a period. State education faces problems with scattered rural communities, but television and radio are being used to combat these. The majority of children receive some basic education, but the literacy rate is still fairly low.

NATIONAL DATA – ALGERIA				
Land area	2,381,740 sq km (919,595 sq mi)			
Climate		Temperatures	Annual	
	Altitude m (ft)	January °C(°F)	July °C(°F)	precipitation mm (in)
Algiers	50 (164)	11 (51)	24 (75)	764 (30)
Major physical features	highest point: Mount Tahat 2,918 m (9,573 feet)			
Population	(2006 est.) 32,930,091			
Form of government	multiparty republic with one legislative house			
Armed forces	army 120,000; navy 7,500; air force 10,000			
Largest cities	Algiers (capital – 1,519,570); Oran (692,516); Qacentina (462,187)			
Official language	Arabic			
Ethnic composition	Arab-Berber 99%; European less than 1%			
Religious affiliations	Sunni Muslim 99%; Christian and Jewish 1%			
Currency	1 Algerian dinar (DA) = 100 centimes			
Gross domestic product	(2006 est.) U.S. $253.4 billion			
Gross domestic product per capita	(2006 est.) U.S. $7,700			
Life expectancy at birth	male 71.68 yr; female 74.92 yr			
Major resources	barley, cattle, citrus fruits, copper, dates, fish, grapes, iron ore, lead, oats, oil and natural gas, olives, phosphates, sheep, timber, uranium, vegetables, wheat, zinc.			

Tunisia

REPUBLIC OF TUNISIA

THE REPUBLIC OF TUNISIA, THE SMALLEST OF the Mediterranean countries of North Africa, is sandwiched between two much larger neighbors: Algeria and Libya. Its long history stretches back 3,000 years.

GEOGRAPHY

Tunisia is bisected by the eastern end of the Atlas Mountains. These form a mountainous plateau called the Dorsale (meaning "backbone" in French), that extends northeast from the Algerian border all the way down to the plains bordering the Gulf of Tunis.

To the south are high steppes and hills, beyond which the land is increasingly flat and dry. The few rivers are seasonal and

NATIONAL DATA - TUNISIA

Land area	155,360 sq km (59,984 sq mi)			
Climate		Temperatures		Annual
	Altitude m (ft)	January °C(°F)	July °C(°F)	precipitation mm (in)
Tunis	3 (10)	12 (54)	26 (79)	660 (26)
Major physical features highest point: Chambi 1,544 m (5,066 ft); lowest point: Chott El Jerid -23 m (-75 ft)				
Population (2006 est.) 10,175,014				
Form of government multiparty republic with one legislative house				
Armed forces army 27,000; navy 4,800; air force 3,500				
Largest cities Tunis (capital - 728,453); Sfax (265,131)				
Official language Arabic				
Ethnic composition Arab 98%, European 1%; Jewish and other 1%				
Religious affiliations Muslim 98%; Christian 1%; Jewish and other 1%				
Currency 1 Tunisian dinar (D) = 1,000 millimes				
Gross domestic product (2006 est.) U.S. $87.88 billion				
Gross domestic product per capita (2006 est.) U.S. $8,600				
Life expectancy at birth male 73.4 yr; female 76.96 yr				
Major resources petroleum, natural gas, phosphates, iron ore, lead, zinc, tourism, fruits, olive oil, barley, cork, dates, fish, grapes, livestock, olives, salt, timber, tomatoes, wheat				

swift-flooding, and the country is dotted with low-lying salt flats known as chott. These are dry and crystalline in summer, but turn to treacherous bogs in winter. The largest, the Chott El Jerid, lies across half the width of this narrow country. Beyond the narrow coastal plains, the southeast fringes of Tunisia swiftly merge into the Sahara.

In the north, by contrast, the main river, the Medjerda, flows down from the Dorsale through a fertile valley formed from ancient lake basins. It enters the Gulf of Tunis via a broad coastal plain covered by rich alluvial silts. To the north of the valley, the Kroumirie mountains run parallel to the coast.

Tunisia enjoys a hot temperate climate, typically Mediterranean in its hot summers and mild winters, during which most of the rain falls. Precipitation is generally low, but the occasional fierce flooding can cause great damage. Equally, rainless summers and the hot sirocco wind from the Sahara can dry out the country, and even the most fertile areas are often hit by droughts.

The steppes south of the Atlas Mountains are carpeted mostly with hardy esparto grass, which gives way gradually to sparse saltbush and acacia shrubland, and finally to desert. Here water drawn from artesian wells supports a few oases, mostly of date palms. In the north, large forested areas, chiefly of evergreens such as the Aleppo pine, grow on the slopes of the Dorsale, giving way to broadleaf woodlands in the northern valleys. The Kroumirie range is covered with a dense expanse of cork oaks.

The forests are still inhabited by lynx and Wild boar, while farther south are found hyenas, jackals and wild goats. Many species of birds flourish, including eagles in the mountains. Small numbers of antelopes and the African ass still survive around the margins of the desert, and hunting here has been prohibited to protect them.

SOCIETY

The capital city of Tunis lies on the site of Carthage, the ancient trading city of the Phoenicians that dominated the western Mediterranean until its destruction by the Romans in the 2nd century BC. From the 7th century AD the surrounding area formed part of the Muslim Arab empire, and in the late 16th century was incorporated into the Ottoman empire as a semi-independent principality. In the mid 19th century economic stagnation put the

country vastly in debt to European bankers, and France used this as an excuse to seize control of the country in 1881. The bey, the local ruler, remained in puppet control, but French colonists took the opportunity to settle on much of the best land.

In 1920 the nationalist Destour Party was formed in opposition to French rule, and in 1934 Habib ibn Ali Bourguiba (b. 1903) broke away to head a less moderate party, the Neo-Destour. During World War II the Neo-Destour leaders set up a pro-Fascist government in Tunisia, but it was toppled in 1943 by the Free French movement led by General Charles de Gaulle (1890–1970) from exile in England.

During the independence struggles of the 1950s, Bourguiba's influence made him the leading figure in negotiations with France, and when Tunisia became an independent republic in 1957, he was its first

Libya and Egypt, and Israel in 1994.

The government was restructured as a multiparty democracy, although the Neo-Destour Party, retitled the Constitutional Democratic Gathering, remains the dominant party. The Islamic fundamentalist party Al Nahda is outlawed. The president can serve no more than three five-year terms and must retire at the age of 70. In 1999 Ben Ali was re-elected to a final term, due to end in 2004.

The Tunisian people are basically Arab, mixed with Berber and a wide range of other Mediterranean and African peoples. Virtually the whole population of the country is Muslim, the once large Jewish community having mostly emigrated.

ECONOMY

Agriculture, though still the largest single employer, has declined in recent years, and now provides around 14 percent of gross national product. The north has been a grain-growing area since Roman times, and now produces wheat and barley as well as olives, sugar beet, citrus fruit, grapes and vegetables. Livestock is limited by poor pastureland, and consists chiefly of goats and sheep, with camels as work animals. Forestry yields important quantities of cork bark, as well as eucalyptus and other oils.

Minerals are now the prime source of national wealth, especially phosphates and petroleum. Both are important exports, together with natural gas, and recent price falls have depressed the economy. Most industrial production is based on these products and on agricultural produce such as olive oil. A large proportion of manufacturing plants are local and small-scale. Textiles are the fastest-growing light industry, though they are also a major import, together with foodstuffs and raw materials. Some heavy manufacturing industry has been set up with foreign assistance. Tunisia signed an association agreement with the European Union (EU) in 1998 which will remove trade barriers. Tourism is also increasingly important, and owes much to the country's climate, the many sandy beaches and the colorful culture.

The country's road and rail networks are extensive, although in need of modernization. International air traffic passes mostly through the main airport near Tunis. Healthcare and social security programs are well established and relatively advanced. Education is free, but suffers from restricted enrollment, which means that the literacy rate is still relatively low.

These eroded terraces (*above*) have been formed by a river cutting its way through layers of rock. The palm trees' long roots enable them to obtain moisture from deep below the ground, even when the river is dried up.

The Grand Mosque (*left*) and the old Medina quarter of Tunis, seen from the roof of the Dar el Bey (the House of the Bey). The mosque was originally founded in 732 AD, but was probably rebuilt in the 9th century.

president. In 1975 he was appointed president for life, ruling a one-party state. Throughout his years of government he checked the rise of Islamic fundamentalism and established rights for women unmatched by any other Arab nation. But Bourguiba's regime became increasingly despotic, and he was removed from power in 1987, at age 84. His successor, General Zine al-Abidine Ben Ali (b. 1936), headed a more moderate regime and introduced a raft of reforms intended to revitalize the stagnant economy. He also renewed diplomatic ties with neighboring states such as

Libya

SOCIALIST PEOPLE'S LIBYAN ARABIC JAMAHIRIYA

L IBYA IS A RICH COUNTRY WITH A REMARK-
ably high standard of living. Both its
wealth and its energy are derived from
vigorous exploitation of its oil reserves.
Its leader, Colonel Muammar Qaddafi (b.
1942), led the overthrow of the monarchy
in 1969, and until the mid 1980s expressed
support for international terrorism.

GEOGRAPHY

More than nine-tenths of Libya's land
area is desert or semidesert, and more
than three-fourths of its population live
along the narrow coastal strip where the
Sahara finally gives way to cultivation.

Tripolitania – a large area in the northwest of the country – extends eastward
from the border with Tunisia. Here the
limestone mountains of the Jabal Nafuṣah
lie between the Saharan plateau and the
sand dunes, salt marshes and steppe of
the Gefara Plain. The Gefara Plain is
Libya's most productive farming area and
contains its largest city, Tripoli. In the
northeast is the former province of
Cyrenaica, where the Al Jabal al Akhdar
(Green Mountains) rise from a narrower
coastal plain, clothed in vegetation that
gradually becomes semidesert grassland.
Eastward lies the Egyptian border.

Southward, in the area known as the
Fezzan, the Sahara rises to the edge of the
North African plateau, bordering Algeria
to the west. Libya's highest mountain,
Bette, is part of the Tibesti Mountains
that span the disputed southern border
with Chad. Libya also shares borders
with Niger (to the southwest) and the
Sudan (to the southeast).

Libya's intensely hot desert climate
(with very cold nights) is moderated only
in the north along the Mediterranean
coast. Annual rainfall is sporadic, especially inland. Drought can occur even

on the coast, where hot, unbearably dry
ghibli winds can bring sand down from
the interior.

Libya has no perennial rivers, so there
is little natural vegetation apart from date
palms near oases, maquis scrubland (predominantly juniper) on the Al Jabal al
Akhdar and overgrazed steppe grassland.
Animals are mostly desert-dwellers such
as jackals, jerboas, the Saharan hare, the
Fennec fox and the skunklike zorilla.

SOCIETY

In the 7th century BC Tripolitania was an
eastern province of the Carthaginian empire, while Cyrenaica was a Greek colony.
In the 1st century BC the Romans conquered both areas, and when the Arabs
invaded in the 7th century AD, the whole
of this part of Northern Africa was part of
the eastern Byzantine empire. It became a
haven for pirates who raided ships in the
Mediterranean. In the 16th century the

Turkish Sultan annexed the whole area
into the Ottoman empire.

Under the Ottomans, Tripolitania,
Cyrenaica and the Fezzan were united
under a single regency in Tripoli. The
Sanusiyya, an Islamic sect, rallied resistance to the Turks in the 19th century and
to the Italians, who invaded the country
in 1911 but failed to win full control of the
interior until 1939.

During World War II the Sanusiyya
helped to secure the expulsion of the
Italians and in 1942 Tripolitania and
Cyrenaica came under British military
control, while the French controlled the
Fezzan. An independent kingdom was
created in 1951, and a national assembly
elected the head of the Sanusiyah as King
Idris I (1890–1983). Idris was generally
pro-British, but in 1953 Libya joined the
Arab League, and in 1956 decided not to
support British attempts to reassert international control over the Suez Canal in
Egypt, which links the Mediterranean to

the Indian Ocean. (This had been brought under national control by the Egyptian president, Gamal Abdel Nasser (1918–70) in 1955.) In 1959 the discovery of oil transformed the Libyan economy. Ten years later Colonel Muammar Qaddafi ousted King Idris, broke Libyan ties with Britain and the United States, and took control of the oil industry.

Today there is one political party, the Arab Socialist Union. Locally elected or appointed bodies supply the 3,000 members of the General People's Congress, or parliament. This body elects a cabinet, known as the General People's Committee, and the "revolutionary leader". Qaddafi has held the latter position since 1969, and retains effective control with the help of a few close advisers.

Qaddafi has openly supported terrorist organizations, most notably extremist Palestinian groups, in the interests of Arab nationalism. Suspected Libyan involvement in several major terrorist

incidents led to an American air raid on Tripoli in 1986. Libya's terrorist activities decreased after 1992 when UN sanctions were imposed. These were suspended in 1999. In return, Libya handed over two suspects for trial in the Netherlands on charges of bombing a Pan Am flight in 1988. The US, however, maintained its embargo.

The vast majority of Libyans are Arabic-speaking Sunni Muslims, mostly of mixed Arab and Berber descent. There is also a minority of Sub-Saharan African origin, the descendants of slaves brought here by the Ottomans in centuries past. Small groups of nomadic Tuareg still exist.

ECONOMY

Petroleum and natural gas account for around one-third of Libya's GDP. The assets of most foreign oil companies were nationalized in 1973, and the state retains a controlling interest in new ventures. Cement production and iron ore extraction are also both state-controlled. Other mineral resources include natron, potash, gypsum and lignite. All Libya's energy comes from domestic oil, though Niger has supplied uranium for a proposed nuclear station.

Libya is not self-sufficient in foodstuffs. Agriculture employs around one-fifth of the workforce and 5 percent of GDP. Arable land is limited and is intensely farmed. Irrigation is a problem. Cereals, especially barley, form the major crop; sorghum is grown in the Fezzan, and wheat, tobacco and olives in the north.

NATIONAL DATA – LIBYAN ARAB JAMAHIRIYA				
Land area 1,759,540 sq km (679,362 sq mi)				
Climate		Temperatures		Annual
	Altitude m (ft)	January °C(°F)	July °C(°F)	precipitation mm (in)
Tripoli	50 (164)	14 (57)	26 (80)	272 (10.7)
Major physical features highest point: Bette 2,286 m (7,500 ft)				
Population (2006 est.) 5,900,754				
Form of government Jamahiriya a "state of the masses"-in practice a one-party state and military dictatorship				
Armed forces army 54,000; navy 8,000; air force 23,000				
Largest cities Tripoli (capital - 1,178,007); Benghazi (950,000); Misurata (300,000); al-Aziziyah (287,407)				
Official language Arabic				
Ethnic composition Berber and Arab 97%; Greeks, Maltese, Italians, Egyptians, Pakistanis, Turks, Indians, Tunisians 3.0%				
Official religion Islam				
Religious affiliations Sunni Muslim 97.0%; others 3.0%				
Currency 1 Libyan dinar (LD) = 1,000 dirhams				
Gross domestic product (2006 est.) U.S. $74.97 billion				
Gross domestic product per capita (2006 est.) U.S. $12,700				
Life expectancy at birth male 74.46 yr; female 79.02 yr				
Major resources petroleum, natural gas, cement, iron ore, gypsum, barley, citrus fruits, dates, figs, groundnuts, livestock, olives, sorghum, tobacco, wheat				

Oasis crops of dates and figs are cultivated, and grapes are also grown on mountain slopes. The government has planted millions of trees, and provides practical and financial help for farmers. Sheep, goats and cattle are reared in the north. Foreign fishing fleets take most of the plentiful fish and sponge stocks.

Manufacturing plays a minor role in the economy. Textiles, footwear and carpets are produced, while petrochemical and construction services are expanding.

There are no railroads, but some two-thirds of the roads are surfaced. Major highways include the coast road to Egypt and the Sabha Road across the desert to Ghat near the Algerian border. The main seaports and airports are at Tripoli in the northwest and Benghazi in the northeast although all international flights were suspended in the 1990s under the UN embargo. Radio, television and the press are under state control.

Libya's extensive social-services program includes a comprehensive pension system, compensation for work injury, and sickness and maternity benefits. Medical care is free through a growing network of hospitals and health centers. Malaria is under control, but diseases such as typhoid, rabies and tetanus are still a serious problem. The three-level educational system is free and compulsory up to the age of 15, and there are two universities. Literacy has improved dramatically in recent years.

The traditional dress (*above*) of these Libyan men keeps them cool in the intense heat of the day and warm during the cold desert nights. The veils are essential to keep out windblown sand.

The theater at Leptis Magna (*left*), built by a wealthy Roman citizen between 1 and 2 AD. The city of Leptis Magna was founded by Phoenician colonists in about 800 BC, and in 46 AD became part of the Roman province of New Africa.

Egypt

THE ARAB REPUBLIC OF EGYPT

EGYPT, IT HAS BEEN SAID, IS THE GIFT OF THE river Nile. For thousands of years the Nile's capricious floods have brought fertile silt to a narrow valley that cuts through the surrounding desert like a blue-green sword. This valley nourished one of the oldest continuous civilizations in the world. The construction of the Suez Canal in 1859–69, linking the Mediterranean and the Indian Ocean, gave Egypt new strategic importance. In the 20th century the Aswan High Dam has transformed Egyptian agriculture.

GEOGRAPHY

Seen from space, Egypt appears divided into a series of north–south strips.

The land

Eastward from the Libyan border the arid limestone of the Western Desert occupies two-thirds of the country's land area. To the southwest outcrops of harder rock rise to some 2,000 m (7,000 ft).

East of the desert lies the Nile valley, a continuous fertile strip measuring only 8–16 km (5–10 mi) wide. This narrow piece of land is occupied by 99 per cent of the population. Lake Nasser, created by the construction of the Aswan High Dam, (completed in 1970 at a cost of $1 million) flooded the first 300 km (200 mi) from the southern border with the Sudan. To the north the Nile delta extends from Cairo (the largest city in Africa) to the coastal towns of Alexandria in the west and Port Said in the east. Running parallel to the

The Nile river, the lifeblood of Egypt, flows past Luxor, famous for its numerous temples and burial grounds. The river is navigable for most of its length, and is used for the transportation of diverse goods, often by narrow, fast sailing boats called feluccas, as well as for extensive year-round irrigation.

Nile and the Red Sea is the Eastern Desert, a series of dissected limestone and sandstone plateaus. Along the coast in the southeast are the Red Sea mountains, an extension of the Ethiopian Highlands.

The northernmost extension of the Red Sea is the Gulf of Suez, linked to Port Said and the Mediterranean by the Suez Canal. East of the canal lies the triangular limestone plateau of the Sinai Peninsula, which extends beyond the Israeli border. To the north it is heavily dissected, and there is a narrow coastal plain running along the Mediterranean seaboard. To the south it rises to Mt. Sinai and Mt. Catherine, Egypt's highest peak.

Climate

Egypt lies mostly within the hot desert zone, and the only moderating influence on its climate is the Mediterranean to the north. Summers (May–October) are hot and dry, but the temperature drops sharply at night. The winter months are cooler, particularly in desert areas. In spring the khamsin wind blowing relentlessly from the Sahara may bring heat, dust and sandstorms.

Plants and animals

Water plants abound along the Nile, and so do birds. Many of the hundreds of migrant and resident bird species feed on the river's varied fish population. Mongooses, egrets and ibises live in the Nile delta. For the most part the deserts are bare expanses of sand, but in the Eastern Desert especially, grasses, tamarisks and mimosas find a foothold. Date or doum palms often grow near surface or underground water. Desert animals include gazelles, hyenas and jackals, with numerous rodents, scorpions and insects.

SOCIETY

Egypt can boast one of the oldest continuous civilizations in the world. Today it is an Arab republic with a distinctive society and culture of its own.

The temple of Abu Simbel (*above*). One of four colossal statues of Ramses II (13th century BC). The entire temple was moved to escape flooding by the Aswan High Dam

The 15th-century fort of Qait Bay (left), now a naval museum, dominates the eastern harbor of Alexandria. The city is built on a narrow strip of land between the Mediterranean and Lake Maryut.

NATIONAL DATA - EGYPT				
Land area 995,450 sq km (384,343 sq mi)				
Climate		Temperatures		Annual
	Altitude m (ft)	January °C(°F)	July °C(°F)	precipitation mm (in)
Cairo	75 (246)	14 (54)	28 (83)	25 (1)
Major physical features highest point: Mount Catherine 2,637 m (8,652 ft); longest river: Nile (part) 6,690 km (4,160 mi)				
Population (2006 est.) 78,887,007				
Form of government single-party republic with one legislative house				
Armed forces army 340,000; navy 18,500; air force 30,000				
Largest cities Cairo (capital - 15,707,992); Alexandria (4,320,129); Giza (1,560,024)				
Official language Arabic				
Ethnic composition Egyptian 98%; Berber, Nubian, Bedouin, and Beja 1%; Greek, Armenian, other European (mainly Italian and French) 1%				
Official religion Islam				
Religious affiliations Muslim (mostly Sunni) 90%; Coptic 9%; other Christian 1%				
Currency 1 Egyptian pound (EGP or LE) = 100 piastres = 1,000 millièmes				
Gross domestic product (2006 est.) U.S. $328.1 billion				
Gross domestic product per capita (2006 est.) U.S. $4,200				
Life expectancy at birth male 68.77 yr; female 73.93 yr				
Major resources petroleum, natural gas, iron ore, phosphates, manganese, gypsum, limestone, talc, asbestos, lead, zinc, tourism, barley, beans, buffaloes, cattle, clover, cotton, dates. goats, gypsum, lentils, maize/corn, millet, rice, sea salt, sugarcane, sheep				

History

Between about 3000 and 525 BC, the united kingdoms of Upper and Lower Egypt were controlled by an unbroken succession of rulers. This period is divided by historians into the Old Kingdom (when the Giza pyramids were built), the Middle Kingdom, the New Kingdom (when the Israelites left Egypt) and the Late Period (when Egypt belonged at times to the Persian empire).

In 332 BC the Macedonian ruler Alexander the Great (356–323 BC) set up a new Greek dynasty. From 30 BC Egypt was a Roman province, which after 395 AD was administered from Constantinople (the capital of the East Roman, later Byzantine, empire). In 642 AD Arab conquerors imposed the Arabic language and the Islamic religion on the country. In 1250 the Mamelukes, formerly royal bodyguards, founded a new royal house, but in 1517 the Ottoman Turks seized control, and Egypt gradually fell into economic and cultural decline.

Following a brief French incursion (1798–1806) power passed to an Ottoman viceroy from Albania called Muhammad Ali (1769–1849). A policy of expansion under Muhammad and his successors left Egypt in debt to the British, who took control in 1882. It became a British protectorate in 1914, achieving independence under a constitutional monarchy in 1922.

In 1952 General Muhammad Naguib (1901–84) seized power, to be overthown by Colonel Gamal Abdel Nasser (1918–70) in 1954. With Soviet funding he began building the Aswan High Dam, and in 1955 he nationalized the Suez Canal, followed by an abortive invasion by Britain, France and Israel who strongly opposed this action. However, the Six-day War with Israel in 1967 cost Egypt the Sinai Peninsula.

Nasser's successor, Anwar Sadat (1918–81), pursued friendlier relations with the West while ending Egyptian dependence on the former Soviet Union. In 1979 a peace treaty was signed between Egypt and Israel which agreed to the phased withdrawal of Israel from Sinai. However, this rapprochement with Israel led to Egypt's isolation in the Arab world, and in 1981 Sadat was assassinated.

Sadat's successor, Hosni Mubarak (b. 1928), continued to support the Israeli–Palestinian peace process, at the same time working to improve Egypt's relations with other Arab countries. Egypt backed the UN coalition against Iraq during the Gulf War in 1991. Islamic fundamentalists opposed to the regime used increasingly militant tactics after 1992, often targeting their attacks against foreign tourists. Mubarak was elected to a fourth term as president in 1999.

Government

Since 1971 Egypt has been a single-party republic. The president is the head of state, nominated by a two-thirds majority of the single-chamber People's assembly.

People

The inhabitants of the Nile Valley are ethnically homogeneous. Small minority

Egypt's presidential system of government The 1971 constitution places great power in the hands of the president, who is nominated by the people's assembly and confirmed by plebiscite for a six-year term. He is supreme commander of the armed forces, and may appoint one or more vice presidents; he also appoints both prime minister and cabinet. The People's assembly has 454 members elected for a five-year term.

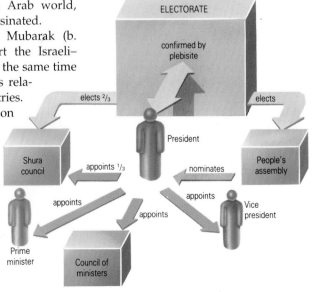

groups include Nubians in the south, the largely nomadic Hamitic Beja in the Eastern Desert, and peoples of mixed Berber and Arab origin in the Western Desert and along the Red Sea coast.

Arabic remains the main language, and Islam the dominant religion, practiced by four-fifths of the population. The Christian minority are mostly Copts.

ECONOMY

Petroleum and its products make up one half of exports and fluctuations in the price of oil had a detrimental effect in the late 1980s and early 1990s. However, Egypt received substantial debt relief as a result of its support of the Gulf War coalition and this, combined with government efforts to restructure the economy and open new markets, led to an improved situation at the end of the 1990s.

Agriculture

Arable farming was traditionally confined to the Nile Valley and its delta, where about 50 percent of the population are fellahin peasant (agricultural laborers) living in mud-brick or stone-built villages. These areas comprise only one-fiftieth of Egypt's total land area. Completion of the Aswan High Dam has extended the area of irrigated agriculture, but has resulted in the loss of fertile silts laid down by the annual floods further down river.

Agriculture is mostly devoted to commercial crops including sugar cane and cotton (Egypt produces about one-third of the world cotton crop). Food crops such as rice, maize, wheat, beans, millet and dates account for only about one-fourth of arable farming, and much of the country's food is imported. Water buffalo and dairy cattle are kept for milk and as draft animals, but there is no pastureland. Fishing thrives (Lake Nasser has been stocked with fish), but the Mediterranean catch has been reduced since the construction of the Aswan High Dam, presumably as a result of changed water-flow patterns.

Industry

Industry contributes a third of GDP and employs nearly one-quarter of the workforce. Petroleum is extracted from the Gulf of Suez and the Sinai Peninsula. There are also two fields in the Western Desert, and natural gas is being extracted

The second Great Pyramid at Giza, one of the largest single structures ever built, dates from the 26th century BC and marks the tomb of the Pharaoh Khafre (reigned c. 2520–2494 BC). In front of it stands the Sphinx. Its face is thought to represent Khafre.

The Suez Canal – Egypt's other waterway

It was the French diplomat and engineer Ferdinand Marie Vicomte de Lesseps (1805–94) who first planned a link between the Mediterranean and the Gulf of Suez. He envisaged a sea-level canal from Port Said to Lake Timsah, the Bitter Lakes and Suez, and between 1859 and 1869 he supervised its construction himself. De Lesseps kept the project under French control, forming an international company that took a 99-year lease from the Ottoman Viceroy of Egypt. The finished canal was about 160 km (100 mi) long and less than 6 m (20 ft) deep; it has since been deepened and widened several times to accommodate ever larger international vessels.

The British government was anxious for a stake in the canal, not least because it cut some 10,000 km (6,000 mi) from the trade route to India and Southeast Asia. In 1875 the British secured the 44 percent share of the company held by the viceroy, and in 1888, an international convention guaranteed the canal's neutrality.

The European interest in the canal was due to lapse in 1968, but in 1956 President Nasser preempted this and seized it for Egypt, closing the Gulf of Aqaba to Israeli shipping. On 29 October, amid international outcry, Israel invaded Egypt, while British and French troops occupied the banks of

The Suez Canal carries vast container ships, but traffic has never fully recovered from the Israeli occupation of the Sinai, and some modern supertankers are now too large to use the canal.

the canal. They were soon replaced by a United Nations force, and Israel withdrew after securing access to the Gulf of Aqaba. The canal, however, remained in Egyptian hands – a vital source of revenue for Nasser's ambitious policies.

In 1967 Nasser again closed the Gulf of Aqaba. A lightning Israeli response destroyed most of the Egyptian air forces on the ground. The resulting Six-day War left the Sinai Peninsula and many other border areas in Israeli hands. The canal was closed – indeed, it marked the front line between the two antagonists. In 1973 a combined Egyptian and Syrian attack was driven back by Israeli forces, who this time even crossed the canal. Even so, the 1974 disengagement agreement in theory returned Sinai to the Egyptians

(though the process was not completed until 1981), and in 1975 the canal was cleared of sunken ships and reopened.

The damage, however, was already done. After eight years many newer ships (especially oil tankers) were too large for the canal, and other routes had become well established. The subsequent war between Iran and Iraq (1980–88) further discouraged traffic. However, the Egyptian government is considering plans to deepen the canal yet again, allowing passage for supertankers of up to 180,000 tonnes, in the hope of restoring its former prosperity.

here as well. Other resources include phosphates, coal, manganese and various other metal ores.

The Aswan High Dam supplies more than half the country's electricity. It could produce much more, but irrigation requirements limit its output. Fossil-fueled power stations supply the rest although solar power use is expanding.

Manufacturing is increasingly important, both for exports and for the domestic market. Nasser's agreements with the former Soviet Union helped Egypt to develop its heavy industrial capacity, and to expand iron, steel and aluminum production and electrification in many of the rural areas.

Egypt's tourist industry contributes vitally to the economy, despite the setbacks suffered in the 1990s as the result of

Islamic terrorist attacks, including the massacre of 60 tourists at Luxor in 1997.

Transportation and communications

Less than half Egypt's roads are surfaced, and the state-owned railroad is comparatively small. Most of the country's extensive waterway network is centered around the Nile and the Suez Canal.

Egypt has 14 daily newspapers; all newspapers and magazines are supervised by the government's Supreme Press Council. The Egyptian Radio and Television Corporation is state-owned

Health and welfare

Government spending on health is steadily increasing, especially in rural areas, where schistosomiasis (a parasitic disease) is common. The battle against

tuberculosis and endemic malaria is not yet won, but urban medical facilities are good, and preventive medicine is actively promoted. Most health services have been nationalized, and health insurance is compulsory.

At present only about half the population over the age of 15 can read or write, but this situation is improving. School education is compulsory from the ages of 6 to 12, with a further nine years available to more successful students. Schools centered on Cairo's Al-Azhar University follow a similar structure to the state system, but place more emphasis on traditional Islamic subjects. There are four state universities as well as the private American University in Cairo and many more specialized institutes of higher learning.

Mauritania

ISLAMIC REPUBLIC OF MAURITANIA

M AURITANIA EXTENDS INLAND FROM THE Atlantic coast of West Africa and northward into the Sahara. It borders Western Sahara and Algeria to the north, with Mali and Senegal to the south.

GEOGRAPHY

Behind the low-lying coastal plains are slightly higher tableland plateaus broken only by occasional isolated peaks and westward scarps. Farther inland, occupying most of the country, are the sand dunes of the Sahara, and in the extreme south the semiarid Sahel scrubland. The climate is hot and dry apart from late summer rains in the south.

Most of the fertile land lies along the coast and in the catchment area of the Senegal river, which forms the southwestern border with Senegal. This is grassy savanna country, with baobab and palmyra trees. In the Sahel vegetation is sparse, and in the desert it is confined to oases and semidry riverbeds, or wadis. Despite conservation measures, hunting and habitat destruction have greatly reduced the numbers of wildlife; this includes antelopes, Senegalese lions, warthogs, African elephants, ostriches, gazelles and crocodiles.

SOCIETY

In medieval times Mauritania was part of the Almoravid empire. It was then overrun by Arabs, or Moors, from the north, who subjugated the various black Sudanic peoples that still live in the south. The Moors (*Maures* in French) gave their name to the country, and the population today is three-fourths Moorish. The Moors are divided into two very distinct groups: the *bidan* or "white Moors", of Arab descent; and the mixed-race *harratin*, or "black Moors", descended from slaves (slavery was not outlawed until 1970, and persists today). Most Moors are nomadic pastoralists, and much of the population is Muslim.

In the 19th century the area was gradually taken over by France, becoming part of French West Africa. In 1960 it won full independence under President Moktar Ould Daddah (b.1924). In 1978 he failed to suppress Polisario guerrillas in Mauritania's Western Sahara territory (now mainly occupied by Morocco), and was deposed by a military coup. Sid'Ahmed Ould Taya, a former prime minister, took over as president. A new constitution to give greater political freedom and allow for multiparty elections was approved by referendum in 1991. However, the opposition parties boycotted elections in 1992 and 1997 on grounds of vote rigging, and on each occasion Taya was reelected president with a substantial majority.

ECONOMY

Agriculture accounts for 25 percent of GDP, and consists mostly of livestock farming, including sheep and goats, with cattle in the south. Arable crops include millet, sorghum, yams, cotton, maize and dates. Some areas are irrigated, but the Sahel droughts have cut production by as much as two-thirds. Fishing accounts for nearly 60 percent of total exports, the waters off Mauritania contain some of the world's richest fishing grounds although these are threatened by overexploitation.

The major industries are fish processing and mining. Iron ore makes up 40 percent of exports and there are also reserves of copper, gold, phosphate and gypsum. Roads are poor, making transportation difficult. There are international airports at Nouâdhibou and Némy. The only railroad runs from Zouérate to the port of Point-Central and is mostly used for the transport of iron ore.

A Mauritanian fisherman carries his nets ashore for repair. Fishing is the backbone of the Mauritanian economy. Foreign investment has developed the industry to the point where it now accounts for almost half the country's exports.

NATIONAL DATA – MAURITANIA				
Land area 1,030,400 sq km (397, 840 sq mi)				
Climate		Temperatures		Annual
	Altitude m (ft)	January °C(°F)	July °C(°F)	precipitation mm (in)
Nouakchott	1 (3)	22 (72)	28 (83)	159 (6.2)
Major physical features highest point: Kediat Idjil 915 m (3,002 ft)				
Population (2006 est.) 3,177,388				
Form of government multiparty republic with two legislative houses				
Armed forces army 15,000; navy 620; air force 250				
Capital city Nouakchott (881,000)				
Official language Arabic				
Ethnic composition Mixed Maur/black 40%; Moor 30%; Black 30%				
Official religion Islam				
Religious affiliations 100% Sunni Muslim				
Currency 1 ouguiya (MRO) = 5 Khoums				
Gross domestic product (2006 est.) U.S. $8.397 billion				
Gross domestic product per capita (2006 est.) U.S. $2,600				
Life expectancy at birth male 50.88 yr; female 55.42 yr				
Major resources iron ore, gypsum, copper, phosphate, diamonds, gold, oil, fish, dates, livestock, millet, rice, salt, sorghum, wheat				

Mali

REPUBLIC OF MALI

A colorful market in the shadow of a mosque at Djenne in central Mali. Mali was a great trading nation in the 13th and 14th centuries, renowned for its gold, but it now has to contend with famine, disease, droughts and the encroaching Sahara.

T HE LANDLOCKED REPUBLIC OF MALI LIES IN the Sahel area of northwest Africa on the fringes of the Sahara.

GEOGRAPHY

Mali's landscape is monotonously flat, consisting mostly of plains and sandstone plateaus or hills. The great northern plains are part of the Sahara, with a hot, almost rainless climate characterized by cold nights. In recent years rainfall has declined in the semiarid Sahel area in the heart of Mali, bringing devastating droughts. Only the south, crossed by the great Niger river and its tributaries, has enough rainfall to support much plant or animal life.

All the arable land is in the south, covering less than one-fiftieth of the total area. Otherwise the south is mostly savanna, with mahogany, kapok and baobab trees, replaced farther north by palm species and eventually by scrub. Animal life includes characteristic African species such as lions, antelopes, hyenas, jackals and many smaller creatures. A national park has been established to protect them.

SOCIETY

Although inhabited since the Stone Age, Mali derives its name from the Muslim empire of the Malinke or Mandingo people that flourished here in the 13th and 14th centuries. The area was a rich center of trade and Islamic scholarship until conquered by the Moors of Morocco. It was colonized by the French in the 19th century as part of French West Africa, and in 1946 became the overseas territory of French Sudan. In 1958 it became autonomous as the Sudanese Republic;

with Senegal it formed the Federation of Mali in 1959 and became fully independent in 1960.

Military coups in 1968 and 1991 were followed by the installation of a multiparty constitution in 1992. Two elections have been held but have been boycotted by opposition parties. Former president Moussa Traore and his wife were sentenced to death in the late 1990s for corruption of state funds.

Mali is home to a variety of ethnic groups, ranging from the Tuareg and Moors in the north to the Bambara, Fulani and Malinke in the south. Each people has its own language, but French and Arabic are widely spoken. Most of the population are Muslim.

ECONOMY

Most people are employed in agriculture, on which the economy largely rests. The country was self-sufficient in food production until the Sahel droughts (1968 onward), which affected cattle and other livestock particularly badly. Staple crops include grains, cotton and groundnuts (peanuts).

Industry is small-scale, and energy is generated by thermal and hydroelectric projects built with foreign aid. Mineral resources include gold, uranium and phosphates, iron and bauxite; however, inadequate transportation services, including poor roads, contributes to their underexploitation. Foreign aid and borrowing have supported the economy, and debt remains a serious handicap despite substantial debt relief. Mali's economy was showing signs of growth following an IMF structural readjustment program.

NATIONAL DATA – MALI

Land area	1,220,000 sq km (471,045 sq mi)			
Climate		Temperatures	Annual	
	Altitude m (ft)	January °C(°F)	July °C(°F)	precipitation mm (in)
Bamako	331 (1,086)	25 (77)	27 (81)	991 (39)

Major physical features highest point: Hombori Tondo 1,155 m (3,789 ft); longest river: Niger (part) 4,200 km (2,600 mi)

Population (2006 est.) 11,716,829

Form of government multiparty republic with one legislative house

Armed forces army 7,350; paramilitary 4,800; inactive militia 3,000

Largest cities Bamako (capital - 1, 690,471); Sikasso (154,823); Mopti (112,248); Koutiala (106,857); Kayes (103,797)

Official language French

Ethnic composition Mande 50% (Bambara, Malinke, Soninke); Peul 17%; Voltaic 12%; Songhai 6%; Tuareg and Moor 10%; other 5%

Religious affiliations Muslim 90%; indigenous beliefs 9%; Christian 1%

Currency 1 Communauté Financière Africaine franc (XOF) = 100 centimes

Gross domestic product (2006 est.) U.S. $14.59 billion

Gross domestic product per capita (2006 est.) U.S. $1,200

Life expectancy at birth male 47.05 yr; female 51.01 yr

Major resources gold, phosphates, kaolin, salt, limestone, uranium, gypsum, granite, hydropower, bauxite, iron ore, manganese, tin, copper

Niger

REPUBLIC OF NIGER

THE REPUBLIC OF NIGER LIES AT THE HEART of northwestern Africa. This land-locked country borders on no fewer than seven other nations. With them it forms the Sahel area, notorious for its repeated droughts since the late 1960s.

GEOGRAPHY

Two-thirds of Niger forms part of the vast Sahara, though the hot, mostly dry climate varies sharply according to the season and the area. The entire north of the country is an arid highland that forms part of a mountain chain stretching from Algeria to Chad. To the east lies desert country, with sand dunes and vast areas of stony ground. Only in the southwest, where the Niger river crosses the country, is there rich arable soil, which becomes flooded in the brief rainy season. In the southeast of the country around Lake Chad lies the Manga wasteland, which mostly supports savanna grassland.

There is a great diversity of plant life in the country, including mahogany, kapok and baobab trees, and date palms near the desert margins. Animal life is rich, but many species are threatened, and conservation is minimal.

SOCIETY

Niger's original settlers were the nomadic Fulani and Tuareg peoples from the north and the Hausa from the east. These remain the chief ethnic groups today, and more than half the population is Hausa. In 1804 the strictly Muslim Fulani subjugated the Hausa – also nominally Muslim – in a holy war (jihad) and established the Sokoto empire.

In the late 19th century France gradually assumed control of the country. In 1922 Niger became part of French West Africa, and in 1946 it became a French overseas territory. After independence in 1960 the first president, Hamani Diori (b. 1916), maintained economic links with France. In 1974 Colonel Seyni Kountché (1931–87) staged a coup and established a military dictatorship. He died in 1987 after a series of coup attempts by his colleagues, and was succeeded by the more moderate and popular General Ali Seybou (b. 1940).

The first multiparty elections (1993) were won by a left-wing coalition, Alliance of the Forces for Change (AFC), with Mahamane Ousmane as president. He was toppled in 1996 in a military coup led by Ibrahim Barre Maïnassara, who was assassinated in 1999. Daouda Mallam Wanké now heads a military dictatorship. Most donor countries suspended aid payments causing Niger economic difficulties.

ECONOMY

Agriculture is Niger's principal employer, and accounts for 39 percent of GDP. The country is virtually self-sufficient in food but suffers from periodic drought. The main crops include grains, with rice in the flooded areas. Groundnuts (peanuts), cotton and cattle are exported.

Most manufacturing is in agricultural products, textiles and chemicals. Energy resources are limited; electricity is generated, but fuels are imported. The main natural resource is uranium ore. It represents over one-third of gross national product, but the economy was badly destabilized by a decline in worldwide uranium prices during the 1980s, and Niger is now burdened with high levels of foreign debt. Surfaced roads have now been constructed across the desert to the mines, and there is an extensive domestic air service.

Social services consist largely of vaccination and disease-elimination campaigns. An education system has been established with French aid, but illiteracy remains high. There is one university, at Niamey, the capital, in the southwest.

Niger has few roads, is subject to repeated droughts, and consists largely of desert. With so few surfaced roads, it is scarcely surprising that the traditional camel train remains a vital means of carrying goods.

NATIONAL DATA - NIGER				
Land area 1,266,700 sq km (489,076 sq mi)				
Climate	Altitude m (ft)	Temperatures January °C(°F)	July °C(°F)	Annual precipitation mm (in)
Niamey	222 (728)	26 (80)	29 (84)	541 (21.2)
Major physical features highest point: Mont Gréboun 1,944 m (6,378 ft); longest river: Niger (part) 4,200 km (2,600 mi)				
Population (2006 est.) 12,525,094				
Form of government multiparty republic with one legislative house				
Armed forces army 5,200; air force 100				
Largest cities Niamey (capital - 707,951); Zinder (170,575); Maradi (148,017); Agadez (78,289); Arlit (69,435)				
Official language French				
Ethnic composition Hausa 56%; Djerma 22%; Fula 8.5%; Tuareg 8%; Beri Beri (Kanouri) 4.3%; Arab, Toubou, and Gourmantche 1.2%				
Religious affiliations Muslim 80%; traditional beliefs and Christian 20%				
Currency 1 Communauté Financière Africaine franc (XOF) = 100 centimes				
Gross domestic product (2006 est.) U.S. $12.23 billion				
Gross domestic product per capita (2006 est.) U.S. $1,000				
Life expectancy at birth male 43.8 yr; female 43.73 yr				
Major resources uranium, coal, iron ore, tin, phosphates, gold, molybdenum, gypsum, salt, petroleum, cotton, groundnuts, livestock, millet, rice, sorghum				

Chad

REPUBLIC OF CHAD

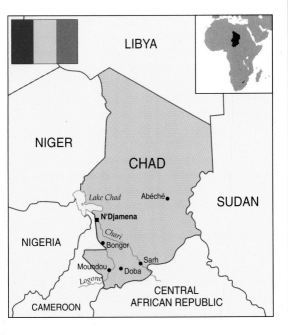

The REPUBLIC OF CHAD STRADDLES THE drought-stricken area of the Sahel. Since the mid-1960s it has been gripped by civil wars, aggravated by conflict with its powerful neighbor Libya.

GEOGRAPHY

Chad is landlocked, and the only large body of water is Lake Chad in the southwest. Shallow and marshy, it once covered a larger area, but is shrinking as the desert advances. The only significant rivers feeding it, the Chari and the Logone, flow from the southeast to meet at the lake. Through it run the country's frontiers with Niger, Nigeria and Cameroon. From the area of the Chad Basin the land gradually rises to the mountains along the opposite frontiers: the volcanic Tibesti mountains to the north; and the Ennedi plateau and Ouaddai mountains that lie to the east.

The climate is generally tropical and hot, but rainfall varies widely. Summer rains support fertile savanna grasslands and broadleaf trees in the south (as much as one-fifth of the land is forested), but rainfall dwindles farther north. The entire northern third of the country consists of arid scrubland fading into barren true desert, with only a few palm oases. The land to the south supports an abundance of wildlife, including African elephants, cheetahs, giraffes, lions and warthogs, in a relatively unspoilt environment.

SOCIETY

Chad is a melting-pot of peoples and cultures, and more than 100 different languages are spoken. Arab settlers from the north founded the Islamic kingdom of Kanem-Bornu, which reached its peak in the 16th century, commanding the main trans-Saharan caravan route to Libya and the Mediterranean. The Muslim slave traders preyed heavily on the people of the south, engendering present-day conflicts. In the 19th century the whole area was conquered by a Sudanese warlord, who in turn was ousted by the French. Chad became part of French Equatorial Africa in 1908, and in 1960 gained independence under President François Tombalbaye (1918–75).

Within several years Libyan-backed guerrilla organizations had raised a revolt in the north. Tombalbaye called in French troops, but in 1975 he was assassinated. Soon afterward Libya annexed the Aouzou Strip along the northern border of Chad. Fighting continued between various guerrilla factions. The most powerful guerrilla leader, Hissen Habré (b. 1942), seized the presidency with French backing in 1984, and in 1986 drove out Libyan incursions. Habré won the 1990 elections, but within six months he was overthrown by Idriss Déby. The approval of a new constitution and the holding of multiparty elections in 1997 and 1998 brought some peace to Chad although further fighting broke out in the north in 1998.

ECONOMY

Agriculture employs about four-fifths of the work force and provides about half of gross domestic product. The country is self-sufficient in food when there are no droughts. The chief food crops are millet, sorghum and rice, with cotton and groundnuts (peanuts) for export. Live-

Villagers take water from a well in Chad. Here, at the very heart of the Sahel drought area, water can never be taken for granted, while war and famine are a constant threat.

stock is the next most important export.

Petroleum is the major resource. Exploitation of oil in the Doba Basin has foundered on environmental objections to a new pipeline requiring World Bank funding. Industry is confined to agricultural processing and minor manufacturing. Energy generation and transportation are extremely limited. Literacy is low as less than half the children are able to attend school. Health services are poor and confined mainly to the south.

NATIONAL DATA – CHAD

Land area	1,259,200 sq km (486,180 sq mi)			
Climate		**Temperatures**		**Annual precipitation**
	Altitude m (ft)	January °C(°F)	July °C(°F)	mm (in)
N'Djamena	295 (968)	21 (70)	26 (79)	510 (20)

Major physical features highest point: Emi Koussi 3,415 m (11,204 ft); longest river: Chari 949 km (590 mi)

Population (2006 est.) 9,944,201

Form of government multiparty republic with one legislative house

Armed forces army 25,000; air force 350

Capital city N'Djamena (753,791)

Official languages Arabic, French

Ethnic composition Sara/Bagirmi/Kreish 30.5%; Sudanic Arab 26.1%; Teda 7.3%; Mbum 6.5%; Masalit/Maba/Mimi 6.3%; Mubi 4.2%; Kanuri 2.3%; Hausa 2.3%; Masa 2.3%; Kotoko 2.1%; others 10.1%

Religious affiliations Muslim 51%; Christian 35%; animist 7%; other 7%

Currency 1 Communauté Financière Africaine franc (XAF) = 100 centimes

Gross domestic product (2006 est.) U.S. $15.26 billion

Gross domestic product per capita (2006 est.) U.S. $1,500

Life expectancy at birth male 45.88 yr; female 49.21 yr

Major resources petroleum, uranium, natron, kaolin, fish (Lake Chad), gold, limestone, sand and gravel, salt, cassava, cattle, cotton, dates, gum Arabic, millet, groundnuts, rice, sorghum, sweet potatoes, yams

Sudan

REPUBLIC OF SUDAN

S UDAN IS AFRICA'S LARGEST STATE, EXTEND-
ing from just south of the Tropic of
Cancer to just north of the Equator. Its
landscape ranges from harsh desert in the
north to dense rainforests in the south.
For thousands of years it formed a trade
bridge between Southern Africa and the
Mediterranean, yet today it is one of the
least-developed countries in the world.

GEOGRAPHY

In southern Sudan the land rises all along
the borders with Kenya, Uganda, Congo
(D.R.O.) and the Central African Republic,
reaching its highest point at mount

Kinyeti on the Ugandan border. The
White Nile enters Sudan at a place called
Nimule, also on the Ugandan border, and
flows northward into a clay plain that
occupies most of southern and central
Sudan. The plain is dotted with isolated
hills (inselbergs), along with the Nuba
Mountains, and in the east it rises to the
foothills of the Ethiopian Highlands. Here
the Blue Nile flows northwestward to
meet the White Nile at the capital,
Khartoum. Between the two rivers is a
huge marshy area known as the Sudd.

On the western side of Sudan, along the
Chad border, the volcanic massif of the
Jebel Marra rises from the sand dunes of
the Qawz. To the north the dunes gradually give way to the rock desert of the
Sahara, which continues to the northern
borders with Libya and Egypt. This barren landscape is interrupted only by a
thin line of cultivation along the Nile. In
the northeast the Red Sea mountains rise
to about 2,000 m (6,500 ft) above a narrow
coastal plain.

The heat in Sudan is moderated only
along the Red Sea coast. Rainfall, vegetation and wildlife decrease gradually
toward the north. The far south has an
equatorial climate with very high rainfall
and a long rainy season. Tropical forests
flourish here, and animals include chimpanzees, hippopotamuses, elephants, zebras, giraffes, leopards, lions and many
species of antelope. The central southern
area supports savanna and semiarid
shrubland; papyrus swamps grow in the
Sudd marshes. The north has a tropical
continental climate with almost no rainfall, creating desert or semidesert.

SOCIETY

The ancient Egyptians knew the Sudan as
Nubia, but from the 11th to the 4th
century BC it was part of the kingdom of
Kush. Coptic Christian missionaries in the
6th century AD found three kingdoms
here: Nobatia, Makurra and Alwah. In the
7th century Makurra, which by then
included Nobatia, stood firm against
Islam and Arab invaders, but in the 14th
century Mameluke Turks devastated the
country and it gradually came under Arab
control. Alwah in the south, the last area
to stand out against the Arabs, fell in
about 1500, but ethnic and religious
conflicts remain here.

In the 16th century the Funj people
carved out a kingdom between the
Ethiopian Highlands and the Blue Nile,
but in 1821 it was shattered when
Muhammad Ali (1769–1849), viceroy of

Egypt under the Ottoman Turks, invaded
the Sudan. In 1881 Muhammad Ahmed
(1848–85), known as the Mahdi or "divine
guide", led a triumphant Islamic revolution to drive out the British and the
Egyptians; this ended in 1885 with the
fall of Khartoum, and the death of the
British General Charles George Gordon
(1833–85), governor-general of Sudan. An
Anglo-Egyptian army under the command of General Horatio Herbert Kitchener (1850–1916) invaded the Sudan in
1897, and the area remained in British
hands until 1951, when the Egyptian
government seized control. In 1953 the
Sudanese were granted self-government.
Three years later they declared independence, but there was continued civil
war because the underprivileged south
resented the Islamic bias of the new
regime in the north.

Parliamentary democracy was short-
lived: in 1959 a bloodless army coup put
General Ibrahim Abbud (b. 1900) in
power. A half-hearted return to democracy in 1964 ended in 1969 with a coup led
by Colonel Gaafar Mohamed el-Nimeiry
(b. 1930), and in 1972 the civil war was
temporarily halted. It erupted again in
1983, leaving the Sudanese Popular
Liberation Army largely in control of the
south. Two years later Nimeiry was overthrown by his defense minister, General
Abd ar-Rahman Siwar ad-Dahab (b.
1934), who once again attempted to restore democracy. But in 1989 Lieutenant-
General Omar Hassan Ahmad al-Bashir

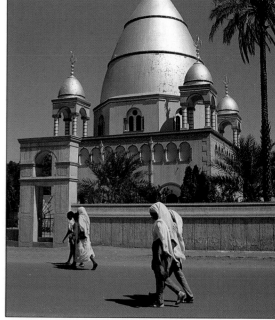

Khartoum has had a troubled history: it was founded in
1823 by an Egyptian ruler, destroyed by the Mahdi, and
rebuilt in 1898 by Lord Kitchener. Fine mosques such
as this one stand near grand colonial buildings, and
there is also a large industrial quarter.

NATIONAL DATA – SUDAN

Land area 2,376,000 sq km (917,379 sq mi)				

Climate		Temperatures		Annual
	Altitude m (ft)	January °C(°F)	July °C(°F)	precipitation mm (in)
Khartoum	380 (1,247)	23 (74)	32 (90)	162 (6.3)

Major physical features highest point: Kinyeti 3,187 m (10,456 ft); longest river: Nile (part) 6,690 km (4,160 mi)

Population (2006 est.) 41,236,378

Form of government multiparty republic with one legislative house

Armed forces army 100,000; navy 1,800; air force 3,000

Largest cities Omdurman (3,127,802); Khartoum (capital - 2,207,794); Khartoum Bahri (1,725,570)

Official language Arabic

Ethnic composition Black 52%; Arab 39%; Beja 6%; other 3%

Religious affiliations Sunni Muslim 70% (in north); traditional beliefs 25%; Christian 5%

Currency 1 Sudanese dinar (SDD) = 10 Sudanese pounds, which are each further subdivided into 100 piastres

Gross domestic product (2006 est.) U.S. $96.01 billion

Gross domestic product per capita (2006 est.) U.S. $2,300

Life expectancy at birth male 57.69 yr; female 60.21 yr

Major resources petroleum, iron ore, copper, chromium ore, zinc, tungsten, mica, silver, gold, hydropower, beans, cotton, natural gas, dates, groundnuts, gum Arabic, livestock, sesame seeds, sorghum, wheat

(b. 1944) seized power and dissolved all political institutions on behalf of the National Islamic Front. Multiparty elections were held in 1996 but the civil war continued. Up to two million people have died in the fighting or through war-induced famine according to estimates made in 1999.

The main internal divisions are between the Arab and the Nubian ethnic groups in the north (mostly Muslim), and the Nilotic and the Sudanic peoples in the south (mostly non-Muslim). During the 1970s and 1980s the southern population has been swelled by vast numbers of refugees from Ethiopia, Eritrea, Uganda and Chad. The official language is Arabic, spoken by about half the population, and the dominant religious group are the Sunni Muslims.

ECONOMY

Most of the work force are involved in agriculture. The area under cultivation varies with the changing rainfall patterns, and much depends on elaborate irrigation schemes. Cotton is a cash crop, usually at the expense of food for the domestic market (agriculture accounts for nearly one-third of the gross national product). Other crops include sorghum (the staple), millet, wheat, barley and groundnuts.

Forests cover about one-fifth of the land area, and most productive woodland is state-owned. A further one-fifth is given over to seasonal pasture for goats, cattle and other livestock such as camels and sheep (both exported). Fishing along the Nile and off the Red Sea coast mostly supplies local demand.

Gold, chromite, salt, manganese and mica are extracted, and oil production started in earnest in 1998. An oil pipeline and refinery were under construction in 1999. Over 50 percent of the Sudan's power production comes from hydroelectricity while the rest of the power comes from fossil fuels. Manufacturing, chiefly of textiles, shoes, beverages and soap, provides just under one-tenth of the gross national product.

A camel train approaches Kassala, 400 km (250 mi) east of the Sudanese capital, Khartoum, and near the border with Ethiopia. Kassala today is a market town noted for its fruit gardens, but its oldest building, constructed in 1840, was an Egyptian fort.

Saudi Arabia is Sudan's chief trading partner, followed by European Community countries. Foreign debt is crippling, and the balance of trade is very unfavorable. Most of the roads are unsurfaced, and impassable after rain. The state-run railroad, with some 5,000 km (3,000 mi) of track, employs one-fifth of the country's work force. Steamer services operate on the White Nile, but Port Sudan in the northeast is the only important outlet to the sea. There are 10 airports operating scheduled flights. Radio, television and both national newspapers are state-controlled.

Social–welfare services are available on a very limited basis, but there are few health care facilities. Infectious diseases (including malaria), malnutrition, poor housing and inadequate sanitation remain serious problems.

Ethiopia

FEDERAL DEMOCRATIC REPUBLIC OF ETHIOPIA

ETHIOPIA IS THE MODERN DESCENDANT OF a powerful and ancient empire, the fabled land of Prester John in medieval European legend. Today it is one of the world's poorest countries, scarred and torn by drought, famine and civil war.

GEOGRAPHY

Ethiopia is a massive plateau cut in half from north to south by the East African Rift Valley. To the west, along the Sudanese border, the land rises as high as 3,700 m (12,000 ft). At the heart of the western highlands lies the capital, Addis Ababa, some 2,500 m (8,000 ft) above sea level. From Lake Tana in the northwest, the Blue Nile gorge cuts a winding course, first to the southeast and then westward. In the north the Simen Mountains rise to Ras Dashan, Ethiopia's highest peak, then drop away to the narrow Red Sea coastal plain.

The Rift Valley averages some 50 km (30 mi) in width between its eastern and western flanks. At the southern end, near the Kenyan border, lakes, lava flows and volcanic cones are characteristic. To the north the valley widens into the broad plains of the Awash river. Northward again, in the Danakil Plains, there are hot springs, active volcanoes, salt lakes and the Kobar Sink, a large desert area below sea level. The border with Djibouti lies to the southeast of this area.

The eastern highlands climb to more than 4,300 m (14,000 ft) along the eastern side of the Rift Valley. Southeastward they descend to the Bale plateau and again to the Harerge plain in the Ogaden, which stretches east and south to the disputed border with Somalia.

The high plateaus enjoy a temperate climate that becomes increasingly hot in lower-lying areas. Rainfall is highest in the south and west. In recent years prolonged drought in the north has brought disease, destitution and famine.

There is an enormous variety of vegetation, from tropical rainforest and savanna grasslands in low-lying areas, to alpine plants in the high mountains. Animals range from ibex and nyala (an antelope) in the mountains to hippopotamuses and crocodiles in lakes and rivers.

SOCIETY

The origins of modern Ethiopia lie in the kingdom of Aksum (2nd century AD), which became Christian in the 4th century. It was cut off from Europe by the Muslim conquests of the 7th century. Not until the middle of the 15th century, when Ethiopian monks seeking to unite their church with Rome were believed to have come from the legendary priest-king, Prester John, was European awareness of Ethiopia resumed.

In 1520 the Portuguese made contact, and gave some aid to Ethiopia's continuing struggle against the Turks and insurrection in its own Muslim provinces in the southeast. However, in 1633 all Jesuit missionaries were expelled. A cultural renaissance followed, but the archaic social system remained unchanged.

The old empire was finally reunified under Menelik II (1844–1913) near the end of the 19th century, though by now the Italians had laid claim to the northern area of Eritrea. Modernization began under Emperor Haile Selassie (1892–1975) in 1930, but five years later the Italians invaded. They were driven out in 1941, and in 1950 Eritrea became a federated state of Ethiopia. Its full incorporation in 1962 sparked a rebellious conflict.

In 1974 a military coup deposed the emperor, triggering a chain of events that ended in civil war. By 1984 the conflict had become so serious that it hampered the international aid agencies bringing much-needed famine relief to the area. In April 1993 Eritrea won its independence from Ethiopia and was recognized as a new state by the United Nations in May. Fighting again erupted along the border between Ethiopia and Eritrea in early 1999. The Ethiopian people are a complex intermixture of ethnic groups, the most powerful are the Amhara, who make up some 30 percent of the population and live mostly in the north and the heartlands. The Oromo constitute about 40 percent of the population, and live mostly in eastern and southern areas. The official language is Amharic, though some 70 other languages are spoken. About half the population are thought to be Muslim. Most people live in rural areas, and many are nomadic or seminomadic.

ECONOMY

Agriculture accounts for some four-fifths of the work force, but output has been severely reduced by war and drought. The highest areas (above 3,000 m; 10,000 ft) support alpine pastures. Below them is a temperate zone where most of the population live. Here maize, wheat, barley and other cereals are grown, and horses, cattle

An Ethiopian woman makes a traditional headdress by stringing together colored beads.

NATIONAL DATA – ETHIOPIA

Land area	1,119,683 sq km (432,312 sq mi)		

Climate		Temperatures		Annual
	Altitude m (ft)	January °C(°F)	July °C(°F)	precipitation mm (in)
Addis Ababa	2,360 (7,741)	20 (67)	20 (67)	1,020 (40.1)

Major physical features highest point: Ras Dashan 4,620 m (15,158 ft)

Population (2006 est.) 74,777,981

Form of government one-party federal republic with one legislative house

Armed forces army 180,000; air force 2,500

Largest cities Addis Ababa (capital – 2,973,000); Dire Dawa (281,800); Nazret (228,600); Bahir Dar (167,300)

Official language Amharic

Ethnic composition Oromo 40%; Amhara and Tigre 32%; Sidamo 9%; Shankella 6%; Somali 6%; Afar 4%; Gurage 2%; other 1%

Religious affiliations Muslim 45%–50%; Ethiopian Orthodox 35%–40%; animist 12%; other 3%–8%

Currency 1 Ethiopian birr (ETB) = 100 cents

Gross domestic product (2006 est.) U.S. $71.63 billion

Gross domestic product per capita (2006 est.) U.S. $1,000

Life expectancy at birth male 47.86 yr; female 50.24 yr

Major resources small reserves of gold, platinum, copper, natural gas, hydropower, barley, beans and peas, cattle, coffee, cotton, maize/corn, millet, potash, salt, sorghum, sugarcane, timber, wheat

and sheep are raised. Below 1,800 m (6,000 ft) the climate is tropical, and high-rainfall areas are suitable for coffee (the main export crop) and sugar cane.

Ethiopia has reserves of gold, platinum and copper, and valuable limestone deposits. Most electricity comes from hydroelectric plants; there is scope for many more, and for geothermal stations along the Rift Valley. Textiles and food processing dominate the small, largely nationalized manufacturing industry.

Famine, drought and civil war drained the economy and disrupted export production in the 1980s, and the border war with Eritrea, which ended in December 2000 under UN auspices, further depressed the economy. Ethiopia maintains the largest standing army in Africa, and its overseas debts are enormous. Major trading partners are Germany, Saudi Arabia, the United States, Italy and Japan.

Surfaced roads link Addis Ababa with provincial capitals. The railroad from Addis Ababa to the port of Djibouti on the Gulf of Aden provides the country's main link to the sea. Waterways have not been developed to any significant extent. There are three international airports. In 1999 there were 25 television stations and one Internet service provider, but only 320,000 televisions.

Few of the population benefit from the four-tiered educational system or from modern health care. Contagious and tropical diseases are rife and average life expectancy is among the world's lowest.

The Tis Isat Falls in full flow during the rainy season at the outflow of the Blue Nile from Lake Tana in northwestern Ethiopia.

Eritrea

STATE OF ERITREA

ON 28 MAY 1993 ERITREA GAINED UNITED Nations recognition as an independent country after a 30-year war with Ethiopia that left the new nation famine-stricken and impoverished. A former Italian colony, and under British rule 1942–52, Eritrea became part of Ethiopia in 1962 against the wishes of its people, whose struggle for independence was won in 1991 when other discontented groups in Ethiopia joined with the Eritrean People's Liberation Front to bring down the Mengistu government.

Eritrea is currently controlled by a transitional government. A constitution, ratified in 1997, has not yet been put into effect, and parliamentary and presidential elections, promised for that year, have been postponed indefinitely following the border war with Ethiopia in 1998.

Eritrea is a small, desperately poor country. Eighty percent of the population are subsistence farmers, and there is a tiny, outmoded industrial sector. Worker remittances from overseas substantially augment the GDP. In the long term Eritrea has plans to develop tourism along the Red Sea as well as an offshore oil industry, and offshore fishing.

NATIONAL DATA - ERITREA	
Land area	121,320 sq km (46,842 sq mi)
Climate	hot, dry desert strip
Major physical features	highest point: Monte Soira 2,989 m (9,807 ft)
Population	(2006 est.) 4,786,994
Form of government	transitional
Armed forces	army 200,000; navy 1,400; air force 350
Capital city	Asmara (500,600)
Official languages	Tigrinya and Arabic
Ethnic composition	ethnic Tigrinya 50%; Tigre and Kunama 40%; others 10%
Religious affiliations	Muslim; Coptic Christian; Roman Catholic; Protestant
Currency	1 nafka (ERN) = 100 cents
Gross domestic product	(2005 est.) U.S. $4.471 billion
Gross domestic product per capita	(2005 est.) U.S. $1,000
Life expectancy at birth	male 57.44 yr; female 60.66 yr
Major resources	gold, potash, zinc, copper, salt, unexploited oil and natural gas, fish, barley, beans and peas, cattle, coffee, cotton, maize/corn, millet, platinum

Djibouti

REPUBLIC OF DJIBOUTI

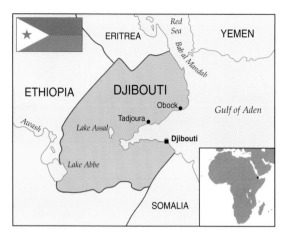

D JIBOUTI – THE SMALLEST COUNTRY IN THE region – is a strategically sited desert republic on the east coast of Africa between the Red Sea and the Indian Ocean. It faces the Gulf of Aden and borders Somalia and Ethiopia.

GEOGRAPHY

The north of Djibouti is moderately mountainous, while the heartland and the south are dominated by volcanic plateaus; faults parallel to the Red Sea have produced deep troughs occupied by saline lakes and the Hanlé Plains. The Gulf of Tadjoura cuts deeply into the eastern coastal plain of Djibouti.

Djibouti is among the hottest countries in the world. Most of its scanty precipitation falls in the northern mountains between late summer and the end of March. The predominant vegetation is desert scrub, with mangroves along the coast and a few wooded areas in the mountains. The animal life is well adapted to the extreme conditions, and includes lynx, jackals and gazelles.

SOCIETY

Djibouti's two main ethnic groups both speak a Hamitic language and share a common religion – Islam. The Afars are descended from 3rd-century settlers, whereas the Issas were later arrivals who came from Somalia.

In 1862 Obock (on the northern side of the Gulf of Tadjoura) was sold to the French by the Sultanate of Tadjoura. Thirty years later the town of Djibouti across the gulf became the capital of French Somaliland (formerly the French Territory of the Afars and Issas). France protected its investment heavily, but a 1977 plebiscite showed 85 percent in favor of independence, and Hassan Gouled Aptidon (b. 1916) became the first president of the new republic. French is the official language (together with Arabic), and French military aid preserves Djibouti's independence and neutrality.

After a referendum, a new constitution was approved in 1992 which permits up to four political parties, each with an ethnic balance of membership. In the subsequent 1992 election the ruling party (Rassemblement Populaire pour le Progrès) won all 65 seats in the Chamber of Deputies and Aptidon was re-elected president. However, other parties boycotted the election.

ECONOMY

Agriculture meets only one-quarter of Djibouti's needs; its development is hampered by the scarcity of water for irrigation. Livestock are raised, but less than one-tenth of the land area is under pasture. Fishing is of increasing importance.

Djibouti's resources are limited and the economy depends on services and external trade, particularly trans-shipment of Ethiopian and Somalian exports and imports. Recession, civil war and high population growth rate have only added to the problems Djibouti faces and the country is heavily dependent on foreign aid. The capital's docks are well developed, and linked by rail with Addis Ababa in Ethiopia. Foreign assistance from countries such as France, Italy and Saudi Arabia is readily forthcoming because of its strategic position. Few of the roads are surfaced. An international airport links the capital with France, Ethiopia, Madagascar, Egypt and other east African countries. There are radio broadcasts in four languages, and one television station.

There are a number of dispensaries and medical posts in rural areas, and two hospitals. Malnutrition, dehydration and poisoning account for almost half the recorded deaths of children up to the age of 10.

Fishermen from Djibouti (*left*) catch tuna in the Bab al Mandab – the strait that links the Red Sea with the Gulf of Aden and the Indian Ocean. A program to develop fishing was started in 1980.

NATIONAL DATA – DJIBOUTI

Land area 22,980 sq km (8,873 sq mi)

Climate	Altitude m (ft)	Temperatures January °C(°F)	July °C(°F)	Annual precipitation mm (in)
Djibouti	8 (26)	25 (74)	36 (97)	164 (6.4)

Major physical features highest point: Musa Ali Terara 2,063 m (6,768 ft); lowest point: Lake Assal -150 m (-492 ft)

Population (2006 est.) 486,530

Form of government multiparty republic with one legislative house

Armed forces army 8,000; navy 200; air force 250

Capital city Djibouti (317,000)

Official languages Arabic, French

Ethnic composition Somali 60%; Afar 35%; French, Arab, Ethiopian, and Italian 5%

Religious affiliations Muslim 94%; Christian 6%

Currency Djiboutian franc (DJF) = 100 centimes

Gross domestic product (2002 est.) U.S. $619 million

Gross domestic product per capita (2005 est.) U.S. $1,000

Life expectancy at birth male 41.86 yr; female 44.52 yr

Major resources geothermal power, gold, clay, granite, limestone, marble, salt, diatomite, gypsum, pumice, petroleum, cattle, dates, fish, fruit and vegetables, goats, sheep

Somalia

SOMALI DEMOCRATIC REPUBLIC

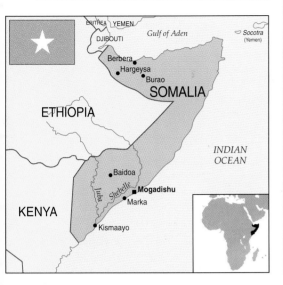

Refugees wait for food at a Somalian camp. In the 1990s continued civil war, floods and drought made an estimated 1.5 million people dependent on foreign aid but movement of food aid only became effective in 1992 when American troops were used to police distribution.

SOMALIA LIES ALONG THE HORN OF AFRICA, a large promontory on Africa's north-eastern coast. In recent years it has been torn apart by civil war and drought.

GEOGRAPHY

In the north lie the Ogo Highlands, where rugged cliffs and mountains of ancient rock extend from neighboring Ethiopia along the Gulf of Aden. The narrow coastal plain is a hot, barren scrubland called the Guban.

South of the mountains lies the Haud Plateau, beyond which the land slopes gradually down toward the Indian Ocean. The two main rivers, the Juba and the Shebelle, flow across the country from Ethiopia, providing a vital water resource in an almost constantly hot and inhospitable arid climate.

Only in the south and on northern slopes is there enough rainfall to support small forests and grassland. Otherwise Somalia is covered by semidesert or dry savanna vegetation. Elephants, zebras, giraffes, antelopes and leopards are among the wildlife typical of the region.

SOCIETY

Somalia was probably the "Land of Punt" referred to in the Bible. Its modern people are descended from Arab or Persian traders along the seacoast and the nomadic Somali tribesmen they came to dominate. More than two-thirds of the population are still nomadic; there are three main tribal groups, but Somali is the dominant language.

In the 19th century the area was colonized by Britain in the north and Italy in the east, though Muslim resistance remained. In 1946 Italian and British Somaliland were united to form a United Nations trust territory, which became independent in 1960.

In 1969 the president was assassinated in a Soviet-backed military coup and replaced by the repressive rule of Major-General Siyad Barre (1921–95). In 1977 he invaded the Ogaden area of Ethiopia, provoking a split with the Soviets and a bloody civil war that was fueled by rivalries between different groups of indigenous peoples. Coup attempts, massacres and growing civil unrest followed, leading in 1991 to the overthrow of Barre's government. Continuing unrest prevented international aid reaching refugee camps in famine areas, and United States troops were sent in December 1992 to assist. They withdrew after a truce in 1994, but clan-based fighting continued, and all UN troops withdrew in 1995.

ECONOMY

Arable farming has traditionally been confined to a small area lying between the two main rivers, which covers less than one-fiftieth of the overall area. Crops include maize, sorghum and groundnuts (peanuts), with bananas and sugar cane along the river banks. About half the land supports the nomadic herding of cattle, sheep and goats and exports consist mainly of livestock, skins, hides and bananas.

Industry accounted for only one-tenth of GDP before civil war and was based mainly on agricultural products and petroleum refining. In the late 1990s, few industries were operating. There are unexploited deposits of chromium, coal, copper, gold and other minerals.

Transportation is limited, with no railroads and limited roads, few of which are surfaced. Port and airport facilities are chiefly at Mogadishu, the capital. State radio stations broadcast in four languages; the television station was destroyed in 1991. State education exists to university level, but is not compulsory and is open to only a few. Literacy levels are low due to the war. Health care, which lacks both staff and basic supplies, is not widely available.

NATIONAL DATA – SOMALIA

Land area	627,337 sq km (242,216 sq mi)			

Climate		Temperatures		Annual
	Altitude m (ft)	January °C(°F)	July °C(°F)	precipitation mm (in)
Mogadishu	17 (56)	27 (81)	26 (79)	427 (17)

Major physical features highest point: Surud Ad 2,406 m (7,894 ft); longest river: Shebelle (part) 2,010 km (1,250 mi)

Population (2006 est.) 8,863,338

Form of government transitional

Armed forces no national armed forces

Largest cities Mogadishu (capital - 2,855,805); Hargeysa (522,508); Marka (350,859); Berbera (263,095)

Official language Somali

Ethnic composition Somali 85%; Bantu and other non-Somali 15% (including Arabs 30,000)

Religious affiliations 100% Sunni Muslim

Currency 1 Somali shilling (SOS) = 100 cents

Gross domestic product (2006 est.) U.S. $5.023 billion

Gross domestic product per capita (2006 est.) U.S. $600

Life expectancy at birth male 46.71 yr; female 50.28 yr

Major resources uranium and largely unexploited reserves of iron ore, tin, gypsum, bauxite, copper, salt, natural gas, possible oil reserves, textiles, bananas, camels, cotton, goats, groundnuts, maize/corn, millet, sheep, sugarcane

Central Africa

COUNTRIES IN THE REGION

SENEGAL · GAMBIA · GUINEA · GUINEA-BISSAU

SIERRA LEONE · LIBERIA · CÔTE D'IVOIRE · BURKINA FASO

GHANA · TOGO · BENIN · NIGERIA · CAMEROON

CENTRAL AFRICAN REPUBLIC · EQUATORIAL GUINEA

GABON · CONGO, Republic of

CONGO, Democratic Republic of (formerly Zaire)

BURUNDI · RWANDA · UGANDA · KENYA · TANZANIA

SEYCHELLES · SÃO TOMÉ AND PRÍNCIPE

CAPE VERDE

Dawn at Lake Naivasha, Kenya The lake is one of many lying in the eastern arm of the Great Rift Valley, the vast geological fault that runs 6,450 km (4,000 mi) down Africa, as it cuts through Kenya's highlands. Its fresh waters are a haven for wildlife, and the lake is much visited by fishermen and birdwatchers.

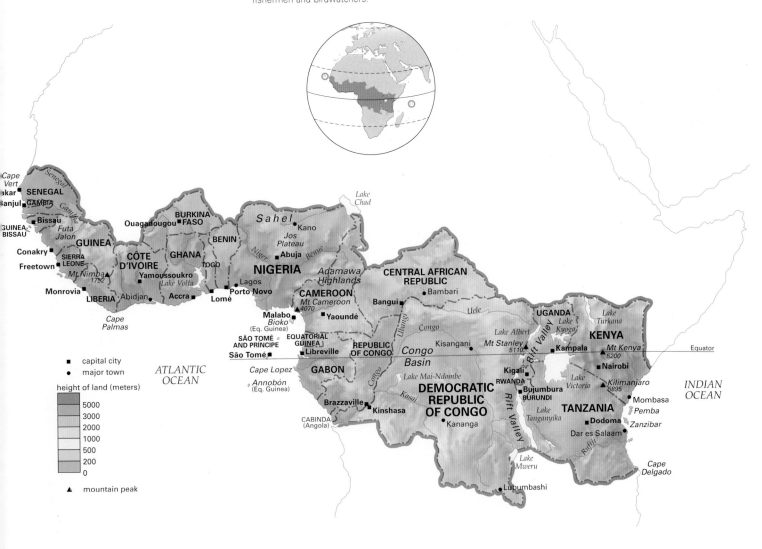

Senegal

REPUBLIC OF SENEGAL

A colorful display of bracelets in beadwork, wood and snakeskin brightens up a busy market stall in Dakar, the capital of Senegal and one of the largest industrial centers in west Africa. Skilled work such as this reflects the country's enduring tradition of folk art.

S ENEGAL LIES ON THE WEST COAST OF WEST Africa, bordered by Mauritania, Mali and Guinea, and surrounding Gambia.

GEOGRAPHY

Senegal's landscape consists mostly of low-lying plains. The only higher ground is in the east, bordering the plateau areas of Mali and Guinea. The plains are crossed by several rivers, including the Gambia, the Casamance and the Senegal, which forms the northern border with Mauritania. The climate is hot with seasonal rains, and the landscape ranges from arid semidesert in the north to lush tropical savanna and rainforest in the south, where there are mangrove swamps along the coast. There is a wide variety of wildlife, including lions, antelopes, monkeys and warthogs, with crocodiles and hippopotamuses in the rivers.

SOCIETY

Senegal is the home of many peoples, notably the Wolof, the Fulani, the Serer and the Tukulor. It is predominantly Muslim. In the 17th and 18th centuries, both the French and the English had a stake in the country. It was fully colonized by France in the 19th century, after the abolition of a prosperous slave trade. In 1958 it gained internal autonomy, and in 1960 it became independent. The first president was the poet Léopold Sédar Senghor (b. 1906), who finally resigned in 1980. In 1981 his successor Abdou Diouf (b. 1935) sent troops into Gambia to help put down a revolt. Diouf was re-elected in 1993, followed by a legislative election. Senegal was ruled by a coalition in 1995.

ECONOMY

Senegal's economy is heavily dependent on agriculture. Groundnuts (peanuts) are the chief crop, followed by millet, beans, rice, sugar cane, sorghum and cotton. Fishing is also important. Large phosphate reserves are mined, and important petroleum and iron ore deposits have been found. Manufacturing, chiefly of agricultural products, especially groundnut (peanut) oil, is under development. A booming service sector based on information technology was providing 60 percent of GDP in the late 1990s. The transportation network is relatively good, especially around the modern capital of Dakar on the Cape Vert peninsula. Outside Dakar, however, health care, welfare and even education are relatively backward. However, the country has a strong tradition of art and literature.

The beach at Ngor, a fishing village on the coast of Senegal. Sea fishing continues to play an important role in the Senegalese economy, but small local boats are giving way to deep-sea trawlers, which now provide the greater part of the catch.

NATIONAL DATA - SENEGAL

Land area	192,000 sq km (74,132 sq mi)			
Climate		**Temperatures**	**Annual**	
	Altitude m (ft)	January °C(°F)	July °C(°F)	precipitation mm (in)
Dakar	23 (75)	22 (72)	26 (79)	514 (20.2)

Major physical features highest point: Futa Jalon (edge) 500 m (1,640 ft); longest river: Senegal (part) 1,633 km (1,015 mi)

Population (2006 est.) 11,987,121

Form of government Multiparty republic with one legislative house

Armed forces army 11,900; navy 950; air force 770

Capital city Dakar (2,564,900)

Official language French

Ethnic composition Wolof 43.3%; Pular 23.8%; Serer 14.7%; Jola 3.7%; Mandinka 3%; Soninke 1.1%; European and Lebanese 1%; other 9.4%

Religious affiliations Muslim 94%, Christian 5% (mainly Roman Catholic), traditional beliefs 1%

Currency 1 Communauté Financière Africaine franc (XOF) = 100 centimes

Gross domestic product (2006 est.) U.S. $22.01 billion

Gross domestic product per capita (2006 est.) U.S. $1,800

Life expectancy at birth male 57.7 yr; female 60.85 yr

Major resources fisheries, phosphates, iron ore, food processing, fertilizer production, petroleum refining, construction materials, textiles, groundnuts, maize/corn, millet, rice, sorghum, timber

Gambia

REPUBLIC OF GAMBIA

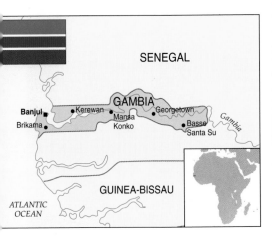

THE TINY WEST AFRICAN STATE OF GAMBIA forms a narrow enclave along the banks and estuary of the Gambia river, surrounded on three sides by Senegal.

GEOGRAPHY

The river dominates the country, whose landscape essentially consists of its wide valley, flanked by low hills and sandy plateaus. The climate is tropical, with heavy seasonal rains. The floodplain supports seasonally flooded grasslands known as *banto faros*, while the river itself is lined with swamps. The hills are largely covered with savanna and open forest, now mostly cleared for agriculture. The river attracts much wildlife, and is particularly rich in birds. Other animals include leopards, eland, antelopes and various species of monkey, together with crocodiles and hippopotamuses in the river.

SOCIETY

The Gambia is inhabited mainly by the Malinke, Fulani, Wolof and Yola peoples, and is distinct from Senegal because it came under British influence from the 17th century onward.

The Gambia river area became a formal colony in 1888. Independence was declared in 1965, and in 1970 the Gambia became a multiparty republic within the British Commonwealth.

An attempted coup in 1981 led to the formation of the federation of Senegambia with Senegal. This ended in 1989 following which the two countries signed a treaty of friendship. A coup in 1994 installed a military council and banned political activity, however, in 1996 a new constitution and presidential elections returned Gambia to civilian rule.

About 80 percent of the population live outside the towns; most are Muslims. A seasonal influx of farm workers from Senegal causes population fluctuations.

At the mouth of the Gambia river, tall fan palms stand atop a cliff that is being rapidly undercut and gullied by the combined action of waves and water runoff. This stretch of coastline was first mapped by Portuguese explorers in the mid 15th century.

ECONOMY

Around 75 percent of the population depend on agriculture. Groundnuts (peanuts) are the major crop far outweighing in importance others which include palm kernels, cotton, millet and rice. Commercial fishing is being developed.

Apart from kaolin and tin, there are no significant mineral resources. Industry is mostly limited to food processing and the chief export market is the European Union. Transportation is mainly by river. Tourism is an important foreign exchange earner.

Health care is limited, but there are programs in place to combat leprosy, malaria and other endemic tropical diseases. Primary education is free but not compulsory, and illiteracy remains extremely high.

NATIONAL DATA - GAMBIA

Land area	10,000 sq km (3,861 sq mi)			
Climate		**Temperatures**		**Annual**
	Altitude m (ft)	January °C(°F)	July °C(°F)	precipitation mm (in)
Banjul	2 (7)	24 (72)	27 (79)	977 (38.4)

Major physical features longest river: Gambia (part) 1,100 km (700 mi)
Population (2006 est.) 1,641,564
Form of government multiparty republic with one legislative house
Armed forces army 800
Capital city Banjul (34,828)
Official language English
Ethnic composition African 99% (Mandinka 42%; Fula 18%; Wolof 16%; Jola 10%; Serahuli 9%; other 4%); non-African 1%
Religious affiliations Muslim 90%; Christian 9%; traditional beliefs 1%
Currency 1 dalasi (GMD) = 100 butut
Gross domestic product (2006 est.) U.S. $3.25 billion
Gross domestic product per capita (2006 est.) U.S. $2,000
Life expectancy at birth male 52.3 yr; female 56.03 yr
Major resources fisheries, titanium (rutile and ilmenite), tin, zircon, silica sand, clay, petroleum, tourism, cassava, cattle, cotton, groundnuts, millet, palm kernels, rice, timber

Guinea

REPUBLIC OF GUINEA

GUINEA LIES ON AFRICA'S WEST COAST, sandwiched between Guinea-Bissau and Sierra Leone, and inland bordering Senegal, Mali, Ivory Coast and Liberia.

GEOGRAPHY

Most of inland Guinea consists of uplands, from the savanna plains around the Niger river in the northeast to the Guinea Highlands in the southeast and the central ranges of the Futa Jalon plateau. The great Niger, Senegal and Gambia rivers all rise here, but flow through many other countries before reaching the Atlantic. From the Futa Jalon the land drops sharply to a relatively short coastline, lined with estuaries and mangrove

A bridge of lianas spans a river near Nzérékoré, in the Guinea Highlands – an ingenious way of exploiting natural resources to solve transportation problems. The country's tropical climate and rich vegetation support a remarkable variety of wildlife.

swamps. The climate is hot, with heavy but variable seasonal rains. The Guinea Highlands are covered with dense tropical rainforest, including hardwood trees such as teak, mahogany and ebony. Much of the rest is savanna grassland. Typical wildlife includes lions, leopards and chimpanzees, with hippopotamuses and crocodiles in the rivers.

SOCIETY

Guinea, once a kingdom of the Islamic Fulani people, was colonized by France in the 19th century. In 1958 French Guinea became independent under President Ahmed Sékou Touré (1922–84), becoming the People's Revolutionary Republic of Guinea. Serious economic difficulties and increased repression led to a military coup in 1984. The new president, Brigadier-General Lansana Conté (b. 1934), instituted economic reform and political liberalization. Despite a number of attempted coups, Conté has remained in power and, in 1993, reintroduced multiparty politics.

He was re-elected in 1998 in a disputed election. There are 24 ethnic groups and most people are Muslim. There are eight official languages apart from French.

ECONOMY

Agriculture employs most of the workforce. Rice, bananas and pineapples, along with oil palms and corn, are grown in the wet coastal southwest, some on reclaimed land. Rice and other grains are grown beside the Niger river. Cattle are raised on the plateau grasslands. The economy, however, rests largely on the mining of immense reserves of bauxite, of which Guinea is a major producer. Otherwise manufacturing is poorly developed, even with foreign aid. Health care is elementary, and although education is compulsory, literacy levels are low.

NATIONAL DATA - GUINEA

Land area	245,857 sq km (94,926 sq mi)			
Climate		**Temperatures**		**Annual**
	Altitude m (ft)	January °C(°F)	July °C(°F)	precipitation mm (in)
Conakry	46 (151)	26 (78)	25 (77)	3,805 (69)

Major physical features highest point: Mount Nimba 1,850 m (6,069 ft); longest river: Niger (part) 4,200 km (2,600 mi)

Population (2006 est.) 9,690,222

Form of government multiparty republic with one legislative assembly

Armed forces army 8,500; navy 400; air force 800

Largest cities Conakry (capital - 2,064,236); Nzerekore (138,862)

Official language French

Ethnic composition Peuhl 40%; Malinke 30%; Soussou 20%; other ethnic groups 10%

Religious affiliations Muslim 85%; Christian 8%; traditional beliefs 7%

Currency 1 Guinean franc (GNF) = 100 centimes

Gross domestic product (2006 est.) U.S. $19.4 billion

Gross domestic product per capita (2006 est.) U.S. $2,000

Life expectancy at birth male 48.34 yr; female 50.7 yr

Major resources bauxite, iron ore, diamonds, gold, uranium, hydropower, fish, salt, tropical fruits, timber

Guinea-Bissau

REPUBLIC OF GUINEA-BISSAU

GUINEA-BISSAU, FORMERLY PORTUGUESE Guinea, lies on the west African coast between Senegal and Guinea. The country also includes the neighboring islands of the Bijagós Archipelago.

GEOGRAPHY

The only area of upland savanna in the country lies on the southeastern borders. Here, deep river valleys cut through the foothills of Guinea's Futa Jalon plateau. The Geba, Corubal and Cacheu rivers meander across the forested central plains to broad estuaries on a coast lined with mangrove swamps. Seasonal rainfall, high everywhere, is torrential here, while temperatures are tropical over most of the country. Riverine wildlife is particularly rich, and includes crocodiles, flamingoes and pelicans.

SOCIETY

From the 15th century the country was exploited by Portugal as a source of slaves. It was colonized as the slave trade declined in the 19th century, and became a Portuguese overseas territory. In 1961 the Partido Africano da Independência da Guiné e Cabo Verde (PAIGC) began a guerrilla campaign, which led to independence in 1974. In 1980, after the failure of a planned merger with the Cape Verde Islands, a military coup installed a Revolutionary Council which eventually gave way to a democratically elected government in 1994. An army uprising in 1998 was followed by a civil war and another coup in 1999. Further elections later that year appointed Koumba Yalla to the presidency.

ECONOMY

Civil war in 1998 damaged Guinea-Bissau's already ailing economy. Most of the population is employed in agriculture which is hampered by periodic droughts and locust swarms. Rice is the major crop and staple food. Cashew and groundnuts form the major cash crop followed by coconuts, palm kernels and timber. Frozen fish and shrimps are also exported.

Food processing and textiles are the only significant industries. Sizeable reserves of bauxite and phosphate remain unexploited as do offshore oil reserves.

NATIONAL DATA - GUINEA-BISSAU

Land area	28,000 sq km (10,811 sq mi)			
Climate		Temperatures		Annual precipitation mm (in)
	Altitude m (ft)	January °C(°F)	July °C(°F)	
Bissau	21 (69)	26 (78)	26 (78)	1,756 (69)
Population	(2006 est.) 1,442,029			
Form of government	multiparty republic with one legislative house			
Armed forces	army 6,800; navy 350; air force 100			
Capital city	Bissau (420,412)			
Official language	Portuguese			
Ethnic composition	African 99% (includes Balanta 30%; Fula 20%; Manjaca 14%; Mandinga 13%; Papel 7%); European and mulatto 1%			
Religious affiliations	traditional beliefs 50%; Muslim 45%, Christian 5%			
Currency	1 Communauté Financière Africaine franc (XOF) = 100 centimes			
Gross domestic product	(2006 est.) U.S. $1.244 billion			
Gross domestic product per capita	(2006 est.) U.S. $900			
Life expectancy at birth	male 45.05 yr; female 48.75 yr			
Major resources	fish, timber, bauxite, clay, granite, limestone, unexploited deposits of petroleum and phosphates, cashew nuts, groundnuts, palm kernels, rice, cereals, coconuts, cotton			

Transportation is mainly by waterway. Health care is limited but around 60 percent of children receive primary education and adult literacy is over 50 percent. There is one television station and three daily newspapers.

A village near Farim in Guinea-Bissau belonging to the Malinke people – one of the forty ethnic groups in the country. Livestock shelters stand alongside the villagers' huts.

Sierra Leone

REPUBLIC OF SIERRA LEONE

SIERRA LEONE LIES ON THE AFRICAN WEST coast bordering Guinea to the north and Liberia to the southeast along the Mano and Moro rivers.

GEOGRAPHY

A narrow coastal belt of swamps and mangroves is interrupted by the thickly wooded mountains of the Sierra Leone peninsula, site of the capital, Freetown. Forest trees include the palm, exploited for its oil and kernels. The southern plains of the interior are wooded, while those farther north are covered with rolling savanna grasslands. These are known as the Bolilands and are subject to seasonal flooding. To the northeast, the Loma Mountains rise to Bintimani Peak, and the Tingi Hills rise to Sankanbiriwa. Many rivers flow through Sierra Leone

from across the Guinean border. Game animals are rare, but monkeys and chimpanzees are common in the forests.

Sierra Leone has a hot, tropical climate with a rainy season from May to October. Rainfall decreases farther north, and temperature ranges are greater in the northeastern highlands. The dry harmattan wind blows from the Sahara in the northeast between November and April.

SOCIETY

Early Portuguese explorers found a cluster of small kingdoms around the Sierra Leone peninsula (the name means "Lion Mountains"). In 1787 homeless freed slaves were settled on the site of Freetown by the British; their descendants are called Creoles. A British protectorate was declared in 1896. In 1961 the country achieved independence under Sir Milton Margai (1895–1964). In 1971 Sierra Leone became a republic, with Siaka Stevens (b. 1905) as its first president. In 1978 a new constitution made the All People's Congress the sole legal party. In 1991 a referendum endorsed multiparty politics but coups in 1992, 1996 and 1997 led to intervention by a Nigerian-led peacekeeping force in 1998. The civil war, which had been raging since 1991, continued with the loss of many lives and the displacement of several million people. Despite the backing of Nigerian troops the government lost control of the diamond fields

in 1999. A peace agreement was signed with the rebels in July 1999 and UN peacekeeping troops were deployed but the situation was far from resolved.

The largest of the 18 ethnic groups are the Mende and the Temne. The official language is English, but most people communicate in Krio, a Creole pidgin language. About half the population hold traditional African beliefs, and about two-fifths are Muslims.

ECONOMY

The majority of the people work on the land. Palm kernels, cocoa and coffee are the main cash crops, and rice is the main staple. Mining is Sierra Leone's key industry; iron ore, diamonds, chromite, rutile, bauxite and gold are exported, mainly to Europe and the United States. Manufacturing centers on Freetown, and is mostly of consumer goods.

The single railroad was finally dismantled in 1975, but the road system is being expanded. The many waterways transport cargo and the main port is at Freetown. The international airport is at Lungi. Radio, television and the press are government-controlled.

Government health-care programs have reduced infant mortality and raised life expectancy, but sleeping sickness and leprosy remain a problem. Primary education is free, but low attendance means that literacy levels are slow to rise.

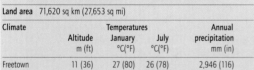

NATIONAL DATA – SIERRA LEONE				
Land area 71,620 sq km (27,653 sq mi)				

Climate	Altitude m (ft)	Temperatures January °C(°F)	July °C(°F)	Annual precipitation mm (in)
Freetown	11 (36)	27 (80)	26 (78)	2,946 (116)

Major physical features highest point: Bintimani Peak 1,948 m (6,391 ft)

Population (2006 est.) 6,005,250

Form of government multiparty republic with one legislative house

Armed forces joint forces 12–13,000

Capital city Freetown (835,108)

Official language English

Ethnic composition Native African tribes 90% (Temne 30%, Mende 30%, other 30%), Creole (Krio) 10%; small numbers of Europeans, Lebanese, Pakistanis, and Indians

Religious affiliations Muslim 60%; traditional beliefs 30%; Christian 10%

Currency 1 leone (SLL) = 100 cents

Gross domestic product (2006 est.) U.S. $5.38 billion

Gross domestic product per capita (2006 est.) U.S. $900

Life expectancy at birth male 38.05 yr; female 42.46 yr

Major resources diamonds, titanium ore, bauxite, iron ore, gold, chromite, cassava, cocoa, coffee, fish, groundnuts, palm nuts, rice, timber

A masked person from the Kanadevil sect (*above*) wears an elaborate headdress. The paleness of the mask suggests it may represent a "white man devil".

Palm-fringed beaches (*right*) are not just a draw for tourists. The palm fruits are highly valued for their kernels and oil, and the trunks make good dug-out canoes.

Liberia

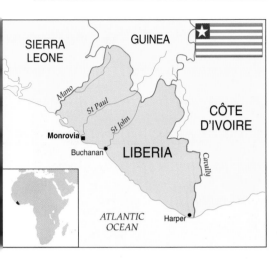

Goods are loaded on board a freighter in Monrovia, Liberia's principal port. Before the civil war in 1990, Liberia had trading links with many countries, but its economy is likely to need time to recover.

LIBERIA, AFRICA'S OLDEST INDEPENDENT republic, borders Sierra Leone to the northwest (along the Mano and Moro rivers), Guinea to the north and Côte d'Ivoire to the east (along the Cavally river). The capital, Monrovia, lies on the coast.

GEOGRAPHY

Swamps and mangroves line Liberia's sandy coast. Inland, rolling hills covered in tropical rainforest rise to a high plateau. The mountains in the north reach the southwestern flanks of Mount Nimba, where the borders with Guinea and Côte d'Ivoire meet. Semideciduous rainforest and elephant grass are the characteristic vegetation in the north. Chimpanzees, pygmy hippopotamuses and crocodiles are found, but game animals are becoming increasingly rare.

Liberia has a hot, humid climate. The rainy season from May to October is interrupted by a short dry season from mid-July to August. Rainfall diminishes inland, and the greatest amount falls on the Cape Mount promontory near the Sierra Leone border. In December the hot, dry harmattan wind blows down from the Sahara.

SOCIETY

In 1822 a group of liberated American slaves landed in Liberia to found a colony that had been negotiated with the indigenous peoples by the American Colonization Society. In 1847 Liberia declared itself an independent republic. The 1920s saw heavy investment by the Firestone Tire and Rubber Company of the United States. Liberian rubber was later vital to the Allied victory in World War II.

The economy suffered when rubber prices fell in the 1960s and 1970s. In 1980 President William Tolbert (1913–80) was shot during a coup led by Master Sergeant Samuel Kanyon Doe (1951–90). In 1990 a civil war broke out, several million people were made homeless and Doe was assassinated by rebel forces. Despite the intervention of a peacekeeping force from the Economic Community of West African States (ECOWAS), the civil war continued until 1996. In 1997 Charles Taylor, leader of the National Patriotic Party, was elected president but fighting between government forces and rebels continued.

Liberia's peoples belong to three different language groups – Mande, Kwa and West Atlantic – but the official language remains English. Many of the population retain traditional African beliefs.

ECONOMY

Liberia's economy has been devastated by civil war, many businessmen fled the country and there are massive international debts. Registration of foreign-owned shipping provides substantial foreign-exchange income. Rubber is the main cash crop. Rice is the main staple, but some food must still be imported. Livestock includes cattle, goats and pigs. Timber is a significant export. The catch from coastal fisheries is supplemented from inland fish farms. Iron ore is Liberia's principal export. Other mineral resources include diamonds and gold. Energy is supplied by hydroelectric and oil-fired stations. Most manufacturing industries serve the local market.

During the civil war, many roads and railroads were destroyed. The two major airports are close to Monrovia. Epidemic diseases are rife and health care and education are poor. Literacy levels are low and infant mortality high.

NATIONAL DATA – LIBERIA

Land area	96,320 sq km (37,189 sq mi)			

Climate		Temperatures		Annual
	Altitude m (ft)	January °C(°F)	July °C(°F)	precipitation mm (in)
Monrovia	23 (75)	27 (81)	25 (79)	5,140 (202.3)

Major physical features highest point: Mount Wutivi 1,380 m (4,528 ft); longest river: Cavally (part) 515 km (320 mi)

Population (2006 est.) 3,042,004

Form of government multiparty republic with two legislative houses

Armed forces Joint forces 15,000

Capital city Monrovia (967,841)

Official language English

Ethnic composition Indigenous African tribes 95% (including Kpelle, Bassa, Gio, Kru, Grebo, Mano, Krahn, Gola, Gbandi, Loma, Kissi, Vai, Dei, Bella, Mandingo, and Mende); Americo-Liberians (descendants of repatriated slaves) 5%

Religious affiliations traditional beliefs 40%; Christian 40%; Muslim 20%

Currency 1 Liberian dollar (LRD) = 100 cents

Gross domestic product (2006 est.) U.S. $2.911 billion

Gross domestic product per capita (2006 est.) U.S. $1,000

Life expectancy at birth male 37.99 yr; female 41.35 yr

Major resources iron ore, timber, diamonds, gold, hydropower, fish, rubber, cocoa, coffee, bananas, cassava, maize/corn, palm kernels, rice

Côte d'Ivoire

REPUBLIC OF CÔTE D'IVOIRE

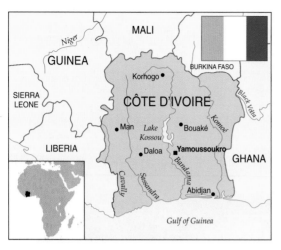

CÔTE D'IVOIRE (IVORY COAST) FACES THE Gulf of Guinea on the west coast of Africa, with Mali and Burkina Faso to the north, and Ghana to the east. It also borders Liberia and Guinea to the west; the three frontiers meet at Mount Nimba.

GEOGRAPHY

The land rises gently northward from the lagoons and sandbars of the coast. Tropical rainforest, much of it cleared for plantations, occupies the southern third of the country, yielding valuable timber. The thinly populated northern plateau is covered in savanna grasslands, while the highlands in the northwest rise to Mount Nimba. The country is dissected by four major north–south rivers: the Cavally, the Sassandra, the Bandama and the Komoé. Typical wildlife includes elephants and pygmy hippopotamuses. There are three major national parks.

The climate is hot and tropical. Along the coast, and to a lesser extent in the forest belt, there are two distinct rainy seasons: May–July and October–November. On the plateau the temperatures are less hot, and rainfall is lower, with a single rainy season. The dry, northeasterly harmattan wind blows here between the months of December and February. The western mountains have no dry season.

SOCIETY

The Portuguese first explored the country in the 15th century, and in the 17th century the French established trading posts here. In the 19th century they concluded protectorate agreements, and in 1893 Louis-Gustave Binger (1856–1936) became the first governor of a new Ivory Coast colony. During the first half of World War II the colony remained under the control of Vichy France (the part of France not occupied by German troops) until 1942. In 1946 it became a territory in the French Union. Upper Volta (Burkina Faso) was separated in 1947. Autonomy came in 1958, and independence in 1960, when Félix Houphouët-Boigny (1905–93) became the first president.

Multiparty politics were reestablished in 1990 and Houphouët-Boigny remained president until his death in 1993. Henri Konan Bedie took over the presidency until deposed in a military coup in 1999.

Fluctuating market prices for coffee and cocoa caused financial problems in the 1990s. There was also an influx of refugees from Liberia. French is the official language but there are 60 ethnic groups each with their own dialect. Most people hold traditional African beliefs often in combination with Muslim and Christian beliefs.

ECONOMY

The majority of the labor force work on the land. Cocoa and coffee are the main cash crops, while yams, manioc, plantains, corn and rice are the chief staples. Timber and fish are important resources; both are exported. Mineral resources include iron ore, copper, nickel, uranium, bauxite, cobalt, diamonds and offshore oil and gas. Power from hydroelectric and oil-fired stations supplies light industries, including textiles, food processing and

Traditional religion In this village near Korhogo in northern Côte d'Ivoire, fetish huts are set aside by the local people for use by ancestral spirits. The villagers leave food offerings inside the huts, and supplicate the spirit for good health and harvest.

vehicle assembly. Tourism is developing.

Communications center on the railroad to Burkina Faso. There is a good road network around the former capital of Abidjan, where the Vridi Canal gives access to the sea. The international airport at Port Bouet also connects with small airfields inland. Health care is poor: malaria, dysentery, yaws and tuberculosis are fairly common. Primary education is compulsory, but not all children are enrolled.

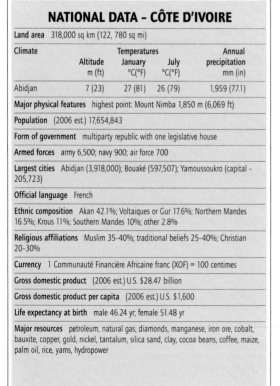

NATIONAL DATA – CÔTE D'IVOIRE

Land area	318,000 sq km (122, 780 sq mi)			
Climate		**Temperatures**		**Annual**
	Altitude m (ft)	January °C(°F)	July °C(°F)	precipitation mm (in)
Abidjan	7 (23)	27 (81)	26 (79)	1,959 (77.1)

Major physical features highest point: Mount Nimba 1,850 m (6,069 ft)

Population (2006 est.) 17,654,843

Form of government multiparty republic with one legislative house

Armed forces army 6,500; navy 900; air force 700

Largest cities Abidjan (3,918,000); Bouaké (597,507); Yamoussoukro (capital – 205,723)

Official language French

Ethnic composition Akan 42.1%; Voltaiques or Gur 17.6%; Northern Mandes 16.5%; Krous 11%; Southern Mandes 10%; other 2.8%

Religious affiliations Muslim 35–40%; traditional beliefs 25–40%; Christian 20–30%

Currency 1 Communauté Financière Africaine franc (XOF) = 100 centimes

Gross domestic product (2006 est.) U.S. $28.47 billion

Gross domestic product per capita (2006 est.) U.S. $1,600

Life expectancy at birth male 46.24 yr; female 51.48 yr

Major resources petroleum, natural gas, diamonds, manganese, iron ore, cobalt, bauxite, copper, gold, nickel, tantalum, silica sand, clay, cocoa beans, coffee, maize, palm oil, rice, yams, hydropower

Burkina Faso

REPUBLIC OF BURKINA FASO

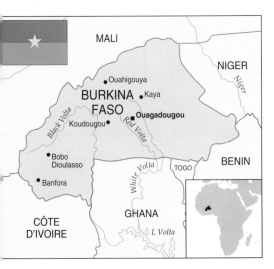

BURKINA FASO, UNTIL 1984 UPPER VOLTA, LIES ON the southern fringe of the Sahara. It is surrounded by Côte d'Ivoire, Ghana, Togo, Benin, Niger and Mali.

GEOGRAPHY

Most of Burkina Faso is a wide plateau, interrupted in the west by an escarpment called the Falaise de Banfora. The three main rivers, the Red, White and Black Volta, have cut deep valleys as they flow southward into Ghana. The climate is hot, with unreliable seasonal rains. The north of the country is typical of the arid Sahel, with scrubland and semidesert. The south is more tropical, ranging from savanna grassland to forest. The country has suffered severely from recurrent drought. Wildlife is abundant, but areas near rivers are plagued by tsetse and simulium flies, carriers of sleeping sickness and river blindness.

SOCIETY

Burkina Faso, once the kingdom of the Mossi people, was colonized by France in the late 19th century. Independence was gained in 1960, but the country was plagued by instability, characterized by frequent and bloody coups in the 1970s and 1980s. Captain Blaise Compaoré has remained president since the coup of 1987. There is an 111-member elected assembly and a 178-member chamber of appointed representatives. There are several ethnic groups, but the official language is French. Most people either follow Islam or hold traditional African beliefs.

ECONOMY

Most people work on the land. Cash crops include groundnuts (peanuts), cotton (the main export) and sesame, while the main food crops are sorghum, millet and sugar cane. Cattle and goats are raised mostly in the north and east. Some gold quartz is mined, along with manganese and limestone. Industry is undeveloped, and is based mostly on agriculture. Power is generated from imported fuels. The transportation networks are good, but health facilities are very poor, with much malnutrition and disease. Education, though compulsory, is available only to a minority, and both press and television are government-controlled.

NATIONAL DATA – BURKINA FASO				
Land area 273,800 sq km (105,715 sq mi)				
Climate		Temperatures		Annual
	Altitude m (ft)	January °C(°F)	July °C(°F)	precipitation mm (in)
Abidjan	304 (997)	25 (77)	28 (79)	895 (35.2)
Major physical features highest point: Téna Kouron 747 m (2,451 ft)				
Population (2006 est.) 13,902,972				
Form of government multiparty republic with two legislative houses				
Armed forces army 6,400; air force 200				
Largest cities Ouagadougou (capital – 1,152,402); Koudougou (91,468); Ouahigouya (63,598)				
Official language French				
Ethnic composition Mossi 47.9%; Mande 8.8%; Fulani 8.3%; Lobi 6.9%; Bobo 6.8%; Senufo 5.3%; Grosi 5.1%; Gurma 4.8%; Tuareg 3.3%; others 2.4%				
Religious affiliations Muslim 50%; traditional beliefs 40%; Christian (mainly Roman Catholic) 10%				
Currency 1 Communauté Financière Africaine franc (XOF) = 100 centimes				
Gross domestic product (2006 est.) U.S. $17.87 billion				
Gross domestic product per capita (2006 est.) U.S. $1,300				
Life expectancy at birth male 47.33 yr; female 50.42 yr				
Major resources manganese, limestone, marble; small deposits of gold, phosphates, pumice, salt, antimony, bauxite, copper, cotton, groundnuts, lead, livestock, maize/corn, millet, nickel, rice, sesame, shea, sorghum, sugarcane				

Colorfully adorned (*above*) this Pau woman lives in Burkina Faso, one of the poorest countries in the world. Her own health seems good, but the prospects for her children are poor in a country plagued by malnutrition and disease.

Rock formations at Fabedougou (*left*) near Banfora in the southwest of Burkina Faso. These spectacular formations are the result of rapid erosion. Floodwaters released during flash storms have eaten away at the soft sedimentary rocks, creating this beautifully sculpted landscape.

Ghana

REPUBLIC OF GHANA

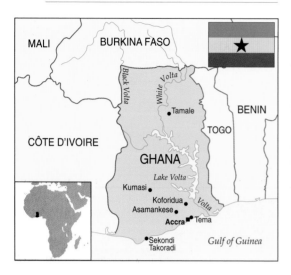

Ghana, formerly the Gold Coast, faces south across the Gulf of Guinea. More than half the country is occupied by the Volta basin.

GEOGRAPHY

In the north the Red and White Volta rivers meet just south of the Burkina Faso border. In the northwest the Black Volta marks the western border with Burkina Faso and part of the border with Côte d'Ivoire, before turning westward. Much of the basin has been flooded, even above the confluence of the Black and White Volta, by the construction of the Akosombo Dam, creating Lake Volta. East of the basin, the Akwapim-Togo Mountains run north–south along the border with Togo, then turn southwest toward the coastal capital, Accra. South of the Akosombo Dam, the Volta crosses a fertile coastal plain before flowing into a series of marshes and lagoons.

Most of Ghana is low-lying, so that the hot, tropical climate remains unmodified by altitude. Temperatures are highest from February to April, and are lowest near the coast, where sea mists have a cooling effect. The south has two rainy seasons (May–June and September–October), whereas in the north there is only one, peaking in August and September.

The coastal vegetation of mangrove swamps rapidly gives way to savanna grassland, and then to tropical forest (much of it cleared for agriculture), covering the southern third of the country. The savanna farther north has fewer trees and shorter grasses. Animal life is still plentiful, and includes lions, leopards, hyenas, antelopes and elephants. The rivers are the home of crocodiles, manatees and hippopotamuses. The largest of the many game reserves is the Mole, near Damongo in the northwest.

SOCIETY

Ghana's interior was settled by many different ethnic groups, but by the 17th century the Akan-speaking Ashanti were dominant. The Portuguese had discovered the Gulf of Guinea in 1471. They built forts along the coast, which was named the Gold Coast on account of the gold to be found there. The 17th century brought other colonists, attracted there by the Ashanti-controlled slave trade, but in the 19th century the British took control. By 1901 the Gold Coast was a British colony, the Ashanti had been subdued and the northern territories had become a British protectorate. After World War I part of neighboring German Togoland also came under British control.

In 1957 the newly renamed state of Ghana won independence – the first state in sub-Saharan Africa to do so – and in 1960 it was declared a republic, with Prime Minister Kwame Nkrumah (1909–72) as its life president. Nkrumah's regime was overthrown by a military coup in 1966. The civilian government elected in 1969 was ousted by a second coup in 1972. Another brief period of civilian rule (1979–81) ended with a coup led by Flight Lieutenant Jerry Rawlings (b. 1947). Rawlings' Provisional National

from Lake Volta) are encouraged, and provide an important supplement to the Ghanaian diet.

Gold is still an important export, as are diamonds, manganese and bauxite. Oil and natural gas provide another significant export. Hydroelectric power from Akosombo supplies domestic needs; the surplus is exported. Manufacturing is limited mainly to supplying local demand, but tourism is increasing.

Road and rail transportation are best in the south, especially in cocoa-growing areas near the coast. About one-fourth of roads are paved, and aid-financed improvements are in hand. The railroad system is largely used for freight. The main port, Tema, and the international airport, Kotoka, are both near Accra. The state airline is Ghana Airways.

Health care has been hindered by rapid population growth, bad sanitation and poor nutrition. Major diseases such as malaria, pneumonia and gastroenteritis have been partly controlled, but infant mortality is high and life expectancy is relatively short. River blindness (onchocerciasis), a disease causing blindness, has had a great impact on local people. Housing is a growing problem, especially in towns. Basic education is free and can be followed by secondary vocational training or preparation for one of the three universities. However, school attendance is still relatively low, which affects literacy levels.

Defense Council created a stable economy although the main exports, gold and cocoa, are highly susceptible to market fluctuations. Democracy returned in 1992. In 2000, Rawlings stepped down and John Kufuor became president. Simultaneous parliamentary elections resulted in victory for the National People's Party ousting the ruling National Democratic Congress.

Ghana has 75 ethnic groups; however, only 10 are of significant size, the largest being the Akan and the Mossi. English is the official language. The majority of people are either Muslims or Christians; a substantial amount though hold traditional African beliefs.

The harbor town of Dixcove on Ghana's Gold Coast. In the 15th century the Portuguese discovered alluvial gold in this area, washed down by the Volta river. However, for hundreds of years the most lucrative trade was in slaves.

ECONOMY

Most people are employed in agriculture, which is the mainstay of the economy. Cocoa and timber are the chief export crops. In the north yams and cereals are grown and cattle are kept, while the forests produce shea and kola nuts. Despite government efforts to increase food production, Ghana is still importing food. Marine and freshwater fishing (notably

---

NATIONAL DATA - GHANA

Land area	230,940 sq km (89,166 sq mi)			

Climate		Temperatures		Annual
	Altitude m (ft)	January °C(°F)	July °C(°F)	precipitation mm (in)
Accra	65 (213)	27 (81)	25 (77)	725 (28.5)

Major physical features highest point: Mount Afadjoto 885 m (2,903 ft); largest lake: Lake Volta 8,482 sq km (3,275 sq mi)

Population (2006 est.) 22,409,572

Form of government republic with multiparty parliament

Armed forces army 5,000; navy 1,000; air force 1,000

Largest cities Accra (capital - 2,096,653); Kumasi (1,604,909); Tamale (390,730); Takoradi (260,651); Ashiaman (228,509); Tema (161,106)

Official language English

Ethnic composition Black African 98.5% (major tribes - Akan 44%; Moshi-Dagomba 16%; Ewe 13%; Ga 8%; Gurma 3%; Yoruba 1%); other 1.5%

Religious affiliations Christian 63%; Muslim 16%; traditional beliefs 21%

Currency 1 cedi (GHC) = 100 pesewas

Gross domestic product (2006 est.) U.S. $59.15 billion

Gross domestic product per capita (2006 est.) U.S. $2,600

Life expectancy at birth male 58.07 yr; female 59.69 yr

Major resources gold, timber, industrial diamonds, bauxite, manganese, fish, rubber, hydropower, petroleum, natural gas, silver, salt, limestone, bananas, cassava, cocoa, maize/corn, sorghum, taro, yams, tourism

Togo

REPUBLIC OF TOGO

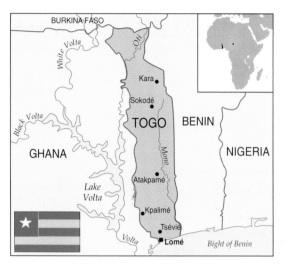

T HE WEST AFRICAN REPUBLIC OF TOGO IS A long, narrow country sandwiched between Ghana to the west and Benin to the east, with Burkina Faso to the north.

GEOGRAPHY

From coastal lagoons the Ouatchi Plateau extends toward a higher tableland. The Togo Mountains, which are covered with tropical forest in the southwest, extend northeast across the Benin border. Savanna dominates the lower-lying areas, where wildlife includes antelopes and elephants. The humid, tropical climate has two rainy seasons (April–June and September–October) in the south and one (June–September) in the north, where the dry and dusty harmattan wind blows for the rest of the year.

NATIONAL DATA - TOGO

Land area	54,385 sq km (20,998 sq mi)			
Climate		Temperatures		Annual
	Altitude m (ft)	January °C(°F)	July °C(°F)	precipitation mm (in)
Lomé	20 (66)	27 (81)	25 (77)	1,577 (62)

Major physical features highest point: Pic Baumann 986 m (3,235 ft); longest river: Mono 400 km (250 mi)

Population (2006 est.) 5,548,702

Form of government Republic under transition to multiparty democratic rule

Armed forces army 8,100; navy 200; air force 250

Capital city Lomé (756,122)

Official language French

Ethnic composition Native African (37 tribes; largest are Ewe, Mina, and Kabre) 99%; European and Syrian-Lebanese less than 1%

Religious affiliations traditional beliefs 51%; Christian 29%; Muslim 20%

Currency 1 Communauté Financière Africaine franc (XOF) = 100 centimes

Gross domestic product (2006 est.) U.S. $9.248 billion

Gross domestic product per capita (2006 est.) U.S. $1,700

Life expectancy at birth male 55.41 yr; female 59.49 yr

Major resources phosphates, limestone, marble, livestock, food processing, cement, textiles, beverages, handicrafts, coffee, copra, cocoa, cotton

SOCIETY

Togo became a German protectorate in 1885. After World War I it was administered by France and Britain. In 1956 the British section was united with the Gold Coast (Ghana), while French Togoland became autonomous, achieving independence in 1960. The first president, Sylvanus Olympio (1902–63), was assassinated in 1963, and the second was overthrown in 1967 by Colonel Gnassingbe Eyadéma (b. 1935). Multiparty rule was instituted in 1992 but the military still dominates the government. Although there are some 30 ethnic groups in the country, French is the official language. There are a large number of Christians, and over half the population retain traditional African beliefs.

ECONOMY

Phosphates account for half Togo's exports, but agriculture is important too. Cash crops include cocoa, coffee and cotton; staples include corn, cassava, rice and yams. Cattle, sheep and pigs are raised in the north. There is an effective road and rail network for carrying food and other resources between the coast and the interior. There are sea ports at Lomé and Kpémé, and an international airport at Lomé. Health conditions and housing are poor. Diseases such as malaria, yellow fever and tuberculosis are still common. Primary education is free, but secondary-school attendance is low, and this is reflected in literacy levels.

Men dancing in traditional costume (*above*) at a village fete in Togo, where the majority of the people hold traditional religious beliefs. The performance of such dances serves to preserve and strengthen the cultural identity of the ethnic group.

Benin

REPUBLIC OF BENIN

BENIN (FORMERLY DAHOMEY) BORDERS THE Gulf of Guinea, and is bounded by Togo to the west, Burkina Faso and Niger to the north, and Nigeria to the east. Though fully independent since 1960, Benin has only recently seen a change toward full multiparty democracy.

GEOGRAPHY

Inland from the coastline are coastal lagoons, behind which the land rises to a fertile, heavily cultivated plateau, and then to the Atakora Mountains – a continuation of the Togo Mountains. The northeastern plains slope down to the valley of the Niger river. Northern areas are covered in mixed forest and savanna, which support a diversity of wildlife including leopards, hyenas, antelopes, elephants and monkeys. The climate is humid and tropical. In the south there are two rainy seasons (March–July and September–November), but in the north there is only one (May–September). The dry harmattan wind blows from the northeast between December and March.

SOCIETY

The French colony of Dahomey won full independence in 1960, but ethnic and religious divisions between north and south contributed to political instability.

NATIONAL DATA – BENIN				
Land area 110,620 sq km (42,710 sq mi)				
Climate		Temperatures		Annual
	Altitude m (ft)	January °C(°F)	July °C(°F)	precipitation mm (in)
Cotonou	5 (16)	27 (81)	26 (79)	1,308 (51.4)
Major physical features highest point: Atakora Massif 641 m (2,103 ft)				
Population (2006 est.) 7,862,944				
Form of government multiparty republic with one legislative house				
Armed forces army 4,300; navy 100; air force 150				
Largest cities Cotonou (711,600); Abomey-Calavi (443,670); Porto Novo (capital – 242,187); Djougou (217,360)				
Official language French				
Ethnic composition African 99% (42 ethnic groups, largest are Fon, Adja, Yoruba, Bariba); Europeans 5,500				
Religious affiliations traditional beliefs 50%; Christian 30%; Muslim 20%				
Currency 1 Communauté Financière Africaine franc (XOF) = 100 centimes				
Gross domestic product (2006 est.) U.S. $8.931 billion				
Gross domestic product per capita (2006 est.) U.S. $1,100				
Life expectancy at birth male 51.9 yr; female 54.22 yr				
Major resources small offshore oil deposits, limestone (cement), marble, timber, beans, cassava, coffee, cotton, groundnuts, maize/corn, palm products, sorghum, yams				

In 1972, after the last of a series of coups, Lieutenant-Colonel Mathieu Kerekou (b. 1933) was named as president. He changed the country's name to Benin in 1975. A new constitution in 1990 paved the way for multiparty elections in March 1991 and again in 1995. French is the official language, but most people speak African languages and retain traditional beliefs.

ECONOMY

About half the population are dependent on agriculture; many are subsistence farmers. Cash crops include cotton, cocoa, coffee and palm kernels, while corn, cassava, yams and beans are the main subsistence crops. Benin has reserves of limestone, gravel, gold and offshore oil, and exports energy from its geothermal stations.

There are two principal roads into Nigeria and Niger, and three railroad lines. The main port and the international airport are at the *de facto* capital, Cotonou, on the coast. Health conditions and housing are poor; major diseases include leprosy, meningitis and malaria. Primary education is compulsory but many children have difficulty in attending and literacy levels are low.

In the marshy coastlands of Benin the best place to build a house is on stilts well above the high-water mark. Fishing nets have been hung out to dry in this quiet southern village.

Nigeria

FEDERAL REPUBLIC OF NIGERIA

Nigeria, bordered by Cameroon, Chad, Niger and Benin, is the largest country on the coast of west Africa, and the most populous in the whole continent.

GEOGRAPHY

At the heart of Nigeria lies the high Jos Plateau, from which the land falls away northward toward the Niger border. There are lower-lying areas to the northwest and to the northeast bordering Lake Chad, where the land becomes swampy in the rainy season. The plateaus become lower farther south, where they are split by shallow river valleys. The broadest of these are formed by the great Niger river and its main tributary, the Benue, which

flow from west and east respectively to meet in the center. The Niger continues southward across the broad coastal plains bordering the Gulf of Guinea, where it spreads out into a vast delta. The plateaus to the west continue into Benin. The mountain chains along the eastern border are a continuation of the volcanic Adamawa Highlands. They include the lofty Shebshi Mountains, and the Mandara Mountains farther north toward Lake Chad.

The climate is tropical with seasonal rains, but varies enormously from south to north. The coast is constantly humid, with an extremely high rainfall. Farther north the wet season decreases in length and intensity, and the far north has the relatively arid climate of the Sahel, with an eight-month dry season and more marked variations in temperature.

In the arid margins of Lake Chad, only hardy scrub vegetation such as acacia and doum palms flourish. Farther south there is savanna grassland, with wildlife such as lions, giraffes and antelopes. Farther south still this becomes savanna woodland, with its distinctive baobab trees. Toward the coast there is a broad rainforest belt, supporting wildlife such as elephants, chimpanzees and the Red river hog. Although a valuable source of timber such as mahogany and iroko, the rainforest has been cleared in many places for cultivation, often of the naturally occurring oil palm. Beyond the cultivated areas, especially along the channels of the Niger delta, lie broad areas of mangrove swamp, giving way to the sandy beaches and lagoons of the coastline.

SOCIETY

Nigeria is home to more than 200 peoples, producing a rich and varied culture. The Hausa and Fulani are dominant in the north, with the Yoruba, Edo and Ibo peoples forming the largest groups in the south. Just under half the population are Muslim. The rest are mostly Christian or follow traditional African beliefs.

There were many ancient city-states in Nigeria, including the Yoruba city of Ife, the mainly Hausa Kanem-Bornu kingdom, and the powerful Benin kingdom in the southwest, famous for its bronze sculptures. Europeans and Americans bought slaves from these kingdoms, but made no attempt to conquer them. In the 19th century the largely Islamic Fulani launched a jihad (holy war), and conquered most of the country.

It was the suppression of the slave trade

that led Britain to take control of the country, formally annexing Lagos in the southwest in 1861. Further areas were later added, and Nigeria became a crown colony and protectorate in 1914. In 1960 it was granted independence as a federal state made up of four regions. In 1961 the northern half of British Cameroon voted to join the federation.

In 1966 Ibo army officers seized the government in a coup. This led to the massacre of thousands of Ibo in the Northern Region. Following a Hausa countercoup, Lieutenant-Colonel Yakubu Gowon (b. 1934) was made president and worked for reconciliation. Nevertheless, in 1967 the Ibo-dominated Eastern Region seceded under Lieutenant-Colonel C. Odumegwu Ojukwu (b. 1933) as the state

---

NATIONAL DATA – NIGERIA

Land area	910,768 sq km (351,649 sq mi)			

Climate		Temperatures		Annual
	Altitude m (ft)	January °C(°F)	July °C(°F)	precipitation mm (in)
Lagos	3 (10)	27 (81)	2 (77)	1,538 (60.5)

Major physical features highest point: Dimiong 2,042 m (6,700 ft); longest river: Niger (part) 4,200 km (2,600 mi)

Population (2006 est.) 131,859,731

Form of government federal republic with military government

Armed forces army 62,000; navy 7,000; air force 9,500

Largest cities Lagos (9,229,944); Kano (3,848,885); Ibadan (3,847,472); Kaduna (1,652,844); Abuja (capital since 1991 - 1,405,201)

Official language English

Ethnic composition Hausa 21.3%; Yoruba 21.3%; Ibo 18.0%; Fulani 11.2%; Ibibio 5.6%; Kanuri 4.2%; Edo 3.4%; Tiv 2.2%; Ijaw 1.8%; others 11%

Religious affiliations Muslim 50%; Christian 40%; traditional beliefs or none 10%

Currency 1 naira (NGN) = 100 kobo

Gross domestic product (2006 est.) U.S. $188.5 billion

Gross domestic product per capita (2006 est.) U.S. $1,400

Life expectancy at birth male 46.52 yr; female 47.66 yr

Major resources natural gas, petroleum, tin, iron ore, lead, zinc, coal, fisheries, palm kernels, cassava, cocoa, cotton, groundnuts, livestock, maize/corn, millet, rice, rubber, sorghum, timber, yams

A Tuareg herdsman (*above*) The Tuareg are traditionally nomadic, ranging as far south as northern Nigeria. They pay no heed to national boundaries, and have a distinct culture of their own.

The long journey to the coast (*left*) In the absence of good roads, the long rivers of Africa, which often flow through several countries, provide the principal means of transporting goods from the heart of the continent to the ports. Here logs have been roped together to be floated down the river Niger to a sawmill on the coast, where they will be prepared for export.

of Biafra. After a bloody war, in which thousands of Ibo died, Biafra collapsed in 1970. Gowon refrained from making reprisals, but was deposed in 1975. Further military coups thwarted subsequent attempts to restore a stable, civilian government.

Multiparty and presidential elections in 1992 and 1993 were followed by the reimposition of military rule. General Sani Abacha dissolved all political parties and established a provisional ruling council headed by himself. Elected president, Chief Moshood Abiola was imprisoned and died a day before his expected release in 1998. Abacha died in June 1998 and was replaced by General Abdulsam Abubakar who released political detainees and paved the way for a return to civilian rule. In 1999 presidential elections resulted in the coming to power of Olusegan Obasanjo. In 1999, the stability of the country looked threatened by the widespread adoption of Islamic sharia law which was seen as underlining the ethnic and religious split between the northern Muslim states and their southern Christian counterparts. Amidst ethnic and religious clashes and violence there were demands for the secession of the south.

ECONOMY

Agriculture still employs the majority of the labor force, but has declined in relation to other sectors. Activities vary between different areas and among the various ethnic groups. The Hausa, for example, tend toward crop growing, whereas the Fulani prefer to raise cattle. Cocoa and rubber, the main export crops, are grown mostly in the forested south, and palm kernels mainly in the east. Groundnuts (peanuts) are another important cash crop. The main staples are cassava, yams and rice in the south, and millet and sorghum in the north, which is also the main livestock area. Production has not kept pace with population growth, and much food must be imported. Fisheries have some potential, but forestry has suffered from overclearance.

Nigeria is rich in minerals including coal, iron ore, lead and zinc. Oil provides 20 percent of GDP and 95 percent of foreign exchange earnings. Manufacturing contributes around 9 percent of GDP and includes paper products, cement, cigarettes and a variety of consumer goods. Nigeria is one of the most indebted of the African countries, in 1997 it owed US$34 billion to the international banks.

The railroads are not as vital as the extensive road network but there are inland waterways and the principal ports are Lagos, Port Harcourt, Warri and Calabar. Lagos, Port Harcourt and Kano have international airports.

Primary education is available to all but female participation is low. There is a growing higher education sector featuring 13 universities and other institutions.

Cameroon

REPUBLIC OF CAMEROON

CAMEROON LIES ON THE GULF OF GUINEA between west and central Africa. It shares borders with Equatorial Guinea, Gabon, Congo (R.O.), the Central African Republic, Chad and Nigeria.

GEOGRAPHY

Inland Cameroon consists mostly of low plateau country, rising northward to the Adamawa Highlands before sloping gradually down toward Lake Chad in the far north. The west is more mountainous volcanic country, dominated near the coast by the active volcano Mount Cameroon, the highest point in West Africa. The southern plateaus are covered with rainforest, and they fall away to a wide band of plains along the coast.

The climate is tropical, with heavy seasonal rainfall and high temperatures that are moderated by altitude. In the southern rainforests many kinds of tree flourish, including ebony, mahogany and sapele. They also support an immense variety of other plant and animal life, including orchids, ferns, chimpanzees, many species of monkey, and a rich bird life. The northern savanna country supports elephants, lions and leopards. A national park has been established here.

SOCIETY

Cameroon was probably originally inhabited by the Pygmies. Later various groups of Bantu and Sudanic peoples moved in. A source of slaves for Dutch and Portuguese traders, the country was colonized by Germany in the 19th century in opposition to Britain. In World War I it was invaded by France and Britain, and was later divided between them under a League of Nations mandate.

In 1960 French Cameroon became independent. In 1961 the southern half of British Cameroon joined it in a federation, while the north joined Nigeria. In 1972 a referendum replaced the federation with a unified government under Ahmadou Ahidjo (b. 1924), the president since independence. In 1982 he resigned and was replaced by Paul Biya (b. 1933), who defeated a military coup in 1984. In 1990 a multiparty constitution was adopted; Biya was re-elected in 1997.

ECONOMY

The economy has benefited from liberal management and the exploitation of oil reserves. It has continued to grow since independence, although it still depends on foreign aid. Agriculture has been less affected by drought than in countries to the north. A wide range of crops such as coffee, bananas and rubber are grown for export, alongside staple crops such as corn, millet, sorghum, groundnuts (peanuts) and cassava. Timber production is now growing.

Industry is also developing fast. Factories range from a large aluminum plant at Edéa, driven by power from the Sanaga river hydroelectric scheme, to a paper mill and tire factory based on local supplies of timber and rubber. Road transportation is often hampered by the rainy season, but rail and air links are well developed, especially to Douala, the main port. Social security is limited, and health care is poor except in the cities. Education is free, and is fast expanding. The media are under government control.

A cattle-breeders' village in Cameroon. The wild and barren landscape around this village seems to offer little in the way of good grazing, but Cameroon has been less affected by periodic drought than many of its neighbors to the north.

NATIONAL DATA –CAMEROON

Land area	469,440 sq km (181,252 sq mi)			

Climate		Temperatures		Annual
	Altitude m (ft)	January °C(°F)	July °C(°F)	precipitation mm (in)
Yaoundé	760 (2,493)	24 (75)	23 (73)	657 (25.9)

Major physical features	highest point: Mount Cameroon 4,070 m (13,353 ft)

Population	(2006 est.) 17,340,702

Form of government	multiparty republic with one legislative house

Armed forces	army 12.500; navy 1,300; air force 300

Largest cities	Douala (1,404,831); Yaoundé (capital - 1,390,516)

Official languages	French, English

Ethnic composition	Cameroon Highlanders 31%; Equatorial Bantu 19%; Kirdi 11%; Fulani 10%; Northwestern Bantu 8%; Eastern Nigritic 7%; other African 13%; non-African 1%

Religious affiliations	traditional beliefs 40%; Christian 40%; Muslim 20%

Currency	1 Communauté Financière Africaine franc (XAF) = 100 centimes

Gross domestic product	(2006 est.) U.S. $42.2 billion

Gross domestic product per capita	(2006 est.) U.S. $2,400

Life expectancy at birth	male 50.98 yr; female 51.34 yr

Major resources	petroleum, bauxite, iron ore, hydropower, cocoa, cassava, coffee, cotton, groundnuts, gold, livestock, maize, millet, natural gas, palm oil, plantains, rubber, sorghum, sweet potatoes, timber, tin, yams

Cent. African Rep.

CENTRAL AFRICAN REPUBLIC

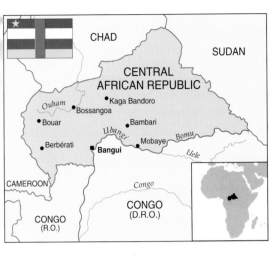

As its name suggests, the Central African Republic is a landlocked country close to the continent's center; it is situated between Cameroon, Congo (R.O.), Congo (D.R.O.), Sudan and Chad.

GEOGRAPHY

Most of the country consists of an undulating plateau, flanked by the higher Bongo Massif in the northeast and the Monts Karre to the west. The plateau forms a major divide between the watershed of Lake Chad, some distance away to the northwest, and that of the Ubangi river, the Republic's southern border. The climate ranges from near-desert in the far north to lush equatorial in the south, with seasonal rainfall (March–October) so heavy that it practically cuts

off the southeast. Here there is dense rainforest – one of the last habitats of the lowland gorilla. However, most of the country is covered with savanna grassland and open forest. Wildlife includes elephants, hyenas, antelopes and buffaloes. Both crocodiles and hippopotamuses are found in the rivers.

SOCIETY

By the 19th century the original Pygmy population had been practically wiped out by slave traders, mostly from Sudan. Bantu and Azande peoples settled in their place. The slave trade was suppressed by the French, who took formal control of the country in 1911, calling it the French Congo or Ubangi-Shari. Exploitation by ruthless commercial concessionary companies caused a major scandal. In 1946 the country became a French overseas territory, and in 1958 it gained full independence under its present name. The national language is Sango, a lingua franca, but French is still used for official communications.

In 1965 the second president, David Dacko (b. 1930), was deposed in a military coup led by Colonel Jean-Bédel Bokassa (1921–96), who ruled as a dictator. In 1975 he renamed the country the Central African Empire, crowning himself emperor. In 1979 he was deposed with French help, and was later tried on charges that included murder and cannibalism. General André Kolingba took power in 1981 but it was not until 1993 that free elections returned a civilian government headed by president Ange-Félix Patassé. In 1996 a government of national unity was formed following an army mutiny and French intervention.

ECONOMY

The Republic's economy is now being rebuilt following social unrest and years of dictatorship. Agriculture is mostly at subsistence level, and less than one-thirtieth of the land is cultivated. Groundnuts (peanuts), cotton, coffee and oil palms are grown for export, while food crops include millet, rice and cassava. Conditions rule out livestock raising.

Diamonds are the Republic's main natural resource, accounting for a substantial proportion of exports. Uranium mining has begun, but is hampered by seasonal flooding. The manufacturing sector is mainly small-scale, but includes among other things an automobile-assembly plant. Most energy comes from

Sunrise over the Ubangi river on the southern border of the Central African Republic. Not far away is the capital, Bangui, which was established by the French in 1889. On the other side of the river are the tropical rainforests of the Democratic Republic of Congo.

the hydroelectric plant at Boali in the southwest of the country.

The Ubangi river provides transportation, with Bangui, the capital, as the main port. A road network remains from colonial days, but is often flooded in the rainy season. A limited social-welfare system exists, but health care is inadequate to cope with sleeping sickness, tuberculosis, leprosy and other endemic diseases. Education is free, and literacy levels are improving. The media are government-controlled.

NATIONAL DATA – CENTRAL AFRICAN REPUBLIC

Land area	622,984 sq km (240,535 sq mi)			
Climate		Temperatures		Annual
	Altitude m (ft)	January °C(°F)	July °C(°F)	precipitation mm (in)
Bangui	381 (1,250)	26 (79)	25 (77)	1,545 (60.8)

Major physical features	highest point: Kayagangiri 1,420 m (4,660 ft)
Population	(2006 est.) 4,303,356
Form of government	multiparty republic with one legislative house
Armed forces	army 1,400; air force 150
Capital city	Bangui (552,904)
Official languages	French, Sango
Ethnic composition	Baya 33%; Banda 27%; Mandjia 13%; Sara 10%; Mboum 7%; M'Baka 4%; Yakoma 4%; other 2%
Religious affiliations	traditional beliefs 35%; Protestant 25%; Roman Catholic 25%; Muslim 15%
Currency	1 Communauté Financière Africaine franc (XAF) = 100 centimes
Gross domestic product	(2006 est.) U.S. $4.913 billion
Gross domestic product per capita	(2006 est.) U.S. $1,100
Life expectancy at birth	male 43.46 yr; female 43.62 yr
Major resources	diamonds, uranium, timber, gold, oil, hydropower, bananas, cassava, coffee, cotton, groundnuts, livestock, maize/corn, millet, plantains, sweet potatoes

Equatorial Guinea

REPUBLIC OF EQUATORIAL GUINEA

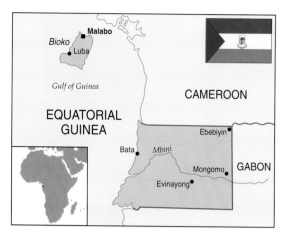

T HE MAINLAND PART OF EQUATORIAL GUINEA faces the Gulf of Guinea. Formerly called Río Muni, it is bounded by Cameroon to the north and Gabon to the south and west. The country also includes five islands, of which volcanic Bioko (the largest) and Pagalu (Annobón) are widely separated across the gulf.

GEOGRAPHY

The mainland rises from a narrow coastal plain to inland plateaus and hills that are extensions of Gabon's Monts de Cristal. Half this area is cloaked in forests, and the animal life is rich and varied. The equatorial climate is hot and humid, with two rainy seasons (February–June and September–December).

NATIONAL DATA – EQUATORIAL GUINEA

Land area	28,051 sq km (10,831 sq mi)			
Climate		Temperatures		Annual
	Altitude m (ft)	January °C(°F)	July °C(°F)	precipitation mm (in)
Malabo	50 (164)	25 (77)	25 (77)	1,799 (70.8)

Major physical features	highest point: Pico de Santa Isabel 3,007 m (9,685 ft); largest island: Bioko 2,017 sq km (779 sq mi)
Population	(2006 est.) 540,109
Form of government	multiparty republic but ruled more as a dictatorship
Armed forces	army 1,100; navy 120; air force 100
Capital city	Malabo (166,535)
Official languages	Spanish, French
Ethnic composition	Fang 72.0%; Bubi 14.7%; Duala 2.7%; Ibibio 1.3%; Maka 1.3%; others 8.0%
Religious affiliations	Roman Catholic 88.8%; traditional beliefs 4.6%; Muslim 0.5%; others 0.2%; none 5.9%
Currency	1 Communauté Financière Africaine franc (XAF) = 100 centimes
Gross domestic product	(2005 est.) U.S. $25.69 billion
Gross domestic product per capita	(2006 est.) U.S. $6,572
Life expectancy at birth	male 48 yr; female 51.13 yr
Major resources	petroleum, natural gas, timber, gold, bauxite, diamonds, tantalum, sand, gravel, clay, bananas, cassava, cocoa, coconuts, coffee, sweet potatoes

SOCIETY

Bioko (formerly Fernando Póo) was discovered by the Portuguese around 1472, and ceded first to Spain and then to Britain. The Spanish reoccupied Fernando Póo in 1858, and began to colonize the mainland. In 1963 Spanish Guinea became Equatorial Guinea, which gained independence in 1968.

A period of military rule lasted from a 1979 coup until the first multiparty elections were held in 1993 as a result of which Brigadier General Obiang Nuguema Mbasogo (b. 1943), who had seized power in 1979, was re-elected president. Obiang was re-elected again in 1996 amidst accusations of fraud.

The Fang form the largest ethnic group. The official language is Spanish and most people are nominally Roman Catholic.

ECONOMY

Oil production started in 1992 and in 2000, production was over 100,000 bbls a day and investment by US-based oil companies was high. There are also unexploited reserves of titanium, iron ore, manganese, uranium and alluvial gold. Agriculture still supports 75 percent of the population in subsistence farming while food processing is really the only industry. Corruption is rife and has led to the suspension of international aid packages. The country is underdeveloped: roads are poor but there are ports at Bata, Luba, and Malabo which also has an international airport. Health care is poor but the literacy rate is the second highest in Africa.

A palm-fringed beach in Equatorial Guinea. The black sand here has been formed by the marine erosion of iron-rich volcanic rock.

Gabon

GABONESE REPUBLIC

GABON SITS ASTRIDE THE EQUATOR ON Africa's west coast, between Equatorial Guinea, Cameroon and Congo (R.O.).

GEOGRAPHY

The country is centered around the great Ogooué river, which rises in western Congo. The Ogooué and its tributaries flow westward, cutting deep valleys through Gabon's uneven plateaus, then spreading out to form a wide delta across the coastal plain. South of the Ogooué the land rises to the Massif du Chaillu. The climate is hot and humid, with heavy rains between October and May. Dense rainforest covers most of the country. Wildlife flourishes here.

SOCIETY

The Pygmies were Gabon's original inhabitants, but Bantu peoples, especially the Fang, later moved into the area. Gabon was discovered by the Portuguese, and for centuries was a source of slaves. It was colonized by France in the late 19th century, and was harshly exploited by concessionary companies. It became independent in 1960. In 1967 Omar Bongo (b. 1935) became the republic's second president. His liberal economic policies heralded a period of stability and some prosperity. In 1990, after over 20 years of one-party rule, there were elections to a new multiparty National Assembly. Multiparty presidential elections were won by Omar Bongo in 1993. Parliamentary elections were held in late 1996.

ECONOMY

Agriculture is still mostly subsistence farming. Coffee, cocoa and oil palms have been encouraged as cash crops, while cassava remains the chief food crop. Forestry has become more important, especially of the gaboon or Gabon mahogany tree. As coastal forests have become depleted, so the TransGabon Railroad has been built to facilitate the exploitation of the interior.

Mining has greatly contributed to the country's prosperity. Coastal and offshore petroleum deposits provide most exports; timber, manganese and uranium provide much of the rest, leaving a substantial surplus. Manufacturing is mostly oil refining and timber processing, and power is provided by the Kinguélé hydroelectric station in the northwest. Social welfare and medical care are fairly good. Education is free and compulsory, and is available to most children.

Lambaréné, in western Gabon, is a small town on an island in the Ogooué river. It is most famous for its hospital, founded by the German theologian, musician and doctor Albert Schweitzer (1875–1965).

NATIONAL DATA – GABON				
Land area 257,667 sq km (99,486 sq mi)				
Climate		Temperatures		Annual
	Altitude m (ft)	January °C(°F)	July °C(°F)	precipitation mm (in)
Libreville	9 (30)	27 (81)	24 (75)	2,509 (98.7)
Major physical features highest point: Mont Milondo 1,020 m (3,346 ft)				
Population (2006 est.) 1,424,906				
Form of government multiparty republic with one legislative house				
Armed forces army 3,200; navy 500; air force 1,000				
Capital city Libreville (604,265)				
Official language French				
Ethnic composition Fang 35.5%; Mpongwe 15.1%; Mbete 14.2%; Punu 1.5%; others 23.7%				
Religious affiliations Roman Catholic 65.2%; Protestant 18.8%; African Christian 12.1%; traditional beliefs 2.9%; Muslim 0.8%; others 0.2%				
Currency 1 Communauté Financière Africaine franc (XAF) = 100 centimes				
Gross domestic product (2006 est.) U.S. $10.21 billion				
Gross domestic product per capita (2006 est.) U.S. $7,200				
Life expectancy at birth male 53.21 yr; female 55.81 yr				
Major resources petroleum, natural gas, diamonds, niobium, manganese, uranium, gold, timber, iron ore, hydropower, bananas, cassava, cocoa, coffee, livestock, maize/corn, palm oil, plantains, rice, sugarcane				

Congo

REPUBLIC OF CONGO

CONGO EXTENDS INTO THE HEART OF AFRICA from the Atlantic coast and shares borders with Congo (D.R.O.), the Central African Republic, Cameroon and Gabon.

GEOGRAPHY

The narrow coastal plain runs between neighboring Gabon and the Angolan enclave of Cabinda. Immediately inland is the Massif du Mayombé, followed by the Niari valley. Farther east the land rises again to a series of plateaus that form the western rim of the great Congo (Zaire) river basin. This covers the northeast of the country – an area crisscrossed by rivers and swamps. The Congo and its great tributary the Ubangi mark Congo's eastern border.

NATIONAL DATA – CONGO (R.O.)

Land area	341,500 sq km (131,854 sq mi)			
Climate		Temperatures		Annual
	Altitude m (ft)	January °C(°F)	July °C(°F)	precipitation mm (in)
Brazzaville	314 (1,030)	26 (79)	23 (73)	1,493 (58.7)

Major physical features	highest point: Monts de la Lékéti 1,040 m (3,410 ft); longest river: Congo (part) 4,630 km (2,880 mi)
Population	(2006 est.) 3,702,314
Form of government	multiparty republic with one legislative house
Armed forces	army 8,000; navy 800; air force 1,200
Largest cities	Brazzaville (capital – 1,370,612); Pointe Noire (708,772)
Official language	French
Ethnic composition	Kongo 48%; Sangha 20%; M'Bochi 12%; Teke 17%; other 3%
Religious affiliations	Christian 50%; animist 48%; Muslim 2%
Currency	1 Communauté Financière Africaine franc (XAF) = 100 centimes
Gross domestic product	(2006 est.) U.S. $4.958 billion
Gross domestic product per capita	(2006 est.) U.S. $1,300
Life expectancy at birth	male 51.65 yr; female 53.98 yr
Major resources	petroleum, timber, potash, lead, zinc, uranium, copper, phosphates, gold, magnesium, natural gas, hydropower, bananas, cassava, cocoa, coffee, groundnuts, palm oil, rice, sugarcane, sweet potatoes

The climate is tropical throughout, but the season of heavy rains varies according to latitude. There is savanna grassland to the east, but most of the country is covered with dense rainforest. Forest animals flourish here, ranging from buffaloes to the rare okapi and gorilla. The savanna areas support other animals such as giraffes and rhinoceroses.

SOCIETY

The Congo area was first settled by the Pygmies. These were later followed by the Kongo peoples, whose kingdoms flourished on the slave trade with Europe. In the late 19th century the area was explored by a French–Italian called Pierre Savorgnan de Brazza (1852–1905). It was colonized by the French, and became French Congo in 1891. Until the 1930s the country was harshly exploited by concessionary companies.

Independence came in 1960, but unrest in 1963 forced the first president to resign. In 1968 his successor was ousted and the new president attempted to remodel the state on Marxist lines. From 1969–91 Congo was renamed the People's Republic of the Congo. In 1979 Colonel Denis Sassou-Nguesso was appointed president. Multiparty elections were held in 1992 and Pascal Lissouba was elected president. The results were strongly disputed and politically-inspired violence erupted into civil war in 1997 and again in 1999. Sassou-Nguesso returned to power and promised to restore full democracy when the conflict ended.

ECONOMY

Agriculture consists mostly of low-yield subsistence farming. Cassava is the main food crop, while sugar cane and coffee are grown for export. Forestry is carried out in the southern rainforests. Most exports are of crude petroleum, which is mined both inland and offshore. Other substantial mineral resources include zinc and lead. Manufacturing includes food processing, textiles, cement, metal goods and chemicals. The civil war had a deleterious effect on industry and infrastructure as well as on the economy as a whole.

Congo has a good rail network in the south. The roads, however, are poor, and are subject to seasonal flooding. Major rivers provide the best transportation in the north. Health care is generally inadequate, especially in rural areas, where tropical diseases are rife. Education is free and compulsory from the ages of 6 to 16.

Congo (formerly Zaire)

DEMOCRATIC REPUBLIC OF CONGO

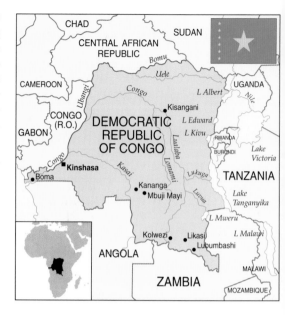

CONGO (D.R.O.) STRADDLES THE EQUATOR IN the very heart of the African continent, and is bordered on all sides by as many as nine different countries. It is landlocked apart from a 50-km (30-mi) strip of coastline, and is the third-largest country in Africa after Algeria and Sudan.

ENVIRONMENT

The great Congo river rises in the west, and flows north and west in a great horseshoe, meeting the Ubangi on the Congo (R.O.) border. The vast Congo Basin occupies central and northwestern Congo, surrounded by higher plateau country. This becomes mountainous along the eastern border, where the Great Rift Valley is marked by a number of lakes such as Lake Tanganyika, and by many ranges of volcanic mountains. To the south rise the peaks of the Shaba plateau. To the west, the Congo river runs between further plateaus to enter the Atlantic.

As well as vast swamps, the Congo Basin contains one of the world's greatest expanses of tropical rainforest, encouraged by torrential rainfall and consistently high temperatures. The forest is one of the last homes of rare and endangered

Traditional farming in Congo (D.R.O.). The country's massive agricultural potential has yet to be fully exploited, and most farmers operate at subsistence level. Livestock raising is an underdeveloped resource.

animals such as the okapi and the mountain gorilla. In the east and south, the forests give way to savanna woodlands and grasslands supported by heavy seasonal rainfall. Wildlife here includes giraffes, lions, antelopes, and the endangered black and white rhinoceroses. Various national parks have been created to conserve wildlife.

SOCIETY

Congo (D.R.O.) is the home of over 200 (mostly Bantu) ethnic groups, many of whom had formed kingdoms before the arrival of Europeans. In the 1870s King Leopold II of Belgium (1835–1909), keen to join the European scramble for African possessions, commissioned the British-born American explorer Henry Morton Stanley (1841–1904) to penetrate the little-known Congo Basin area. Leopold established the so-called Congo Free State at his own private expense, and ran it as his own domain. Widespread atrocities practiced by concessionary companies aroused international protests, and in 1908 Leopold was forced to grant a proper colonial status to the Belgian Congo.

After many decades of centralized rule from Belgium, growing nationalist parties demanded independence. It was granted in 1960, and Patrice Lumumba (1925–61) was the first prime minister. But an army mutiny broke out, led by Colonel Joseph-Désiré Mobutu (b. 1930). The southeastern province of Katanga (later Shaba) seceded under the leadership of Moise Tshombe (1919–69). The government called in United Nations peacekeeping forces, but Lumumba was murdered and the United Nations Secretary-General, Dag Hammarskjöld (1905–61), was killed in a plane crash while visiting the country.

When United Nations forces left in 1964, further revolts were suppressed by Tshombe, now president of the whole country. In 1965 Mobutu seized the presidency, promoting intensely nationalist policies. The country's name was changed to Zaire, and its citizens had to take African names; the president himself became Mobutu Sese Seko.

Mobutu's administration brought stability to Zaire and multiparty elections were promised for 1992. However, economic problems led to rioting and mutiny of the army in 1991. Political and economic problems continued exacerbated by an influx of half a million refugees from Rwanda. In 1997 Mobutu was ousted by rebel forces headed by Laurent Kabila who declared himself president and

NATIONAL DATA – CONGO (D.R.O.)				
Land area 2,267,600 sq km (875,525 sq mi)				

Climate		Temperatures		Annual
	Altitude m (ft)	January °C(°F)	July °C(°F)	precipitation mm (in)
Kinshasa	311 (1,020)	26 (79)	23 (73)	1,358 (53.4)

Major physical features highest point: Mount Stanley 5,109 m (16,763 ft); largest lake: :Lake Tanganyika (part) 32,900 sq km (12,700 sq mi)

Population (2006 est.) 62,660,551

Form of government transitional multiparty republic

Armed forces army 60,000; navy 1,800; air force 3,000

Capital city Kinshasa (8,418,819)

Official language French

Ethnic composition Luba 18.0%; Kongo 16.1%; Mongo 13.5%; Rwanda 10.3%; Azande 6.1%; Bangi/Ngale 5.8%; Rundi 3.8%; Teke 2.7%; Boa 2.3%; Chokwe 1.8%; Lugbara 1.6%; Banda 1.4%; others 16.6%

Religious affiliations Roman Catholic 50%; Protestant 20%; Kimbanguist 10%; Muslim 10%; other syncretic sects and traditional beliefs 10%

Currency 1 Congo franc (CDF) = 100 centimes

Gross domestic product (2006 est.) U.S. $44.6 billion

Gross domestic product per capita (2006 est.) U.S. $700

Life expectancy at birth male 50.01 yr; female 52.94 yr

Major resources cobalt, copper, niobium, tantalum, petroleum, industrial and gem diamonds, gold, silver, zinc, manganese, tin, uranium, coal, hydropower, timber, bananas, cassava, cocoa, coffee, cotton, groundnuts, maize/corn, millet, natural gas, plantains, palm oil and kernels, rice, rubber, sugarcane, tea

renamed Zaire the Democratic Republic of Congo. Violence continued in 1998 and 1999, underlain by ethnic rivalry; in 2001 Kabila was assassinated by his own bodyguard and his son took over the country.

ECONOMY

The civil war had a serious impact on the economy: in 1998 GDP was estimated at 65 percent lower than in 1960, and inflation reached 240 percent. Congo's (D.R.O.) agriculture is potentially rich, but apart from some large plantations it remains mostly at subsistence level. Cash crops include coffee, cocoa, rubber and cotton. The main staples are cassava, corn, rice and millet. The vast forests are rapidly disappearing.

Congo (D.R.O.) enjoys great mineral wealth. A major copper belt extends across the southern border into Congo's southeastern province of Shaba, with secondary deposits of cobalt, both among the world's largest reserves. Uranium, gold and silver ores are extracted and there are petroleum reserves off the coast. Ore smelting and petroleum refining are the main industries. Congo (D.R.O.) also produces over half the world output of low-grade industrial diamonds.

Transportation depends heavily on the river and railroad networks. Social welfare and health facilities are limited. Primary education is good and literacy levels are relatively high.

Burundi

REPUBLIC OF BURUNDI

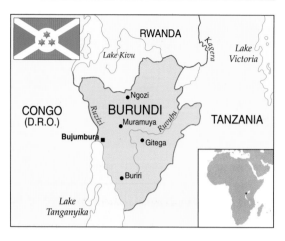

BURUNDI IS A SMALL, LANDLOCKED AND densely populated country in Central Africa on the northeastern shore of Lake Tanganyika. It is bordered by Congo (D.R.O.), Tanzania and Rwanda, with which it was formerly federated.

GEOGRAPHY

Along Burundi's western border, the Ruzizi river flows along part of the Great Rift Valley into Lake Tanganyika. Above here the land rises steeply to a mountainous ridge that extends into Rwanda, forming the ultimate divide between the great Congo and Nile river systems. Beyond the mountains the land falls away into an irregular plateau, cut through by the valley of the Ruvubu river. The tropical climate is moderated by the high altitude, and the seasonal rainfall pattern leads to both floods and droughts. Forests in the mountains give way to wooded country and then grassland savanna.

SOCIETY

The Twa, the original Pygmy inhabitants of the area, and the Hutu peoples that later settled there, were subjugated by the less numerous Tutsi, who ruled as a feudal aristocracy through their control of cattle, the primary source of wealth. Burundi, a former German colony mandated to Belgium, was granted independence in 1962 as a Tutsi kingdom, but became a one-party republic after two separate coups in 1966. A series of coups followed involving both Tutsi and Hutu.

Political and ethnic unrest between Hutu and Tutsi has continued unceasingly up to the present day, displacing hundreds of thousands of people and killing an estimated quarter of a million more between 1972 and 1999. In the 1990s a number of attempted coups and political assassinations led to a further coup in 1996 which installed Major Pierre Buyoya as president. Kurundi and French are the two official languages of the country.

ECONOMY

The economy rests largely on subsistence agriculture and has suffered from the effects of continued ethnic violence. Coffee is by far the most important export but cotton is exported as well. Staple crops include cassava, yams and grains such as corn and millet. Livestock includes cattle – reared in quantity for status and wealth – together with pigs, goats and sheep. Cattle hides are also a significant export.

Mining is small scale and manufacturing is limited. There are no railroads, and few good roads across the mountainous terrain. Social welfare and health care provide a very limited service; malaria and tuberculosis are widespread. Education reaches only a minority of the population.

A Tutsi man (*below*) sits within a house compound in the Burundi village of Gisozo. The Tutsi, who are thought to be of Nilotic origin, are only a minority in Burundi, but have dominated the country for many years. Most Burundians belong to the Hutu group – an agricultural people of Bantu origin.

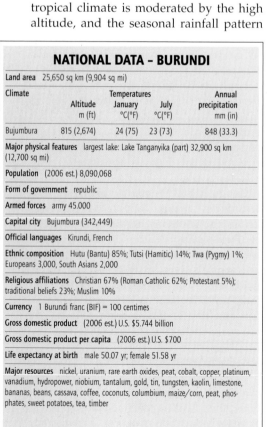

NATIONAL DATA – BURUNDI

Land area	25,650 sq km (9,904 sq mi)			
Climate		Temperatures		Annual precipitation
	Altitude m (ft)	January °C(°F)	July °C(°F)	mm (in)
Bujumbura	815 (2,674)	24 (75)	23 (73)	848 (33.3)
Major physical features	largest lake: Lake Tanganyika (part) 32,900 sq km (12,700 sq mi)			
Population	(2006 est.) 8,090,068			
Form of government	republic			
Armed forces	army 45.000			
Capital city	Bujumbura (342,449)			
Official languages	Kirundi, French			
Ethnic composition	Hutu (Bantu) 85%; Tutsi (Hamitic) 14%; Twa (Pygmy) 1%; Europeans 3,000, South Asians 2,000			
Religious affiliations	Christian 67% (Roman Catholic 62%; Protestant 5%); traditional beliefs 23%; Muslim 10%			
Currency	1 Burundi franc (BIF) = 100 centimes			
Gross domestic product	(2006 est.) U.S. $5.744 billion			
Gross domestic product per capita	(2006 est.) U.S. $700			
Life expectancy at birth	male 50.07 yr; female 51.58 yr			
Major resources	nickel, uranium, rare earth oxides, peat, cobalt, copper, platinum, vanadium, hydropower, niobium, tantalum, gold, tin, tungsten, kaolin, limestone, bananas, beans, cassava, coffee, coconuts, columbium, maize/corn, peat, phosphates, sweet potatoes, tea, timber			

Rwanda

REPUBLIC OF RWANDA

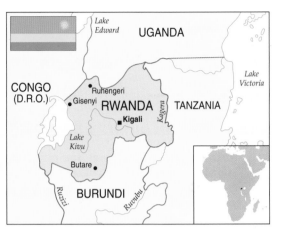

T HE MOUNTAINOUS, LANDLOCKED COUNTRY OF Rwanda in Central Africa shares its borders with Congo (D.R.O.), Uganda and Tanzania, and to the south with Burundi, with which it was formerly federated.

GEOGRAPHY

In the west of Rwanda, Lake Kivu on the Congo (D.R.O.) border is part of the Great Rift Valley. East of here the land rises steeply to the mountains of the Congo-Nile divide. The Virunga Mountains to the north include two active volcanoes. The high plateaus to the east fall away toward the Tanzanian border. The tropical climate is moderated by altitude, with heavy seasonal rains. The mountain forests are the last home of the mountain gorilla. The plateaus are covered with woodland and savanna.

A smallholding in Rwanda Subsistence agriculture like this forms the backbone of the Rwandan economy. Bananas and other fruits are being grown, and food crops are likely to include cassava, yams and grains.

SOCIETY

The original inhabitants were Twa Pygmies. The Hutu settled later in the area, and both were conquered by invading Tutsi. The country was claimed by Germany, and was mandated to Belgium after World War I. In 1959 the Hutu rebelled, ousting the Tutsi king and many of his people. In 1962 Rwanda became an independent republic. Interethnic clashes between Hutu and Tutsi began almost immediately, leading eventually to a military coup in 1973. A period of relative stability ended with the breaking out of civil war in 1990. The war culminated, in April 1994, in a genocide in which around one million Tutsis and moderate Hutus were killed. International intervention halted the genocide but approximately two million Hutu refugees – many fearing Tutsi retribution – fled the country. Civil strife continued into the 2000s.

ECONOMY

The economy is based on subsistence agriculture. Cash crops include coffee (the main export), tea and sugar cane. Cassava, corn, plantains and vegetables are the main food crops. The Tutsi raise cattle, which confer social status, whereas the Hutu raise sheep and goats. Some tin ore is mined. Social welfare is very limited.

NATIONAL DATA – RWANDA

Land area	24,948 sq km (9,632 sq mi)			
Climate		Temperatures		Annual
	Altitude m (ft)	January °C(°F)	July °C(°F)	precipitation mm (in)
Kigali	1,545 (5,069)	19 (67)	21 (70)	1,028 (40.4)

Major physical features highest point: Mount Karisimbi 4,507 m (14,787 ft); largest lake: Lake Kivu (part) 2,699 sq km (1,042 sq mi)

Population (2006 est.) 8,648,248

Form of government presidential multiparty republic. New Constitution adopted in 2003.

Armed forces army 40,000; air force 1,000

Capital city Kigali (857,719)

Official languages Kinyarwanda, French, English

Ethnic composition Hutu 84%; Tutsi 15%; Twa (Pygmoid) 1%

Religious affiliations Roman Catholic 56.5%; Protestant 26%; Adventist 11.1%; Muslim 4.6%; traditional beliefs 0.1%; none 1.7%

Currency 1 Rwandan franc (RWF) = 100 centimes

Gross domestic product (2006) U.S. $13.54 billion

Gross domestic product per capita (2006) U.S. $1,600

Life expectancy at birth male 46.26 yr; female 48.38 yr

Major resources gold, cassiterite (tin ore), wolframite (tungsten ore), methane, hydropower, coffee, tea, sugarcane, bananas, beans and peas, cassava, cattle, groundnuts, goats, potatoes, pyrethrum, pigs, sheep, sorghum, sweet potatoes

Uganda

REPUBLIC OF UGANDA

T HE EAST AFRICAN REPUBLIC OF UGANDA HAS been torn apart by war in recent decades, but since 1986 it has achieved a slow return to stability.

GEOGRAPHY

The country is landlocked, bordering Rwanda and Tanzania to the south and Kenya to the east, with Sudan to the north and Congo (D.R.O.) to the west. Most of it occupies a northward-sloping plateau north and west of Lake Victoria. The capital, Kampala, lies near Lake Victoria's northern shore. In the far southwest, the border with Rwanda runs through the volcanic Virunga mountains. The western border with Congo runs parallel to the western branch of the Great Rift Valley,

features of which include Lake Edward, the Ruwenzori range (rising to Mount Stanley) and Lake Albert. The river Nile flows north from Lake Victoria through Lake Kyoga, then north and west into Lake Albert, and finally north into Sudan.

Uganda's equatorial climate is moderated by its altitude, and by large areas of lake and swamp that occupy one-sixth of the total area. The north has a single wet season from April to November, whereas the south has two rainy seasons (April–May and October–November). Rainfall patterns are reflected in the vegetation, with modified tropical forest around Lake Victoria and wooded savanna parkland in the center and the north. Uganda's national parks and game reserves protect numerous animals and habitats.

SOCIETY

Oral tradition describes several African kingdoms in the area, including the Nilotic state of Bunyoro-Kitara. Buganda, a former satellite of Bunyoro, rose to dominance in the 18th century. In 1894 it became a British protectorate, and in 1896 it was extended to include Bunyoro and neighboring kingdoms. Independence under a federal constitution was granted in 1963, with Mutesa II (1924–69), *kabaka* (king) of Buganda, as the first president. In 1966 Mutesa was ousted by prime minister Milton Obote (b. 1924), who declared a republic. Obote himself was deposed by Idi Amin (b. 1925) in 1971.

Under Amin's dictatorship Asians and other skilled professionals were expelled, the majority Christian population was persecuted, and at least 300,000 people were murdered or disappeared. The economy began to collapse. In 1978 Amin invaded Tanzania, only to be overthrown by a combined force of Tanzanians and Ugandan exiles. In 1980 Milton Obote became president of a shattered country. However, after coups in 1985 and 1986 Yoweri Museveni (b. 1944) took the presidency, facing civil war, troubled relations with Kenya and the burden of some 100,000 refugees.

In 1994 a new constitution created a directly-elected 278-member single assembly and multiparty democracy is planned early in the 2000s. The monarchy was reinstated in 1993 with Ronald Muwenda Mutebi II (b. 1955). Television and radio broadcasting is state-controlled, but some press freedom has been reintroduced.

Uganda has some 40 different ethnic groups, the largest being the Bantu-speaking Ganda (about one-fifth of the

population). Other large groups include the Teso, Nkole, Soga and Gisu. English is the official language, but Swahili is widely spoken.

ECONOMY

During the 1990s the economy increasingly stabilized, inflation reduced and there was sizeable investment in roads and railroads and in projects which will increase agricultural production and manufacturing. Around 80 percent of the workforce is involved with agriculture which accounted for almost half of GDP in 1996. Coffee, cotton, tea and tobacco are the principal exports. Spain, France and Germany are the main export markets. Uganda is rich in minerals including copper, gold and cobalt. Power is chiefly supplied by the Owen Falls Dam on the river Nile.

The road network is undergoing an internationally-funded 10-year improvement plan and there is a project to establish a direct rail link between Kampala and Johannesburg, South Africa. There is an international airport at Entebbe.

An emphasis on preventive medicine has had some success, and there are some rural clinics as well as centralized medical facilities. However, malaria, hookworm, AIDS and gastric disorders are common and housing is in short supply. There is a contributory social security system. Secondary education is limited, and is closely linked to economic needs.

Boys at the pump (*above*) in the Ugandan village of Bugiri, not far from Lake Victoria. Only one-eighth of the population has access to clean drinking water. One-tenth of Ugandan children die in infancy, and AIDS has become a serious problem in the region.

NATIONAL DATA - UGANDA

Land area	199,710 sq km (77,108 sq mi)			
Climate		Temperatures		Annual
	Altitude m (ft)	January °C(°F)	July °C(°F)	precipitation mm (in)
Kampala	1,150 (3,773)	23 (67)	21 (70)	1,150 (46.2)

Major physical features highest point: Mount Stanley 5,109 m (16,763 ft); largest lake: Lake Victoria (part) 62,940 sq km (24,300 sq mi)	
Population (2006 est.) 28,195,754	
Form of government republic with one legislative house part-elected by popular vote	
Armed forces joint forces 45,000	
Largest cities Kampala (1,455,027)	
Official languages English, Swahili	
Ethnic composition Baganda 17%; Ankole 8%; Basoga 8%; Iteso 8%; Bakiga 7%; Langi 6%; Rwanda 6%; Bagisu 5%; Acholi 4%; Lugbara 4%; Batoro 3%; Bunyoro 3%; Alur 2%; Bagwere 2%; Bakonjo 2%; Jopodhola 2%; Karamojong 2%; Rundi 2%; non-African (European, Asian, Arab) 1%; other 8%	
Religious affiliations Roman Catholic 33%; Protestant 33%; Muslim 16%; traditional beliefs 18%	
Currency 1 Ugandan shilling (UGX) = 100 cents	
Gross domestic product (2006 est.) U.S. $51.89 billion	
Gross domestic product per capita (2006 est.) U.S. $1,800	
Life expectancy at birth male 51.68 yr; female 53.69 yr	
Major resources copper, cobalt, gold, hydropower, limestone, salt, coffee, cotton, tea, tobacco, bananas, fish, livestock, maize/corn, millet, phosphates, sorghum, sugarcane, timber, yams, tourism	

Kenya

REPUBLIC OF KENYA

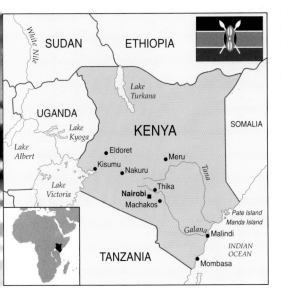

Masai herders in the arid landscape of Kenya's Rift Valley. Livestock raising is mainly confined to the drier parts of the country. Most agricultural land is in the wetter areas, where it is communally owned and used to grow cash crops or staples.

Kenya straddles the equator on the eastern coast of Africa. Like many African countries, it is a melting pot of different peoples and cultures.

GEOGRAPHY

The coastal plain is narrow in the south, but broadens northeastward toward the Somali border. Inland, the Nyika Plain rises northwestward from the coast. The volcanic western highlands are split in two by the north–south line of the Great Rift Valley. On its western side, the Mau and Elgeyo Escarpments stretch from the Tanzanian border in the south to the Cherangany Hills near Mount Elgon on the Ugandan border. East of the rift are the Laikipia Escarpment and the Aberdare Range, at the southern end of which lies the capital, Nairobi. East of the Aberdare Range is the country's highest mountain, Mount Kenya. The two major rivers, the Tana and the Galana, drain southeastward from the highlands across the Nyika Plain. West of the highlands are grasslands bordering Lake Victoria.

The climate is equatorial and temperatures are similar throughout the year. Large areas have been developed as game reserves, notably the Nairobi and Tsavo National Parks. However, ivory poachers are a continuing threat, especially along the border with Somalia.

SOCIETY

Arab traders settled along the Kenyan coast from about the 9th century onward. Masai invaders moved south in the 18th century, but in about 1830 their progress was checked by resistance from the Kikuyu. In 1887 the Sultan of Zanzibar leased part of the coast to the British, and by 1901 a railroad had been built from Mombasa to Lake Victoria. White settlers found the highlands congenial, and in 1920 Kenya became a British colony.

After World War II rising African nationalism in the 1950s led to a campaign of violence by the Mau Mau, a Kikuyu anticolonial organization. Following several years of unrest, Kenya finally gained independence in 1963, and in 1964 a republic was declared, with the Kikuyu leader Jomo Kenyatta (1889–1978) as its first president. In 1992 the first multiparty elections were held after a series of violent demonstrations which legalized opposition parties. During the last few decades, however, growth has been marred by inflation, droughts, and a rapidly rising population leading to unemployment.

Most Kenyans are Africans of Bantu, Nilotic and Kushitic origin. Out of the 70 or more distinct ethnic groups, the largest and most influential are the Kikuyu and the Luo. Apart from Nairobi and the port of Mombasa, the few large settlements are in the highlands and along the railroad.

ECONOMY

Most Kenyans are still agriculturalists living and working on communally owned land. Agriculture is the chief export earner, but farming is largely restricted to high-rainfall areas of the highlands, central Rift Valley and around Lake Victoria. Coffee and tea are the main cash crops. Staples include corn, potatoes, pulses, millet and bananas. Horticulture is a major growth area. Tourism, an invaluable source of foreign exchange, has been damaged by political violence.

Wood from very limited forests is the only natural source of fuel. Major industrial activities include cement production and petroleum processing based on refineries at Mombasa. Hydroelectricity from the Tana river was supplemented by a geothermal station in 1981.

Road connections are good in the more populated regions, and a highway links Nairobi with Addis Ababa (Ethiopia). Mombasa is the principal port. Nairobi's airport is important in East Africa.

A variety of health-care services is provided by central and local government, voluntary agencies and private doctors. Diseases include malaria, sexually transmitted diseases (particularly AIDS), dysentery and trachoma. Self-help welfare activities are encouraged. Education is not compulsory, but the facilities are improving. Seven years of primary education can lead, after an examination, to four years of secondary education.

NATIONAL DATA - KENYA

Land area	569,250 sq km (219,789 sq mi)			
Climate		**Temperatures**		**Annual**
	Altitude m (ft)	January °C(°F)	July °C(°F)	precipitation mm (in)
Nairobi	1,616 (5,302)	18 (67)	15 (70)	1,024 (40.3)

Major physical features highest point: Mount Kenya 5,199 m (17,058 ft); longest river: Tana 708 km (440 mi)

Population (2006 est.) 34,707,817

Form of government multiparty republic with one legislative house

Armed forces army 20,000; navy 1,620; air force 2,500

Largest cities Nairobi (capital – 2,982,226); Mombasa (847,626); Nakuru (273,039); Eldoret (237,398)

Official languages Swahili, English

Ethnic composition Kikuyu 22%; Luhya 14%; Luo 13%; Kalenjin 12%; Kamba 11%; Kisii 6%; Meru 6%; other African 15%; non-African 1%

Religious affiliations Protestant 45%; Roman Catholic 33%; traditional beliefs 10%; Muslim 10%; other 2%

Currency 1 Kenyan shilling (KES) = 100 cents

Gross domestic product (2006 est.) U.S. $40.77 billion

Gross domestic product per capita (2006 est.) U.S. $1,200

Life expectancy at birth male 49.78 yr; female 48.07 yr

Major resources limestone, soda ash, salt, gemstones, fluorspar, zinc, diatomite, gypsum, wildlife, hydropower, tourism, bananas, beans, cattle, cassava, coffee, cotton, maize/corn, millet, potatoes, pyrethrum, sisal, sugarcane, sweet potatoes, tea

Tanzania

UNITED REPUBLIC OF TANZANIA

Tanzania's abundant wildlife is protected in its many national parks, where lions such as this one generally live in family groups. The females are usually the hunters, though the long-maned males are perfectly able to kill for themselves.

THE EAST AFRICAN REPUBLIC OF TANZANIA was created in 1964 by the union of Tanganyika and Zanzibar, which consists of the islands of Zanzibar and Pemba.

GEOGRAPHY

Most of Tanzania consists of a plateau traversed by a sprawling network of mountains and depressions. A narrow coastal plain runs south from the Kenyan border toward Mozambique.

In the southwest Lake Malawi marks the Malawian border. From here the Zambian frontier runs northwestward to the southern end of Lake Tanganyika. The northwestern corner of the country borders Burundi, Rwanda and Uganda. The Ugandan border cuts across Lake Victoria, meeting the Kenyan border near its eastern shore. In the many scattered mountain ranges are several volcanoes flanking the Great Rift Valley. These include the Ngorongoro Crater and also mount Kilimanjaro, which is Africa's highest peak.

The climate is equatorial, with a fairly narrow range of temperatures; rainfall depending largely on altitude and aspect. There are two rainy seasons in the north, whereas the south has a single, much longer season.

Much of the country is covered in grassland with scattered woodland but tsetse-fly infestation makes a great deal of this land unsuitable for cattle or human settlement. Alpine desert is found at higher altitudes with semidesert on some north-facing slopes. Coastal vegetation ranges from mangroves to reedy swampland. The abundant wildlife includes elephants, rhinoceroses and lions, and there are many fine national parks, notably the Serengeti in the north.

SOCIETY

Excavation of the Olduvai Gorge in the northeast has revealed some of the oldest known human remains, dating back 2 million years. Coastal trade with Arabia and India began in about the 1st century AD. An Arab civilization flourished on the coast from the 8th century until the Portuguese took over in the 16th century. Arabs later regained control, and their trade in slaves and ivory spread inland. In 1886 Germany and Britain both established claims in the area. The Germans controlled Tanganyika, and the British made Zanzibar a protectorate.

During World War I Britain occupied Tanganyika, but in 1947 it was placed under United Nations trusteeship. The 1950s saw the emergence of nationalism, with the foundation of the Tanganyika African National Union (TANU), leading to independence in 1961. In 1962 a republic was declared, with Julius Nyerere (1922–99) as president. Zanzibar gained independence in December 1963. A month later a violent revolution replaced the mainly Arab government with a republic controlled by the Afro-Shirazi Party (ASP). A similar revolution in Tanganyika was quelled with British help. In 1964 Tanganyika and Zanzibar united.

TANU and the ASP united in 1977 to form the Revolutionary Party of Tanzania (Chama Cha Mapinduzi or CCM), which became Tanzania's only political party. In 1992 the CCM promised to abolish one-party rule; first multiparty elections were held late in 1995. The president is the head of state, and his cabinet includes a vice-president and prime minister. Zanzibar's internal affairs are administered by its own elected council and president.

Tanzania is one of the least urbanized countries in Africa. There are some 120 ethnic groups, the largest being the Nyamwezi and the Sukuma. Swahili is Zanzibar's principal language. Christianity and Islam each claim about one-third of the population, but Islam dominates in Zanzibar.

ECONOMY

Agricultural products (notably coffee and cotton) are the major export earners, but low selling prices create a massive trade gap, and overseas aid is vital. Cattle are the main livestock, and staple crops include grains, fruit and vegetables. Production is susceptible to drought and fuel shortages. Many mineral resources are unexploited, but salt and gems are exported. Energy comes from hydroelectric and geothermal sources, and petroleum products are exported to Zambia. Other industries include food processing, textiles and tourism.

The infrastructure is slowly improving, with a new highway being built into Zambia. Dar es Salaam is the main port, and steamers operate on the Tanganyika and Victoria Lakes. The main airports are at Dar es Salaam and Kilimanjaro. Malaria and sleeping sickness are being eradicated, but malnutrition is a problem. Primary education is free and compulsory, and secondary education is oriented toward practical training.

NATIONAL DATA - TANZANIA

Land area	886,037 sq km (342,101 sq mi)			

Climate		Temperatures		Annual
	Altitude m (ft)	January °C(°F)	July °C(°F)	precipitation mm (in)
Salaam	58 (190)	27 (81)	23 (74)	1,056 (41.6)

Major physical features highest point: Kilimanjaro 5,895 m (19,341 ft); largest lake: Victoria (part) 62,940 sq km (24,300 sq mi)

Population (2006 est.) 37,445,392

Form of government multiparty republic with one legislative house

Armed forces army 23,000; navy 1,000; air force 3,000

Largest cities Dar es Salaam (2,915,878); Mwanza (480,279); Dodoma (capital - 196,006)

Official languages Swahili, English

Ethnic composition African 99% (of which 95% are Bantu consisting of more than 130 tribes); other 1%

Religious affiliations mainland - Christian 30%, Muslim 35%, traditional beliefs 35%; Zanzibar more than 99% Muslim

Currency 1 Tanzanian shilling (TZS) = 100 cents

Gross domestic product (2006 est.) U.S. $29.25 billion

Gross domestic product per capita (2006 est.) U.S. $800

Life expectancy at birth male 44.93 yr; female 46.37 yr

Major resources hydropower, tin, phosphates, iron ore, coal, diamonds, gemstones, gold, natural gas, nickel, beans, cassava, cloves, coconuts, coffee, cotton, maize/corn, sisal, tobacco

Seychelles

REPUBLIC OF SEYCHELLES

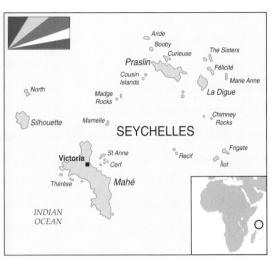

THE SEYCHELLES FORM A WIDELY SCATTERED archipelago of 100 or so islands, plus innumerable islets and cays, in the Indian Ocean northeast of Madagascar.

GEOGRAPHY

The islands fall into two main physical types. The islands of the central group, including the main island of Mahé, are founded on granite, and consist of a mountainous heart surrounded by a flat coastal strip. The outer islands are made up of coral accretions at various stages of formation, from reefs to atolls. These are generally smaller and almost entirely flat, only a few meters above sea level. Most of these islands have virtually no water, and few are inhabited.

The main islands enjoy high humidity and rainfall, feeding the remaining lush tropical rainforest of hardwoods and palms such as coco-de-mer that once totally covered them. Wildlife includes a rare giant land turtle and colorful reef animals such as the Green sea turtle. This island paradise has suffered from human incursions, but is now valued as a major tourist attraction; some coral islands are bird sanctuaries.

SOCIETY

Occupying a strategic position on the Europe–India sea route, the Seychelles were annexed by France in 1756 to support its colony on Mauritius. It passed to Britain after the Napoleonic Wars. Most of the people are of mixed colonial descent and speak a Creole dialect. The Seychelles became independent in 1976 with James R. Mancham (b. 1939) as president and Albert René (b. 1935) as prime minister. In 1979, while Mancham was abroad, René seized the presidency in a coup. In 1993 he was re-elected in the first multiparty elections.

ECONOMY

The mainstay of the economy is tourism, served by an international airport on Mahé and many hotels. The islands' agriculture no longer supports the expanding population. Copra, tobacco, cinnamon, vanilla and coconuts are grown for export, and much food is imported.

Other important exports include fish (especially canned tuna), guano and re-exported petroleum products. Industry is largely confined to food processing. Health care is excellent, and education is available to most children.

La Digue, the third-most populous island of the Seychelles archipelago after Mahé and Praslin, has an area of only 15 sq km (6 sq mi). High temperatures and rainfall have eroded this granite outcrop to create an isolated dome or inselberg.

NATIONAL DATA – SEYCHELLES				
Land area 455 sq km (176 sq mi)				
Climate		Temperatures		Annual
	Altitude m (ft)	January °C(°F)	July °C(°F)	precipitation mm (in)
Victoria	3 (10)	27 (80)	26 (56)	2,172 (85.5)
Major physical features largest island: Mahé 153 sq km (59 sq mi)				
Population (2006 est.) 81,541				
Form of government multiparty republic with one legislative house				
Armed forces army 200				
Capital city Victoria (22,336)				
Official languages English, French, Creole				
Ethnic composition Seychellois Creole (Asian/African/European) 89.1%; Indian 4.7%; Malagasy 3.1%; Chinese 1.6%; English 1.5%				
Religious affiliations Roman Catholic 82.3%, Anglican 6.4%, Seventh Day Adventist 1.1%, other Christian 3.4%, Hindu 2.1%, Muslim 1.1%, other non-Christian 1.5%, unspecified 1.5%, none 0.6%				
Currency 1 Seychelles rupee (SCR) = 100 cents				
Gross domestic product (2002 est.) U.S. $626 million				
Gross domestic product per capita (2002 est.) U.S. $7,800				
Life expectancy at birth male 66.69 yr; female 77.63 yr				
Major resources fish, coconuts, copra, cinnamon, tobacco, vanilla, tourism, fruit				

São Tomé & Príncipe

DEMOCRATIC REPUBLIC OF SÃO TOMÉ AND PRÍNCIPE

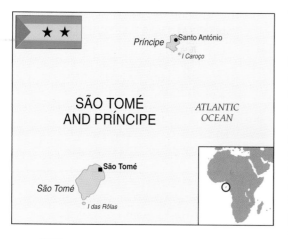

THE VOLCANIC ISLAND GROUP OF SÃO TOMÉ and Príncipe forms an independent republic in the Gulf of Guinea off the west coast of Africa.

GEOGRAPHY

São Tomé is the largest and most populous island, rising at the center to the Pico de Tomé. Rainforest clothes the upper slopes of the mountains. The smaller rectangular island of Príncipe lies some 150 km (100 mi) to the northeast; it is equally mountainous. Other islets in the group include the Pedras Tinhosas and the Ilha das Rôlas, which lies directly astride the Equator. The hot, humid equatorial climate is moderated by the surrounding ocean and the cold Benguela Current. The mountain slopes enjoy somewhat cooler temperatures.

NATIONAL DATA – SÃO TOMÉ AND PRäNCIPE

Land area 1,001 sq km (386 sq mi)

Climate	Altitude m (ft)	Temperatures January °C(°F)	July °C(°F)	Annual precipitation mm (in)
São Tomé	8 (26)	27 (81)	25 (77)	966 (38.1)

Major physical features highest point: Pico de Tomé 2,024 m (6,640 ft); largest island: São Tomé 855 sq km (332 sq mi)

Population (2006 est.) 193,413

Form of government multiparty republic with two legislative houses

Armed forces no armed forces

Capital city São Tomé (65,416)

Official language Portuguese

Ethnic composition African (mainly Fang) 90%; Portuguese and Creole 10%

Religious affiliations Catholic 70.3%; Evangelical 3.4%; New Apostolic 2%; Adventist 1.8%; other 3.1%; none 19.4%

Currency 1 dobra (STD) = 100 cêntimos

Gross domestic product (2003 est.) U.S. $214 million

Gross domestic product per capita (2003 est.) U.S. $1,200

Life expectancy at birth male 65.73 yr; female 68.95 yr

Major resources cocoa, copra, coffee, bananas, palm oil, fisheries, tourism, fish, hydropower, coconuts, timber

SOCIETY

São Tomé was discovered by European explorers in the 1470s. It was used as a settlement for convicts and Jewish refugees banished to the island by the Portuguese, who later brought slaves from the mainland to work on their sugar plantations. The islands became independent from Portugal in 1975. The Movement for the Liberation of São Tomé and Príncipe (MLSTP) became the sole legal party in the 51-member National People's Assembly. A plan for full democracy was launched in 1988, and in 1990 the MLSTP was defeated in multiparty elections, but returned to power in 1994, and again in 1995 after a military coup failed. The inhabitants are largely Portuguese-speaking Africans, including immigrant workers from Angola and Mozambique.

ECONOMY

Agriculture, fishing and tourism form the backbone of the economy, and cocoa is the main export crop. About one-third of the land is cultivated, mostly for export, leaving little opportunity for staples; dependence on imported food has led to widespread malnutrition. São Tomé has one railroad and several roads. It is linked by air to Príncipe and Luanda (Angola), and by boat to the other islands of the group. Education and health care are free.

Cape Verde

REPUBLIC OF CAPE VERDE

Cape Verde consists of two small archipelagos lying some 600 km (400 mi) off the coast of west Africa.

GEOGRAPHY

All the islands are volcanic, ringed by steep rocky cliffs and reefs. Most have been eroded into jagged shapes by wind-blown sand. The island of Fogo has an active volcano. The climate is stable and moderately warm, but extremely arid. Agriculture is only possible where sea mists linger, and periodic droughts have devastated the islands.

A fishing village on one of the many volcanic islands of Cape Verde – one of the smallest nations in the world. The white painted church and buildings in the background are evidence of the country's long association with Portugal.

SOCIETY

The islands were settled by Portugal as early as the 15th century. They became a slave-trading center, and in the 19th century they were an important staging post for transatlantic sea traffic. The majority of the people are mulattos, descended from West African slaves and speaking a Creole dialect called Crioulo. In 1951 the islands became an overseas province of Portugal. In 1975 they became independent under President Aristides Pereira (b. 1924). Moves to unite the country with Guinea-Bissau did not come to fruition. In 1990 the ruling party was defeated in the first multiparty elections.

ECONOMY

The economy rests mainly on agriculture. The main crops include bananas, coconuts and sugar cane; fishing is being developed. Industries are based mainly on local food produce. Other resources include salt and a volcanic rock used in cement. Transportation is mainly by sea and air, and tourism is increasing. There are limited health and welfare programs. Education is freely available, but jobs are scarce and many islanders are forced to emigrate in order to find work. Most of the population is under the age of 20.

NATIONAL DATA – CAPE VERDE				
Land area 4,033 sq km (159 sq mi)				
Climate		Temperatures		Annual
	Altitude m (ft)	January °C(°F)	July °C(°F)	precipitation mm (in)
Praia	27 (89)	23 (73)	26 (79)	2,172 (85.5)
Major physical features highest point: Pico do Cano (Fogo) 2,829 m (9,281 ft); largest island: São Tiago 991 sq km (383 sq mi)				
Population (2006 est.) 420,979				
Form of government multiparty republic with one legislative house				
Armed forces army 1.000; coast guard 100; air force 100				
Capital city Praia (121,393)				
Official language Portuguese				
Ethnic composition Creole (mulatto) 71%; African 28%; European 1%				
Religious affiliations Roman Catholic 97.8%; others 2.2%				
Currency 1 Cape Verdean escudo (CVE) = 100 centavos				
Gross domestic product (2006 est.) U.S. $3.129 billion				
Gross domestic product per capita (2006 est.) U.S. $6,000				
Life expectancy at birth male 67.41 yr; female 74.15 yr				
Major resources salt, basalt rock, limestone, kaolin, fisheries, clay, gypsum, bananas, coffee, coconuts				

Southern Africa

COUNTRIES IN THE REGION
ANGOLA · ZAMBIA · MALAWI · MOZAMBIQUE
NAMIBIA · BOTSWANA · ZIMBABWE · SOUTH AFRICA
LESOTHO · SWAZILAND · MADAGASCAR
MAURITIUS · COMOROS

DEPENDENCIES IN THE REGION
ST HELENA · MAYOTTE · REUNION

Mountain kingdom The people of Lesotho inhabit a
landlocked country of snowcapped peaks and
windswept uplands cut by deep gorges. Thick, colorful
blankets are worn to withstand cold winter temperatures
and frequent summer storms.

- ■ capital city
- ● major town

height of land (meters)
- 3000
- 2000
- 1000
- 500
- 200
- 0 (sea level)

▲ mountain peak

Angola

REPUBLIC OF ANGOLA

ANGOLA IS A LARGE STATE ON AFRICA'S southwest coast, flanked by Namibia to the south, Zambia to the east and Congo (D.R.O.) to the north. The coastal enclave of Cabinda is divided from the rest of Angola by a strip of Congo (D.R.O.), and borders Congo (R.O.) to the north, Until 1974 Angola was an overseas colony of Portugal, and since independence in 1975 it has suffered civil war.

GEOGRAPHY

From the narrow coastal plains, the land rises to an extensive tableland that covers most of the country. The uplands are fairly level in the south, but become mountainous in southwestern and central Angola, where the highest point is Mount Moco. In the north the plateau becomes more fragmented; the valleys and coastal plains are broader toward the mouth of the Congo river and in the Cabinda enclave beyond. A network of rivers fans out from the central highlands; some flow into the Atlantic, while others feed the rivers of adjoining countries.

The climate is generally hot except in the central highlands and along the coast, where temperatures are moderated by the cold Benguela Current that flows north from the colder southern oceans. Rainfall varies widely across the country and according to the season. The north receives heavy rains over a seven-month period, whereas the west and south are subject to long droughts.

The vegetation varies according to climate, from dense tropical rainforest in Cabinda and the north to savanna woodland and then more arid grassland, dotted with euphorbia, acacia and baobab trees. Along the southwest coast a strip of pure desert extends north from Namibia.

Angola has a rich variety of African wildlife: cheetahs, leopards, elephants and rhinoceroses. Many of these are threatened by the destruction of their habitats for farming land, by the disruptions of civil war and by the illicit trade in ivory and rhinoceros horns. National parks exist, but conservation measures have so far been ineffective.

SOCIETY

The earliest inhabitants of Angola, hunter–gatherers related to the Khoisan-speaking San (Bushmen) of present-day Namibia, were displaced by the influx of Bantu peoples from the north, who introduced cereal agriculture and ironworking to the area. By the time the Portuguese arrived in the 16th century, the Bantu had established a number of kingdoms, some of which were highly organized. The largest, that of the Kongo, included the northwest of modern Angola, while the Ndongo kingdom occupied the south.

The Portuguese called the country Angola after the name Ngola, which the Ndongo gave to their kings. At first the Bantu kingdoms enjoyed friendly relations with the Portuguese, exploiting the slave trade in order to expand their own power. Nevertheless, the Kongo kingdom was overrun by a neighboring group in 1568. Meanwhile, the Ndongo thrived until the Portuguese decided on a policy of subjugation. Despite this they were not finally conquered until a century later.

Angola became the slave center for the main European powers, leaving the area largely depopulated and with no chance of thriving even as a colony. Slavery was abolished in the early 19th century, only to be replaced by indentured labor. Portugal established its right to colonize the whole area, and initiated projects such as

railroads and diamond mines, but the borders of Angola were not formally drawn until as late as 1926. No rights were granted to indigenous Africans, who continued to remain impoverished, often as forced laborers.

Growing nationalist feeling in the 1960s gave rise to three different guerrilla campaigns: the primarily Marxist-oriented Angolan People's Liberation Movement (Movimento Popular de Libertação de Angola; MPLA), the Kongo-led Angolan National Liberation Front (Frente Nacional de Libertação de Angola; FNLA), and in the south the National Union for Total Angolan Independence (União Nacional para a Independência Total de Angola; UNITA). All three were fiercely resisted by the Portuguese. In 1974, however, the new regime in Lisbon abruptly withdrew the Portuguese troops, paving the way for full independence the following year.

Luanda, capital of Angola (*left*), was founded as a trading post by the Portuguese in 1576. It remains a busy port, and today modern towers of concrete and glass rise up alongside the spacious buildings of the colonial past.

Village life (*below*) Women carry out the traditional tasks of food preparation in a small village in the countryside south of Luanda. Agriculture suffered during the years of fighting, which disrupted the lives of Angola's peasant farmers.

War immediately broke out. UNITA and FNLA, with South African backing, confronted the Soviet-backed MPLA, supported by Cuban troops. The MPLA took over the government, creating a one-party republic. However, the war continued for many years, with large areas under UNITA control. The withdrawal of Cuban troops (and of Soviet backing) in 1989 opened the way for peace negotiations, which were concluded in 1991. But hopes for peace were dashed when fighting resumed in November 1992 after UNITA lost an election. Further peace agreements in 1994 and 1998 provided temporary respite from the fighting which resumed again in 1999.

Portuguese is the official language, although spoken only by a minority; the status of the native Bantu languages is being improved. The largest of the various ethnic groups are the Ovimbundu in the central highlands, the Mbundu along the coast around the capital, Luanda, and the Kongo in the north. Most people are Roman Catholic.

ECONOMY

Despite many years of incessant civil war, Angola's economy has begun to improve in some respects, thanks partly to abundant natural resources. A wide variety of crops are grown, especially in the more fertile north. The main cash crops are coffee, sugar cane, cotton and oil palm. Maize is the main staple, together with cassava, yams and millet. The hardwoods of the Cabinda forests are also an important source of revenue, and the coastal fisheries bring in mackerel and sardines.

Up until the 1960s diamonds were the only mineral resource to be exploited. Since then substantial deposits of iron and other metal ores have been opened up. Extensive petroleum and natural-gas fields, mostly off the coast of Cabinda, now provide the main source of foreign income, and are the only energy source apart from hydroelectricity.

The main manufacturing industries are petroleum and steel processing. Other concerns are on a smaller scale, and are mostly for domestic markets; products include food, textiles, building materials and plastics. Petroleum products account for the vast majority of exports, together with diamonds and food products; the main imports include machinery and chemicals. The road system remains poor, and railroads are used mainly for transporting freight and raw materials.

Health and welfare services are in theory available to all, but there is little provision in rural areas, where disease and malnutrition are rife. Primary education is free, but has been hampered by the civil war. However, literacy levels are rising thanks to the greater use of African languages.

NATIONAL DATA - ANGOLA

Land area	1,246,700 sq km (481,354 sq mi)			

Climate		Temperatures		Annual
	Altitude m (ft)	January °C(°F)	July °C(°F)	precipitation mm (in)
Luanda	44 (144)	26 (67)	21 (70)	323 (12.7)

Major physical features	highest point: Mount Moco 2,559 m (8,397 ft)

Population	(2006 est.) 12,127,071

Form of government	presidential multiparty republic

Armed forces	army 100,000; navy 1,000; air force 6,000

Capital city	Luanda (2,977,212)

Official language	Portuguese

Ethnic composition	Ovimbundu 37%; Kimbundu 25%; Bakongo 13%; Mestico (mixed European and native African) 2%; European 1%; other 22%

Religious affiliations	traditional beliefs 47%; Roman Catholic 38%; Protestant 15%

Currency	1 kwanza (AOA) = 100 centimos

Gross domestic product	(2006 est.) U.S. $51.95 billion

Gross domestic product per capita	(2006 est.) U.S. $4,300

Life expectancy at birth	male 37.47 yr; female 39.83 yr

Major resources	petroleum, natural gas, diamonds, iron ore, phosphates, copper, feldspar, gold, bauxite, uranium, asphalt, bananas, cassava, citrus fruits, coffee, cotton, fish, maize/corn, palm oil, salt, sisal, sugar beet, sweet potatoes, timber

Zambia

REPUBLIC OF ZAMBIA

ZAMBIA IS A LANDLOCKED COUNTRY IN south-central Africa. Congo (DRO) to the north almost divides it in two. It also shares borders with many other states, and its history has been closely linked with Zimbabwe to the south.

GEOGRAPHY

Zambia occupies a broad, rolling plateau, broken only by scattered mountains and deep valleys. The Zambezi river rises in Zambia, flows through part of Angola and back into Zambia forming the country's southern frontier with Namibia and Zimbabwe. It has been dammed at Kariba to form the great Lake Kariba. The Kafue river rises farther east, spreading out into wide floodplains to the south of the capital, Lusaka, before joining the Zambezi river via a gorge to the north of Kariba. The Luangwa river rises in the eastern mountains on the border with Malawi, then runs southwest along a broad rift valley to join the Zambezi river at the Mozambique border. North of here, across the Muchinga Escarpment, is a swampy basin around Lake Bangweulu. This is drained by the Luapula river, which runs north along the Congo border to enter Lake Mweru in another deep rift valley. The southern end of Lake Tanganyika occupies the far northeast of Zambia.

The climate is subequatorial, but is moderated by the country's altitude and location. The warm wet season (from November to April) is characterized by high humidity and heavy rain, often with thunderstorms. The dry season is cooler, with night frosts in mountainous areas, but becomes very hot soon before the next rains begin.

Most of the country is covered by "bush" – a mixture of woodland and savanna grassland. Larger forests grow in the southwest, consisting chiefly of valuable Rhodesian teak. The floodplains

In Lusaka a striking monument commemorates the 1970 Conference of Non-Aligned Countries, opposed to colonialism and the dominance of the major world powers. It stands in sharp contrast to the pillared colonial facade of the High Court building.

are covered with richer grassland, and the lake and swamp areas are thick with papyrus reeds. Wildlife is varied, including characteristic African grassland species such as giraffes, elephants, leopards and lions. However, it has been decimated by hunting, and destruction of habitat for livestock grazing. Game reserves have been established to preserve the wildlife, including the Kafue and Luangwa National Parks, but there are wide variations in the protection and conservation measures that are used.

SOCIETY

Humankind's early ancestors lived in the Zambia area millions of years ago, and it has been inhabited by various peoples ever since. Most of the present ethnic groups, however, arrived from the north, probably in the 17th century, not long before the first Europeans. Two small groups, the Ngoni and the Kololo, arrived as late as 1835, fleeing from the Zulus in the far south.

In the late 19th century Arab slave traders began to raid the extreme east of what is now Zambia. The country first came prominently to European notice with the explorations of the Scottish missionary David Livingstone (1813–73) in the Upper Zambezi area, where he discovered and named the Victoria Falls in 1855. Later the British South Africa Company, led by the British-born South African politician Cecil Rhodes (1853–1902), began to expand into the area from the south, concluding treaties with many local peoples. In 1911 the country was renamed Northern Rhodesia, with its capital at Livingstone. European settlement began to increase with the establishment of copper mines, and in 1924 the country was formally made a British protectorate.

After World War II a nationalist movement grew up, headed by the United National Independence Party (UNIP) and

David Livingstone's statue stands gazing toward the Victoria Falls on the Zambezi river, bordering Zambia and Zimbabwe. A Scottish missionary and explorer, Livingstone was the first European to see these falls, and it was he who gave them their present name.

retained its legislative function. After unrest in 1990 multiparty politics returned, and Frederick Chiluba, a former trade unionist, was elected president in democratic elections in 1991 and 1996.

English is the official language, but the majority speak a variety of Bantu languages, the most important being Bemba in the far northeast, Nyanja in the east and Tonga in the south. Although most people are nominally Christian, many also hold to their traditional (animistic) African beliefs.

ECONOMY

Agriculture employs a majority of the work force, but most production is absorbed by the domestic market. Most export crops are grown on large European-run farms. They include tobacco, sugar cane, cotton and groundnuts (peanuts). Most farmers are subsistence producers. Maize is the main food crop, along with millet, cassava, sorghum and various vegetables. Cattle are the main livestock, but are limited by the local incidence of tsetse fly. The teak forests in the southwest are extensively exploited, and there are newer plantations in the Copperbelt along the border with Congo (DRO). The many lakes support a flourishing fishing industry.

Zambia has effectively a single-commodity economy, based on copper. The extensive ore deposits in the Copperbelt have made Zambia the world's fourth-largest producer. This has funded much of the country's development, but has also made it vulnerable to world price fluctuations. Other metal ores and precious stones are also mined. Some coal is extracted, but hydroelectricity is the main energy source. Manufacturing benefited somewhat from the break in Rhodesian trade, which encouraged local enterprise. Among the major industries are copper fabrication and the refining of imported petroleum, obtained via a Tanzanian pipeline. Other products include textiles, tires, fertilizers and explosives.

Rhodesia's UDI cut the main transportation links, so a new northeastern railroad was built through Tanzania. The road system is still mostly unpaved, but is gradually being improved. Aero Zambia is the national carrier and there is an international airport at Lusaka. Health provision has improved despite staff shortages, and headway has been made in combating endemic diseases. Primary education is available to most children, and further education is being developed.

its rival, the African National Congress (ANC). In 1953 the country formed a federation with Southern Rhodesia and Nyasaland (now Zimbabwe and Malawi), but in 1963 an African majority gained control of the government and dissolved the federation. In 1964 full independence was granted, and the UNIP leader, Kenneth Kaunda (b. 1924), became Zambia's first president.

Zambia relied heavily on its southern neighbor Rhodesia (now Zimbabwe) for trade and communications. Consequently Rhodesia's unilateral declaration of independence (UDI) in 1965 severely disrupted the Zambian economy until 1980, when links were restored with the new Zimbabwean state. From the 1960s until 1989 Zambia provided a refuge for guerrilla forces operating in Rhodesia, Angola, Mozambique and finally South Africa. Meanwhile inter-party violence, mostly reflecting differences between the ethnic groups, led to the creation in 1972 of a one-party state. A new constitution in 1973 gave more power to the president, although the single-chamber parliament

NATIONAL DATA – ZAMBIA				
Land area 740,724 sq km (285,995 sq mi)				

Climate		Temperatures		Annual
	Altitude m (ft)	January °C(°F)	July °C(°F)	precipitation mm (in)
Lusaka	1,297 (4,195)	22 (71	16 (61)	843 (33.1)

Major physical features highest point: Nyika Plateau (part) 2,164 m (7,100 ft)

Population (2006 est.) 11,502,010

Form of government multiparty republic with one legislative house

Armed forces army 13,500; navy 1,600; air force 1,400

Capital city Lusaka (1,346,522)

Official language English

Ethnic composition Bemba 36.2%; Maravi 17.6%; Tonga 15.1%; Northwestern tribes 10.1%; Barotze 8.2%; Mambwe 4.6%; Tumbuka 4.6%; others 3.6%

Religious affiliations Christian 50%–75%, Muslim and Hindu 24%–49%, indigenous beliefs 1%

Currency 1 Zambian kwacha (ZMK) = 100 ngwee

Gross domestic product (2006 est.) U.S. $11.51 billion

Gross domestic product per capita (2006 est.) U.S. $1,000

Life expectancy at birth male 39.76 yr; female 40.31 yr

Major resources copper, cobalt, zinc, lead, coal, emeralds, gold, silver, uranium, hydropower, fisheries, cassava, cattle, groundnuts, maize/corn, tobacco, sugarcane

Malawi

REPUBLIC OF MALAWI

M ALAWI LIES AT THE SOUTHERN END OF THE East African Rift Valley, bordering Zambia to the west, Tanzania to the northeast, and Mozambique to the east and to the southwest.

GEOGRAPHY

Northeastern Malawi is occupied by Lake Malawi, which covers nearly one-fifth of the total area. Its southern outflow, the Shire river, runs through a deep, often swampy valley flanked by mountains to the east. Central and western Malawi consists mostly of plateaus, with high mountains in the north.

The tropical climate is modified by the country's landlocked position and by variations in relief. Temperatures are highest just before the summer rainy season in the Rift Valley, while rainfall is greatest in the northern highlands.

Savanna grasslands in the valleys give way to open woodland higher up, and then to grassland and evergreen forest in the mountains. Much of the original woodland has been cleared for cultivation. Larger animals are mainly confined to the reserves, although hippopotamuses are found in Lake Malawi.

SOCIETY

Bantu speakers first settled in the area many centuries ago. Muslim slave traders arrived in the early 19th century, closely followed by Christian missionaries. In 1891 Britain created the Nyasaland Districts Protectorate, which became Nyasaland in 1907. In 1953 the country joined a federation with Northern and Southern Rhodesia, where it remained until 1963.

Malawi achieved full independence in 1964, becoming a single-party republic two years later. In 1971 its first president,

Hastings Banda (1905–1997), was proclaimed president for life but this was repealed in 1993 following a referendum which introduced a multiparty democracy. In the first multiparty election held in 1994 the United Democratic Front led by Bakili Muluzi won an overwhelming victory over Banda's Malawi Congress Party. Malawi has had differences with its African neighbors but during Mozambique's civil war, it took in refugees.

Almost all the population are Bantu-speaking; English, though an official language, is spoken by a minority. About half the population are Christian, but traditional beliefs are also widely held, and there is a strong Muslim minority.

ECONOMY

Maize is the chief staple, while cash crops, notably tobacco, sugar cane and

Happy faces (*above*) Young girls at Nkhotakota on the shore of Lake Malawi. The spectacular woodland scenery, glimmering lakes and the dramatic waterfalls in the south make Malawi a tourist paradise.

The Nyika Plateau (*left*) in northern Malawi. Flowers enliven the bleak highland landscape, which is one of the wildest in the country. Much of this area on the Zambian border has been declared a nature reserve.

The road system is limited, and most roads are unsurfaced. Railroad links with Mozambique give access to the sea at Beira and Nacala, but in the 1980s these were effectively cut off by the civil war in Mozambique. Malawi now spends over 40 percent of gross national product on transportation by truck through Zambia, Botswana and Zimbabwe to South African ports. The Shire river is navigable, and there are passenger and cargo services on Lake Malawi. The national airline, Air Malawi, operates an international service from Lilongwe, the capital city since 1975, in the center of the country. Press and broadcasting are controlled by the government.

Some free medical treatment is provided, but preventive health care is limited. Major diseases include malaria, bilharzia, gastrointestinal diseases and leprosy. Housing is in short supply, but welfare help is available for children, the handicapped and the poor. Limited resources mean that less than half the school-age population have access to education, and literacy is correspondingly low.

tea, account for the majority of export revenue. New forest plantations have created a surplus for export. Fishing, mostly from Lake Malawi, is also an important industry. Mineral resources are scarce, but there is abundant water and hydroelectric potential. Most power comes from this source, reducing import bills for fuel. Many consumer products are now manufactured, but basic products and machinery are imported.

NATIONAL DATA - MALAWI

Land area 94,080 sq km (36,324 sq mi)

Climate		Temperatures		Annual
	Altitude m (ft)	January °C(°F)	July °C(°F)	precipitation mm (in)
Lilongwe	1,095 (3,593)	22 (71)	15 (59)	775 (30.5)

Major physical features highest point: Mount Mulanje 3,000 m (9,843 ft); longest river: Shire (part) 400 km (250 mi); largest lake: Lake Malawi 29,600 sq km (11,400 sq mi)

Population (2006 est.) 13,013,926

Form of government multiparty republic with one legislative house

Armed forces army 5,300

Largest cities Blantyre (618,153); Lilongwe (capital - 721,988); Mzuzu (143,320); Zomba (86.217)

Official language English

Ethnic composition Maravi 58.3%; Lomwe 18.4%; Yao 13.2%; Ngoni 6.7%; others 3.4%

Religious affiliations Christian 79.9%; Muslim 12.8%; other 3%; none 4.3%

Currency 1 Malawian kwacha (MWK) = 100 tambala

Gross domestic product (2006 est.) U.S. $8.038 billion

Gross domestic product per capita (2006 est.) U.S. $600

Life expectancy at birth male 41.93 yr; female 41.45 yr

Major resources limestone (marble), hydropower, unexploited uranium, coal and bauxite, beans, cassava, cotton, groundnuts, maize/corn, millet, sugarcane, sorghum, tobacco, tea

Mozambique

REPUBLIC OF MOZAMBIQUE

MOZAMBIQUE IS A LARGE, IRREGULARLY shaped country on Africa's southeast coast. It is bordered by South Africa and Swaziland to the southwest, Zimbabwe and Zambia to the west, and Tanzania to the north; southern Malawi is sandwiched between western and northern Mozambique. The east coast faces the island of Madagascar across the Mozambique Channel.

GEOGRAPHY

The wide coastal plain occupies most of the south, but becomes narrower farther north. Away from the sea the land rises in a series of steps to the tablelands of the High Veld. The granite plateaus of the north, beyond the Zambezi river, are higher than the sandstone plateaus of the south. In the far northwest the land drops sharply to the shores of Lake Malawi.

The coastal plain is crossed by numerous rivers. These include the great Zambezi, the Lugenda, the Save and the Limpopo. The capital, Maputo (formerly Lourenço Marques), lies in the far south on Delagoa Bay, which is the finest natural harbor on Africa's east coast. North of here the coastline features sandbars and mangrove swamps, but beyond the Zambezi river it is lined by sandy beaches interspersed with rocky cliffs and small islands.

The climate is subtropical, with a very marked division between rainy and dry seasons. However, it is surprisingly dry in view of the prevailing onshore winds and the warm offshore Mozambique Current. Many areas suffer from drought, especially south of the Zambezi river, and high temperatures increase water loss. Plateau areas enjoy cooler temperatures and higher rainfall, especially in the north.

Savanna woodland and grassland are the dominant vegetation types, with mangroves and palms (especially coconut palms) along the coast. Typical trees include ironwood, ebony and khaya, with mopani and baobab on the dry savanna. Bamboo and spear grass are common in river and swamp areas. Wildlife is still abundant despite excessive hunting. Mammals include White rhinoceroses, zebras, warthogs and various species of monkey. Bird life is similarly varied, with flamingoes, cranes, ibis and many other species.

SOCIETY

When the Portuguese navigator Vasco da Gama (1460–1524) reached Mozambique in 1498, he found Arab and Swahili traders already active there. The Portuguese founded several settlements, both along the coast and up the course of the Zambezi river, where they sought gold from the nearby empire of the Mwene Matapa. However, they administered the colony from Goa in India until 1752. Slave trading then became a bulwark of its economy, and continued, albeit illegally, until the 1850s. At first the inland areas were controlled by powerful landholders, but these were eventually replaced by private companies. From the late 19th century the government increasingly took over the administration.

In 1930 limited autonomy was replaced by centralized government, and in 1951 Mozambique was declared a Portuguese overseas province. A nationalist movement led by the Mozambique Liberation Front (Frente de Libertação de Moçambique, or Frelimo) erupted into violence in 1964. The war was long and costly, and contributed to the fall of the government in Portugal in 1973. Independence in 1975 brought Frelimo at last to power.

The resulting one-party state was headed by a president, who in turn was chosen by the People's Assembly. Election to this body was by delegation from regional and local levels of government.

The government's authority was very quickly challenged by rebel movements, resulting in a prolonged civil war. The main armed opposition was from the Mozambique National Resistance (MNR), also known as Renamo (Resistência Nacional Moçambicana), which was backed by South Africa. The late 1980s saw an improvement in relations between the Frelimo government and South Africa, with promises of cooperation and joint development in return for greater democracy. However, the Nkomati Accords – a non-aggression pact signed between South Africa and Mozambique – were violated soon after the signing ceremony. In 1992 the United Nations agreed to provide troops to supervise demobilization of warring armies and oversee future elections. First troops arrived in 1993. Presidential and multiparty legislative elections were finally held in 1994.

Most Mozambiquans belong to one of ten major ethnic groups, the largest of which is the Makua-lomwe, accounting for almost half the population. There are tiny minorities of Europeans and Asians. The official language is Portuguese, but few can actually speak it. Several distinct Bantu languages are spoken in different parts of the country, and many people speak Swahili. Almost half the population hold traditional African beliefs. The two other main religious groups are Muslims, mostly in the north, and Christians, mostly in the south.

ECONOMY

Mozambique is one of the poorest countries in southern Africa. It suffered disastrous floods in 1998 and again in 2000. The Mozambiquan economy is centrally planned, and chiefly depends on agriculture, fishing and light industry. Cash crops include cashew nuts, cotton, sugar cane, copra and citrus fruits. Shrimps and lobsters are also exported. Maize is the

NATIONAL DATA – MOZAMBIQUE

Land area 784,090 sq km (302,739 sq mi)

Climate	Altitude m (ft)	Temperatures January °C(°F)	July °C(°F)	Annual precipitation mm (in)
Maputo	60 (197)	27 (80)	19 (56)	814 (32)

Major physical features highest point: Monte Binga 2,436 m (7,992 ft)

Population (2006 est.) 19,686,505

Form of government multiparty republic with one legislative house

Armed forces army 10,000; navy 200; air force 1,000

Largest cities Maputo (capital – 1,249,309); Matola (570,242); Beira (561,151); Nampula (407,338)

Official language Portuguese

Ethnic composition Traditional tribal groups 99.66% (Makhuwa, Tsonga, Lomwe, Sena, and others); Europeans 0.06%; Euro-Africans 0.2%; Indians 0.08%

Religious affiliations Catholic 23.8%; Muslim 17.8%; Zionist Christian 17.5%; other 17.8%; none 23.1%

Currency 1 metical (MZM) = 100 centavos

Gross domestic product (2006 est.) U.S. $29.32 billion

Gross domestic product per capita (2006 est.) U.S. $1,500

Life expectancy at birth male 39.53 yr; female 40.13 yr

Major resources coal, titanium, natural gas, hydropower, tantalum, graphite, clothing, bananas, cashew nuts, cassava, cereals, coconuts, sisal, tea

Colorful Mozambiquans (*above*) The introduction of modern dyes has transformed the dress of African women by making colored fabrics – until the 20th century a luxury only for the rich – easily available.

Rural tranquility (*left*) in the Niassa province of northern Mozambique. The children running down this country track are more fortunate than those thousands, mainly in the south, who have been injured or made homeless in the civil war.

chief staple, but production cannot fully meet demand, and food remains a major import. Tsetse flies, which transmit sleeping sickness to cattle, prevent livestock rearing in many areas.

Mozambique's great mineral potential has so far remained largely untapped, and production has mostly been limited to coal, limestone and salt. However, natural gas and iron ore are now being exploited, together with small amounts of tantalite (a rare mineral). Manufacturing industry is mainly based on the processing of local raw materials. Machinery, chemicals and spare parts have to be imported. Electricity is generated both from thermal stations and from hydroelectric plants, most notably the Cabora Bassa Dam on the Zambezi river.

Mozambique also supplies thousands of migrant workers to South Africa in return for payment in gold at the official rate. It then sells the gold at the much higher free-market rates in order to reduce the overall economic deficit.

The transportation system was originally designed for colonial exploitation rather than to meet local needs. The three major ports of Maputo in the south, Beira in the center and Nacala in the north are the focal points for both road and rail. Much of the country is still without roads, and only a small part of the existing network is surfaced. Zambia and Zimbabwe are working with Mozambique to improve and expand the railroad network, which provides a link for these two countries to the coast, reducing their reliance on South Africa. There are international airports at Maputo, Beira and Nampula in the north, as well as many smaller airports and rural airstrips.

Health care has been nationalized and medicine is free, although private healthcare was introduced in 1992. Government efforts have so far failed to combat major diseases such as malaria, tuberculosis, hepatitis and pneumonia. Education too has been nationalized, and is compulsory up to the age of 14. An adult literacy scheme has been established but enrollment is low, and this is reflected in the lower literacy levels.

Namibia

REPUBLIC OF NAMIBIA

THE NEWLY INDEPENDENT COUNTRY OF Namibia on the southwest coast of Africa borders South Africa to the south, Botswana to the east and Angola to the north. From its northeastern corner, the narrow Caprivi Strip extends its territory eastward as far as Zambia. The main port, Walvis Bay, was South African territory until 1994.

GEOGRAPHY

The sand dunes of the Namib Desert extend from the coast toward the foot of the Great Escarpment that rims the great central plateau. The mountains of the escarpment are highest in the central

NATIONAL DATA - NAMIBIA				
Land area 825,418 sq km (318,696 sq mi)				
Climate		Temperatures		Annual
	Altitude m (ft)	January °C(°F)	July °C(°F)	precipitation mm (in)
Windhoek	1,738 (5,669)	12 (54)	13 (56)	362 (14.2)
Major physical features highest point: Mount Brand 2,574 m (8,445 ft)				
Population (2006 est.) 2,044,147				
Form of government multiparty republic with two legislative houses				
Armed forces army 9,000; navy 200				
Capital city Windhoek (286,762)				
Official language English				
Ethnic composition Ovambo 49.8%; Kavango 9.3%; Herero 7.5%; Damara 7.5%; European 6.4%; Nama (Hottentots) 4.8%; others 14.7%				
Religious affiliations Christian 80%–90% (Lutheran 50% at least); traditional beliefs 10%–20%				
Currency 1 Namibian dollar (NAD) = 100 cents and 1 South African rand (ZAR) = 100 cents				
Gross domestic product (2006 est.) U.S. $15.04 billion				
Gross domestic product per capita (2006 est.) U.S. $7,400				
Life expectancy at birth male 44.46 yr; female 42.29 yr				
Major resources diamonds, copper, uranium, gold, silver, lead, tin, tungsten, lithium, cadmium, zinc, salt, hydropower, fish, possible deposits of oil, coal, and iron ore, cattle, maize/corn, millet, sorghum				

portion around the capital, Windhoek. The plateau itself occupies the eastern part of the country, and is covered by the dry scrub of the Kalahari Desert.

The climate is generally hot and very dry. Seasonal rainfall is largely confined to the north. The Namib Desert supports little vegetation apart from succulents and the extraordinary welwitschia plant, which absorbs moisture from the fog that rolls in from the sea. The welwitschia, often described as a living fossil, can survive to a great age; some specimens are thought to be up to 2,000 years old.

Inland areas are less barren, however, with many drought-resistant grasses and shrubs, and savanna woodland in the extreme north. The rich wildlife is typical of southern Africa, and includes baboons, kudu antelopes, elephants, giraffes, zebras, lions and flamingoes. The Etosha National Park is among the world's largest game parks.

SOCIETY

South West Africa, as Namibia was formerly known, became a German protectorate in 1884. During World War I it was occupied by South Africa, to which it was mandated in 1920 by the League of Nations as a trust territory. In 1946 it came under the United Nations trusteeship system, despite South Africa's stated intention to annex the country. In 1966 the United Nations called for total South African withdrawal, and the South West Africa People's Organization (SWAPO) began a lengthy guerrilla war against the continuing South African occupation. Free elections in 1989 brought SWAPO to power, and independence followed soon afterward in 1990.

The multiparty government is headed by a president, who is accountable to a two-chamber parliament. Both the president and the lower house are directly elected every five years. Namibia is still largely economically dependent on South Africa, with which it remains in customs union; but it is gradually becoming more independent.

The two oldest racial groups are the desert-dwelling San (Bushmen) and Nama (Hottentots). Bantu peoples include the Kavango and the Ovambo – the largest ethnic group, who occupy the more populated northern areas. Europeans are mainly confined to the towns; Afrikaans and English are the two official languages, but German is spoken by a large number of people. Christianity is the dominant religion.

ECONOMY

Most people work on the land, but the soils and climate do not favor agriculture. About half Namibia's food is imported, largely from South Africa. Livestock farming is the chief activity, producing mainly beef, mutton and karakul (Persian lamb) pelts. The once-prosperous offshore fishing industry is gradually recovering from overfishing in the 1970s.

Mineral resources include diamonds, gold and various metal ores, including uranium dioxide. Copper and lead are smelted at Tsumeb in the north of the country, but other minerals are exported unprocessed. Namibia depends on South Africa for fuel and manufactured goods. Its own industries are limited to food processing. The road and rail networks, though expanding, are still largely confined to the south, so that local air services are vital; there is an international airport at Windhoek.

Healthcare services are concentrated in a few large hospitals. Major diseases include tuberculosis, measles and malaria. Primary education is compulsory, but literacy is low outside the white community; skilled training must be sought in South Africa.

Dressed up to the nines The traditional costume of these Herero women was copied from 19th-century German Lutheran missionaries. The long skirts and puff-sleeved bodices have been colorfully adapted.

Botswana

REPUBLIC OF BOTSWANA

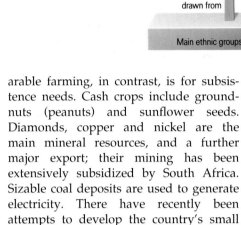

An African democracy (*right*)
Botswana's national assembly elects a president to head the government. He is advised by a house of chiefs, elected from eight ethnic groups. There are two main parties.

Red lechwe antelopes (*below*)
are specially adapted to living in the swampy Okavango Basin of northern Botswana, an area that constitutes one of the richest wildlife reserves in Africa.

Botswana is a landlocked state that borders Namibia to the west and north, South Africa to the south, and Zimbabwe and Zambia to the northeast. It is in customs union with South Africa, on which its economy depends.

GEOGRAPHY

The country is part of the southern African plateau. It is covered in the west by the dry scrub of the Kalahari Desert, and in the east by savanna grassland. Surface water is confined to the north and east, in the river basins of the Okavango, the Chobe and the Limpopo. The Okavango river flows southeast from Angola via Namibia and into the Okavango Basin, from where there is an eastward seasonal flow toward Lake Dow and the Makgadikgadi Pan. The climate varies from semiarid to subtropical, with a hot season from October to March. Rainfall is scanty and unpredictable, especially in the south, and droughts can last for years.

Savanna grassland is the dominant vegetation type, while palms grow in the Okavango Delta and deciduous forests border the Chobe river on the northern border. Wildlife is rich and plentiful in the many game reserves and national parks. Typical animals include lions, elephants, giraffes and many species of poisonous snake.

SOCIETY

The original inhabitants of present-day Botswana were nomads called the San (Bushmen), of whom few remain. Eastern parts were mainly settled by Bantu peoples called the Tswana. The first European arrivals were British missionaries from the south in the early 19th century. From the 1820s onward there were raids by the Ndebele and other refugees escaping from the northward advance of the Zulus. In 1835 the Boers – descendants of the 17th-century Dutch settlers in South Africa – reached the southern borders of the country. The discovery of gold in the east led to several territorial conflicts. In 1884 the British occupied the area at the instigation of the British-born South African politician Cecil Rhodes (1853–1902), and in 1885 it became the British Bechuanaland Protectorate. It remained so until full independence was achieved in 1966.

Legislative power rests with the 40-member parliament, most of whom are directly elected for a five-year term; there are many different political parties. The 15-member House of Chiefs advises on ethnic and constitutional matters. There are eight major ethnic groups, but the Tswana account for three-fourths of the population. The two official languages are Tswana and English. Christianity is the dominant religion, though traditional beliefs are also widely held.

ECONOMY

Agriculture is dominated by cattle farming, which provides a major export. Most arable farming, in contrast, is for subsistence needs. Cash crops include groundnuts (peanuts) and sunflower seeds. Diamonds, copper and nickel are the main mineral resources, and a further major export; their mining has been extensively subsidized by South Africa. Sizable coal deposits are used to generate electricity. There have recently been attempts to develop the country's small manufacturing base, and to exploit its possibilities for tourism.

The main railroad was formerly operated by Zimbabwe, but is now under Botswanan management. Most roads are of dirt or gravel, but there is an all-weather north–south artery. Domestic and international flights are provided by Air Botswana. Health care is widely available through mobile clinics. Major health problems include tuberculosis, gastroenteritis and malnutrition. There is free primary education, and literacy levels are relatively high.

NATIONAL DATA – BOTSWANA

Land area	585,370 sq km (226,013 sq mi)			
Climate	Altitude m (ft)	Temperatures January °C(°F)	July °C(°F)	Annual precipitation mm (in)
Francistown	983 (3,225)	25 (77)	15 (59)	448 (17.6)

Major physical features	highest point: Tsodilo Hills 1,375 m (4,511 ft)
Population	(2006 est.) 1,639,833
Form of government	multiparty republic with one legislative house
Armed forces	army 8,500; air force 500
Capital city	Gaborone (220,558)
Official language	English
Ethnic composition	Tswana (or Setswana) 79%; Kalanga 11%; Basarwa 3%; other, including Bushmen, Kgalagadi, and White 7%
Religious affiliations	Christian 71.6%; Badimo 6%; other 1.4%; unspecified 0.4%; none 20.6%
Currency	1 pula (BWP) = 100 thebe
Gross domestic product	(2006 est.) U.S. $18.72 billion
Gross domestic product per capita	(2006 est.) U.S. $11,400
Life expectancy at birth	male 33.9 yr; female 33.56 yr
Major resources	diamonds, copper, nickel, salt, soda ash, potash, coal, iron ore, silver, asbestos, beans, cattle, meat processing, sorghum, manganese, maize/corn, tourism

Zimbabwe

REPUBLIC OF ZIMBABWE

Members of the ZANU PF District Choir (*above*) at the national Heroes' Day celebrations in the Zimbabwean capital, Harare. The image of Robert Mugabe on their costumes proclaims the leading role that he played in their liberation struggle.

ZIMBABWE, FORMERLY SOUTHERN RHODESIA, lies in southeastern Africa between the Zambezi river, which forms its northern border with Zambia, and the Limpopo river, which forms the border with South Africa. Botswana lies to the west and Mozambique to the east.

GEOGRAPHY

Zimbabwe is dominated by a broad mountainous ridge known as the High Veld, which spans the country from the southwest to the northeast. To the southeast of the capital, Harare (formerly Salisbury), is mount Inyangani, the highest point of the escarpment that borders Mozambique. The mountains are flanked to the north and south by the rolling plateaus of the Middle Veld. These in turn fall away to the Low Veld of the two great river basins. The Zambezi river on the northern border plunges over the Victoria Falls into a series of gorges. Farther downstream the valley is overlooked by steep escarpments, and has been dammed and then flooded to create Lake Kariba. The Limpopo river in the far south flows through much gentler country. Both rivers flow out through Mozambique into the Indian Ocean.

Zimbabwe lies within the tropics, but its climate is modified by altitude and aspect. Winters are dry and often cool, and rain is mostly confined to the hotter summer months. The rains are heaviest in the eastern mountains, which support areas of evergreen forest. The land to the south around the Limpopo river receives the least rain, and is arid in places. Over most of the country the natural vegetation is savanna grassland with extensive woodlands, where native trees include mohobohobo and baobab.

Large areas have been cleared for cultivation, limiting the habitat of the local wildlife: baboons, elephants, aardvarks and gorillas, with hippopotamuses and crocodiles in the rivers. Some conservation measures have been adopted, mostly for the benefit of tourism.

SOCIETY

The name Zimbabwe (meaning "house of the chief" in the language of the Shona people) was taken from the Great Ruins, massive fortifications in the south of the country. Once thought to be the legendary King Solomon's Mines, the oldest parts date from about the 9th century, and were built by Bantu invaders who had driven out the original inhabitants.

Some of the modern Bantu peoples arrived as late as the mid-19th century, about the same time as the Europeans, fleeing the expanding Zulu kingdom of Shaka to the south. One such group, the Ndebele, settled in the southeast (now called Matabeleland), dominating the more peaceable Shona people. In the 1890s the British South Africa Company, founded by the British-born South African politician Cecil Rhodes (1853–1902), began the settlement of the country. Both ethnic groups rebelled, but to little effect. The country was run by the British South Africa Company until 1923, when it rejected union with South Africa to become the autonomous British colony of Southern Rhodesia.

In 1957 limited political rights were extended to the majority African population. Their National Democratic Party (NDP) opposed the federation with Northern Rhodesia and Nyasaland (now Zambia and Malawi respectively), which had been established in 1953. The NDP was banned following a number of riots, and the African movement split into broadly ethnic divisions. The Ndebele people supported the Zimbabwe African People's Union (ZAPU), led by Joshua Nkomo (b. 1917), while the majority Shona peoples formed the Zimbabwe African National Union (ZANU) under Ndabaningi Sithole (b. 1920) and later Robert Gabriel Mugabe (b. 1925).

When the federation collapsed in 1963, the European-dominated Rhodesian Front party, led by the white Rhodesian politician Ian Douglas Smith (b. 1919), pressed for immediate independence under their own minority rule. When Britain failed to grant this, the Front made a unilateral declaration of independence (UDI) in 1965. Few countries recognized Rhodesia,

tion. Agriculture, for example, became more diversified. The larger farms, most of which are still European-owned, produce most of the cash crops, including tobacco, cotton and sugar cane. An accelerated land reclamation program in 2000 added to the economic destabilization. African farming is mainly small scale, producing staples such as maize and cassava, to which wheat has now been added. Beef cattle are the main livestock, and the country is self-sufficient in dairy products.

Manufacturing was greatly encouraged by sanctions pressure, and is now a major part of the economy. Industries are diverse, including textiles, light machinery and the processing of various agricultural products. Major exports include tobacco and gold, while imports include machinery, manufac-tured goods and fuel. Fuel is needed for electricity generation, although some hydroelectricity is available.

The rail system is excellent, with links to neighboring countries, including the nearest ports in Mozambique. The poor road system is being improved. The main airport is at Harare in the northwest. Health care and education were well provided for in the 1990s until foreign debt crippled government budgets.

The mighty stone walls of the Great Zimbabwe ruins, part of a vast complex of buildings that were occupied from about the 9th to the 17th century. At its height it was the center of an extensive trading empire, whose wealth was based on gold.

and the United Nations imposed economic sanctions. However, South Africa and other states continued to allow vital supplies into the country. ZANU and ZAPU launched a guerrilla campaign against Ian Smith and his party from inside Zambia and later Mozambique. Eventually the parties united to form the Patriotic Front.

The war continued, and mounting international pressure forced the prime minister, Ian Smith, to make concessions toward black majority rule, allowing the formation of the more moderate United African National Council (UANC) under Bishop Abel Muzorewa (b. 1925). An internal settlement in 1979 led to multiracial elections and a UANC government. Eventually, a new constitution led to elections in 1980 which the Patriotic Front won, and Robert Mugabe became president of a fully independent Zimbabwe.

The Patriotic Front split, and Nkomo was ousted from government. However, in 1987 ZANU and ZAPU were again reunited. In 1990 the 1979 constitution lapsed, and a new constitution created a single-chamber parliament in what had effectively become a one-party republic. In the late 1990s as Zimbabwe plunged into economic crisis, an opposition Movement for Democratic Change (MDC) arose. The MDC won significant support in national elections in 2000 despite intimidation at the polls. There were calls for Mugabe to resign. After the March 2002 presidential elections, Zimbabwe was suspended from the Commonwealth for one year as observers declared the elections flawed by violence and intimidation. Mugabe then charged MDC leader, Morgan Tsvangirai, with treason.

The official language is English, although a majority speak Chishona as their first language. Most indigenous Africans hold traditional beliefs, but a large number are Christian.

ECONOMY

The economy suffered during the war and sanctions of the 1970s, but some sectors benefited from the lack of outside competi-

NATIONAL DATA – ZIMBABWE

Land area	386,670 sq km (149,294 sq mi)			

Climate		Temperatures		Annual
	Altitude m (ft)	January °C(°F)	July °C(°F)	precipitation mm (in)
Harare	1,479 (4,852)	21 (70)	14 (57)	841 (33.1)

Major physical features	highest point: Inyangani 2,592 m (8,504 ft)

Population	(2006 est.) 12,236,805

Form of government	one-party republic with one legislative house

Armed forces	army 25,000; navy 4,000; air force 21,800

Largest cities	Harare (capital - 1,607,022); Bulawayo (713,340); Chitungwiza (352,204); Mutare (193,629); Gweru (148,935)

Official language	English

Ethnic composition	Shona 82%; Ndebele 14%; other Africans 2.0%; Asian 1.0%;

Religious affiliations	syncretic (Christianity and indigenous beliefs) 50%; Christian 25%; indigenous 24%; others 1%

Currency	1 Zimbabwean dollar (ZWD) = 100 cents

Gross domestic product	(2006 est.) U.S. $25.05 billion

Gross domestic product per capita	(2006 est.) U.S. $2,000

Life expectancy at birth	male 40.39 yr; female 38.16 yr

Major resources	coal, chromium ore, asbestos, gold, nickel, iron ore, vanadium, lithium, tin, copper, platinum group metals, maize/corn, millet, sugarcane, wheat, tobacco

South Africa

REPUBLIC OF SOUTH AFRICA

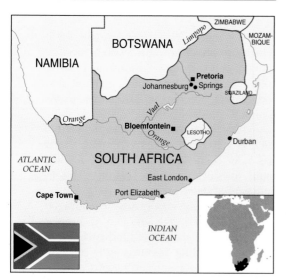

SOUTH AFRICA IS A LARGE COUNTRY THAT occupies the southern tip of the African continent, bordered by the Indian Ocean to the southeast and by the Atlantic Ocean to the southwest. In the north it borders (from west to east) Namibia, Botswana, Zimbabwe, Mozambique and Swaziland, while in the east it totally surrounds the independent state of Lesotho. Walvis Bay, the main port on the coast of Namibia, was South African territory until 1994.

From 1948 to 1991 South African politics were dominated by its policy of apartheid which involved the almost complete segregation of the white minority community from the rest of the population. Despite strong international reaction against this policy, the country remained very prosperous in a hostile economic climate, thanks to its enormous wealth of natural resources, notably its minerals, and its strategic position.

GEOGRAPHY

South Africa can be divided into two main topographic areas: the vast central interior plateau, and the lands that surround it. These areas are separated by the Great Escarpment.

The land
The Great Escarpment or plateau edge varies in altitude, structure and steepness of slope. It forms a rim around the central plateau from the Transvaal Drakensberg in the northeast to Namibia in the northwest. The highest peaks are found along the Lesotho border in the Natal Drakensberg.

The central plateau is also varied in altitude, being highest along the edges of the Natal Drakensberg and lowest in the northwest on the sandy plateau of the

Kalahari Desert. The judicial capital, Bloemfontein, lies in the center, just to the west of Lesotho. In the northeast the Witwatersrand ridge divides the basins of the Vaal and the Limpopo rivers. The Witwatersrand area is the site of South Africa's largest conurbations, including the administrative capital of Pretoria, and Johannesburg, the largest city. The Witwatersrand gold deposits occur in some of the world's oldest rocks. The central plateau is largely drained by the Orange river, of which the Vaal is a tributary.

The lands surrounding the central plateau include the dissected slopes below the plateau edge, and in the south the lower-lying plateau of the Great Karoo, which is bounded by the Great Escarpment and by the ranges of the Cape fold mountains, such as the Great Swartberge. Table Mountain, part of the fold-mountain belt, overlooks the legislative capital, Cape Town, on the southwest coast. The coastal plain is generally quite narrow except immediately south of Mozambique, and contains no good natural harbors.

Climate
Most of South Africa lies within the warm temperate zone, and the climate is predominantly dry. The higher altitudes across the vast inland plateau tend to cancel out the effects of changing latitude, so that temperatures are generally moderate over most of the country.

Off the west coast the cold Benguela Current flows northward from the southern oceans, restricting rainfall in the west of the country. Off the southeast coast is the warm Agulhas Current, which brings greater rainfall to the eastern side of the country. On the east coast most of the rain falls during the summer months of November to April. The winter months, May to October, are generally fairly cool, bringing rain or even snow to the far southwest. However, on the east coast even the winters are warm.

Uneven distribution of rainfall means that more than half the country is arid or semiarid. Only about one-twentieth of the land area receives more than 1,000 mm (40 in) of rain per year.

Vegetation and wildlife
Higher rainfall in the east creates a characteristic landscape of savanna grassland. In the drier country to the west this gives way to the semiarid scrubland of the Karoo. In the north there is dry savanna bushland. Along the west coast the effects of the cold Benguela Current

A Ngoma Zulu dance team performs in the traditional dress of their warrior elite. The Zulus are proud of their warrior traditions and their distinctive spear and shield, but performances like this are usually put on for the entertainment of tourists.

have helped to create the sand desert of the Namib. In the far southwest the vegetation is Mediterranean in character. Along the warmer east coast, small forested areas can be found in the ravines that line the Great Escarpment.

South Africa boasts a great variety of wildlife, but much of it is under threat of extinction, and the majority is confined to game reserves. Lions, leopards, rhinoceroses, elephants, hippopotamuses, antelopes and giraffes are all found here. Altogether there are eight game reserves, including the famous Kruger National Park, the Kalahari Gemsbok National Park and the Addo Elephant National Park near Port Elizabeth on the south coast. All contribute to animal conservation.

SOCIETY

South Africa's policy of apartheid produced international condemnation, economic sanctions, arms-sale embargoes and exclusion from international sporting events. There had also been violent internal opposition from black African movements such as the African National Congress (ANC). Only in recent years have there been any significant changes within the system.

History
The Portuguese explorer Bartolomeo Diaz (c. 1450–1500) rounded the Cape of Good Hope in 1488, but it was not until 1652 that the Dutch East India Company founded a post there for supplying provisions. Dutch settlers, later called Afrikaners or Boers ("farmers"), moved inland from the Cape, encountering Bantu

agriculturalists (Xhosa) who were extending their territories westward. Disputes with the Bantus developed into war, while the indigenous San (Bushmen) and Hottentot (Khoi) populations were either subjugated or driven away.

At the end of the 18th century the British twice captured the Cape Colony before purchasing it (for £6 million) in 1814. They systematically began to conquer Bantu territories to the northeast. In the 1830s disgruntled Boers left the colony to begin the Great Trek, which took them across the Orange and Vaal rivers to found the South African Republic (later known as the Transvaal) in 1838 and the Orange Free State in 1854. Britain acknowledged the independence of both states in the 1850s, but after 1870 the discovery of diamonds and gold there brought new waves of settlers. The British fought two wars with the Boers, eventually defeating them in 1902. Afterward they forged South Africa into a single colony, establishing the Union of South Africa in 1910. In the 1920s South Africa's dominion status was recognized.

During World War I South African forces drove the Germans out of South West Africa (present-day Namibia). In World War II, however, the Afrikaners

Voting for the first time, this South African woman embraces the opportunity to have a say in her own future. The huge distances that many had to travel did not deter black citizens from coming to the polling stations in their millions.

were critical of the government's decision to join what they saw as Britain's war against Germany. In 1948 the Afrikaner-backed National Party came to power, and began to implement the policy of apartheid or separate development.

Black South Africans were segregated in ten different "homelands" or Black States, which occupied only about one-eighth of the land area; no similar provision was made for other nonwhite South Africans (mainly Asians brought by the British as indentured laborers on the sugar plantations of Natal). Despite the limited resources of the homelands and their effective dependence on South Africa, the government announced that they would eventually become separate states. By 1981 four had become so, but none was recognized outside South Africa. In practice racial separation was never complete. Black laborers frequently worked alongside whites, especially in towns and in the mineral industries, but were denied skilled employment and were paid minimal wages. In cities such as Johannesburg, white residential areas are close to the center, while black areas such as Soweto lie much farther out.

In 1961 South Africa became a republic and seceded from the British Commonwealth. Anti-apartheid protest was repressed with increasing force, and in 1976 the response to a riot by unemployed black youths in Soweto brought death and injury on an unprecedented scale.

In 1990 the African National Congress (ANC) leader Nelson Mandela (b. 1918) was released after 27 years of imprisonment, signaling a change of heart on the government's part. The ANC suspended its 30-year guerrilla war against the government, while president Frederik Willem de Klerk (b. 1936) began to dismantle apartheid. A 1992 referendum approved a new constitution for multi-racial majority rule despite the fact that only whites were allowed to vote in the referendum. The country's first multi-racial elections were held in 1994. The new government announced plans to establish a "Truth and Reconciliation Commission"(TRC) to investigate violations of human rights perpetrated under the apartheid regime and to award reparations to the families of those who had suffered abuse. The TRC concluded its report in 1998 stating that the apartheid regime had been a crime against humanity. It also reported abuses by the ANC and Inkatha Freedom Party (IFP). Many public figures including Winnie Madikizela-Mandela and former president Pieter Willem Botha caused controversy.

Government

Nelson Mandela became the first president of a newly democratic South Africa in May 1994. In 1996 a new constitution was approved. One of its stated aims being to heal the divisions of the past and establish a society based on democratic

NATIONAL DATA – SOUTH AFRICA

Land area	1,219,912 sq km (471,011 sq mi)			
Climate	Altitude m (ft)	Temperatures January °C(°F)	July °C(°F)	Annual precipitation mm (in)
Cape Town	12 (39)	21 (70)	12 (59)	515 (20.2)

Population	(2006 est.) 44,187,637

Form of government multiparty republic with two legislative houses

Armed forces army 36,000; navy 4.500; air force 9,250

Capital cities Cape Town (legislative – 3,660,197); Pretoria (administrative – 1,757,505); Bloemfontein (judicial – 498,404)

Official languages Afrikaans, Ndebele, Northern Sotho, Southern Sotho, Swati, Tsonga, Tswana, Venda, Xhosa, Zulu, English

Ethnic composition Black African 79%; White 9.6%; Mixed race 8.9%; Indian/Asian 2.5%

Religious affiliations Zion Christian 11.1%; Pentecostal/Charismatic 8.2%; Catholic 7.1%; Methodist 6.8%; Dutch Reformed 6.7%; Anglican 3.8%; other Christian 36%; Islam 1.5%; other 2.3%; unspecified 1.4%; none 15.1%

Currency 1 South African rand (ZAR) = 100 cents

Gross domestic product (2006 est) U.S. $576.4 billion

Gross domestic product per capita (2006 est.) U.S. $13,000

Life expectancy at birth male 43.25 yr; female 42.19 yr

Major resources gold, chromium, antimony, coal, iron ore, manganese, nickel, phosphates, tin, uranium, gem diamonds, platinum, copper, vanadium, salt, natural gas, fruits, maize/corn, cotton, potatoes, tobacco, livestock, tourism

values, social justice and fundamental human rights. The new constitution provides for an executive president elected by a 400-seat democratically elected National Assembly and a National Council of Provinces (NCOP) which has 54 permanent members and 35 special delegates. The NCOP represents provincial interests and considers legislation passed by the National Assembly. The nine new provinces which were established in 1994, and in which the former homelands were reintegrated, have their own legislatures.

Until the early 1990s, South Africa had troubled relations with its neighbors and funded dissenting groups in order to destabilize their regimes. Under Mandela, South Africa has pursued an independent foreign policy, establishing diplomatic relations with China but also strengthening ties with the United States. South Africa showed its commitment to future regional integration by joining the Southern African Development Community (SADC) in 1994.

People

Pre-1994, the South African government distinguished four racial groups, but the true picture is far more complex. Almost three-fourths of the population are blacks, generally of Bantu ancestry, the largest groups being the Zulu, the Xhosa and the Sotho. The whites make up less than one-fifth of the population; they are the descendants of European settlers, and the majority of them are Afrikaners. The so-called coloreds are of mixed racial origin, and form about one-tenth of the population. There is also a small Asian minority, mostly of Indian descent.

Afrikaans and English are the official languages, but at least eight others are widely spoken. More than three-quarters of the people are Christian (religious practice remains very strong among rural Afrikaners, members of the Dutch Reformed Church). Some traditional beliefs survive among black Africans, and there are small Hindu and Muslim minorities.

ECONOMY

Although South Africa is rich in mineral resources, it has no petroleum deposits, and in the light of the world reaction to apartheid, this has had a profound effect on the organization of its economy.

Agriculture

Although almost half the country's population lives in rural areas, agriculture employs only a small portion of the work force, and provides an even smaller percentage of gross domestic product. Maize is the chief staple, while export crops include sugar cane, citrus fruits, grapes, cotton and tobacco. Livestock is varied, and produces a number of export products. Land shortage among the traditional subsistence farmers has forced many black South Africans to become migrant laborers.

There is good coastal and trawler fishing, mainly off the west coast, where the rich plankton in the Benguela Current attract pilchards, mackerel, sardines and snoek. Although forestry potential is relatively small, the existing stands produce the greater part of South Africa's domestic needs.

Industry and resources

It was South Africa's gold and diamond resources that helped to make the country so attractive to European settlers. Gold and precious stones still account for up to half the country's total exports. Other important minerals include asbestos, uranium and other valuable metal ores. Ore-based industries include three iron-and-steel plants and an aluminum smelter.

Energy conservation is a necessary policy, as all petroleum except that produced from coal has to be imported, and many oil-producing countries have imposed an embargo on exports to South Africa. Low rainfall reduces the potential for hydroelectricity, but several schemes have been initiated, including a large installation on the Orange river. The country's substantial coal output has been increased to supply the electricity industry, and a nuclear power station in the Western Cape has been operating since 1982. The result is that South Africa generates about half the electricity used in the African continent, and the government's energy-conservation policy allows much of this to be exported. Mozambique and Namibia are among the many countries that are linked to the South African national grid.

The development of South Africa's manufacturing industries has been exceptionally rapid, and they now provide almost one-fourth of export earnings. The principal manufactured goods include beverages, chemicals, textiles, transportation equipment and electrical goods.

Trade and commerce

South Africa's chief trading partners include the United States, Germany, Japan, Switzerland and the United Kingdom.

Foreign investment declined after the Soweto riots of 1976, and most industrial finance now comes from within the country, helped by South Africa's wealth in gold. Tariff barriers protect local manufacturing industries and reduce imports.

Transportation and communications

The backbone of South Africa's transportation system is the state-owned railroad network. At least one-third of it has been electrified, and diesel locomotives service the rest of the system. The same adminis-

A well-stocked vegetable stall (*above*). The country's subsistence farmers grow a wide variety of fruit and vegetables for their own use and for sale. Maize is the staple crop.

The imposing facade of City Hall, Cape Town (*left*) is dwarfed by the timeless bulk of Table Mountain behind. In the 1980s the ANC rally in front of it would have seemed almost unthinkable, but now the police look on without interfering.

Health and welfare

Health care is better in urban areas than rural and while the majority of white South Africans contribute to private medical schemes black people are still largely dependent on state health care. Facilities are poor in the areas which were previously segregated, the hospitals are overcrowded while poverty and malnutrition make people susceptible to infection and disease. In the late 1990s, an estimated 20 percent of the South African population was HIV positive with the highest rates among young women aged 15-25 and young men aged 20-29. In 1994, the government introduced a new national health plan which included free health care for children under 6 years and pregnant women as well as the establishment of additional health clinics.

The government is also taking action against poverty and the chronic housing shortage, providing additional grants and pensions to those in need.

A new multi-racial education system was established in 1996 providing free, compulsory education for children aged 7-15. In the 1980s the universities were formally segregated but later began to admit students of all races. By 2000, the number of African students at former white universities was around 60 percent.

tration is also responsible for bus services on the extensive road network. However, few of the small, local roads in rural areas have been surfaced.

Water transportation is confined to the coast, as there are no navigable rivers. The harbors are state-owned; the main passenger port is at Cape Town in the southwest, while freight ports include Durban and the new complex at Richard's Bay, both on the east coast. Internal air services are excellent, and there are international airports at Cape Town, Johan-nesburg and Durban. There are many private airline companies in addition to the state-owned South African Airlines.

An independent broadcasting authority was set up in 1994 to establish a system free from state control. There are three television channels broadcasting in 11 languages as well as cable and satellite services. Newspapers are published in English and Afrikaans along with other black languages and black empowerment of the press has been encouraged. Black readership is subsequently increasing.

Lesotho

KINGDOM OF LESOTHO

LESOTHO, FORMERLY BASUTOLAND, IS A small kingdom near the southern tip of Africa. It is completely surrounded by South Africa.

GEOGRAPHY

Lesotho is the only country in the world where all the land is above 1,000 m (3,300 ft). Along its southeastern border the Drakensberg range rises to Thabana Ntlenyana, the highest point in southern Africa. The Orange, South Africa's longest river, rises in the Maloti Mountains in central-northern Lesotho. To the west the land descends to a more fertile plateau. The most populous area is the valley of the Caledon river, which forms the northwestern border.

The climate is warm-temperate, with more precipitation and greater ranges of temperature at higher altitudes. The low-lying areas have warm, wet summers, while the highlands have bitterly cold winters with snow on the peaks. The dominant vegetation is grassland scrub, giving way to better grassland on the mountain slopes and to alpine vegetation on high ground. Scattered woodland of wild olive and willow trees grows in the remoter valleys. Antelopes, hares and reptiles are common.

SOCIETY

In the early 19th century a Sotho chief, Mshweshwe (c. 1786–1870), secured his people against the double threat of the advancing Zulus and the colonizing European nations. After many disappointments he eventually won British protection. In 1871 the country was annexed to the Cape Colony, but became a separate crown colony in 1884.

Lesotho became an independent kingdom in 1966. In 1970, when the opposition Congress Party won the election, the outgoing prime minister, Chief Leabua Jonathan (1914–87) declared a state of emergency, during which the king, Mshweshwe II (1938–96) was deposed. The king was later allowed to resume a figurehead role. In 1986 a military coup restored the full monarchy, and political activity was banned. However, in 1990 the king's powers were again curtailed, as Major-General Justin Metsing Lekhanya, chairman of the 14-member Military Council, placed the country under military rule. Later the same year the king was once more deposed and replaced by his son (Letsie III). Free elections in 1993 ended military rule, but the government was dismissed in 1994. There is a National Assembly made up of an 80-member lower house and a Senate of 22 chiefs and 11 nominated members.

Almost all Lesotho's population belong to the Sotho peoples, who are Bantu in origin. Sotho and English are the official languages. Most people are Christian.

ECONOMY

Economically, Lesotho is mostly dependent on South Africa, with which it is in customs union. However, it is also a member of the Southern African Development Coordination Conference (SADCC), created in 1980 with the aim of lessening its member states' dependence on South Africa's dominant economy.

Agricultural land is very limited; sub-

Mohair spinning in Lesotho is largely a cottage industry using treadle wheels. Mohair is one of the country's most prestigious products, both as yarn and in finely woven materials such as those proudly displayed on the wall behind.

sistence crops include maize, sorghum, wheat and beans, but yields are low. Cattle-rearing is uncommercial, but wool and mohair from sheep and goats are a vital export. Mineral resources are scanty; diamond production is controlled by De Beers Consolidated Mines of South Africa. Industry is limited to small-scale manufacturing. However, water is exported to South Africa, and major hydroelectric schemes are planned. Earnings from migrant laborers in South Africa help to balance the economy, and a growing tourist industry brings welcome hard currency.

Roads are few except in the west, and only a minority of them are surfaced. A short railroad links Maseru to the South African network. There is a commercial airport at Maseru, but no international services. There are several hospitals and clinics, mostly in towns, but health conditions are poor; the most common ailments are those associated with diet deficiencies. Education is free but not compulsory. Enrolment is high in primary schools and literacy is increasing.

NATIONAL DATA - LESOTHO

Land area	30,355 sq km (11,720 sq mi)			

Climate	Altitude m (ft)	Temperatures January °C(°F)	July °C(°F)	Annual precipitation mm (in)
Maseru	1,528 (5,013)	21 (70)	8 (46)	725 (28.5)

Major physical features	highest point: Thabana Ntlenyana 3,482 m (11,425 ft)
Population	(2006 est.) 2,022,331
Form of government	constitutional monarchy
Armed forces	army 2,000
Capital city	Maseru (115,559)
Official languages	Sesotho, English
Ethnic composition	Sotho 99.7%; Europeans, Asians, and other 0.3%
Religious affiliations	Christian 80%; traditional beliefs 20%
Currency	1 loti (LSL) = 100 lisente; 1 South African rand (ZAR) = 100 cents
Gross domestic product	(2006 est.) U.S. $5.195 billion
Gross domestic product per capita	(2006 est.) U.S. $2,600
Life expectancy at birth	male 35.55 yr; female 33.21 yr
Major resources	water, diamonds, sand, clay, building stone, clothing, asparagus, beans, livestock, maize/corn, pulses, wheat, mohair, wool, leather, jute

Swaziland

KINGDOM OF SWAZILAND

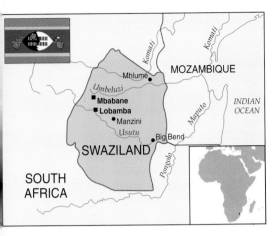

THE KINGDOM OF SWAZILAND BORDERS Mozambique to the east, and is otherwise entirely surrounded by South Africa, with which it is in customs union.

GEOGRAPHY

The landscape descends in a series of steps from the mountainous High Veld in the west to the Lebombo Escarpment in the east. Numerous rivers, including the Umbeluzi and the Great Usutu, have cut deep gorges as they flow eastward. Mbabane, the administrative capital, and Lobamba, the royal and legislative capital, are in the western half of the country.

The climate is subtropical with a wet summer season, but becomes cooler and wetter with increasing height. Grasslands flourish on the High Veld, as do plantations of pine and eucalyptus. On lower land woodland savanna is typical. The wildlife includes lions, leopards, hippopotamuses, antelopes and giraffes.

SOCIETY

In 1893 Swaziland was effectively placed under South African administration, having been granted almost entirely to concessionary companies. However, after the Boer War (1899–1902) between the Dutch settlers and the British, it came under British control. Limited self-government was introduced in 1963; four years later the Kingdom of Swaziland was proclaimed, and full independence from Britain followed in 1968. In 1973 and 1977 the king dismissed parliament and a new constitution was adopted.

The reigning King is Mswati III (b. 1968). The House Assembly has 55 elected members and 10 appointed by the King. There is also a House of Senators consisting of 30 members. Ten of these are elected by the House Assembly while 20 are appointed by the King who retains much of his personal power. There are no political parties. The Swazi national Council is a consultancy body for Swazi men.

Swazis form the vast majority of the population, but there are also other ethnic groups, and a small European community. There are some 70 clans, including the royal Dlamini clan. The official languages are English and the royal (Siswati) dialect of Swazi.

ECONOMY

The great majority of the labor force work on the land, mostly as subsistence farmers. Cash crops such as sugar cane, citrus fruits and tobacco are produced by European-controlled companies and by Swazis on resettlement schemes. The Low Veld is particularly suitable for livestock, a traditional sign of wealth. Forestry is a major export earner, and there is a wood-pulp factory near Bunya to the southwest of Lobamba.

Production of local mineral resources, principally asbestos and iron ore, has declined. Power is generated from coal and a series of hydroelectric schemes. Manufacturing is varied but small scale. The economy depends heavily on the wages of migrant Swazi laborers in South Africa, and on the income from tourists from neighboring countries.

Good roads link the principal towns, but most of the system is still unsurfaced. There is one railroad, linking Swaziland to Mozambique. Aircraft from South Africa and Mozambique fly into an airport at Matsapa near Lobamba, and there are many small local airstrips.

Health facilities to combat malnutrition and endemic diseases are provided by the government. The welfare system provides disability, retirement and survivor pensions. Education is free but not compulsory; literacy levels are rising.

A Swazi woman (*above*) leads a colorful processional dance, waving a sheaf of long grass. Her particular traditional costume indicates that she is a married woman.

A Swazi craft stall (*left*) in the open-air market at Mbabane, the administrative capital. Tourists flock here to choose from the wide selection of traditional mats, baskets and bead curtains that are on offer. If customers cannot find exactly what they want, the local craftworkers may make goods on the spot to their requirements.

NATIONAL DATA - SWAZILAND

Land area	17,203 sq km (6,642 sq mi)			
Climate		**Temperatures**		**Annual**
	Altitude m (ft)	January °C(°F)	July °C(°F)	precipitation mm (in)
Mbabane	1,145 (3,757)	20 (68)	12 (54)	1,442 (56.7)

Major physical features	highest point: Emlembe 1,862 m (6,109 ft)
Population	(2006 est.) 1,136,334
Form of government	monarchy with two legislative houses
Armed forces	no armed forces
Largest cities	Mbabane (administrative – 81,312); Lobamba (legislative – 4,819)
Official languages	Swazi, English
Ethnic composition	Swazi 84.3%; Zulu 9.9%; Tsonga 2.5%; Indian 1.6%; others 1.7%
Religious affiliations	Zionist 40% (a combination of Christianity and traditional ancestor worship), Roman Catholic 20%; Muslim 10%; Anglican, Bahai, Methodist, Mormon, Jewish and other 30%
Currency	1 lilangeni (SZL) = 100 cents
Gross domestic product	(2006 est.) U.S. $5.91 billion
Gross domestic product per capita	(2006 est.) U.S. $5,200
Life expectancy at birth	male 32.1 yr; female 33.17 yr
Major resources	asbestos, coal, clay, cassiterite, hydropower, forests, small gold and diamond deposits, quarry stone, citrus fruits, maize/corn, sugarcane, timber, tobacco

Madagascar

DEMOCRATIC REPUBLIC OF MADAGASCAR

MADAGASCAR IS A LARGE ISLAND STATE lying off the southeast coast of Africa, opposite Mozambique. It remains very distinct from the African mainland, not least in its people and their culture, and in its unique and diverse wildlife.

GEOGRAPHY

Madagascar is the fourth largest island in the world. Geologically, the island was formed when it broke away from the landmass of Africa approximately 50 million years ago.

The east coast is lined with coral beaches and lagoons, and from here the land rises sharply to the great central plateau that covers most of the island. This is ridged and crossed by many rivers, and is interrupted by mountainous massifs. The highest of these is the Massif du Tsaratanana in the north, which culminates in the peak of Maromokotro. There are volcanic features such as crater lakes, and steep sea cliffs in the far north and south. On the western side the mountains fall away more gradually. While the eastern rivers plunge steeply to the Indian Ocean, those to the west such as the Ikopa and Mangoky meander through gentler landscapes, washing down rich sediments to the broad coastal plains.

The climate is greatly varied, from temperate in the highlands to tropical along the coasts; there are heavy seasonal rains in the north and east, but conditions are relatively dry in the central highlands and semiarid in the south. Until recently most of the island was covered in forests, ranging from tropical rainforest to cooler evergreen and deciduous woodlands. However, these forests have been devastated by slash-and-burn cultivation so that only small areas are left. The exposed soil is now suffering drastically from severe erosion.

Madagascar's isolation has allowed a unique and rich diversity of wildlife to evolve. This includes the small, spiny, insectivorous tenrec, and many remarkable species of lemur. There are also crocodiles, chameleons, butterflies and great numbers of birds. Sea fish native to the area include the spiny globefish and a famous "living fossil" called the coelacanth. The wildlife has, however, also been devastated, both by deliberate killing and by the destruction of habitats. Several lemur species have become extinct, and those remaining are endangered. Conservation efforts are small-scale, and though the government is attempting to limit the destruction, it may already be too late.

SOCIETY

Madagascar was probably first settled some time before the 10th century by seafarers from Indonesia. Then in about the 14th century settlers arrived from the African mainland, together with Arab and Asian traders, who introduced Muslim influences. The people of Madagascar, the Malagasy, thus show considerable ethnic mixing, but their language is still clearly Indonesian in origin.

In the 16th century a French trading post was established, but the coast became a haunt of Indian Ocean pirates, and so the Malagasy began to settle the island's interior. They established powerful kingdoms, often with the help of European weapons. The central kingdom of the Merina people eventually emerged as the most dominant, founded by King Andrianampoinimerina (reigned 1787–1810) with guns obtained from the French in exchange for slaves. After 1810 his son Radama I (reigned 1810–28) joined with the British rulers of the neighboring island of Mauritius to end the slave trade and to expel the French. However, Radama's successors invited foreigners back in, and French Roman Catholics vied for influence with British Protestants. Matters came to a head with the Franco-Merina Wars of the 1880s.

In 1890 Britain recognized Madagascar as a French protectorate, and by 1895 the Merina had been subjugated by the French. Malagasy resistance was suppressed, and French culture came to dominate the prevailing mix of indigenous and Protestant custom. New economic development and health care programs were introduced. During World War II Madagascar was occupied briefly by British and South African troops, but became a French overseas territory in 1945. In 1947 there was a strong nationalist revolt, and in 1958 the people voted in a referendum for an autonomous Malagasy Republic. Full independence followed in 1960 under President Philibert Tsiranana (b. 1912).

In 1975 his appointed successor, Major-General Gabriel Ramanantsoa (b. 1906), broke ties with France and sought to align the country with the Soviet bloc. In 1976, following a series of coups, Lieutenant-

Commander Didier Ratsiraka (b. 1936) became the president of the newly constituted Democratic Republic of Madagascar, in which the Supreme Revolutionary Council was the most powerful body. A new constitution was approved in 1992 and multi-party elections were held the following year. The prime minister is appointed by the president.

ECONOMY

The Madagascan economy suffered greatly from the abrupt transition to state planning that followed the 1975 revolution. Anti-government strikes and demonstrations have hindered economic reform. Foreign debt was over $4 billion in 1996. France remains the country's chief trading partner.

Madagascar still relies largely on traditional agriculture. The majority of its exports are of cash crops such as coffee, cloves, vanilla, sisal and sugar. Cattle are the main livestock, and various fruits are grown, largely for local consumption. The main staple crops are rice and cassava, but the island is not self-sufficient in either, because production cannot keep pace with the fast-growing population; the introduction of collective farming after 1975 caused much resentment without improving efficiency. Fisheries have so far remained undeveloped. There has been considerable reforestation, usually with non-native species.

Chromite and graphite are mined for export, together with small amounts of gold and precious stones. There are no energy resources except hydroelectricity, which fulfills half of the island's requirements; the rest is generated with imported fuels. Manufacturing industry is based mainly on agriculture; typical products include soap, textiles, sisal rope, sugar and other processed foods. There is a printing works and a vehicle-assembly plant in the capital, Antananarivo, roughly in the center of the country.

Only a few major roads are surfaced, but the few railroad and port facilities are good, and there are excellent internal and international air services. Primary education is free and compulsory, and is provided both by the state and by Protestant and Roman Catholic churches. However, literacy levels are still relatively low. Health services are available mostly to more affluent city-dwellers, while the majority live in poor conditions. There has, however, been considerable success in the eradication of endemic diseases.

Outrigger canoes glide back to shore after a day's fishing at Nosy Komba, Madagascar, watched by the village children. Many of the smaller communities live by fishing, which is mainly a subsistence occupation.

NATIONAL DATA – MADAGASCAR

Land area 581,540 sq km (224,534 sq mi)

Climate	Altitude m (ft)	Temperatures January °C(°F)	July °C(°F)	Annual precipitation mm (in)
Antananarivo	1,310 (4,297)	22 (72)	15 (59)	1,365 (25.9)

Major physical features highest point: Tsaratanana Massif 2,876 m (9,436 ft); longest river: Mangoky 560 km (350 mi)

Population (2006 est.) 18,595,469

Form of government multiparty republic with one legislative house

Armed forces army 12,500; navy 500; air force 500

Capital city Antananarivo (1,538,349)

Official languages French, Malagasy

Ethnic composition Malagasy 98.9%; Comorian 0.3%; Indian 0.2%; French 0.2%; others 0.4%

Religious affiliations traditional beliefs 52%; Christian 41%; Muslim 7%

Currency 1 Madagascar ariary (MGA) = 5 iraimbilanja

Gross domestic product (2006 est.) U.S. $17.27 billion

Gross domestic product per capita (2006 est.) U.S. $900

Life expectancy at birth male 54.93 yr; female 59.82 yr

Major resources graphite, chromite, gold, coal, bauxite, salt, quartz, tar sands, precious stones, mica, fish, hydropower, tourism, bananas, cassava, cloves, coffee, maize/corn, peppers, potatoes, rice, sisal, sugarcane, vanilla, zirconia

Mauritius

REPUBLIC OF MAURITIUS

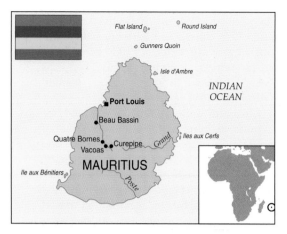

MAURITIUS, ONE OF THE MASCARENE Islands, is an independent island state in the Indian Ocean some 800 km (500 mi) east of Madagascar. Its territory also includes Rodrigues to the east and the tiny dependency of the Agalegals to the north.

GEOGRAPHY

Mauritius is a rugged volcanic outcrop fringed with coral reefs. From the broad coastal plains in the north, it rises to a plateau surrounded by mountain peaks – the remnants of a giant volcano. The highest point is Piton de la Petite Rivière Noire in the south. Rodrigues is similarly volcanic and rugged in character. The tropical climate is moderated by the southeasterly trade winds, which bring ample rainfall and the occasional cyclone.

NATIONAL DATA - MAURITIUS

Land area	2,030 sq km (784 sq mi)			
Climate		Temperatures		Annual
	Altitude m (ft)	January °C(°F)	July °C(°F)	precipitation mm (in)
Port Louis	55 (180)	28 (72)	23 (59)	711 (27.9)

Major physical features highest point: Piton de la Petite Rivière Noire 826 m (2,711 ft); largest island: Mauritius 1,865 sq km (720 sq mi)	
Population (2006 est.) 1,240,827	
Form of government multiparty republic with one legislative house	
Armed forces Paramilitary 2,000	
Capital city Port Louis (158,196)	
Official language English	
Ethnic composition Indo-Mauritian 68%; Creole 27%; Sino-Mauritian 3%; Franco-Mauritian 2%	
Religious affiliations Hindu 48%; Roman Catholic 23.6%; other Christian 8.6%; Muslim 16.6%, other 2.5%; unspecified 0.3%; none 0.4%	
Currency 1 Mauritian rupee (MUR) = 100 cents	
Gross domestic product (2006 est.) U.S. $16.72 billion	
Gross domestic product per capita (2006 est.) U.S. $13,500	
Life expectancy at birth male 68.66 yr; female 76.66 yr	
Major resources sugarcane, tea, bananas, fish, tourism, potatoes, tobacco	

The dense forest that once covered Mauritius has largely been cleared, to the detriment of local wildlife. This includes some remarkable species, notably the introduced hedgehog-like tenrec, and a fruit bat. Mauritius was also once the home of the now extinct dodo. Conservation projects have been set up with British assistance.

SOCIETY

Mauritius was named in 1598 by Dutch navigators after Prince Maurice of Nassau (1567–1625), but remained uninhabited until it was colonized by the French in the 18th century. They called it Ile de France, and imported African slaves to work on their sugar plantations, both here and on neighboring Rodrigues. In 1810 both islands were ceded to Britain, but the French and mixed-race Creole communities each retained their own language and culture. Following the abolition of slavery, large numbers of Indian laborers were brought in, and Chinese traders also settled on Mauritius.

After World War II economic reforms and increasing local self-government were introduced. In 1968 Mauritius was granted independence within the British Commonwealth. Its parliamentary government endured despite political turbulence until the declaration of a republic in 1992. English is the official language, but French, Creole, Hindi and Chinese are in everyday use. The main religious groups are Hindus, Christians (mainly Roman Catholics) and Muslims, along with a smaller number of Buddhists.

ECONOMY

Mauritius still remains heavily dependent on sugar cane, although great efforts are being made to diversify. Other crops include tea, bananas, tobacco and various food crops, and fishing methods are also being modernized.

The manufacturing sector is developing fast, thanks to the establishment of free-trade zones, within which imported raw materials are processed and reexported; successful products include textiles and clothing, plastics and electronics. The island's natural beauty and lively cultural mix have encouraged a flourishing tourist industry, which is now the country's second-largest revenue earner. Exports are still heavily outweighed by imports of food, fuel, machinery and manufactured goods. Hydroelectricity and even sugar-derived fuels are being developed to reduce the country's reliance on imported energy sources.

The roads on Mauritius are good and mostly surfaced, and buses have replaced the old railroad system. The capital, Port Louis, on the northwest coast, has very good dock facilities, and there is an international airport at Plaisance in the southwest of the island. The Mauritius Broadcasting Corporation operates radio and television services. Health and welfare services are good despite problems of overpopulation. The vast majority of children receive primary education, and secondary education is readily available. The University of Mauritius provides mainly technical subjects.

Sugar cane fields in Mauritius spread out beneath jagged peaks of volcanic stone – an image that unites the island's two great assets: the exotic beauty of its landscape – a draw for tourists – and the crop on which the country depends.

Comoros

FEDERAL ISLAMIC REPUBLIC OF THE COMOROS

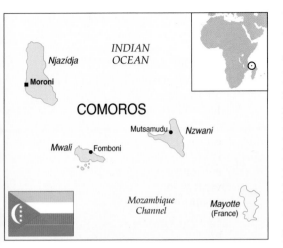

THE COMOROS ISLANDS LIE IN THE INDIAN Ocean between northern Madagascar and the African coast. Three out of the four islands form the independent republic. The southeasternmost island, Mayotte, has remained a French dependency.

GEOGRAPHY

The archipelago is volcanic in origin. Njazidja (formerly Great Comoro) in the northwest is mostly a rough plateau that rises southward to the active crater of Mount Kartala. Moroni, the national capital, lies on the west coast. The climate is hot with seasonal rains, but the ground is porous, and only the upper slopes support rainforests. The island of Nzwani (formerly Anjouan) to the southeast is very similar in character, but although its soils are better, much of the topsoil has been eroded after land clearance. Mwali (formerly Mohéli), the smallest island, has well-watered river valleys and densely forested hillsides. Wildlife includes several species of land bird and a lemur that are endemic to the islands.

SOCIETY

The Comoro Islands were settled by many different peoples at various times, including mainland Africans, Malagasy, and Arab and Iranian traders. Arabs were the dominant group when the first Europeans arrived in the 16th century. In 1843 Mayotte was annexed by France, and in 1886 the other islands became French protectorates. Arabic and French are the official languages in this predominantly Muslim state, but the lingua franca is Comorian, which resembles Swahili.

After World War II the islands became a French overseas territory, and in 1961 they were granted internal autonomy. In 1975 President Ahmed Abdallah (1919–89) declared independence unilaterally. Abdallah was deposed later that year, but in 1978 he was restored after a coup, and became president of the newly constituted Federal Islamic Republic. In 1989 Abdallah was assassinated but 1990 saw the first multiparty elections since independence and Said Mohamed Djohar became president. Further coups in 1995 and 1999 and attempts at secession by the islands of Nzwani and Mwali left Comoros in a troubled state as the century closed.

ECONOMY

The Comorian economy has traditionally depended on subsistence agriculture. Goats and cattle are the main livestock, and food crops include cassava, sweet potatoes and mountain rice, but much food is imported. Exports consist mainly of cash crops such as vanilla and cloves.

Fishing boats lie in the harbor of Moroni, capital of the Comoros. Behind is the ornate white facade of the Great Mosque, witness to the islands' long Arab connections. Although the population is today very mixed, Islam is the dominant religion and culture.

There are a few hydroelectric plants, but most fuel has to be imported. Manufacturing industry consists mostly of small–scale processing of vanilla, cloves and other spices. Tourism also provides an important source of revenue. France remains the chief trading partner and the main source of foreign aid.

The roads are often impassable in the rainy season, but good air and ferry services operate between the islands and to Madagascar and East Africa. There is an international airport on Njazidja. The three French-built hospitals suffer from a lack of medical staff. Education is provided both by Islamic schools and by French-style lycées.

NATIONAL DATA – COMOROS

Land area	2,170 sq km (837 sq mi)			

Climate		Temperatures		Annual
	Altitude m (ft)	January °C(°F)	July °C(°F)	precipitation mm (in)
Moroni	12 (39)	27 (81)	23 (73)	2,542 (100)

Major physical features highest point: Mount Karthala (Njazidja) 2,361 m (7,746 ft); largest island: Njazidja 1,148 sq km (443 sq mi)

Population (2006 est.) 690,948

Form of government federal multiparty republic with one legislative house

Armed forces army 700 (reorganized in 1995); navy 200

Capital city Moroni (44,518)

Official languages Arabic, French

Ethnic composition Comorian (Bantu/Arab/Malagasy) 96.9%; Makua 1.6%; French 0.4%; others 1.1%

Official religion Islam

Religious affiliations Sunni Muslim 98%; Roman Catholic 2%

Currency 1 Comoran franc (KMF) = 100 centimes

Gross domestic product (2002) U.S. $441 million

Gross domestic product per capita (2005) U.S. $600

Life expectancy at birth male 60 yr; female 64.72 yr

Major resources negligible apart from cash crops such as vanilla, copra, cloves, ylang-ylang, coffee, sweet potatoes, rice, coconut, maize/corn, yams

Dependencies in the region

MAYOTTE
FRANCE

The southeasternmost island of the Comoros archipelago is a French dependency with the official status of a *collectivité territoriale*. It is mountainous, largely volcanic and fringed by coral reefs. The capital and port, Dzaoudzi, is perched on a rocky outcrop. The wet, tropical climate supports lush forest vegetation.

The people are mainly Malagasy in origin and speak Comorian, which is closely related to Swahili. Most are Muslim (Islam was introduced by Arab traders), but there is a strong Roman Catholic minority. In 1975, when Comoros declared its independence from France, Mayotte chose by plebiscite to remain French. Comoros still claims the island, supported by a 1979 United Nations resolution. However, in 1984 the French abandoned plans to return Mayotte to Comoros on account of local opposition. The island has its own elected assembly, but is also represented in the French parliament.

The economy is largely agricultural; the main cash crops are vanilla, ylang-ylang (used in perfumes), coconuts and coffee. Manufacturing is mainly limited to food processing. The education system includes both Islamic and French schools.

REUNION
FRANCE

Réunion, the most southerly of the Mascarene Islands, lies in the Indian Ocean off the east coast of Madagascar and southwest of Mauritius. It is an overseas department of France.

The island is mountainous and volcanic in origin; the highest peak, Piton de la Fournaise, is still active. The tropical climate is cooled by southeasterly trade winds, which bring ample rainfall and occasional severe cyclones. The rich rainforest that once covered the island is all but gone, except in uninhabitable upland areas, where the tamarind tree is still found. Much of the native wildlife has now disappeared, but that which remains includes several species of bat, and among the birds the Bourbon flycatcher and the Mascarene bulbul.

The island, claimed by France in 1638, was originally known as Bourbon. It was first settled in the 17th century by French colonists, who brought African slaves to work their sugar plantations. These were

later followed by Malay, Chinese and Indian laborers. The present population includes all these peoples, but is predominantly mixed race, or Creole. The everyday speech is Creole, but French is the official language, and most people are Roman Catholic. The islanders are French citizens, and return five deputies and three senators to the French parliament. Internal government is by an elected council of 44 members, headed by a French-appointed prefect.

Réunion's economy is predominantly agricultural, and is largely dependent on sugar cane. Other cash crops include tobacco, vanilla, ornamental and perfume plants, and various fruits and vegetables. Sugar, together with its derivatives, rum and molasses, accounts for the great majority of exports, but tourism is an increasing source of revenue. Imports

include food, fuel and manufactured goods. Manufacturing at home is mostly confined to sugar refining and other food-processing industries. Health facilities are being developed to combat indigenous diseases, and the French-based education program is contributing to the increase in the literacy figures.

ST HELENA
UNITED KINGDOM

St Helena – a remote volcanic island lying some 1,930 km (1,200 mi) off the southwest coast of Africa – is a British colony which itself has two dependencies: Ascension Island to the northwest and Tristan da Cunha to the southwest.

St Helena covers some 120 sq km (47 sq mi) of mostly rugged and mountainous

A striking aerial view of Dzaoudzi (*above*) The capital of Mayotte, it occupies a rocky outcrop connected by a causeway to the islet of Pamanzi, part of the coral reef that surrounds the main island of Mayotte. Water to supply the town is drawn from reservoirs that are filled in the rainy season.

Towering mountain peaks (*left*) on Réunion, eroded into spectacular shapes, betray the island's volcanic origins. Volcanic activity continues in the eastern part of the island, where there is one active volcano and nine dormant cones.

terrain, but with enough fertile land to support a small farming economy for its 6,000–7,000 inhabitants. The main port and capital is Jamestown, which also acts as an administrative center for Ascension Island and Tristan Da Cunha.

The island was discovered on 21 May 1502 by the Portuguese navigator Jõao da Nova (d.1509) and named after St Helena (c. 248–329), the mother of Constantine the Great, whose feast day it was. It was subsequently settled by the East India Company and became a British colony in the mid-19th century. Following his abdication as emperor, Napoleon Bonaparte (1769–1821) was exiled there by the British from 1815 until his death in 1821.

Ascension Island is a barren, rocky peak, almost totally lacking in vegetation and covering only 88 sq km (34 sq mi). It was uninhabited until Bonaparte was

exiled to St Helena, when a small British garrison was established on Ascension as a precautionary measure. Today it has a tiny agricultural community, most of whom are St Helenians. Ascension Island is perhaps most famous for its green turtles, which come to the island between December and May each year to lay their eggs in the sand.

Strategically, however, Ascension Island has become a vital link across the South Atlantic, and many of its 1,000 inhabitants are service personnel or communications workers. It came into the public spotlight briefly during the British recapture of the Falkland Islands from Argentina in 1982, when it acted as a critical air–sea staging post for British and US armed forces. It also provided a base for US tracking equipment during the Apollo space program. In addition, the

island is a relay station for telecommunications between South Africa and Europe, and links international broadcasting via satellite and submarine cables. It was given its name by a Portuguese expedition – led by Jõao da Nova – which landed there on Ascension Day, 1501.

The Tristan da Cunha group of islands – Tristan da Cunha, Inaccessible Island, Gough Island and the three Nightingale Islands (Nightingale, Stoltenhoff and Middle) – lies about 2,550 km (1,600 mi) off the coast of Cape Town. Tristan da Cunha itself is a volcanic cone, 100 sq km (40 sq mi) in area, which last erupted in 1961 and was settled again in 1963. It has only a few hundred residents. An administrator who represents the British government heads the local council on Tristan da Cunha and acts as the magistrate for the whole group. The other islands in the group are largely uninhabited, but the low cliffs of Nightingale Island and its two rocky islets support millions of nesting seabirds.

Tristan da Cunha first came to prominence as a garrison in the early 19th century; now three main activities support the islands' economy. The group's main source of revenue comes from selling postage stamps to philatelists. It also lies at the center of a small fishing industry run from Cape Town, which exports goods to Japan, the United States and Britain. In addition, Gough Island is the home of a small weather station.

The Nordic Countries

A tranquil fishing village (*left*) in the Lofoten islands, off the northwest coast of Norway. The sea has always played an important role in the region's economic and cultural life, and most of the population is found along the coasts. Fishing and sailing are popular leisure pursuits even with city dwellers.

- ■ capital city
- ● major town

height of land (meters)

	2000
	1000
	500
	200

☐ permanent ice

▲ mountain peak

211

Norway

KINGDOM OF NORWAY

NORWAY LIES ALONG THE WESTERN COAST of the Scandinavian peninsula. Its spectacular rugged land supports a relatively small population.

GEOGRAPHY

Long and thin – only 450 km (280 mi) across at its widest point – Norway narrows in places to 80km (50 mi).

The land

Norway's ancient granite bedrock was eroded by ice age glaciers and shaped into high mountains and deep valleys. There are more than 160,000 lakes; the largest is lake Mjøsa in the southeast. Fjords (highwalled sea lakes in glacial valleys) turn the coastline into a maze flanked by some 50,000 islands.

Oslo and the southeast are in the lowlands. The Glama river and others run down from the mountains, creating fertile valleys and lakes in the main agricultural area. Other important lowland farming is in the Trøndelag in central Norway, and near Stavanger in the southwest.

Glacial erosion has flattened some mountains, creating immense plateaus such as the Hardangervidda in southern central Norway. Farther north lies the Jotunheimen mountain range. Toward the coast is the Jostedalsbreen icefield with its flowing glaciers. Northern Norway, lying mostly above the Arctic Circle, ends in the Finnmark Plateau. Norway also includes the bleak Arctic islands of Svalbard and Jan Mayen.

Climate, plants and animals

Below the Arctic Circle, the North Atlantic

A peaceful scene in summer Vetle Fjord is a northern arm of southern Norway's Sogne Fjord, the longest – at 184 km (115 mi) – of the country's great coastal inlets. It was carved out by the grinding action of glaciers during the last ice age, and flooded by the sea.

Drift gives western Norway a relatively temperate climate. Eastern Norway has colder winters, warmer summers and lower rainfall. About one-quarter of the country is forested. The richly varied vegetation includes blueberries, cranberries and cloudberries – the fruits of which are Scandinavian delicacies. Recently, large areas have been damaged by acid rain from industrial air pollution originating in neighboring countries. Large predatory animals such as wolves, bears and lynx have been hunted almost to extinction in the southern forests.

SOCIETY

Norway's immense mountains and deep fjords have always made land travel and communications extremely difficult, so it is not surprising that the country has a long maritime and seafaring history

History

Norway's earliest settlers, Central European hunter–gatherers, gradually moved north as the last ice age ended. By the 9th century AD settled farming was well established. As the population expanded, settlers began trading and raiding in Scotland, Ireland, northern England and parts of northern France (Normandy). They also settled in Iceland and Greenland, and crossed the Atlantic to North America. The name Viking comes from Skagerrak, the strait between Norway and Denmark, then called the Vik.

King Harald Fairhair (c. 860–940) united Norway about 900. Christianity and the eventual decline of the Viking era strengthened the monarchy, but the country was devastated by the bubonic plague, which reached Norway in 1350.

The Kalmar Union of 1397 combined the royal lines of Norway, Sweden and

Filigree of ice Snow lies on the ground on many days, and winters are long and dark. But the moderating influence of the warm North Atlantic Drift and westerly winds means that coastal waters are rarely frozen over even when the land is snowbound.

Denmark. In 1814 Norway was forced into a union with Sweden but did gain limited internal self-government. After a series of popular protests, Sweden recognized Norway's independence in 1905.

Norway sought to remain neutral in World War II but was occupied by the German army from 1940 to 1945. After the war, Norway joined the North Atlantic Treaty Organization (NATO) and the European Free Trade Association (EFTA).

Government

The constitutional monarchy adopted in 1814 remains today. The parliament, or Storting, is elected by universal adult suffrage for a 4-year term and in turn selects a prime minister and cabinet. The king acts as head of state but his functions are mainly ceremonial.

People

Most Norwegians share common ethnic and linguistic origins with their Scandinavian neighbors. Bokmal, the urban educated form of Norwegian, and Nynorsk (New Norsk), based on rural western dialect, are gradually being combined into Samnorsk. Most Norwegians belong to the national Lutheran Church.

NATIONAL DATA – NORWAY

Land area	307,442 sq km (118,673 sq mi)			
Climate		**Temperatures**		**Annual**
	Altitude m (ft)	January °C(°F)	July °C(°F)	precipitation mm (in)
Oslo	96 (314)	-4 (24)	17 (62)	655 (26)
Bergen	39 (127)	1 (34)	14 (58)	2,074 (82)

Major physical features highest point: Galdhøpiggen 2,469 m (8,100 ft); longest river: Glåma 611 km (380 mi)

Population (2006 est.) 4,610,820

Form of government multiparty constitutional monarchy with one legislative house

Armed forces army 14,700; navy 5,310; air force 5,000

Largest cities Oslo (capital - 521,886); Bergen (237,430); Trondheim (154,351)

Official language Norwegian

Ethnic composition Norwegian 95.8%; other Scandinavians 1.0%; others 3.2%

Religious affiliations Church of Norway 85.7%; Pentecostal 1%; Roman Catholic 1%; other Christian 2.4%; Muslim 1.8%; other 8.1%

Currency 1 Norwegian krone (NOK) = 100 øre

Gross domestic product (2006) U.S. $207.3 billion

Gross domestic product per capita (2006) U.S. $47,800

Life expectancy at birth male 76.91 yr; female 82.31 yr

Major resources petroleum, natural gas, iron ore, copper, lead, zinc, pyrites, nickel, fish, timber, hydropower, apples, barley, coal, livestock, oats, potatoes, silicon, titanium

Ski jumping (*above*) at Holmenkollen, the biggest of Norway's many ski festivals. Norwegians excel at crosscountry skiing, which developed as a means of traveling across frozen terrain, but alpine events are very popular as well.

Oil-led economy (*below*) A huge drilling rig under construction. The discovery of rich oil and gas reserves in the North Sea in the 1960s gave an enormous boost to Norway's economy, and energy products now form its single largest export.

Industry and trade

Metallic ores are the only substantial mineral resources on the mainland. Iron and steel production, once the most important industries, have been replaced by aluminum production from imported bauxite. Norway is still a leading exporter of nickel, copper and zinc.

During the 1970s, deposits of natural gas and oil were found in the Norwegian North Sea sector. Norway became a significant exporter of oil and gas, while other industries suffered from lack of investment. The oil-dependent economy was hit very hard by fluctuating oil prices in the late 1970s and 1980s. Today much of the country's energy is produced by hydroelectric plants on the many fast-flowing rivers.

Norway's large merchant fleet, though declining in size, still makes a large contribution to the balance of payments. Trade has been affected by Norway's decision in 1994 not to join the European Union, but free-trade agreements have been concluded. The scenery and outdoor sports facilities draw tourists year-round.

Some 20,000 Sami (Lapps) and 10,000 Finns make up the main ethnic minorities with their own languages. They live mainly in traditional homelands in the far north. Some 2,000 Sami on the Finnmark Plateau are seminomadic herders of reindeer.

Norway's cultural history is long and distinguished. A large number of figures have become world-famous, including the composer Edvard Grieg (1843–1907); the playwright Henrik Ibsen (1828–1906); and the painter Edvard Munch (1863–1944). A notable line of famous scientist-explorers includes Fridtjof Nansen (1861–1930), Roald Amundsen (1872–1928) and Thor Heyerdahl (b. 1914).

ECONOMY

The difficult climate and terrain (only 5 percent of the land area is farmed) restrict agricultural development, and Norway's economy is chiefly industrial.

Agriculture and fisheries

Nearly half of the small southern coastal agricultural area provides pasture; the rest is used to grow cereals, potatoes and animal fodder. Most holdings are small. Forestry and fishing, the major agricultural earners, are heavily subsidized by the government. However, Norway's fishing fleet has been reduced by a quarter as a result of international overfishing.

Transportation and communications

Coastal shipping is still the main means of communication; overland transportation is heavily subsidized. The rail network links Oslo with Stavanger and Bergen. A line extends to Trondheim but the rail link to Narvik in the north is through Sweden. About two-thirds of the road system is surfaced. Bus transport (with linked ferry services) is important in rural areas. There are over 40 local airfields and international airports at Oslo, Stavanger and Bergen. It is a partner in the Scandinavian Airlines System.

Broadcasting is controlled by an independent national body with two radio networks and one television network.

Health and welfare

The Norwegian welfare system offers compulsory national insurance for healthcare and a comprehensive pension system. Education is free and compulsory for 10 years. Further secondary and higher education are available at vocational schools, colleges or at one of Norway's four universities.

Norway's Arctic island territories

A trapper's cabin in the Svalbard Islands. Most of the islands' 3,000 residents are Russian and Norwegian coal miners who live there only during the short summer months.

Svalbard is the name given to a group of Arctic islands due north of mainland Norway between latitude 74° and 81° north, and between longitude 10° and 35° east; it was recognized as a Norwegian possession by international treaty in 1920, and contains the Spitsbergen group of islands as well as White Island, King Charles Land, and the southerly Bear Island. The name "Svalbard" means "the cold coast" – in winter temperatures can fall as low as –40°C (–40°F), and gale force winds and pack ice floes almost completely cut the islands off from each other.

Well known to the Vikings, the islands were only rediscovered in 1596. At first they were a center for whaling, and later for hunting and trapping. Since 1773 they have been used as a base for explorations to the North Pole. But it was the discovery of coal in the early 20th century that brought the islands to international prominence. The 1920 treaty that recognized Norwegian sovereignty over the islands shared mineral rights between several of the signatory nations, but by the 1990s only Norway and Russia continued to mine coal there. Mining, the main activity on the islands, takes place in summer when the population increases to 3,000. The extreme winter weather conditions mean that the population is entirely seasonal; few if any would wish to stay permanently. In the brief summer months, however, the North Atlantic Drift can raise the temperature to 15°C (59°F).

Vegetation consists of lichens, mosses, liverworts and a few very hardy flowering plants. Many migratory birds use the islands as a rest stop on their seasonal flights – birdwatching and related tourism is becoming a major industry. Seals, walruses and whales breed and live off the coasts of the island group.

Jan Mayen island off the northeast coast of Greenland is named after a 17th-century Dutch seafarer. A bleak volcanic island, it is dominated by the extinct volcano Beerenberg. Annexed by Norway in 1929, it is used as a radio and navigational station. It supports scant wildlife.

Sweden

KINGDOM OF SWEDEN

S WEDEN IS THE FOURTH LARGEST COUNTRY in Europe, occupying the southeastern part of the Scandinavian peninsula. Its people enjoy a very high standard of living, a highly developed welfare system and one of the highest average incomes in the world. In the 20th century the country has always taken a strongly independent approach to world affairs.

GEOGRAPHY

Sweden can be divided into two parts. The mountains and plateaus of the north account for two-thirds of the land area; they are thickly forested and rich in mineral resources. The southern lowlands contain four-fifths of the population, most of the agricultural land and the majority of the manufacturing industries.

The land

The mountainous Norrland region in the north and center of the country includes the Kjolen range along the Norwegian border and the Kebnekaise, Sweden's highest mountain peak at 2,111 m (6,926 ft). A series of plateaus, the source of most of Sweden's largest rivers, drops gently away southeastward toward the coastal plains along the Gulf of Bothnia.

Farther south is Svealand, a lowland area with many lakes – including the country's largest, Lake Vänern. Lakes are a major feature of the country, covering almost one-twelfth of the total area. The Göta is the only major river in the south, flowing southwest from Lake Vänern to the North Sea at Göteborg. Eastward the lowland area extends to the Baltic, with

Uppsala and the capital city Stockholm near its western edge. The Baltic coastline is rocky and there are numerous small islands lying just off the coast.

South of the lakes, in Småland, the land rises again to 300 m (1,000 ft) above sea level before dropping to the densely populated Skåne plain at the southern tip of the country.

Climate

Almost 15 percent of Sweden's land area lies within the Arctic Circle, while its southern tip is on the same latitude as Copenhagen, Denmark. This long north–south extent gives the country a varied climate. Southwestern Sweden benefits from the warm winds associated with the North Atlantic Drift. These bring mild, wet winters and cool summers. The north and east, areas affected by cold air masses from Siberia, are significantly cooler than the south. The northern Baltic is usually icebound throughout the winter.

Differences in altitude also have a marked effect. In the Kjolen range snow can lie for as long as eight months, but in the lowlands groundsnow lasts for a month or less.

Plants and animals

Vegetation, like climate, is affected by latitude and altitude. The plains of Skåne and the southwest are characterized by woodlands of beech, oak and hornbeam. Farther north and at higher elevation these give way to pine and spruce. The plateaus of Norrland and Småland support much of the coniferous forests that cover almost 60 percent of the country. On the mountains – from 900 m (3,000 ft) in the south and 500 m (1,600 ft) in the north – the vegetation becomes alpine. Mosses and lichens predominate on the highest ground, birch and dwarf shrubs grow lower down.

Northern Norrland (Swedish Lapland) is the only area that preserves Sweden's ancient forests. Many of the country's remaining trees have been planted by forest farmers: three-quarters of Sweden's forests are controlled by corporations and private owners.

Bears and lynx, once common, are now restricted to the forests of the north. Like the increasingly rare wolf they are protected species. There are no wild reindeer, though in the north the Sami (Lapps) keep herds of domesticated rein-

Northern twilight (*left*) A pink sunset glows over one of the numerous lakes in central Sweden. The country's landscapes of undulating granite plateaus, ice-scoured lakes and dark coniferous forests provided the setting for Sweden's rich legacy of ancient myths and legends.

In need of protection (*right*) The Brown bear of Sweden, once common, is now a protected species, living among Sweden's million hectares of nature reserves. Sweden's first national parks were established in 1909 – among the earliest in Europe. Many are located in remote mountainous areas, free from public intrusion.

deer. Many other common species are hunted for sport, notably the abundant elk and the Roe deer, common in central and southern areas. Other game animals include foxes, hares and otters.

Sweden is rich in bird life, with gulls, terns and eider ducks on the coast, birds of prey in the forests and cranes in northern marshlands. Golden eagles and cranes are protected by law. There are no poisonous snakes other than the viper, and freshwater and marine fish are plentiful. However, in 1988 sea pollution along the west coast caused largescale destruction of marine fish stocks. The population of Harbor seals in Sweden has also been greatly reduced by a lethal virus infection, which quickly spread to other colonies as far away as the Netherlands, Germany and the British Isles.

SOCIETY

Sweden's political, social and cultural history is closely linked to the other countries of Scandinavia. Since World War II, Sweden's predominately socialist governments have provided ample welfare provisions for its citizens.

History

The early history of Sweden, apart from legends and mythology, is not well documented, but even before the Viking age (9th to 11th centuries) Swedish chieftains had exerted overlordship around the Baltic and beyond. Their Viking descendants opened up trade routes along rivers across Europe to Byzantium (modern Istanbul) and Baghdad, and established new centers of power in Russia. Christianity fared badly in this strongly pagan society, and the church was not established until King Olof Skötkonung (d. 1022) was baptized in the 11th century.

A period of instability and civil war came to an end in the mid 13th century, but the struggle for power between Sweden's nobles and kings continued into the 14th century. Then, in 1397, Erik of Pomerania (c. 1381–1459) was crowned king of Norway, Denmark and Sweden at Kalmar, in southwestern Sweden. The Kalmar Union sought to counter the might of the Hanseatic League, a confederation of north German towns and merchants. This tripartite union was short-lived; the Swedes effectively left it in 1448 by electing their own king, Karl Knutsson (1408–70). Danish efforts to regain control ended in 1523, when Gustav I Vasa (1496–1560) established a new royal dynasty in Sweden that lasted until 1720. Its most notable member was Gustav II Adolf (1594–1632), a skilled military leader who played a decisive role in the religious and political conflicts of the Thirty Years War (1620–48) in Germany, leading the Protestant forces in an overwhelming victory over the Roman Catholic army at Breitenfeld in 1631.

In the long term this was an expensive triumph, demanding a continuous Swedish presence in Germany and the Baltic. In 1700 Charles XII (1682–1718) gained a string of initial victories against the united armies of Denmark, Poland and Russia in the Great Northern War. However, after 1709 the fortunes of war went against Sweden and the country continuously lost territorial possessions. Charles was killed at the siege of Frederiksten, and three years later in 1721 Sweden ceded its Baltic provinces, part of Karelia and the city of Vyborg (near St Petersburg) to Tsar Peter the Great (1672–1725) of Russia.

Parliamentary power became stronger after Charles's death, but a second war

NATIONAL DATA - SWEDEN

Land area	410,934 sq km (158,663 sq mi)			

Climate		Temperatures		Annual
	Altitude m (ft)	January °C(°F)	July °C(°F)	precipitation mm (in)
Stockholm	52 (170)	-3.5 (25.7)	17 (63)	502 (20)

Major physical features highest point: Kebnekaise 2,111 m (6,926 ft); longest river: Göta-Klar 715 km (447 mi); largest lake: Lake Vänern 5,584 sq km (2,156 sq mi)

Population (2006 est.) 9,016,596

Form of government multiparty constitutional monarchy with one legislative house

Armed forces army 13,800; navy 7,900; air force 5,900

Largest cities Stockholm (capital - 1,889,945); Göteborg (481,410); Malmö (271,271)

Official language Swedish

Ethnic composition Swedish 88.5%; Finnish 2.2%; other Europeans 4.7%; Asians 3%; other 1.6%

Religious affiliations Lutheran 86.5% (non practicing 30.0%); Muslim 2.3% Roman Catholic 1.8%; Pentecostal 1.1%; others 8.3%

Currency 1 Swedish krona (SEK) = 100 öre

Gross domestic product (2006) U.S. \$285.1 billion

Gross domestic product per capita (2006) U.S. \$31,600

Life expectancy at birth male 78.29 yr; female 82.87 yr

Major resources iron ore, copper, lead, zinc, gold, silver, tungsten, coal, uranium, arsenic, feldspar, timber, hydropower, cereals, dairy products, potatoes, rape seed, sugar beet

with Russia (1741–43) left Finland, then a duchy of Sweden, occupied by Russian troops, while war with Prussia (1757–62), the rising power in northern Germany, put a heavy strain on the country's finances. Gustav III (1746–92) restored absolute rule and attempted major social reforms. His efforts to regain the Finnish provinces were foiled by treason in 1788, and four years later he was assassinated.

During the Napoleonic wars at the beginning of the 19th century, Sweden aligned itself with Britain, but its ally could offer little support against the combined forces of France, Russia and Denmark. In 1809 Finland and the Åland Islands at the southern end of the Gulf of Bothnia were ceded to Russia, and in 1810 one of the emperor Napoleon's marshals, Jean-Baptiste Bernadotte (1763–1844), was chosen as crown prince, taking the name of Charles John. He failed to negotiate a return of the ceded territories, but in 1813 took a combined Prussian, Russian and Swedish army into Denmark, forcing its king, Frederik VI (1768–1839), to surrender Norway under the Treaty of Kiel in 1814. Shortly afterward, following a brief rebellion, Norway was forced into political union with Sweden.

Liberal reforms in the 19th century included free trade (1846) and a two-chamber parliament (1865–66). In 1905 Oscar II (1829–1907) dissolved the union of the Swedish and Norwegian crowns.

Sweden remained neutral throughout both World Wars. It later joined the United Nations (UN), but stayed outside the North Atlantic Treaty Organization (NATO). From 1946 onward the Social Democratic Party developed a comprehensive welfare system. The murder of prime minister Olaf Palme (1927–86) deprived the Social Democrats of their most charismatic and persuasive leader.

Government

Sweden is a constitutional monarchy, though today the monarch's role is almost entirely ceremonial. The Riksdag or parliament consists of a single chamber with 349 members elected for a four-year term. Everyone aged 18 or over can vote. Some 310 seats are allocated directly, and the rest are divided between parties in order to achieve proportionality with the accumulated votes throughout the country. This proportional representational system is designed to avoid having an overabundance of small or fringe parties.

The principal organs of government lie within 14 committees, and committee

Seasonal festivities (*far right*) Communities in Sweden still celebrate Midsummer's Eve – the longest day – in the traditional way with children dancing around a cross of flowers and leaves, and with all-night parties. Candles on St Lucy's day (13 December) light up the winter gloom, and spring is greeted on Walpurgis night (the last day of April) with the lighting of bonfires.

Symbol of Nordic political union (*right*) The castle stronghold of the historic city of Kalmar overlooks the narrow straits lying between the mainland and Öland, a long narrow island off the southeast coast. It was here that the Union of Kalmar was sealed in 1397 with the coronation of Eric of Pomerania as king of Denmark, Norway and Sweden. Queen Margrethe I (1353–1412) who was regent of Denmark (from 1375), of Norway (from 1380) and of Sweden (from 1389) had been instrumental in making her nephew Erik undisputed king of Scandinavia. After his coronation Margrethe remained regent and effectively Scandinavia's ruler until her death. Her diplomacy that had led to the Union of Kalmar united Denmark, Norway and Sweden until 1523 and Denmark and Norway until 1814.

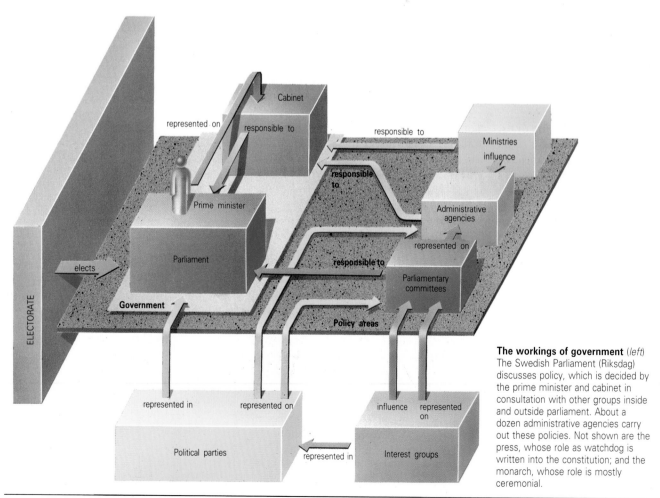

The workings of government (*left*) The Swedish Parliament (Riksdag) discusses policy, which is decided by the prime minister and cabinet in consultation with other groups inside and outside parliament. About a dozen administrative agencies carry out these policies. Not shown are the press, whose role as watchdog is written into the constitution; and the monarch, whose role is mostly ceremonial.

members are elected by proportional representation. Executive control belongs to an 18-member cabinet led by the prime minister, who is appointed by the speaker of the Riksdag and must be approved by its members. The cabinet, or an individual minister, can be dismissed by a simple majority vote in the Riksdag. At local level there are 24 counties (each headed by a governor) and 23 county councils (the island county of Gotland in the Baltic does not have one). There are in addition 284 town councils.

Environmental awareness brought the Green Ecology Party enough support to make its first appearance in the Riksdag in 1988. The victory of the conservative Moderate Party (1991) turned Sweden from socialism towards a free market economy. The Social Democratic Labor Party was returned to power in 1994 but had to continue with unpopular levels of taxation and welfare cuts until the economy improved in 1999.

People

The people of Sweden share common ethnic and cultural origins. The Swedish language is closely related to Danish, Norwegian, Icelandic and Faeroese, and all belong to the same branch of the Indo-European family of languages as German and English. The most sizable ethnic minority are the Sami (Lapps), who have a distinctive culture and language. However, most Samis also speak Swedish, and in the north both languages are taught in schools. Recent immigration has created other minority groups including Finns, Greeks and Turks.

Most Swedes belong to the Lutheran Church whose organization and finances are closely linked with the state, but it retains religious independence. Swedish culture is lively and productive, particularly in literature, art and film-making.

ECONOMY

The prime natural resources of Sweden are lumber, iron ore and water, which is harnessed for hydroelectric power. All three are largely confined to Norrland. Forestry and engineering are the chief export earners since the decline of the shipbuilding industry.

Agriculture

The lumber industry is carefully regulated by the government. Some 60 percent of Sweden's total land area is devoted to productive forestry, which yields turpentine, resins, dyes, rayon and plastics as well as traditional timber products such as pulp for paper. The rivers of Norrland supply timber and hydroelectric power to sawmills and pulp mills on the Bothnian

The Vikings: founders of Russia?

In the 8th century a strong kingdom based around Uppsala in Sweden made the first of many forays into the eastern Baltic. Within 50 years two Swedish settlements had developed in what is now Latvia, with another at Apulia in modern Lithuania. Although these towns were probably established as trading posts, they became the starting point for Viking expansion and conquest farther south.

In the early 9th century dense forest separated the Baltic coast from the Black Sea and Central Asia, which at that time were dominated by two great trading empires. The Arab Caliphate, with its capital at Baghdad, controlled everything south of the Caucasus, including the caravan routes to China and the Far East. The Byzantine empire, with its capital at Constantinople (modern Istanbul), controlled the Black Sea and southeastern Europe. For enterprising and adventurous traders there were two possible routes southward from the Baltic coasts – down the Dnieper river to Constantinople, and down the Volga river to the Caspian Sea and Baghdad. Archaeological and written evidence indicates Swedish Vikings were using both routes by the end of the 9th century.

The Russian Primary Chronicle, also known as the Nestor Chronicle, describes the origins and growth of Russia up to the 12th century. It cites a tradition that the royal families of Novgorod and Kiev were founded by Scandinavians – and there is tangible evidence to support this idea. Archeologists digging at Staraya Ladoga in northern Russia have found the remains of a Swedish colony. Their excavations have shown that Swedes

lived there from the early 9th century to the middle of the 11th. The town was ideally placed for access to both major routes south, and another written source proves the existence of a northern kingdom of "Rus" as early as 839. It also records that these "Rus", although not Swedish, were "of Swedish origin". Inscriptions and archeological finds from Sweden confirm strong links with Kiev and Constantinople, and Scandinavian names began to appear in the Slavic regions throughout the 9th century: Helgi changed to Oleg, Valdemar became Vladimir, Ingvar became Igor and so on.

So was Russia founded by Swedish Vikings? Probably not as such – but the Vikings almost certainly did found a great number of towns in Russia and the Ukraine, including Novgorod, Izborsk, Polotsk, Smolensk, Murom, Rostov, Chernigov and Kiev.

Viking hoard A cache, found in an 11th-century burial site in Sweden, consists principally of Byzantine and Islamic coins, and provides striking evidence of the great distances traveled by the Viking adventurers.

coast. From there the finished products are shipped abroad or sent to the industrial south for further processing.

Less than 7 percent of Sweden's land area is farmed, and agriculture is protected by government levies; however, it does meet four-fifths of the domestic requirement. Most cultivated land is in Skåne and the central lowlands, where arable crops include sugar beet, barley and wheat. Elsewhere farms are smaller and livestock (cattle, sheep and pigs) is more important; only hardy varieties of grass for fodder, potatoes, barley and rye can be grown successfully in the north's shorter growing season. Fur-bearing animals such as mink and fox are also raised for their pelts. Increasing mechanization and rationalization of farms is causing a rural exodus. Some 83 percent of the population now live in towns and cities, and farming employs less than 3 percent of the workforce.

Transportation and communications

State involvement in transportation and communications is quite high. The trucking industry, dominated by two government-run companies, is the chief goods carrier. Buses are the backbone of most local public transportation services. Air services are dominated by the Scandinavian Airlines System (SAS), a joint venture of the Danish, Norwegian and Swedish governments.

Since the railroads were built in the 19th century internal shipping services have declined in importance, though international shipping traffic remains very active. Recently the state-owned railroads have also suffered low profits and shutdowns. Most people now travel in cars, increasing demand for improved roads.

There are 168 daily newspapers, but many of them are state-subsidized and the number has been falling steadily since World War II. Sveriges Radio operates a comprehensive radio network and two television channels. There are four commercial television channels.

Health and welfare

Health insurance is compulsory and comprehensive. The government provides virtually everyone with child allowances, pensions for old age and disability, and benefits for sickness, maternity and unemployment. Medicines, doctors' fees and hospital charges are largely paid for by the state. As a result, both direct and indirect taxation is very high. Postwar governments have always been committed to full employment. In the 1990s unemployment figures averaged 7.5 percent.

Education

Sweden has a completely unified education system that coordinates university and vocational studies with secondary-school programs. Everyone has to attend comprehensive schools between the ages of 7 and 16, and special provision is made for the disabled. English is taught from the third year on, and German or French can also be selected in later years. In the seventh year pupils choose their own subject groups.

Everyone is entitled to continue their studies after leaving school, to prepare for university or for vocational training. There are 13 universities, of which the oldest is Uppsala (founded 1477). Adult education is available through daytime and evening classes, or at the folk high schools. Distance or home learning is available with support from radio, television or correspondence courses.

Unlike Norway, fishing – mostly from harbors on the west coast – makes a marginal contribution to the economy. The catch is dominated by fish such as herring, mackerel and cod.

Industry and trade

Kiruna in northern Norrland has one of the world's richest deposits of iron ore, and a processing plant has been built at the port of Luleå on the Bothnian coast. Other mineral resources include a range of precious metals as well as copper, zinc, lead and uranium. There are low-quality coal deposits in the south.

Sweden is one of the largest users of hydroelectric power in the world. A series of plants, mostly along the Norrland rivers, provides a large percentage of the country's electricity requirements. Coal and nuclear fuel are imported, along with all of Sweden's oil. In 1988, after the nuclear disaster at Chernobyl in the

Seaport city Göteborg, overlooking the straits between the North and Baltic seas, is Sweden's largest port and second largest city. Its industries include Volvo, textiles and the SKF steel giant. It is also a large freight and passenger ferry center.

Ukraine (then part of the Soviet Union), the decision was taken to phase out the country's 12 nuclear power stations, which supplied half of Sweden's energy requirements, by 2010.

The manufacturing industry in Sweden benefits from the country's rich resources. Products include electronic and telecommunications hardware as well as transportation equipment and machinery.

Close to half of the country's heavy industrial production is sold abroad, making the economy increasingly dependent on foreign trade. Sweden had been a full member of the European Free Trade Association (EFTA) since 1960, but, after a referendum in favor, became a member of the European Union (EU) in 1995.

Finland

REPUBLIC OF FINLAND

FINLAND'S CULTURE AND LANGUAGE ARE quite distinct from those of the other Nordic Countries. For long periods of its history it was ruled by Sweden and Russia, but neither was able to subdue the Finnish spirit.

GEOGRAPHY

Finland is the sixth largest country in Europe and one of the most northerly. Its climate and topography reflect the fact that one-third of the country's area lies within the Arctic Circle.

NATIONAL DATA - FINLAND

Land area 338,145 sq km (130,559 sq mi)				

Climate	Altitude m (ft)	Temperatures January °C(°F)	July °C(°F)	Annual precipitation mm (in)
Helsinki	51 (167)	-6 (21)	17 (63)	635.4 (25)

Major physical features highest point: Haltiatunturi (northern Finland) 1,328 m (4,357 ft); longest river: Kemi 548 km (343 mi); largest lake: Lake Saimaa 1,147 sq km (443 sq mi)

Population (2006 est.) 5,231,372

Form of government multiparty republic with one legislative house

Armed forces army 1,920; navy 5,000; air force 2,800

Largest cities Helsinki (capital - 559,046); Espoo (231,704); Tampere (202,932)

Official language Finnish, Swedish

Ethnic composition Finn 93.4%; Swede 5.7%; Russian 0.4%; Estonian 0.2%; Roma 0.2%; Sami 0.1%

Religious affiliations Lutheran National Church 84.2%; Greek Orthodox in Finland 1.1%; other Christian 1.1%; other 0.1%; none 13.5%

Currency 1 euro (EUR) = 100 euro cents

Gross domestic product (2006) U.S. $171.7 billion

Gross domestic product per capita (2006) U.S. $32,800

Life expectancy at birth male 74.99 yr; female 82.17 yr

Major resources timber, iron ore, copper, lead, zinc, chromite, nickel, gold, silver, limestone, cereals, dairy products, fish, livestock, potatoes

"Land of lakes and marshes" is the literal meaning of Suomi, the name the Finns give their country, and this view of the Lapland landscape in summer shows how well it fits the description. Birch trees flourish on the peaty soils of this part of the country.

The land

Finland is landlocked to the north and east; to the south lies the Gulf of Finland and to the west the Gulf of Bothnia. The Åland Islands, lying at the entrance to the Gulf a few miles off the southwestern coast of Finland, are an extension of the coastal lowland. Some 80 of the 3,000 islands in the archipelago are inhabited.

Apart from a small hilly area in the northwest, Finland is lowlying, its ancient granite rocks ground down by the glacial icesheets. The ice remained until only a few thousand years ago, leaving an undulating landscape dotted with lakes. The Finnish name for the country, Suomi, means "land of lakes and marshes". There are over 60,000 lakes covering at least one-tenth of its area, and they are linked by extensive river systems. Lake Saimaa, the largest, covers 4,400 sq km (1,700 sq mi). Most of the lakes, however, extend no more than 25 sq km (10 sq mi). The

longest rivers are the Kemi and the Oulu, both of which drain west into the northern end of the Gulf of Bothnia.

Climate

Separated from the Atlantic by the Scandinavian peninsula, Finland's climate is influenced by the continental weather system. Above the Arctic Circle the weather is typically extreme; very cold winters are matched by short summers in which temperatures often soar during the long days. Farther south the climate is moderated by the proximity of the sea. The mild summers allow Finland's agricultural area to extend farther north than in the other Nordic countries. However, from mid November until April the whole country is snow-covered and the sea is frozen over.

Plants and animals

Most of the country is heavily forested, chiefly with pine, spruce and other conifers. Hardy broadleaf trees grow in the south. In the extreme north the forests give way to swampy tundra, a landscape of few trees but rich in lichens and berry-

producing plants, especially cloudberries.

The woodlands are home to many mammal species, including bears, wolves and lynx; elk and reindeer are relatively common. The rivers are rich in salmon and trout, and they also support ducks and other waterfowl.

As elsewhere in the Nordic Countries, the forests, lakes and rivers show signs of environmental damage from acid rain, caused by windblown air pollution originating outside the region. In addition, radiation from the Chernobyl nuclear disaster in the Ukraine in 1986 still affects Sami (Lapp) reindeer herds.

SOCIETY

Although Finland has been independent only since the early 20th century, its people retain a proud and enduring sense of national identity.

History

Finland was first inhabited by neolithic nomadic hunter–gatherers, probably ancestors of the present-day Sami. Later settlers came from around the Baltic area,

speaking a Finno-Ugaric language that is closely related to both Estonian and Lithuanian but is completely unrelated to the Scandinavian languages, which are Indo-European in origin.

Finland remained outside the cultural orbit of Europe until the 12th century, when the Swedish kings launched a crusade to subdue and convert its pagan inhabitants. The Swedes' English-born bishop, Henry, murdered by a Finnish peasant, later became Finland's patron saint. Between the 13th and 19th centuries Finland was effectively a territory of Sweden, and its culture and education were entirely Swedish-dominated. Swedish became the language of government and the aristocracy.

Russia, however, continually eroded parts of Finnish territory, and in 1809, when Sweden failed to defend Finland, the Finns accepted an offer by Tsar Alexander I (1777–1825) to make the country an autonomous grand duchy. Finland enjoyed considerable political freedom, and prospered under the favorable rule of Alexander II (1818–81). He encouraged wider use of Finnish, which had survived only as the language of the working people; this was partly to weaken links with Sweden, and partly to foster a sense of nationalism and new belief in a Finnish cultural identity. Harsh moves by his successor, Nicholas II (1868–1918), to Russify the country only encouraged a growing nationalist movement that successfully united all levels of society.

At the outbreak of the Russian revolution (1917–18) Finland seized the opportunity to declare its independence; a communist coup was defeated and a republic established in 1919. The Åland islanders in 1917 sought to become part of Sweden, with which they had long economic and cultural links dating back to the 12th century. Finland granted them autonomy in 1920, but refused to allow them to secede; today they remain an autonomous part of the state of Finland.

In 1939 the Soviet Union took advantage of its pact with Nazi Germany to seize much of Karelia in eastern Finland. With the Allies unable to help, Finland had to turn to Germany, for which the Soviet Union imposed heavy postwar reparations. During the Cold War era Finland managed to steer a delicate course, maintaining its democratic tradition and mixed economy without offending the Soviet Union. The term "Finlandization" came to describe the situation of a small, neutral country coexisting beside a communist superpower.

Government

The president is elected for six years by popular vote with a second round of voting if there is no absolute majority. The president appoints the administration, the prime minister and cabinet. Legislative authority is vested in a single-chamber parliament, the Eduskunta with 200 directly elected members. The autonomous province of the Åland Islands has its own elected congress and land councillor, or prime minister, but the provincial governor, appointed by the Finnish government, has the right to veto the decisions made by congress.

Proportional representation in Finland has allowed a mass of major and minor parties to develop, and most governments are center-left coalitions. The largest parties are the Social Democrats, the Center Party and the National Coalition Party; the once-popular Communist Party has suffered from declining membership.

People

Finland is more linguistically complex than the other Nordic Countries. In most of the country the Finnish-speaking majority is dominant, including regional groups with their own dialects, such as the Tavastlanders. Most of the country's Sami inhabitants, who speak their own distinct language and now number only about 2,500, live as seminomadic reindeer herders in the north. In the coastal south and the Åland Islands there is a Swedish-speaking minority. Swedish is still an official language alongside Finnish, but is in declining use.

The Lutheran Church is the national church, to which about 90 percent of the population belong. A tiny proportion (about 1 percent) belong to the Finnish Orthodox church.

Finland's international cultural contribution belies its size and its linguistic isolation. The arts, particularly the music of Jean Sibelius (1865–1957) and the folk-epic *Kalevala* compiled by Elias Lönnrot (1802–84), played an important part in establishing Finland's distinctive identity and, latterly, its nationhood. In the field of town-planning, the architects Alvar Aalto (1898–1976) and Eliel and Eero Saarinen (1873–1950 and 1910–61) have been influential in shaping the modern cityscape. Finnish design skills have become world-famous, particularly in glasswork and the bright Marimekko fabrics. Also internationally famous, particularly with younger readers, are the novels of Tove Jansson (b. 1914) about the Moomintroll family and their adventures.

ECONOMY

Finland enjoys its share of the Nordic Countries' prosperity, and its national income is currently growing at a much faster rate than its population.

Agriculture and forestry

Timber is Finland's main natural resource, and the country's forestry and lumber industries are among the most important in Europe. Agriculture is confined to the south by climate; only about 9 percent of the land is cultivated. Most farmers own smallholdings and combine agriculture with forestry. Cereals are grown, mainly as fodder, with potatoes and sugar beet and, in the extreme south, rapeseed and some fruit. Livestock production traditionally concentrated on the raising of dairy cattle, but this has declined significantly; there is also some pig and poultry farming.

Industry

Industry – primarily manufacturing – accounts for one-quarter of the country's income. Mining industries are small but significant, including important European supplies of copper, nickel and vanadium, and other rare metals such as zinc, chromium and titanium. Iron ore is no longer mined.

Manufacturing growth after World War II was spurred on both by the Soviet Union's demands for reparations and by favorable trade agreements soon after, whereby Russian oil was exchanged for Finnish manufactured goods. This was a considerable advantage for Finland, as the country was largely dependent on peat for its energy. One-third of the country's energy requirements is derived from hydroelectric plants, and more than one-third from nuclear plants.

Heavy industries include specialized shipbuilding, particularly of icebreakers, and the manufacture of machinery and cement, together with building and construction, chemicals and petroleum refining. Important lighter industries include electronics, textiles, printing and publishing, and all types of timber products, including wood and paper. Tourism is increasingly important. Most major industries have some degree of state ownership or government investment.

Floating assets Forestry is the mainstay of the Finnish economy. The cut logs are floated down the rivers to huge timber mills on the coast. Wood products, including paper, wood pulp, board and plywoods, account for over one-third of exports.

Trade and commerce

Following a referendum in favor, Finland became a member of the EU in 1995.

Timber related products, such as paper, dominate exports, while petroleum, machinery and transport equipment are the major imports.

Transportation and communications

The road network is extensive, though only about half the roads are surfaced. The rail network was privatized in 1995. It is extensively developed in the southwest, and is undergoing modernization and electrification elsewhere. Air services operate between various domestic airports all around the country, and internationally via Malmi airport at Helsinki. The inland waterway network of lakes and rivers is an important transportation route for the timber industry, and its scenic beauty also attracts tourists.

Health and welfare

Finland's state welfare system offers a wide range of benefits, including some subsidized housing in the towns to counter an acute national shortage. The population is aging as life expectancy increases under Finland's excellent health service. Elderly people enjoy good pensions and generous care provisions. The nine-year comprehensive education system is free and compulsory.

Finland Awakes!

Perhaps more than any other single person, it was the composer Jean (or Jan) Sibelius who helped to put his struggling new nation on the international map.

Sibelius was born in 1865 in the small town of Hämeenlinna in the south of Finland. Like most educated Finns at that time, his family were Swedish-speaking, but their forward-looking views led them to Finnicize their Swedish name, Sibbe, and Jean was sent to the Finnish-speaking grammar school. It was there that he first showed unusual talent for the violin, and composed his first piece at the age of 10. He was sent to university in Helsinki to study law, but soon abandoned this for music, going to Berlin and Vienna to study.

His time abroad reinforced his love of Finnish, as did his secret engagement to Aino Järnefelt, daughter of a fervently nationalist aristocratic family. Finnish influences became pervasive in his music. His first major work, the Kullervo Symphony, was a massive choral symphony based on a tragic legend from the *Kalevala* folk epic. First performed in 1892, it won him immediate recognition.

In 1899 Sibelius, by now beginning to establish an international reputation and devoting himself fulltime to composing, wrote the first of his seven symphonies. However, the political upheavals of that year were having an effect on Finnish society. Under Tsar Alexander II (1818–81), Finland had enjoyed freedoms unknown in Russia itself, but following the accession of Nicholas II (1868–1918) a Moscow-appointed governorship was imposed, and Russian was made the legal state language. Major Finnish newspapers were closed down, and protest meetings and political plays were banned. To find a way around the ban a series of allegorical tableaux was put on. Sibelius wrote the music for these tableaux and some of them, including the *Karelia Suite*, became the basis of his most popular works. For the last tableau he wrote *Finland Awakes!* – a strikingly dramatic finale.

Renaming the piece *Finlandia*, Sibelius redrafted the finale as a concert overture for the Helsinki Orchestra to play at the Paris World Fair in 1900, where Finland was given a pavilion of its own on a par with other countries. It is from this World Fair that Finland usually dates its acceptance as a nation among nations. The pavilion, designed by Finnish architects including the young Eliel Saarinen, and filled with original Finnish works of art and design, made an immense impression, as did *Finlandia*. Soon the piece was being played all over the world. Even today *Finlandia* is considered the country's unofficial national anthem.

Monument to a national hero The Sibelius monument in Helsinki honors the composer whose music captured the growing sense of Finnish identity. His *Finlandia* overture was performed in 1900, 17 years before Finnish independence.

Denmark

KINGDOM OF DENMARK

DENMARK, THE SMALLEST SCANDINAVIAN country, enjoys possibly the highest standard of living in the European Union. Until recently it also had the highest income tax and foreign debt.

GEOGRAPHY

Mainland Denmark, the Jutland peninsula, extends 300km (200mi) north from the border with Germany. Dunes, lagoons

NATIONAL DATA - DENMARK

Land area	42,394 sq km (16,368 sq mi)			
Climate		**Temperatures**		**Annual**
	Altitude m (ft)	**January** °C(°F)	**July** °C(°F)	**precipitation** mm (in)
Copenhagen	9 (30)	0 (32)	17 (63)	587 (23)

Major physical features	highest point: Yding Skovhøj (central Jutland) 173 m (568 ft); largest island: Sjaelland 7,016 sq km (2,709 sq mi)

Population	(2006 est.) 5,450,661

Form of government	multiparty constitutional monarchy with one legislative house

Armed forces	army 12,500; navy 3,800; air force 4,200

Largest cities	Copenhagen (capital - 1,086,762); Arhus (228,547); Odense (145,554); Alborg (121,549)

Official language	Danish

Ethnic composition	Danish 95.1%; other European 1.5%; Asian 1.3%; other 2.1%

Religious affiliations	Evangelical Lutheran 95%; other Protestant and Roman Catholic 3%; Muslim 2%

Currency	1 Danish krone (DKK) = 100 øre

Gross domestic product	(2006) U.S. $198.5 billion

Gross domestic product per capita	(2006) U.S. $37,000

Life expectancy at birth	male 75.49 yr; female 80.22 yr

Major resources	petroleum, natural gas, fish, salt, limestone, chalk, stone, gravel, sand, electronics, textiles, cattle, cereals, coal, cryolite, fodder, iron ore, lead, molybdenum, natural gas, pigs, potatoes, poultry, uranium, zinc

and sandbars shelter the west coast from North Sea storms; the eastern plains face the Baltic sea and an archipelago of 483 islands, of which 97 are inhabited.

The land

Denmark averages only about 30m (100 ft) above sea level. Its landscapes are the product of ice age glaciation; glacial debris divides eastern and western Jutland. Denmark's largest lagoon, the Lim Fjord, cuts north across the country. In the extreme north, the long sandspit known as the Skaw points toward Sweden.

East of Jutland, across the Little Belt strait, is the island of Fyn and, to the southeast, the smaller island of Langeland. Across the Great Belt is the island of Sjaelland. Copenhagen, the capital city, is on its eastern coast. Suburbs now cover most of the coast from Copenhagen to Helsingor (Elsinore). Far out in the Baltic lies the granite island of Bornholm.

Climate, plants and animals

Denmark's cool and temperate climate is warmed by the North Atlantic Drift. Summers are warm, with high rainfall.

Conifer plantations have replaced the natural deciduous woods, and trees have been planted as windbreaks on the North Sea coast. Few large mammals have survived in the wild, but there are red deer in Jutland's forests and plantations, and a profusion of bird life. Pollution has affected marine and freshwater animal life.

SOCIETY

Progressive, but deeply interested in their past, the Danes have achieved a healthy and equitable society.

History

Reliable historical accounts begin in the 9th century, when the Danes resisted the northward expansion of the Franks. King

Early morning in Copenhagen Dawn brushes the bronze statue of Hans Christian Andersen's *Little Mermaid* in Copenhagen harbor, a symbol of Denmark's close association with the sea, as are the wharves and dockyards in the background.

Harald Bluetooth of Jutland (c.910–c.985) brought Christianity to his people. His son Sweyn Forkbeard (d.1014) conquered England, creating an empire that lasted for two generations.

Norway and Sweden were united under the crown of Denmark by the Kalmar Union (1396–1523), which tried to counter the commercial power of the German-based Hanseatic League. Civil war in 1534–36 resulted in the formation of the new Lutheran Church of Denmark, and many Catholic properties were seized. King Christian IV (1577–1648) tried and failed to reestablish Danish authority over the Baltic, and modern national borders were established by 1645. Norway, still subject to the Danish crown,

The Faroes – islands of sheep

The Faeroes are a group of 22 islands lying in the Atlantic between Scotland and Iceland, formed by basaltic lavas similar to those found in Iceland. Some 18 of the islands are inhabited. The characteristic landscape is harsh and rugged with high coastal cliffs and treacherous straits between islands.

The climate, with mild winters and cool summers, benefits from the warm North Atlantic Drift. Trees survive only in plantations that are sheltered from the frequent strong winds. Native wildlife consists mainly of seals and a rich variety of seabirds, including puffins and eider, valued for their feathers. There are no native land mammals: rats and mice were introduced by early settlers, and Viking colonizers in the 9th century brought sheep to the islands. An old proverb says that "sheep's wool is Faeroe gold", and the islands' name even means "sheep islands".

In about 1000 AD the Faeroese were converted to Christianity. The islands came under Norwegian rule in 1035, and passed to Denmark in 1380. From 1709 to 1856 a Danish royal trade monopoly held back economic development and impoverished the population. In the 19th century a growing nationalist movement led to the creation of the written Faeroese language, based on Old Norse. Demand for home rule strengthened, and in 1948 the Faeroes became a self-governing dependency of Denmark. Today the Faeroes have their own 32-member parliament, the Lagting, which elects the Landsstÿri, an administrative committee. The Faeroese government has been pressing hard for independence since the early 1990s and wants to phase out Danish subsidies by 2012.

Nearly one-third of the inhabitants live in or around the main harbor and capital, Thorshavn. Faeroese is the main language, though Danish is still taught in all schools. The islands have their own radio broadcasting system and there is also a limited television service. Most people are members of the Danish Lutheran Church.

The islanders enjoy many distinct cultural traditions. Ballad-dances, less common than they used to be, take place between Christmas and Lent in the Christian calendar. Ballads concerning legend, folklore or Faeroese history are sung aloud to provide the music for a ring-dance. The wording of the ballads has remained unchanged from one generation to the next.

Woolen yarns and knitted garments are produced by the islanders, but today fishing and associated industries are far more important, providing some 95 percent of exports. Whaling, for meat and oil, has always been part of the Faeroese economy and continues today despite international concern over the depletion of whale stocks.

Less than one-twentieth of the land is cultivated, mostly producing grass for fodder and vegetables, especially potatoes. Coal is mined on Sudhuroy, but hydroelectric power meets all the islands' needs. There are regular inter-island ferries and shipping communications with Denmark, Britain, Norway and Iceland throughout the year. In summer, passenger and vehicle ferries serve a small tourist industry. The airport on Vágar offers regular flights to Iceland and Norway and daily flights to Copenhagen.

Living in isolation A turf-covered wooden cabin on Strømø, the largest of the Faeroe Islands, is silhouetted against a silver, wind-ruffled sea. Fishing and farming communities are scattered around the islands' deeply indented coasts.

was relinquished in 1813 after Swedish forces defeated Denmark following the Napoleonic wars.

In 1864 war broke out with Prussia and Austria over long-disputed Schleswig and Holstein at the southern tip of Jutland. Denmark lost, but in 1920 a plebiscite returned Schleswig to Danish rule.

Although neutral in World War II, Denmark was occupied by German troops from 1940 and had an active resistance movement. As a member of the North Atlantic Treaty Organization (NATO) from 1949, Denmark refused to allow United States' air bases on Danish soil.

In 1973 Denmark joined the European Community (EC). In 1999 Denmark ratified the Treaty of Amsterdam, the EU's new constitution, setting the agenda for enlargement of the Union including relaxation of border controls.

Government

Denmark has been a constitutional monarchy since 1849. The monarch appoints the prime minister and the cabinet, who are accountable to the 179-member single-chamber parliament or Folketing.

Parliamentary elections based on proportional representation are held every four years. All citizens aged 18 and over may vote. Greenland and the Faeroes each have two representatives in the Folketing but follow Danish foreign affairs, defense, financial and legal policies.

Denmark's non-nuclear policy has led to debates over the presence of nuclear weapons on visiting warships.

People

The Danes are closely related to the other Scandinavian peoples, and there is a significant German minority. Both German and English are taught in Danish schools.

More than 90 percent of Danes belong to the national Lutheran church. The country has produced internationally renowned writers including the philosopher Soren Kierkegaard (1815–55) and Hans Christian Andersen (1805–75). The Royal Danish Ballet preserves 19th-century choreographic traditions.

ECONOMY

Denmark today is an industrial society, but agriculture still plays a vital role.

Agriculture

Farming accounts for some two-thirds of the land area, but occupies only one-sixteenth of the labor force. Small farms and agricultural cooperatives form an efficient system of intensive farming, emphasizing livestock and dairy products. The major crop is barley. About half the country's farm produce is exported, mostly to the EU. The rest supplies Denmark's food-processing industry, its second-largest employer. Denmark's fishing industry is among the world's top 15; herring, cod and flatfish account for half of the total catch.

ing in the 1980s caused a severe balance-of-payments problem. The government responded with tax cuts and reductions in welfare expenditure.

Transportation and communications

The sea has always been central to Denmark's main lines of communication. In addition to sea routes there are about 420km (260 mi) of inland waterways. Two bridges connect Jutland with Fyn, and in 1989 a massive project was launched to make a permanent link between Fyn and Sjaelland, formerly accessible only by ferry. The Øresund bridge connecting Copenhagen with Malmö, Sweden opened in July 2000. Road and railroad networks are extensive and well maintained.

Denmark has 13 airports. Kastrup, near Copenhagen, serves Jutland and Bornholm as well as international traffic, and is a major base for the co-owned SAS (Scandinavian Airlines System).

Until the 1980s Danish broadcasting was controlled by Danmarks Radio, an independent public body funded by license fees. A second television channel with commercial advertising was launch-ed in 1988, and Danes can receive broadcasts from Sweden, Norway and Germany. There are about 60 newspapers.

Health and welfare

Denmark has one of the oldest and most extensive welfare programs in the world. It offers universal benefits including pensions and free health checks for children. Compulsory health insurance provides free hospital treatment and home nursing. Unemployment insurance is voluntary but widespread.

Education is compulsory from age 7 to 16. Most schools are free; private schools receive state support. Youth schools offer both vocational and other training. The traditional folk high schools, widely imitated abroad, provide a broad-based curriculum of adult education. There are five universities and a number of professional institutions, commercial and technical colleges.

Model attractions A model of Copenhagen harbor on display at Legoland, a purpose-built park at Billund in central Jutland – home of Lego building blocks. The displays, made from 30 million blocks, attract tourists from all over Europe and beyond.

The Øresund Bridge (*above*) opened in July 2000 and links Scandinavia with the European continent via a 10-mile railway and freeway that runs between the Danish capital, Copenhagen, and the Swedish city, Malmö. A combined bridge and tunnel, its budget was $2 billion.

Industry

The country's mineral resources are heavily exploited. They include granite, boulder clay, white chalk and kaolin.

North Sea gas and oil supply about half of Denmark's energy. Surplus energy is sold to Germany and Sweden. Wind turbines are increasingly common in flat, windy Jutland, and newer houses may have solar or geothermal heating.

Industrial products include cement, crude steel, machinery and transportation equipment, processed foods, chemicals, paper and furniture. Shipping and tourism are also important.

Trade and commerce

Denmark trades almost entirely within Europe, particularly with Germany, Britain and Sweden. High consumer spend-

Iceland

REPUBLIC OF ICELAND

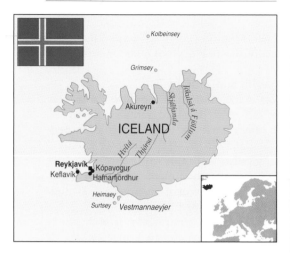

ICELAND IS AN ISLAND REPUBLIC JUST SOUTH OF the Arctic Circle. Its culture derives from the original 9th-century Viking settlers. The country and its people have grown up in a harsh, volcanically active environment unmatched anywhere else in Europe.

GEOGRAPHY

Geologically Iceland is very young. It has been built up by a series of volcanic eruptions from the mid-Atlantic ridge – the boundary of two of the Earth's tectonic plates. In 1973, the island of Heimaey off the south coast had to be evacuated after the eruption of the volcano Helgafell. Earthtremors are frequent, and hot springs are common in volcanic areas, most notably at Geysir in the southwest. Similar springs gave the capital its name – Reykjavik, or "Smoky Bay" – and today they provide most of the country's hot water and heating.

At the heart of the island is a high plateau partly covered by a glacier and surrounded by mountains. Glaciers account for over one-tenth of Iceland's total land area. The largest is Vatnajökull in the southeast, covering an area of 8,300 sq km (3,200 sq mi), which, in late 1996, split open due to volcanic activity beneath. Most of the population live on the narrow coastal plain around Reykjavik in the southwest. In the north steep valleys penetrate deep into the interior.

The northwest of the island tends to be cooler than the southeast, and snowfall there is more frequent. Temperatures are well above the average for this latitude, due to the influence of warm southwesterly winds from the tropics associated with the North Atlantic Drift, but gales are common, especially in the mild, humid winters. Summers are generally cool, and rainfall is high throughout the year. Aurora borealis, or the northern lights, can often be seen, especially in fall and early winter.

Woodland once covered large areas of the island, but has long since been cleared by settlers. The characteristic vegetation is mosses and lichens with a few shrubs. Sparse grassland is limited to valleys and the coastal plain.

Foxes are Iceland's only native mammal. Mice and rats came with the Vikings, as did domestic animals such as the sturdy Icelandic pony. Reindeer were introduced in the 18th century. Bird life is rich and varied, especially on sea cliffs and at Lake Mývatn in the north. Many migratory birds visit the island.

Water and rock A waterfall plummets down a cliff of basalt rock at Seljalands Foss in southern Iceland. The island's dramatic scenery includes lunar-like lava fields, vast glaciers and snowfields, active volcanoes and geysers spouting steam high into the air.

SOCIETY

The first settlers in Iceland were probably Irish monks. They were followed in about 870 AD by Vikings from Norway, Ireland and Scotland. The chieftains in each district set up a "thing" or local assembly to settle quarrels and feuds common among landholders, and about 930 a national assembly, the Althing, was established – the basis of Europe's oldest parliament. It was through a decision reached in the Althing that the population accepted Christianity in 999–1000.

During the 13th century Iceland's chieftains accepted Norwegian rule in order to preserve domestic peace, but in 1380 Norway itself fell under Danish sovereignty, and Iceland entered a long period of decline, exacerbated by the Danish crown's commercial monopoly, which crippled trade.

In the 19th century the demand for independence and constitutional reform grew. Trading restrictions were lifted, and the Althing was granted a degree of legislative authority. Conditions remained harsh, however, and there was massive emigration to Canada until the development of the fishing industry brought increased prosperity at the beginning of the 20th century; the island achieved full self-government in 1918.

During World War II British and American forces used Iceland as a base. The German invasion of Denmark in 1940 had severed communications with the island, and in 1944 the Althing cut the last ties with Denmark by declaring Iceland an independent republic.

Since 1951 there has been an American military base at Keflavík; this, and Iceland's membership of the North Atlantic Treaty Organization (NATO), have been hotly debated political issues. Fishery protection is another major concern. In 1975 the government declared a 370-km (200-mi) fishing limit – a decision that led to one of several "cod wars" with the United Kingdom. However, in 1976 the British government recognized the new international limit.

Icelanders are proud of their Viking roots, and many can trace their ancestry back to the earliest settlers, whose names are recorded in the *Landnama Bok*, or Book of Settlement. The modern Icelandic language is directly descended from Old Norse, the language of the Vikings. The medieval sagas, which mingle imagination and folklore with historical facts about the early settlers, can be easily read by modern Icelanders. Most of the islanders belong to the Lutheran Church, which was established in Iceland in the 16th century.

ECONOMY

Fishing is the cornerstone of Iceland's economy, accounting for three-quarters of all exports. Much of the fishing fleet is government-owned. Attempts are being made to tackle the problems of over-

The sun at midnight illuminates the harbor of Höfn, in southeastern Iceland. Fishing, once a smallscale family activity, is now big business and the huge trawlers and processing plants are owned by a few large companies or cooperatives.

fishing and declining fish stocks of cod, capelin and herring. The Marine Research Institute regulates operations with a system of quotas and bans. Fish farming is proving increasingly successful. Concern over loss of fishing rights is a major reason for Iceland's reluctance to join the European Union.

Until the beginning of this century sheep farming was the main occupation of all Icelanders. There are still some 4,500 farms, but only about 1 percent of the total land area is cultivated, producing fodder crops, potatoes and turnips. Grazing for livestock (sheep, cattle and ponies) occupies about 20 percent of the land. Fresh vegetables and flowers are available all year round from greenhouses heated by geothermal power, a cheap and abundant energy source.

Several major industrial installations are also run on geothermal power. Minor industries include fish processing, skins and furs. In the 1990s Iceland has diversified into software production, biotechnology and financial services. Tourism is also growing steadily, especially in whale-watching, ecotourism and cycle touring.

Iceland has about 11,000 km (7,000 mi) of roads, but many of them are impassable in winter. There are no railroads on the island and in many places Icelandic ponies still provide an important means of local transportation. Icelandair operates domestic and international flights from Reykjavik. The merchant fleet carries up to half of Iceland's imports and more than half of its exports.

Icelanders are among the healthiest people in the world. Compulsory health insurance allows many medical services to be offered at minimal cost, and hospital care is free. All education is also free, and compulsory from the ages of 7 to 16. Further education is available and there is a university at Reykjavik.

NATIONAL DATA – ICELAND				
Land area 100,250 sq km (38,707 sq mi)				

Climate		Temperatures		Annual
	Altitude m (ft)	January °C(°F)	July °C(°F)	precipitation mm (in)
Reykjavik	52 (170)	0 (32)	11 (52)	817 (32)

Major physical features highest point: Hvannadalshnúkur 2,119 m (6,952 ft); longest river: Thjórsà 230 km (143 mi)

Population (2006 est.) 299,388

Form of government multiparty republic with one legislative house

Armed forces paramilitary 130

Capital city Reykjavik (184,244)

Official language Icelandic

Ethnic composition Icelandic 96.4%; other Europeans 2.5%; Asian 0.7%; others 0.4%

Religious affiliations Lutheran 85.5%; Reykjavik Free Church 2.1%; Roman Catholic 2%; Hafnarfjoorour Free Church 1.5%; other Christian 2.7%; other or unspecified 3.8%; unaffiliated 2.4%

Currency 1 Icelandic króna (ISK) = 100 aurar

Gross domestic product (2006) U.S. $11.4 billion

Gross domestic product per capita (2006) U.S. $38,100

Life expectancy at birth male 78.23 yr; female 82.48 yr

Major resources cattle, dairy products, diatomite, fish, fodder, hay, hydroelectric and geothermal power, potatoes, poultry, sheep

The British Isles

COUNTRIES IN THE REGION

UNITED KINGDOM · IRELAND

The traditional English pub (*left*), or public house, is found on countless street corners in towns and villages throughout the region – the meeting place of people from all walks of life, both young and old. *The Sherlock Holmes*, in London, named after the celebrated 19th-century fictional detective, is a typical example.

Unst
Yell
Shetland Islands
Mainland
Lerwick

Sanday
Mainland
Orkney Islands
Kirkwall
Hoy
John O'Groats

Butt of Lewis
Lewis
Outer Hebrides
Harris
N Uist
Skye
S Uist
Barra
Rhum
Mull
Jura
Islay
Inner Hebrides

North West Highlands
Inverness
Moray Firth
Loch Ness
Cairngorm Mts
Aberdeen
Ben Nevis ▲ 1344
Grampian Mountains
Dundee
Firth of Tay
Perth
Firth of Forth
Edinburgh
Clyde
Glasgow
Irvine
Arran
Southern Uplands
Tweed
Cheviot Hills
Dumfries
Newcastle upon Tyne

ATLANTIC OCEAN

North Sea

Malin Head
Mts of Antrim
Londonderry
Lough Neagh
Belfast
Donegal Bay
Lough Erne
Mourne Mts
Sligo
Douglas
Isle of Man
Cumbrian Mts
▲ Scafell Pike 977
Pennines
Middlesbrough
North York Moors
Ouse
Leeds
York
Hull

Achill I
Lough Mask
Lough Corrib
Lough Ree
Shannon
Dublin
Anglesey
Liverpool
Manchester
Sheffield
Irish Sea

Galway Bay
Galway
IRELAND
Aran Is
Burren
Lough Derg
Barrow
Wicklow Mts
Snowdon ▲ 1085
The Wash
Norwich

Limerick
Aberystwyth
Cambrian Mountains
Birmingham
Coventry
The Fens
Ipswich
Suir
Wexford
Waterford
St George's Channel
Severn
Trent
Gt Ouse

Macgillycuddy's Reeks
Cork
Brecon Beacons
Swansea
Newport
Cardiff
Cotswolds
Oxford
Chiltern Hills
Thames
London
North Downs
Strait of Dover

Bristol Channel
Bristol
Mendip Hills
Exmoor
Southampton
South Downs
Brighton
Isle of Wight

Dartmoor
Exeter
Plymouth
English Channel

Land's End
Isles of Scilly

UNITED KINGDOM

■ capital city
● major town

height of land (meters)
1000
500
200
0 (sea level)

▲ mountain peak

United Kingdom

THE UNITED KINGDOM OF GREAT BRITAIN AND
NORTHERN IRELAND

THE UNITED KINGDOM LIES OFF THE ATLANTIC
northwest coast of Europe and consists of the union of England, Scotland
and Wales (Great Britain) with the six
counties of Northern Ireland.

GEOGRAPHY

A fairly small land area contains great
diversity of landscape and geology. Much
of southern and central England has gently rolling countryside, but northern
England, Scotland, Wales and Northern
Ireland have a high proportion of rugged
terrain and mountains.

The land

Scotland and its offshore islands – the
Outer Hebrides, the Orkneys and the
Shetlands – are the northernmost part of
Britain. The Scottish mainland divides
into the Highlands in the north, the
Lowlands in central Scotland, and the
Southern Uplands to the south. The
highest peak in the British isles is Ben
Nevis in the western Scottish Highlands
at 1,344 m (4,408 ft). The Cheviot Hills to
the far south form a natural border with
England.

In England, the Pennine chain rises
south of the Cheviots and extends into
central England. In the northwest are
the Cumbrian mountains and the Lake
District; in the northeast are the North
Yorkshire moors. South of the Pennines,
much of central England has rolling hills,
while the east tends to be flat. Britain's
longest river, the Thames, rises in the
Cotswold hills in the south central region

The Channel
Islands

The Channel Islands, which lie in the
English Channel off the northwest
coast of France, are a self-governing
group of islands. They have been
dependencies of the British crown
since the Norman invasion of 1066,
when they belonged to the Duchy of
Normandy. There are four main islands: Jersey, Guernsey, Alderney and
Sark. Jersey has its own constitution
and a separate system of government.

During World War II the Channel
Islands were the only British territory
to be occupied by Nazi Germany.
English is now the dominant language,
though many people speak French, the
official language of government on

Fertile arable land (*above*) is common in southern England. Five percent of Britain is now forested, with enclosed fields such as these being the rule. The use of traditional hedgerows is being revived to preserve wildlife habitat.

Natural harbors and rich farmland (*left*) characterize the county of Devon in the extreme southwest of England. Tourism and fishing are the major industries along the coast, and the area's mild winters make it a popular retirement area.

Jersey. The distinctive Jersey cattle are a main export, together with potatoes, flowers, fruit and a much diminished crop of greenhouse tomatoes. The Battle of Flowers is an annual tourist event that also helps to promote the island's successful cut-flower industry. Their southerly location has made the islands a popular resort area, while low taxes have attracted a number of very wealthy businesses to register there.

The vexed question of sovereignty over some of the very small islands was revived in the 1970s, when France and the United Kingdom disputed the right to explore for oil nearby.

Albert Harbor marina, St Helier, Jersey Lying only 19 km (12 mi) west of France, Jersey is a popular port of call for both British and French tourists, particularly yacht owners. Most place-names on the island are French

and flows southeast into the heart of London (the capital city) and out into the North Sea.

Wales, on the western border of England, is bounded by the Irish Sea to the north, St George's Channel to the west, and the Bristol Channel to the south. The Cambrian Mountains fill most of the center of the country. The second longest river in Britain, the Severn, rises in the Cambrian mountains and flows into the Bristol Channel.

Across the Irish Sea is the island of Ireland. The Republic of Ireland occupies some five-sixths of the land area, with the British province of Northern Ireland in the northeastern corner. North of Belfast, the provincial capital, are the Mountains of Antrim; one slope forms spectacular chalk and basalt cliffs facing Scotland's southwestern coast.

Climate

Despite the northerly latitude, the climate is generally mild. Warm air derived from

NATIONAL DATA - UNITED KINGDOM

Land area 241,590 sq km (93,278 sq mi)

Climate	Altitude m (ft)	Temperatures January °C(°F)	July °C(°F)	Annual precipitation mm (in)
London	7 (22)	4 (39)	17 (63)	611 (24)

Major physical features highest point: Ben Nevis 1,343 m (4,406 ft); lowest point: Holme Fen (The Fens) -4 m (-13 ft); longest river: Severn 290 km (180 mi)

Population (2006 est.) 60,609,153

Form of government multiparty constitutional monarchy with two legislative houses

Armed forces army 117,300; navy 40,700; air force 48,585

Largest cities London (capital - 7,172,091); Birmingham (989,141); Leeds (715,402); Glasgow (629,501); Sheffield (513,234); Edinburgh (452,194); Liverpool (439,473); Manchester (395,323); Bristol (380,615); Cardiff (305,353); Belfast (277,391)

Official language English

Ethnic composition White (English 83.6%; Scottish 8.6%; Welsh 4.9%; Northern Irish 2.9%) 92.1%; Black 2%; Indian 1.8%; Pakistani 1.3%; Mixed 1.2%; other 1.6%

Religious affiliations Christian (Anglican, Roman Catholic, Presbyterian, Methodist) 71.6%; Muslim 2.7%; Hindu 1%; other 1.1%; unspecified or none 23.6%

Currency 1 pound sterling (GBP) = 100 pence

Gross domestic product (2006) U.S. $1.903 trillion

Gross domestic product per capita (2006) U.S. $31,400

Life expectancy at birth male 76.09 yr; female 81.13 yr

Major resources coal, petroleum, natural gas, iron ore, lead, zinc, tin, limestone, salt, clay, chalk, gypsum, potash, silica sand, slate, barley, fruit, vegetables, fish, potatoes, sugar beet, wheat, tourism

the North Atlantic Drift and from western Africa often combines with cold, wet air from Iceland and the Arctic, producing high winds and heavy rainfall.

There are strong regional variations. Western Scotland has heavy snow in the mountains. Northern Ireland is humid, but there is little snow in winter. Wales has cool summers and mild winters, with high rainfall in the uplands. Southern England has the highest summer temperatures and the widest annual range; winters are often cold and frosty.

Plants and animals

Over 5,000 years ago much of Britain was covered in dense deciduous woodland, but today hardly any survives. Moorland is common to Scotland, Northern Ireland and the uplands of northern England and Wales. Most valleys and woodlands have been cleared to make way for farmland. Commercial tree plantations are common.

Large mammals such as wolves were long ago hunted to extinction or died out as their habitats were destroyed. The largest surviving mammal is the Red deer, found in the Scottish Highlands. Smaller species include the Red fox, the badger, the otter and the stoat. Bird life is

The rugged mountains of Snowdonia (*right*) in northwestern Wales support slate quarrying, hydroelectricity generation and aluminum smelting, as well as attracting outdoor sports enthusiasts.

The Isle of Man

The Isle of Man lies in the Irish Sea midway between the northwest coast of England, the south coast of Scotland and the east coast of Northern Ireland. It is not part of the United Kingdom, but a crown possession that is virtually self-governing.

The island is about 48 km (30 mi) long by 16 km (10 mi) wide, with a rocky coastline enclosing a central highland area rising to 620 m (2,034 ft) at Snaefell. Low-lying land on either side produces cereals and potatoes.

A number of Neolithic monuments on the island indicates that there were settlers as early as 2000 to 1500 BC. The next wave of settlers were the Celts, and their language, Gaelic – known on the island as Manx – remained in daily use until the 19th century. In the 5th century AD the island was settled by Irish missionaries, and in the 9th cen-

tury it was invaded by Vikings from Norway. It remained a Norwegian dependency until it was sold to Scotland in 1266. It came under English control in the early 15th century.

Since 1828 the island has been a crown possession with a governor appointed by the monarch. It has its own two-chamber parliament, which dates from the 9th century. The legislative council, or upper house, and the elected lower house, the House of Keys, sometimes assemble together as the Court of Tynwald, but they normally operate as separate bodies.

The Isle's chief industry is tourism, and its Tourist Trophy (TT) motorcycle races attract many visitors. It is also famous for its tailless Manx cats.

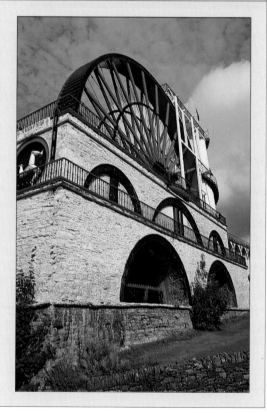

The Laxey Wheel (or Lady Isabella Wheel), on the Isle of Man, is the largest of its kind in the world. Built in 1854, it stands 22 m (72 ft) high, and was used to pump water from the Laxey iron ore mines in the Agneash Valley. Now it is a tourist attraction.

rich and varied, and freshwater fish are plentiful.

SOCIETY

The British Isles were the birthplace of modern parliamentary democracy, of the Industrial Revolution and of the welfare state. An equally great contribution is the English language, spoken by some 350 million native speakers worldwide, and by another billion as a second language.

History

Britain's first settlers came from mainland Europe across an Ice Age land bridge that later disappeared. There is evidence of Neolithic settlement as early as 7000 BC and Bronze Age metalwork in about 1800 BC. Celtic peoples from Central Europe arrived in about 450 BC.

Rome occupied most of what is now Wales and England from the 1st to the 5th centuries AD, and introduced Christianity and the Latin language. As the empire collapsed, the Celtic Britons were driven into the far north and southwest of Wales and England by waves of settlers – the Angles, Saxons and Jutes (Germanic peoples from northern Europe). By the 7th century, when missionaries reintroduced Christianity, Anglo-Saxon kingdoms had emerged.

During the 8th and 9th centuries the northeast and east of England came under Viking rule from Norway and Denmark. Regained by the mid 10th century, England, now a single kingdom, was again invaded and ruled by the Danes from 1016–42.

In 1066 the English king Harold II (b. 1022) was killed in battle by William, Duke of Normandy, who claimed the English throne and established a royal dynasty. The new feudal structure of English society gave his barons considerable power. Royal power was limited in 1215 by the Magna Carta or Great Charter, which led to the establishment of Great Councils, later known as parliaments. The first recognizable parliament met during the reign of Edward I (1239–1307).

Ireland had been conquered in the late 12th century. In 1284 Wales was made subject to the English crown, but only became fully united with England in 1536. In 1314 Edward II (1284–1327) was forced to recognize Robert the Bruce (1274–1329) as king of Scotland. Territorial quarrels with France led to the Hundred Years War (1337–1453), and English forces were eventually expelled from all French land.

The Wars of the Roses (1455–85) saw a bloody internal struggle for the throne, ending with the accession of Henry VII (1457–1509), who founded the Tudor dynasty. Henry VIII (1491–1547) broke away from the Roman Catholic Church in 1533 and proclaimed himself Supreme Head of the Church of England. Under his daughter Elizabeth I (1533–1603), England enjoyed a golden age of prosperity, literature and exploration. The navy defeated a Spanish invasion by sea, and colonies were founded in North America.

Elizabeth's successor, James VI of Scotland (1566–1625), became James I of England and united the two kingdoms. During his reign thousands of Scottish and English Protestants settled on Catholic owned lands in northern Ireland, thereby sowing the seeds of conflict between Catholics and Protestants that has lasted to this day. James's son Charles I (1600–49) quarrelled with parliament over the extent of the monarch's power, leading to civil war and his execution in 1649. The parliamentary general Oliver Cromwell (1599–1658) was appointed Lord Protector of the Commonwealth, but two years after his death, Charles II (1630–85) was invited to return to the throne, though with reduced powers.

James II (1633–1701), a Catholic, alienated large numbers of his subjects and was forced into exile in 1688 (the

The huge stone circle at Stonehenge in central southern England is the country's most fascinating prehistoric puzzle. Built between 1800 and 1300 BC, it was possibly used for astronomical purposes and worship, but the religious practices of the people it served are not known.

Henry VIII (*above*) by the German artist Hans Holbein the Younger (1497–1543), English court painter from 1536. His astute portraits of Henry and his wives provide an excellent pictorial record of the Tudor court.

The Spanish Armada (*top right*) In 1588 Philip II of Spain attempted to invade England with an armada of 130 ships against England's 197 ships with more advanced firepower. The battle took only 100 English lives but cost Spain over half of its 27,000 troops.

The Great Screen at Winchester Cathedral (*left*) a magnificent Norman church built in the late 11th century in central southern England on the site of the Saxon church of St Swithun. The cathedral is the home of one of the most famous choral schools in Britain and is a center of excellence for church music.

"Glorious Revolution"). His Protestant daughter Mary II (1662–94) ascended the throne with her husband, the Dutch prince William of Orange (1650–1702). England, Wales and Scotland were formally united into the kingdom of Great Britain in 1707. Under a new dynasty, the Hanovers, a Scottish rebellion to restore the exiled Stuart line was defeated in 1715 and a second one bloodily suppressed at the Battle of Culloden in 1746.

In the 1760s Britain defeated France over rival claims in India and North America, becoming a leading colonial power. Explorations of Australia and New Zealand in 1770 by Captain Cook (1728–79) were the basis for later settlements. But in 1775, angered by taxes, the American colonies rebelled. Winning the resulting war, they became independent in 1783. Britain had better results against the emperor Napoleon I of France (1769–1821): in 1805 with a naval victory at Trafalgar off the Atlantic coast of Spain, and on land at the Battle of Waterloo in 1815 which finally defeated Napoleon.

Britain's rise as an overseas trading nation coincided with the Industrial Revolution, which, based on local iron

William Shakespeare

William Shakespeare (1564–1616) is regarded by many literary experts as the greatest writer of all time, and certainly the greatest dramatist and poet to have written in the English language. A genius with words and images, his poetry and plays explore the complete range of human emotions and conflicts. His versatility was extraordinary, with works ranging from histories to comedies, tragedies and romances.

Shakespeare was born and educated at Stratford-upon-Avon in the English Midlands, where he married Anne Hathaway (1556–1623) at the age of 18. Later he moved to London, where he became an actor with the city's leading theater company, and by 1584 he had emerged as a rising playwright.

The first play attributed to his authorship, based on the life of the English king Henry VI (1421–71), was produced in 1590–92. The precise sequence in which the rest of the plays were written and performed is hard to establish, though the early historical plays were written as a tribute to Queen Elizabeth I to support the Tudor claim to the throne. By 1597 he had earned enough money from his writings to buy New Place, one of the largest houses in Stratford, still standing today. In 1598 he became involved

in the building and management of a new London theater, the Globe, whose foundations were rediscovered in 1989. Shakespeare's famous love sonnets, 154 in all, were mostly written between 1593 and 1596, but were not published until 1609.

In the early years of the 17th century, under the patronage of King James I, Shakespeare produced his great tragedies: *Hamlet*, *Othello*, *King Lear* and *Macbeth*, all profound explorations of human character.

Publication in 17th-century England was a chancy business. Copyright laws were virtually nonexistent, and copyists would often write down the text of a play during its performance and then publish it for their own profit. Shakespeare's name was well established, and some unscrupulous publishers applied it to the writings of others to increase their sales. The first collected edition of Shakespeare's plays, the *First Folio*, was published in 1623, just seven years after the playwright's death, and contained all but one of what we now take to be his plays.

In the 18th century scholars began to suggest that Shakespeare, the product of a provincial schooling, could not have written all the plays published under his name. A number of apparent cyphers and cryptograms in the plays were presented as evidence that the philosopher and essayist Francis Bacon (1561–1626) had written them. Modern

Shakespeare's original Globe Theater, built in Southwark, London, in 1599. It burned down in 1613 during a performance of "Henry VIII", was rebuilt in 1614, and was demolished in 1644 during the English Civil War. It has now been rebuilt.

scholars, however, have demonstrated that this theory cannot be supported. Although Shakespeare's surname occurs in a number of different spellings in various documents, this was not at all unusual at the time. In fact, the documentary evidence for Shakespeare's life is remarkably comprehensive for a man of his station in life at this period.

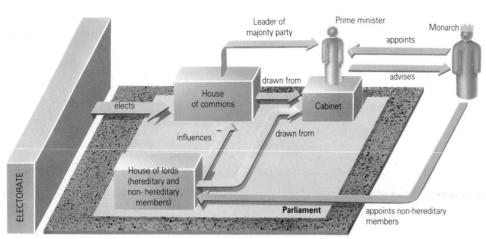

ore and coal deposits, established a new economic base in manufacturing. During the reign (1837–1901) of Queen Victoria (1819–1901) the British empire expanded to cover one-quarter of the world's land surface, and on its behalf the British army fought in various regions almost continuously from 1851 to 1902.

By the end of the 19th century reforms improved the welfare of the working class, while in Ireland there was a growing demand for Home Rule. In 1914 the United Kingdom entered World War I in alliance with France and Russia against Germany and Austria, suffering enormous casualties. Southern Ireland (later the Republic of Ireland) won independence in 1922 as the Irish Free State, but the province of Northern Ireland elected to remain part of the United Kingdom.

In 1931 the United Kingdom's self-governing overseas dominions of Australia, Canada, Ireland, New Zealand and South Africa became known as the Commonwealth of Nations. At home there was rising unemployment and economic depression well into the 1930s. London withstood heavy bombing by German aircraft in World War II; British casualties in the war were again very high. Rationing lasted into the 1950s.

A socialist Labor government elected immediately after the war created the first welfare state and placed major industries such as transportation, coal mining, electricity and gas under state ownership or control. Later Conservative governments, notably in the 1980s and 1990s, returned most of these to the private sector.

By the 1970s the British Commonwealth was almost entirely an association of independent states. The 1960s saw an increase in the conflict in Northern Ireland between the Protestant majority, who wished to remain in the United Kingdom, and the Catholic minority seeking union with the Republic of Ireland.

Britain's constitutional monarchy (*above*) Britain has no written constitution: the relationship between the monarch as head of state and the prime minister as head of government is based on rigid convention. The 650 members of the Commons are elected locally for a maximum five-year term by voters aged 18 or over. Executive power rests with the prime minister (leader of the majority party in the Commons) and a cabinet of ministers.

The Houses of Parliament (*right*) are built on the site of the 14th-century royal Palace of Westminster, which was destroyed by fire in 1834. The present buildings are the seat of government for the United Kingdom and house both the upper and lower chambers. Big Ben, one of London's most famous features, is the name of the clocktower's 13-tonne bell. It was named after Sir Benjamin Hall, the commissioner of works at the time of its installation.

Violence and terrorism escalated in the 1970s and 1980s, but a ceasefire came into force in 1994. In February 1996 the Irish Republican Army (IRA) renewed its bombing campaign. A fragile peace was brokered with multi-party talks held at Stormont in 1998. It was agreed that a new Northern Ireland Assembly would be set up, but the disarmament of the paramilitary forces is a major stumbling block.

The United Kingdom has a permanent seat on the Security Council of the United Nations, and is a founder member of the North Atlantic Treaty Organization (NATO). In 1973 it joined the European Community (EC). In 1982 it recaptured the Falkland Islands in the South Atlantic after an Argentinian invasion.

Government

The United Kingdom is a monarchy governed by parliamentary democracy but without a written constitution. The law is based on parliamentary statutes, on the interpretation of statutes in the law courts, and (except in Scotland) on common law. The parliament consists of the House of Commons and the House of Lords; members of the former are elected while the latter inherit their positions or are appointed. The monarch's right to veto legislation has not been used since the 18th century. London became the first city to have an elected mayor in 2000.

At the local level, counties (regions in Scotland) and districts have their own elected councils. After referendums in 1998 the government agreed to set up a

locally elected parliament in Scotland and assembly in Wales, with administration powers being transfered from London.

People

The English, Irish, Scottish and Welsh are all centuries-old mixes of peoples. The most numerous are the English. English is the main language, though Gaelic is still spoken in parts of Scotland, while in Wales the use of Welsh has increased since its introduction into local schools.

Since 1945 immigration from Commonwealth countries, notably the West Indies, India and Pakistan, created a multiracial and multicultural society. There are growing numbers of Muslims and other faiths. Among the 90 percent of the population that is nominally Christian, church attendance is low. Less than 10 percent of people now live in rural areas.

241

Animal pyres (*above*) such as this one in Cumbria, England were a familiar site in 2001. Drastic measures were taken to eradicate an epidemic of hoof and mouth which brought terrible losses to farming and tourism.

Lloyd's of London (*right*) A giant in the world of international insurance, the Corporation of Lloyd's moved into its dramatic modern premises in 1986 in the heart of London's financial district. All around it the major banking and financial services of the city are packed into the region known as the "Square Mile".

ECONOMY

Undermined by increasing competition in the 20th century, Britain's early industrial superiority was diminished by the breakup of its empire and the economic destruction of two world wars. Since 1945 the most significant developments have been the exploitation of North Sea oil and gas reserves; the changing patterns of trade within the European Union (EU); and the shift to a services economy.

Agriculture and fisheries
Agriculture supplies about half the demand for food but employs only 1 percent of the workforce. Eastern areas tend to be arable, western areas pastoral. Sheep, cattle, pigs and poultry are the main livestock, while wheat, barley, potatoes and sugar beet are the main crops. Farming is highly mechanized.

Fishing is also highly mechanized but is declining as a result of overfishing and foreign competition. There have been disputes with Iceland and Denmark over territorial waters. Fish farming, especially of salmon, is increasing. Productive woodland covers a little more than 5 percent of the country. Forest management is divided between the government-run Forestry Commission and private landowners.

Industry
Britain's limited mineral resources have been heavily depleted. By the 1990s nearly all coal mines in southern Wales –

once the mainstay of the economy – had been closed. Most of the coal consumed in Britain now comes from cheaper imports.

Natural gas first came ashore in 1962. North Sea oil followed in 1975, making Britain a net oil exporter. It is now self-sufficient in energy production. Nuclear power provides up to a quarter of the energy supply. Smallscale hydro-electricity is generated in Scotland.

The chief manufacturing products include steel, motor vehicles, machinery, electrical goods, aerospace equipment, leather, textiles and clothing. New indus-

tries such as chemicals and electronics have replaced declining traditional industries such as shipbuilding. Britain's wealth of historic sites, diverse landscapes, and numerous cultural events have encouraged the growth of the tourist industry. Britain remains one of the largest trading nations, but its trade imbalance has grown steadily.

Transportation
Canals and railroads were built in the 19th century to meet the urgent transportation needs of the Industrial Revolution.

Oxford and Cambridge

In termtime, the narrow cobbled lanes of Oxford and Cambridge – England's two most ancient universities – are crowded with students hurrying between colleges and lecture halls and libraries as they have done for the last 800 years. The modern universities of today evolved from corporations, or private societies, of students and masters that developed in Europe in the 11th and 12th centuries to provide a place of education and learning outside the narrow confines of the monastic and cathedral schools. The two oldest were at Bologna, in Italy, and Paris. Oxford, England's first such school of learning, was set up at the beginning of the 12th century after English students were banned from the University of Paris.

At first the university had no buildings of its own. Students and masters lived in lodgings in the town, and met in any suitable room or open space they could find. Disputes between "town" and "gown" were frequent and often very violent. Quarrels broke out between rival groups of scholars, and

in 1209 a dissident group left to set up a sister university in Cambridge.

In the 13th century the first colleges in both Oxford and Cambridge were built, founded by wealthy patrons as places where masters and scholars could live and work as a self-contained community, in sets of rooms surrounding a quadrangle or court. Each college had a chapel for common worship, and a large hall where members could dine together every day.

About 1500 the practice began of admitting boys (often as young as 12 or 13) to courses of study. Then they were trained for careers in the church, the law, or the civil service. Today students are taught in every branch of learning from computer technology to Egyptology. The first women's colleges were established in 1871 (in Cambridge) and 1878–79 (in Oxford), though in Oxford they had to wait till 1920, and in Cambridge till 1947, to be admitted to full membership of the university. In the 1970s and 1980s most colleges became coeducational.

At both Oxford and Cambridge, college buildings of all periods from the 13th to the 20th centuries are scattered through and around a basically medieval town center. As a result

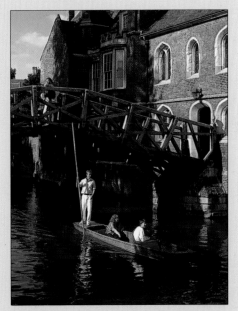

Punting on the river Cam, tourists float past historic Queens' College, Cambridge. Members of the university have included the internationally renowned scientists Isaac Newton (1642–1727) and Charles Darwin (1809–82).

the visitor can explore a superb display of architecture from every period, with an almost unparalleled richness of historical association. But anyone expecting to find a "university campus" among the cluster of streets and buildings will be disappointed.

Gateway to Europe A high speed train leaves Waterloo Station, London, bound for France via the Channel Tunnel. Although the Tunnel has improved links between Britain and Europe, it has also opened up another route for illegal immigrants looking for a better life in Britain.

The entire rail network was nationalized in 1947. Investment has been concentrated on mainline routes between cities, along with containerized freight traffic. An extensive network links the suburbs with central London. In the late 1990s the railroad system was privatized.

Since 1960 the motorway or expressway system has been extended and has become overburdened with traffic, especially in the heavily populated southeast. Yet government attempts to reduce car use have met with resistance. The bus network is well developed.

British Airways, a privatized company, has one of the world's most comprehensive international air networks. Other airlines operate from London's two major airports, Heathrow and Gatwick, and from regional airports.

A road and rail tunnel under the English Channel, under construction since the late 1980s, opened in 1994 linking England to France and the main European transportation network.

Communications and media

The telecommunications industry is also privatized. All British newspapers are pri-

vately owned, mostly by two big groups. There is no formal censorship. The British Broadcasting Corporation (BBC) operates as an independent public corporation controlling a substantial share of local and national radio and television. Wales has a number of Welsh language channels. The Independent Television Commission (ITC) licences and regulates commercial television and the Radio Authority, commercial radio. Television broadcasting is now set to provide up to 500 channels along with high speed Internet access.

Health and welfare

The national health service, introduced in 1948, offers free healthcare and hospital treatment. Eye tests, dental treatment and prescriptions are free for children, the elderly and those with low incomes. There are 2,500 national health hospitals and many private feepaying hospitals.

Education is both free and compulsory between the ages of 5 and 16. A national curriculum was introduced in the late 1980s. There are also many hundreds of fee-paying or public schools outside the state system. Almost two-thirds of students go on to some form of higher education.

Ireland

REPUBLIC OF IRELAND (EIRE)

THE REPUBLIC OF IRELAND OCCUPIES SOME five-sixths of the island of Ireland, about 80 km (50 mi) off the west coast of Great Britain. Ireland is renowned for natural beauty and turbulent history.

GEOGRAPHY

Ireland's lush green landscapes have earned it the title "the Emerald Isle."

The land

Ireland is surrounded by a broken rim of mountains. The central plain is low-lying – less than one-sixth of the country rises above 200 m (700 ft) – with hills and ridges protruding from the limestone layers beneath. Ice-age glaciers deposited fertile, rock-scattered soils. Small lakes, known as loughs, and bogs are common. The main rivers – the Liffey, the Blackwater, the Clare and the Shannon (the longest river in the British Isles) – are broad and slow moving.

Shorter, fast-flowing rivers fall to the coasts from the highland fringes. The largest ranges are the granite Wicklow Mountains in the east and Macgillicuddy's Reeks in the extreme southwest.

The coastline is most dramatic in the west, where it is rocky and deeply indented, with many spectacular cliffs. In the far southwest the mountains strike out into the Atlantic along a ragged row of broad peninsulas separated by deep inlets such as Dingle Bay and Bantry Bay.

Climate

The warm waters of the North Atlantic Drift create moderate temperatures, and mild Atlantic winds bring rain on two days out of three. Annual average rainfall ranges from 2,540 mm (100 in) in the mountains of the southwest to 762 mm (30 in) in the east. Summers are cool and

Lough Leane, Co. Kerry is the largest of a series of picturesque lakes in the far southwest. Lough Leane (or Lower Lake) covers 2,024 ha (5,001 acres) and encompasses 30 islands, one of which – Inisfallen – has a 9th-century abbey.

winters mild, with very little snow except in the mountains.

Plants and animals

Flourishing grasslands cover much of the country, but plant life is less rich than Britain's and there are fewer trees. Under one-twentieth of the country is forested; little forest has been restored even for commercial purposes. Peat bogs – habitats for many rare plants – carpet large areas of central Ireland. Long exploited for domestic fuel, the bogs are now threatened by the use of peat in commercial power stations and for export for garden and horticultural use. Along the west coast, where soils are poorer, heath and scrub vegetation predominate.

At the end of the last ice age Ireland was colonized by plants and animals migrating overland from Europe. However, many common European and British animals such as weasels, moles and snakes never reached the island, which

was cut off from Britain some 11,000 years ago when the glaciers melted and the sea level rose. According to folklore, St Patrick – the patron saint – banished the snakes.

SOCIETY

Ireland's history is one of turmoil, bloodshed, religious conflict and the fight for independence. Today it is a stable democracy with a strong literary tradition.

History

The earliest known occupants of Ireland date back to about 7000 BC, and Bronze Age artifacts from Ireland were traded all over Europe. In about 300 BC, Celtic ironworkers arrived from Europe and gradually became organized into five tribal groups or kingdoms.

The Irish Celts were unaffected by the Roman conquest of Britain in the 1st century AD. Converted to Christianity in the 5th century, they remained a center of faith despite the collapse of the Roman empire and several centuries of raids by Vikings, who established a fortified base at Dublin before they were finally de-

A Celtic cross at Drumcliffe in the northwest of Ireland, erected by early Christians in the 10th century on the site of a former monastery. The Anglo–Irish poet and mythologist W. B. Yeats (1865–1939) is buried in Drumcliffe churchyard.

feated by Brian Boru (941–1014), the first king of all Ireland, in 1014.

In 1172 Ireland was invaded by Anglo-Norman knights sent by England's king Henry II (1133–89). His sovereignty was recognized by the Church and many of the Irish knights. Over the next 300 years the Anglo–Norman barons became assimilated and Ireland effectively came to be ruled by the earls of Kildare.

England's break with the Roman Catholic Church in 1533 under King Henry VIII (1491–1547) increased the resentment of the Irish, who remained firmly Catholic. Repeated uprisings took place, often aided by Catholic Spanish or French troops. Many Irish Catholics emigrated to Europe. When the earls of Ulster in the north chose this course, James I of England (1566–1625) encouraged Scottish Protestants to settle on the vacated lands. In 1641 an uprising among the Irish of Ulster led to the massacre of thousands of colonists. An equally savage massacre of the Irish was at the hands of Parliamentary forces fighting in Ireland under Oliver Cromwell (1599–1658).

In 1690 an uprising led by England's deposed Roman Catholic king James II (1633–1701) was defeated at the Battle of the Boyne by the Protestant forces of King William III (1650–1702). The English government imposed civil disabilities and land confiscations (known as the penal laws) on Irish Catholics, but uprisings continued. In 1801 an Act of Union abolished Ireland's parliament, integrating it with that of Britain.

Under English rule, Ireland was divided into large estates owned by absentee landlords and farmed by local Irish peasants. From 1846 to 1851 the potato crop – the Irish staple food – was destroyed by a fungal disease, leading to the Great Famine. More than a million people died; many more emigrated to the United States, halving the population in Ireland.

From about 1870, the movement for Irish Home Rule gathered momentum, led in the British House of Commons by the Irish Member of Parliament Charles Stewart Parnell (1846–91). Home rule was finally granted in 1914, though violently opposed by Ulster's majority Protestant population. World War I delayed its implementation, fueling discontent. In 1916 an armed insurrection was suppressed by the British army; 14 leaders were executed. Eamon de Valera (1882–1975), a survivor, formed the Sinn Fein party and in 1919 set up a provisional Irish government. After a bitter civil war, Britain agreed to partition the island. The six northeastern counties became Northern Ireland, a province of the United Kingdom. The self-governing Irish Free State was created in the south.

In 1937 the Irish Free State renamed itself Eire and adopted a new constitution with an elected president, severing its links with the British crown. In 1949, on leaving the British Commonwealth, it declared itself an independent republic. Economic development was slow, and large numbers continued to emigrate in search of work, particularly to Britain. In 1973 Ireland joined the European Community (EC) now the European Union (EU).

In the 1960s the fragile stability in Northern Ireland began to break down as the minority Catholic population began campaigning for civil rights. In the 1970s the situation became increasingly violent, British troops were sent onto the streets and direct rule from Britain was imposed. Various attempts at reconciliation floundered during the 1980s and it was not until 1998 that the Good Friday Agreement set up a new constitution for Northern Ireland that was ratified by referendum held in both southern and Northern Ireland. In 1999 the Republic of

NATIONAL DATA - IRELAND

Land area	68,890 sq km (26,599 sq mi)			

Climate	Altitude m (ft)	Temperatures January °C(°F)	Annual July °C(°F)	precipitation mm (in)
Dublin	81 (266)	5 (41)	15 (59)	733 (28.8)

Major physical features highest point: Carrauntoohill (Macgillicuddy's Reeks) 1,041 m (3,414 ft); longest river: Shannon 370 km (230 mi)

Population (2006 est.) 4,062,235

Form of government multiparty republic with two legislative houses

Armed forces army 8,500; navy 1,100; air force 860

Largest cities Dublin (capital - 1,186,159); Cork (119,143); Galway (65,832); Limerick (54,023); Waterford (44,594)

Official language Irish, English

Ethnic composition Irish 95%; British 2.7%; others 2.3%

Religious affiliations Roman Catholic 88.4%; Church of Ireland 3%; other Christian 1.6%; other 1.5%; unspecified 2%; none 3.5%

Currency 1 Euro (EUR) = 100 euro cents

Gross domestic product (2006) U.S. $177.2 billion

Gross domestic product per capita (2006) U.S. $43,600

Life expectancy at birth male 75.11 yr; female 80.52 yr

Major resources natural gas, peat, copper, lead, zinc, silver, barite, gypsum, limestone, dolomite, barytes, cereals, fish, livestock, potatoes, sugar beet, vegetables, tourism

Ireland gave up its territorial claims to the whole island and in 2000 the Northern Ireland Assembly was established.

Government

By its constitution, adopted in 1937, Eire is a parliamentary democracy. The head of state is the president, who is elected for up to two seven-year terms. The National Parliament or Oireachtas consists of the Dail, whose 166 members are appointed by the Taoiseach (prime minister) or elected by various public bodies. The Dail initiates all legislation, the Seanad can revise or delay its enactment and the President may refer proposed legislation to the Supreme Court if it might infringe the Constitution. Voting in parliamentary elections is through a system of proportional representation.

People

The distinction between English-speaking and Celtic-speaking Irish largely vanished as a result of the mass emigrations of the 19th century. Although Irish Gaelic is taught in all schools, it is now the first language only in some rural areas. The Irish are noted for their love of good talk, and Ireland has produced many famous writers, such as Jonathan Swift (1667–1745), Oliver Goldsmith (1730–74), Oscar Wilde (1854–1900), George Bernard Shaw (1856–1950), William Butler Yeats (1865–1939), James Joyce (1882–1841) and Samuel Beckett (1906–1991).

The story of the Irish people has been one of continued emigration. However, Ireland's economic renaissance in the late 20th century led to net immigration for the first time and in 2000 Ireland found itself in the strange position of needing to attract thousands of immigrants to deal with skills shortages.

The vast majority – over 90 percent – are Roman Catholic; other groups include Church of Ireland Episcopalians, Presbyterians, Methodists and Jews. The influence of the church remains strong over social issues such as divorce, abortion and birth control, though the 1990s have seen a demand for greater flexibility on these moral issues. Freedom of conscience and religious practice are guaranteed under the Irish constitution. In 1995 a referendum was narrowly in favor of divorce.

ECONOMY

Ireland both encourages private initiative and undertakes the administration of essential services and development.

Agriculture

Agriculture still accounts for employment among ten percent of the workforce. Most Irish farms are owner-occupied small-holdings. The limited area of arable land is highly productive. Cereals, sugar beet and potatoes are the main crops. Large areas of rich grassland support livestock almost all the year round. Cattle and pigs are kept in great numbers, while sheep graze on the upland pastures. Meat, milk and dairy products account for nearly one-sixth of Irish exports.

Forestry is limited, and concentrates mostly on fast-growing conifers. Sea fisheries were modernized with government aid in the 1970s. Although potentially rich, they are subject to overfishing controls. Salmon farming in the country's rivers, a potential source of earnings, has recently been endangered by pollution.

A son of Dublin

Every year on "Bloomsday" (June 16th) visitors and residents in Dublin retrace the footsteps of the city's best-known fictional character, Leopold Bloom. He was the creation of another Dubliner, James Joyce (1882–1941), a man who dropped out of school for two years at the age of nine due to his family's financial difficulties, but went on to university and published his first critical essay at the age of 18. Subsequently he became one of the most influential writers in the history of modern English literature.

Joyce was a quiet, mild-mannered, but deeply unconventional man who was plagued by poor health, including extremely poor sight, and a precarious financial situation. Given a scholarship to a Jesuit grammar school as a teenager, he performed well academically but lost favor with his teachers as he questioned his Roman Catholic faith. In 1904 he met Nora Barnacle, with whom he lived for the rest of his life. He refused to marry her until 1931, when he agreed to his daughter's wish that her parents should get married.

Joyce spent his adult life abroad, chiefly in France and Switzerland, but wrote exclusively about the people of Dublin, whose hospitality and poetic speech he loved as much as he hated their rigid, oppressed and oppressive social institutions. *Dubliners*, published in 1914, was the title of his first work of fiction; the publication date was delayed because the original publisher feared a libel suit and destroyed the printed galleys.

Ulysses, his best-known novel, first appeared in serial form in New York in 1917, but several episodes were confiscated and burned by the New York post office in 1919–20 on the grounds of obscenity. In 1921 the New York Society for the Prevention of Vice took Joyce's publishers to court and won a ban on further publication. The decision was finally reversed by a higher court in 1936, and a complete edition appeared that year. But Ulysses had long since assured James Joyce's international reputation. The novel records a day in Dublin – 16 June 1904 – with incidents based on scenes from Homer's ancient Greek epic, the *Odyssey*, and from Joyce's own life.

James Joyce is probably Dublin's most famous citizen, yet he wrote much of his work in exile in Zurich or Trieste, and was banned from publication at home. Ireland is now a writer's tax haven and home to an international literary set.

The date was deliberately chosen by Joyce in remembrance of the day he first met Nora Barnacle. Now it is celebrated by Joyce's admirers all over the world.

A horsefair (*above*) in a village southwest of Cork. The Irish thoroughbred is prized throughout the racing world and horse breeding contributes significantly to agricultural earnings, but trading in more humble horses remains an Irish passion.

Adjoining pubs in the same street of Allihies, Co. Cork. The pub is so much a part of Irish culture and social life that many small towns boast a dozen or more licensed establishments. Many ordinary retailers provide "bar" facilities for thirsty shoppers.

Industry and commerce

Ireland's mineral resources are not great, and only gypsum, zinc, and lead ores are mined in any quantity. There is little coal but large reserves of peat which are being developed for home and commercial use. Irish offshore fields supply natural gas meeting 20 percent of national requirements. Oil and gas exploration continues.

The 1980s and 1990s saw changes in output and employment. The industrial sector now accounts for 39 percent of GDP and manufacturing has expanded strongly, especially in engineering, pharmaceuticals, computers and computer software, telemarketing and financial services. In 1998 Ireland was the second biggest exporter of software in the world. Foreign investment has been important in this expansion, in 1998 there were over 1,000 foreign-owned companies. The services sector accounts for slightly over half of GDP and for almost 61 percent of employment. In the late 1990s, Ireland had become one of the most trade-dependent countries in the world. Exports include machinery, transport equipment, chemicals, and food products. A quarter of exports were to the UK. At the end of the 1990s the economy had an average annual growth of over 8 percent.

Transportation and communications

Ireland's road system is extensive, even in the most rural areas, and its major roads have generally been modernized in recent years. By contrast, the railroad network has shrunk, with many branch lines closed in the 1960s and 1970s. The railway and bus services are state-supported and the extensive inland waterways are maintained by the government as a public amenity. Ireland has several seaports, with the most important for trade being Dublin and Cork. Air traffic is provided through Aer Lingus with competition from eighteen independent operators. There are international airports at Dublin, Shannon and Cork.

Radio Telefis Eireann is the state-owned broadcasting company. The Independent Radio and Television Commission controls an independent television channel and the many commercial radio stations. There are Irish language radio stations and a TV channel. Cable television and British television channels are also available. Publishing and printing are major industries, there are eight national dailies and five Sunday newspapers. Freedom of the press is guaranteed by the constitution, but the publication of matter deemed to undermine Roman Catholic moral standards is forbidden.

The telecommunications industry is privatized. There are three mobile phone operators and expansion of this sector is a high priority. Ireland intends to become a leader in the field of electronic commerce and internet-based industry.

Health and welfare

Eire operates a social insurance scheme providing all major pensions, unemployment, and health benefits, with added charges that are calculated according to income. Health conditions and facilities are generally high, and hospitals raise extra income through the internationally famous Irish Hospitals' Sweepstake.

Education is free and compulsory between the ages of 6 and 15, and schools are almost entirely run by religious organizations. Further education is provided by secondary schools, secular vocational schools, and at a higher level by technical colleges and the major universities. The most famous of these is Trinity College, Dublin, founded in 1591. The National University, was founded in 1908.

France and its neighbors

COUNTRIES IN THE REGION
FRANCE · ANDORRA · MONACO

The Arc de Triomphe de l'Etoile (*left*), hung with the tricolored flag of France, at the Place Charles de Gaulle. Commissioned in 1806 by Napoleon, it looks down the Avenue Champs-Elysees and across the Place de la Concorde to its twin, the Arc de Triomphe du Carrousel at the Louvre.

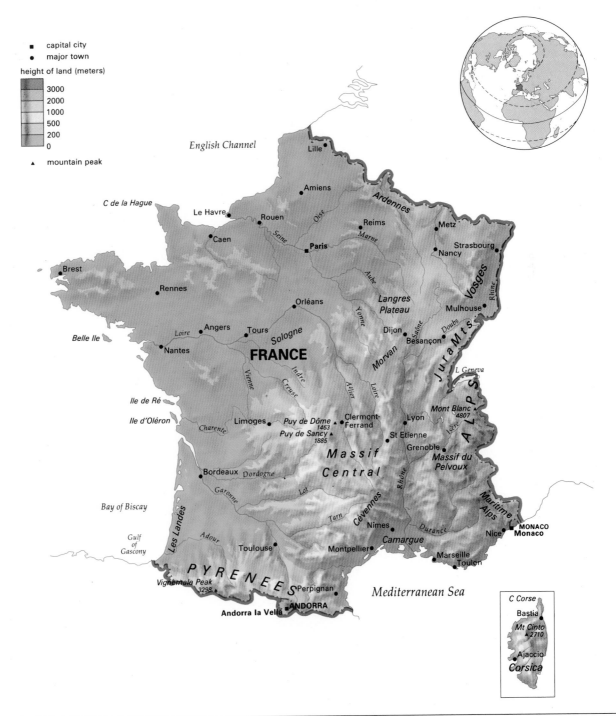

■ capital city
● major town

height of land (meters)

3000
2000
1000
500
200
0

▲ mountain peak

English Channel

Lille
Amiens
Ardennes
C de la Hague
Le Havre
Rouen
Reims
Metz
Caen
Oise
Seine
Marne
Paris
Strasbourg
Nancy
Vosges
Rhine
Brest
Aube
Rennes
Langres Plateau
Mulhouse
Orléans
Yonne
Dijon
Doubs
Belle Ile
Angers
Tours
Sologne
Saône
Besançon
Jura Mts
Loire
Nantes
FRANCE
Indre
Creuse
Allier
Loire
Morvan
L Geneva
Ile de Ré
Vienne
ALPS
Mont Blanc
4807
Ile d'Oléron
Limoges
Puy de Dôme
1463
Clermont-Ferrand
Lyon
Isère
Charente
Puy de Sancy
1885
St Etienne
Grenoble
Massif du Pelvoux
Massif Central
Bordeaux
Dordogne
Rhône
Garonne
Lot
Bay of Biscay
Les Landes
Tarn
Cévennes
Maritime Alps
Durance
Nîmes
MONACO
Monaco
Gulf of Gascony
Adour
Toulouse
Montpellier
Camargue
Nice
Marseille
Toulon
PYRENEES
Mediterranean Sea
Vignemale Peak
3298
Perpignan
Andorra la Vella
ANDORRA

C Corse
Bastia
Mt Cinto
2710
Ajaccio
Corsica

France

FRENCH REPUBLIC

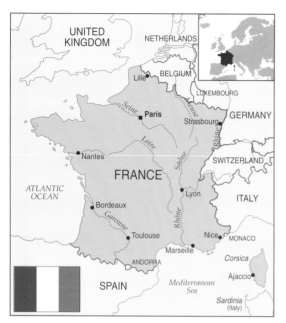

FRANCE, IS THE LARGEST COUNTRY IN Western Europe, and has a long history of cultural influence both within Europe and across the world as a result of its former colonial power.

GEOGRAPHY

France is bordered on the northwest and west by the English Channel and Atlantic respectively, on the southwest by the Pyrenees, on the south by the Mediterranean, and in the southeast and east by the Alps. In the north and northeast are low-lying plains and hills.

The land

The steep mountain range of the Pyrenees divides France from Spain in the southwest. At the eastern end of the Pyrenees mountains the Mediterranean coast curves down to the Rhône delta. From the interior the mountains of the Massif des Maures, the Massif d'Esterel and the Maritime Alps stretch to meet the sea. Here on the Côte d'Azure lie the wealthy resort towns of the French Riviera.

The Alps form a barrier along the Italian border. Farther north, the high limestone ridges of the Jura form the border with Switzerland. North of these the Vosges mountains overlook the Rhine valley and mark the boundary with Germany. The northeastern border meets Germany, Luxembourg and Belgium. Cliffs and beaches on the north coast give way to the English Channel. West of the peninsula of Cotentin in Normandy, Brittany juts out into the Atlantic. The low plateaus of the Armorican Massif stretch inland, and the sandy plains run along the west coast to the Pyrenees.

Volcanic wonderland (*above*) The Massif Central, seen from the Puy Mary, is a huge granite plateau covering much of south-central France. It is dotted with extinct volcanoes and hot volcanic springs, around which spa towns such as Vichy have grown up.

Wedded to the sea (*left*) The deeply indented south Brittany coast in the northwest of France provides excellent natural harbors for the traditional fishing fleets of the area and the pleasure boats of vacationers from Paris and abroad.

The four regions of the mountainous Massif Central dominate central and southern France. In the south are the Causses – a high, arid limestone plateau dissected by deep river canyons. To the east, steep granite massifs stretch north toward Burgundy, dropping sharply into the Rhône and Saône valleys.

Around Limousin, in the northwest of the Massif, are stepped plateaus cut by more river gorges. The higher central plain of the Auvergne rises to the Puy de Sancy, the highest peak in the Massif Central. Many local features here were formed by volcanoes.

The lowlands are defined by river systems. The largest rivers of the northeast are the Meuse and the Moselle, both tributaries of the Rhine. The Saône river rises in the Vosges and meets the Rhône at Lyon; the Rhône flows through Lake Geneva between the Jura and the Alps.

In the northwest, the Seine drains the Ile-de-France – the low-lying region around Paris – and flows northwest to the English Channel. Beyond the plains to the southwest is the longest river, the Loire, extending some 1,020 km (630 mi) and draining much of western France. In the far southwest the river Garonne rises in the Pyrenees. It is fed by the Tarn and Lot rivers, and meets the Dordogne north of Bordeaux, where the two rivers flow into the broad Gironde estuary.

Climate

The climate of France is temperate, with three regional variations. In the north and west, Atlantic winds bring high humidity and unpredictable weather, especially in Brittany. Summers are cool and winters mild; farther south, summers are warmer and winters colder.

The Paris Basin has a more continental climate, with considerable rain in the spring and fall, and thunderstorms in the summer. In eastern France and the Massif Central, winters are colder, with heavy snow in the mountains. Strasbourg, near the German border, has the biggest temperature range in the country.

The Mediterranean climate of Provence in the southeast has only a few days of frost each year. Summers are hot and dry. The mistral, a cold dry northerly wind, is funneled down from the Alps mainly in the winter months.

Plants and animals

The fertile plains of the north and west rise to hills covered in gorse and heather, with oak forests growing in sheltered

NATIONAL DATA – FRANCE

Land area 545,630 sq km (210,669 sq mi)

Climate	Altitude m (ft)	Temperatures January °C(°F)	July °C(°F)	Annual precipitation mm (in)
Brest	103 (338)	7 (45)	17 (63)	1,145 (45)
Paris	53 (174)	5 (41)	20 (68)	650 (25.5)
Marseille	8 (26)	7 (45)	24 (75)	555 (21.8)

Major physical features highest point: Mont Blanc 4,807 m (15,771 ft); longest river: Loire 1,020 km (634 mi)

Population (2006 est.) 60,876,136

Form of government multiparty republic with two legislative houses

Armed forces army 137,000; navy 44,300; air force 64,000

Largest cities Paris (capital – 2,142,800); Marseille (795,600); Lyon (468,300); Toulouse (426,700); Nice (339,000); Nantes (276,200); Strasbourg (273,100); Montpellier (244,700); Bordeaux (229,500); Lille (222,400)

Official language French

Ethnic composition French 90.6% (including Occitan 2.7%; Alsatian 2.3%; Breton 1%; Catalan 0.4%); Algerian 1.5%; Portuguese 1.4%; Moroccan 0.8%; Spanish 0.6%; Italian 0.6%; others 4.5%

Religious affiliations Roman Catholic 83%–88%; Protestant 2%; Jewish 1%; Muslim 5%–10%; unaffiliated 4%

Currency 1 euro (EUR) = 100 euro cents

Gross domestic product (2006) U.S. $1.871 trillion

Gross domestic product per capita (2006) U.S. $30,100

Life expectancy at birth male 76.1 yr; female 83.54 yr

Major resources coal, iron ore, bauxite, zinc, uranium, antimony, arsenic, potash, feldspar, fluorospar, gypsum, timber, fish, barley, cattle, fruit, vegetables, grapes/wine, maize, oats, oil, natural gas, pigs, potatoes, poultry, salt, sheep, sugar beet, timber, uranium, wheat, tourism

Corsica

The island of Corsica, which lies off the northwest coast of Italy and is separated by a narrow strait from Sardinia, has been part of France for 200 years. Since 1975 it has been a single French region divided into two *départements*. It is a popular holiday resort, and many French people have second homes there.

Corsica's coastline has been compared to a fist with one finger extended toward Italy. The east coast is flat and marshy, while the west coast rises steeply to a mountainous interior. A chain of peaks runs like a backbone down the center of the island. The climate is Mediterranean, cooled by altitude over much of the island. Substantial rainfall feeds the island's many rivers and lush vegetation. More than one-fifth of Corsica is forested; chestnut trees grow in abundance, with tall pines on higher slopes. Much of the rest is covered with maquis, whose aromatic plants give the island a distinctive fragrance. Wildlife includes mouflon, wild goats and large numbers of feral pigs.

Corsica was conquered by the Romans in the 3rd century BC (there are many Roman ruins on the island). As the western Roman empire collapsed during the 5th century AD, Corsica was overrun many times by invading Germanic peoples. The island was eventually claimed by the Pope, but in 1077 it was ceded to Pisa in Italy. After a long struggle the Pisans were finally ousted by the Genoese in 1299. The Genoese ruled the island for about four centuries, putting down rebellions by Corsican nationalists. The island was sold to France in 1769 – the year in which its most famous native, Napoleon Bonaparte, was born. After brief periods of British occupation during the Napoleonic Wars, it finally became a *département* of France. In recent decades a nationalist movement has been active in demanding separation from France, often using terrorism.

The Corsican people, over one-third of whom live in the main towns of Bastia and Ajaccio, are full French citizens. They speak French as their official language, but most also speak one of the island's many dialects – some of which are similar to the Tuscan dialect of Italian. The Corsicans also resemble Italians in appearance, traditional dress and way of life. The island's population is declining and aging, largely because so many young people emigrate. However, it has also seen a huge influx of immigrants, especially from the former colony of Algeria after it gained its independence in 1963.

The main reason for the high level of emigration is the lack of opportunity in Corsica's mainly agricultural economy. Wheat is the main cereal, and sheep the main livestock animal. Major exports include citrus fruits, wines and cheeses. Fuel and machinery have to be imported, and there is little industry apart from tourism – a major source of income. The road network has been improved as tourism has increased, and the main towns are linked by rail. Regular sea and air services operate to the Italian mainland.

The citadel of Calvi in the northwest of Corsica was fortified by the Genoese in the 16th century. Corsica itself is a natural fortress of 20 mountain peaks. It was the first part of France to free itself from German occupation in World War II.

areas. The northeastern regions and much of the Paris Basin are covered in loess soils. There are lush pastures and slopes on the western and central slopes of the Pyrenees and the Massif Central. The vegetation in the southeast is typically Mediterranean, with olive, box, thyme, rosemary and cork oak. Maquis scrubland occurs in drier parts. Unique wildlife includes feral white horses, bulls and flamingoes in the Camargue at the Rhône delta on the southern coast.

The river Lot (*right*) at Estaing, near Espalion. The Lot rises west of the Massif Central and flows in a deep valley to join the Garonne river south of Bordeaux. Along its banks can be seen many ancient castles, quiet villages and terraced vineyards.

Jagged peaks and icy pools (*below*) in the French Alps, part of a massive mountain range that extends across Switzerland, northern Italy and Austria. The range is between 65 and 2 million years old. Mont Blanc in France is the highest peak at 4,807 m (15,770 ft).

SOCIETY

After centuries of cultural and military power, France remains very influential as a founding member of the European Community.

History

The 30,000-year-old remains of the first modern human, Cro-Magnon man, were discovered in France and named for the caves in the Dordogne where they were found. The history of France as a country began with settlement by the Gauls, a Celtic people who had spread west from the Rhineland about 800 BC. By the 5th century BC they had conquered large areas of Europe and were rivals to Rome, which they failed to conquer in 390 BC.

Rome conquered the whole of France between 58 and 50 BC, and the region remained stable for nearly 500 years. The Christian Frankish kings (the Merovingian dynasty) helped keep Gaul's Roman culture alive after the empire disintegrated, but were gradually displaced by the Pepins. One of these, Charles Martel (c. 688–741), defeated Muslim raiders from Spain (the Moors) at Poitiers in 732 and founded the Carolingian dynasty. His grandson, Charlemagne (742–814), spread Frankish rule back into Germany and to the borders of Scandinavia and Spain, thus establishing the major Christian kingdom of western Europe with his capital at Aix-la-Chapelle (now Aachen in Germany).

The Capetian kings from 987 began as figureheads but gradually extended their authority. Philip VI (1293–1350), the first of the Valois kings, had much of present-day France under his control. But large areas came under English control in the Hundred Years War (1337–1453), while the whole country suffered from high taxes and the Black Death (1347–51). Charles VII (1403–61), aided by Joan of Arc (c. 1412–31), drove out the English and had reconquered his realm by 1453.

France in the 16th century was overshadowed by religious conflict between Catholics and French Protestants or Huguenots. The Wars of Religion ended in 1589 when the Protestant prince Henry of Navarre (1553–1610) became a Catholic in order to inherit the throne as Henry IV, founding the Bourbon dynasty and granting tolerance to the Huguenots. After his death, however, the old divisions reappeared. Under Louis XIII (1601–43) the ruthless chief minister Cardinal Richelieu (1585–1642) built up French power abroad and suppressed all dissent at home. A series of revolts known as the Frondes followed under his successor, Cardinal Mazarin (1602–66).

Louis XIV (1638–1715), the "Sun King," built the most opulent court in Europe to distract the nobility from politics. Government became even more centralized and the Huguenots were persecuted. Louis sought to dominate Europe and strengthen France's territorial possessions. France colonized parts of North America, India and the Caribbean, where plantations used African slaves to produce vast wealth in coffee and sugar.

During the 18th century, expensive wars lost France both Canada and India, and ruined its financial system. Under Louis XV (1710–74) the monarchy's popularity waned. The economy approached breaking point under the well-meaning but inept Louis XVI (1754–93) and his hated queen Marie Antoinette (1755–93). The French Revolution, which began in 1789, finally toppled the monarchy. Seizure of the government by fanatics such as Danton (1759–94) and Robespierre (1758–94) led to the Terror, a period of mass murder and anarchy. Louis XVI and Marie Antoinette were both executed.

From 1799 a brilliant young Corsican general, Napoleon Bonaparte (1769–1821), assumed power, proclaiming himself Emperor in 1804. He instituted many domestic reforms, among them the Napoleonic Code, which is still central to French law. But his chief concern was expanding his empire by foreign conquest, and he overextended himself and his army. Defeated in 1814, he was exiled from France but returned in 1815 to face final defeat at Waterloo in Belgium.

The unpopular restored Bourbon kings were followed by the unstable Second Republic, and Napoleon's charismatic nephew, Louis-Napoleon Bonaparte (1808–73), was elected president. In 1852 he staged a coup and declared himself Napoleon III of the Second Empire. During his reign the Industrial Revolution spread throughout France, bringing peaceful social change and unprecedented prosperity – along with political and financial corruption.

In 1870 Otto von Bismarck (1815–96), architect of the rising German empire, goaded France into war with Prussia. The French armed forces collapsed and Napoleon III was defeated and captured. The new Third Republic sued for peace, and the Germans annexed Alsace-Lorraine. A radical uprising established a Commune in Paris, which was savagely defeated: more than 20,000 Communards died.

Under the Third Republic France regained domestic stability, consolidated its presence in Algeria, and established more colonies in West Africa and Indochina. Much of World War I took place on French soil, and France lost almost 1.4

Symbolic victory On 14 July 1789 a Paris mob stormed the Bastille – the state prison – and freed its 7 prisoners in an enraged gesture against Bourbon despotism. The Bastille was an 8-tower fortress built in the 14th century to protect Paris from English attack.

million men. After the war France recovered Alsace-Lorraine and occupied the German Rhineland from 1923 to 1930.

Northern France came under German rule in the invasion of 1940, while the south was governed by the collaborationist Vichy regime under Marshal Philippe Pétain (1856–1951). Both were opposed by the British-based Free France movement led by General Charles de Gaulle (1890–1970). In 1947 the Fourth Republic was established, but it lost public confidence during disastrous wars in French Indochina (Vietnam) and Algeria.

In 1958 General de Gaulle returned to serve 10 years as president of the Fifth Republic. French prosperity revived in the newly established European Economic Community (EEC), but in 1968 the country was shaken by violent student riots in Paris. At the same time de Gaulle was forced to grant independence to Algeria and other colonies, but he pursued nationalist policies, resisting US influence in Europe and repeatedly vetoing British entrance into the EEC.

De Gaulle's successors, Georges Pompidou (1911–74) and Valéry Giscard d'Estaing (b. 1926), both encouraged industrial development but faced growing economic problems. In 1981 de Gaulle's old rival, François-Maurice Mitterrand (1916–96), was elected France's first socialist president and was re-elected in 1988. Jacques Chirac (b. 1932) was elected president in 1995. In January 1996 he announced the end of French nuclear testing in the Pacific, and revealed plans to phase out military conscription by 2001.

Government

The constitution of the Fifth Republic, adopted in 1958, made France a multi-party democracy with the president as head of state elected by popular vote for a 7-year term. Appointed ministers, led by the prime minister, carry out administrative business. The president can also summon or dissolve parliament, hold a referendum and assume emergency powers.

The two parliamentary chambers are the Senate and the National Assembly. The Senate's 321 members have limited legislative powers. The National Assembly's 577 deputies are the main legislative body, but finance, defense and other national issues lie in the hands of the president, who has more power than is usual in Western democracies – a legacy of the de Gaulle era and the political instability that preceded it.

Local government wields little power apart from administration and public

Hero or tyrant?

Nearly 200 years after his death, Napoleon (1769–1821) is a controversial figure. The impact he had is unquestioned: within a few years he changed the history of France and of Europe. What is still debated is whether or not his regime was beneficial to France.

Napoleon was born at Ajaccio on Corsica in 1769, only shortly after the Italian city-state of Genoa had sold the island to France. His name Napoleone Buonaparte – is an indication of his Italian ancestry. Corsicans were then Italian in both language and culture, a feeling that Napoleon himself never lost, although he was educated in France and graduated from the military academy in Paris before becoming an artillery lieutenant. Briefly involved in the Corsican nationalist movement, he then joined the Jacobin revolutionaries in France. He distinguished himself against the British at the siege of Toulon becoming a brigadier general.

His career prospered in spite of the political turmoil of the early 1790s and by 1795 he was given the command of the army of the interior. In 1796 he married his mistress, Josephine de Beauharnais (1763–1814), widow of a general who had been guillotined during the Revolution. That same year he led a successful invasion of Italy, and forced Austria to yield valuable territory. He also organized an antiroyalist coup in the French government. At the zenith of his power he declared himself Emperor of France and King of Italy; married the daughter of the Emperor of Austria; and controlled Austria, Prussia, Spain and Portugal.

His defeats were equally spectacular. Clashing with Britain, he invaded Egypt in 1798 to cut off wealth derived from colonial India. He occupied the country successfully and introduced many European political institutions, administration and technical skills there. But his fleet was destroyed by the British under the command of Admiral Horatio Nelson (1758–1805) in the Battle of the Nile (1798). An attempted invasion of Britain ended in another naval defeat at Nelson's hands at Trafalgar (1805). In 1812 Napoleon turned on Russia, the other great power not yet under his control, which had become hostile to French ambitions in the Baltic. He defeated the

Russian army at Borodino and entered Moscow. But the Russians' scorched-earth tactics deprived his army of essential supplies, and the savage Russian winter forced him into a disastrous retreat. Defeat followed defeat, forcing the French army back within the borders of France. Invaded by combined Russian, Prussian, Austrian and British forces, Napoleon was forced to abdicate in 1814, but returned within a year from exile on the Mediterranean island of Elba. His final defeat by British and Prussian forces at Waterloo in Belgium came in 1815. By then he was so unpopular in France that he was forced to put himself under British protection. He died after six years in exile on the remote Atlantic island of St Helena.

Napoleon crossing the Alps An idealized portrait by Jacques-Louis David (1748–1825). Born in Corsica and educated in France, Napoleon had a spectacular career: brigadier-general at the age of 24, military dictator at 30, emperor at 35.

Was he a far-seeing statesman or merely a ruthless tyrant? Much that he did for France was good – he restored its prestige abroad and laid the foundation of the modern French state by establishing the prefecture system of administration, the Code Napoleon, the Bank of France and other financial institutions. But – like so many dictators before and after him – he was unable to distinguish between self-interest and the national interest, and his ambitions led to his ruin. However, he was a hero to many of his countrymen.

services. There are 22 regions, 96 *dé-partements* and 4 overseas *départements*; the smallest local government body is the commune, responsible for municipal services and establishments such as schools, hospitals and fire stations. The commune is headed by a mayor and deputies, who are important figures in local life.

People and language

In addition to physical differences and regional dialects, linguistic minorities reflect the mixed ancestries of the French people. The northern accents are close to standard French; those of the south have highly distinctive nasal vowel sounds. The Occitan dialect of the south is derived from the Provençal language of the Middle Ages, and a small written tradition has been preserved. The most marked language differences are in outlying areas: Catalan, spoken in the eastern Pyrenees; Breton, a Celtic language of Brittany; the Germanic dialect of Alsace; and Flemish (similar to Dutch) along the border with Belgium. The Basque language of the western Pyrenees (also spoken in northern Spain) is linguistically unique among European languages.

The Académie Française – established in 1635 to uphold the highest standards in French language and literature – insists on the use of French-derived terms, newly coined if necessary, to replace adopted English language phrases. "Le computer" has become "le ordinateur."

Culture and religion

French culture has enriched the world from the time of Charlemagne onward. Playwrights, painters, philosophers, writers, composers, film directors and sculptors are included in the long list of influential French cultural figures. But perhaps France's most lasting contribution has been its cooking. From traditional peasant dishes to elegant nouvelle cuisine, it has come to dominate the tables of the world.

There is no state church in France. The majority of the people belong to the Roman Catholic Church. There are large Protestant and Jewish minorities, and a large number of Muslims among the immigrant-based communities from Northern Africa.

ECONOMY

One of the most highly developed in the European Union (EU), the French economy has enjoyed a high growth rate since the 1950s.

Strategic landmark (*above*) The Eiffel Tower, during the Universal Exposition of 1900. Named after its French designer, it was derided by the critics, but its importance in World War I as a communications tower saved it from demolition.

Impressions of light and shadow (*right*) Edouard Manet's 1874 painting "Monet Working on his Boat in Argenteuil". These two French masters shared a passion for painting outdoors – *en plein-air* – and for exploring the transient effects of light on color.

Agriculture, forestry and fisheries

France, the largest agricultural producer in western Europe, has benefited enormously from EU subsidies, though the agricultural sector is shrinking. Small farmers, who form the majority, are offered incentives to consolidate farms, modernize techniques and form agricultural cooperatives. Overproduction of wheat, dairy produce and wine is causing problems in the EU and overseas.

Wheat, barley, oats, rye, maize, rice, sunflower seeds and root crops, chiefly sugar beet and potatoes, are the main crops. Normandy and Brittany are famous for apples and the making of cider and calvados (apple brandy). Aquitaine in the southwest is equally known for plums. Peaches and apricots are grown in the warm Rhône valley.

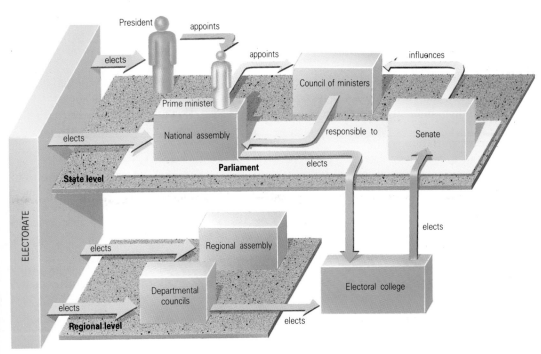

France is one of the world's largest producers of wine. Languedoc in the south produces the most wine, but the largest share of high-quality wines comes from Bordeaux, followed by Burgundy and some of the regions in the southeast.

Livestock farming accounts for an increasing proportion of agricultural production; the government has encouraged farmers to shift to meat production to reduce surpluses of cereals and butter. Cattle are the main stock animals throughout the country. Pigs are also important. Sheep and goats are raised principally in the less fertile highlands of central and southern France.

About one-fifth of France is forested, and it is Europe's third largest producer of raw timber. Large areas of forest are state-managed and there are extensive reforestation programs.

Fisheries are confined mainly to the Atlantic coast, as fish stocks are poor along the Mediterranean coast. La Rochelle, Concarneau, Boulogne and Lorient are the major fishing ports.

Industry, trade and commerce

Industry was devastated in World War II. Its recovery was initially financed by American aid under the Marshall Plan, which sought to improve Europe's war-torn economies with cash input, and coordinated by a government-led recovery plan. France is now the world's fifth greatest industrial power, after the United States, Japan, Germany and China.

Compared with other European countries, there are few large companies. There is much public and relatively little

France's system of government Under the constitution of the Fifth Republic the president appoints the prime minister, who then advises him on appointments to serve on the council of ministers. This is responsible to the national assembly, which can censure its handling of the administration.

private investment. Industry suffered during the oil-based recessions of the 1970s and 1980s, but there has since been some recovery. France's mineral resources are now limited. Lorraine's supplies of iron ore and coal have dwindled, and most is now imported. Bauxite, potash, rock salt and uranium ore are mined in smaller quantities.

Almost three-quarters of electrical power is now generated by state-owned nuclear plants. Many hydroelectric plants have been built along the extensive river systems. Petroleum and natural gas reserves are found chiefly in Les Landes and in the central part of the Paris Basin, but most has to be imported. France nevertheless has one of Europe's largest refining and petrochemical industries.

France is also a leading steel and automobile manufacturer, as well as a major producer of both civil and military aircraft. It is involved in joint European aircraft projects such as Concorde and the Airbus. Other major industries are armaments, rubber and chemicals. The textile industry still thrives, and food processing is a quickly growing sector. Prestige industries such as perfume and fashion are world market leaders.

Manufactured goods, and machinery in particular, constitute France's largest category of exports. Other important exports include automobiles, agricultural

products and chemicals. The country's major trade is mostly within the EU.

The state is heavily involved in banking and finance. The major clearing banks are state-owned, while the main stock market, the Paris Bourse, plays a smaller role than in other Western countries.

Transportation

France has developed a comprehensive transportation system, headed by Europe's most extensive road network, but there are fewer urban expressways than in other European countries. The streets of provincial cities tend to be choked with traffic.

The state-owned railroad system has its heart in Paris. In recent years it has been modernized and considerably improved. New tracks accommodate the new high-speed trains known as TGVs (*trains à grande vitesse*), which run at speeds of up to 300 km/h (186 mph) on the main routes between Paris and Lyon, the French Riviera and Switzerland. A newly opened route links Paris with Nantes and Bordeaux. The network extended to Britain via the Channel Tunnel, which opened for business in 1994.

Paris, the hub of air transport, has three main airports: Charles de Gaulle and Le Bourget to the north, and Orly to the south. Internal air services have grown rapidly and are mostly operated by Air France's subsidiary, Air Inter. France boasts the longest inland waterway system in Europe. Many of its rivers have

High fashion France has long been the world leader in women's fashion, its most exclusive designers based in Paris's Faubourg Saint Honoré. Chanel, Dior, Cardin and Saint Laurent are all household names, although few can afford their prices.

Champagne – king of all wines

Wine has been grown on the chalky soils of the Champagne region, southeast of Paris, at least since Roman times. But when exactly the *méthode champenoise*, the process that adds the sparkle to the world famous white wine, was developed is uncertain. Legend ascribes its invention to a certain Dom Pierre Pérignon, a local cellarer and Benedictine monk, in the 17th century.

The method consists basically of bottling the wine after its first fermentation and allowing it to ferment a second time within the bottle. It is not easy to achieve this. The bottles have to be securely corked and placed in sloping racks, then turned and shaken a little every day. This process, known as *rémuage*, can last for anything from six months to several years. It was originally done by men walking along the rows – one of the labor-intensive aspects of production that made champagne so expensive. These days huge, specially engineered racks are used, which can be shaken as a unit and will turn the bottles automatically. Some companies have introduced completely automated systems controlled by computer that carry out *rémuage* continually – all day, every day – speeding up the maturing process without, apparently, affecting the quality of the wine produced.

The *rémuage* finishes with the bottles upended, so that the deposit created by the second fermentation collects on the cork. The bottles are then passed through a bath of freezing brine. This encases the deposit in a plug of solid ice that is removed with the cork. The bottle is then topped up with a solution of cane-sugar syrup in old champagne, and restoppered. If little or no syrup is added, the resulting champagne becomes *brut*, or extra dry; more, and it becomes *sec* (dry), or *demi-sec* (medium dry – acutally quite sweet). There are rarer extremes: sugary *doux* is reminiscent of cough syrup, and sugarless *brut sauvage* is positively arid.

The wine is produced in three areas to the south of the ancient city of Reims. The northernmost area is the Montagne de Reims, where one of the principal varieties of champagne

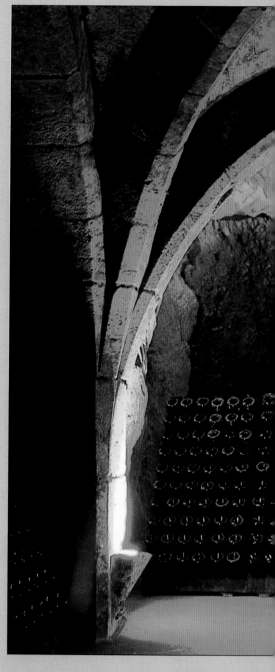

grape, the black Pinot Noir, is grown. To the south are the vineyards along the Marne Valley from Château-Thierry to Epernay – home of another black grape variety, the Meunier. South of Epernay is the district known as the Côte des Blancs, where the white Chardonnay grape is grown.

Champagne is almost always blended by its producers from the products of particular vineyards. The less common varieties known as *blanc des blancs* and *blanc des noirs* are made exclusively from white and black grapes respectively. The best champagne will be blended only from the wines of a particularly good year, and hence known as vintage champagne. Otherwise different years will be blended to produce the best flavor; from skillful

Drink of celebration Champagne, perhaps the world's most copied wine, comes from the region of chalky soil around Reims, where summers are hot and winters cold. No other region may legally call its wine champagne

hands these nonvintage champagnes may be as good or better than the vintage varieties.

Many other regions in France make their own sparkling wine by the *méthode champenoise*; much of it is excellent. Other sparkling wines, including most of the equally good German varieties, are produced by the much cheaper *cuve close* technique. But only the true champagne can boast the distinctive crisp, uncloying flavor – derived, apparently, from the chalky soil of its native slopes – that has earned it its truly regal title: king of all the wines.

Ten percent of the French labor force earn their living from the automobile industry. The firm PSA Peugeot-Citroën has operations in 140 countries and sold 1.5 million automobiles in 1999.

been canalized, though only the most modern parts of the system in the industrial northeast can carry large international freight barges. Pleasure boating is also popular.

Media and communications

France has a vigorous publishing tradition. Many of its newspapers and magazines – among them *Le Monde*, *Le Figaro*, *Paris-Match* and the leading women's magazines *Elle* and *Marie Claire* – are read internationally.

Broadcasting, until recently, was much less diverse, partly because it was a state monopoly. In 1975 the central controlling body, the Office de Radiodiffusion-Télévision Française (ORTF), was broken up into separate companies controlled by representatives of the state, parliament, the press and the broadcasters themselves. The number of television channels increased to seven. Since the advent of cable and satellite broadcasting, the viewers' choice has become even wider.

Health and welfare

The French social-security system was founded in the 1930s and expanded after World War II. It is mostly financed by the state, with compulsory worker and employer contributions.

Medical expenses are dealt with by reimbursement, using a standard fee

scale to allow private medicine to coexist with public services. Medical facilities are excellent by world standards, but there are shortages in the number of hospital beds, and severely limited special care facilities such as mental institutions and nursing homes for the aged.

Education

France has an excellent tradition of public education, which is free at all levels and compulsory between the ages of 6 and 16. The Ministry of National Education draws up curricula and administers the system. After primary education up to age 11, *lycées* or grammar schools prepare some students for the national *baccalauréat* examinations; others go on to secondary colleges providing general or technical education.

Higher education was reformed after the student unrest of 1968 and now consists of multidisciplinary universities, polytechnics and professional colleges such as medical schools. The quality of public education makes private education less popular; most private schools are Roman Catholic, but some receive limited government assistance.

Andorra

PRINCIPALITY OF ANDORRA

THE TINY AND ISOLATED PRINCIPALITY OF Andorra lies in the heart of the eastern Pyrenees, bounded by France to the north and Spain to the south. It is centered around the triple valley complex of the Valira river, which flows south into Spain and forms the only breach in the surrounding mountains. The landscape is extremely rugged, with high peaks towering above deep valleys and gorges that were originally carved out by glaciers. No part of the country lies below an altitude of 900 m (3,000 ft).

The climate is typically alpine, but with variable rainfall. Winters are cold, with heavy snow usually blocking the one pass into France. Spring and summer are mild and sunny. The plants and animals are also alpine in character. Hardy trees such as pine, fir and birch have been largely cleared, giving way to extensive mountain pastures on the slopes, where the chamois and Spanish goat roam wild.

Andorra first gained its independence from Spanish Muslim rule in 803 AD when Charlemagne (742–814) recovered the area from the Muslims. In 819 his son Louis the Pious (778–840) gave the region to the Spanish Bishop of Urgel in northeastern Spain. In 1278 an agreement was drawn up to divide dominion of Andorra between Spain and the French Counts of Foix, and apart from a short period following the French Revolution, it has remained as a co-principality up to the present day.

The heads of state are the President of France and the Bishop of Urgel, but since the new constitution of 1993 they only have power to veto matters that affect the borders of Spain and France. The Andorran legislature (General Council of Andorran Valleys) has 28 elected members.

Strong cultural links are maintained with the neighboring Spanish province of Catalonia. Catalan is the main language, though French and Spanish are also spoken. Most of the population are Roman Catholic, and Andorra belongs to the diocese of Urgel across the border in Spain. The only large town is the capital, Andorra la Vella. In recent decades the population has been greatly swelled by immigrants, mostly from Catalonia, lured by the growing tourist industry and the promise of jobs. Immigrants now make up some two-thirds of the population.

The economy is traditionally based on agriculture. In the winter cattle and sheep are kept in the villages or in lower-lying pastures in Spain, and in summer they are moved up the mountains to graze on the pastures; many of the villagers move with their livestock to temporary summer settlements. There is some forestry, and the tiny area of cultivable land is devoted to tobacco growing. Industry is mostly limited to timber and tobacco processing and hydroelectric power generation.

Tourism is now the main source of revenue. Visitors are attracted by the magnificent scenery, the local architecture and folklore, and the good winter skiing. Roads have been greatly improved, but there is no railroad and there is only one through road from Spain into France.

Beneficial tax and customs laws have encouraged the development of Andorra as a place of international commerce. Both French and Spanish currencies are used. Education, television, postal and telephone services are provided jointly by the French and Spanish governments. The Council General is responsible for other public services.

Mountain hideaway Andorra, high in the eastern Pyrenees, encompasses a cluster of mountain valleys that provides tourists with excellent skiing. They also come for the scenic beauty and duty free goods. The official language is Catalan.

NATIONAL DATA - ANDORRA

Land area	464 sq km (179 sq mi)			

Climate	Altitude m (ft)	Temperatures January °C(°F)	July °C(°F)	Annual precipitation mm (in)
Andorra la Vella	1,080 (3,543)	2 (36)	19 (67)	808 (31.8)

Major physical features	highest point: Coma Pedrosa 2,946 m (9,665 ft)

Population	(2006 est.) 71,201

Form of government	parliamentary democracy with one legislative house

Armed forces	no armed forces

Largest cities	Andorra la Vella (22,035)

Official language	Catalan

Ethnic composition	Spanish 38%; Andorran 37%; Portuguese 11.5; French 6%; others 13.5%

Religious affiliations	Roman Catholic 89.1%; other Christian 4.3%; other 1.6%; nonreligious 5%

Currency	1 euro (EUR) = 100 euro cents

Gross domestic product	(2004) U.S. $1.84 billion

Gross domestic product per capita	(2004) U.S. $24,000

Life expectancy at birth	male 80.61 yr; female 86.61 yr

Major resources	hydropower, mineral water, timber, iron ore, lead, barley, cattle, potatoes, rye, sheep, tobacco, vegetables, tourism

Monaco

PRINCIPALITY OF MONACO

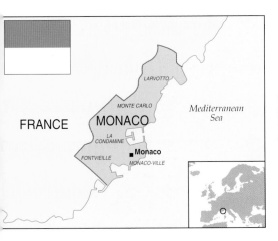

THE TINY INDEPENDENT ENCLAVE OF Monaco is situated on the French Mediterranean coast just to the east of Nice and very close to the Italian border. It enjoys a Mediterranean climate, with hot, dry summers and mild winters; and its sheltered location on the lower slopes of the rugged, maquis-covered Maritime Alps affords protection against the cold northerly wind known as the mistral.

The whole area – only 1.95 sq km (0.75 sq mi) – is densely populated, and there are four main sectors or quartiers. To the northeast is fashionable Monte Carlo, with its hotels, villas and high-rise apartments climbing up the hillside. The commercial center of La Condamine looks across the modern harbor to the old town of Monaco proper, which is perched on a rocky headland. Across the old harbor (now a marina for luxury yachts) to the southwest is the newer industrial quarter of Fontvieille, part of which has recently been reclaimed from the sea.

From 1297 until 1792, Monaco was ruled by the princes of the Grimaldi family from Genoa. In 1792, however, the local people declared the principality a republic, and in 1793 revolutionary France took control of Monaco. In 1815, at the Congress of Vienna, Monaco regained its independence, but placed itself under the protection of the kings of Sardinia. The Grimaldis were restored to power in 1861, but by then the territory had been considerably reduced. A treaty with France in 1918 ensured Monaco's continuing autonomy even if the Grimaldi line were to become extinct. In 1962, following a disagreement with France over taxes, Prince Rainier III (b. 1923) introduced a new constitution, whereby legislative power is shared between the monarch and the 18-member elected National Assembly.

Only a small minority of the population are native Monégasques – more than half are French citizens. There are also a number of Italians and a significant international community. The majority of the population are Roman Catholic, and the official language is French, though the local Monegasque dialect is also spoken.

The principality's chief industry is tourism. Its location and excellent climate have made Monaco one of Europe's most prestigious resorts. Monte Carlo contains a world-famous casino and a theater that is renowned for its opera and ballet. Monaco is also home to two international motor sport events: the Monte Carlo Rally and the Monaco Grand Prix round the streets of the principality.

Monaco is in customs and monetary union with France. Up until 1962 its tax-exempt status provided a haven for the world's rich, fueling the town's prosperity. Taxes have since been introduced, though native Monegasques are exempt from income tax. Monaco is a major international business center, drawing revenue from taxes, from radio and television franchises, and from state-operated businesses such as tobacco, postage stamps and the casino. Some light industries have developed in Fontvieille.

Grand Monaco The Monaco Grand Prix, held each May, is one of the most glamorous venues on the circuit. Monaco is a tiny principality with some big drawcards, from motor racing and gambling at the famous casino to its splendid Oceanographic Museum and old harbor.

NATIONAL DATA – MONACO

Land area	1.95 sq km (0.75 sq mi)			
Climate		Temperatures		Annual
	Altitude m (ft)	January °C(°F)	July °C(°F)	precipitation mm (in)
Monaco	55 (180)	10 (50)	24 (75)	816 (32.1)

Major physical features	Mont Agel 140m (459ft)
Population	(2006 est.) 32,543
Form of government	nonparty constitutional monarchy with two legislative houses
Armed forces	no armed forces
Capital city	Monaco (15,500)
Official language	French
Ethnic composition	French 47%; Monegasque 16%; Italian 16%; other 21%
Religious affiliations	Roman Catholic 90%; others 10%
Currency	1 euro (EUR) = 100 euro cents
Gross domestic product	(2000) U.S. $870 million
Gross domestic product per capita	(2000) U.S. $27,000
Life expectancy at birth	male 75.85 yr; female 83.74 yr
Major resources	tourism, financial services, property revenue, ceramics, metal works, textiles, instruments, plastics

The Low Countries

COUNTRIES IN THE REGION

NETHERLANDS · BELGIUM · LUXEMBOURG

Mist lies in the valley (*left*) in Luxembourg's most fertile region, the Gutland; throughout the Low Countries mixed farming predominates. But in this densely populated region intense agricultural production is never far from highly developed manufacturing, agricultural and commercial industries.

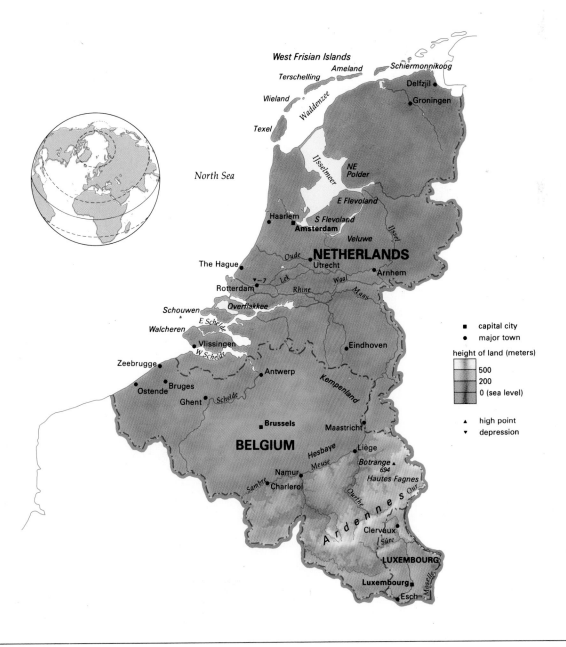

Netherlands

KINGDOM OF THE NETHERLANDS

T HE SMALL, DENSELY POPULATED NETHER-lands is remarkable not only for its predominantly flat landscape – one-third of which lies below sea level – but also for the vast tracts of land that have been reclaimed from the sea. The country is bordered by the North Sea to the north and west, Germany to the east and Belgium to the south. It controls access to the North Sea from three major European rivers – the Schelde, Meuse and Rhine – making the Netherlands an important international commercial center.

NATIONAL DATA - NETHERLANDS

Land area 33,883 sq km (13,082 sq mi)				
Climate		Temperatures		Annual
	Altitude m (ft)	January °C(°F)	July °C(°F)	precipitation mm (in)
Amsterdam	-4 (-13)	3 (37)	17 (63)	831 (33)

Major physical features highest point: Vaalserberg (in far southeast) 321 m (1,053 ft); lowest point: (in the west) -7 m (-22 ft); longest rivers: Rhine (part) 1,319 km (820 mi) and Meuse (part) 933 km (580 mi)
Population (2006 est.) 16,491,461
Form of government multiparty constitutional monarchy with two legislative houses
Armed forces army 23,150; navy 12,130; air force 11,050
Largest cities Amsterdam (capital - 1,002,868); Rotterdam (989,956); The Hague (610,245); Utrecht (366,186); Eindhoven (302,274)\
Official language Dutch
Ethnic composition Dutch 83%; other 17% (mainly Germans, Turks, Moroccans, Antilleans, Surinamese, and Indonesians)
Religious affiliations Roman Catholic 31%; Dutch Reformed 13%; Calvinist 7%; Muslim 5.5%; other 2.5%; none 41%
Currency 1 euro (EUR) = 100 euro cents
Gross domestic product (2006) U.S. $512 billion
Gross domestic product per capita (2006) U.S. $31,7000
Life expectancy at birth male 76.39 yr; female 81.67 yr
Major resources natural gas, petroleum, peat, limestone, salt, sand, gravel, arable land, cereals, fish, fruit, vegetables, livestock, potatoes, poultry, sugar beet, horticultural plants

As early as 1000 AD, simple sluices were being used in low-lying areas of the Netherlands to improve drainage. Four hundred years later the clay flats of the west and north were crisscrossed by an intricate network of dikes and drains. From the 15th to the 17th century, windmills were built on a large scale to pump water from low-lying land behind the dikes, creating "polders" – areas where the water level can be controlled – for farming. Steam pumps were introduced in the 19th century, making it possible to reclaim the Haarlemmermeer in 1852. Schiphol airport lies in the northeastern corner of the Haarlemmermeer Polder 4 m (13 ft) below sea level.

The land

The north coast of the Netherlands is fringed by the long arc of the West Frisian Islands. These are formed from dunes and sandbars at the outer limits of the Waddenzee – a stretch of shallow salt-water lying between the islands and the mainland. The Waddenzee is separated from the freshwater IJsselmeer by a broad 32 km (20 mi) long dike.

The dike was the first stage of a massive scheme, conceived as early as 1891, to reclaim land from what was then the Zuider Zee, and to link the province of North Holland with the northwestern province of Friesland. While the dike was under construction the shallow Wiering-ermeer, to the southwest, was drained. This became the first of four large polders to be reclaimed, and today they cover an area of more than 2,000 sq km (800 sq mi).

The western provinces of North and South Holland are older reclaimed lands.

In the southwestern province of Zee-land, the rivers Rhine, Schelde and Meuse flow out into the North Sea by way of a delta made up of estuaries and islands. After disastrous flooding in 1953 most of the estuaries and sea-arms were sealed off by 32 km (20 mi) of new dikes, and one inlet was spanned by a tidal barrage. Two major shipping lanes remain open, however: the New Waterway linking Rotterdam with the sea, and the Western Schelde, which gives access to the Belgian port of Antwerp. In early 1995, the government declared a state of emergency during severe flooding.

Inland, the flat, clay lowlands of the north give way to sandy soils in the eastern, central and southern Netherlands. In the extreme southernmost province of Limburg, a chalk plateau rises toward Vaalserberg, which is the highest point in the Netherlands.

Climate

Summers are relatively cool, though there can be very warm days, and winters are generally mild and wet. Winds from the north may bring cold weather in the spring, especially to the north and east. Rainfall is frequent, but often light and

Low-lying coastal marshes *(right)* are part of the distinctive landscape of the Netherlands. Flat as well as low, they offer little resistance to sea winds, and from the 15th century windmills were used extensively to drain the land for farming.

Behind the protective coastal sand dunes *(below)* a scrubland habitat supports a range of plants adapted to tolerate strong winds and high levels of salt. Lichens, mosses, low shrubs and deep-rooted grasses bind and stabilize the sands.

evenly distributed thoughout the year. In the fall, heavy cloud cover, fog and mist are common.

Plants and animals

Compared with prehistoric times, today both forest and marshland are much reduced. Most of the oak and birch woodlands in eastern areas have been planted, as have the large coniferous and deciduous forests on the central hills of the Veluwe. Large areas of agricultural land are now occupied by commercial nurseries.

In the north the Waddenzee, with its sand dunes, mudflats and marshes, is the most important wetland in western Europe, supporting many species of native and migrating birds, as well as fishes and crustaceans. Marram and other coarse grasses cover the coastal dunes, while the tidal mudflats attract salt-loving plants such as glasswort; otters are found on the polders.

SOCIETY

The histories of the Netherlands, Belgium and Luxembourg are all intimately connected. They are known collectively as the Low Countries, and many attempts have been made to unite them.

History

In the 1st century BC the Celtic and Germanic peoples of the Low Countries were invaded by Rome's legions. Roman occupation came to an end some 450 years later with the rise to power of the Germanic Franks in the early 5th century AD.

The Franks founded a kingdom that covered the area between the Rhine and the Loire rivers. Later they extended it to the coastal area of what is now the Netherlands, which was occupied by another Germanic people, the Frisians.

After the death of the Carolingian emperor Charlemagne (742–814), the greatest of the Frankish rulers, there was a gradual decline in Frankish power. In 888 the Carolingian empire finally broke up and from its ruins the kingdoms of France and Germany emerged.

From the 10th century the Low Countries were effectively ruled by the vassals of the German Holy Roman Emperor. In the late 14th and early 15th centuries, marriages, inheritance and annexation brought many of these territories under the control of the French dukes of Bur-

finally recognized the provinces' independence by the Peace of Westphalia agreement in 1648.

The 17th century was a golden age of scientific and cultural achievements for the Dutch Republic. It produced among others, Antonie van Leeuwenhoek (1632–1723), the first person to study microscopic organisms; the greatest Dutch painter Rembrandt van Rijn (1606–69) and the Jewish philosopher Baruch Spinoza (1632–77).

In 1794 the French invaded and when Napoleon I (1769–1821) became emperor of France in 1806 he appointed his younger brother, Louis Bonaparte (1778–1846), as king of the Netherlands. After Napoleon's defeat in 1814, hereditary stadtholder (governor), William VI of Orange (1772–1843), returned as King William I to create a constitu-

The Dutch East India Company

By the end of the 15th century Dutch commercial explorers from the newly independent United Provinces were keen to challenge Portugal for a share of the spice trade.

To divide the risks – as well as the profits – involved in trading with the East Indies, rich Amsterdam merchants invested in a joint-stock company called the Dutch East India Company. The government granted the company a trading monopoly between the Cape of Good Hope and the Straits of Magellan, with the right to draw up treaties with neighboring provinces, maintain armed forces and appoint officials in new territories.

In 1603 the Dutch took the island of Amboina from the Portuguese, in the first of many Portuguese losses.

Malacca fell to the Dutch in 1641, and control of Ceylon, with its precious stones, pearls and cinnamon trees, was finally wrested from the Portuguese in 1656. By 1755, most of the island of Java was in Dutch hands.

In the 18th century the company changed from a commercial shipping venture to a territorial administration interested in agricultural produce. It began purchasing local produce. The produce went to Europe, and the profits remained in Dutch hands. But

gundy. The Low Countries then passed to the powerful Habsburg dynasty through the marriage of Mary (1457–82), daughter of the last Duke of Burgundy, to the Austrian archduke Maximilian (1459–1519), heir to the Holy Roman Empire. In 1504 their son, Philip the Handsome (1478–1506), and his wife inherited the Spanish crown.

The Spanish king who succeeded Philip, Charles V (1500–58), regarded the Spanish Netherlands (as the Low Countries were now called) as a single unit, and attempted to rule them as such. His fiercely Catholic son Philip II (1527–98), who became King of Spain in 1556, was determined to wipe out what he saw as the Protestant heresy in the Netherlands. A Burgundian was appointed as the Catholic Primate of the Netherlands, provoking outright resistance and fueling Protestant opposition movements. Alarmed, Philip II sent Fernando Álvarez de Toledo, third Duke of Alba (1508–82), with an army of 10,000 to the Netherlands to enforce order and deal with all heretics.

The Duke's repressive rule quickly brought open rebellion led by Prince William the Silent, Prince of Orange

De Haar castle, to the northwest of Utrecht. Built in medieval times, its strong defenses afforded protection over the centuries from the warring French, German and Spanish forces in the region.

(1533–84), a Germanborn nobleman living at the Burgundian court at Brussels who came to side with the Protestants. By 1573, most of the provinces of Holland and Zeeland were rebel-held. In 1576 William marched into Flanders, took the town of Ghent and signed a treaty – called The Pacification of Ghent – allowing freedom of worship in the Low Countries. Three years later the seven northern Dutch provinces of the Netherlands signed the Union of Utrecht, creating a loose federation known as the United Provinces. The 1581 Act of Abandonment declared the independence of all the Low Countries from Spain, but it was only truly effective in the north.

In the years that followed, the United Provinces forged and protected their independence with a well-equipped and efficient army. Commercially, the Dutch became Europe's carriers, and burgeoning trade – especially with the East and West Indies – ensured the success of their banking and financial institutions. Spain

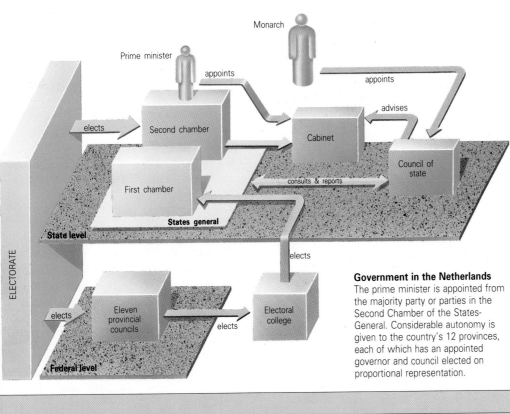

Government in the Netherlands
The prime minister is appointed from the majority party or parties in the Second Chamber of the States-General. Considerable autonomy is given to the country's 12 provinces, each of which has an appointed governor and council elected on proportional representation.

tional monarchy. His kingdom included modern Belgium and Luxembourg, but they seceded to become independent countries in 1831 and 1867 respectively.

The Netherlands remained neutral during World War I, but in May 1940, soon after the outbreak of World War II, the country was invaded by Nazi Germany.

After the war, American economic aid – part of the Marshall Plan – helped the Dutch to create a powerful industrial base and neutrality was abandoned in favor of full participation in Western European affairs. Overseas, a nationalist revolution in the Dutch East Indies (modern Indonesia) eventually led to full independence from the Netherlands in 1949. Independence came to Dutch New Guinea (Irian Jaya) in 1962, though it was annexed by Indonesia the next year, and to Suriname in South America in 1975.

Building an empire In the early 17th century the Dutch East India Company built its headquarters at Batavia (Djakarta) on the Indonesian island of Java. Fleets carrying precious spices returned each year to Holland, to the profit of the shareholders. In 1625 thousands of clove trees were destroyed in order to raise the price of cloves in Europe.

these measures could not halt the company's slow decline. Its monopoly was broken, first by smuggling, and then, after the Treaty of Paris in 1784, by its old rivals the British, who were granted the right to trade in the East

Indies. In 1799, while the Netherlands lay under the dominion of the French Napoleonic empire, the company was finally wound up by the Dutch government, which acquired all its debts and possessions.

Government

The monarch is the head of state, but most real power is vested in the democratically elected parliament, or States-General, which meets in The Hague. The constitution provides for two parliamentary chambers. The 150 members of the Second Chamber are directly elected by the people for four years and only they can propose and amend laws.

The 12 provinces are administered by elected councils and an appointed commissioner. Both national and local elections involve a proportional system in which a large number of parties and opinions are represented.

The Netherlands belongs to the North Atlantic Treaty Organization (NATO), and is a full member of the European Union (EU).

People

Most of the 15 million people speak standard Dutch, though West Frisian is still spoken in the northern province of Friesland. Some Saxon dialects can still be heard in the northeastern area bordering Germany. More than half the population is Christian – Roman Catholics comprise 31 percent and various Protestant groups 21 percent. The Netherlands also has many new immigrant communities, including people from former Dutch colonies, as well as small Turkish and Moroccan minorities, mostly made up of so-called "guest workers".

The country maintains a strong cultural tradition, particularly in music and art, boosted by state subsidy. Drama and literature – seldom translated for a global audience – also remain vigorously creative in the Netherlands.

Wheels of yellow cheese (*above*) are carried on traditional "sledges" at the cheese market at Alkmaar near Amsterdam. Highly productive dairy farming provides milk for processing into cheese and butter, much of which is exported.

Flowers under glass (*right*) A flower exhibition at Zoetermeer, near Rotterdam. Flowers have been an important Dutch industry since the 17th century: in the 1630s tulip speculation ran wild and many people were ruined with the crash of the market.

ECONOMY

Benefiting from its key location at the mouth of the Rhine, the Netherlands has become one of Europe's most prosperous nations.

Agriculture

Farms are large and highly mechanized, but they employ only a tiny percentage of the workforce. About one-fifth of the land is devoted to agriculture and almost two-thirds of this is pasture. The principal cattle-farming regions are Friesland and North Holland, the two coastal provinces on either side of the IJsselmeer. The country remains a major exporter of cheese and butter, and meat products, bacon and eggs are also significant exports. Agricultural produce represents nearly one-quarter of the value of the country's exports.

Crop farming (sugar beet, potatoes and vegetables) is often combined with cattle farming, but is generally less important. The Netherlands grows a considerable amount of salad vegetables – for example cucumbers, tomatoes and lettuces – under glass.

Another unique feature of the landscape is the bulbfields. In addition to the famous export trade in bulbs, cut flowers – tulips, roses, chrysanthemums and lilies – are exported to florists and distributors all over the world.

Forestry and fisheries

Forests account for a little under one-tenth of the land area of the Netherlands. Many smaller woods have been planted as recreational areas, or to prevent the movement of sand dunes.

The small fishing industry has tended to concentrate on shellfish, but stocks have been greatly reduced by the coastal barriers that were erected as flood control measures. The North Sea herring fisheries have also declined, largely due to falling herring populations and the extension of fishing limits by neighboring countries.

Industry

During the 1950s substantial natural gas reserves were discovered, both offshore and in the northeastern province of Groningen. Much of the gas is exported. The Netherlands produces its own crude petroleum from wells in the provinces of Drenthe (in the northeast) and South Holland. Output is too small to meet domestic demand, but much of the crude petroleum that is imported into the Netherlands is reexported after passing through Europoort – a large storage,

The industrial complex of Europoort is an extension of the inland port of Rotterdam some 28 km (17 mi) away. Building started on the complex in 1957, and the port can now accommodate the world's largest bulk carriers of oil, ore and grain.

refining and petrochemical complex outside Rotterdam. At the ports of Delfzijl in the far northeast and Vlissingen in the extreme southwest, smelting works fueled by natural gas have been set up to extract aluminum.

About one-seventh of the labor force works in manufacturing, chiefly located in a densely populated arc known as the Randstad, which runs from Dordrecht in the south, via The Hague, Leiden, Haarlem, Amsterdam and Hilversum, to Utrecht. Important manufacturing industries include food processing, chemicals and petrochemicals, motor vehicles, commercial aircraft, electrical equipment and computers and microelectronics.

Trade, commerce and transportation
Exports from the Netherlands amount to about half its income. More than half the country's external trade is within the EU, in particular with the other Benelux countries of Belgium and Luxembourg and with Germany. The dense road system is extensive and well maintained, with Dutch trucking companies handling a large share of the EU road freight. Utrecht is the hub of the state-owned rail network.

Rotterdam-Europoort, the largest port in Europe, handles the biggest tonnage of any harbor in the world. Nearly 5,000 km (3,000 mi) of canals provide the main arteries for freight transportation, linking the major coastal ports with inland areas and main European rivers. The national carrier Royal Dutch Airlines (KLM) provides passenger services to most major international destinations.

Media and communications
Post, Telegraph and Telephone (PTT), the national mail and telephone service, was privatized in 1989 and divided into PTT-Post and PTT-Telecom. The press includes more than 40 national and local daily newspapers. Broadcasting, which is controlled by the Netherlands Broadcasting Foundation (NOS), is free of government interference.

Health, welfare and education
All citizens are entitled to retirement, disability and survivors' pensions, and there are also children's allowances. For people in the labor market, incapacity and unemployment insurance are compulsory, and most have health insurance.

School attendance is compulsory for children between the age of 6 and 15. Education is provided in both state and private schools, all of them government funded. Universities, colleges and technical institutions are of a high standard.

Belgium

KINGDOM OF BELGIUM

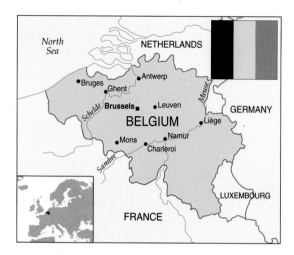

BELGIUM LIES ON THE NORTH SEA COAST OF Europe, with the Netherlands to the northeast, Germany to the east, the Grand Duchy of Luxembourg to the southeast and France to the south and west.

GEOGRAPHY

Belgium's landscape naturally divides along the central valley of the Meuse river, with lowlands to the north and uplands to the south.

The land
The short Belgian coastline resembles the coast of the Netherlands, with long sandy beaches bordered by a belt of sand dunes that broadens out toward the French frontier. Behind the dunes runs a strip of flat polder land 16 km (10 mi) across in places. This low-lying area was once marshy, but it has been drained to create a characteristic landscape of dikes and canals. Farther inland, the plain of Flanders in the northwest is drained by the river Schelde and its tributaries.

In the northeast the land rises gently to an area of heath, grasslands and coniferous forests. Farther west, central Belgium forms a low, fertile, loam plateau, rising in the south to about 200 m (650 ft). Brussels, the capital, lies to the north of this region. Farther south, beyond · the Famenne depression, lies the high plateau of the Ardennes, whose flat-topped mountains are cut by the deep, wooded river valleys of the Meuse and its tributaries. In the east, near the German border, are the highlands of the Hautes Fagnes, and the Botrange, the highest point in Belgium at 694 m (2,276 ft).

In the extreme south of the country is the area known as the Belgian Lorraine, geographically very similar to the adjoining French Lorraine. Here, the landscape is hilly and partly wooded, with north-facing scarps.

Climate
Belgium has a temperate climate moderated by the North Atlantic Drift and by prevailing westerlies from the sea. The weather is changeable, but rainfall can be both heavy and frequent. Winters are mild but damp, and fog and mist are common. Summers tend to be cool.

The Ardennes – the area farthest inland and highest above sea level – experiences lower winter temperatures and more rain and snow than anywhere else in the country.

Plants and animals
Half the land has been cleared for agriculture or pasture, though some oak and beech forests have survived. Belgium's largest forests are in the Ardennes, where the native trees include beech, oak and hornbeam.

Heathland is typical of the sand and gravel areas of the northeast. Marram and other coarse grasses grow on the dunes along the coast.

In the highland peat bogs of Haute Fagnes a reserve has been established, providing a protected habitat for wild boar, red deer, wildcats, woodpeckers and owls.

SOCIETY

The history of Belgium is closely bound up with those of its neighbors Luxembourg and the Netherlands. Like them it has suffered from having Germany and France as its nearest large neighbors.

History
In the 1st century BC the Romans invaded the region now known as the Low Countries. To the south of the Rhine they created a province that they called Belgica after a Celtic people, the Belgae, who lived there, and on whom the Romans imposed their Latin language and culture.

In the 5th century AD Belgica was occupied by Germanic invaders known as the Franks. The Frankish settlement was densest in the north, and here their Franconian dialect, which became the basis of modern Dutch, replaced the native Latin vernacular which formed the basis of the French language.

The kingdom of the Franks became the heart of the Carolingian empire, which reached its greatest extent during the reign of Charlemagne (742–814). However, by 925 the territory of modern Belgium was divided between German and French control. Most of the French-speaking south, forming part of the

Duchy of Lorraine, came under the control of the German Holy Roman Empire. Most of the Dutch-speaking north came under French rule.

Between 1363 and 1477 the Low Countries fell into the hands of the powerful dukes of Burgundy, and so became embroiled in the long and bitter struggle between France and England known as the Hundred Years War (1331–1453).

Following the death of the last Burgundian duke, Charles the Rash (1433–77), at the battle of Nancy, the Low Countries passed to the Habsburg line through the marriage of Charles' daughter Mary (1457–82) to Maximilian of Austria (1459–1519), the heir to the Holy Roman Empire. In 1504 their son, Philip the Handsome (1478–1506), received the crown of Spain,

The forested uplands of the Ardennes in southern Belgium. A number of rivers, including the Semois, cut through this sparsely populated area where farmsteads, for the most part, have been converted into holiday homes for Belgians and foreign tourists.

After Napoleon's defeat in 1814, the Congress of Vienna united Belgium with the Netherlands. However, the two regions had conflicting interests and religions. Belgium proclaimed national independence in 1831, when Prince Leopold of Saxe-Coburg (1790–1865), was chosen as the first ruler of the new kingdom.

In 1885, in a bid to make Belgium a colonial power, King Leopold II (1835–1909) acquired the vast territory of the Congo in Central Africa; this became the Belgian Congo in 1908. The Congo brought Leopold great wealth from rubber production, but the African laborers suffered atrocities against them by overseers, causing an international scandal.

In July 1914 Germany invaded Luxembourg and demanded passage through Belgium for its troops. When the demand was refused, Germany invaded Belgium immediately. Most of the country was occupied, and many of the bloodiest battles of World War I (1914–18) were fought on Belgian soil. During the war, Belgian forces from the Congo invaded Ruanda-Urundi in German East Africa.

In 1930 the Belgian government, in recognition of the growth of the Flemish nationalist movement, divided the

forging a link between Austria, Spain and the Low Countries.

The Protestant leader Prince William of Orange (1533–84) hoped to win independence for the whole of the Spanish Netherlands (as the Low Countries were known), but the southern provinces remained loyal to Roman Catholicism and to Spanish rule. In 1579 the Protestant northern provinces responded with the Union of Utrecht to form the independent United Provinces of the Netherlands under William's leadership. The Spanish Netherlands lost further territory in the

Thirty Years War (1618–48). In 1700, when Charles II (1661–1700), the last of the Spanish Habsburgs, died, the remaining territories fell briefly into French hands. They were then occupied by the Dutch and the English during the War of the Spanish Succession (1702–13) before Austrian rule was restored in 1748.

Under the Habsburg empress Maria Theresa (1717–80) the modernization of agriculture brought new prosperity and stimulated new industries in the region. Roads and waterways were built during this period. But in 1792 war broke out between revolutionary France and Austria. The French defeated the Austrians in June 1794, a prelude to the French annexation of Belgium, ending the region's autonomy.

NATIONAL DATA – BELGIUM

Land area 30,278 sq km (11,690 sq mi)

Climate	Altitude m (ft)	Temperatures January °C(°F)	July °C(°F)	Annual precipitation mm (in)
Brussels	55 (180)	3 (37)	18 (64)	822 (32.3)

Major physical features highest point: Botrange 694 m (2,276 ft); longest rivers: Meuse (part) 933 km (580 mi) and Schelde 432 km (270 mi)

Population (2006 est.) 10,379,067

Form of government multiparty federal monarchy with two legislative houses

Armed forces army 24,800; navy 2,450; air force 10,240

Largest cities Brussels (capital – 1,031,925); Antwerp (463,256); Gent (233,111); Charleroi (199,898); Liege (181,715); Brugge (116,618); Namur (106,539)

Official language Dutch, French, German

Ethnic composition Fleming 58%; Walloon 31%; mixed or other 11%

Religious affiliations Roman Catholic 81%; Protestant or other Christian 7.3%; Muslim 3.6%; other 0.6%; nonreligious 7.5%

Currency 1 euro (EUR) = 100 euro cents

Gross domestic product (2006) U.S. $330 billion

Gross domestic product per capita (2005) U.S. $31,800

Life expectancy at birth male 75.59 yr; female 82.09 yr

Major resources construction materials, silica sand, carbonates, barley, coal, flax, hay, livestock, oats, potatoes, sugar beet, timber, vegetables, wheat, textiles, oil refining, diamond processing

country into French-speaking and Dutch-speaking areas, each of which was to be administered in its own language. Germany once again invaded Belgium in May 1940 during World War II.

When Allied forces finally reached Belgium in August 1944, resistance fighters prevented the Germans from destroying the port of Antwerp, which served as a vital base for the continuing Allied advance into the German heartland.

Unpopular after the war, King Leopold III abdicated in 1951 in favor of his son, Baudouin I (1930–93). In 1960 Belgium granted independence to the Congo (formerly Zaire), and in 1962 to Ruanda-Urundi (now Rwanda and Burundi). In 1980 the country reorganized into French-speaking Wallonia and Dutch-speaking Flanders, while a distinct bilingual area was established around Brussels.

Government

The 1993 constitutional amendments state that Belgium is a federal state and devolved many powers from the national to community and regional governments. The Senate (71 members, 40 of which are directly, 21 indirectly elected from community councils) lost most of its power to the Chamber of Representatives (150 elected members). The king lost his power to dissolve parliament. There are four levels of sub-national government: community (Flemish, Francophone, Brussels, Germanophone), regional (Walloonia, Brussels, Flanders), provincial (10 provinces), and at the lowest level communal (589 communes).

Belgium belongs to the North Atlantic Treaty Organization (NATO), and is a full member of the European Union (EU), the main headquarters of which are in Brussels.

People

More than half of the population of 10 million are Flemings; they live mostly in the northern provinces and speak Flemish, a strongly localized version of Dutch. The French-speaking Walloons in the southern and eastern provinces make up most of the rest. A small number of German-speakers live in the sparsely populated eastern districts (the Cantons de l'Est) of Liège province. Nearly one million people live in or around Brussels.

The vast majority of Belgians are Roman Catholic. However, there is a tiny Protestant minority, most of them Walloons, and a growing Muslim minority reflecting the presence of a sizable immigrant community. The Belgians have excelled in art, music and architecture.

Notable figures include the Flemish artists Pieter Breughel the Elder (c.1525–69) and Pieter Paul Rubens (1577–1640), the surrealist René Magritte (1898–1967), the composer César Franck (1822–90) and the architect Henri van de Velde (1863–1957).

ECONOMY

The Belgian economy is highly diversified, and the country is an important center of industry and commerce. Belgium's principal trading partners are the other members of the European Union which by 1995 totaled 15.

Agriculture, forestry and fishing

Farming in Belgium concentrates on the production of livestock; much of the arable land is used for pasture and forage crops. Less than 3 percent of the labor force is employed on the land, and agriculture accounts for only a tiny percentage of the country's income. About one-fifth of Belgium is forested, and the lumber industry has grown rapidly. Hardwoods and pulp are both produced.

Belgium has a small fleet of trawlers operating in the North Sea. The catch, chiefly herring and flatfish (turbot and plaice), meets about one-third of local demand.

Industry, trade and energy

Manufacturing industries in Belgium account for almost half the total value of the country's exports. Most raw materials have to be imported.

Much heavy industry can still be found near the old coalfields of the Sambre-Meuse valley and north of Ghent. This includes glassmaking, metal refining, and the manufacture of heavy machinery. The traditional arms manufacturing is still important. Belgium is also a major producer of rarer metals such as cobalt, radium and germanium. Steel production, an important export industry, relies on imports of ore and coal. Imported petroleum from the North Sea, the Middle East and the former Soviet Union is refined in Antwerp, and some is used to make plastics, part of the city's thriving petrochemical industry.

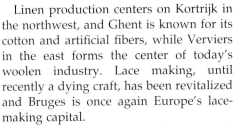

Lace making (*above*) in Belgium dates from the mid 17th century. Made from high quality local linen, the lace was exported to Britain, where it was a popular adornment to court dress. The craft has recently been revived in Bruges, a traditional center of lace making.

At the heart of Europe (*left*) In 1958 Brussels played host to the World's Fair, symbolized by the futuristic Atomium structure. The city is also headquarters of the European Commission, whose emblem is a ring of 12 yellow stars on a blue background, as seen on the front of the large creature.

Ostende (*below*) is Belgium's main fishing port, as well as being a major shipbuilding center and resort with a popular marina.

Linen production centers on Kortrijk in the northwest, and Ghent is known for its cotton and artificial fibers, while Verviers in the east forms the center of today's woolen industry. Lace making, until recently a dying craft, has been revitalized and Bruges is once again Europe's lace-making capital.

Belgian chocolates are world renowned and supply a large international export market. Antwerp's diamond-cutting industry has been established for 500 years, and carries out 70 percent of the world's diamond dealing.

About one-half of Belgium's electricity is generated by nuclear power. In 1999, the government announced that it would phase out nuclear power by 2015.

Transportation

Road and rail networks are extensive and mostly radiate from Brussels. A system of expressways links all the main centers of population, both within Belgium and with neighboring France and the Netherlands. The railroad network is state owned; fast and frequent services connect all the main towns and cities.

Antwerp is the world's fifth largest port, and handles most of the overseas trade. The canals are fully modernized to cater for standard European-size barges. Belgium has an international airport at Brussels and Antwerp.

Media

Belgium has more than 35 daily newspapers controlled by various consortia. The circulation is limited by the size of the readership, which is either Flemish- or French-speaking. (There is also one German-language newspaper.)

The principal broadcasting organizations are the French-language Belgian Television Broadcasting (RTBF) and the Dutch-language Belgian Radio and Television (BRT). Both are paid for by a tax on audiences.

Welfare, health and education

Social insurance provides health and unemployment insurance, family allowances and pensions, and compensation for injuries sustained at work.

Education is compulsory between the ages of 6 and 16. In general, children are taught in either French or Flemish depending on where they live. In Brussels, however, parents can choose their children's language of instruction. In the many institutes of higher education, which include several universities, tuition is in one or other of the main languages.

Luxembourg

GRAND DUCHY OF LUXEMBOURG

L UXEMBOURG IS ONE OF EUROPE'S SMALLEST states, landlocked between France, Germany and Belgium.

GEOGRAPHY

Luxembourg's territory consists mainly of the basins of the rivers Sûre and Alzette. The eastern border with Germany is defined by the course of the rivers Our, Sûre and Moselle.

Geographically the northern region, known as the Oesling, occupies about one-third of the country, and consists of a dissected plateau. Here, the soils are thin, acidic and barren.

The rest of the country is known as the Gutland or "Good Land". This region of hills and plains is covered by a layer of much more fertile soil underlain by beds of iron ore. The river Alzette flows north to meet the Sûre, while smaller rivers to the southeast drain into the Moselle. The climate is wet and temperate but with relatively long winters and temperatures low enough to bring heavy snowfalls to the northern slopes. The lower-lying Gutland is milder in winter and warmer in summer.

Coniferous trees grow on a few of the northern slopes, with some broadleaf species in the valleys. The south is more heavily wooded. Toward the extreme east, in the Müellerthal, lies a great beech forest. Wild boar and a few species of deer survive in the forests.

SOCIETY

Luxembourg, whose history is closely intertwined with those of its more powerful neighbors – France, Belgium and Germany – retains a keen sense of its cultural identity and independence.

History
When the Romans colonized the region in about 50 BC it was inhabited by Celtic peoples. Some 450 years later, the region became part of the Frankish kingdom of Austrasia. Originally called Lucilinburhuc, or "Little Fortress" Luxembourg became a county within the Holy Roman Empire in 963. In the 14th century Henry IV, Count of Luxembourg (1269–1313), became Holy Roman Emperor as Henry VII. His descendant, Emperor Charles IV (1316–78), made Luxembourg an independent duchy in 1354. Subsequently it became a political pawn in the power struggle between Europe's ruling houses.

In the 16th century Luxembourg passed from the Austrian Habsburgs to the rulers of Spain, before being returned, together with Belgium, to the Austrian Habsburgs in 1713. Luxembourg was later under the rule of revolutionary France, but with the fall of Napoleon I (1769–1821) in 1814, it came under the control of the Dutch king William I (1772–1843).

The Sûre river in northern Luxembourg cuts its way through winding wooded valleys, past ancient towns such as Esch-sur-Sûre. The country's scenic beauty and small size belie its status as an urban industrial nation.

After a revolt in 1830, Luxembourg was split into the French-speaking Belgian province of Luxembourg and the smaller Grand Duchy, centering on the capital, under William. In 1867, the Grand Duchy was given to the House of Nassau, who remain in power to this day.

Despite its neutrality, Luxembourg was occupied by Germany in both World Wars. In 1949 Luxembourg abandoned its neutral status to become a founder member of the North Atlantic Treaty Organization (NATO) and later of the European Union (EU). The Grand Duchy is the seat of the European Court of Justice created in 1958.

Luxembourg is a constitutional mon-

demand has led to considerable diversification. American-based multinational corporations, for example, have set up major tire, chemical and metal fabrication plants. Local industries, such as tanning, textiles and brewing, operate on a much smaller scale. Tourism is important.

Increasingly, finance is the lifeblood of Luxembourg. Drawn by tax advantages, many financial institutions, including the European Investment Bank, have established their headquarters there. Today Luxembourg boasts more than 200 different banks. A shipping register offers a flag of convenience to other European Union shipowners. The European free market may challenge some aspects of Luxembourg's status as a financial center, but is just as likely to boost others.

Luxembourg's road and rail networks provide major links with neighboring countries. Barge traffic on the river Moselle carries import and export freight. The national airline, Luxair, operates from Findel airport near the capital.

Health and social security schemes, which were first introduced after World War I, are comparable with others in Western Europe. Education is compulsory from the ages of 6 to 15. After secondary education many young people go abroad to study.

archy with a democratically elected government headed by the Grand Duke. His executive power is exercised through a cabinet of ministers headed by a prime minister. These in turn are responsible to the 60-member Chamber of Deputies, who are directly elected, under a system of proportional representation, for a five-year term. In Luxembourg voting is a compulsory duty.

Luxembourgers speak their native language, Letzeburgesch, and also French and German, the other two official languages. There are also Italian- and Portuguese-speaking minorities. Most of the population is Roman Catholic.

Luxembourgers enjoy close cultural ties with other European countries. The multilingual Radio Luxembourg has become an institution throughout Europe. Luxembourg has no university, but the arts, culture and sciences are promoted by

the Institut Grand-Ducal, the state library and museums.

ECONOMY

Although Luxembourg is still strongly agricultural, its prosperity depends almost entirely on heavy industry and international finance.

Farm output fulfills most domestic needs. The Gutland provides the richest pasture for both beef and dairy cattle, and the best land for grains and fodder crops. Fruit is grown here, too, especially grapes for the wine industry.

Luxembourg's main mineral resource is iron ore from deposits in the southwest, but most of these have been depleted. For energy, the country depends heavily on imported fuels and hydroelectricity.

The iron and steel industries were once the mainstay of the economy, but falling

NATIONAL DATA – LUXEMBOURG

Land area 2,586 sq km (998 sq mi)

Climate		Temperatures		Annual
	Altitude m (ft)	January °C(°F)	July °C(°F)	precipitation mm (in)
Luxembourg	330 (1,082)	0 (32)	17 (63)	874 (34.4)

Major physical features highest point: Wemperhardt 559 m (1,843 ft); longest river: Moselle (part) 545 km (340 mi)

Population (2006 est.) 474,413

Form of government multiparty constitutional monarchy with one legislative house

Armed forces army 900

Capital city Luxembourg (76,420)

Official language French, German, Letzeburgesch

Ethnic composition Luxembourger 72.5%; Portuguese 9%; Italian 5.4%; French 3.4%; Belgian 2.5%; German 2.4%; others 4.8%

Religious affiliations 87% Roman Catholic; Protestants, Jews, and Muslims 13%

Currency 1 euro (EUR) = 100 euro cents

Gross domestic product (2006) U.S. $32.6 billion

Gross domestic product per capita (2006) U.S. $68,800

Life expectancy at birth male 75.6 yr; female 82.38 yr

Major resources iron ore (no longer exploited), cattle, cereals, dairy, grapes, potatoes

Spain and Portugal

COUNTRIES IN THE REGION

SPAIN · PORTUGAL

DEPENDENCIES IN THE REGION

GIBRALTAR

Wealth and grandeur of a bygone era (*left*) Both Spain and Portugal derived vast wealth from their global colonial empires, building magnificent private and public buildings. The Vila Real in the village of Mateus near Oporto, Portugal, built in the mid 18th century, is the home of a rosé wine now sold worldwide.

ATLANTIC OCEAN

Bay of Biscay

La Coruña

Santander

San Sebastián

Cape Finisterre

Oviedo

Bilbao

Cantabrian Mountains

Pamplona

PYRENEES

Aneto 3404

León

Vigo

Minho

Burgos

Sierra de la Demanda

Esla

Douro

Zaragoza

Ebro

Lérida

Segre

Barcelona

Valladolid

Tarragona

Oporto

Douro

PORTUGAL

Salamanca

Sierra de Guadarrama

SPAIN

Serrania de Cuenca

Sierra de San Just

Estrela 1991

Coimbra

Serra da Estrela

Sierra de Gata

Sierra dos Gredos

Madrid

Meseta

Castellón de la Plana

Minorca

Mahón

Majorca

Palma

Tagus

Tagus

I b e r i a n   P e n i n s u l a

Júcar

Valencia

Ibiza

Balearic Islands

Ibiza

Lisbon

Badajoz

Guadiana

Albacete

Formentera

Guadiana

S i e r r a   M o r e n a

Sierra de Segura

Murcia

Alicante

Mediterranean Sea

Guadalquivir

Córdoba

Cartagena

Huelva

Seville

Genil

Sierra Nevada

Faro

Granada

Mulhacén 3482

Almería

capital city

Cape São Vicente

Cádiz

Málaga

major town

height of land (meters)

Gibraltar (U.K.)

3000
2000
1000
500
200
0

Strait of Gibraltar

mountain peak

Spain

KINGDOM OF SPAIN

S PAIN OCCUPIES THE IBERIAN PENINSULA between the Atlantic Ocean and the Mediterranean Sea. France is across its border to the north, Portugal to its west. The Balearic Islands in the Mediterranean and the Canary Islands off the Atlantic coast of Africa are administered by Spain.

GEOGRAPHY

Spain's Meseta, a huge central plateau of ancient rock, covers more than half the country. It is surrounded by mountains and has an average elevation about 600 m (2,000 ft).

The land

The Pyrenees mountains form a chain from the Mediterranean to the Bay of Biscay in the northeast. The Cantabrian Mountains run along the north coast before turning southwest toward northern Portugal and the Atlantic Ocean. The Sistema Ibérico runs from the Cantabrian mountains north of the Meseta toward the Mediterranean coast. Between it and the Pyrenees lies the Ebro River valley. The Meseta is drained by the rivers Douro and Tagus, which run toward the Atlantic on either side of the mountains of the Sistema Central.

The Sierra Morena range at the southern edge of the Meseta overlooks the Guadalquivir river, which enters the Atlantic at Cadiz. Farther south are the coastal ranges of the Sistema Penibética. Mulhacén, at 3,482 m (11,408 ft), is the highest peak in mainland Spain.

Climate

Spain's Mediterranean climate is modified by westerly North Atlantic winds and occasional warm dry air from the Sahara.

Nestling in the foothills (*above*) Small isolated villages, maize fields and lush pasture characterize the lower slopes of the rugged Cantabrian Mountains, a snowcapped range that extends across northwestern Spain for some 500 km (300 mi). This area is one of the most fertile in Spain.

Bleak mountain refuge (*right*) The Pyrenees on the northeastern border of Spain provide a safe haven for many animal species including the Spanish goat or ibex. The long distinctive horns of the male can grow up to 75 cm (30 in) long.

The northern zone, running from Galicia to the western Pyrenees, has a humid maritime climate. Winters are mild, with relatively cool summers. The central zone has a continental climate with very cold winters and extremely hot summers. Rainfall is low and sporadic. Along the eastern and southern coasts a Mediterranean climate predominates, with mild winters and warm, dry summers. The

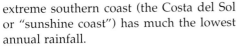

Parched and barren (*above*) The Sierra de los Filabres in Almeria in southeastern Spain is extremely arid. Lying in the rainshadow of the Sierra Nevada, it receives an average of only 114 mm (4.5 in) of rain a year, in sudden storms that cause devastating erosion.

extreme southern coast (the Costa del Sol or "sunshine coast") has much the lowest annual rainfall.

Plants and animals

There is deciduous forest in the humid north, and conifers in the south. Parts of the southern coast support subtropical vegetation. After many centuries of use, most of the Meseta is covered in a sparse scrubland of thorny trees and aromatic shrubs. In the arid southeast there are areas of grass steppe known as esparto. Isolated from Europe by both the Pyrenees and the sea, Spain retains some indigenous animals and reptiles now extinct elsewhere in Europe, and some typical of Africa, such as the cat-like genet. Spain lies on a major bird migration route between Europe and Africa.

SOCIETY

Spain has sharp regional differences in language, local custom and culinary tastes in spite of ethnic uniformity. The Basque country in the northwest and Catalonia in the northeast are most notably distinct.

History

The Iberian peninsula was inhabited as long as 500,000 years ago. Cave paintings of bison, horses and other animals – of which those at Altamira in Cantabria are the most famous – survive from between 25,000 and 10,000 years ago. Spain – the Tarshish of the Bible – was a center of trade about 1000 BC. Its metal deposits attracted Phoenicians, Greeks and Carthaginians to the eastern coast, and from the 9th century BC Celts to the north.

The Romans gradually conquered the peninsula after defeating the Carthaginian general Hannibal (247–182 BC). For nearly 600 years Spain was an integral part of the Roman empire. Cities, roads and aqueducts were built, and Latin became the dominant language throughout the country. In the 5th century AD, as the Roman empire began to weaken, Germanic peoples invaded Spain from the

---

NATIONAL DATA – SPAIN

Land area   499,542 sq km (192,874 sq mi)

Climate	Altitude m (ft)	Temperatures January °C(°F)	July °C(°F)	Annual precipitation mm (in)
Santander	64 (210)	10 (50)	19 (66)	1,095 (43)
Madrid	609 (1.998)	5 (42)	25 (76)	437 (17.2)
Seville	13 (43)	11 (52)	27 (81)	533 (20.9)

Major physical features   highest point: (mainland) Mulhacén 3,481 m (11,421 ft); (Canaries) Pico de Teide 3,715 m (12,188 ft); longest river: Tagus (part) 1,007 km (626 mi)

Population   (2006 est.) 40,397,842

Form of government   multiparty republic with two legislative houses

Armed forces   army 95,600; navy 19,455; air force 22,750

Largest cities   Madrid (capital – 3,092,759); Barcelona (1,578,546); Valencia (785,732); Sevilla (704,154); Zaragoza (638,799); Malaga (558,287)

Official language   Castillian Spanish

Ethnic composition   Spanish 44.9%; Catalan 28%; Galician 8.2%; Basque 5.5%; Aragonese 5%; Roma 2%; others 6.4%

Religious affiliations   Roman Catholic 92%; Muslim 0.5%; Protestant 0.3%; other 7.2%

Currency   1 euro (EUR) = 100 euro cents

Gross domestic product   (2006) U.S. $1.07 trillion

Gross domestic product per capita   (2006) U.S. $27,000

Life expectancy at birth   male 76.32 yr; female 83.2 yr

Major resources   coal, lignite, iron ore, copper, lead, zinc, uranium, pyrites, mercury, tungsten, magnesite, fluorspar, gypsum, sepiolite, kaólin, potash, hydropower, almonds, cereals, cork, cotton, fruit, fish, grapes/wine, olives, sugar cane, leather, timber, tin, tobacco, vegetables, tourism

the north. The Visigoths, the strongest of these, were converted to Christianity.

In 711 the Muslim governor of Tangier on the north coast of Africa conquered Spain, which became part of the Umayyad caliphate. Between the 9th and 13th centuries, Spain was a flourishing center of Islamic art, literature and science, especially astrology. Charlemagne (741–814), who expelled the Moors from France, led the recovery of Catalonia, which became an independent kingdom by the 9th century, though subject to invasions from Córdoba, the Muslim capital in Spain. A number of rival Christian kingdoms developed in the north. Castile and Aragón eventually emerged as leaders, and by the end of the 13th century only Granada in the south remained in Muslim hands. In 1479 the accession of Ferdinand of Aragon (1452–1506), who had married Isabella of Castile (1451–1504) 10 years earlier, united the two Christian kingdoms. With the fall of Granada in 1492 the country was unified. In that same year Christopher Columbus (1451–1506) crossed the Atlantic to the Americas and 170,000 Jews were expelled from Spain on the orders of the Spanish Inquisition, a religious court.

The Canary Islands

The Canary Islands, in the Atlantic Ocean about 95 km (60 mi) off the northwest coast of Africa, are an autonomous region of Spain, and are divided into two provinces. The western province of Tenerife includes the islands of Hierro, La Palma, Gomera and Tenerife itself. The eastern province of Gran Canaria is made up of Gran Canaria, Fuerteventura, Lanzarote and a number of islets.

The islands are the cones of extinct volcanoes and are mostly very steep, those of Tenerife being higher than those of Gran Canaria. The Pico de Teide, on Tenerife, at 3,718 m (12,195 ft) is the highest mountain on Spanish territory, and its summit is snow-covered from November to April. The climate is generally warm and dry. Rainfall is scanty, occurring mostly in the winter, and is heavier in the west than in the east.

The vegetation is often sparse, especially on the drier eastern islands, but the porous volcanic soils support a wide variety of unusual plants such as the dragon tree (*Dracaena draco*) and succulents such as *Euphorbia* and *Sempervivum*. A range of subtropical and temperate plants is found on the higher mountains of the west, with bananas, coffee, sugar cane, oranges and dates on the lower slopes giving way to wheat, barley, corn (maize), potatoes and grapes. Trees such as holly, olive and various species of laurel are succeeded by Canary pine and tree heaths near the cloud layer, and brooms and other low-growing shrubs dot the lava slopes above the treeline, about 1,830 m (6,000 ft).

The bird life is extremely rich, and includes the wild canaries that take their name from the islands. There are, however, no indigenous mammals,

though the Roman writer Pliny the Elder (23–79 AD) records that the islands (then the most westerly known point in the world) derived their name Canaria from the large dogs (*canes*) that were found there. These, like the sheep, pigs and goats that were present when the Spanish colonized the islands in the 15th century, must have been introduced by the original settlers from Africa, known as Guanches.

Although known to the Romans (who also called them the Fortunate Islands), the islands disappeared from history until the Portuguese rediscovered them in 1341. By the time Christopher Columbus (1451–1506) used them as a staging post for his first transatlantic voyage in 1492, they had been placed under Spanish control, and the Guanches subdued. The islands continued to act as a vital link

Tenerife's Pico de Teide (*above*), protected within a national park, rises to 3,718 m (12,195 ft). Although the upper slopes of pumice and lava are barren, the rich volcanic soil of the lower slopes supports many exotic plant species.

between Spain and its American possessions. In July 1936 Francisco Franco (1892–1975) launched the Nationalist uprising in a broadcast from the Canary Islands before flying to Spain.

Year-round visitors have made tourism the Canary Islands' chief source of revenue, though agriculture remains important. Bananas, potatoes and tomatoes are all grown for export. The main seaports and both international airports are located at the two provincial capitals: Santa Cruz on Tenerife and Las Palmas on Gran Canaria. There are universities at La Laguna on Tenerife and at Las Palmas.

The Moors of Granada were later given the choice of conversion or expulsion.

Columbus' voyages heralded a period of conquest and exploitation that established a Spanish empire in Central and South America and in Southeast Asia. Court life, art and literature flourished on the wealth that flowed in from the Americas over the next 200 years. The Spanish crown became increasingly involved in European affairs, acquiring parts of Italy and the vast Habsburg territories of Austria, Burgundy and the Netherlands.

Philip II (1527–98) acquired Portugal in 1580, but his attempt to invade England in 1588 ended in disaster with the defeat of the Spanish Armada. In 1648 Spain lost the United Provinces of the Netherlands after an 80-year war. Increasing resentment against Spain's European wars led the Catalans and the Portuguese to revolt against Spanish rule in 1640. Portugal formally regained its independence in 1668. Continuing war against France drained Spanish resources still further, and in 1714 Spain conceded the remainder of its European territories.

Spain lost a war against revolutionary France in 1794, and in 1808 the French emperor Napoleon I (1769–1821) installed his brother Joseph Bonaparte (1768–1844) as king. In the War of Independence (1808–14) the French were driven out and the Spanish Bourbon monarchy (founded in 1700 by Philip V [1684–1746], grandson of Louis XIV of France) was restored.

By 1826 the spread of nationalism cost Spain all of its mainland colonies in Central and South America. At home, political quarrels stifled effective government, and defeat in the Spanish–American War of 1898 resulted in the loss of Cuba, Puerto Rico and the Philippines to the United States. Alarm at a fast-growing popular movement calling for workers' rule led General Prima de Rivera (1870–1930) to seize power in 1921. His rule became increasingly authoritarian, and in 1930 Alfonso XIII (1886–1931) forced de Rivera to resign. A year later Alfonso was himself forced to abdicate after a landslide electoral victory in favor of the Republicans.

In 1936, after the Republicans were reelected, fears of a socialist revolution prompted a military uprising backed by the right-wing Nationalists. They then

Olive trees and wild flowers cover the plains of Spain's vast central plateau of La Mancha. It was here that the mythical knight Don Quixote set out on his chivalric adventures in the 17th-century novel by Miguel de Cervantes (1547–1616).

The Balearic Islands

The Balearic Islands lie in the Mediterranean Sea off the east coast of Spain. They form an autonomous region consisting of a single province. The larger island group to the east includes Majorca, Minorca and Cabrera, while Formentera and Ibiza make up the smaller group to the west.

Geologically, the islands – with their rich and varied landscapes – are a continuation of the Andalusian mountain chain in southwestern Spain. The highest mountains are in the northwest of Majorca, and the coastline, like that of Ibiza, is rugged. Southern Majorca, together with Minorca and Formentera, is much hillier. Winters are mild and summers generally hot. Rainfall is restricted to spring and autumn, with the mountains of Majorca receiving the greatest amounts. There is a marked drought between May and August. Most of the natural vegetation has long been cleared for farming, and garigue scrub is now predominant. A number of reptiles, including the Majorcan midwife toad, are unique to the islands.

The Balearics have a long history of occupation, and are rich in prehistoric remains. Phoenicians, Carthaginians and Romans all came to the Balearics, which derive their name from the Latin *balearii* meaning stone slingers –

the islanders were renowned for their skill with this weapon. The islands were later occupied by the Vandals and then by Moors. Annexed by the kingdom of Aragon in the 13th century, they became a great center of medieval cartography and navigation. In 1708 the British seized and occupied the port of Mahón on Minorca; it remained in British, and then French, hands until 1802.

Today Majorca is the most heavily populated of the islands, with most people living in the capital, Palma. The people show great ethnic and cultural diversity, reflecting the islands' long history of invasion. Castilian Spanish is commonly spoken in the towns, but Catalan remains the dialect of country districts.

The traditional agriculture of the islands mixed the growing of fodder crops and cereals with fruit such as almonds, apricots and figs in orchards. There is also some pastoralism, especially in Minorca. Farming remains an important activity, as does fishing. Sea salt is processed on Minorca, and there are some local manufacturers. However, tourism provides most of the islands' income. There are good sea links with Barcelona and international airports at Palma and Mahon.

Fiesta time (*below*) in Minorca, one of the Balearic Islands and a popular tourist resort. Boisterous crowds celebrate at Alayor, a city founded by the Spanish in 1304 after the islands had been seized from the Moors.

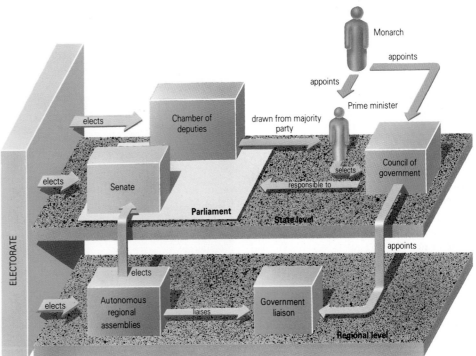

The Spanish system of government The parliament, called the Cortes Generales, consists of two chambers. The Congress of Deputies, the lower and more powerful chamber, has 350 members who are directly elected every four years by proportional representation. The 252 members of the Senate are elected (208) or appointed (44) from the autonomous regional and provincial councils. The prime minister is appointed by the king.

The Güell Park (*left*), built by the Catalan architect Antonio Gaudi (1852–1926) on a hillside in Barcelona, is now a park-playground with mosaic pagodas.

won the bitter civil war (1936–39), led by General Francisco Franco (1892–1975) and supported by the fascist regimes in Germany and Italy. Despite this support Spain remained neutral during World War II. General Franco was made head of state for life in 1947.

Under Franco individual, political, economic and cultural freedoms were severely constrained. Minority languages and customs were banned, fueling the growth of separatist movements in Catalonia and the Basque country. These became increasingly violent in demanding cultural and political freedom. In 1969 Juan Carlos de Borbón y Borbón (b. 1938), Alfonso XIII's grandson, was named as Franco's successor; he became king in 1975. Full democratic elections followed two years later, and a new democratic constitution in 1978. Spain joined the North Atlantic Treaty Organization (NATO) in 1982 and is a member of the European Union (EU).

Government

Under the 1978 constitution, the king is head of state and of the armed forces. He appoints, nominates and may dismiss the prime minister, who must also be approved by the two-chamber parliament, the Cortes Generales.

Spain, including the Balearic Islands and the Canary Islands, is divided into 17 autonomous regions. The Basque Country and Catalonia were given this status in 1980; the others were created by 1983. Each region has its own parliament and executive, with varying degrees of self-government. Below the regions, each of Spain's 47 provinces elects a delegated body of local municipal officers to liaise with central government. The municipalities, the smallest local governments, elect individual governing councils.

People

Castilian Spanish, a Romance language descended from Latin, is spoken throughout the country. In addition to regional dialects are three distinct languages – Galician (Gallego), in the rural northwest; Catalan, in Catalonia; and Basque – which are taught in local schools. Basque is closely associated with the independence movement and now has official status, with Spanish, in the Basque country.

The only sizable ethnic minority are the Gypsies, originally from the Indian subcontinent, who reached Spain in the 15th century. They retain their cultural distinctiveness and rarely marry outside the Gypsy community, but their music and flamenco dancing have strongly influenced Spanish culture.

Roman Catholicism is the predominant religion in Spain, but formal religious practice is declining, especially in the cities. Most non-Catholics in Spain are members of Protestant churches. Jews are the only other significant religious minority.

Spain's rich cultural heritage owes much to its diverse influences. Its Arab period can still be seen in architectural styles, ceramic decorative motifs and traditional music. The guitar, much featured in Spanish classical and folk music, was originally Arab.

From the 15th to 17th centuries Spain enjoyed a great flowering of art and literature. Its writers included the playwright Lope de Vega (1562–1635) and Miguel de Cervantes (1547–1616), author of *Don Quixote*, one of the masterpieces of European literature. Among Spain's great artists were El Greco (1541–1614), Diego Velázquez (1599–1660), Francisco de Goya (1746–1828), Pablo Picasso (1880–1973), Joan Miró (1893–1983) and Salvador Dali (1904–91).

ECONOMY

Spain's predominantly agricultural economy was transformed by very rapid economic growth after the mid 1950s – stimulated by the rationalization of major industries. Even so, imports normally exceed exports and tourism is vital to maintain Spain's balance of trade.

Agriculture
Spain's agricultural output has increased in recent years though there are ever fewer people working on the land. Many rural dwellers have migrated to the cities. In the south, traditional large estates owned by absentee landlords have now been broken up into smaller units or been mechanized for commercial farming. In some areas of the north, small inherited family farms have been combined into more profitable units. Soil erosion as a result of deforestation and overgrazing is a serious problem in some regions. Most arable land is on the Meseta.

Barley and wheat are the chief cereal crops, and fruit growing is widespread. Spain is the third largest producer of wine in Europe. Pulses are a staple crop on parts of the Meseta, and rice is cultivated along the east coast. Tobacco and cotton are also grown. Irrigated cultivation of vegetables and salad crops has increased enormously, especially along the south and east coasts. Olives are grown mostly for oil production. Almost one-fifth of Spain's land area is pasture. Cattle and dairying are limited to the north, while sheep are found throughout the Meseta,

with goats in areas of poorer pasture and pigs in the holm-oak woodlands of the southwest. Draft animals, particularly mules and donkeys, are being replaced by modern farm machinery.

Fisheries and forestry
Fishing is largely confined to the Atlantic coasts, with the main ports at Vigo and La Coruna in Galicia. The catch includes mainly sardines, anchovies, cod and hake but stocks are declining.

Spain is second only to Portugal as a producer of cork, but forestry is generally not well developed. There has been considerable reforestation in recent years to check soil erosion.

Intensively farmed family smallholdings (*above*) on green and verdant hillsides in the province of Galicia in northwestern Spain. These lush hills produce a wide variety of grains, fruit and vegetables, as well as 25 percent of Spain's timber.

Casks of sherry (*right*) piled high in the historic Domecq cellars at Jerez de la Frontera, the world-famous center of the sherry trade in Andalucia from which the drink takes its name. Sherry has been made here since Roman times.

Industry
Spain's coal-mining industry, once one of the largest in western Europe, experienced major cutbacks in the 1980s. Other mineral resources include mercury, iron ore and uranium. Spain has insignificant oil reserves and relies heavily on

imported petroleum. Nuclear power provides about a third of its energy. Periodic droughts make hydroelectricity an erratic source. Natural gas is extracted at Jacca in the northeast and in the Gulf of Cadiz. This supplies about 10 percent of Spain's energy needs.

Spain's industrial growth from the mid 1950s until the oil price crisis of the late 1970s was among the fastest in the world. Investment in new technology was centered in the industrialized area around Madrid. Heavy industry is concentrated in the north. Spain's largest steelworks is at Avilés in Asturias. Like the shipyards along the north coast and at Cadiz in the southwest, the steel industry suffered from the recession of the 1980s.

Barcelona is a major center for chemicals, especially fertilizers and dyeing, textiles and manufacturing. In the 1970s multinational companies such as Ford, General Motors, Peugeot, and Renault opened automobile plants in provincial towns in northern and central Spain. A large oil-refining industry is based at Algeciras in the far south, and Valencia has both ceramic and paper industries that date back to Moorish times.

Trade and tourism

Spain's chief trading partners are fellow EU countries, the United States and Saudi Arabia: its most important exports are food and agricultural products, transportation equipment, chemicals, plastics, textiles and clothing. Heavy reliance on imported petroleum has led to a significant visible trade deficit. However, this is offset by the revenue from the tourist industry. Tourism is now the main industry along the Mediterranean coast, and supports the construction industry as well as Spain's traditional handicraft industries, such as leather crafts, ceramics, glass and metalwork.

Transportation

Spain's road system, which radiates outward from the center at Madrid, dates from the 18th century. Since the 1960s the busiest highways have been widened and resurfaced, and a network of modern expressways has been constructed. But congestion remains a problem, particularly on the summer tourist routes.

The railroad network, constructed in the late 19th century, was extensively damaged in the civil war of 1936–39. In 1941 the government took over the railroads in order to rebuild them. Since modernization dating from 1964 many lines have been electrified and new rolling stock introduced. Madrid, Seville and Barcelona have subway systems.

The main shipping lines to the Balearic Islands, the Canary Islands, Ceuta and Melilla are state-owned, and there are important international sea links with South America and the United States. There are four major ports: Barcelona, Valencia, Bilbao and Huelva. Algeciras handles international container traffic.

There are two national airlines – Iberia and Aviaco – and a number of small regional airlines handling both freight and passengers. International airports include Madrid, Barcelona, Palma in the Balearic Islands and Las Palmas in the Canary Islands.

Media and communications

The 1978 constitution guarantees freedom of the press. With the end of censorship there was a great increase in magazines and newspapers: there are more than 100 daily newspapers, including a number of regional dailies.

Televisión Española, the national TV service, broadcasts two channels and there are three commercial TV channels as well as cable TV and numerous radio stations. The whole country, including the major islands, is covered by the modern telecommunications system.

Welfare and education

Healthcare is provided by state-funded hospitals and clinics, and by specialist surgical, maternity and psychiatric units. There are also hundreds of other hospitals run by the church, the Red Cross and private institutions. Life expectancies have risen considerably in the latter part of the 20th century.

A comprehensive welfare system offers sickness and maternity benefits, help with hospital and medical costs, and financial assistance in the event of disability, unemployment, retirement or death. Funds are mostly provided by the government and employers. The self-employed can also contribute to the system, or can opt to pay into private schemes.

The Spanish education system was reformed in 1970 and 1991. Pre-school for children age 3 to 5 is followed by compulsory basic education. After qualifying examinations, a three-year *bachillerato* program leads to vocational schools or to a one-year preparation course for university or advanced technical school. Spain has 40 state universities, four open universities and other universities which are privately owned.

Portugal

REPUBLIC OF PORTUGAL

PORTUGAL OCCUPIES THE WESTERN SIDE OF the Iberian Peninsula in southwest Europe, between the Atlantic Ocean and Spain. Its territory includes the Atlantic islands of Madeira and the Azores.

GEOGRAPHY

The country falls into two main geographical regions, roughly divided by the river Tagus. The lusher, more mountainous north contrasts with the drier lowland areas of the south.

NATIONAL DATA - PORTUGAL

Land area	91,951 sq km (35,493 sq mi)			

Climate	Altitude m (ft)	Temperatures January °C(°F)	July °C(°F)	Annual precipitation mm (in)
Lisbon	77 (252)	11 (51)	22 (71)	702 (27)

Major physical features highest point: (mainland) Estrela (Serra de Estrela) 1,991 m (6,532 ft); (Azores) Ponta do Pico 2,351 m (7,713 ft); longest river: Tagus (part) 1,007 km (626 mi)

Population (2006 est.) 10,605,870

Form of government multiparty republic with one legislative house

Armed forces army 26,700; navy 10,950; air force 7,250

Largest cities Lisbon (capital - 1,977,000); Porto (1,303,000)

Official language Portuguese

Ethnic composition Portuguese 91.9%; African 1.6%; Brazilian 1.4%; others 5.1%

Religious affiliations Roman Catholic 87.4%; other Christian 3.7%; Protestant 1.3%; others 1.1%; nonreligious 6.5%

Currency 1 euro (EUR) = 1 euro cent

Gross domestic product (2006) U.S. $203.1 billion

Gross domestic product per capita (2006) U.S. $19,100

Life expectancy at birth male 74.43 yr; female 81.2 yr

Major resources fish, cork, iron ore, copper, zinc, tin, tungsten, silver, gold, uranium, marble, clay, gypsum, salt, hydropower, cereals, citrus fruits, figs, vegetables, fish, grapes/wine, olives, rice, timber, tourism

The land

Portugal's northeast plateaus are an extension of the Spanish Meseta, divided by mountains, river valleys and deep gorges. Farther west, the land drops away toward the hills bordering the Atlantic coast. The inland plateaus extend south beyond the valley of the river Douro, falling away westward to a broad coastal plain.

At the southeast border of the plateau region, mountain chains form a barrier between northern and southern Portugal, becoming lower as they approach the coast. To their south the river Tagus runs from east to west, entering the Atlantic at the capital, Lisbon.

South of the Tagus are undulating hills and plains. The coastal lowlands spread inland into the uplands of Alto Alentejo in the east, which dwindle southward to form the low rolling hills of Baixo Alentejo. In the far south, the hills and scarps of the Algarve run from the Spanish border across to Cabo de São Vicente at Portugal's southwestern tip.

Climate, vegetation and wildlife

The Mediterranean climate is heavily influenced by sea and landscape. Winters are cool and wet, and summers on the coast are humid. Summers in the interior are hot and dry; drought is common in the south and east. The mountains of the northwest are barriers to clouds from the Atlantic and receive more than 1,000 mm (40 in) of rain a year. The plateaus of the northeast are drier and cold in winter.

The higher rainfall in the north supports forests. The south is less wooded, but the southern Algarve region is famous for its groves of carob, fig and almond trees. Olive trees are found on lower slopes throughout the country, and Alpine vegetation in the higher mountains. Portugal has a rich variety of mammal life, including wolves, wild boar and lynxes. Reptiles are also plentiful, especially lizards and chameleons in the south. Bird life, which includes some species found only on the Iberian Peninsula, is enriched by seasonal migrants en route to and from Africa.

Gulls circle the cliffs near Albufeira (*above*) one of many resort towns in the Algarve, Portugal's busiest tourist area. Once the Moorish kingdom of Al Gharb, the province offers holidaymakers a wonderful climate and good beaches.

SOCIETY

Portugal's history includes an overseas empire in which the Portuguese language and culture have left a significant mark. Macao, its last colony, was returned to China in 1999.

History

Little is known about Portugal's earliest inhabitants, the Iberians. The Celts began to enter through the Pyrenees about 1000 BC. The Lusitani were a Celticized people who gave their name to the Roman province in central Portugal. The modern name comes from the Roman fortress Portus Cale (now Oporto).

Following the breakup of the Roman empire in the 5th century AD, Portugal was contested between the Suebi, invading from the north, and the Visigoths

The Azores

The Azores are a scattered achipelago in the mid Atlantic made up of nine islands and many smaller islets. There are three main groups. The largest, São Miguel, is also the nearest to Portugal, lying about 1,200 km (750 mi) due west of Lisbon. Most of the population lives here and on the neighboring island of Santa Maria. Other islands include Terceira and Pico in the central group, with Flores and Corvo in the far west.

All the islands are volcanic in origin, and the archipelago remains volcanically active. The resulting landscape is varied and often spectacularly beautiful, with many crater lakes. The mild Atlantic climate produces almost no rain during the summer months. The vegetation is rich, but there are almost

An island archipelago of volcanic origin in the mid Atlantic, the Azores were first settled by the Portuguese in 1431. Two crater lakes – Verde and Azur – on the largest island of São Miguel typify the spectacular landscape.

no native trees, and wildlife consists mostly of introduced species.

The uninhabited Azores became a Portuguese colony in 1431. A part of metropolitan Portugal, the islands constitute three of its 22 administrative districts, with a greater degree of autonomy than those on the mainland.

The islands' main crops include tea, tobacco and a variety of tropical fruits. Wine, fish and handicrafts are the main exports. Tourism is an increasingly important source of revenue. There is little industry, but the main airport on Santa Maria is an important transatlantic refueling station.

already present as Roman auxiliaries. The country remained fragmented until it was overrun in 711 by Muslim Moors from northern Africa. Only the north remained in Christian hands. Reconquest was begun by Afonso Henriques (1109–85), who became king. His successors recaptured Portugal from the Moors, until in 1270 the present Portuguese frontier was defined; it is the oldest surviving border in Europe.

The new kingdom soon became involved in power struggles with the Spanish kingdom of Castile. In 1385 John

I of Castile (1358–90) invaded, but was routed at Aljubarotta. Under their own king John I, Portugal established political and trading links with England and Flanders, and conquered Ceuta in northern Africa.

From the 15th century, under the patronage of Prince Henry the Navigator (1394–1460), Portuguese sailors explored the Atlantic coast of Africa and sailed east to India, Indonesia and China. They also crossed the Atlantic to the Americas, where in 1500 Pedro Alvares Cabral (1467–1520) claimed Brazil for Portugal.

Madeira

Madeira is a small archipelago lying in the Atlantic about 1,000 km (600 mi) southwest of Portugal off the coast of Northern Africa. Formed by volcanoes, it consists of two islands and several uninhabited islets.

The island of Madeira is steep and mountainous, with deep valleys. There is little flat land except on the south coast around the capital, Funchal. Porto Santo to the northeast is smaller and flatter, and has only a tiny population. Madeira has a mild climate, with moderate rainfall in the winter, supporting lush vegetation. There are no native mammals, though the rare Mediterranean monk seal may be seen off the coast. There are 40 bird species and numerous insects.

Madeira was uninhabited until it was colonized by the Portuguese in the 15th century. The islands were discovered by João Gonçalves Zarco on an expedition sponsored by Prince Henry the Navigator. The name Madeira, meaning timber, refers to the dense forest that covered the island until it was cleared to provide fertile farmland for settlers.

The islands today constitute one of Portugal's 18 national administrative districts, but with a greater degree of autonomy. The economy depends mostly on the export of handicrafts, bananas, sugar and the famous Madeira wine, which is fortified with cane sugar and has a distinctive smoky taste. Tourism is increasingly important. Visitors are attracted by the spectacular scenery and relaxed lifestyle.

Traditional farming methods are practiced throughout much of northern Portugal. Maize, seen growing here, has been one of Portugal's chief crops since the late 15th century when it was brought back from the Americas by Christopher Columbus (1451–1506)

Due to intermarriage with the Castilian royal family, the throne passed to the Spanish king Philip II (1512–80) in 1580. Spain treated Portugal like a minor colony, fit only for exploitation. Portuguese South American possessions, threatened by the Dutch, were poorly defended. In 1640 the Portuguese revolted, supported by the French. John, Duke of Bragança (1603–56), became King John IV, but Portugal's independence was only recognized in 1668. Portugal's recovery was speeded by a new influx of wealth from its colonies, especially by Brazil's gold and diamond reserves.

The French emperor Napoleon I invaded Portugal in 1807. Portugal's ally Britain evacuated the Portuguese royal family to Brazil, and after a three-year campaign, British troops finally drove the French out of Portugal in 1811. The royal family remained in Brazil, which became a joint kingdom with Portugal. After a democratic revolt in 1820 a liberal constitution was introduced.

Two years later Brazil seceded from Portugal. An inconclusive civil war, known as the War of the Two Brothers (1826–34), marked the beginning of a long period of turmoil and decline. By the end of the 19th century, republicanism was gaining the upper hand. In 1908 King Charles (1863–1908) and his heir were assassinated. His successor Manuel II (1889–1932) was overthrown by a revolution in 1910. Portugal became a republic, initially under the presidency of the

writer Téofilo Braga (1843–1924), then in 1911, after an election, under Manuel José de Arriaga (1842–1917).

After continuing unrest, a right-wing army coup in 1926 brought Antonio de Oliveira Salazar (1889–1970), a professor of economics, to prominence. He restored the country's devastated finances, turning Portugal into a semi-fascist "corporative republic" under his personal, if discreet, dictatorship. Salazar refused to allow any progress toward independence in any of Portugal's African colonies. In 1968 Dr Marcelo Caetano (1906–80) replaced Salazar.

The African independence movements became increasingly hostile to Portugal, leading to guerrilla war. In 1974 General António de Spinola (b. 1910) was sacked for criticizing Portugal's colonial policy. His dismissal led to a left-wing coup. Spinola was made president, acceptable to all sides, with an equally broadbased cabinet. Within a year all the Portuguese colonies (excluding only tiny Macao in Asia) had gained independence. The Marxist-led Armed Forces Movement (MFA) became increasingly dominant. The minority Communist Party won control of labor organizations and began expropriating private land. Spinola was ousted in 1974. In the following year the MFA mounted a military coup, but this was suppressed, and a transitional constituent assembly was elected in 1975.

A new democratic constitution in 1976 led to national elections. General António

Ramalho Eanes (b. 1935) was chosen as president with Dr Mario Soares (b. 1924) as prime minister. He began to repair the economy, badly affected by the revolution, and Portugal gradually achieved greater stability. In January 1986 Portugal joined the European Community, now the European Union (EU), and later that year Dr Soares was elected Portugal's first civilian president since the 1926 coup.

Government

Portugal has been a republic since 1910, but has been truly democratic only since 1976. The president, as head of state, is chief of the armed forces. He is advised by the Council of the Revolution, which he leads, and by the Constitutional Commission; both bodies are less powerful under civilian rule. The president also

appoints a prime minister to form the Council of Ministers, or cabinet. The legislative parliament, the Assembly of the Republic, is elected every four years by proportional representation.

Portugal's traditional provinces have no official status. The 18 administrative districts each have their own governor. The Azores and Madeira districts have a greater degree of autonomy.

People
The Portuguese people are a homogeneous group of typically Mediterranean appearance. Immigrants from the former colonies live mainly in Lisbon and Oporto. Recently many Portuguese have emigrated, particularly from rural areas, to other western European countries.

ECONOMY

Portugal's economy suffered badly from the 1974 revolution, the African guerrilla wars and the loss of income from the colonies. Recovery, now underway, has been further encouraged by Portugal's membership of the EU.

Agriculture and fisheries
Portugal's predominantly agricultural economy has a relatively low output. In the north, small farm holdings limit the possibilities for modern methods. The large estates (*latifundia*) of the south were nationalized in 1975, but have not yet been transformed by modernization.

Large quantities of wheat and maize, the chief crops, still have to be imported to meet domestic needs. Fruit crops include figs and olives. Wine is the major agricultural export, along with tomatoes and cork. Timber and woodpulp are also exported. A commercial fishing industry has developed along the Atlantic coast.

Industry and trade
Portugal has few energy reserves. Gas and petroleum are imported; about one-third of electricity comes from hydroelectric plants. Small amounts of tungsten, copper pyrites and kaolin are exported.

Manufacturing is increasingly important. Heavy industry is concentrated in Lisbon's shipyards and has suffered from falling demand. Fish canning, tobacco processing, cork products, footwear and textiles, and paper are the main light industries and exports. Portugal's chief trading partners are within the EU. Revenue from tourism is growing fast, particularly in the Algarve.

Transportation and communications
The road network covers most of the country and is now being improved. The railroad system, linking all the main towns, has been extensively electrified.

Airlines serve the main airports at Portela de Savacem near Lisbon, Oporto, and Faro in the Algarve. Portugal has a large merchant fleet. Radio and television were nationalized in 1975; satellite television is received from elsewhere in Europe. Most daily newspapers are either owned or subsidized by the state. Censorship ended with the 1976 constitution.

Welfare and education
The state provides an extensive social welfare system. Further schemes are run by trade unions, professional associations and major private employers. There are both public and private hospitals some run by charitable and religious organizations. Free education is available in state schools, and there are some private schools. There are five universities.

Dependencies in the region

GIBRALTAR
UNITED KINGDOM

Gibraltar is a small, self-governing British colony, 6.5 sq km (2.5 sq mi) in extent, occupying the southern tip of Spain. Facing it across the Straits of Gibraltar is the Spanish enclave of Ceuta on the coast of northern Africa.

Gibraltar consists of a single steep limestone ridge of rock shaped like an arrowhead that rises to 425 m (1,394 ft). From the Mediterranean side it is almost unassailable. A low, sandy isthmus joins it to Spain. The Rock has many caves and fortifications from all periods of its history. There are no natural springs or rivers, so rainwater catchment reservoirs occupy a large area of the summit. There is almost no rainfall in the hot summer.

The name Gibraltar derives from the Arabic Jabr al-Tariq, or "Mount Tariq," after the Moorish emir Tariq ibn Ziyad, who captured it in 711. It remained under Muslim rule until 1462. Captured by the British in 1704, during the War of the Spanish Succession, in the 19th century it was developed as a major British naval base and dockyard. Gibraltar played an important strategic role during World Wars I and II, and today it remains an airbase and communications center.

Spain has never ceased to lay claim to it, but in 1967 Gibraltarians voted overwhelmingly to remain British. In early 2002 talks were held between Britain and Spain to resolve the issue and decided to conclude a deal by summer 2002. Joint sovereignty is one of the issues under discussion but any decision made will be determined by a Gibraltarian referendum.

The government of the colony is headed by a governor, as representative of the British sovereign, who appoints a council of ministers from the majority party in the 18-member assembly. Gibraltarians are a mixture of Moorish, Maltese, Spanish, Genoese, British and Jewish descent. Most speak Spanish and English, and most are Roman Catholic. They possess full British citizenship. About one-third of the population are resident foreigners, including Asians, Moroccans and British military personnel. Spaniards cross daily to work on the Rock.

The colony's economy depends largely on the extensive port facilities and on light manufacturing, banking and finance. All food is imported. Tourism is becoming increasingly important, and is served by the airport close to the Spanish border. There is good education and healthcare.

Italy and Greece

COUNTRIES IN THE REGION

ITALY · VATICAN CITY · SAN MARINO

GREECE · MALTA · CYPRUS

Bathed in Mediterranean sun (*left*), a Greek windmill on Antiparos in the Cyclades, a group of islands off the southeast coast of Greece. The harsh, arid, sun-baked landscape and whitewashed buildings are common to both Italy and Greece, where traditional ways of life are still followed in many rural areas.

capital city
major town

height of land (meters)

3000
2000
1000
500
200
0

▲ mountain peak

Italy

ITALIAN REPUBLIC

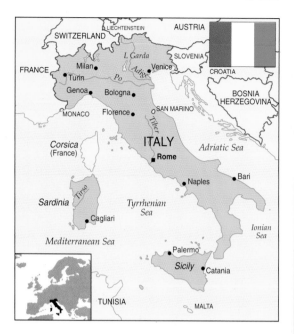

TALY HAS ONE OF THE OLDEST HISTORIES in Europe, second only to Greece. It is also one of Europe's youngest unified states. The rural landscape is spectacular, and the country's long cultural history has left a legacy of ruined temples and classical sculptures, medieval castles and fine Renaissance churches and painting.

NATIONAL DATA – ITALY

Land area	294,020 sq km (181,252 sq mi)			
Climate		**Temperatures**		**Annual**
	Altitude m (ft)	January °C(°F)	July °C(°F)	precipitation mm (in)
Rome	17 (56)	8 (46)	23 (73)	724 (28.5)

Major physical features highest point: Monte Rosa 3,505 m (11,500 ft); longest river: Po 652 km (405 mi); largest lake: Lake Garda 370 sq km (143 sq mi)

Population (2006 est.) 58,133,509

Form of government multiparty republic with two legislative houses

Armed forces army 112,000; navy 34,000; air force 45,152

Largest cities Rome (capital - 2,547,677); Milan (1,308,735); Naples (984,242); Turin (900,608); Palermo (670,820); Genoa (620,316); Bologna (373,473); Florence (366,901)

Official language Italian

Ethnic composition Italian 94.1%; Sardinian 2.7%; Rhaetian 1.3%; others 1.9%

Religious affiliations Roman Catholic 90%; Protestant and Jewish communities and growing Muslim immigrant community

Currency 1 euro (EUR) = 100 euro cents

Gross domestic product (2006) U.S. $1.727 trillion

Gross domestic product per capita (2006) U.S. $29,700

Life expectancy at birth male 76.88 yr; female 82.94 yr

Major resources coal, hydroelectric power, mercury, zinc, potash, marble, barite, asbestos, pumice, fluorospar, feldspar, pyrite (sulfur), natural gas and crude oil reserves, fish, cereals, citrus fruits, grapes/wine, iron ore, livestock, olives, potatoes, sugar, vegetables, tourism

GEOGRAPHY

To the north Italy is separated from France, Switzerland, Austria and Slovenia by the formidable barrier of the Alps. Southward, beyond the fertile expanse of the great northern plain, are the Apennines, which form a rocky backbone running from the Alps in the northwest down to Calabria in the toe of the peninsula. The islands of Sicily to the south and Sardinia to the west each have a distinct character of their own.

The land

The western and central Alps reach their highest points in Italy at Mont Blanc – more than 4,800 m (15,700 ft) – on the border with France, and at Monte Rosa – 4,634 m (15,203 ft) – on the Swiss border.

The northern Italian lakes – Maggiore, Como, Iseo and Garda, noted for their beauty – are water-filled valleys scooped out by ice age glaciers. Northeast of Lake Garda, the Dolomites, a breath-taking series of limestone peaks that are part of the eastern Alps, rise to the Marmolada – 3,342 m (10,965 ft).

The fertile soils of the north Italian plain were washed down by the melting of ice age glaciers. Over 320 km (200 mi) long and up to 200 km (125 mi) wide, more than half of its 70,000 sq km (27,000 sq mi) area is drained by the Po, Italy's longest river.

The long, rocky spine of the Apennines, south of the plain, begins in the Ligurian Alps. The mountains hug the northwest coast of the Italian Riviera, and then sweep in a bow shape down the

The Amalfi coast (*left*) as seen from the Villa Rufolo at Ravello, south of Naples. Many small resort towns such as Ravello shelter between the steep cliffs of this spectacular peninsula, where earth tremors are still a frequent occurrence.

The towering cliffs of Volterra (*above*), to the southeast of Pisa, form part of the Colline Metallifere, or Ore Mountains, rich in minerals and rising to a height of 1,050 m (3,400 ft). Farther north, the Apuan Alps are a famous source of high-quality marble.

length of the peninsula, reaching their highest point at Monte Corno in the Gran Sasso d'Italia northeast of Rome. On the western side, rivers such as the Arno and the Tiber run through a complex landscape of hills and plateaus. To the east, dozens of rivers create a landscape of parallel ridges and valleys running down to a narrow coastal plain. In the southeast a limestone platform forms the promontory of Gargano, and in the extreme south the Gulf of Taranto divides the "heel" (Apulia) from the "toe" (Calabria) of Italy.

Across the Strait of Messina lies the island of Sicily, the largest island in the Mediterranean. Almost entirely mountainous, with the highest ranges in the north, the eastern end is dominated by the active volcano of Etna. Italy lies on the boundary between the Eurasian and the African tectonic plates where earth tremors occur frequently. Other volcanoes in the region include Stromboli in the Lipari islands to the north of Sicily, and Vesuvius on the Bay of Naples, made famous by the eruption of 79 AD, which destroyed the city of Pompeii. The nearby Phlegraean Fields bubble with volcanic gases. Near Rome, the uplands between Monte Amiata and the Alban Hills are clearly volcanic: there are mud springs at Viterbo, and the area is dotted with crater lakes such as Bolsena and Nemi. Thermal springs rise as far north as the Euganean Hills near Padua. The whole region is subject to occasional severe earthquakes.

Off the western coast of Italy, the rugged mountains of the island of Sardinia are generally formed of older igneous and metamorphic rocks than those of the mainland.

Climate

In the Alps, the winter months are usually cold and can bring heavy snow, while summers are short and cool; but there is considerable variation. Lower temperatures on the higher mountain slopes tend to be offset by intense sunshine.

The northern plain, which is shielded from the Ligurian Sea by the Ligurian Alps, enjoys long, warm summers ideal for cultivation, though winters are short and often bitter. Farther south the climate is characteristically Mediterranean, with warm and sunny summers and mild winters. The high Apennines receive up to three times as much rainfall as the lowlands, where there may be prolonged summer drought.

Strong winter winds blow off the Ligurian Sea. The island of Sardinia endures the cold northerly maestrale, while the Italian Riviera is swept by the westerly *libeccio*. In spring the hot *sirocco* from Northern Africa brings hazy, stifling conditions to southern Italy.

Plants

Like much of Europe, the Italian lowlands, were once clothed in dense woodland, but little remains in the northern plains after 22 centuries of intensive farming. Cultivated crops dominate the landscape; even the poplars on the river margins have been planted. Farther south, cleared and abandoned woodland has reverted to a dense scrubland of aromatic plants (macchia) or, in stony areas, to garigue – open ground scattered with low shrubs and flowering annuals. In some areas, such as Calabria in southern Italy, the environment has been reduced to near desert.

Distinctive buildings of Pisa. The leaning tower, one of the world's most famous buildings, now tilts more than 5.2 m (17 ft) in contrast to the cathedral baptistery (foreground) and the Duomo. Signs of subsidence began appearing soon after construction began in 1174.

Rainfall in the Apennines encourages deciduous oak, chestnut and beech trees. On the western side and in Sardinia, evergreen oaks cling to lower mountain slopes up to about 600 m (2,000 ft). In the Alps very few plants can survive above 2,250 m (7,500 ft). Forest management has preserved stands of oak and sweet chestnut, which have also survived in the foothills. Around the lakes olive, cypress and laurel flourish.

Animals

Centuries of hunting and other human activities have reduced the number of animal species to be found in Italy, though the Alps shelter animals such as ermine, Alpine rabbits and mountain partridge; larger Alpine mammals include ibex and the increasingly rare Brown bear (both protected species) as well as chamois and roe deer. In the south of the country foxes and even wolves live in the mountains; on Sardinia there are fallow deer, mouflon sheep and wild boar. The variety of fish around the coasts include bluefin tuna, swordfish and white shark.

Pollution is a massive and growing problem. Along the Adriatic coast seasonal blooms of algae, nourished by the nutrients in the industrial and agricultural chemicals washed down by the rivers of the northern plain into the sea, have deprived other marine life of oxygen, killing thousands of fish, and covering tourist beaches in a stinking slime. However, new water purification plants, brought into operation in the 1990s, have begun to improve conditions in the sea.

SOCIETY

In the course of its long history Italy was at one time the power base of the mighty Roman empire, the birthplace of the Roman Catholic Church and the cradle of the Renaissance. It has a literary tradition going back almost unbroken to Roman times, and a long history of political disunity that continues to this day.

History

As early as the 8th century BC Italy was dominated by two major civilizations: the Greeks in Sicily and the south, and the

Etruscans in the center. Under Etruscan rule the small city of Rome became increasingly important as a power base and trading center, but in the late 6th century BC the native Latin-speaking people of the city – the Romans – overthrew the Etruscans and established Rome as a republic. Over the next five centuries the Romans waged war almost continuously on the neighboring peoples, imposing centralized power across the whole peninsula. Under Rome's first emperor, Augustus, they were able to gain control of the entire Mediterranean area.

At its height, in the 1st and 2nd centuries AD, the Roman empire stretched from Britain in the northwest to Egypt in the southeast, creating the cultural boundaries of modern Europe. However, by the 3rd century the western half of the empire had begun to decline economically, and the emperor Constantine established a new capital, Constantinople, in the east. The western empire became progressively weaker as waves of invaders from central Europe swept into Italy, and in 476 the last Roman emperor was deposed.

During the mid 6th century the country was invaded again, this time by the Eastern empire of Byzantium, which claimed Italy as an outlying province. Further Germanic invaders, the Lombards, conquered much of Italy in the late 5th and early 6th centuries, but were themselves overthrown by the Frankish king, Pepin. He gave the lands around Rome and Ravenna to the pope, the head of the Roman Catholic Church, marking the beginning of the Papal States.

In the 9th and 10th centuries Italy was nominally unified within the Holy Roman Empire, but the country remained fragmented into powerful city-states. Among the greatest in late medieval times were Florence and the maritime states of Genoa and Venice. From the 15th century the north came increasingly under the rule of Spain, and later of the Austrian empire. The south, particularly Sicily and Naples, also came under Spanish rule, but then changed hands several times between Spain, Austria and France. The country did not break free from foreign domination until the bloody wars of the Risorgimento, a cultural and political move-

ment for unification led by figures such as Giuseppe Garibaldi (1807–82) and the Piedmontese statesman Conte Camillo Benso di Cavour (1810–61). Victor Emmanuel II (1820–78), king of Sardinia and Piedmont, was installed (1861) as the first king of a reunited Italy. Venice joined in 1866, and Rome in 1870. However, the pope, having been granted sovereignty over the Vatican City, refused to recognize the new order. San Marino also remains an independent enclave.

In the late 19th and early 20th centuries Italy became a colonial power in its own right, occupying Libya, Italian Somaliland and Eritrea. In 1922 Benito Mussolini (1883–1945) seized power. Under his fascist dictatorship Italy expanded its empire to include the whole of Ethiopia between 1936 and 1945, and became an ally of Nazi Germany during World War II. In 1943 Mussolini lost support of the Fascist Council and was arrested by royal command. Though rescued and reinstated by the Germans as a puppet, he was murdered in 1945 trying to escape. The monarchy was discredited, and in 1946 it was abolished and a republic declared.

Florence – city of the Renaissance

During the Middle Ages, Florence – a town on the river Arno at the foot of the Apennines – emerged as one of the most powerful city-states in Italy, its wealth based on woolen cloth and international banking. Its merchant aristocracy supported social improvements ranging from street paving to schools, hospitals and one of Europe's foremost universities. This turbulent but exciting time produced many talents, notably the painter and architect Giotto (1266–1337) and two of the world's greatest writers, Dante Alighieri (1265–1321) and Giovanni Boccaccio (1313–75). The city entered a period of financial eclipse in the mid 14th century, but recovered under the powerful merchant princes of the Medici family, Cosimo (1389–1464), Piero (1416–69) and Lorenzo the Magnificent (1449–92), who ruled Florence by dominating its senate, the Signoria.

Directly or indirectly, the Medici were the patrons of a remarkable generation of artists, writers and

thinkers who have come to epitomize "Renaissance man" – widely talented creators who were deeply interested in all aspects of humanity and human knowledge. Leonardo da Vinci (1452–1519) – artist, inventor, writer – exemplified Renaissance man, as did Lorenzo de' Medici himself, a notable poet and naturalist. Other Renaissance artists included the architect Filippo Brunelleschi (1377–1446) and innumerable painters, among them the Della Robbia family; Tommaso Masaccio (1401–28); Sandro Botticelli (1445–1510); Raffaello Sanzio, known as Raphael (1483–1520); and Michelangelo Buonarroti (1475–1564).

Under the later Medici, Florentine arts declined, and by the 17th century they had stagnated altogether. Their influence, however, had spread far beyond Italy, and stagnation preserved many Renaissance treasures intact – to the benefit of the thriving tourist trade, which began in the 18th century. Modern Florence still owes its rich character and its distinctive skyline to the Renaissance churches, palaces and public buildings that are its living inheritance from the Medici.

Detail from "La Primavera (Spring)" by Sandro Botticelli (1445–1510), tempera on wood. Botticelli used line to depict both contour and movement, showing typical Renaissance passion for sculpturally solid classical figures bathed in light.

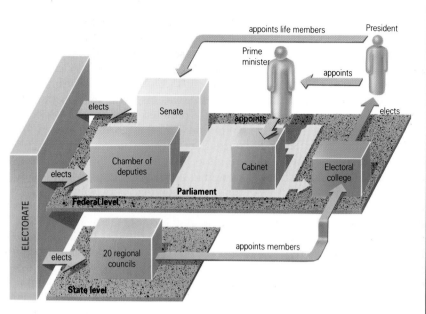

People

Centuries of foreign domination have left a distinctive mark on the Italian people and their dialects. In the south and Sicily, Greek and Arab influences are strong, while in the north people tend to be taller and fair-skinned. In the postwar period these distinctions became more blurred as many people from the rural areas of the south migrated to the industrialized cities of the north, such as Milan and Turin.

In the Valle d'Aosta and Trentino-Alto Adige (South Tyrol) regions – isolated enclaves in the high Alps on the northern borders – there are long-established French- and German-speaking communities. The Alto Adige only became part of Italy after World War I, as a reward for victory over Austria, and many of its inhabitants resented the transfer of authority. There are also significant ethnic minorities from neighboring states such as Albania and Slovenia.

Each of the former independent states boasts a distinctive character and dialect, but the Tuscan form of the Italian language, which was used by the 14th-century poet Dante Alighieri in his masterpiece the *Divine Comedy*, has become the literary standard. The Roman Catholic Church has played a major part in shaping the country. Until 1870 part of the country was under direct papal rule. More than 90 percent of Italians are Catholic, though in the postwar period the authority of the church has declined, particularly over controversial moral issues such as divorce, contraception and abortion. Religious regard for the sanctity of family life is still important, especially in rural communities.

The Italian Renaissance of the 14th to 16th centuries created an astonishing

Italy's government (*above*) The president, elected by parliament and the regional councils, has little actual power. A 1997 parliamentary commission recommended great changes to the constitution including a directly elected president.

Mount Etna (*right*) Europe's highest active volcano at 3,323 m (10,902 ft) looms over eastern Sicily. It remains snow-capped for six months each year and at night the glow of molten lava from its crater can be seen for many miles. It has been active for at least the last 2,500 years.

wealth of painting and sculpture, as well as a school of music that dominated Europe throughout this time. The artists of Florence in particular turned to the art of classical Greece and Rome for inspiration. It was also a period that saw the growth of skepticism and freethinking, belief in the individual and new discoveries in science and mathematics.

Opera is an Italian invention: it is still a national passion today and, like the rest of the country's magnificent heritage, is a potent tourist attraction. The richness of Italian cooking has been equally influential in more recent years. Widespread emigration has established Italian cooking throughout the world. In the growing fast-food culture the Italian pizza, introduced into the United States by Italian immigrants, has been the hamburger's greatest rival.

Government

The present structure of Italian government was adopted in the Republican constitution of 1947. The country is divided into 20 autonomous regions, each with its own legislature and executive council headed by a regional president.

The national parliament consists of a lower chamber of deputies with 630 members elected for a five-year term, and an upper house of 315 senators elected by region for five-year terms. There are also 11 life senators appointed by the president, who is elected by both chambers for a seven-year term.

There were many changes among political parties in the 1990s: the Christian Democrats were devastated by scandal and reformed into the Italian Popular Party (PPI); the Communists split into the Democratic Party of the Left (DS) and the Reconstructed Communism Party (RC); and the centrist Democratic Union for the Republic (UDR) was formed in 1994. Three new parties, Forza Italia, the Northern League and the National Alliance caused ripples of alarm across Europe when they took a majority of seats at the 1994 general election, espousing neo-fascist sentiments. Proportional representation and fierce regional self-interest have contributed to a splintered political scene with many coalition governments – 49 between 1945 and 1991. There were two attempts to abolish proportional representation in the 1990s by referendum,

but low turnouts led to no mandate.

Activities of the Mafia have been a major challenge to governments. A determined effort was made to combat the Mafia in the 1990s, following the murders of some members of the judiciary.

ECONOMY

Italy's dependence on imports for food and energy has created a large and almost permanent balance-of-trade deficit and a weak currency. However, despite a huge public deficit, industrial unrest, high wage settlements and high inflation, the country had experienced a boom economy and a thriving manufacturing industry until the beginning of the 1990s, when it was hit hard by the world recession. Recovery was slow and tough economic policies were pursued by the government in order to meet European Union criteria that would allow Italy to achieve European monetary union in 1999.

Agriculture

Despite a long farming tradition, Italy is a massive net importer of agricultural pro-

Small family-run farms (*below*) are a feature of the Italian landscape, and they employ 6.6 percent of the workforce. Italy is one of Europe's major producers of fruit, vegetables, wine and olive oil. About 40 percent of the land is arable.

ducts. Although some 65 percent of available land is still farmed, much of it is in hilly or mountainous country. The fertile northern plain accounts for roughly 40 percent of total production, almost as much as the whole of the south. Of more than 3 million farms, 94 percent are very small; one family will own several holdings, and part-time farming is common. Typical livestock includes cattle, pigs, chickens, sheep and goats. Traditional exports (wine, olives and fruit) remain important despite growing competition within Europe.

Forests account for some 25 percent of usable land, but the quality of the timber is generally low and none is exported. The small fishing industry relies largely on individual enterprise. Pollution of the Adriatic from industrial agricultural chemicals is now a serious threat to the fishing industry.

Industry

There are few mineral resources in Italy; only mercury, sulfur, rock salt and marble are present in significant quantities. However, other minerals are also extracted, notably in Sardinia and Sicily. Oil deposits in Sicily provide some 6 percent of Italy's energy needs. Reserves of iron ore, coal and natural gas are too small to meet national requirements, and up to 80 percent of the country's energy is imported. However, Italy earns valuable revenue by processing imported metals and refining imported oils.

The 1981 National Energy Plan proposed energy conservation, a nuclear power program, increased use of gas and the conversion of oil-fired power stations to coal. Implementing these ideas has proved slow and laborious, particularly as the construction of nuclear power complexes (in Piedmont, Lombardy, Latium and Apulia) has been blocked by both local governments and environmental protection movements.

In this mountainous country hydroelectricity is an obvious potential source of power. More than half the power requirements of northern Italy are met by installations near lakes Como, Maggiore and Garda, and there are similar schemes in Sardinia. Geothermal power sources are also being tapped while a major natural gas pipeline runs from Algeria.

Manufacturing industry, now involving more than one-fifth of Italy's labor force, is largely concentrated in the north. FIAT (for Fabbrica Italiani Automobile Torino) in Turin is one of Europe's largest manufacturers of automobiles. Fashion

items, especially shoes and clothes, are important exports, as are chemicals for agriculture and industry. Until the late 1990s state involvement in industry was the highest in the European Union, with three state holding companies. The Instituto per la Ricostruzione Industriale (IRI) is chiefly involved in mechanical products, steel, banking and airlines; the Ente Nazionale Idrocarburi (ENI) in hydrocarbons; and the Ente Finanzario per l'Industria Meccanica (EFIM) in light engineering.

Tourism is an important source of revenue, providing employment and an influx of foreign currency. Each year, Italy receives over 20 million tourists.

Trade and commerce
Trade and manufacturing industry are major contributors to the economy, closely followed by services. Italy's chief trading partners are Germany, France, the United States, the United Kingdom, the Netherlands, Switzerland and Belgium. Imports include machinery and transportation equipment, precision machinery, crude petroleum, chemicals and chemical products, food and live animals, and metal and semiprocessed metal.

Transportation and communications
Italy's road network, the origins of which date back to Roman times, is the heart of its communication system, covering some 300,000 km (186,000 mi). To date the IRI has built some 7,000 km (4,500 mi) of autostradas (toll expressways). Other roads are well maintained but cannot

The Galleria in Milan (*below*) is a bustling arcade of shops and cafés. Spurred by a huge postwar influx of immigrants from the impoverished south, Milan developed into Italy's financial, manufacturing and commercial capital.

support fast-moving traffic. There are many road passes over the Alps, but most are closed during the winter; the all-weather St Bernard and Mont Blanc tunnels keep traffic flowing at these times.

The part state-run railroad, Ferrovie dello Stato, is being modernized. More than half the 16,000 km (10,000 mi) of track are now electrified, and more investment is being made in the system. Main routes run parallel to the autostradas, and in the Alps railroad traffic passes through the Mont Cenis, Simplon, St Gotthard and Brenner passes. Rome has a clean and efficient subway system.

Apart from the Po, most of Italy's rivers are unsuitable for navigation. The Italian merchant fleet carries less than one-third

A mecca for sunlovers Rows of orderly sun umbrellas such as these at Sorrento, south of Naples, are a common sight on Italy's long, sandy beaches. Each year some 20 million tourists visit the country, providing a vital contribution to the economy.

of the country's external trade, creating a massive balance-of-payments deficit on maritime freight. The state, a major shareholder in the merchant fleet, has responded with a restructuring program that includes the purchase of new vessels. Principal services are within the Mediterranean and to North America and Australia. Genoa, Naples, Venice, Trieste and Leghorn are the chief ports.

The national airline, Alitalia, flies to most major cities worldwide. Italy has over 30 airports, the largest being Leonardo da Vinci, 24 km (15 mi) southwest of Rome.

There are nearly 100 daily newspapers; the most influential are *Corriere della Sera*, published in Milan, and *Repubblica*, published in Rome. In addition to the two state TV channels there are several hundred private radio and television channels appealing to local audiences.

Health and welfare

Since 1980 Italians have enjoyed a comprehensive national health service that aims to provide free medical care for all citizens. However, minimum charges are made for essential medicines, medical examinations and hospital treatment. A unified medical insurance plan is also being put into operation. Welfare includes unemployment and disability benefits, as well as retirement pensions and family allowances, and there is a scheme for industrial injuries.

Education is compulsory between the ages of 6 and 14, and there is a standardized national curriculum. Private schools, often run by the church, flourish alongside the public elementary schools (ages 6–11) and secondary schools (ages 11–14). Preschool facilities (ages 3–6) are excellent. Further education and preparation for university is available at a wide variety of institutions, and although scholarships or grants are hard to obtain, recent legislation aims to improve this situation. Standards of literacy are high.

Vatican City

ALTHOUGH COMPLETELY SURROUNDED BY the city of Rome, the Vatican City is an independent state – the smallest in the world. It lies on the west bank of the river Tiber, its medieval walls cutting it off from the city beyond – except at St Peter's Square on its southeastern flank.

The Vatican City is the last remnant of the Papal States, the great temporal domain that successive popes had built up around Rome over the centuries. The papacy refused to recognize the incorporation of these lands into the newly independent Italian state in 1860 and 1870. In 1929, however, it finally concluded the Lateran Treaty with the government of Mussolini, ceding all claims to the papal territories in return for sovereignty over the Vatican City.

Within its walls the pope has absolute legislative, executive and judicial powers, exercised in practice through a commission of five cardinals and an appointed governor and council. The Vatican also has its own army, the Swiss Guard – about a hundred soldiers, whose 16th-century parade uniforms belie their responsibility for papal security.

The Vatican operates its own radio station, telephone and postal systems independently of the Italian state. It has its own banking system, issuing its own coinage and having no exchange controls. This is a powerful advantage in the Vatican's complex involvements in international finance, carried out chiefly through its merchant bank, L'Istituto per le Opere di Religione.

All essential supplies and energy are imported and the Vatican's extensive publishing operations are the nearest thing to an export industry. Its products are sold all over the world, and its daily newspaper, *L'Osservatore Romano*, has influence far beyond the city boundaries.

Tourism is also important. The city attracts great numbers of visitors from all over the world, both for its religious significance and for its rich cultural legacy. Pilgrims flock to St Peter's basilica, until recently the world's largest church, to celebrate the great festivals of Christmas and Easter. The church dominates the main public entrance to the city. A church stood on this site in the 4th century, but the building took its present shape in the 16th and 17th centuries when it was embellished by some of the greatest artists and sculptors of the Renaissance, among them Raphael, Michelangelo and Gian Lorenzo Bernini (1598–1680). The most famous works of art in the Vatican City are probably the frescoes on the walls and ceiling of the Sistine Chapel, the personal chapel of the popes in the Vatican Palace. These include priceless works by both Michelangelo and Raphael.

The Vatican also has jurisdiction over a number of buildings elsewhere in Rome, including the churches of St John Lateran and St Paul Outside the Walls, the largest church in the city after St Peter's; and the pope's summer palace at Castel Gandolfo, about 20 km (12 mi) southeast of Rome.

St Peter's Square, the magnificent keyhole-shaped colonnade in the Vatican, seen from the dome of St Peter's basilica. Many great artists contributed to rebuilding St Peter's in the 16th century, a grandiose project designed to be a marvel of Christendom. The cost was so great that it led to abuses in the sale of indulgences to raise funds, precipitating an attack on the papal system by the German reformer Martin Luther (1488–1546) and leading to the Reformation.

NATIONAL DATA – VATICAN CITY			
Land area 0.44 sq km (0.17 sq mi)			
Climate	Temperatures		Annual
	January °C(°F)	July °C(°F)	precipitation mm (in)
	8 (46)	23 (73)	724 (28.5)
Major physical features none			
Population (2004 est.) 780			
Form of government city-state with one appointed absolute ruler – the pope			
Armed forces Swiss Guard 100			
Official language Italian			
Official religion Roman Catholicism			
Religious affiliations Roman Catholic 100%			
Currency 1 Vatican lira (VL) = 1 euro (EUR) = 100 euro cents			
Major resources tourism			

San Marino

REPUBLIC OF SAN MARINO

THE MOST SERENE REPUBLIC OF SAN MARINO, smallest republic in Europe, second smallest in the world, is an enclave in the Apennine mountains of northern Italy. The towers and battlements of the Rocca Fortress in San Marino, the capital, are perched on the rugged western slope of Mount Titano, while woods and pasture-land cover the foothills. This ancient state largely owes its survival to its rugged and isolated terrain, which helped preserve it from predatory neighbors throughout the turbulent Renaissance and after.

Legends trace San Marino's origins to the 4th century AD, when St Marinus, a Dalmatian stonemason, was thought to have settled there with his Christian followers. In recorded history, it first appears in the 12th century as a self-governing commune.

In 1600 a constitution was adopted that remains broadly in force today (although in the 18th century government was shared between a few leading families). The Congress of Vienna (1815), which met to decide the future shape of Europe after the Napoleonic wars, agreed to preserve San Marino's independence. In the mid 19th century the leaders of the Italian unification movement, among them Giuseppe Garibaldi (1807–82) took refuge in San Marino, and this helped it to conclude favorable treaties with the new Kingdom of Italy after 1861. This newfound security, along with the development of modern roads and communications, helped reduce San Marino's isolation. Although many of its people have emigrated, chiefly to Italy and the United States, a large number of Italians have also moved to San Marino and settled there. Italian is the official language, though a much older dialect is also still spoken.

The mountainous landscape, ancient fortifications and unspoilt countryside now attract increasing numbers of tourists, which enables the republic to survive as an independent state. Until this century agriculture, well suited to the mild Mediterranean climate, was the main occupation. Cattle are reared on mountain pastures, and the lower slopes are planted with vines and cereals. Until the 1940s there was no industry except stone quarries (now exhausted) on Mount Titano. Since World War II various light manufacturing industries have been introduced, though tourism is quickly overtaking them as a source of revenue. San Marino is now investing in its own resort hotel facilities. Related industries, ranging from traditional handicrafts to the republic's own finely printed postage stamps, have also benefited.

An extensive welfare system is provided; many San Marino citizens are employed on state concerns, and assisted with home ownership. Free education is provided to the age of 14, and there are grants for further study outside the republic. Government is headed by a two-house parliament. The Great and General Council are elected by universal suffrage at five-year intervals; every six months the Council elects the State Congress, a council of executive ministers, from among its own members, and two captains-regent to head it. The largest political party is traditionally communist, but moderate in the Italian tradition.

The Rocca Tower fortress dominates San Marino, the triple-walled capital of Europe's smallest republic, self-governed since the 12th century. Its first known settlers were a group of Christians led by St Marinus seeking refuge from persecution in the 4th century.

NATIONAL DATA – SAN MARINO

Land area	61 sq km (24 sq mi)
Climate	mediterranean
Major physical features	highest point: Monte Titano (capital on its slopes) 739 m (2,424 ft)
Population	(2006 est.) 29,251
Form of government	multiparty republic with one legislative house
Capital city	San Marino (2,294)
Official language	Italian
Ethnic composition	Sammarinesi 84.9%; Italian 14.6%; others 0.5%
Religious affiliations	Roman Catholic 95.2%; nonreligious 3%; others 1.8%
Currency	1 euro (EUR) = 100 euro cents
Gross domestic product	(2001) U.S. $940 million
Gross domestic product per capita	(2001) U.S. $34,600
Life expectancy at birth	male 78.23 yr; female 85.5 yr
Major resources	building stone, cattle, dairy products, olives, vines, wheat, tourism

Greece

HELLENIC REPUBLIC

THE GREAT FLOWERING OF ART, LITERATURE and ideas that took place in the city-states of Greece between the 8th and 5th centuries BC has been a major influence on the development of European culture and civilization. Today the country is renowned throughout the world for its rich cultural heritage – particularly its architecture and sculpture – as well as its great natural beauty.

NATIONAL DATA - GREECE

Land area	130,800 sq km (50,502 sq mi)			
Climate		Temperatures		Annual
	Altitude m (ft)	January °C(°F)	July °C(°F)	precipitation mm (in)
Athens	107 (351)	9 (48)	28 (82)	424 (16.6)

Major physical features highest point: Olympus 2,917 m (9,570 ft); longest rivers: Vardar (part) 388 km (241 mi) and Aliàkmon 314 km (195 mi)

Population (2006 est.) 10,688,058

Form of government multiparty republic with one legislative house

Armed forces army 110,000; navy 19,250; air force 23,000

Largest cities Athens (capital - 745,514); Salonika (363,987); Piraeus (175,697); Pátrai (161,114)

Official language Greek

Ethnic composition Greek 98%; other 2%

Religious affiliations Greek Orthodox 98%; Muslim 1.3%; other 0.7%

Currency 1 euro (EUR) = 100 euro cents

Gross domestic product (2005) U.S. $236.8 billion

Gross domestic product per capita (2005) U.S. $22,200

Life expectancy at birth male 76.72 yr; female 81.91 yr

Major resources lignite, petroleum, iron ore, bauxite, lead, zinc, nickel, magnesite, marble, salt, hydropower potential, barytes, chrome, cotton, fruit, vegetables, grapes/wine, livestock, olives, sugar beet, tobacco, wheat, tourism

GEOGRAPHY

Greece is a land of peninsulas and island chains formed by parallel mountain ranges that were flooded by the rising levels of the Mediterranean. Its coastline is deeply indented with bays and inlets. The whole region is geologically unstable and experiences frequent tremors.

The land
Mainland Greece forms the southern end of the Balkan peninsula. The Pindus mountains in the northwest extend from the Dinaric Alps of the Former Yugoslav Republic of Macedonia and Albania. In the far north, Greek Macedonia and Thrace are bounded by the Rhodope

Stripped to the bone The thin, rocky soils of the Eurotas valley and Taygetus mountains in the southern Peloponnese are typical of much of Greece. Clearance of the natural vegetation and overgrazing by sheep and goats have caused widespread soil erosion

mountains on the Bulgarian border and the Maritsa river on the Turkish border.

East of the Pindus the lowland plain of Thessaly is bounded on the north by mountains. The Othrys range runs south-east into the Aegean Sea and reappears to form the northern part of Euboea, a long, narrow island parallel to the coast.

To the south is the Peloponnese, separated from the northern mainland by the Corinth Canal, which links the Aegean and Ionian seas. The mountainous plateau of the Peloponnese ends

Peaceful scene on the tip of a volcano Thira, the southernmost island of the Cyclades in the Aegean, is the remnant of a huge volcanic eruption in about 1500 BC. It consists largely of lava and pumice, the latter being the island's major export.

Balkans. An icy wind, the *boreas*, brings snow to the highest mountains. In the south, warm winds make mild winters. Rainfall is most frequent during fall and winter, and heaviest on the western coastal ranges. Summers are hot and dry with heatwaves. Sea breezes may fan the coast, but hot dry winds called *meltemi* intensify the heat in Crete and the southern islands.

Plants and animals
Thousands of years of clearing and grazing by sheep and goats have reduced the original forest cover to one-fifth of the land area, mostly in the north. Maquis scrubland predominates in cleared lowland areas and throughout the Peloponnese. Flowering plants, including crocuses and anemones, make a brilliant display in spring. The Ionian islands of Corfu and Levkas receive more rain and have lush vegetation.

The northern forests shelter some large mammals that have now virtually vanished from the rest of Europe, such as wolves, brown bears and lynxes. There is a wide variety of reptiles in the south including chameleons. Turtles are endangered by tourism and pollution on the beaches where they lay their eggs.

in four rocky peninsulas, the longest of which, the Taygetus, arcs southeast to form the island chain of Kithira, Crete and Rhodes.

More than 2,000 islands (less than 200 of which are inhabited) account for one-fifth of the total land area. Crete, in the southern Aegean, is the largest. The Ionian islands, west of the mainland, include Corfu, Levkas, Cephalonia and Zakinthos. The two largest clusters are the Cyclades in the south Aegean and, to their east, the 12 Dodecanese.

Climate
The Mediterranean climate has marked regional variations. In the northeast, cold winter air is drawn in from the eastern

SOCIETY

Greeks have a strong national pride and identity derived from their classical heritage. Influenced by later Mediterranean civilizations, present-day Greek life is particularly rich and vital.

History
Between 2000 and 1500 BC Crete was the center of the Minoan civilization, usually considered Europe's earliest. By the 16th century BC its influence passed to Mycenae in the Peloponnese, which in turn was overrun about 1200 BC by Dorian warriors from the north, who had developed the use of iron weapons. The Greek city-states may have originated at this time as villages joined for self-defense within the walls of a citadel (*acropolis*).

These city-states became more prosperous; among them Athens established trading links and colonies around the Mediterranean. Literature, architecture, politics, art and philosophy were vigorously practiced. This rich civilization reached its zenith in the 5th century BC, but rivalry between the leading city-states eventually weakened them, and they were taken over by Philip II of Macedonia (382–336 BC). His son, Alexander the Great (356–323 BC), spread Greek culture across Asia to northern India, but after his death his empire was divided. Internal struggles continued to weaken the Greek city-states and by the end of the 3rd century BC Greece had succumbed to the rising power of Rome.

The Romans admired Greek art and scholarship but allowed Greece to fall into economic decline. After the 4th century AD, Greece was part of the eastern Byzantine empire ruled from Constantinople. The Greek language and Eastern Orthodox Church remained unifying forces in spite of successive invasions. From the 11th century the Byzantine emperors were unable to retain control of the Peloponnese, where Norman and French adventurers established feudal kingdoms. Venice controlled many Ionian and Aegean islands. By 1460, following the fall of Constantinople, the Turkish Ottoman empire absorbed nearly all of Greece. The Ionian islands remained in Venetian hands until 1797. During the 18th and 19th centuries, the Orthodox Church became a focus for growing nationalism. In 1830, after a bitter struggle, an independent Greek state was established in the Peloponnese, central Greece and the Aegean. Added to these were the Ionian islands in 1863, Thessaly in 1881, Epirus and Macedonia in 1913, and Thrace in 1918. After an unsuccessful attempt to annex Anatolia from Turkey after World War I, the Greek community in Turkey was expelled.

Occupation by German forces between 1941 and 1944 was followed by civil war. By the time monarchist forces defeated the communists in 1949, 10 percent of the Greek population had died in battle or of starvation. A military coup in 1967 overthrew the monarchy; a new republic proclaimed in 1973 lasted only a year. The ruling military clique was almost certainly involved in the overthrow of the government of Cyprus in 1974. Under the constitution of 1975, democratic elections were restored and Andreas Papandreou (1919–96) was elected in 1981 as Greece's first socialist prime minister. His popularity waned during the 1980s, and he fell from office in 1989. The socialist PASOK party won the 1993 general elections and were re-elected in 1996 and 2000.

People

The Greek people form a uniform majority. There are also significant Macedonian, Turkish and Albanian minorities. Almost all Greeks are active members of the Greek Orthodox Church. Although Greece is rapidly becoming an urban society, most Greeks feel strong attachment to their rural roots. Strict marriage and divorce laws and the position of women have been improved by recent legislation.

Government

Greece is governed by a single-chamber parliament of 300 members. The leader of the majority party becomes prime minister and head of government. The president is elected by parliament for a five-year term and is the supreme commander of the armed forces. Governments are elected for a four-year term, and all Greeks who are over the age of 18 are required to vote.

Greece's 13 administrative regions are divided into 51 departments. Mayors and councils are elected locally, but are funded by the national government.

Greece is a member of the North Atlantic Treaty Organization (NATO) and also a member of the European Union (EU). Greece's continuing involvement in Cyprus continues to cause difficult relations with neighboring Turkey.

ECONOMY

Greece is the poorest country in the EU. Services (especially tourism, banking and shipping) are the most profitable sector and do much to relieve a continuing balance-of-payments problem.

Agriculture

Agriculture employs 18 percent of the workforce, but productivity is poor. Only about one-quarter of the land can be cultivated. Many farmers still rely on draft animals. A wide variety of crops includes cotton, tobacco, olives and citrus fruits. Sheep and goats are raised throughout the country. Fishing, though modernized, remains a minor industry. Greece has benefited from EU subsidies to agriculture.

Industry

Greece has some mineral resources, but most raw materials must be imported, including 75 percent of its fuel. New plans for increased energy production focus on hydroelectricity and converting oil-burning power stations to locally

available lignite and peat. Geothermal, solar and wind power are also utilized. Most of the country's industry is based in or near Athens; Salonika in the northeast is the only other significant industrial city. Manufactured products include cigarettes, textiles, finished clothes, footwear, cement and fertilizer. The Greek shipping industry is one of the largest in the world and accounts for more than 30 percent of the income derived from services. However, rising fuel prices and demands for higher wages are affecting profits. Tourism provides foreign exchange as well as jobs in the service and construction industries.

Trade and commerce

Greek trade has expanded within the EU, but low-value exports and high-value imports have created a steadily growing

A hard living (*above*) Some 90 percent of Greece's farms are small to medium sized, typically 5 ha (12 acres), due to fragmentation by inheritance laws, poor farming methods and the loss of male workers to the cities. In addition, soils are rocky and rainfall is sparse.

trade deficit. Apart from crude oil, major imports include meat, milk, iron, steel, automobiles and pharmaceutical products. Important exports include textiles and handicrafts, tobacco, olive oil, tomato puree, petroleum products and minerals.

Transportation and communications

Greece's geography accounts for the importance of shipping and for the lack of surfaced roads well into the 20th century. Modern highways now link Athens with other centers. There is an extensive network of rural bus routes on the mainland and in Crete, and ferries between the mainland and the islands. The railroad system, though modernized, is confined to the mainland. The state-owned Olympic Airways operates both domestic and international flights. There are nine international airports, including Heraklion (Crete) and Rhodes, mainly servicing the major tourist industry.

Broadcasting is state-controlled and, like newspapers, subject to some censorship. There are 130 newspapers.

Health and welfare

Malaria, typhoid, dysentery and tuberculosis have been brought under control by improved medical care and water supplies. Life expectancy is now comparable with most European countries. Most hospitals are in the larger cities, but there are local health centers in rural areas and the islands. Care is free for those unable to afford it. A social security system is supplemented by the close-knit family. The state school system offers 9 years of free compulsory education between the ages of 6 and 15. Entry to the national universities (private universities are banned by the constitution) is controlled in order to limit unemployment among graduates.

A tourist haven (*above*) Cruise ships lie at anchor offshore from Mykonos ("white island" in Greek), one of the smallest islands in the eastern Cyclades. With poor soil and little fresh water, Mykonos depends heavily on tourism for income. As a result, it is well connected to mainland Greece by air from Athens and to many islands by boat. It supports a population of 5,500 and hosts a branch of the Athens School of Fine Art.

Table talk in Athens (*left*) One of the most striking features of modern Greek society is its great uniformity. Most of the population are Greek and belong to the Greek Orthodox Church. They also share a magnificent heritage, a strong sense of equality and a passion for politics. Thanks to a rise in the birthrate after World War II, Greece is now a young and increasingly urbanized country – one-third of its people live in Greater Athens.

Malta

REPUBLIC OF MALTA

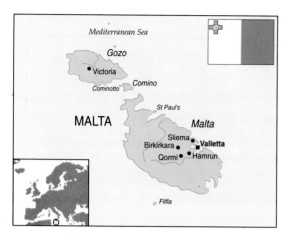

THE REPUBLIC OF MALTA CONSISTS OF THE islands of Malta, Gozo and Comino in the Mediterranean between Sicily and north Africa. The commanding position they hold in the relatively narrow straits between Sicily and Tunisia makes them strategically important, and they have changed hands many times throughout their history.

GEOGRAPHY

The islands are the peaks of a massive limestone outcrop stretching south from Sicily, dissected by deep bays and valleys. With no fresh water apart from low rainfall, desalinization plants are

NATIONAL DATA - MALTA

Land area	316 sq km (122 sq mi)			
Climate		Temperatures		Annual
	Altitude m (ft)	January °C(°F)	July °C(°F)	precipitation mm (in)
Valletta	70 (230)	12 (54)	26 (79)	519 (20.4)

Major physical features largest island: Malta 246 sq km (95 sq mi); highest point: southwest cliffs 253 m (830 ft)	
Population (2006 est.) 400,214	
Form of government multiparty republic with one legislative house	
Armed forces joint forces 2,237	
Capital city Valletta (7,137)	
Official language Maltese, English	
Ethnic composition Maltese 95.7%; British 2.1%; others 2.2%	
Religious affiliations Roman Catholic 98%; other 2%	
Currency 1 Maltese lira (MTL) = 100 cents	
Gross domestic product (2006) U.S. $8.122 billion	
Gross domestic product per capita (2006) U.S. $20,300	
Life expectancy at birth male 76.83 yr; female 81.31 yr	
Major resources limestone, salt, tourism, port facilities, fruit, goats, pigs, poultry, tobacco, vegetables, wheat	

necessary for a stable water supply.

The natural vegetation is sparse due to centuries of overgrazing and erosion. The remains of prehistoric animal species have been found in extensive caves, but today only a few small mammals (such as hedgehogs, shrews and bats) are found. Birds are more numerous, Malta lying on an important migratory route between Africa and Europe.

SOCIETY

Malta, possibly a former colony of Carthage (in north Africa), was part of first the Roman and then the Byzantine empires. Seized by the Arabs in 870, it was reconquered and ruled by Christian feudal lords until 1530 when it was ceded to the Knights of St John, a religious and military order. Malta was seized by Napoleon in 1798 but returned to the Knights in 1802. However, after a rebellion the island became a British colony in 1814. It became an important staging post on the sea route to India via the new Suez Canal (1869) and developed a dockyard industry.

In World War II Malta withstood repeated attacks by Germany and its allies, earning the entire population the British George Cross for bravery.

Full independence was granted in 1946. Good relations with Britain were maintained until the Maltese Labor Party came to power in 1971, swinging toward the Eastern Bloc, ousting NATO forces and sharply increasing Britain's rental fees for the dockyards. In 1979 the dockyards were nationalized and all British forces expelled. An attempt to replace Britain with Libya as mentor and partner proved stormy. In 1987 the Labor Party was ousted by the Nationalist Party, which sought to restore relations with the West while maintaining links with Libya.

Prehistoric Malta was inhabited by people from Sicily or Italy, and there has been a strong Italian and north African influence. The Maltese language is derived from Arabic with Sicilian elements. English is the second language, and Italian is widely spoken.

Christianity survived the Muslim occupation, and most Maltese are Roman Catholics. The church is a major property owner, wielding considerable influence.

At independence Malta's constitution set up a two-chamber parliamentary system. In 1974 this was amended to make Malta a republic, with a single chamber that is elected by proportional representation. This House of Represen-

Passing the time A passerby chats with a priest in Valletta, capital of Malta. Locals speak Maltese, a distinct language blended from North African Arab and Sicilian Italian, but written in Latin script. Schools, however, teach in English.

tatives (65 seats) elects the president for a five-year term. Malta is divided into six administrative regions.

ECONOMY

Agriculture is restricted to terraced hillsides and irrigated lowlands. Vegetables, cereals and citrus fruit are the chief crops; cattle and goats the main livestock. Intense cultivation has brought near self-sufficiency, but agriculture accounts for only 4 percent of gross national product. Fishing is local and small-scale.

The economy depends largely on shipping and port facilities. Light industry has become increasingly important. The textile and garment industries, which are among the major exports, have suffered from European Union (EU) import controls. Other products include machinery and electronics. The major source of foreign exchange is tourism. In 1995 it was agreed that EU accession negotiations would start in 1997.

Roads are extensive and well surfaced. Ferries run between Malta and Gozo, and Valletta's modern airport is served internationally. Publishing and the media are active and are generally bilingual, as is television. Communications systems are excellent, and strong social and health services reflect a high standard of living. Education is compulsory and free between the ages of 5 and 16.

Cyprus

REPUBLIC OF CYPRUS

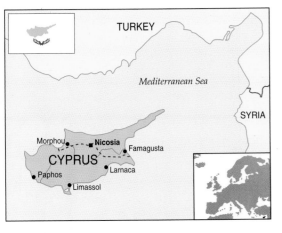

CYPRUS, THE THIRD LARGEST ISLAND IN THE Mediterranean, is divided by a long-running conflict between the majority Greek and minority Turkish population.

GEOGRAPHY

On the north coast are the Kyrenia mountains; in the south, the Troodos, from which the Pedias and Kariyoti rivers flow toward the bays of Famagusta and Morphou. The fertile Mesaoria plain in the center is the site of Nicosia, the capital. Winter is the rainy season, and summers are hot and dry. Scrubland predominates; the formerly extensive woodlands now remain only in the mountains. Many animals have been overhunted, and fishing is poor. Pasture used by sheep, goats and pigs covers one-tenth of the island. Crops include wheat, barley, citrus fruits, potatoes, tobacco, grapes, olives, almonds and vegetables.

SOCIETY

Once a colony of the Greeks, Phoenicians and Romans, Cyprus spent more than 700 years as part of the Byzantine empire before passing to the Ottomans in 1573 and to Britain in 1878. It became an independent republic in 1960. By 1963 the Turkish minority had withdrawn from the government. Amid a terrorist campaign in favor of union with Greece, UN forces were sent to keep the peace.

In 1974 a coup ousted and exiled the president. Turkey invaded Cyprus five days later, seizing the north of the island. Some 200,000 Greek Cypriots fled south. The Turkish Republic of Northern Cyprus, proclaimed in 1983, is recognized only by Turkey. UN forces remain, but negotiations for Cyprus to join the EU in the 2000s have prompted the two sides to hold open-ended talks to resolve the "Cyprus problem" once and for all.

Four-fifths of Cypriots are of Greek descent. The Turkish-speaking minority arrived from 1573. Almost all Greek Cypriots belong to the Greek (Eastern) Orthodox Church while the Turkish

Kykko monastery in the eastern Troodos mountains, decorated in classical Byzantine fashion. Greek Cypriots are members of the Church of Cyprus, an independent Eastern Orthodox Christian church, while Turkish Cypriots are mainly Sunni Muslim.

Cypriot community is mostly Sunni Muslim. English is widely spoken.

The Republic of Cyprus still claims legitimacy under the constitution of 1960. Many constitutional provisions for Turkish Cypriot representation have been suspended including the vice-presidency and seats in the House of Representatives. The breakaway Turkish state introduced a new constitution in 1985 and now elect their own government.

ECONOMY

Cyprus' economic expansion – funded by foreign aid – was severely disrupted in the Turkish invasion of 1974. Although most factories came under Turkish control, the Turkish sector remains dependent on agriculture and support from Turkey. The Greek sector has enjoyed impressive growth in industry and tourism. The Republic offers six compulsory years of free primary education and three years of free secondary education.

NATIONAL DATA – CYPRUS				
Land area 9,240 sq km (3,568 sq mi)				
Climate		Temperatures		Annual
	Altitude m (ft)	January °C(°F)	July °C(°F)	precipitation mm (in)
Nicosia	175 (574)	10 (50)	29 (84)	355 (13.9)
Major physical features highest point: Mount Olympus 1,951 m (6,403 ft)				
Population (2006 est.) 784,301				
Form of government multiparty republic with one legislative house (Turkish republic not recognized internationally)				
Armed forces national guard 10,000				
Largest cities Nicosia (capital – 202,488); Limassol (156,286); Larnaka (70,000); Famagusta (43,975)				
Official language Greek, Turkish				
Ethnic composition Greek 77%; Turkish 18%; other 5%				
Religious affiliations Greek Orthodox 78%; Muslim 18%; Maronite, Armenian Apostolic, and other 4%				
Currency 1 Cypriot pound (CYP) = 100 cents; 1 Turkish New lira (YTL) = 100 kurus				
Gross domestic product (2006) Republic of Cyprus: U.S. $17.79 billion; north Cyprus: U.S. $4.54 billion				
Gross domestic product per capita (2005) Republic of Cyprus: U.S. $22,700; (2004) north Cyprus: U.S. $7,135				
Life expectancy at birth male 75.44 yr; female 80.31 yr				
Major resources copper, pyrites, asbestos, gypsum, timber, salt, marble, tourism, almonds, beans, carob, citrus fruits, chromium, clay, grapes/wine, iron ore, olives, potatoes, vegetables, barley, wheat				

Central Europe

COUNTRIES IN THE REGION

GERMANY · SWITZERLAND

LIECHTENSTEIN · AUSTRIA

The city of Frankfurt (*left*) on the Main river in western Germany has hosted international trade fairs since 1240, and every year holds the largest Book Fair in the world. It is also a leading financial center, housing the principal stock exchange and headquarters of the national bank.

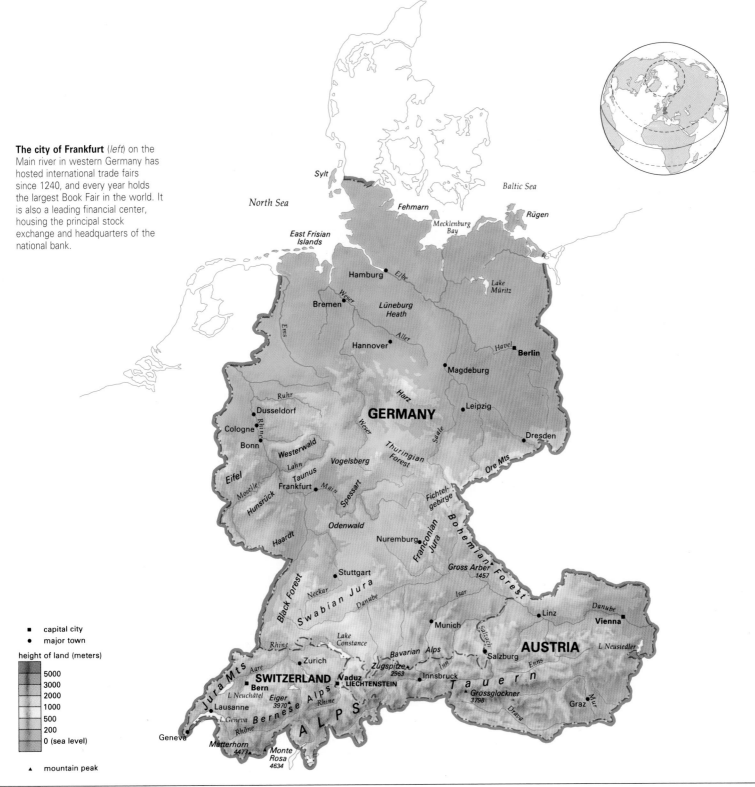

North Sea

Sylt

Baltic Sea

Fehmarn

East Frisian Islands

Mecklenburg Bay

Rügen

Hamburg

Elbe

Lüneburg Heath

Lake Müritz

Bremen

Weser

Ems

Aller

Hannover

Havel

Berlin

Magdeburg

Ruhr

Harz

GERMANY

Leipzig

Dusseldorf

Rhine

Weser

Saale

Cologne

Bonn

Dresden

Westerwald

Thuringian Forest

Eifel

Lahn

Vogelsberg

Ore Mts

Moselle

Taunus

Frankfurt

Main

Spessart

Fichtel-gebirge

Hunsrück

Odenwald

Haardt

Nuremburg

Franconian Jura

Bohemian Forest

Gross Arber 1457

Black Forest

Stuttgart

Isar

Danube

Neckar

Swabian Jura

Danube

Linz

Vienna

Rhine

Lake Constance

Munich

Salzach

AUSTRIA

L Neusiedler

Bavarian Alps

Salzburg

Enns

Zurich

Zugspitze 2963

Inn

Innsbruck

SWITZERLAND

Vaduz

LIECHTENSTEIN

Bern

Eiger 3970

Rhine

Tauern

Mur

Aare

Grossglockner 3798

Jura Mts

Lausanne

Bernese Alps

Graz

L Neuchâtel

L Geneva

Rhône

A L P S

Drava

Geneva

Matterhorn 4477

Monte Rosa 4634

■ capital city

● major town

height of land (meters)

5000
3000
2000
1000
500
200
0 (sea level)

▲ mountain peak

Germany

FEDERAL REPUBLIC OF GERMANY

G ERMANY IS EUROPE'S SECOND MOST populous country, with nine other European countries at its borders. Historically, Germany has been a major power since the 8th century, and its economic and political influence has become a major force in late 20th-century Europe.

GEOGRAPHY

Germany's varied landscapes reflect western, central and eastern Europe. Its northern coast faces the North Sea on the west and the Baltic on the east.

The land
Germany can be divided into four areas: the North German Plain; the central uplands; southern Germany; and the Bavarian plateau bordering the Alps along the Austrian border. In many places the Rhine, Elbe and smaller rivers have carved out a landscape of their own.

Germany's two coasts are quite different. The Baltic coast is generally flat, with some low chalk cliffs and long, white, sandy beaches. The shallow North Sea coast has dikes protecting farmland reclaimed from former marshes. Beyond its extensive tidal flats lie the East Frisian Islands and farther out to sea the rocky island of Helgoland. The German state of Schleswig-Holstein, at the base of the Jutland peninsula, has coasts facing both seas, with the tidal western coast flanked by the North Frisian Islands.

The northern German plains are part of a massive lowland belt that stretches from northern France into Russia. Ice-age glaciers left ridges and depressions, notably in Mecklenburg in the northeast, which has numerous small lakes amid

In mountainous Bavaria, rolling pasturelands are framed by the wooded foothills and jagged peaks of the Bavarian Alps, close to the Austrian border. Bavaria's picturesque Alpine scenery attracts tourists, particularly hikers, from all over the world.

rolling farmland and wooded hills. Farther south there is another hilly stretch running from the sandy Lüneburg Heath in central–northern Germany and continuing southeast toward the Polish border. Northwestern Germany is more level, alternating between sandy heathland and marshy peatlands.

Some of the country's greatest rivers flow through the North German Plain. The Rhine in the west, the Ems, the Weser and the Elbe all flow into the North Sea; in the east, the Oder and the Neisse mark Germany's frontier with Poland.

Along the southern margins of the Plain is more rolling countryside called the Borde. The irregular division between the country's lowland and mountain areas includes several broad basins partly divided by uplands: the Leipzig basin in the east, and the Cologne basin and the Munsterland in the west.

Germany's central uplands form part of a broad mountainous arc that sweeps across Europe from central France in the west to the Carpathians in the east. In Germany these mountains rarely rise

above 1,000 m (3,300 ft), producing forested slopes separated by fertile river valleys. The rugged Harz Mountains, in the northern uplands, straddle the former East–West German border and contain some of the highest peaks.

Farther south, the main river valleys divide the upland belt into ranges. In the far west the Eifel and Hunsrück mountains are separated by the deep river valley of the Moselle. The Rhine valley divides both these ranges from the Sauerland and Westerwald massifs. South of the ancient city of Koblenz, where the Moselle joins the Rhine, the spectacular scenery of the Rhine gorge separates the Hunsrück from the Taunus range.

In the center of Germany are two massive volcanic plateaus – the Vogelsberg and the Rhön. Beyond, to the east, lies the Thuringian Forest, part of a longer mountain arc that runs southeast via the Franconian Forest, the Fichtelgebirge and the Ore Mountains along the Czech border in the province of Saxony.

Much of central–southern Germany is a low plateau divided by deep, winding river valleys. Most of the rivers, notably the Main and the Neckar, flow westward to enter the Rhine. The Rhine itself flows north and extends from the Swiss border as far north as Frankfurt. In the south-

The Lüneburg Heath (*above*) is a vast heather-covered expanse in central-northern Germany between the Elbe and the Aller rivers. Punctuated by solitary trees, birch woods and timber-framed farmhouses, it has a haunting beauty of its own.

west, where the Rhine forms the border with France, the rounded peaks of the Black Forest look across to their French counterparts, the Vosges. The Swabian and Franconian Jura, east of the Black Forest, is a continuation of Switzerland's Jura Mountains. Farther east are the parallel ranges of the Bavarian Forest and the Bohemian Forest.

Germany's most dramatic scenery lies in the far south, on the border with Austria, where the Bavarian plateau rises toward the northern Alps. The area contains numerous high peaks, such as the Zugspitze, the highest of Germany's mountains at 2,963 m (9,720 ft). Lake Constance, bordered by Germany as well as Switzerland and Austria, is where the Rhine begins its long journey through Germany. The great river Danube (twice as long as the Rhine) rises on the east slope of the Black Forest.

Climate

Germany's climate becomes more continental, with colder winters and warmer summers, toward the south and east of the country. Rain falls throughout the year, and there is frequent snow in the winter; the driest areas are in the northern lowlands. The Bavarian plateau has cold winters but early springs, partly

as a result of the *föhn* – a warm, dry wind that blows down from the Alps. The Rhine valley, protected by mountains, enjoys warm summers, relatively dry winters and plenty of sunshine.

Plants and animals

Almost 30 percent of Germany is still wooded. Although lowland areas have been largely cleared for agriculture, open woodlands of oak and birch still survive, and pine plantations have been planted on sandy heathland.

Forests cover most of the wilder uplands, growing almost to the summits of the Black Forest – named after its lofty, dark firs – and as high as 1,500 m (5,000 ft) in the Bavarian Alps. Many of Germany's forests have suffered badly from the effects of acid rain, but is in the forefront of those countries trying to stem the damage by reducing pollution.

Germany's wildlife is typical of continental Europe, while the country's central position makes it the destination for many migratory birds particularly in the forests. There is a wide variety of mammals throughout the country although few are unique to the region. Nature parks and reserves have been set up across the country to protect a range of species and environments.

NATIONAL DATA – GERMANY

Land area 349,223 sq km (89,166 sq mi)

Climate	Altitude m (ft)	Temperatures January °C(°F)	July °C(°F)	Annual precipitation mm (in)
Berlin	55 (180)	–1 (30)	19 (66)	580 (23)
Hamburg	14 (46)	1 (34)	17 (63)	772 (30.3)
Munich	528 (1,732)	–2 (29)	17 (63)	928 (37)

Major physical features highest point: Zugspitze 2,963 m (9,720 ft); longest rivers: Danube (part) 2,850 km (1,771 mi), Rhine (part) 1,319 km (820 mi); Elbe (part) 1,165 km (724 mi); largest lake: Lake Constance (part) 544 sq km (210 sq mi)

Population (2006 est.) 82,422,299

Form of government multiparty republic with two legislative houses

Armed forces army 191,500; navy 25,600; air force 67,500

Largest cities Berlin (capital - 3,388,477); Hamburg (1,734,083); Munich (1,247,873); Cologne (965,954); Frankfurt (643,432); Dortmund (589,661); Essen (589,499); Stuttgart (589,161); Düsseldorf (572,511); Bremen (544,853)

Official language German

Ethnic composition German 88.2%; Turkish 3.4%; other 8.4% (mainly Greek, Italian, Polish, Russian, Serbo-Croatian, Spanish)

Religious affiliations Protestant 34%; Roman Catholic 34%; Muslim 3.7%; unaffiliated or other 28.3%

Currency 1 euro (EUR) = 100 euro cents

Gross domestic product (2006) U.S. $2.585 trillion

Gross domestic product per capita (2005) U.S. $31,400

Life expectancy at birth male 75.81 yr; female 81.96 yr

Major resources coal, lignite, natural gas, iron ore, copper, nickel, uranium, potash, salt, construction materials, timber, antimony, arsenic, barley, bismuth, cobalt, fish, fruit, vegetables, grapes/wine, beer, lead, livestock, dairy products, oats, potatoes, rye, sugar beet, wheat, zinc, tourism

SOCIETY

For centuries Germany was a patchwork of rival city-states, principalities and kingdoms that only slowly coalesced into a loose confederation. The price of nationalism was high. At the end of World War II Germany lay in ruins, divided between the Eastern and Western blocs. From 1945 until 1990 the gulf between the two Germanys widened until the collapse of communism in Eastern Europe made reunification possible.

History

Germanic peoples living along the shores of the Baltic gradually migrated down into the lands east of the Rhine, the Carpathians and the territories north of the Black Sea. In the 4th century AD the Germanic peoples were displaced westward by nomadic Huns invading from central Asia. In 410 AD the Germanic Visigoths from the Black Sea region sacked Rome. The collapse of the western Roman empire had begun, and by 500 a patchwork of Germanic kingdoms stood in its place.

Most of these kingdoms were short-lived until the Frankish kingdom, comprising most of modern France and much of Germany. Charlemagne (c. 742–814), the greatest Frankish ruler, built an empire that stretched from Rome to the North Sea and from the Elbe to beyond the Pyrenees. The empire survived until 888, when it was divided into the separate kingdoms of East Francia (modern Germany), West Francia (modern France), Italy, Burgundy and Provence. When Charlemagne's line died out in the East Frankish kingdom, the crown passed, in 919, to Henry Duke of Saxony (c. 876–936) who became, in effect, the first true German ruler. His son Otto I (912–973) is regarded as the founder of the Holy Roman Empire which united Germany, Burgundy and Italy under a single ruler, but by 1254 the empire was fragmenting. In 1273 Rudolf I (1218–91), the first of the Austrian Habsburg dynasty, was elected to the imperial throne. With one exception his descendants retained the title of Holy Roman Emperor until 1806 – but the empire was never totally unified under their control.

In the 16th century Emperor Charles V

Light of the Renaissance (*above*) A self-portrait by the German graphic artist and painter Albrecht Dürer (1471–1528), court painter to Charles V. Dürer is often regarded as the greatest Renaissance artist to have emerged from northern Europe.

Imperial splendor (*right*) The Charlottenberg Palace in west Berlin, built in 1695–99, is a sumptuous monument to centuries of Prussian dominance and imperial rule. Prussian rulers used the palace as their city residence.

The Legacy of Johannes Gutenberg

In the German city of Mainz on the west bank of the Rhine river, an obscure pensioner was found dead in February 1468. For some years this destitute, half-blind old man had received a yearly allowance of cloth, grain and wine from the local ruler who held court just across the Rhine at Weisbaden. But few people remembered who he was, or what he had achieved. His name was Johannes Gutenberg (c.1395–1468), and he was the father of modern printing.

Johannes Gensfleisch zur Laden zum Gutenberg was born into a patrician family in Mainz, where he became a goldsmith and metalworker. He later joined a goldsmiths' guild in Strassburg (now Strasbourg, France), where he began to pursue an obsessive and very costly dream: a method of printing painstakingly hand-lettered medieval manuscripts without sacrificing their elaborate ornamental design.

After twenty years of experimentation Gutenberg devised a number of innovations that allowed him to realize

his ambition of mass production: these included quickly made movable type; an efficient printing press, allegedly inspired by watching someone operate a winepress; oil-based ink for the press; and the use of cheap rag paper. His techniques were largely unknown in existing Chinese and European printing, and were not surpassed until the 19th century.

In 1455 Gutenberg produced his first printed book – a beautifully illustrated copy of the Bible in an edition of 200 copies. However, by this time he was heavily in debt to his business partner, Johann Fust (c.1400–66), who had invested in the venture and was impatient for some return. Fust took Gutenberg to court and won his suit, taking control of all Gutenberg's printing equipment – including the type for the famous 42–line (or Mazarin) Bible. Gutenberg faced financial ruin. But the effects of his genius were to alter the course of history, and laid the foundations for the first largescale information revolution.

Spreading the word A page from Johannes Gutenberg's magnificent 15th-century Bible. His revolutionary printing techniques – including movable metal type – allowed books to be mass-produced for the first time. By 1500 some 250 towns, including Nuremburg, were equipped with printing presses.

(1500–58) inherited Luxembourg, Burgundy and Spain. But the many territories were scattered, disparate and often rebellious. Instability intensified as a growing movement for religious change – the Protestant Reformation – gripped Europe. An increasing sense of national awareness made Protestantism the focus of opposition both to the Roman Catholic Church and its champion Charles V. To oppose the emperor the Protestant princes united in 1530 to form the Schmalkaldic League under the leadership of John Frederick, Elector of Saxony (1503–54). The Peace of Augsburg in 1555 fully recognized Lutheranism in those states with a Protestant leader. Among the peasants and urban poor, Protestantism inspired social revolutionary movements, as in 1524–25, that were denounced by Luther and brutally suppressed by the princes.

In 1618 Emperor Ferdinand II (1578–1637) started a long campaign against the Bohemian Protestants, plunging Europe into the Thirty Years War (1618–48). Over the next three decades the region's population declined from 21 million to 13.5 million. The Peace of Westphalia (1648) recognized the sovereignty of some 300 German bishoprics, principalities and free cities – preserving the fragmented character of the empire.

The weakening of the Habsburg empire allowed other leaders to increase their influence and power. The Hohenzollern rulers of Brandenburg-Prussia in the northeast made territorial gains along the Baltic coast of Poland. Under Frederick the Great (1712–86) Prussia began to accelerate its remarkable rise to military power. During his reign, Prussian territory extended from modern Lithuania to Magdeburg in the west.

Prussia's transformation into a major European power was checked temporarily by the invasion of Napoleon's armies in 1806. Austria was defeated by the French, and Francis II of Austria (1768–1835) was forced to abandon the title of Holy Roman Emperor. In the following year, most of the minor German states became members of the French-dominated Rhine Confederation.

One of the results of the Congress of Vienna (1814–15) was the formation of a loose German Confederation of 39 independent states dominated by Austria. In 1834, a German "customs union" was formed, creating a unified inland market. Prussia, meanwhile, grew in strength. The Austro-Prussian War of 1866 led to Austria's expulsion from the Confedera-

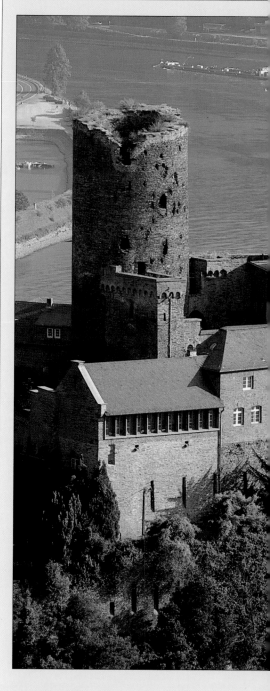

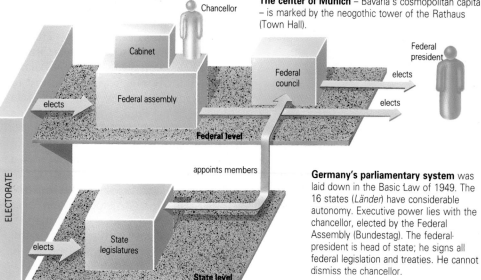

The center of Munich – Bavaria's cosmopolitan capital – is marked by the neogothic tower of the Rathaus (Town Hall).

Chancellor

Cabinet

Federal president

Federal council

elects

elects

Federal assembly

elects

Federal level

ELECTORATE

appoints members

elects

State legislatures

State level

Germany's parliamentary system was laid down in the Basic Law of 1949. The 16 states (*Länder*) have considerable autonomy. Executive power lies with the chancellor, elected by the Federal Assembly (Bundestag). The federal president is head of state; he signs all federal legislation and treaties. He cannot dismiss the chancellor.

Valley of legend

The river Rhine winds northward for 140 km (90 mi) between Bingen and Bonn, through a spectacular valley of gorges with the Taunus mountains to the east and the brooding peaks of the Hunsrück to the west. This valley is the home of some of Germany's most famous legends, sources of inspiration to its poets, novelists and composers for hundreds of years.

The most famous of these tales is the "Song of the Nibelungs," a 13th-century epic poem by an unknown writer. A combination of Nordic myth and early medieval German history, it tells of the hero Siegfried, who performs magical feats such as winning

Passion for pretzels (*above*) A street stall displays a mouth-watering variety of pretzels – glazed and salted crackers – during *Oktoberfest*, Munich's 16-day beer festival. A love of beer and pretzels unites Germany's regional appetites.

Romance of the Rhine The ruined tower of a medieval castle near St Goarshausen. Not far from the castle is the notorious Lorelei Rock, where a water nymph is said to lure sailors to their death in the treacherous currents below.

the gold of the Nibelungs – a race of dwarfs – and killing a monster that guards the Dragon's Rock (near the modern city of Bonn). Later married to Kriemhild, the sister of King Gunther of Worms, Siegfried helps Gunther wed Queen Brunhild of Iceland. A complicated plot involving trickery, jealousy and revenge leads to Siegfried's murder and the casting of the Nibelungs' gold treasure into the Rhine. The composer Richard Wagner (1813–83) used a version of the story as the core of his operatic tetralogy "The Ring of the Nibelung," composed between 1853 and 1874. Two of his other operas, "Tannhauser" (1845) and "Lohengrin" (1850), were based on other legends of the Rhine. His contemporary Friederich Hebbel (1813–63) wrote a dramatic trilogy which was based directly on passages from the Nibelung epic itself.

Other stories are of more recent origin. The Lorelei Rock, which towers 122 m (400 ft) over a bank above St Goarshausen, is said to have been the home of a water spirit who lured unwary travelers to their death. This story, echoing the ancient Greek legend of the Sirens, was conceived by the German Romantic poet Clemens Brentano (1778–1842) and was retold in a famous version by another poet, Heinrich Heine (1797–1856).

tion, and the subsequent Franco-Prussian War (1870–71) finally established Prussian domination of Europe. By this time the Prussian chancellor, Otto von Bismarck (1815–96), had forged a unified Germany, which began to build an overseas empire and a powerful navy to defend it.

By 1907, years of uneasy peace had left Germany, Italy and Austria–Hungary aligned against Britain, France and Russia. The assassination of the Austrian Archduke Franz Ferdinand (1863–1914) in Sarajevo by a Serbian nationalist finally tipped the two alliances into war in 1914. Germany's defeat in 1918 led to the abolition of its monarchy and the French occupation of the Rhineland, its industrial heart. In February 1919 a new constitution established the democratic Weimar republic. Its problems were considerable, compounded by massive reparations and boundary changes imposed on Germany. By November 1923, rocketing inflation had bred extremist parties. A brief economic recovery collapsed with the Depression, and in 1933 Adolf Hitler (1889–1945), leader of the National Socialist (Nazi) Party, became chancellor. Hitler's Third Reich soon threw Europe into a second world war. Dissidents were ruthlessly suppressed, and millions of Jews and other minorities killed in extermination camps.

At the end of World War II (1939–45) – which cost some 50 million lives – Germany lost all territories east of the Oder and Neisse rivers, and was divided into zones of occupation by Britain, France, the United States and the Soviet Union. In 1949, the British, French and American zones were consolidated to form the Federal Republic of Germany (West Germany). The Soviet Union responded by creating the communist German Demo-

cratic Republic (East Germany). In 1952, the Soviet Union established a heavily guarded cordon of land along the whole border. West Berlin – an isolated enclave 190 km (120 mi) within East Germany – became the favored destination for refugees. An uprising against the communist regime in June 1953 was crushed by Soviet tanks.

In 1955 the Warsaw Treaty Organization unified the military command structure of the Eastern bloc countries. West Germany achieved full sovereignty, and became a member of the North Atlantic Treaty Organization (NATO).

In order to stem the exodus of East Berliners to West Berlin, East Germany built the massive Berlin Wall in 1961. In 1968, a thaw in relations between East and West was interrupted by the Soviet invasion of Czechoslovakia, in which East German troops played a token role. Chilly relations lasted until 1970, when a treaty confirmed existing German frontiers and renounced the use of force for political purposes. Two years later, a pact between East and West Germany opened a door to cooperation.

The 1980s saw a slowly increasing wave of pressure for reform in the Soviet bloc. In 1989 a wave of demonstrations was followed by largescale resignations in both national and local governments. In November the border with West Germany was opened, releasing a flood of emigrants. East Germany's communist government finally collapsed in the spring of 1990. On 3 October 1990, East and West Germany were formally reunited.

Government

West Germany's constitution of 1949 created a federal government based on 10 *Länder*, or autonomous states. The 1990 Unification Treaty admitted the five reconstituted East German *Länder* to the federation; the reunited city of Berlin became the 16th *Länd* (state) and the reunited Germany's capital city.

The 669 members of the Federal Assembly (Bundestag) are elected every four years by popular vote; they in turn elect the chancellor, who heads the government. The 69 members of the Federal Council (Bundesrat) are appointed by the 16 *Länder*. The president, who is the formal head of state, is elected every five years by an electoral college made up of delegates from both the federal and state assemblies. The government structure within each *Länd* is generally similar to that of the federal government.

People

The German people have a common history going back many centuries, but a unified language began only in the 16th century. Martin Luther's Bible translation of 1522 laid the foundations of today's written language.

Ethnic Germans make up most of the country's population, and include many communities who fled former German territory east of the Oder river. Germany is also home to numerous minorities, notably Turks, Yugoslavs and Italians, who came as "guest workers" to West Germany in the postwar years. Since 1990 the reunited country has taken in great numbers of refugees from Africa, Asia and the poorer parts of Eastern Europe, stoking violent right-wing opposition and a neo-Nazi resurgence, especially among the poorer youth of eastern Germany who resent the foreigners' claims on the heavily burdened resources of the unified government.

Religious practice reflects a long history of divided loyalties since the Reformation. Generally Lutherans are strongest in the north and east, Roman Catholics in the south and in the Rhineland.

Germany has a rich legacy of art and literature. Its renowned composers include Johann Sebastian Bach (1685–1750), Ludwig van Beethoven (1770–1827) and Richard Wagner (1813–83).

ECONOMY

Reunification has had a dramatic impact on the German economy. The infrastructure of former East Germany – its housing, transportation networks, telecommunications and industrial base – is very poor in comparison with the west. Its industries also caused largescale environmental damage. Addressing these problems is stretching state resources.

Agriculture and industry

Since 1990 increased mechanization has boosted output, but agriculture remains second to industry in economic importance. The former East German cooperatives have little place in the free-market economy. Just over one-third of Germany's land is under cultivation. Germany is a

The huge steel plants at Duisburg, on the junction of the Rhine and Ruhr rivers, are supplied with iron ore and coal from the canals of the Ruhr valley. This is the most heavily industrialized part of Germany, and pollution is a major problem.

leading producer of hops, which supply more than 5,000 varieties of beer. The Rhine and Moselle valleys produce grapes for wine-making, while dairy cattle supply milk, cheese and butter.

Germany's extensive forests are a well-developed resource, though East Germany imported lumber until 1990. A large modern fishing fleet operates in the North Sea, the Baltic, the Atlantic, the Pacific and even the Antarctic.

Mining and manufacturing account for about one-quarter of the nation's income, despite relatively poor mineral resources. Most electricity is generated through the burning of fossil fuels. There are 21 nuclear and hydroelectric plants.

German manufacturing is widely diversified, but the east is less developed, and uncompetitive products can no longer find a market in the new Germany. The main industrial area extends southwest from the Ruhr to Cologne. Manufactured goods include chemicals, machinery, iron and steel products, automobiles and trucks, textiles and clothing, and modern electronic and precision equipment. Construction is another important industry, especially since reunification.

In western Germany, tourism is a profitable industry – though German travelers abroad spend four times what the industry earns at home. Tourism is less developed in the east, though the many resorts and historic sites have great potential to attract visitors.

Transportation and communications

The contrast between eastern and western Germany is more marked in this sector than many others. The west has a comprehensive network of surfaced roads, including a sophisticated system of *autobahns* (expressways) linked to neighboring countries. In the east, by contrast, many roads were still unsurfaced in 1990, and the autobahns had changed surprisingly little since they were first built, just before World War II. The state of the railroads is very similar.

The German waterways form a complex network. The Rhine is at the center of a vast canal system from the heart of Europe to the North Sea. Further canals link the Rhine to the Elbe, the Oder and the Danube, creating a doorway to Eastern Europe. Germany's most important port, Hamburg, is easily accessible from the North Sea, while the Kiel Canal allows shipping through Schleswig–Holstein to the Baltic.

Lufthansa, the federal state airline, was an elite international carrier from the 1950s onward. Domestic air traffic is light in the west, largely because the road and rail networks are so efficient.

Telecommunications throughout Germany have needed vast capital investment. The media, too, faced problems of readjustment. West Germany's press had always been in private hands, operating without government interference; in the east, the press was tightly controlled. Radio and television were state-owned in both East and West Germany, but West German broadcasting preserved an independent voice in the divided country.

Welfare and education

Healthcare and medical provision throughout Germany are generally very good, though the west has significantly more resources. The west also has a well-developed private health network.

Welfare systems were well developed in both countries before reunification, with a comprehensive range of benefits. However, the East German system was strongly tied to employment, and included extra provisions such as free nursery care for working mothers. After 1990 many in the east resented the loss of these services. Unemployment benefits paid to the many citizens in the east who lost their jobs as a result of reunification have been a drain.

Education is free and compulsory, and standards are generally high. An examination known as the *Abitur* is the passport to higher education at college or university. However, the system is very different in east and west, and there are discrepancies in the west between individual *Länder*. Germany's universities have a history of excellence, notably in theology, philosophy, philology and chemistry. Among the most historic institutions are Heidelberg, founded in 1386, and Tübingen, 1477. Marburg, founded in 1527 in central Germany, is the oldest Protestant educational establishment.

River of wine A barge glides past vineyards reflected in the peaceful waters of the Moselle river in western Germany. The steeply terraced vineyards along the sheltered Rhine and Moselle river valleys produce Germany's world-famous white wines.

Switzerland

SWISS CONFEDERATION

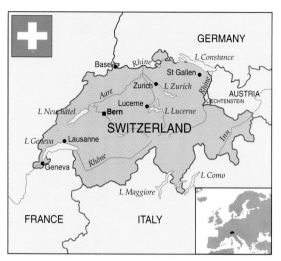

S WITZERLAND IS A LANDLOCKED COUNTRY IN the mountainous heart of Europe. Politically a federal state made up of 26 separate cantons – many of them tiny, and speaking as many as four different languages – the Swiss nation still preserves a remarkable sense of unity.

GEOGRAPHY

Switzerland is renowned for its spectacular Alpine scenery. In no other nation does it so dominate the landscape.

Topographically Switzerland comprises three distinct areas. The plateaus of the Jura Mountains in the northwest give way to long, grass-covered mountainous ridges divided by deep wooded valleys and gorges; the southern edge overlooks Lake Neuchâtel and the Aare valley.

The Mittelland plateau is the least mountainous part of Switzerland, and holds most of the population.

The southern margins of the Mittelland merge into the foothills of the Alps, which occupy well over half Switzerland's total land area. The two ranges of the Alps almost meet at the St Gotthard massif, where the Rhône and Rhine rivers rise; the rivers form parts of Switzerland's two largest lakes, Lake Geneva and Lake Constance respectively. The highest peaks are in the Pennine Alps in the southwest along the Italian frontier, and include Monte Rosa at 4,634 m (15,203 ft) and the Matterhorn at 4,477 m (14,688 ft). The Bernese Alps to the north of the Rhône form the northern rim of Europe's largest glacier, the 24 km (15 mi) long Aletsch. Alpine rivers have carved several gorges here, and there are a number of spectacular lakes, such as Lake Lucerne and Lake Lugano.

Switzerland's climate is temperate continental, but the weather varies depending on terrain and altitude. Midwinter

temperatures normally fall below freezing everywhere. The highest Alps lie under perpetual snow and ice. Summers in the Mittelland, however, are warm, and in the southern valleys they are often quite hot. Rain falls throughout the year.

Over the centuries, grazing sheep and goats have stripped much of the country's original plant cover, although rare plants, and flowers such as the edelweiss, can still be found above 2,000 m (6,500 ft). The lower slopes support woodland; the southern valleys contain trees, such as cypresses and figs. As well as the typically Alpine animals – marmots, chamois and ibex (reintroduced) – Switzerland harbors many species familiar in other parts of Central Europe.

SOCIETY

In the 13th, 14th and 15th centuries the Swiss cantons struggled to achieve independence. Since then they have kept that freedom, forging disparate cultures, religions and languages into a strong national identity. In 58 BC the Romans placed permanent garrisons in the Alpine passes of Helvetia – modern Switzerland. In 259 AD, the Alemanni were the first of several successive waves of Germanic invaders to arrive. The area later came under Frankish rule, and by 1033 it had become a part of the Holy Roman Empire.

In 1291 three cantons around Lake Lucerne– Uri, Schwyz and Nidwalden –

On the rooftop of Europe (*left*) evening sunlight is reflected from the snowy peaks of the Bernese Alps, which form the northern rim of Europe's largest glacier. The small landlocked country of Switzerland is the most mountainous in Europe.

Lake Lucerne (*above*), the historic heart of the nation, is the ancient terminal basin of one of Switzerland's great glaciers. In 1291 the confederation of the three Forest cantons around Lake Lucerne sowed the seed of the independent Swiss nation.

formed a confederation to break free of Habsburg rule. In 1332 Lucerne joined; Zurich joined in 1351, followed by neighboring Glarus, Zug and Bern. By 1388 the Swiss Confederation was firmly established. During the next 150 years the Confederation grew in power and influence, while continuing to struggle for its independence against the threat of warring neighboring states.

The rise of Protestantism led to open conflict between cantons. Yet, while religious wars ravaged Europe, the Swiss remained neutral, patching up their domestic differences. The Peace of Westphalia (1648) finally acknowledged the cantons' freedom from imperial rule. Switzerland maintained its neutrality

well into the 18th century. Then in 1794 revolutionary France occupied much of Switzerland, establishing the puppet Helvetic Republic four years later. In 1815 the last remaining cantons were admitted to the confederation, and Switzerland regained its neutrality.

In the mid 19th century, following the forced secularization of Jesuit monasteries, the Catholic cantons formed a protective union known as the Sonderbund. However, this alliance was nullified in 1847, and in 1848 a new constitution created the modern federal state. Bern became the capital of the confederation. Swiss neutrality made the country a haven for many international bodies, such as the International Red Cross and the short-lived League of Nations. The Geneva Convention of 1864 established laws for the conduct of war. Today many United Nations' and other international agencies have offices there. Even so, a referendum in 1986 rejected United Nations' membership as a threat to Swiss neutrality.

Switzerland is a federal republic made up of 20 full cantons and six half-cantons. National policy is decided by the Federal Assembly: the 200-member National Council, elected every four years; and the Council of States, whose 46 members are appointed by the cantons. Federal

NATIONAL DATA - SWITZERLAND

Land area	39,770 sq km (15,355 sq mi)			
Climate	Altitude m (ft)	**Temperatures** January °C(°F)	July °C(°F)	Annual precipitation mm (in)
Zurich	540 (1,772)	0 (32)	18 (79)	1,086 (42.7)

Major physical features highest point: Monte Rosa 3,505 m (11,500 ft); longest river: Rhine (part) 1,319 km (820 mi), Rhone (part) 813 km (505 mi); largest lake: Lake Constance (part) 544 sq km (210 sq mi)

Population (2006 est.) 7,523,934

Form of government federal multiparty republic with two legislative houses

Armed forces joint forces 4,300

Largest cities Zurich (342,518); Geneva (177,535); Basel (165,051); Bern (capital - 122,707); Lausanne (116,332); Lucerne (116,332)

Official language German, French, Italian, Romansch

Ethnic composition German 65%; French 18%; Italian 10%; Romansch 1%; other 6%

Religious affiliations Roman Catholic 41.8%; Protestant 35.3%; Orthodox 1.8%; other Christian 0.4%; Muslim 4.3%; other 1%; unspecified 4.3%; none 11.1%

Currency 1 Swiss franc (CHF) = 100 centimes

Gross domestic product (2005) U.S. $241.8 billion

Gross domestic product per capita (2005) U.S. $32,300

Life expectancy at birth male 77.69 yr; female 83.48 yr

Major resources hydropower, timber, salt, apples, barley, building stone, grapes, livestock, potatoes, salt, wheat, tourism

Chocolate, a Swiss specialty, was first made in Zurich in 1845 by the master confectioner Rudolf Sprüngli (shown in the portrait). The company, now Lindt-Sprüngli, exports highest quality chocolates to over 80 countries around the world.

government is administered by the Federal Council, consisting of seven members, each from a different canton, whom the Federal Assembly elects for four-year terms. Each year a member of the council is elected federal president. Major issues, such as constitutional changes, are decided by popular referendum. Women were denied voting rights until 1971 (1989 in the two Appenzell half-cantons).

Switzerland has four different linguistic communities. About 65 percent of the population speak Swiss German (*Schwyzerdütsch*); they live in the central, northern and eastern cantons and the local dialects differ greatly. French speakers, who constitute about 18 percent, live in the west, and Italian-speakers live in the southern canton of Ticino. Romansh, a descendant of Latin, is spoken by only one percent of Swiss. Switzerland is evenly divided between Roman Catholics (49 percent) and Protestants (48 percent). Most Protestants are either Lutherans or Calvinists.

ECONOMY

Switzerland is one of the most prosperous nations in the world. Its strategic location at the crossroads of Europe, combined with its political stability and tradition of confidentiality, has boosted its development as a center of international finance. Switzerland's few flat areas are heavily populated, leaving little fertile land for farming. Nearly half the nation's food has to be imported. However, Swiss cheese and chocolates are exported all over the world. Mountain slopes provide pasture for beef and dairy cattle. Forests – protection against avalanches and floods – cover

The Matterhorn, one of the most spectacular mountain peaks in the world and one of the ultimate challenges to skilled mountaineers, seen from the terrace of an Alpine hostel. Tourists are attracted to Switzerland all year round and provide the main source of income for many Swiss communities.

many valley slopes, and clearance is forbidden by law.

Hydroelectric power meets almost all the country's needs; nuclear power stations supply the rest. A 10-year ban on expansion was imposed in 1990. High-grade manufactured goods for export are the mainstay of the Swiss economy. They include chemicals, heavy machinery and precision instruments such as clocks and watches. Trade tariffs provide the greater part of federal government revenue, while tourism is an all-year-round source of foreign income.

Several important trans-European routes pass through the Swiss Alps. The road system is very well developed; tunnels keep traffic flowing in winter. There are hundreds more tunnels on the country's extensive electrified railroads.

The northern city of Basel, on the Rhine, is Switzerland's only freight port.

Swissair is among the world's major commercial carriers, with principal airports at Zurich and Geneva.

Broadcasts and media are in three languages. The Swiss Broadcasting Corporation, a nongovernmental body, supervises all television and radio output.

Healthcare is free or subsidized. The welfare system provides child allowances and pensions; all salaried workers must take out unemployment insurance.

Education is free and compulsory from the age of 7 to 14, and secondary education is either academic or vocational. Higher education attracts an unusually large number of foreign students.

Liechtenstein

PRINCIPALITY OF LIECHTENSTEIN

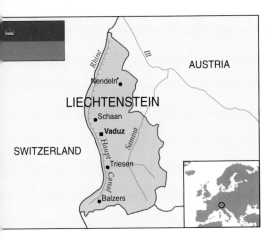

Liechtenstein's decorative postage stamps, famous around the world, contribute about 10 percent of the national income.

LIECHTENSTEIN IS A VERY SMALL CENTRAL European principality on the east bank of the upper Rhine, sandwiched between Austria to the east and Switzerland to the west and south. Although it uses Swiss currency and belongs to the Swiss customs union, it is otherwise a fully independent state.

GEOGRAPHY

Liechtenstein measures about 25 km (16 mi) from north to south, and is 6 km (3.7 mi) wide. To the north and west the land is part of a broad floodplain bordering the Rhine. Once marshy, it was drained in the 1930s for agricultural purposes. To the east, the foothills of the Rhaetian Alps rise to snow-capped peaks along the southern border. The climate is surprisingly mild considering its continental location, and rainfall is plentiful. The valley suffers from sharp frosts, but the *föhn* – a warm, dry wind that blows from the south – brings an early spring.

These conditions allow a rich variety of plant life, with Alpine plants at higher altitudes, and forests of beech, maple, elm and ash on the lower slopes. The forests cover two-fifths of the land area, and are rigorously protected. Animal life includes several species of deer, as well as marmots, foxes, badgers and weasels.

SOCIETY

The area was first settled by the Alemanni, a Germanic people, in about 500 AD, and Liechtensteiners still speak an Alemannic dialect of German. Under Charlemagne (742–814) it became part of the Carolingian empire, and for centuries the land was divided between two independent counties based at Vaduz (the present-day capital) and Schellenberg in the north. In 1719 these two counties were amalgamated to form the principality of Liechtenstein, which remained a part of the empire. In 1806 the Austrian king abandoned the title of Holy Roman Emperor, and Liechtenstein became part of the Rhine Confederation. In 1815 it joined the German Confederation, and achieved full independence in 1866 when the Confederation was dissolved.

Liechtenstein is a constitutional monarchy with a single-chamber parliament, whose 25 members are elected every four years. There are 11 local authorities, each controlled by an elected mayor and council. Women may vote only in national elections. The official language is German, and most people are Roman Catholics, but fewer than two-thirds of the population are native Liechtensteiners; most of the rest are foreign industrial workers. No town in the country has more than 5,000 inhabitants.

ECONOMY

Since 1921 Swiss currency has been used in Liechtenstein, and its advantageous tax and company legislation have made the principality an important banking center. Revenue also comes from the registration of foreign firms, and from a thriving government-sponsored tourist industry. Liechtenstein is particularly famous for its decorative postage stamps, which add to the government revenue. Good roads provide excellent access from neighboring countries, and the Paris–Vienna railroad crosses its territory.

In 1930 the great majority of Liechtensteiners were farmers. Today, fewer than one in 40 of the population works on the land, and farms are being enlarged to improve efficiency, notably in stock breeding and dairy farming. The principal crops are corn and potatoes.

Liechtenstein has no raw materials or fuels, so all these items have to be imported. There is no heavy industry either, and consequently little pollution. Smallscale manufacturing produces a variety of goods, including optical, electronic and precision equipment, and various consumer goods.

Healthcare in the principality is excellent, and the Liechtenstein government contributes to facilities in neighboring Switzerland, so that they are made available to Liechtensteiners too. Compulsory insurance schemes financed by workers, employers and the government provide social security. Educational standards and literacy levels are both extremely high: Liechtensteiners receive 11 years of compulsory education, which may be followed by study at a technical, vocational, academic or other institution.

NATIONAL DATA - LIECHTENSTEIN	
Land area 160 sq km (62 sq mi)	
Climate temperate continental	
Major physical features highest point: Grauspitz 2,599 m (8,527 ft)	
Population (2006 est.) 33,987	
Form of government Hereditary constitutional monarchy on a democratic and parliamentary basis	
Capital city Vaduz (5,038)	
Official language German	
Ethnic composition Alemannic 86%; Italian, Turkish, and other 14%	
Religious affiliations Roman Catholic 76.2%; Protestant 7%; unknown 10.6%; other 6.2%	
Currency 1 Swiss franc (CHF) = 100 centimes	
Gross domestic product (2001) U.S. $1.786 billion	
Gross domestic product per capita (1999) U.S. $25,000	
Life expectancy at birth male 76.1 yr; female 83.28 yr	
Major resources hydroelectric potential, grapes, livestock, maize (corn), potatoes, timber, vegetables, wheat, tourism	

Austria

REPUBLIC OF AUSTRIA

Today Austria is a small federal republic, but it was formerly one of the greatest empires in the history of Europe. Although its power has declined, its geographical situation on the frontier of Western and Eastern Europe ensures Austria's continuing influence on the continent's political and economic life. The magnificent Alpine scenery, together with the cosmopolitan atmosphere and cultural attractions of its cities, also make Austria a major center of tourism.

NATIONAL DATA – AUSTRIA

Land area	82,444 sq km (31,832 sq mi)			

Climate		Temperatures	Annual	
	Altitude m (ft)	January °C(°F)	July °C(°F)	precipitation mm (in)
Vienna	212 (695)	-3 (27)	20 (68)	620 (24.4)

Major physical features highest point: Grossglockner 3,801 m (12,470 ft); longest rivers: Danube (part) 2,850 km (1,770 mi), Inn (part) 510 km (317 mi)

Population (2006 est.) 8,192,880

Form of government federal multiparty republic with two legislative houses

Armed forces army 33,200; air force 6,700

Largest cities Vienna (capital - 1,651,365); Graz (244,537); Linz (188,407); Salzburg (148,549); Innsbruck (116,881)

Official language German

Ethnic composition Austrians 91.1%; former Yugoslavs 4% (includes Croatians, Slovenes, Serbs, and Bosniaks); Turks 1.6%; German 0.9%; other or unspecified 2.4%

Religious affiliations Roman Catholic 73.6%; Protestant 4.7%; Muslim 4.2%; other 3.5%; unspecified 2%; none 12%

Currency 1 euro (EUR) = 100 euro cents

Gross domestic product (2006) U.S. $279.5 billion

Gross domestic product per capita (2006) U.S. $34,100

Life expectancy at birth male 76.17 yr; female 82.11 yr

Major resources oil, coal, lignite, timber, iron ore, copper, zinc, antimony, magnesite, tungsten, graphite, salt, hydropower, anhydrous gypsum, barley, cattle, clay, hay, fodder, kaolin, maize (corn), natural gas, potatoes, sheep, sugar beet, talc, temperate fruits, vines/wine, wheat, tourism

GEOGRAPHY

Mountainous and landlocked, Austria is bordered by Switzerland and Liechtenstein to the west; Germany to the northwest; the Czech Republic and Slovakia to the north and northeast; Hungary to the east; Slovenia to the southeast; and Italy to the south.

The land
Austria can be divided into three main topographical areas: the Alps; the lowlands in the east; and the hill country along the borders with the Czech Republic and Slovakia.

The Alps cover nearly two-thirds of the country, stretching from west to east in several major ranges. The central mountain chain, the Tauern, includes some of the highest peaks. On either side of the Tauern, separated by deep river valleys, their streams fed by snowfields and glaciers, are two younger, more rugged limestone ranges. Some of the most majestic scenery is in the Ötztaler Alps near the borders with Switzerland and Italy.

The mountains gradually fall away into the Danube valley in the north and to the Vienna basin in the east. These lowlands are Austria's main agricultural area, with vineyards on the valley slopes. The forested hills north of the Danube become more mountainous along the frontier with Slovakia, the Czech Republic and Germany.

The Danube, Europe's second longest river, flows from west to east across northern Austria, its tributaries draining much of the country. Most of Austria's lakes occur around the flanks of the Alpine ranges. But the largest lakes are at either end of the country: the slightly saline, shallow Lake Neusiedler on the border with Hungary and Lake Constance in the far west, lying mainly in Germany and Switzerland.

Climate
The southern and western mountains have a typically Alpine climate, with high rainfall throughout the year and frequent summer thunderstorms; the northeast of the country has a drier, more continental

Tirolean landscape Coniferous forests clothe the lower slopes of the Alpine province of Tirol, where snow often remains on the ground until well into the spring. Higher up, the permanent snow cover has given rise to numerous popular ski resorts.

climate. In general, Austria has warm, humid summers and bitingly cold winters, when temperatures often plunge below freezing. In the more remote Alpine valleys snow may remain well into spring, while higher mountains are snow-covered all year round.

Plants and animals

Austria is largely covered with either forests or meadows. The Alpine areas support coniferous forests on the lower slopes and pastureland above, with Alpine plants at higher levels, including the rare edelweiss. In the lower valleys, the conifers give way to deciduous woodland and wild fruit trees. Bears, wolves and wild boars are now virtually extinct. Strict conservation laws are in force to protect the remaining rare ibex and chamois in the high mountains, and birds such as eagles, falcons and owls.

SOCIETY

Situated at the heart of Europe, Austria has a long history of wars, invasions and territorial struggles. The modern federal republic, however, is a stable and neutral country, with a reputation as an international center of culture.

History

The area has been inhabited since the early Stone Age. From about 1000 BC the Illyrian people from the Adriatic coast established a thriving Iron Age culture at Hallstatt in Upper Austria. The Celts founded a kingdom in the eastern Alps in about 400 BC. Some 200 years later the Romans arrived: roads were built and vineyards planted. Early invasion by Germanic peoples from the north were repulsed, but by the 5th century Germanic Bavarians and Slavs occupied a large part of the country.

Conquest by the Frankish emperor Charlemagne (742–814) led to increasing Christianization. However, his empire soon fragmented, and Magyars invaded from present-day Hungary in the east. The Magyars were eventually repelled, and in 976 a distinctive Austrian domain had emerged within the German Holy Roman Empire. The Babenberg rulers of Austria extended the country's boundaries, and presided over its cultural and economic development.

The Babenberg dynasty was followed by that of the Germanic Habsburgs, who were to rule Austria until 1918. By the 16th century, the Habsburgs had extended their empire to include Burgundy, Spain, the Netherlands and parts of Italy, and gained control of Spain's American colonies. However, Habsburg claims over neighboring Hungary and Bohemia were strongly resisted. The 16th and 17th centuries were marked by battles against Turkish incursions from the southeast and by the repression of Protestant reformers at home. The Thirty Years War (1618–48) brought the Habsburgs into conflict with Bohemia, Sweden and Denmark. By 1648, however, the empire's grip on Hungary and Bohemia was firm. The 18th century saw conflicts with France (1701–14) and with Prussia (1740–48), the latter over who was to inherit Habsburg territories on the death of the last male Habsburg, Charles VI (1711–40). The War of Austrian Succession lasted eight years (1740–48).

Though weakened by almost continuous war, Austria remained a great power until the Napoleonic Wars (1792–1815). At this stage the Habsburg empire included Croatians, Hungarians, Czechs, Italians, Slovaks and Poles. The map of Europe was redrawn in 1815 at the Congress of Vienna: Austria presided over the German Confederation – a grouping of 39 independent German states.

The restoration of Austria's power in Europe had been mainly due to the political skills of Prince Metternich (1773–1859), the Austrian foreign minister. But his policies provoked major unrest and reinforced Austria's reputation for conservatism and repression. Opposition to the government's denial of liberal and nationalist rights led to the 1848 revolution. Metternich was forced to resign, and there followed a brief period of reform. Austria was permanently weakened; unrest among the empire's ethnic minorities increased, as did Prussian domination over the German states.

After the Austro–Prussian War of 1866, Austria was expelled from the German Confederation. A new constitutional arrangement created the Austro–Hungarian empire. Ethnic unrest continued, and the assassination of Archduke Franz Ferdinand (1863–1914), the heir to the throne, by a Serb nationalist at Sarajevo (in present-day Bosnia) triggered World War I. After the war Austria was proclaimed a republic within its present borders.

With so much of its land and resources lost, Austria could not hold off economic depression or German domination. In 1938, Adolf Hitler (1889–1945) annexed Austria and the country was incorporated into the German Third Reich. At the end of World War II Austria was occupied by Allied forces. On 13 May 1955, the Austrian State Treaty established Austria as a sovereign independent republic committed to permanent neutrality, and all foreign soldiers left Austrian land. Since then, the country has been politically and economically stable. In 1989, Austria became the first destination for thousands of refugees as the borders of Eastern Europe were opened up.

Government

Austria is made up of nine semi-autonomous *Länder*, or provinces. The president and head of state is elected for a six-year term; he appoints the chancellor and cabinet, who must command a majority in the National Council (183 members), the more influential of the two legislative houses, elected every four years by a system of proportional representation. The Federal Council represents the

Länder, and its 64 members are elected by the provincial assemblies.

The *Länder* are responsible for implementing local legislation; each has its own administration led by a governor elected by the provincial assembly, which is itself elected by popular vote.

The Socialist Party of Austria (SPO) was for many years the largest single party in government. The main opposition party, the center-right Austrian People's Party (VPO), regularly featured in coalition governments with the Socialist Party. In 1999, the right-wing Freedom Party of Austria (FPO), led by Joerg Haider, made substantial electoral gains, taking 6 out of 10 ministerial posts in the government formed early in 2000.

People

Postwar Austria has become increasingly urbanized, with nearly one-fifth of the population living in the capital, Vienna. The vast majority are German-speaking Austrians, with a few minorities, mainly of neighboring nationalities such as Hungarians, Croats and Slovenes. German is spoken in a variety of dialects; those of Tirol, Carinthia and Vorarlberg are particularly strong. Although most Austrians are Roman Catholic, the Lutheran Church has a significant membership; the 12,000 members of the Jewish community, much diminished since World War II, are concentrated in Vienna.

The country has long been famed for its cultural riches in many fields, especially music: Austrian composers have included Joseph Haydn (1732–1809), Wolfgang Amadeus Mozart (1756–91) Franz Schubert (1797–1828), Johann Strauss the Elder 1804–49) and the Younger (1825–99), and Anton Bruckner (1824–96). The German-born composer Ludwig van Beethoven (1770–1827) also spent virtually all his working life in Vienna. Among later Austrian figures were Arnold Schoenberg (1874–1951), founder of the modern 12-tone school of composition, and Alban Berg (1885–1935). Today the Vienna Philharmonic Orchestra and Vienna State Opera – whose directors have included Gustav Mahler (1860–1911) and Herbert von Karajan (1908–92) – have outstanding international reputations. Austrian painters have included Gustav Klimt (1862–1918), Egon Schiele (1890–1918) and Oscar Kokoschka (1886–1980); and in the realm of psychoanalysis, the psychiatrist Sigmund Freud (1856–1939). Folk culture thrives in the Tirol in the west, where it receives support from the provincial government.

Fairytale architecture The Tirolean tradition of painting architectural features, floral wreaths and cherubs on their buildings has been lavishly upheld on this expensive craft shop in Seefeld, a fashionable winter sports resort to the northwest of Innsbruck.

ECONOMY

Austria's position between Western and Eastern Europe has helped to give the country an important trading role. The republic has successfully transformed its economy from an agricultural to a modern industrial one. Combining market principles with a high level of state control, Austria has continued to expand its economy while managing to keep inflation and unemployment down.

Agriculture and forestry

Agriculture no longer dominates the economy as it once did. Austrian farmers are generally conservative, and smallholdings predominate. Although only 20 percent of the land is arable, the intensive modernization of farming practices has made the country self-sufficient in 75 percent of its cereal and vegetable needs. Most of the arable land is in the eastern lowlands, where barley, wheat, maize and oats are grown together with fruit and vegetables. Vineyards, mostly in the provinces of Lower Austria and Burgenland – and in Vienna itself – supply a flourishing wine industry. Livestock, mainly dairy cattle and pigs, are raised on open pasture in the east, and on Alpine pasture in the west. The fish catch is limited to freshwater species.

Forests cover about 40 percent of the land, a greater proportion than in any other European country, apart from Sweden and Finland. Forestry is divided between smallholdings and large estates, some of them state-owned. The forests provide timber for domestic use as building material, fuel and matches, and paper is manufactured in the southeast at Graz. Untreated lumber is exported.

Trade and commerce

Monetary policy is largely controlled by the state; the Austrian National Bank is partly owned by the government, though other banks and financial institutions operate freely. Austria's dominant trading partner is Germany, followed at a distance by Italy and Switzerland. In 1960, Austria became a founder member of the European Free Trade Association (EFTA), and after 1977 it had a free-trade agreement with the European Community, now the European Union (EU), by which trade in certain industrial goods was duty-free. However, in 1995, Austria became a full member of the EU.

Although Austria has a balance-of-payments deficit, this has so far been largely offset by income from tourism, Austria's biggest invisible export. Millions of foreign visitors are drawn to Austria throughout the year by the beautiful mountain and lakeland scenery, the excellent skiing facilities, and the architectural and cultural attractions of both Vienna and Salzburg.

Transportation and communications

Although Austria lies at the center of the European transportation routes, its mountainous terrain imposes restrictions on communications; road and rail routes have to follow the river valleys. This has been partly overcome by the construction of several major tunnels. The roads are excellent, with an extensive autobahn (expressway) system between major cities that also links up with the highways of Germany and Italy. Austrian railroads are state-owned; more than half the tracks, including all main lines, are electrified.

The Danube, which flows through eastern Austria for some 358 km (224

Festival of flowers Farmers in Austria decorate their cattle with sprays of Alpine flowers in celebration of the spring thaw. Along with tourism, smallscale dairying and beef production are the main sources of income in Alpine areas.

miles), is the main navigable river, and the federally owned Danube Steamship Company carries freight and passengers along the route between Germany and the Black Sea. Austrian Airlines runs domestic flights between Vienna, Linz, Salzburg, Graz and Klagenfurt (in the south of Carinthia), as well as flights to European and other destinations from the main airport at Schwechat, about 19 km (12 miles) southeast of Vienna.

Radio, television and telephone services are well developed. All the broadcasting media are run by the state-owned Austrian Broadcasting Company, but a specific broadcasting law prohibits censorship or interference by the state. There are 17 daily newspapers, the most important – including the *Neue Kronen–Zeitung*, with two-and-a-half million readers – being based in Vienna.

Welfare and education

The state provides a full social welfare system, including unemployment, old age, disability and maternity benefits. National health insurance covers all medical and hospital treatment, and the standard and provision of care is excellent.

Education in Austria is free and compulsory, with ample provision for further, higher and adult education; literacy levels are extremely high. Austria has some 15 universities, including technical and other specialized institutions, and several academies for music and the arts. The University of Vienna, founded in 1365, is one of Europe's oldest.

Industry

Austria is rich in mineral and energy resources, and both mining and energy production are nationalized industries. The country is one of the world's largest producers of magnesite, used in the chemical industry. Other materials mined include iron ore, lignite (brown coal), graphite, salt, quartz and clay. Austria is one of Europe's leading oil producers, although reserves of petroleum and natural gas are diminishing. Hydroelectricity supplies most of the country's power.

The nationalized iron and steel industry is a model of efficiency, and nearly half of the annual production of 2 million tonnes of pig iron and 4.9 million tonnes of crude steel is exported. Austria also has one of the world's largest aluminum plants. There are important petroleum and chemical industries, as well as plants manufacturing plastics and electrical goods. Optical instruments for export are assembled near Salzburg.

Eastern Europe

COUNTRIES IN THE REGION

POLAND · CZECH REPUBLIC · SLOVAKIA

HUNGARY · ROMANIA · BULGARIA

YUGOSLAVIA (SERBIA AND MONTENEGRO)

SLOVENIA · CROATIA · BOSNIA HERZEGOVINA

FORMER YUGOSLAV REPUBLIC OF MACEDONIA

ALBANIA

■ capital city
● major town

height of land (meters)

2000
1000
500
200
0

▲ mountain peak

Baltic Sea

Gulf of Gdansk

Gdansk
Elblag
Olsztyn
Szczecin
Bydgoszcz
Bialystok
Bug
Poznan
Warta
Vistula
Warsaw
POLAND
Lodz
Wroclaw
Lublin
Oder
Silesian Plain
Vistula
Krakow
Ore Mts
Sudetic Mts
Mt Snezka ▲ 1603
Prague
Plzen
Bohemian Forest
CZECH REPUBLIC
Ostrava
Vltava
Brno
Carpathian Mountains
Tatra Mts
Gerlach Peak ▲ 2665
Slovakian Ore Mts
Kosice
SLOVAKIA
Vah
Bratislava
Miskolc
Kékes ▲ 1015
Tisza
Alföld
Györ
Budapest
HUNGARY
Oradea
Cluj-Napoca
Iasi
Somes
Carpathian Mts
Triglav ▲ 2864
Ljubljana
SLOVENIA
Kras
Lake Balaton
Danube
Great Alföld
Pécs
Subotica
Timisoara
Bihor Mts
ROMANIA
Brasov
Mures
Stret
Galati
Zagreb
CROATIA
Istria
Rijeka
Krk
Drava
Osijek
Novi Sad
Transylvanian Alps
Mt Negoiu ▲ 2548
Cres
Sava
Banja Luka
Oli
Pag
Zadar
Dinaric Alps
BOSNIA HERZEGOVINA
Sarajevo
Drina
SERBIA
Morava
Craiova
Bucharest
Constanta
Split
Dalmatia
Mostar
YUGOSLAVIA
Ruse
Black Sea
Brac
Hvar
Durmitor ▲ 2522
MONTENEGRO
Nis
Pleven
Varna
Danube
Korcula
Mljet
Dubrovnik
Lake Shkoder
Daravica ▲ 2656
Sofia
Balkan Mountains
BULGARIA
Burgas
Stara Zagora
Musala ▲ 2925
Maritsa
Adriatic Sea
Shkodër
Skopje
Rhodope Mts
Plovdiv
Durrës
Tiranë
FORMER YUGOSLAV REPUBLIC OF MACEDONIA
Pirin
L Ohrid
L Prespa
ALBANIA

Count Dracula's abode (*left*) High in the Transylvanian plateau of Romania, encircled by mountains, is the town of Sighisoara, birthplace in about 1430 of the bloodthirsty Prince Vlad Tepes, or Vlad the Impaler, inspiration for the fictional character Count Dracula.

Poland

THE REPUBLIC OF POLAND

T HE EAST CENTRAL EUROPEAN STATE OF Poland lies on the southern shore of the Baltic Sea, with Germany to the west, the Czech Republic and Slovakia to the south, and the Ukraine and Belarus to the east. To the northeast it adjoins Lithuania and a small enclave of the Russian Federation.

GEOGRAPHY

Poland is mainly low lying. The only mountainous areas are the Sudetic Mountains in the southwest, and the Tatra and Carpathian Mountains in the far south along the border with Slovakia.

The land

From north to south Poland is made up of five main geographical belts. In the north-west, the lagoons and bays of the Baltic coast are fringed with broad, sandy beaches and dunes rising to some 30 m (100 ft). South of this plain, but reaching the coast near the port of Gdansk, is a range of hills known as the Baltic Heights. Glacial in origin, they enclose the sandy lake-strewn hinterland of Pomerania.

The central lowlands to the south, which cover one-third of the country, are Poland's richest agricultural area. Here the broad plains are covered with fertile loess soil and are crossed by marshy river valleys. In the east the capital, Warsaw, lies on the banks of Poland's longest river, the Vistula.

A series of uplands marks the southern boundary of the central plains. The two main areas here are the Lublin plateau lying to the east of the Vistula valley and the uplands of Little Poland in the center. Little Poland includes two higher ranges: the Jura, running north from the central southern city of Kracow, and the thickly forested Holy Cross Mountains a little farther to the northeast. Poland's highest mountains lie in the far south straddling

Rolling green pastures (*above*) extend across a vast area of central and southern Poland. In contrast to the north, the soil here is fertile and the climate favorable. The gently undulating southern uplands also support wheat and sugar beet.

The blue-gray waters of the Baltic (*right*) along the northern coast of Poland moderate the cold northern climate. During World War II boundary changes settled by the Allies restored 700 km (435 mi) of coastline to Poland, including the port of Gdansk,

the Slovakian border, broken in the center by the Moravian Gate where the river Oder enters from the south. The southwestern province of Silesia includes the northern slopes of the Sudetic Mountains. In the southeast the massive central ridge of the Carpathian Mountains enters Poland where the border loops southward into the spectacular glaciated scenery of the Tatra Mountains.

Climate

The lack of mountain barriers within Poland makes for few climatic contrasts. The south and east of the country have a continental climate, with cold dryish winters and warm wetter summers. In the

northwest the moderating effect of the sea and of westerly winds means that summers are cooler and the winters milder, though wetter, than elsewhere. To the south, in the Sudetic and Carpathian Mountains, winter snows may be heavy.

Plants and animals

Much of Poland, once a densely forested area, is now farmland, but large areas of forest remain. Good forest management has preserved some areas of primeval forest, most notably the Bialowieza National Park on the Belorussian border in the east. This protected area also shelters the last remaining European bison, once common throughout the continent.

SOCIETY

The history of Poland is a long and complex one. The modern republic was formed after World War II. In 1989 Poland was among the first of several European countries to break free of the communist system.

History

Tradition dates the foundation of Poland to 966 AD, when Mieszko I (c. 930–92), prince of the Polanie (the plain dwellers), was converted to Christianity. In 1386 Jagiello (c. 1351–1434), Grand Duke of Lithuania, married Jadwiga (1373–99), Queen of Poland, creating a new Jagiellon dynasty. In 1466 the Jagiellons subjugated the Teutonic Knights, the German rulers of Prussia in the north, opening a Polish gateway to the Baltic.

When the Jagiellon line died out Poland became a royal republic, whose monarchs were elected by a council of nobles called the Sejm. In 1587 Sigismund III Vasa (1566–1632) was elected King of Poland. Five years later he inherited the Swedish throne, and his attempts to unite the mainly Roman Catholic Poland with Protestant Sweden embroiled the countries in a war that lasted 35 years (1599–1635).

In the mid 17th century the political fortunes of Poland went into decline. The 20-year reign of John II Casimir Vasa (1609–72) saw the loss of eastern Poland to Russia and temporary occupation by Sweden. By the late 18th century, after losing the southeastern territories (including much of the Ukraine) to the Ottoman Turks, Poland was severely weakened and at the mercy of the military might of Prussia, Austria and Russia. After successive partitions in 1772, 1793 and 1795, all Polish territories were divided between these powers, and Poland temporarily disappeared from the map.

During the Napoleonic Wars (1803–15), rebellious Poles helped the Napoleonic armies to capture Warsaw, resulting in the creation of the Duchy of Warsaw. By 1813 this too had come under Russian control. Two years later Alexander I of Russia (1777–1825) established the Congress Kingdom of Poland in an effort to win Polish loyalty, but he failed to do so. A succession of Polish rebellions in 1830, 1846 and 1863 led to ever more repressive legislation by Russia.

During World War I the Congress Kingdom of Poland was invaded by Austria and Germany. After the Russian Revolution of 1917, the new Bolshevik government recognized Poland's right to independence, and in 1918, as the war ended, a second Polish republic was established. In 1926 Jósef Pilsudski (1867–1935), Poland's first president, staged a military coup. In 1939, only four years after Pilsudski's death, Poland was invaded by Nazi Germany, triggering the outbreak of World War II. When Germany attacked the Soviet Union in 1941, many Poles joined the Soviet forces. They continued to fight alongside the Soviet Union

---

NATIONAL DATA - POLAND

Land area 304,465 sq km (117,555 sq mi)				

Climate

	Altitude m (ft)	Temperatures January °C(°F)	July °C(°F)	Annual precipitation mm (in)
Gdansk	12 (39)	-2 (28)	17 (63)	497 (19.5)
Warsaw	110 (361)	-2 (28)	18 (64)	520 (20.4)

Major physical features highest point: Rysy (Tatra Mountains) 2,498 m (8,197 ft); lowest point: near Gulf of Gdansk -10 m (-33 ft); longest river: Vistula 1,086 km (675 mi)

Population (2006 est.) 38,536,869

Form of government multiparty republic with two legislative houses

Armed forces army 89,000; navy 14,300; air force 30,000

Largest cities Warsaw (capital - 1,692,854); Lodz (774,004); Krakow (757,430); Wroclaw (636,268); Poznan (570,828); Gdansk (459,072); Szczecin (411,900); Bydgoszcz (368,235); Lublin (355,998)

Official language Polish

Ethnic composition Polish 96.7%; German 0.4%; Belarusian 0.1%; Ukrainian 0.1%; other and unspecified 2.7%

Religious affiliations Roman Catholic 89.8% (75% practicing); Eastern Orthodox 1.3%; Protestant 0.3%; other 0.3%; unspecified 8.3%

Currency 1 zloty (PLN) = 100 groszy

Gross domestic product (2006) U.S. $542.6 billion

Gross domestic product per capita (2006) U.S. $14,100

Life expectancy at birth male 70.95 yr; female 79.23 yr

Major resources coal, sulfur, copper, natural gas, oil, silver, lead, salt, amber, cereals, fish, iron ore, livestock, oil seed, potatoes, sugar beet, timber, zinc

Pope John Paul II (*above*) on his visit to Poland in 1979. Born Karol Wojtyla, he was Archbishop of Cracow before being elected pope in 1978. The Roman Catholic faith thrived in Poland in spite of communist repression, and John Paul's election was seen as a sign of hope for the nation.

until the Germans were finally driven out of Poland in the winter of 1945. Devastated by war, and with a population reduced by almost 11 million, Poland became a Soviet satellite; the constitution of 1952 inaugurated a communist state.

Protests and strike action against the communist regime in 1956 and 1970 were brutally suppressed. Nothing, however, could stifle the depth of religious feeling among Poland's predominantly Roman Catholic population. Hopes were renewed when, in 1978, a Polish cardinal, Karol Wojtyla (b. 1920), was elected pope, taking the name John Paul II. His triumphant visit to Poland in 1979 highlighted the strength of the Roman Catholic Church and boosted the morale of the people. Within a year the shipyards at Gdansk were paralyzed by a series of strikes organized by the charismatic Lech Walesa (b. 1943), leader of the then illegal trade union *Solidarnosc* (Solidarity). Some concessions were made by government, and the party leader was replaced, ultimately by General Wojciech Jaruzelski (b. 1923).

Eight years of further Soviet repression (1981–89) failed to hold back the tide of dissent. Additional strikes, many led by the now officially recognized Solidarity union, forced Jaruzelski's government to negotiate. The results were startling, and 1989 saw major moves toward a free-market economy. In a partially free democratic election, Solidarity candidates took all the seats in the Sejm reserved for the opposition (35 percent). Jaruzelski found himself forced to accept Tadeusz Mazowiecki (b. 1927), a noncommunist, as the new prime minister. Stringent austerity measures, needed to adapt and

Birthplace of Solidarity (*left*) The Baltic port of Gdansk was a pawn in Eastern European politics for 700 years. After it was returned to Poland in 1945, its shipyard workers founded the Solidarity movement that toppled communist rule.

A costumed dancing troupe (*right*) performs traditional Polish dances such as the polonaise and the mazurka. Strong cultural traditions helped to keep the Polish national identity intact during its long history of foreign domination.

to revive the economy, led to split loyalties within Solidarity, and Mazowiecki was forced to resign in 1990 amid a bitterly fought election. The same election made Lech Walesa state president in December 1990, a position he lost in elections in November 1995.

Government

The constitution of 1997 provides for a two-chamber parliament: the lower house, the Sejm, has 460 members; and the upper house, the Senate, has 100 members, with power of veto over measures proposed by the Sejm. Representatives of both houses are directly elected for four-year terms, while the head of state is a freely elected executive president.

People

World War II drastically reduced Poland's former ethnic diversity. The deportation of Jews or their massacre at the hands of the Nazis, the mass exodus of many Germans and major boundary changes have meant that the vast majority of today's population are Poles. Despite centuries of foreign domination, Poles have a rich cultural heritage, especially in music, literature and film-making, that has often been encouraged rather than eclipsed by attempts to suppress it. Figures of renown include the astronomer Copernicus (1473–1543), the composer Frédéric Chopin (1810–49), the novelist Henryk Sienkiewicz (1846–1916) and the film director Andrzej Wajda (b. 1926).

ECONOMY

In 1989 Poland dismantled its centralized, state-controlled economy in favor of a free-market system. The major industries were privatized, and price and wage subsidies abolished; foreign investment was encouraged, while interest and exchange rates were set at commercial levels. By the late 1990s the economy was one of the fastest-growing ex-communist countries with a vibrant private sector.

Agriculture, industry and commerce

Almost half of Poland's land area is given over to arable farming: the main crops are potatoes, sugar beet and cereals. Extensive areas are assigned to pastureland, supporting pigs and cattle. Forestry produces sawn timber for export.

The rich reserves of bituminous coal in Upper Silesia have provided Poland with a valuable export income for many years. Other mineral resources include sulfur in the southeast, and zinc and lead in the south near Katowice; in 1985 petroleum was discovered off the coast. Most of Poland's electricity comes from coal-fired stations. The communist years emphasized engineering and heavy industry at the expense of consumer manufacturing, and adapting to a free-market economy is requiring extensive change.

Transportation and communications

Poland has extensive road and railroad networks. Many of the railroads were electrified after World War II. The road network, once an important link between Soviet bloc countries for freight transportation, is also well developed. The main ports at Szczecin, Gdynia and Gdansk have busy international shipping schedules. The state airline, Polskie Linie Lotnicze (Lot), provides travel to internal and international destinations.

Polish Radio and Television (PRT) has three radio and two television channels, and also satellite links with western European networks.

Welfare and education

Healthcare is free for all workers and their families, and also extends to farmers, pensioners, students and invalids. Welfare services are also provided. However, facilities are often stretched and foreign support is needed to maintain an adequate level of care. Education is free and compulsory from the age of 7 to 16; free higher education is available at general or vocational schools, polytechnics and universities.

The Czech Republic

IN 1993 THE FORMER EASTERN BLOC STATE OF Czechoslovakia split into two new republics: the Czech Republic and Slovakia. Most of the Czech Republic occupies the territory of the ancient Kingdom of Bohemia; Moravia and parts of historical Silesia (the rest of which is in modern Poland) make up the remainder of this new state. The Czech Republic is bordered by Slovakia to the east, Poland to the northeast, Germany to the north and west and Austria to the south.

GEOGRAPHY

In the west of the Czech Republic the Bohemian massif and plateau occupy the greater part of the former Kingdom of Bohemia. This high broken plateau is encircled by a number of mountain chains: the Ore Mountains in the northwest, and the Bohemian Forest in the southwest, both bordering Germany; the Sudetic Mountains to the northeast along the border with Poland; and the much lower Bohemian–Moravian Heights in the southeast. The Carpathian Mountains separate the Czech Republic from Slovakia. The central plateau is crossed by the Elbe (Labe) and Vltava rivers, which merge near the capital, Prague. Southeast of the Bohemian–Moravian Heights lies the Morava river valley in the heart of Moravia; here hills and valleys surround the ancient Moravian capital of Brno.

The Czech Republic lies in the middle of a climatic transition zone. To the west of the country the climate is mostly maritime, but to the east it is increasingly continental. Precipitation is generally low on the Bohemian plateau and in the plains, both of which are sheltered from winds bringing rain and snow by the mountains of Slovakia.

SOCIETY

Among the early inhabitants in the present Czech region were the Celtic Boii people, from whom Bohemia took its name. Successive waves of westward moving peoples overran this part of Europe, notably the Slavs in the 5th century AD, who lived as herdsmen and hunters. The Slavic Kingdom of Bohemia was set up by the legendary Přhemysl dynasty in about 800, after the Frankish emperor Charlemagne (c. 742–814) had suppressed the powerful neighboring Avars. After conquering Moravia, a dukedom created by Charlemagne, the Přhemysls became independent kings within the Holy Roman Empire in 1198.

The austere beauty of Prague, capital of the Czech Republic. Founded about 800 AD, the city straddles the Vltava river. The Old and Lesser towns are linked by a series of bridges including the Charles Bridge, built in 1357 and seen here crowded with people.

On the extinction of the Přhemysl line in 1306, Hungary's nobility gave the throne to John, prince of Luxembourg (1296–1346), whose son Charles I (1316–78) became Holy Roman Emperor in 1355, making Prague his imperial capital. During the reign of the Bohemian king Wenceslas IV (1361–1419), the preacher Jan Hus (1372–1415) led a popular campaign for religious reform among the Czechs. When Hus was betrayed and burned at the stake for heresy, his followers, the Hussites, began a series of

The Bohemians were soundly defeated, heralding a long period during which the Czechs came under harsh imperial rule from Austria, and the German language was imposed as the medium of authority. In the late 18th and early 19th centuries, however, a new sense of national identity was fostered with a revival of Czech language and culture. In 1918, following the collapse of the Austro–Hungarian empire in World War I, the republic of Czechoslovakia was created, with Tomaš Masaryk (1850–1937) as its first president. Following invasion by Germany during World War II the communists seized power in 1948, and in 1960 the "people's republic" was changed to "socialist republic". The republic lasted until 1 January 1993 when the two main regions split to form fully independent countries.

ECONOMY

With the advent of communism in the 1950s, agriculture was collectivized. Since then, industry has encroached heavily on farmland, and the country is no longer self-sufficient in food. Major crops include wheat, sugar beet, barley and potatoes, but yields are declining. Hops are grown around the western city of Plzen, home of Pilsner beers. There are also vineyards along the Vltava river and in southern Moravia. Pigs, poultry and cattle are the main livestock animals.

Mineral resources are limited, the most important being coking coal. Poor quality brown coal (lignite) is also mined, mostly for power generation. The use of lignite as a fuel is controversial since it causes heavy industrial pollution. Small quantities of petroleum and natural gas are also produced domestically, but most of these fuels have to be imported. Local and abundant sources of coal and iron ore have made iron and steelworking the major industries in the Czech Republic. In former Czechoslovakia other sectors also operated successfully, producing a range of manufactured goods from heavy machinery to highly exportable products such as Bohemian glass and Pilsner beer.

Until 1990, trade was mainly with the Soviet Union and Czech goods were barely competitive in the world markets. In the early 1990s a steadily increasing budget deficit – nearly 10 percent of GDP in 1996 – led to investment failures and a currency crisis. Since then the government has introduced austerity packages and a revitalization program aimed at attracting foreign investment and privatizing state enterprises.

"Good King Wenceslas", patron saint of the Czech Republic. He helped convert the ancient kingdom of Bohemia to Christianity, but in 929 was murdered by his heathen brother. His statue stands in Wenceslas Square in the heart of Prague.

violent revolts against the empire, plunging Bohemia into near anarchy.

The Czech and Slovak peoples were united under one ruler in 1471, when Bohemia became part of a joint kingdom with Poland and Hungary. In 1526 the Bohemian crown passed to Ferdinand of Habsburg (1503–64), the Austrian heir to the Holy Roman Empire. The Habsburgs, although obliged to tolerate Protestantism, gave power to a minority of pro-Catholic officials. In 1618, the so-called Defenestration of Prague (in which the Protestant nobility had imperial governors thrown out of a high window into a conveniently placed dungcart) sparked off a wholesale rebellion that developed into the Thirty Years War (1618–48).

NATIONAL DATA – CZECH REPUBLIC				
Land area 77,276 sq km (29,836 sq mi)				
Climate		Temperatures		Annual precipitation
	Altitude m (ft)	January °C(°F)	July °C(°F)	mm (in)
Prague	374 (1,227)	-3 (27)	18 (64)	526 (20.7)
Major physical features highest point: Mt Snezka (Sudetic Mountains) 1,602 m (5,256 ft); longest rivers: Elbe (part) 1,165 km (724 mi), Vltava 430 km (267 mi)				
Population (2006 est.) 10,235,455				
Form of government multiparty parliamentary democracy with two legislative houses				
Armed forces army 16,663; air force 5,609				
Largest cities Prague (capital - 1,181,610); Brno (366,757); Ostrava (310,078); Plzen (162,759)				
Official language Czech				
Ethnic composition Czech 90.4%; Moravian 3.7%; Slovak 1.9%; other 4%				
Religious affiliations Roman Catholic 26.8%; Protestant 2.1%; other 3.3%; unspecified 8.8%; unaffilfated 59%				
Currency 1 Czech koruna (CZK) = 100 halura				
Gross domestic product (2006) U.S. $221.4 billion				
Gross domestic product per capita (2006) U.S. $21,600				
Life expectancy at birth male 72.94 yr; female 79.69 yr				
Major resources hard coal, lignite, kaolin, clay, graphite, timber, antimony, cereals, iron ore, livestock, magnesium, mercury, potatoes, sugar beet, uranium, beer, tourism				

Slovakia

REPUBLIC OF SLOVAKIA

SLOVAKIA – UNTIL 1993 PART OF FORMER Czechoslovakia – is a newly created state at the center of Europe. The "Velvet Divorce" from the Czech Republic in January 1993 gave Slovakia complete independence for the first time in over a thousand years.

GEOGRAPHY

Slovakia is mostly mountainous, with lowland plains around the basin of the Danube on the southern border and on the eastern border with the Ukraine. To the west, the western Carpathian Mountains rise sharply from the border with the Czech Republic. Central Slovakia contains the parallel ranges of the Tatra and Slovakian Ore Mountains.

Slovakia's climate is continental, with cold winters, and warm summers, often with thunderstorms. In the far east and in the mountains winters can be very severe with heavy snow or rain. Precipitation is lower on the sheltered plains.

SOCIETY

Illyrian, Celtic and German peoples were the first settlers in modern Slovakia, and in the 6th century AD Slavic peoples arrived from Russia and the east. After the collapse in the 10th century of the Moravian empire, Slovakia began a long period under Hungarian control. Following the breakup of the Austro-Hungarian empire in 1918, the republic of Czechoslovakia was created, with Tomaš Masaryk (1850–1937) as its first president.

In the early 1990s, following the dissolution of the Soviet Union and the subsequent political upheaval in Eastern Europe, the Slovaks and the Czechs agreed to separate. From 1 January 1993 Slovakia gained full independence as a republic in its own right.

Czechoslovakia 1918–92

When the Austro-Hungarian empire collapsed at the end of World War I, the reorganization of its former territories and disparate peoples within new borders took no account of the cultural and nationalist aspirations of the peoples involved. Although obvious linguistic and geographic links between Czechs and Slovaks made the formation of the Republic of Czechoslovakia in 1918 seem a logical step, the union of these two peoples within one state was to be an uneasy one.

In 1918, Tomaš Masaryk (1850–1937) was elected president of the Republic of Czechoslovakia. He was faced with the task of uniting the powerful Czech kingdom of Bohemia and Moravia with Slovak provinces that had been under Hungarian rule for nearly a thousand years. He also faced the additional problem of appeasing several other minority groups including ethnic Germans, Poles and Ukrainians.

Under Masaryk's influence Czechoslovakia managed to sustain a parliamentary democracy while other democracies in Eastern Europe were falling

Alexander Dubček, in the "Prague Spring" of 1968, the year he was removed from office by the Soviets. In 1989 he returned to guide Czechoslovakia to democracy, but by 1993 ethnic conflicts had led to Slovakia's independence.

apart. But growing unrest between two of Czechoslovakia's major ethnic groups soon led to political tension. The Slovaks resented Czech domination of political and economic life, finding it difficult to maintain their national identity within the new republic. A great deal more threatening, however, was the disaffection of the Sudeten Germans in the Sudetic Mountains on the border with Poland, who looked to the emergent Nazi party in Germany for support in their nationalist aspirations. In 1938, under the notorious Munich Agreement between Britain, France, Italy and Germany, the Sudeten territories were ceded to Germany. Hitler subsequently used this as a pretext for invading and seizing large areas of Bohemia and Moravia. In 1939 both provinces were incorporated into a German Protectorate within the Third Reich, while a puppet government under the pro-fascist Josef Tiso (1887–1947) was set up in Slovakia.

After World War II the Soviet Union capitalized on its liberation of most of Czechoslovakia from German occupation and rapidly established a power base there. In 1948 the communists declared Czechoslovakia a People's Democracy. In the years that followed, political repression manifested itself in a series of Stalinist show trials, in which the communist leadership turned on its own ranks. Thirteen prominent communists, all of them Jews,

and including the vice premier Rudolf Slansky (1901–52) were arrested, and all but three executed. The hardline regime in Czechoslovakia became ever more entrenched and by the time Antonin Novotny (1904–75) became president in 1957, Czechoslovakia's status had been reduced to nothing more than that of a Soviet satellite.

The brutalities of the old Stalinist regime were slow to recede in Czechoslovakia in spite of the political "thaw" in the Soviet Union under the premier Nikita Khrushchev (1894–1971). But eventually, in the 1960s, growing pressure for political and economic reforms resulted in Novotny's dismissal from office. A Slovak, Alexander Dubček (1921–92), took over as Czechoslovakia's First Secretary in 1967. The period of political tolerance and reform that followed in 1968 became known as the "Prague Spring".

Alarmed that the Czech example might be followed by other Eastern European satellite states, the Soviet Union launched an invasion of Czechoslovakia by Warsaw Pact troops in August 1968, on the pretext of "defending socialism". Resistance was put down, Dubček was removed from office and a puppet government installed under Gustav Husak (1913–91).

During the late 1980s – the Soviet Union's years of *perestroika* under the leadership of Mikhail Gorbachev (b.1931) – Husak scorned the reforms that were sweeping across the Soviet Union. But nothing could suppress the continuing calls for reform in Czechoslovakia and, after massive demonstrations, the communists were finally driven from power at the end of 1989. Alexander Dubček returned from obscurity as chairman of a new Federal Assembly with the former dissident playwright Vaclav Havel (b. 1936) as president. However, the transition to a parliamentary democracy was dogged by the age-old problem of Czech–Slovak relations, as each nation tried to look after its own best economic and political interests. A series of political crises followed in late 1990 and 1991 as divergences between Czechs and Slovaks deepened. On 1 January 1993 the two countries formally separated, and the states of Slovakia and the Czech Republic came into being.

An elaborate green patchwork Strip farming and coniferous forests cover the rolling foothills of the Tatra Mountains – a range of the Carpathians – in central Slovakia. The extensive forests supply the country's lumber and paper industries.

ECONOMY

Although only one third of Slovakia is cultivated, agriculture produced 5 percent of GDP in 1996. Major crops include wheat, sugar beet, barley and potatoes. Sheep, pigs and cattle are the main livestock animals. Forestry is also important. Iron, copper, lead, manganese and zinc are mined from various sites, but in small quantities. Coal is mainly poor-quality lignite, used mostly for power generation, though the main source of power is hydroelectricity.

Postwar industrialization was concentrated in the Slovakian cities of Kosice and Bratislava; manufacturing in these areas is still the biggest sector of the economy. By 1995 extensive privatization had taken place and the economy had stabilized.

NATIONAL DATA – SLOVAKIA

Land area	48,800 sq km (18,842 sq mi)			
Climate		**Temperatures**		**Annual precipitation mm (in)**
	Altitude m (ft)	January °C(°F)	July °C(°F)	
Bratislava	132 (433)	-1 (30)	21 (70)	557 (21.9)

Major physical features highest point: Gerlach Peak 2,655 m (8,711 ft); longest rivers: Danube (part) 2,850 km (1,771 mi), Vah 394 km (245 mi)

Population (2006 est.) 5,439,448

Form of government multiparty parliamentary democracy with one legislative house

Armed forces army 12,860; navy 5,160; joint forces 2,175

Largest cities Bratislava (capital - 425,533); Kosice (235,281); Presov (92,147); Zilina (85,278); Nitra (86,138)

Official language Slovak

Ethnic composition Slovak 85.8%; Hungarian 9.7%; Roma 1.7%; Ruthenian/Ukrainian 1%; other and unspecified 1.8%

Religious affiliations Roman Catholic 68.9%; Protestant 10.8%; Greek Catholic 4.1%; other or unspecified 3.2%; none 13%

Currency 1 Slovak koruna (SKK) = 100 hellers

Gross domestic product (2006) U.S. $96.35 billion

Gross domestic product per capita (2006) U.S. $17,700

Life expectancy at birth male 70.76 yr; female 78.89 yr

Major resources brown coal and lignite; small amounts of iron ore, copper, manganese ore, zinc, salt, antimony, cereals, livestock, magnesium, mercury, potatoes, sugar beet, barley, timber, uranium

Hungary

REPUBLIC OF HUNGARY

HUNGARY IS ONE OF THE OLDEST STATES IN Eastern Europe, with an ancient Magyar culture. Landlocked, it is bordered by Slovakia and the Czech Republic to the north, Austria to the west, Slovenia to the southwest, Croatia and Serbia to the south and Romania to the east; in the far northeast it shares a short frontier with the Ukraine.

GEOGRAPHY

Hungary is mainly low lying, the broad fertile plains of the central Danube basin extending across much of the country. Low hills and mountains dominate the north and west.

A rich lowland area – the Great Hungarian Plain – extends eastward from the river Danube, which flows south across the center of the country. To the north it is rimmed by mountains. West of the river Danube lies hilly Transdanubia rising to several mountain chains. To the northwest of these uplands is the Little Hungarian Plain, a lowland area bounded to the north by the Danube.

Hungary's climate is continental, with cold winters and warm summers. Droughts are common in the summer months, though thunderstorms can also produce heavy rainfall. High winter snowfalls in the mountains of neighboring countries (including the Alps to the west) cause a spring thaw that swells Hungary's rivers.

Hungary's once extensive natural forests were cleared long ago. Some small broadleaf forests have survived, and there has also been some reafforestation in mountain areas. Here deer and Wild boar are still abundant. Some natural marshlands remain. Intensive industrialization has caused pollution; effluent controls and conservation programs have yet to make much impact. The marshlands are a

Vineyards in the Mecsek hills (*above*) near Pecs in the far south of Hungary, where summer temperatures are warm and rainfall can be heavy. Local wines such as Tokaj and "Bull's Blood" are extremely popular in Hungary, and are also exported to the West.

paradise for waterfowl, including exotic migratory species such as flamingoes and spoonbills.

SOCIETY

Hungary has been settled by many different peoples. However, native Hungarians – descendants of the "Magyars", originally nomadic horsemen – have maintained a strong national identity.

Hungary was part of the extensive Bulgarian empire before it fell to the Byzantines and was incorporated into the eastern Roman empire. During the 9th century the area was overrun by the Magyars from the Volga steppes. Under their supreme chieftain Árpád (d. 907), the Magyars became a menace to the German kingdom. Árpád's descendant, King Stephen I (977–1038), established Hungary as a kingdom in 1001 and began

to build it into a feudal monarchy in line with much of Europe.

Extending its dominion, Hungary became one of Europe's largest kingdoms, but in 1241 a Mongol horde overran the country. From 1301 Hungary was ruled by foreign monarchs. Sigismund (1368–1437) neglected Hungary in favor of his other kingdoms in Germany and Bohemia (now the Czech Republic). The Turkish rulers of the nearby Ottoman Empire quickly seized this opportunity to make serious encroachments on Hungarian territory.

In the 16th century much of southeast Hungary was annexed to the Ottoman empire; the rest became part of the Holy Roman empire in 1558. But Turkish attempts at gaining supremacy failed, and in 1699 the Ottoman empire ceded Hungary to the Holy Roman Emperor.

Magyar nationalism grew under leaders such as Lajos Kossuth (1802–94), uniting Hungarian society against Austrian rule. The 1848 uprising in Hungary forced the Austrian Emperor Ferdinand (1793–1875) to abdicate in favor of his nephew Franz

Joseph (1830–1916). The rebellion was only put down with Russian help. In 1867, compromises on both sides won Hungary self-government as one half of the dual monarchy of the Austro-Hungarian empire. World War I shattered the Austro-Hungarian empire, and in the chaos that followed more than two-thirds of Hungary was annexed by surrounding countries and the rest fell to a communist coup. However, in 1920 Admiral Miklos Horthy (1868–1957) restored Hungary as a monarchy with himself as regent.

During World War II Hungary sided with Nazi Germany. As the Soviet army pushed westward in 1945 Hungary was occupied and the inevitable communist takeover followed in 1949. In 1956 Imre Nagy (1896–1958) led a nationalist rebellion which was savagely suppressed by the Soviet Union. A pro-Moscow regime under the leadership of János Kádár (1912–89) was then imposed.

While adhering closely to Soviet policies abroad, Kádár allowed a degree of political and economic liberalization at home. But the 1980s saw the collapse of the Hungarian economy, and in 1989 Kádár was ousted.

In 1989 the new constitution paved the way for multiparty democracy. The 386 seats of the one-chamber assembly are now freely elected for a five-year term. After the first multiparty elections in 1990, the writer and former dissident Árpád Goncz (b. 1922) was elected president. The election of 1994 was won by the former ruling Communist Party, as voters showed their dissatisfaction at the fall in living standards. Goncz was re-elected president in 1995.

ECONOMY

During the 1990s, the former communist controlled economy is slowly gave way to free-market capitalism. By 2000 over 85 percent of the economy was privatized.

As much as three-fifths of the country is rich arable land, and Hungary is practically self-sufficient in food. Since 1990 land has been restored to precollectivation owners.

The main crops include cereals, sugar beet and sunflowers; pigs and poultry are the main livestock. Vineyards, mostly in upland areas, yield some excellent wines. Commercial forestry along the Tisza river supplies the cellulose industry.

Hungary's main mineral resources include low-grade coal (lignite), its main fuel for energy. Only small deposits of petroleum and natural gas have been tapped. Manufacturing industries are extensive, particularly around Budapest. Major products include chemicals and engineering, automobiles and transportation equipment. Lighter manufacturing has grown in importance.

A massive program of privatization is underway. Trade with the European Union (EU) has greatly increased, with Germany, Austria and Italy as the major trading partners. Hungary is hoping to gain EU membership in the early 2000s.

Hungary's road and rail networks, both centered on Budapest, are highly developed but increasingly crowded. The Danube carries a vast amount of freight by water, especially from abroad. The national airline, Malév, is a significant international carrier.

Political changes have brought restructuring of the comprehensive welfare benefits provided by the communist state, and there is economic hardship for some Hungarians. Primary and secondary education is free and compulsory, and further education is available at a variety of colleges and universities.

Baroque splendor The Fertöd Palace, built in the mid 18th century by the aristocratic Esterhazy family, leading Magyar landowners of the period. Between 1761 and 1790 the Austrian composer Joseph Haydn (1732–1809) was court musician to the Esterhazys.

NATIONAL DATA – HUNGARY

Land area	92,340 sq km (35,653 sq mi)			
Climate		**Temperatures**		**Annual**
	Altitude m (ft)	January °C(°F)	July °C(°F)	precipitation mm (in)
Budapest	139 (456)	-1 (30)	21 (70)	516 (20.3)

Major physical features highest point: Kékes 1,014 m (3,327 ft); longest river: Danube (part) 2,850 km (1,771 mi); largest lake: Lake Balaton 601 sq km (232 sq mi)

Population (2006 est.) 9,981,334

Form of government multiparty republic with one legislative house

Armed forces army 23,950; navy 7,500; joint forces 850

Largest cities Budapest (capital - 1,697,343); Debrecen (204,297); Miskolc (175,701); Szeged (162,889); Pécs (156,567)

Official language Hungarian

Ethnic composition Hungarian 92.3%; Roma 1.9%; other or unknown 5.8%

Religious affiliations Roman Catholic 51.9%; Calvinist 15.9%; Lutheran 3%; Greek Catholic 2.6%; other Christian 1%; other or unspecified 11.1%; unaffiliated 14.5%

Currency 1 forint (HUF) = 100 filler

Gross domestic product (2006) U.S. $172.7 billion

Gross domestic product per capita (2006) U.S. $17,300

Life expectancy at birth male 68.45 yr; female 77.14 yr

Major resources bauxite, lignite, natural gas, tourism, cereals, fruit, vegetables, grapes/wine, livestock, oil, potatoes, sugar beet, sunflowers

Romania

REPUBLIC OF ROMANIA

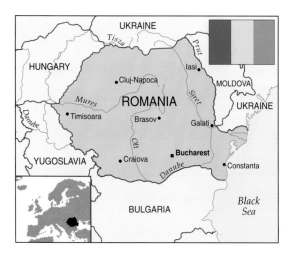

ROMANIA – THE "LAND OF THE ROMANS" – was founded as part of the ancient Roman empire. In its southeastern corner it has a short coastline along the Black Sea. Apart from this port access Romania is landlocked. The river Danube forms its southern border with Bulgaria. Serbia lies to the southwest, Hungary to the west, and the Ukraine to the north and east, sandwiching Moldova to the northeast.

GEOGRAPHY

Romania is dominated by the great mountain barrier of the Carpathians that extends from the Ukraine in the north and sweeps through Romania in a broad arc, first running southeastward before turning sharply west toward Serbia. The highest and most spectacular peaks are in the southern part of the main range, known as the Transylvanian Alps. Here the mountains are separated into massifs by deep defiles created by rivers such as the Olt and the Buzău.

North and west of the Carpathians, the upland basin of Transylvania is bounded by the Bihor Mountains, which are also cut by major rivers: the Somes running northwest into Hungary, and the Mures flowing due west into the Banat lowlands bordering Hungary. The northeast of Romania is occupied by the low plateau of Moldova, which continues into the newly independent republic of Moldova, formerly part of the Soviet Union.

South of the Transylvanian Alps, the river Danube forms the southern edge of the flat, fertile plains of Walachia, in the center of which lies the capital, Bucharest. To the east, the uplands of Dobruja give way to a series of marshes and a broad, swampy delta leading to the Black Sea.

The climate of Romania is largely continental, with cold, relatively dry winters and warm summers, which can be prone to thunderstorms. The mountains modify the climate considerably, receiving most of the country's rain and snow, and protecting western and southern areas from the bitterly cold winter winds that blow from the northeast.

SOCIETY

In 106 AD Romania was conquered by the Romans, becoming the Roman province of Dacia. The native Thracian peoples were quickly Romanized, and their culture survived wave after wave of incursions by the Germanic Goths, the Huns, the Slavs, the Avars, the Bulgars and finally, in the 10th century, the Magyars.

In 1365, two independent Romanian principalities emerged: Walachia in the Danube plains of the south, and Moldova to the northeast of the Carpathian Mountains. Their independence was brief: in 1396 Walachia became a vassal state of the burgeoning empire of the Ottoman Turks, and Moldova followed in 1455.

With the wane of Ottoman power at the end of the 18th century, Romania came increasingly under Russian influence. This period fueled an intense and long-lasting dislike of the Russians. At the end of the Crimean War in 1856, the principalities of Walachia and Moldova were united, and the Treaty of Berlin of 1878 recognized the new Kingdom of Romania as an independent state.

Romania gained much new territory as a result of the peace treaty ending World War I; Bessarabia, Bukovina and Transylvania were all incorporated into the country. It also inherited new problems and tensions with the absorption of a diverse ethnic mix of peoples: Magyars, Jews, Germans and Ukrainians.

In 1940 Soviet troops occupied Bessarabia (the modern Republic of Moldova). Endeavoring to recover the lost territory, Romania allied with Nazi Germany and in 1941 joined the German invasion of the Soviet Union – with disastrous results. In 1944 the country was occupied by the Soviets who installed a puppet government. In 1947 the monarchy was abolished and Romania was proclaimed a republic.

In 1965 Romania entered a dark and repressive period of its history as Nicolae Ceauşescu (1918–89) became secretary general of the Romanian Communist Party. He assumed the presidency in 1967, and under his dictatorial rule corruption at high level became rife. Resentment erupted into popular revolt in December 1989. Ceauşescu's attempt to reassert his authority failed when the army turned against him and the dissident National Salvation Front (NSF) proclaimed a provisional government. Ceauşescu and his wife were captured, tried and executed on 25 December 1989.

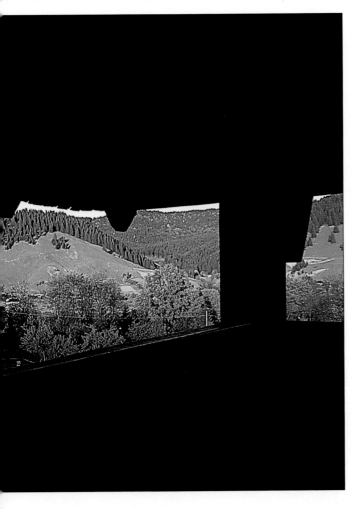

Ethnic Romanians (*above*) make up 89 percent of the population. They can be distinguished from the Hungarian, German, Turkish and Gypsy minorities by their costume and language. Ceauşescu's regime strongly emphasized mass culture.

A typical Romanian Orthodox church (*left*) in the village of Humor. Villages and the Church both suffered under Ceauşescu, who suppressed worship and destroyed some 7,000 villages to make way for agro-industrial complexes.

Ion Iliescu (b. 1930) remained president through a period of political and economic troubles in the early 1990s. In 1996 the Democrats were voted into power, and new president Emil Constantinescu promised to deal with the corruption and organized crime which had hampered economic development. In the late 1990s reform was slow as a result of disagreements between the ruling coalition.

ECONOMY

Repeated attempts have been made to transform Romania into a free-market economy: state subsidies have been cut; price controls removed; and a program of mass privatization has been undertaken. Efforts at reform have been hampered by political and popular dissent, especially in areas where the closure of unprofitable heavy industries and coal mines would have an adverse effect on the local population. The IMF granted a large loan to Romania to assist with restructuring in the late 1990s. In 1998 Romania began negotiating to join the EU but it must first control inflation and unemployment.

Farming, collectivized under communist rule, is still a cornerstone of the Romanian economy. Ceauşescu's policy of destroying Romanian villages and replacing them with huge industrial complexes has done much to damage the centuries-old way of life of the rural population. But the traditional crops grown in these areas, such as cereals and sugar beet, fruit and vegetables, are essential to the economy.

Sheep rearing is widespread, and there are many mixed farms in Transylvania and Banat. The Romanian forests are heavily exploited, less for sawn timber than for export products such as furniture and chipboard.

Although Romania is well supplied with coal, petroleum and natural gas, some fuel still has to be imported, together with the majority of raw materials for industry. A number of hydroelectric projects supply some of Romania's electricity, but most comes from thermal stations, many burning imported fuel. One nuclear power plant began operation in 1996. Since the 1960s much emphasis has been placed on the development of the manufacturing sector, dominated by heavy industry. The two main activities – machine building and metal processing – cause pollution. Much of the wildlife of the Danube is threatened by industrial effluent. The tourist industry is as yet underdeveloped. Romania has much to offer: ski resorts, historic towns and castles and stunning mountain scenery.

Romania's standard of living is lower than that of most other European countries; less than one percent of the population own a car. Medical treatment is free and more hospitals are being built and more doctors trained to cope with the high demand. But the costly transition to a market economy has left little funding to support the social welfare system. Ceausescu's policy of denying Romanian women access to contraception in order to raise the birthrate has left the country with the problem of providing for large impoverished families that cannot support themselves.

NATIONAL DATA - ROMANIA

Land area	230,340 sq km (88,935 sq mi)			
Climate		Temperatures		Annual
	Altitude m (ft)	January °C(°F)	July °C(°F)	precipitation mm (in)
Bucharest	82 (269)	-2 (28)	22 (72)	595 (23.4)

Major physical features highest point: Mount Negoiu 2,535 m (8,317 ft); longest rivers: Danube (part) 2,850 km (1,771 mi) and Prut (part) 909 km (565 mi)

Population (2006 est.) 22,303,552

Form of government multiparty republic with two legislative houses

Armed forces army 66,000; navy 7,200; air force 14,000

Largest cities Bucharest (capital - 1,921,751); Iasi (321,580); Cluj-Napoca (318,027); Timisoara (317,651); Constanta (310,526); Craiova (298,622); Galati (298,584)

Official language Romanian

Ethnic composition Romanian 89.5%; Hungarian 6.6%; Roma 2.5%; Ukrainian 0.3%; German 0.3%; Russian 0.2%; Turkish 0.2%; other 0.4%

Religious affiliations Eastern Orthodox 86.8%; Protestant (denominations including Reformate and Pentecostal) 7.5%; Roman Catholic 4.7%; other (mainly Muslim) and unspecified 0.9%; none 0.1%

Currency 1 leu, plural lei (RON) = 100 bani

Gross domestic product (2006) U.S. $197.3 billion

Gross domestic product per capita (2006) U.S. $8,800

Life expectancy at birth male 68.14 yr; female 75.34 yr

Major resources petroleum, timber, natural gas, coal, lignite, iron ore, salt, hydropower, cereals, copper, fish, fruit, vegetables, grapes, gold, livestock, oil, potatoes, silver, sugar, uranium

Bulgaria

REPUBLIC OF BULGARIA

BULGARIA LIES ON THE EASTERN SIDE OF THE Balkan peninsula bordering the Black Sea. It is bounded by Romania to the north (across the river Danube), Serbia and Macedonia to the west and Greece and Turkey to the south. The country was one of the most acquiescent of the Soviet satellite states. Since the quiet revolution of 1989 it has been undergoing rapid political and economic changes.

GEOGRAPHY

Two large mountain ranges divide the country into four distinct areas: the northern plateau, the east–west range of the Balkan Mountains, the central and coastal lowlands and the southwestern Rhodope Mountains.

Across the north of the country the steep banks of the Danube (Dunar) rise to the low, fertile Danubian plain. South of this the land rises gently to the fragmented ridge of the Balkan Mountains, and then falls sharply to a complex series of river basins in the central lowlands. Here the river Maritsa drains southward into Greece. Rivers in the eastern lowlands drain to the Black Sea coast. Southern and western Bulgaria is dominated by the great massifs of the Vitocha, Rila, Rhodope and Pirin Mountains, which include Musala, the highest peak in the Balkans at 2,925 m (9,594 ft). The capital city, Sofia, lies in a high basin at the conjunction of the Balkan and Vitocha Mountains.

The influence of the Mediterranean Sea creates a temperate climate in the southeast with warm dry summers and mild wet winters. In the north and west the country is subject to greater extremes of climate with colder winters and hotter summers. The mountains are covered by heavy snowfalls in winter.

SOCIETY

The Bulgars were originally a Turkic people, who first crossed the river Danube from central Asia in the 7th century AD. Gradually assimilated by the Slavs that had settled there a century or so earlier, the Bulgarians soon created a large empire in the Balkans. This first empire fell to its southern neighbor Byzantium in 1018. A second empire, wrested from Byzantium in 1185, survived until 1396, when the country was

Alexander Nevsky Cathedral (*above*) in Sofia was built as a symbol of gratitude to the Russian troops who liberated Bulgaria in 1878 from nearly 500 years of Ottoman rule. The following year the city was designated Bulgaria's capital.

taken by the Ottoman Turks. For the next 500 years Bulgaria was subjected to Turkish rule.

In the 19th century, growing nationalist movements throughout the Ottoman empire fueled the decline of Turkish power. The Congress of Berlin, called in 1878 to address the territorial disputes in the

Balkans, gave Bulgaria the status of an autonomous principality. Continuing upheavals in the Ottoman empire led Ferdinand of Saxe-Coburg-Gotha (1861–1948), Bulgaria's elected prince, to proclaim himself king in 1908. Bulgaria at this time was fully independent of Turkey. In 1912 Bulgaria joined forces with its neighbors Serbia, Greece and Montenegro to extract more territorial concessions from the declining Ottoman empire. However, in 1913 it turned on these allies and lost further territory to Greece and Serbia after a humiliating defeat. Balkan politics encouraged Bulgaria to align with Germany in World War I, and again in World War II. It was unable to realign before Soviet troops "liberated" Bulgaria from German occupation in 1944. Two years later the monarchy was abolished and so were all opposition parties. A Soviet-style communist republic was established in 1947, with the government in the hands of the Fatherland Front.

Bulgaria, like other Eastern European countries, was eventually caught up in the wave of change that swept through the communist bloc in the late 1980s. Following a series of strikes and demonstrations, Bulgaria's president of 27 years, Todor Zhivkov (b. 1911), was ousted in 1989. The constitution was amended and the Communist Party's sole right to govern was abolished. The next two years saw the decentralization of parts of the economy, the recognition of Turkish minority rights, and government liberalization on an unprecedented scale. A democratic coalition, led by Zhelyu Zhelev (b.1935) as president, came to power in 1992 and altered the process of election to Bulgaria's 240-member National Assembly whereby all are elected by proportional representation. American-style primary elections were introduced in 1996 and in June 2001 former king Simeon II became prime minister. He is the first exiled monarch in post-communist eastern Europe to return to power.

ECONOMY

The reforms of 1989 effectively abolished Bulgaria's centralized economy, allowing existing businesses to transform themselves into independent companies with the right to issue shares. The process of transition to a market economy has been difficult. Hyperinflation in the mid 1990s and the near collapse of the banking system prompted a restructuring program aided by a US$850 million IMF loan.

Cereals and vegetables are the main staples, while cash crops include sunflower seeds, tobacco and grapes; Bulgaria is famous for its fine wines. Livestock and forestry are less significant, though of growing importance. Fishing is a major industry; besides local fisheries in the Black Sea and inland waters, Bulgarian trawlers travel as far as the Mediterranean and the Atlantic.

Bulgaria has deposits of copper and some iron ore. There are also fossil fuel reserves of coal, but little petroleum or natural gas. Hydroelectric plants generate up to one-tenth of the country's power; one-half comes from thermal plants, and the rest from Soviet-designed nuclear stations. Bulgaria has become a major distribution center for energy in the Black Sea region. The communist regime invested heavily in manufacturing industry, and its products – notably machinery and equipment, metals and petrochemicals – earn the bulk of Bulgaria's export revenue. Tourism, particularly to the Black Sea coast, also brings useful foreign exchange.

Plovdiv, Bulgaria's second city, has a charming old quarter with well-preserved 19th-century houses, as well as some impressive Roman remains. The city saw a sixfold population increase between 1946 and 1985, and is now a major industrial center and market town.

NATIONAL DATA - BULGARIA

Land area	110,550 sq km (42,684 sq mi)			
Climate		Temperatures		Annual
	Altitude m (ft)	January °C(°F)	July °C(°F)	precipitation mm (in)
Sofia	550 (1,804)	-1 (28)	20 (68)	571 (39)

Major physical features highest point: Musala 2,925 m (9,596 ft); longest river: Danube (part) 2,850 km (1,771 mi)

Population (2006 est.) 7,385,367

Form of government multiparty republic with two legislative houses

Armed forces army 25,000; navy 4,370; air force 13,100

Largest cities Sofia (capital - 1,138,950); Plovdiv (341,464); Varna (312,026); Burgas (189,529)

Official language Bulgarian

Ethnic composition Bulgarian 83.9%; Turk 9.4%; Roma 4.7%; other 2% (including Macedonian, Armenian, Tatar, Circassian)

Religious affiliations Bulgarian Orthodox 82.6%; Muslim 12.2%; other Christian 1.2%; other 4%

Currency 1 lev (BGL) = 100 stotinka

Gross domestic product (2006) U.S. $77.13 billion

Gross domestic product per capita (2006) U.S. $10,400

Life expectancy at birth male 68.68 yr; female 76.13 yr

Major resources bauxite, copper, lead, zinc, coal, timber, apples, barley, cattle, grapes, lignite, limestone, maize (corn), manganese, oil and natural gas, pigs, poultry, sheep, sunflowers, tobacco, uranium, wheat, tourism

Yugoslavia

FEDERAL REPUBLIC OF YUGOSLAVIA
(SERBIA) AND MONTENEGRO

I N 1990 THE FEDERATION OF THE SIX REPUBLICS comprising Yugoslavia (Slovenia, Croatia, Bosnia Herzegovina, Serbia, Montenegro and Macedonia) began to dissolve with the secession of Slovenia and Croatia. This prompted Serbia to declare war on Croatia. Meanwhile Bosnia Herzegovina and Macedonia also declared independence, leaving Serbia and Montenegro in the new rump state of Yugoslavia.

SERBIA

Serbia, the largest of the former Yugoslav republics, covers some 88,361 sq km (34,116 sq mi) and incorporates the previously independent regions of

Vojvodina in the north and Kosovo in the south. It has international borders with eight states: clockwise from Hungary in the north these are Romania, Bulgaria, the Former Yugoslav Republic of Macedonia, Albania, Montenegro, Bosnia Herzegovina and Croatia.

Geography

Excluding the two provinces of Vojvodina and Kosovo, Serbia is almost completely mountainous. The most fertile land is found on the Pannonian plain, which runs from Hungary and Croatia across into Vojvodina and northern Serbia. A number of rivers crisscross this plain, including the Sava, which eventually flows into the Danube. In the southwestern corner of the country bordering Albania and Montenegro, the province of Kosovo rises out of the Sar and Kopaonik Mountains. Serbia's highest peak, Daravica, at 2,656 m (8,712 ft), is at the meeting point of these three states, amid the most rugged mountain terrain on the Balkan peninsula.

Eastern Serbia is also rugged and wild. Here the Balkan Mountains, which form the border between Serbia, Bulgaria and Romania, rise to 1,800 m (6,000 ft). The climate is almost totally continental. Temperatures are generally mild, but can range from very hot in summer to very cold in winter. Precipitation is steady throughout the year, mostly taking the form of snow on higher ground.

Society

Slavonic in origin, the Serbs first settled in the Balkans during the 6th century AD. Divided into warring clans, the Serbs were dominated by the neighboring Bulgarian and Byzantine empires until the 13th century. The Byzantine influence is still evident: most Serbs belong to the Greek Orthodox church, setting them apart from the Bosnian Muslims and Catholic Croats.

As Byzantium declined the Serbs briefly came to dominate the Balkans under their king Stephen Dušan (c. 1308–55). The rise of the Ottoman Turks, however, signaled an end to Serbian expansionism. After its crushing defeat at the Battle of Kosovo ("the field of black birds") in 1389, Serbia became an Ottoman vassal state. In 1537 the Ottoman empire took complete control. The Serbs continued to wage guerrilla attacks against the Turks, allying themselves by turn with Hungary, Austria, the Holy Roman Empire and Russia. In 1817 Serbia eventually gained

Concrete apartment blocks and offices dominate the skyline of modern Belgrade, capital of Eastern Orthodox Serbia. Built as cheap accommodation for workers, they are a legacy of the rapid process of industrialization that took place after World War II.

recognition as an autonomous principality with its capital at Belgrade. Total independence came in 1878, granted under the Treaty of Berlin concluding the Russo-Turko War. However, Serbia's aggressive forays into neighboring territory continued. By the outbreak of World War I, the Serbian state still lacked stable borders. After the collapse of the Habsburg monarchy in 1918 Serbia was instrumental in setting up the forerunner to the Yugoslav federation, the Kingdom of Serbs, Croats and Slovenes. Serbian domination of the new state became complete in 1921 with the adoption of the Serbian regent as King Alexander I (1888–1934).

After World War II Serbia became part

NATIONAL DATA – SERBIA

Land area	88,361 sq km (34,116 sq mi)			
Climate		Temperatures		Annual
	Altitude m (ft)	January °C(°F)	July °C(°F)	precipitation mm (in)
Belgrade	132 (433)	1 (45)	22 (63)	1,145 (45)

Major physical features highest point: Daravica 2,656 m (8,714 ft) in Prokletija Range on Albanian border

Population 9,396,411

Form of government republic

Armed forces no figures

Largest cities Belgrade (capital – 1,120,092); Novi Sad (191,405); Nis (173,724)

Official language Serbian

Ethnic composition Serb 66%; Albanian 17%; Hungarian 3.5%; other 13.5%

Religious affiliations Serbian Orthodox, Muslim, Roman Catholic, Protestant

Currency 1 new Yugoslav dinar (YUM) = 100 euro cents

Gross domestic product (2006) U.S. $44.83 billion

Gross domestic product per capita (2005) U.S. $4,400

Life expectancy at birth male 71 yr; female 76 yr

Major resources oil, gas, coal, iron ore, copper, lead, zinc, antimony, chromite, nickel, gold, silver, magnesium, pyrite, limestone, marble, salt, agricultural produce

Tito, Marshal of Yugoslavia

When Josip Broz - better known as Marshal Tito - died on 4 May 1980, Yugoslavia lost its most powerful symbol of unity. Born in Croatia on 7 May 1892, Broz was the seventh of 15 children. During World War I, while serving in the Austro-Hungarian army, he was captured on the Russian front. After his return to Croatia in 1920 (then part of the Kingdom of the Serbs, Croats and Slovenes) he joined the illegal Communist Party. He was arrested three times for communist activities, and in 1928 was sentenced to five years' imprisonment for conspiring against the regime in the state that was about to be renamed Yugoslavia. Following his release in 1934 he rose to prominence in the Communist Party of Yugoslavia (CPY), becoming its secretary-general in 1940.

Tito's leadership was quickly put to the test. In April 1941 German forces invaded and occupied Yugoslavia. Two months later Serbian peasants in Herzegovina rose against savage massacres perpetrated by the Ustase – fascist Croatian separatists who supported German rule. With the German invasion of the Soviet Union, Tito called a general uprising and soon half the country was in the hands of the Yugoslav resistance. Wherever his partisans were in control, Tito created revolutionary councils in preparation for the united Yugoslavia of his dreams – a policy that antagonized both the government in exile and the Soviet Union. The partisans survived a series of increasingly powerful German offensives and civil conflict with the royalist chetniks. In 1943 Tito established a new government with himself as marshal, and the king was deposed.

Despite a growing rift between Tito and the allies – Great Britain, the United States and the Soviet Union – Yugoslavia was given military assistance by them in 1944. By the time the war had ended in 1945 Yugoslavia was in ruins and more than one-tenth of its population was dead. Tito's new 1946 constitution was modeled on that of the Soviet Union, but important differences swiftly became apparent. In 1948

Marshal Tito held together the disparate peoples of the region, with their cultural and religious differences, from 1945 until his death in 1980. His brand of communism allowed a higher standard of living than other communist countries.

Tito broke with his Stalinist masters and the Soviet Union effectively isolated Yugoslavia in the hope of bringing Tito down. They succeeded only in unifying the country behind its leader as he built his own independent style of communism (popularly known as "Titoism"). Five years later, in 1953, Tito became Yugoslavia's first president, supervising the enactment of a new federal constitution.

The same spirit of independence also showed itself in Tito's determinedly nonaligned foreign policy. While denouncing Western colonialism on the one hand, he was swift to condemn the 1968 Soviet invasion of Czechoslovakia, and responded by strongly reinforcing his own country's defenses. The 1974 constitution included the idea of a collective presidency – an idea that became reality on Tito's death in 1980. In the event, government by committee proved a poor substitute for his effective dictatorship. In 1990 the peoples of Yugoslavia had a brief tantalizing glimpse of a freer, more open society until rising ethnic tensions finally plunged the country into civil war.

Tito's dream of a federal Yugoslavia, where all its nationalities would be equal, had occupied much of his energy and skill for more than 40 years. Within 11 years of his death it had become apparent that his vision of harmony and mutual cooperation between the disparate peoples of Yugoslavia had not been realized.

of the federal Yugoslav state. Its role in the federation began to change significantly, however, when Slobodan Milosevic (b.1941) was elected president of Serbia in 1989 and again, under a new constitution, in 1990. This removed the term "socialist" from the official name of the Republic of Serbia, and revoked the special rights and autonomous privileges of Kosovo and Vojvodina.

During this period Serbia continually vetoed federal assembly motions by Croatia and Slovenia to devolve to a looser federation of Yugoslav republics. As the federation began to fragment, Serbia was able to hold onto majority control of the Yugoslav People's Army (YPA). The YPA failed to prevent the independence of Slovenia in 1991, but in Croatia and in Bosnia Herzegovina it intervened in support of ethnic Serbian minorities who were opposed to independence, plunging both republics

Yugoslavia 1918–95: the rise and fall of a country

The complex and troubled history of the Balkan peninsula has created an enduring legacy of hostility and mistrust between its varied ethnic and religious communities. Only one man effectively held the Yugoslav federation together: Josip Broz (1892–1980), better known as Marshal Tito. His leadership of Yugoslavia as a communist state between 1945 and 1980 was the only cohesive period in the region's turbulent history. Little more than 10 years after his death, the countries of former Yugoslavia were plunged into a bloody civil war.

The early days of the Yugoslav federation were steeped in violence. The Balkan Wars of 1912–13 had given extensive territory to both Serbia and Montenegro and had destroyed the last remnants of Turkish power on the peninsula. Relations between the Austro-Hungarian empire and Serbia, built on years of rivalry and suspicion, grew increasingly bitter as the two countries vied for influence in the Balkans. Austro-Hungary annexed Bosnia in 1908. On 28 June 1914 Archduke Franz Ferdinand (1863–1914), heir to the Austro-Hungarian throne, was assassinated by a Serb nationalist at Sarajevo, the capital of Bosnia. A month later Austria–Hungary declared war on Serbia. Russia, France and Great Britain rose to Serbia's defense, while Germany and the Turks sided with Austria–Hungary.

Though defeated in 1915, the Serbs reaped a rich reward when, in 1918, at the end of World War I, the Kingdom of the Serbs, Croats and Slovenes was established following Austria-Hungary's defeat. Although not in the majority, they were the largest single ethnic group, and as such dominated the government. In 1929 King Alexander I (1888–1934) renamed the kingdom Yugoslavia (meaning "the country of the south Slavs"), and divided it into nine counties based on geography rather than race. During World War II Yugoslavia was invaded by German troops and divided between Germany, Italy and Bulgaria. Bitter fighting took place in the civil war of 1941–45. The seeds of many hatreds between Serb

and Croat were sown in the period surrounding World War II, which saw the extermination of many Serbs, Jews, Gypsies and other ethnic groups.

In 1944 the Soviet army liberated Yugoslavia from German occupation, and in 1945 the Croat communist Josip Broz (known as Tito) established a communist republic. The ensuing alliance between Yugoslavia and the Soviet Union proved an uneasy one. Tito's efforts in the late 1940s to build a Balkan federation independent of Soviet influence were seen by the Soviet Union as a threat to its power-base in the region. The Soviets withdrew economic support in the hope of toppling Tito, but he remained in power until his death in 1980.

The adoption of a new constitution in 1974 was a step toward a shared and equal leadership. A nine-member state presidency (one from each of the six republics and two autonomous provinces with Tito at its head) failed to satisfy Serbia's demand for a greater share in the balance of power.

The introduction of political and economic reforms after Tito's death in 1980 was hampered by the decentralization of government. Unemployment and inflation escalated in the late 1980s. Slovenia and Croatia, alarmed at the prospective loss of their economic

stability, held multiparty elections in 1990 and voted in noncommunist republican assemblies. Serbia, under its president Slobodan Milosevic (b.1941), strongly resisted this action, fearing that change would threaten the 700,000-strong Serbian communities in Croatia and Bosnia Herzegovina. When Slovenia and Croatia declared independence in June 1991 civil war broke out – the first one in Europe for 46 years.

A succession of abortive ceasefires failed to halt the rising tide of violence, until United Nations' (UN) intervention and a peacekeeping force secured a formal ceasefire agreement in January 1992. However, this did not prevent the spread of the civil war into neighboring Bosnia Herzegovina – regarded as the powder keg of the Balkans because of its volatile ethnic mix. When Bosnia Herzegovina and Macedonia declared their intention of seceding from the Yugoslav state, the Serbs unleashed fierce attacks on the cities of Sarajevo and Mostar. Serbia and Montenegro continued to fight for the existence of the Yugoslavian state in the face of trade sanctions and worldwide condemnation for Serbian atrocities. These included the systematic "ethnic cleansing" of Muslims from Bosnian territory.

International sanctions and pressure

into bloody civil war. This led to accusations that Milosevic's government was planning to create a "Greater Serbia" out of the ruins of Yugoslavia. Despite Serbia's expulsion from the General Assembly of the United Nations (UN) and the imposition of economic sanctions by the European Union and the United States, the continuing strength of Serbian support for Milosevic was confirmed by his re-election as president in 1993.

UN trade sanctions continued in 1996, even though Serbia had announced earlier it would not supply Bosnian Serbs.

Serbia has a complex ethnic mix. Excluding Kosovo and Vojvodina, the population is about 90 percent Serb. As a group, they account for 60 percent of all the ethnic Serbs in the former Yugoslav Republic. Kosovo is peopled mainly by Muslim Albanians (they form 90 percent of the local population – the remaining 10 percent are Serbs) who have always resisted Serb domination. Fighting broke out between the separatist Kosovo Liberation Army and Yugoslav security forces in early 1999. Widespread outrage at the slaughter of Albanian civilians spurred the NATO alliance to begin bombing military and economic targets in Yugoslavia in April. In 2000 President Slobodan Milosevic of Yugoslavia lost an election to Vojislav Kostunica but only left office after mass protests. In June 2001 he was handed over by the Serbian authorities to the United Nations War Crimes Tribunal in The Hague for prosecution.

Economy

The rich Pannonian plain of northern Serbia and Vojvodina was the breadbasket of the former Yugoslav Republic. Agriculture, especially cereal growing, remains important, employing around a quarter of the population. In Kosovo the farming is mostly subsistence-oriented. Here, two-fifths of the population work on small family-run farms. Throughout the country, vines are grown for wine-making and there is a healthy forestry industry in the mountainous areas.

Civil war and international sanctions have severely damaged Serbia's economy. The NATO air strikes of 1999 destroyed much of the transportation infrastructure and industrial areas, and an embargo on petroleum imports damaged further attempts at economic recovery. In the late 1990s, over half the population lived on less than US$60 per month and unemployment was near 50 percent. There are, however, considerable natural deposits of fossil fuels, including petroleum and coal.

Sarajevo's parliament building burns after a Serbian rocket attack in 1992. As the capital of Bosnia Herzegovina, it has been at the heart of the civil war. Much of the city has been destroyed, and food shortages in 1993 were common.

from the UN forced Serbia and Montenegro to the negotiating table. As part of the US-brokered Dayton peace accord (November 1995) Serbia agreed to recognize the separate existence and independence of Croatia and Bosnia Herzegovina. The question of the autonomy of the province of Kosovo was not tackled as part of this agreement. Conflict erupted in 1998 as the Serbs retaliated against increasing unrest in Kosovo by launching an offensive against the separatist Kosovo Liberation Army (KLA). The reimposition of sanctions had little effect, and by March 1999 250,000 people had been displaced by "ethnic cleansing."

In frustration, NATO launched a series of air strikes which eventually ended hostilities and forced Serbian troops to withdraw from Kosovo. A large UN peacekeeping force oversaw elections in which the question of Kosovan independence was paramount. At the same time the electoral defeat and departure of Slobodan Milosevic as president of Yugoslavia gave hope that future negotiations regarding the status of Kosovo would be resolved peacefully.

MONTENEGRO

The Italian dialect words for the country's mountains gave Montenegro its name; it is also known by the Serbo-Croatian for black mountains – *Crna Gora*. This small country on the Adriatic Sea is bordered by Albania to the south, Serbia to the east, Bosnia and Hercegovina to the north and the Adriatic Sea to the west. With Serbia it comprises the much reduced state of Yugoslavia following the breakup in 1990 of the federation of the six republics that made up the former Yugoslavia (Slovenia, Croatia, Bosnia Herzegovina, Serbia, Montenegro and Macedonia).

Society

For centuries Montenegro's rocky landscape has been a mountain stronghold resistant to conquest. It is one of the few Balkan countries never to have been

NATIONAL DATA - MONTENEGRO

Land area	13,812 sq km (5,333 sq mi)			
Climate		Temperatures		Annual
	Altitude m (ft)	January °C(°F)	July °C(°F)	precipitation mm (in)
Podgorica	44 (144)	5 (41)	26 (79)	1,361 (53.5)
Major physical features	highest point: Bobotov Peak 2,523 m (8,277 ft)			
Population	(2006 est.) 630,548			
Form of government	republic			
Armed forces	no figures			
Capital city	Podgorica (136,473)			
Ethnic composition	Montenegrin 43%; Serbian 32%; Bosniak (Muslim) 8%; Albanian 5%; other (Muslims, Croats, Roma) 12%			
Religious affiliations	Orthodox, Muslim, Roman Catholic			
Currency	1 euro (EUR) = 100 euro cents			
Gross domestic product	(2006) U.S. $3.394 billion			
Gross domestic product per capita	(2005) U.S. $3,800			
Major resources	bauxite, hydroelectricity			

Deceptively tranquil The harbor at Kotor, a small port in Montenegro, is only a short distance from the devastated port of Dubrovnik in Croatia, attacked early in the civil war. All regions have lost valuable revenue through the disruption to tourism.

overrun by the Ottoman Turks. Although parts of the principality of Zeta, the early precursor to Montenegro, were captured by the Serbs in the 12th century and by Venice in the 15th, the Montenegrins retreated to the rugged Black Mountains near the coast. The mountain village of Cetinje became the new capital of an independent state, ruled by prince-bishops from 1516. Until 1697 these rulers or *vladike* were elected by popular assemblies. Montenegro was ruled by the Petrovich Njegosh family from 1697 until it became part of Yugoslavia in 1918.

Almost 200 years of continuous war with the Turks and Albanians turned the Montenegrins into a highly militaristic society. Peter I (1747–1830), doubled the size of the country through successful military campaigns against the Venetians

and Albanians. In 1878 the Congress of Berlin, called to resolve territorial claims in the Balkans at the end of the Russo-Turkish War (1877–78), ceded yet more territory to Montenegro and recognized its independence. The Montenegrins' traditional solidarity with the Serbian Orthodox church was reinforced by their support of Serbia in the Balkan Wars (1912–13) and World War I (1914–18).

Overrun by Austro-Hungarian troops in 1916, Montenegro was occupied by the Serbian army and formally annexed by Serbia in 1919, and its monarchy replaced. It has remained allied to Serbia ever since and suffered from the sanctions imposed against Yugoslavia during the 1990s as well as from the NATO air strikes of 1999. The close economic and political ties with Serbia looked increasingly threatened as Montenegro replaced the dinar with the deutschmark in 1999, sought diplomatic links with the other Balkan nations and Italy, and publicly discussed the possibility of a referendum on independence.

Economy

Yugoslavia's federal government spent considerable sums of money developing Montenegro's industrial base before its dissolution in 1990, especially in iron and electrical power. Tourism brought in a significant amount of much needed foreign currency before it was interrupted by the civil wars of the 1990s. By 2000 Montenegro was in tatters, with living standards only a tenth of what they were in 1990, and rising levels of organized crime involving the smuggling of cigarettes, stolen cars and drugs. Montenegro remains a largely agricultural society dominated by the rearing of sheep, goats and pigs. There is also some cereal and tobacco production. The only international airport is in the capital, Podgorica (formerly Titograd). Although there are extensive rail links between Montenegro and Serbia, its mountainous terrain has prevented largescale industrial development and hampered the introduction of an efficient communications system

Slovenia

REPUBLIC OF SLOVENIA

SLOVENIA WAS THE FIRST OF THE FORMER Yugoslav republics to declare independence in 1991; it was formally recognized as a new state in January 1992. An alpine country at the northern end of the Balkan peninsula, it is bordered by Italy to the west (where it also has a very short Adriatic coastline), Austria to the north, Hungary to the northeast and Croatia to the south.

GEOGRAPHY

Most of Slovenia is mountainous, with the exception of the lowlands near the coast. The Alps extend into Slovenia from Italy and Austria, and include the country's highest point, Triglav, some 2,864 m (9,393 ft) above sea level. From the Alps, the landscape descends through dense woodland into deep and fertile glacial valleys, including the basin of the Sava river near the capital, Ljubljana. In the east, the Drava river flows into Croatia along the Pannonian plain, a rich agricultural area. The western part of the country has a Mediterranean climate, with warm summers and heavy rainfall in late summer and early fall. Eastern Slovenia's climate is much more continental.

SOCIETY

The first Slovenes settled here in the 5th and 6th centuries AD, when the region was part of the eastern Roman empire. In the 9th century Slovenia was annexed to the Frankish empire by Charlemagne (742–814). In the years that followed, the country was ruled by a succession of Austrian dukes as part of the Holy Roman Empire. Unlike many neighboring Balkan states, it escaped occupation by the Turks during the period of Ottoman expansion in the 15th and 16th centuries, remaining under Austrian control until the collapse of the Austro-Hungarian empire at the end of World War I. In 1918, anxious to preserve its national identity against possible future incursions by Italy and Austria, Slovenia became a founder member of the Kingdom of the Serbs, Croats and Slovenes, the precursor of the modern Yugoslav Republic.

During World War II Slovenia lost some of its territory to Italy and Germany. Part of that territory was given back to it in 1954, but Italy retained possession of the important coastal city of Trieste. Slovenian nationalism resurfaced in the 1980s after the death of the Yugoslav leader, Tito (1892–1980). In 1989 the Slovenian Assembly adopted a resolution of sovereignty that brought it into confrontation with Serbia. The country formally broke away from the former state of Yugoslavia in June 1991.

ECONOMY

Before the disintegration of Yugoslavia, Slovenia was the most prosperous of its constituent republics, and generated 20 percent of the country's income. The effects of the civil war with Serbia and of the sanctions leveled against that republic took their toll on Slovenia's formerly healthy economy. However, the economy

The clear waters of Lake Bled in northern Slovenia, a glacial lake in the Julian Alps. Bled itself – a ski resort in winter – is situated on the north shore, and is dominated by an impressive Baroque castle that clings to the top of a rocky crag.

had largely recovered by the mid-1990s, and Slovenia was hopeful of achieving membership in the European Union in the first decade of the 2000s. Agriculture is concentrated on livestock raising, with substantial cereal, potato and sugar beet cultivation, and wine production is well established around the Drava river region. Iron and steel are major industries in Ljubljana, and textile factories are found throughout the country. Slovenia also has abundant resources of brown coal (lignite), lead and mercury.

NATIONAL DATA - SLOVENIA

| Land area | 20,151 sq km (7,780 sq mi) |

Climate	Altitude m (ft)	Temperatures January °C(°F)	July °C(°F)	Annual precipitation mm (in)
Ljubljana	448 (1,469)	0 (32)	21 (70)	1,266 (49.8)

Major physical features highest point: Triglav 2,864 m (9,396 ft); longest river: Sava (part) 940 km (584 mi)

Population (2006 est.) 2,010,347

Form of government multiparty republic with two legislative houses

Armed forces army 6,550

Largest cities Ljubljana (capital - 247,772); Maribor (91,540); Celje (36,369)

Official language Slovene

Ethnic composition Slovene 83.1%; Serb 2%; Croat 1.8%; others 13.1%

Religious affiliations Catholic 57.8%; Orthodox 2.3%; Muslim 2.4%; other 37.5%

Currency 1 tolar (SIT) = 100 stotins

Gross domestic product (2006) U.S. $43.08 billion

Gross domestic product per capita (2006) U.S. $22,900

Life expectancy at birth male 72.63 yr; female 80.29 yr

Major resources lignite, lead, zinc, mercury, uranium, silver, hydropower, forests, apples, barley, bauxite, grapes/wine, hops, livestock, mercury, oats, pears, potatoes

Croatia

REPUBLIC OF CROATIA

THE BEAUTIFUL AND RUGGED DALMATIAN coastline runs the length of western Croatia. Inland the country is bounded by Slovenia to the northwest, Hungary to the northeast, Serbia to the east and Bosnia Herzegovina to the south and east. The historic medieval city of Dubrovnik is separated from the rest of Croatia by a band of territory that gives Bosnia Herzegovina access to the sea. South of here, a narrow strip of Croatia has a further border with Montenegro.

NATIONAL DATA - CROATIA

Land area 56,414 sq km (21,782 sq mi)				
Climate		Temperatures		Annual
	Altitude m (ft)	January °C(°F)	July °C(°F)	precipitation mm (in)
Dubrovnik	49 (161)	9 (48)	25 (77)	1,037 (40.8)

Major physical features highest point: Bobotov Kuk 2,522 m (8,274 ft); longest river: Sava (part) 940 km (584 mi)

Population (2006 est.) 4,494,749

Form of government multiparty republic with one legislative house

Armed forces army 14,050; navy 2,500; air force 2,300

Capital city Zagreb (691,724)

Official language Croatian

Ethnic composition Croat 89.6%; Serb 4.5%; other 5.9% (including Bosniak, Hungarian, Slovene, Czech, and Roma)

Religious affiliations Roman Catholic 87.8%; Orthodox 4.4%; other Christian 0.4%; Muslim 1.3%; other and unspecified 0.9%; none 5.2%

Currency 1 kuna (HRK) = 100 lipa

Gross domestic product (2006) U.S. $59.41 billion

Gross domestic product per capita (2006) U.S. $13,200

Life expectancy at birth male 71.03 yr; female 78.53 yr

Major resources oil, some coal, bauxite, low-grade iron ore, gypsum, natural asphalt, silica, mica, clays, salt, hydropower, corn, flax, fruits, grapes, oats, olives, potatoes, timber, wheat, tourism

GEOGRAPHY

This boomerang-shaped country can be divided into three distinct geographical and historic regions. The long, mountainous expanse of the Dalmatian coast and the Dinaric Alps are extensions of the continental alpine system. Two other ranges, the Velebit and Velika Kapela, run parallel to the Dinaric Alps. Along the Dalmatian coast changes in sea level have drowned alpine valleys to create a landscape of steep islands and rocky peninsulas. A short distance inland from the northeastern end of the Adriatic Sea is the Kras plateau: an area of limestone scenery that has been eroded into a complex of caves, pavements and potholes. The central northern part of Croatia around the capital Zagreb is dominated by rolling hills and fertile inland plains formed by the Sava river. In northern Croatia the Pannonian plain extends from Slovenia.

Croatia has two separate climatic patterns: the Mediterranean climate of the coast; and a colder continental climate on the Pannonian plain.

SOCIETY

In the 6th century AD what was then the Roman province of Illyricum was overrun by settlers from White Croatia (now in the Ukraine). As the Roman empire in the east disintegrated, the Croats set up a kingdom that stretched into Bosnia, and were able to exist independently of any foreign power, except for a short time in the 9th century. Civil war led to the decline of the kingdom and in 1091 the country was conquered by Hungary. Following the collapse of Hungary in 1526 Croatia was divided between the Ottoman and the Austrian Habsburg empires. By 1699 the Ottomans had been driven out and the whole country came under Austrian control.

Following the collapse of the Austro-Hungarian empire in 1918 the Croats, together with the Serbs and Slovenes, were instrumental in setting up the precursor to the Yugoslav state. Their alliance lasted until 1991, despite a five-year period during World War II when the fascist movement in Croatia led it to declare itself an independent state allied with Germany. Multi-party elections in 1990 brought to power the right-wing nationalist Croatian Democratic Union (CDU). Following the Croatian assembly's call for secession from Yugoslavia in 1991, Serb militants in the region of Krajina declared independence.

The historic city of Dubrovnik before the outbreak of war. Its double ramparts and bastions enclosing the Baroque cathedral and numerous fine churches did little to save the city from devastation during the fighting between Serbs and Croats in the early 1990s, which aroused worldwide concern.

Civil war broke out, with the Serb-dominated Yugoslav army joining forces with the ethnic Serbs. Despite a number of ceasefires and the presence of UN peace-keeping forces, fighting continued through the early 1990s. In a series of offensives during 1993–94 Croatia regained almost all the territory it had lost to Serbia. In late 1995 the Croatian Serbs agreed to hand back East Slavonia, the last Serb enclave in Croatia. Diplomatic relations between Croatia and Yugoslavia were restored in the late 1990s, although conflict over the Prevlaka peninsula in southern Croatia continued and international observers remained in the area to oversee security.

ECONOMY

Civil war virtually wiped out Croatia's tourist trade overnight and it was not until the late 1990s that the industry showed signs of significant recovery. The war caused tremendous structural damage, and there were large-scale movements of refugees from Bosnia and also within Croatia. International aid helped stabilize the economy, but in the early 2000s Croatia was still struggling to modernize its industry, transform its economy and escape the effects of war. Rich natural resources should boost future prosperity: sizable deposits of oil, coal and bauxite. There is also some wine making and other agricultural activity in the fertile Pannonian plain.

Bosnia Herzegovina

REPUBLIC OF BOSNIA HERZEGOVINA

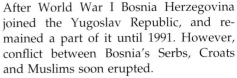

BOSNIA HERZEGOVINA DECLARED ITS INDE-pendence from former Yugoslavia in October 1991 and was invaded by the Serbian-dominated Federal Army, which has polarized its population and destroyed its cities. It is bordered by Croatia to the north and west, Serbia to the east and Montenegro to the south.

GEOGRAPHY

Like most of the Balkan peninsula, Bosnia Herzegovina is mountainous and rugged. It has only 19 km (12 mi) of coastal access through a narrow strip that separates the Croatian city of Dubrovnik from the rest of Croatia. Along the northern part of its border with Croatia, the Sava river forms a broad and fertile plain. The capital, Sarajevo, lies roughly in the center of the country near the headwaters of the river Bosna, from which Bosnia takes its name. The southwest is dominated by a karst landscape of eroded limestone plateaus. Cold winds from the Dinaric Alps can make winters bitterly cold; summers are generally mild.

SOCIETY

Southern Slavs settled in the area about the 6th century AD. Croatia fell under the aegis of the successors of the western Roman empire, and Serbia was controlled by the Byzantine empire, leaving Bosnia and Herzegovina as a buffer zone between the two. In the 12th century it was ruled by Hungary, and was subsequently fought over by several powers, including Venice, Serbia and Croatia.

Despite this continued warfare, Bosnia was able to assert a brief period of independence from 1353. Reconquered by the Ottoman empire in 1463, it was joined by Herzegovina in 1482, and the two have been linked together ever since. Under Ottoman control many of Bosnia's former Christians converted to Islam, the dominant religion today.

In 1875 the Bosnians successfully rebelled against Ottoman rule and became a protectorate of the Austro-Hungarian empire, by which it was annexed in 1908.

The Turkish bridge at Mostar in Herzegovina, built in 1566 when the region was under Ottoman rule, was a familiar landmark until its destruction in 1993–94, when the town was the center of fierce fighting between Bosnian Serbs and Croats.

After World War I Bosnia Herzegovina joined the Yugoslav Republic, and remained a part of it until 1991. However, conflict between Bosnia's Serbs, Croats and Muslims soon erupted.

The move to independence, recognized in 1991, prompted Serbian fears of the creation of an "Islamic state" by Bosnia's Muslim majority. Serbia, together with Croatia, used the pretext of protecting its Christian peoples there to send in paramilitary groups, and civil war broke out in 1992. However, late in 1995 a peace plan was signed by the presidents of Serbia, Croatia and Bosnia Herzegovina in which the unified state called the Union of Bosnia Herzegovina was to consist of a Muslim–Croat federation and Bosnian Serb republic. NATO peace-implementation troops arrived late in 1995 and early 1996. UN estimates are that 2.8 million people were displaced during the war.

ECONOMY

Agriculture is crucial to the economy, with most of the fertile land in the north; fruit, olives, rice, tobacco and grapes for wine are all grown. Sheep rearing is also important. Vast timber reserves in the mountains are the basis of an important forestry industry. The country is also rich in mineral resources, and coal, iron, copper, salt and zinc are mined extensively.

Despite substantial international aid, in 2001 Bosnia Herzegovina was still struggling to overcome the huge postwar problems of homelessness, refugees and mass unemployment that hindered plans for modernization and industrial growth.

NATIONAL DATA - BOSNIA AND HERZEGOVINA

Land area 51,129 sq km (19,741 sq mi)

Climate	Altitude m (ft)	Temperatures January °C(°F)	July °C(°F)	Annual precipitation mm (in)
Sarajevo	630 (2,067)	-1 (28)	19 (90)	931 (36.6)

Major physical features highest point: MaMaglic 2,386 m (7,828 ft); longest rivers: Sava (part) 940 km (584 mi), Bosna 241 km (150 mi)

Population (2006 est.) 4,498,976

Form of government multiparty republic with two legislative houses

Armed forces state army in transition 11,825

Largest cities Sarajevo (capital - 380,000); Banja Luka (165,100)

Official language Serbo-Croat

Ethnic composition Bosniak (Muslim) 48%; Serb 37.1%; Croat 14.3%; other 0.6%

Religious affiliations Muslim 40%; Orthodox 31%; Roman Catholic 15%; other 14%

Currency 1 marka (BAM or KM) = 100 fening

Gross domestic product (2006) U.S. $24.8 billion

Gross domestic product per capita (2006) U.S. $5,500

Life expectancy at birth male 74.39 yr; female 81.88 yr

Major resources coal, lignite, iron ore, bauxite, copper, lead, zinc, chromite, cobalt, manganese, nickel, clay, gypsum, salt, sand, forests, hydropower, apples, corn, grapes

Macedonia

THE FORMER YUGOSLAV REPUBLIC OF Macedonia (FYROM) is a small state in the middle of the Balkan peninsula. It shares borders with Greece to the south, Bulgaria to the east, Serbia to the north and Albania to the west. It gained United Nations membership in 1993 following a compromise with Greece by which it is known as FYROM, whereupon Greece reopened its border to Macedonian trade. It reimposed the economic blockade in 1994 after the majority of European Union states recognized FYROM.

GEOGRAPHY

A large part of Macedonia is a high plateau 600–900 m (2,000–3,000 ft) above sea level; the rest of the roughly rectangular country is mountainous. The Kozuf and Nidze ranges separate it from the

---

NATIONAL DATA - MACEDONIA

Land area	24,856 sq km (9,597 sq mi)
Climate	continental with cold winters and hot summers
Major physical features	highest point: within Korab Mountains 4,634 m (15,203 ft); longest river: Po 620 km (380 mi); largest lake: Lake Garda 370 sq km (140 sq mi)
Population	(2006 est.) 2,050,554
Form of government	multiparty republic with one legislative house
Armed forces	army 9,760; air force 1,130
Capital city	Skopje (467,257)
Official language	Macedonian
Ethnic composition	Macedonian 64.2%; Albanian 25.2%; Turkish 3.9%; Roma 2.7%; Serb 1.8%; other 2.2%
Religious affiliations	Macedonian Orthodox 64.7%; other Christian 0.37%; Muslim 33.3%; other and unspecified 1.63%
Currency	1 Macedonian denar (MKD) = 100 deni
Gross domestic product	(2006) U.S. $16.91 billion
Gross domestic product per capita	(2006) U.S. $8,200
Life expectancy at birth	male 71.51 yr; female 76.62 yr
Major resources	iron ore, copper, lead, zinc, chromite, manganese, nickel, tungsten, gold, silver, asbestos, gypsum, timber, antimony, arsenic, barley, maize (corn), cotton

---

Greek region of Macedonia. The highest point at 2,764 m (9,066 ft) is in the Korab Mountains in the far northeast of the country on the border with Serbia.

The climate is continental; rain is heaviest in the summer and there are significant snow falls in winter. Less traveled than the other more accessible republics of former Yugoslavia, it has large areas around its three great inland lakes – Ohrid, Prespa and Doiran – that remain unspoilt by human activity.

SOCIETY

Modern Macedonia was a part of the ancient Empire of Macedonia, which partly covered territory now in southwestern Bulgaria and northern Greece. Founded in the 6th century BC as the Kingdom of Macedon, its most famous ruler was Alexander the Great (356–323 BC). After his death, Macedon, like Alexander's wider empire, was split between warring factions. It was eventually defeated by the Romans and became a Roman province in 148 BC.

The Romans were quick to exploit the strategic location of this region, which controlled the fastest land route from the Adriatic to the Bosporus (the narrow strait linking the Black Sea to the Aegean) and Asia Minor. Following the decline of Roman power, in the 6th century AD the region was settled by the Slav peoples from whom the majority of modern Macedonians are descended. Macedonia was fiercely fought over by the Byzantines, Bulgars and Serbs until it was conquered by the Ottomans in 1371. With the demise of the Ottoman empire in the

The medieval monastery of St Naum near Lake Ohrid is typical of many such churches throughout the country. Most Macedonians belong to the Eastern Orthodox church, which has been independent from the Serbian Orthodox church since 1967.

19th century, each of these three countries claimed that the Macedonians were their ethnic descendants and the region was therefore their legitimate territory. The Balkan Wars of 1912–13 resulted in the division of the country between all three claimants. The largest Serbian part of the region (Vardar Macedonia) was brought by Serbia into the newly formed Kingdom of the Serbs, Croats and Slovenes after World War I.

Macedonia declared independence in 1992 following a referendum overwhelmingly in favor. In 1993 Macedonia joined the UN but Greece blocked formal recognition and imposed a trade embargo when the European Commission began legal proceedings against it. In 1995, after Macedonia agreed to redesign its flag and renounce any present or future claims to the Greek region of Macedonia, Greece capitulated. UN peacekeeping troops remained in Macedonia through 2000.

ECONOMY

Civil war in the Balkans has destabilized Macedonia since its independence. The Greek trade embargo in 1994 caused energy and food shortages, and in 1999 large numbers of refugees from Kosovo arrived. The economy nevertheless showed signs of recovery. The country is rich in minerals. Agriculture is a key sector, with livestock, cereals and tobacco. Manufacturing includes steel, chemicals and machinery.

Albania

REPUBLIC OF ALBANIA

A LBANIA IS A MOUNTAINOUS COUNTRY ON the western (Adriatic) coast of the Balkan peninsula, bordered by Montenegro to the northwest and Serbia and the Former Yugoslav Republic of Macedonia to the east. To the southeast is Greece.

GEOGRAPHY

Albania's mountainous topography has contributed to its isolation from Europe, and much of the broad coastal lowlands are swampy and inaccessible. The climate is largely Mediterranean, though winters in the mountains are often severe.

SOCIETY

The Albanians are descended from the Indo-European Illyrians. In addition to strong regional differences between the northern and southern dialect groups, there are numerous ethnic minorities, including Greeks and Macedonians.

Conquest by the Romans in the 1st century BC was followed by incursions by Germanic Goths, Greek Byzantines and Slavic Serbs, until in the late 14th century Albania succumbed to Ottoman rule.

Independence, won from the Turks in 1912, ended with invasion by fascist Italy in 1939. In 1946, after a bitter civil war, Albania reemerged as a single-party communist republic under the isolationist rule of General Enver Hoxha (1908–85).

Albania's hardline government was the last communist regime in Europe to give way to popular pressure for reform in the early 1990s. The collapse of the economy that followed was exacerbated by the arrival of ethnic Albanian Muslims fleeing former Yugoslavia, and Kosovan refugees in 1999. Civil violence erupted in 1997, after the collapse of investment schemes.

ECONOMY

The government faces massive problems in its attempts both to modernize Albania's underdeveloped economy and to provide for a population that is still primarily rural. The limited export market relies heavily on minerals such as chromite, iron ore, petroleum and natural gas. However, Albania still lacks the infrastructure necessary to promote the growth of a successful market economy.

NATIONAL DATA – ALBANIA				
Land area 27,398 sq km (10,578 sq mi)				

Climate		Temperatures		Annual
	Altitude m (ft)	January °C(°F)	July °C(°F)	precipitation mm (in)
Tirana	89 (292)	7 (45)	24 (75)	1,353 (53.2)

Major physical features highest point: Mount Korabit 2,751 m (9,026 ft); longest river: Drin (part) 282 km (175 mi); largest lake: Lake Scutari 370 sq km (143 sq mi)

Population (2006 est.) 3,581,655

Form of government emergent democracy

Armed forces army 16,000; navy 2,000; air force 3,500

Capital city Tiranë (376,642); Elbasan (123,270); Durrës (116,275); Shkodër (109,604); Vlorë (104,118)

Official language Albanian

Ethnic composition Albanian 95%; Greek 3%; other 2% (Vlach, Roma, Serb, Macedonian, Bulgarian)

Religious affiliations Muslim 70%; Albanian Orthodox 20%; Roman Catholic 10%

Currency 1 lek (ALL) = 100 qintars

Gross domestic product (2006) U.S. $20.21 billion

Gross domestic product per capita (2006) U.S. $5,600

Life expectancy at birth male 74.78 yr; female 80.34 yr

Major resources petroleum, natural gas, coal, bauxite, chromite, copper, iron ore, nickel, salt, timber, hydropower, asphalt, bitumen, cereals, cotton, grapes, olives, potatoes, tobacco

The Vijosë river in southern Albania cuts its way through rugged mountain peaks typical of 70 percent of the country's landscape. Albania's geographical isolation was reinforced by the policies of the communist leader Enver Hoxha (1908–85).

Russia and Northern Eurasia

COUNTRIES IN THE REGION

ESTONIA · LATVIA · LITHUANIA · MOLDOVA

BELARUS · UKRAINE · RUSSIA · GEORGIA

ARMENIA · AZERBAIJAN · TURKMENISTAN

UZBEKISTAN · TAJIKISTAN · KYRGYZSTAN

KAZAKHSTAN · MONGOLIA

Domes of victory (*left*) The multicolored, onion-shaped cupolas of St Basil's Cathedral in Moscow have been a characteristic feature of the city skyline ever since they were built by Tsar Ivan IV in 1555. The cathedral commemorates his defeat of the Tartars – one of many peoples who have invaded Russia through the centuries.

capital city
major town

height of land (meters)

5000
3000
2000
1000
500
200
0 (sea level)

▲ mountain peak
▼ depression

Estonia

REPUBLIC OF ESTONIA

THE NORTHERNMOST OF THE THREE SMALL Baltic republics, Estonia lies on the eastern shore of the Baltic Sea, with Latvia to the south and Russia to the east. Finland lies to the north, across the Gulf of Finland. In 1991 Estonia regained its independence after half a century of Soviet rule that had never been officially recognized by some Western governments. The Estonian people, who speak a language similar to Finnish and share a common descent with them, have a strong cultural heritage that has survived many centuries of foreign domination.

NATIONAL DATA - ESTONIA

Land area 43,211 sq km (16,684 sq mi)

Climate		Temperatures		Annual
	Altitude m (ft)	January °C(°F)	July °C(°F)	precipitation mm (in)
Tallinn	44 (144)	-5 (23)	17 (68)	676 (26.6)

Major physical features highest point: Munamägi 318 m (1,043 ft); largest lake: Lake Peipus (part) 3,548 sq km (1,370 sq mi)

Population (2006 est.) 1,324,333

Form of government multiparty republic with one legislative house

Armed forces army 3,429; navy 331; air force 193

Largest cities Tallinn (capital – 396,193); Tartu (101,740); Narva (66,936); Kohtla-Järve (45,740); Pärnu (44,198)

Official language Estonian

Ethnic composition Estonian 67.9%; Russian 25.6%; Ukrainian 2.1%; Belarusian 1.3%; Finn 0.9%; other 2.2%

Religious affiliations Evangelical Lutheran 13.6%; Orthodox 12.8%; other Christian (including Methodist, Seventh-Day Adventist, Roman Catholic, Pentecostal) 1.4%; unaffiliated 34.1%; other and unspecified 32%; none 6.1%

Currency 1 Estonian kroon (EEK) = 100 cents

Gross domestic product (2006) U.S. $26 billion

Gross domestic product per capita (2006) U.S. $19,600

Life expectancy at birth male 66.58 yr; female 77.83 yr

Major resources oil shale, peat, phosphorite, clay, limestone, sand, dolomite, sea mud, engineering, electronics, wood and wood products, textiles, tourism, cotton, fish, fruits, grains, livestock, potatoes, vegetables

GEOGRAPHY

Estonia's low and undulating landscape bears the marks of extensive glacial activity. Notable features are the many islands off the coast including Hiiumaa and Saaremaa to the west (across the mouth of the Gulf of Riga), and Lake Peipus along the Russian border.

The climate is cool and northerly. Winds from the Baltic make the winters damp and raw, but provide relief from the much colder weather to the east. Although more than half the land is given over to pasture and fields, much of the natural vegetation remains. This includes mixed forest, with marsh and swamp in low-lying areas. Wildlife is still abundant in these parts, including deer and lynx.

SOCIETY

Estonians have inhabited this area for at least 2,000 years. Their first experience of foreign invasion came with the Vikings in the 9th century AD. In the 13th century Estonia came under the Christianizing influence of Germanic knights, who joined the Teutonic Order in 1237. German domination later gave way to Swedish and finally Russian rule in the 18th century. A growing awareness of Estonian culture and identity developed during the 19th century, and in 1918, following the upheavals of the Russian Revolution, the country at last became independent.

The new democratic regime proved fragile, however, and during World War II Estonia was little more than a pawn in the struggles between its more powerful neighbors. After being formally annexed by the Soviet Union in 1940 the country was under Nazi German occupation until 1944. The reimposition of communist rule after World War II brought with it enforced collectivization and an influx of Russian and Slav immigrants. In the late 1980s, however, the Soviet grip on the republic relaxed. Baltic nationalism revived, and independence came in 1991 following the breakup of the Soviet Union. Initially, the civil rights of nonethnic Estonians, who represent about 40 percent of the population, were reduced to favor the native population. However, the majority of the ethnic Russian population was able to vote in 1995 elections.

Estonia has a strong cultural tradition, particularly distinctive in its music and literature. The Lutheran Church has a long history of choral singing, reflected in the music festivals held.

Wooded marshland and bog islands – such as these in the Endla State National Reserve – and boulder-strewn lowlands cover much of Estonia, the smallest of the Baltic states, making the land difficult to farm. Estonia also includes some 800 Baltic islands.

ECONOMY

Estonia had an efficient and prosperous economy, but still suffered much hardship after 1991, as other former Soviet Republics. By the early 2000s the transition to a market economy was showing signs of success and Estonia hoped to join the European Union in the near future.

Keeping livestock is the main agricultural activity, in particular raising beef and dairy cattle. Forest cover has been severely reduced in recent years and woodworking industries have suffered as a result, though a concerted reforestation program is now underway. The country has few natural mineral resources, apart from bituminous shale, which supplies power for the industrial sector and numerous byproducts for the chemical industry. Most other industries depend on imported raw materials.

Estonia has a well-developed communications infrastructure and a good welfare support network. There are some 40 Estonian newspapers, and since 1955 the republic has had its own Estonian-language television channel. Educational standards and literacy levels are high, and Estonia's 17th-century university at Tartu remains one of the great universities of the former Soviet Union.

Latvia

REPUBLIC OF LATVIA

L ATVIA IS A SMALL, NEWLY INDEPENDENT state that has much in common with the two neighboring republics of Estonia to the north and Lithuania to the south. Situated on the eastern shore of the Baltic Sea, it also shares borders with the former Soviet republics of Russia to the east and Belarus to the southeast.

GEOGRAPHY

The country has a long, sandy coastline that faces the Baltic to the west, while to the northwest, around the Gulf of Riga, it forms a broad arc with the capital, Riga, in the center. The landscape consists of undulating plains, rising in the east to a low plateau dotted with lakes and bogs. The cool, damp climate with moderate rainfall brings some very cold spells in winter. Oak and pine forests interspersed with pasture and meadows cover large areas of the country. The abundant wildlife includes foxes, lynx and squirrels, and the rare European beaver has been reintroduced to woodland areas.

SOCIETY

Like their Lithuanian neighbors, the Latvians (or Letts) are an ancient Baltic people. In the 13th century Latvia and Estonia came under the control of the German crusading order of the Livonian Knights. Latvia was later divided between Poland and Sweden, but by the end of the 18th century it had been absorbed into the vast Russian empire of Catherine the Great (1729–96).

The Russian Revolution of 1917 provided the opportunity for an upsurge in Baltic nationalism, and by 1920 Latvia was recognized as an independent nation. In 1940 Latvia was annexed as a constituent republic of the Soviet Union under the terms of the Nazi–Soviet Non-Aggression Pact. From 1941 until its reconquest by Soviet forces in 1944 Latvia was under German occupation.

A rapid industrialization program instituted in Latvia by the Soviets in the postwar period led to massive immigration into Latvia from other parts of the Soviet Union. The Latvians have, however, maintained a strong and vibrant culture and in 1988 their language replaced Russian as the official language; a prelude to independence from the Soviet Union, which was achieved in 1991. A liberalization of citizenship laws in 1998 eased the way for Russians to gain legitimate citizenship.

ECONOMY

Industrial production, once a major contributor to gross national product, was in decline for the first half of the 1990s. Once a leading Soviet producer of motor vehicles, consumer goods and telecommunications equipment, Latvia's main trading partner by the late 1990s was the EU. Latvia established free trade regimes with EFTA and the EU in 1996.

The capital city, Riga, has a long tradition as a seaport and trading center, and is one of the busiest of the Baltic cities. Elsewhere, the country is well served by rail, road and sea links, which have helped to encourage the growth of tourism. The welfare support network is highly developed. Educational standards are good and literacy levels high.

NATIONAL DATA – LATVIA				
Land area 63,589 sq km (24,552 sq mi)				
Climate		Temperatures		Annual
	Altitude m (ft)	January °C(°F)	July °C(°F)	precipitation mm (in)
Riga	3 (10)	-5 (23)	17 (63)	633 (24.9)
Major physical features highest point: Vidzeme 311 m (1,020 ft); longest river: Western Dvina = Daugava (part) 1,060 km (635 mi)				
Population (2006 est.) 2,274,735				
Form of government multiparty republic with one legislative house				
Armed forces army 1,817; navy 685; air force 255				
Largest cities Riga (capital - 727,578); Daugavpils (109,482); Liepāja (85,915); Jelgava (66,087)				
Official language Lettish				
Ethnic composition Latvian 57.7%; Russian 29.6%; Belarusian 4.1%; Ukrainian 2.7%; Polish 2.5%; Lithuanian 1.4%; other 2%				
Religious affiliations Lutheran, Roman Catholic, Russian Orthodox				
Currency 1 Latvian lat (LVL) = 100 santims				
Gross domestic product (2006) U.S. $35.08 billion				
Gross domestic product per capita (2005) U.S. $15,400				
Life expectancy at birth male 66.08 yr; female 76.85 yr				
Major resources peat, limestone, dolomite, amber, hydropower, arable land, timber, paper-making industry, fish, flax, hydroelectric power, livestock, potatoes, rye, sugar beet, vegetables, wheat				

Riga's 13th-century Doma cathedral dominates the Baltic city, which spans the Western Dvina river. The city was founded in 1201 by Bishop Albert I, whose crusaders, the Brothers of the Sword, Christianized the whole of Livonia (modern Latvia and southern Estonia).

Lithuania

LITHUANIAN REPUBLIC

L ITHUANIA LIES IN THE SOUTHEASTERN corner of the Baltic Sea, with Latvia to the north and Belarus to the east. To the southwest it shares borders with Poland and the Kaliningrad enclave that is part of the Russian Federation.

Although Lithuania is a small country now, in medieval times, as part of the Grand Duchy of Lithuania, it was one of the most powerful states in Eastern Europe. A strong sense of national identity going back some 2,000 years was vital to the struggle for independence from Soviet domination, which began in 1940 during World War II.

NATIONAL DATA - LITHUANIA

Land area	65,200 sq km (25,200 sq mi)			
Climate		Temperatures		Annual
	Altitude m (ft)	January °C(°F)	July °C(°F)	precipitation mm (in)
Rome	17 (56)	7 (45)	25 (77)	657 (25.9)

Major physical features	highest point: Juozapine 292 m (958 ft); longest river: Neman (part) 963 km (582 mi)

Population	(2006 est.) 3,585,906

Form of government	multiparty republic with one legislative house

Armed forces	army 10,100; navy 710; air force 1,200

Largest cities	Vilnius (capital – 541,824); Kaunas (360,637); Klaipeda (187,316); Siauliai (129,037); Panevezys (115,315)

Official language	Lithuanian

Ethnic composition	Lithuanian 83.4%; Polish 6.7%; Russian 6.3%; other or unspecified 3.6%

Religious affiliations	Roman Catholic 79%; Russian Orthodox 4.1%; Protestant (including Lutheran and Evangelical Christian Baptist) 1.9%; other or unspecified 5.5%; none 9.5%

Currency	1 litas (LTL) = 100 cents

Gross domestic product	(2006) U.S. $54.03 billion

Gross domestic product per capita	(2006) U.S. $15,100

Life expectancy at birth	male 69.2 yr; female 79.49 yr

Major resources	peat, amber, barley, fish, flax, legumes, livestock, potatoes, rye, sugar beet, timber, wheat

GEOGRAPHY

The terrain is mainly low-lying glaciated plains, with sand dunes along the coast and lake-strewn hills in the southeast. The marshy plains of the center are crossed by numerous rivers, including the Neman, which flows from the south into the Baltic Sea. The capital, Vilnius, is on the Neris river in the southeast.

Warm summers and cold winters, moderated by damp westerly winds from the Baltic, characterize the mainly continental climate; rainfall is highest in late summer. The plentiful natural vegetation ranges from forests of pine and oak to meadows and marshes. Wildlife includes wolves, badgers, Wild boars and numerous birds.

SOCIETY

The Lithuanians, unlike their neighbors the Latvians, resisted conquest by the German crusading order of the Livonian Knights in the 13th century and, as a result, were the last European people to be converted to Christianity. For 300 years the Lithuanians were united under the powerful rulers of the Gediminas dynasty, who established the Grand Duchy of Lithuania incorporating modern Lithuania, Belarus and western Ukraine. In the 15th century, alliance with Poland and later full union extended Lithuanian territory southeastward as far as the Black Sea. However, the expansionist aspirations of the nearby state of Muscovy were a constant threat. In the late 18th century Lithuania finally came under Russian domination, following the partition of Poland.

A precarious independence was gained in the aftermath of the Russian Revolution of 1917, but it was repeatedly threatened by border disputes with Poland. In 1940, under the terms of the Nazi–Soviet Non-Aggression Pact, Lithuania was annexed by the Soviet Union as a constituent republic, but in 1941 the country was occupied by German forces. The expulsion of the Germans by the Soviet Army in 1944 reinstated the process of Sovietization.

With the collapse of central Soviet communist power in 1991, Lithuania finally won full independence. Lithuanian, which preserves more archaic features than almost any other language in Europe today, replaced Russian as the official language. The local cultural tradition was also kept alive during the Soviet era through music festivals and folk art.

Festive crowds (*above*) gather in Gediminas Square, Vilnius, capital of Lithuania, to celebrate their national identity. Lithuanians are mostly Roman Catholic, with a distinct language and culture. They finally became independent of the Soviet Union in 1991.

Lithuania received far fewer Russian immigrants under Soviet rule than Latvia and Estonia, but it has always had a significantly large Polish community. Consequently it has maintained a strong Roman Catholic tradition, unlike its mainly Protestant Baltic neighbors.

ECONOMY

Lithuania's economy was disrupted following its liberation from Soviet rule. By the late 1990s, 30 percent of exports were going to the European Union and trade agreements were in place with neighboring countries. Growth was slow in the 1990s and inflation and unemployment rates remained high.

Agriculture, traditionally strong in meat and dairy farming, has diversified into grain and vegetable crops through marshland reclamation schemes. Forestry provides local woodworking and other light manufacturing with raw materials. Mineral resources include some petroleum as well as limestone and other construction materials.

Railroads are gradually giving way to roads, as motor vehicles become the chief means of transportation. The Soviet regime encouraged the development of generous medical, welfare and educational facilities, but all three areas are now suffering from poor financial support and lack of resources.

Moldova

REPUBLIC OF MOLDOVA

MOLDOVA, THE FORMER SOVIET REPUBLIC of Moldavia, is a small country near the north coast of the Black Sea. It borders Romania to the west, and is otherwise surrounded by the Republic of Ukraine.

GEOGRAPHY

Moldova is a hilly country much of which lies between the river Prut, on its border with Romania, and the river Dniester, close to the Ukrainian frontier. It forms part of an ancient eroded plateau covered in rolling steppelands that is crisscrossed by deep valleys and ravines with heavily forested slopes. The climate is generally temperate with short, sharp winters and high rainfall in summer.

SOCIETY

Once known as Bessarabia, the land between the Prut and the Dniester has had a checkered history. The name "Moldova" dates from 1359, when the principality of that name was formed. However, Moldova was soon dominated by Hungary, Poland and, from the 16th century, by the Turkish Ottoman empire, which finally ceded part of it (Bessarabia) to Russia in 1812. In 1859, when the rest of Moldova united with Walachia to form Romania, Bessarabia was retained by Russia. But in 1918, when the Russian empire collapsed, its people chose to become part of Romania too, and the Dniester became the border with what later became the Soviet Union.

Rural life Wearing bright headscarves, two Moldovan housewives exchange news, while their compatriots relax on a bench. By culture and language, Moldovans are indistinguishable from Romanians, their western neighbors from whom they parted, politically, in 1940.

During World War II Bessarabia changed hands more than once and in 1944 it was taken over by the Soviet Union. The Soviets quickly instituted a system aimed at erasing all historical links with Romania. The Russian (Cyrillic), alphabet was imposed on the Moldovan language, which is otherwise little different from Romanian. During the breakup of the Soviet Union in 1991, immigrant Ukrainians and Russians, who together outnumbered Moldovans in the east, opposed Moldovan moves toward closer ties with Romania by declaring a Trans-Dniestrian Moldovan Republic. Since 1992 a ceasefire has held and in 1994 the Moldovan government promised a high degree of autonomy. The idea of reunification with Romania was defeated in a referendum in 1994.

ECONOMY

Since Moldova has no mineral resources, agriculture is the mainstay of the economy. Sunflower seeds are an important crop, along with a variety of fruit, nuts, tobacco and vegetables. Much of the steppe has been turned over to growing wheat and other cereals. Food processing, engineering and the production of agricultural machinery also form a major part of the industrial sector. But it is in wine producing that Moldova has the greatest potential for international trade. The wine industry supplied the former Soviet Union annually with 320 million bottles of sparkling wine alone. Moldova is now developing the industry commercially.

A casualty of the old communist system of centralism, Moldova has now become heavily reliant on other members of the Commonwealth of Independent States, initially for medicine, petrol and electricity and now as trading partners, especially Russia. At the end of the last century, Moldova was struggling with negative economic growth, a balance of payments deficit and high inflation.

NATIONAL DATA – MOLDOVA

Land area	33,371 sq km (12,885 sq mi)
Climate	moderately continental
Major physical features	highest point: Balaneshty 430 m (1,409 ft); longest river: Dniester (part) 1,420 km (880 mi)
Population	(2006 est.) 4,466,706
Form of government	multiparty parliamentary republic with one legislative house
Armed forces	army 5,710; air force 1,040
Largest cities	Chisinau (capital - 593,800); Tiraspol (157,000); Beltsy (122,700); Tighina (96,000); Ribnita (53,00)
Official language	Moldovan
Ethnic composition	Moldovan/Romanian 78.2%; Ukrainian 8.4%; Russian 5.8%; Gagauz 4.4%; Bulgarian 1.9%; other 1.3%
Religious affiliations	Eastern Orthodox 98%; Jewish 1.5%; Baptist and other 0.5%
Currency	1 Moldovan leu (MDL) = 100 bani
Gross domestic product	(2006) U.S. $8.971 billion
Gross domestic product per capita	(2006) U.S. $2,000
Life expectancy at birth	male 61.61 yr; female 69.88 yr
Major resources	lignite, phosphorites, gypsum, limestone, barley, clay, corn, fruit, grapes, livestock, oil and natural gas, quartz sands, rye, soybeans, sugar beet, sunflowers, tobacco, vegetables, wheat

Belarus

T HE FORMER SOVIET REPUBLIC OF BELORUSSIA, now officially known by its native Belorussian name of Belarus, stretches eastward from the borders of Poland to Russia. Ukraine lies to the south, and Lithuania and Latvia to the northwest. Often referred to as White Russia (a literal translation), the country has had a long history of domination by neighboring powers. In recent years it has also suffered significant land contamination and its people have been exposed to increased health risks, both associated with the after-effects of the Chernobyl nuclear disaster in Ukraine in 1986, when radioactive fallout was carried across the country.

GEOGRAPHY

The landscape of Belarus consists largely of rolling plains produced by ice-age glaciation. The north is crossed by a series of sandy hills and long ridges made up of glacial debris, while in the south the Pripyat and Dnieper rivers flow to Ukraine through marshlands that have been mostly drained for their fertile soils. A temperate climate, with moderate precipitation, varies between the more equable conditions of central Europe and the extremes of continental Russia.

A third of the land is covered by forests of birch, pine and fir in the north, and oak, ash, maple and beech farther south. In the west, the Belovezhskaya Pushcha – a remnant of the primeval forest that once covered much of Europe – stretches across the Polish frontier. Today it is a nature reserve maintained by both Belarus and Poland, rich in rare wildlife such as elk (moose), Wild boar and beaver. It also offers a refuge to the last of the European bison, descendants of zoo stocks introduced into the wild.

SOCIETY

The country was settled by eastern Slavic peoples about the 6th century AD. Gradually, their cluster of small princedoms came under the sovereignty of the Viking state of Rus, with its capital Kiev to the south. After the overthrow of Kiev in the 13th century, the area became part of the Grand Duchy of Lithuania, which, after two centuries of close alliance, finally entered a union with Poland in 1569.

In the 18th century, when Poland was repeatedly partitioned between Austria, Prussia and Russia, Belarus came under the jurisdiction of Russia and became a political and economic backwater. Under tsarist rule this area, together with Ukraine, became the homeland for most of Russia's Jewish population. Many knew it as the "Pale of Settlement". Forced to live within strictly defined areas, the Jews were also denied many of the basic freedoms of Russia's other subjects, including higher educa-tion and freedom of travel.

World War I left the country devastated, and the subsequent Russo-Polish War

Production line pottery (*right*) in a Minsk porcelain factory that produces teapots. Since being rebuilt after its near total destruction during World War II, Minsk has become a major manufacturing center producing a variety of products ranging from household goods, chemicals and foodstuffs to radios, trucks and tractors.

The Victory obelisk (*below*) in Minsk commemorates the dead of World War II, particularly almost the entire Jewish population. Minsk has been virtually rebuilt since 1944 when its recapture by the Russians left hardly a building standing. Its population has tripled in the past 30 years.

NATIONAL DATA – BELARUS				
Land area 207,600 sq km (80,155 sq mi)				

Climate		Temperatures		Annual
	Altitude m (ft)	January °C(°F)	July °C(°F)	precipitation mm (in)
Minsk	234 (767)	-7 (19)	13 (55)	782 (30.7)

Major physical features highest point: Dzyarzhynsk Mountain 345 m (1,135 ft); longest river: Dnieper (part) 2,280 m (1,420 ft)

Population (2006 est.) 10,293,011

Form of government republic in name, although in fact a dictatorship

Armed forces army 29,600; navy 18,170; joint forces 25,170

Largest cities Minsk (capital - 1,765,800); Homyel (481,197); Mahilyow (366,900); Vitsyebsk (342,700); Hrodna (316,700); Brest (298,329)

Official language Belarusian, Russian

Ethnic composition Belarusian 81.2%; Russian 11.4%; Polish 3.9%; Ukrainian 2.4%; other 1.1%

Religious affiliations Eastern Orthodox 80%; other (including Roman Catholic, Protestant, Jewish, and Muslim) 20%

Currency 1 Belarusian ruble (BYB/BYR) = 100 kopecks

Gross domestic product (2006) U.S. $80.74 billion

Gross domestic product per capita (2006) U.S. $7,800

Life expectancy at birth male 63.47yr; female 74.98 yr

Major resources timber/forests, peat deposits, oil, natural gas, granite, dolomitic limestone, marl, chalk, sand, gravel, clay, buckwheat, chloride, potassium, rye, sodium chloride, sugar beet, tobacco, wheat

of 1919–21 resulted in Belarus being partitioned between Poland and the newly formed Soviet Union. At the outbreak of World War II, Soviet forces seized the Polish sector, which was then overrun by the Germans. The war years took a terrible toll on the population of Belarus; some 1.3 million people died, including most of the Jewish population. In addition, about 75 percent of all urban housing was destroyed, along with almost all of the country's industrial infrastructure. After long and bitter fighting, the Soviet Union took over the entire country, deporting its Polish population and establishing a nominally independent republic with a seat at the United Nations.

Increased freedoms in the late 1980s led to a resurgence of anti-Soviet feeling in Belarus, as elsewhere in the Soviet Union, culminating in the declaration of a fully independent republic of Belarus in September 1991. With the final breakup of the Soviet Union in December 1991, the capital of Belarus, Minsk, became the headquarters of the new Commonwealth of Independent States.

ECONOMY

There was much hardship after 1991; the average inflation rate for 1990–96 was 715 percent. Belarus has opted for close economic and cultural ties with Russia.

The traditional basis of the Belarussian economy is agriculture, which benefits from fertile soils and a favorable climate. The main crops are oats, rye and other grains, along with potatoes, a mainstay of the local diet, and sugar beet. Meat and dairy cattle are the main livestock. Forestry is significant, as are traditional beekeeping and fur-farming activities. However, all these activities, as well as the health of the people, have been severely affected by the nuclear accident in 1986 at

Chernobyl, in neighboring Ukraine.

Industry has increased in importance since the discovery of new mineral resources, including coal, oil shale, petroleum, potash, limestone, metal ores and enormous salt deposits. During the years of the Soviet regime, the Belarussian capital, Minsk, rebuilt its industrial sector to the extent that it became the industrial hub of the western Soviet Union. Today, manufactured goods range from heavy machinery to tools and instruments, furniture, textiles, television sets and electronics components.

Transportation is good, as the country is crossed by several major routes between Russia, the Ukraine and Central Europe. Its many lakes and waterways are also used to float timber to the factories. Welfare and education were highly developed under the Soviet regime. Literacy became virtually universal and education was free and compulsory between the ages of 7 and 17. For those who were able to continue into higher education, there were some 30 institutions to choose from. These good education facilities helped to sustain a strong Belarussian literary and musical heritage. After World War II, the republic built up a conservatory of music, a state opera and ballet company, two state theaters and folk music companies.

Ukraine

REPUBLIC OF UKRAINE

T HE EASTERN EUROPEAN REPUBLIC OF THE
Ukraine lies on the northern side of
the Black Sea, bordering Russia to the east
and northeast, Belarus to the north,
Poland, Slovakia and Hungary to the
west, and Romania and Moldova to the
southwest. The second most populous of
the former Soviet republics (after Russia),
it also ranks second in terms of its
economic resources. Economic reforms
following the collapse of the Soviet Union
caused production to decline.
Unemployment and hyperinflation were
persistent problems in the late 1990s.

NATIONAL DATA – UKRAINE

Land area	603,700 sq km (233,100 sq mi)			
Climate		Temperatures		Annual
	Altitude m (ft)	January °C(°F)	July °C(°F)	precipitation mm (in)
Kiev	179 (587)	–5 (23)	20 (68)	643 (25.3)

Major physical features highest point: Hoverla 2,061 m (6,762 ft); longest river: Dnieper (part) 2,280 km (1,419 mi)

Population (2006 est.) 46,710,816

Form of government multiparty republic with one legislative house

Armed forces army 125,000; navy 13,500; air force 49,100

Largest cities Kiev (capital – 2,660,401); Kharkov (1,464,740); Dnepropetrovsk (1,056,497); ; Odessa (1,007,131); Donetsk (999,975); Zaporozhye (799,348)

Official language Ukrainian

Ethnic composition Ukrainian 77.8%; Russian 17.3%; Belarusian 0.6%; Moldovan 0.5%; Crimean Tatar 0.5%; Bulgarian 0.4%; Hungarian 0.3%; Romanian 0.3%; Polish 0.3%; Jewish 0.2%; other 1.8%

Religious affiliations Ukrainian Orthodox - Kiev Patriarchate 19%; Orthodox (no particular jurisdiction) 16%; Ukrainian Orthodox - Moscow Patriarchate 9%; Ukrainian Greek Catholic 6%; Ukrainian Autocephalous Orthodox 1.7%; Protestant, Jewish, none 38%

Currency 1 hryvnia (UAH) = 100 kopiykas

Gross domestic product (2006) U.S. $355.8 billion

Gross domestic product per capita (2006) U.S. $7,600

Life expectancy at birth male 64.71 yr; female 75.59 yr

Major resources iron ore, coal, manganese, natural gas, oil, salt, sulfur, graphite, titanium, magnesium, kaolin, nickel, mercury, timber, food processing, consumer goods, heavy engineering, chemicals and chemical equipment, asbestos, barley, corn, flax, gypsum, limestone, livestock, marble, ozocerite, peat, rye, sugar beet, tobacco, wheat

GEOGRAPHY

Ukraine is famous for its steppe landscape: mile after mile of level, rolling
terrain broken only by the Carpathian
Mountains, which cross the far southwest
between the Slovak and Romanian borders. Equally dramatic, though less high,
are the Crimean Mountains that overlook
the south coast of the Crimean Peninsula
in the far south, dividing the northwestern part of the Black Sea from the
smaller Sea of Azov.

Western Ukraine consists mainly of the
Dnieper plateau. This low tableland, cut
by deep valleys, descends to the Pripyat
marshes in the north, to the Dnieper river
valley in the east and to broad coastal
lowlands in the south. East of the
Dnieper, the land rises once again to low
hills along the northeastern border, and
to the Azov Hills and the Donets Ridge in
the southeast.

The climate varies from temperate continental in the northeast to Mediterranean on the south coast of the Crimea.
Average January temperatures are below
freezing point and precipitation is variable. The hot, dry summers of the southern Crimea have for a long time made it
attractive to tourists.

Most of the forests are to be found in
the south and southwest – in the Carpathian and Crimean Mountains. Farther
north, mixed woodland alternates with
bog and marshland, while a central band
of woodland and steppe is largely given
over to farmland. Some areas of virgin

Berries and cherries are weighed by a proud, bemedaled veteran of World War II at a market stall in Kiev, the Ukrainian capital. Under Soviet rule, all food marketing was handled by the state; but today, small-scale private enterprise is making a return.

steppe in the south are designated as
nature reserves, the oldest being As-
kaniya Nova, immediately north of the
Crimean Peninsula. The south coast of
the Crimea is renowned for its particularly lush vegetation of deciduous and
evergreen grasses and exotic flowers and
shrubs. Ukrainian wildlife is varied, including woodland animals – foxes, wildcats and roe deer – and steppe-dwelling
hamsters and jerboas.

SOCIETY

The Ukrainian capital, Kiev, was once the heart of Kievan Rus, a trading empire founded in the 9th century AD by the Varangians – Viking adventurers from the north. This empire prospered until the 13th century, when it was overrun by Mongol and Tartar invaders from the east. In 1240 these invaders razed the city of Kiev to the ground. During the 14th century most of the area of modern Ukraine came under the sway of Poland and the expanding Lithuanian empire, while the south remained under Mongol control. In 1386 the Polish and Lithuanian crowns were united when Jagiello of Lithuania married Jadwiga (daughter of Louis I of Poland) and became Ladislaus II of Poland.

As Poland–Lithuania pushed its frontier eastward, across the *ukraina*, or "borderlands", newly installed Polish landowners came into conflict with the Cossacks – Tartar warriors whom the Poles had originally recruited to defend their borders. In 1648 the Cossacks declared a semi-independent state, but faced with hostility from the Poles they turned to Russia for support. In 1654 a union of Russia and the Ukraine was negotiated. This proved unacceptable to Poland, and a Russo-Polish war broke out, ending in

Swirl of skirts (*below*) A colorful folk ensemble puts on an exciting display of popular Ukrainian dance and song. Ukraine has a vital cultural tradition, reflected especially in folk music, but also in classical music, the theater, literature and the cinema.

1667 with Kiev, and territory east of the Dnieper being taken by Russia. In 1772 Poland was partitioned by Russia, Prussia and Austria, with the far west of the Ukraine going to Austria. By a second partition in 1793 areas to the west and south of Kiev went to Russia.

In the 19th century a Ukrainian nationalist movement was born; this enjoyed some success in the civil war that followed the Russian Revolutions of 1917. However, in 1922 the country became one of the four founder-republics of the Soviet Union. Although the 1920s saw a period of improved living conditions and the encouragement of the Ukrainian language, the situation changed dramatically in the 1930s. Stalin's policy of enforced collectivization of agriculture resulted in a terrible famine that claimed millions of lives. Harsh political repression followed, as Stalin systematically crushed the thriving Ukrainian national revival. In addition, millions died during World War II and much of the country's historic architecture and valuable art treasures were destroyed. After the war, Ukrainian territory in the far west was reunited under Soviet control. The expanded and influential Soviet Ukrainian Republic became a founder member of the United Nations.

Ukrainian nationalism, forced underground in the 1960s and 1970s, reemerged strongly in the late 1980s. Building a campaign on this new nationalism, the Rukh (or People's Movement) received powerful support in free elections held in March 1990. The attempted Soviet coup of August 1991 won no support in Ukraine, which declared independence only weeks later. In December 1991, the Ukraine joined the Commonwealth of Independent States, and began the long process of reshaping its political and economic structures as a fully independent multiparty democracy. Since then, there have been differences with the Russian Federation over former Soviet military installations, and over the future of the Crimea, which was transferred from Russia to Ukraine. In 1995 the Supreme Council of the Ukraine voted to suppress the Crimean constitution and abolish the Crimean presidency.

Out of more than a hundred nationalities represented in the Ukraine, the great majority are ethnic Ukrainians, and most of the others speak closely related Slavic languages such as Russian or Polish. Russians form the largest minority, notably in the Crimea, whose original Tartar population was deported to central Asia in 1944 amid much suffering. New-

found religious freedoms have allowed historical differences to re-emerge: Christians are divided between Eastern Orthodox in the east, Roman Catholics in the far west, and the majority Uniate or Eastern Catholics, who under 16th-century Polish rule acknowledged papal supremacy while retaining Eastern rites.

ECONOMY

A favorable climate, and the fertile *chernozem* or black soil of the steppe, made Ukraine the breadbasket of the former Soviet Union. At the height of production Ukraine was responsible for almost half of the Soviet Union's total agricultural output. The country is a major producer of grain, sugar beet, cotton, tobacco, wine, meat and dairy products. The Black Sea is an important source of fish, especially sturgeon for caviar.

Mineral resources include vast iron-ore reserves and the world's richest concentration of manganese-bearing ores. The Donets basin has large coal reserves, while oil is extracted near the Carpathians, in Crimea, and in the Donets area. The leading industries are the production of various metals and food processing. A large manufacturing sector also produces heavy machinery, chemicals and fertilizers. Most energy comes from fossil fuels, though a small proportion comes from hydroelectric installations that have been built along the Dnieper. The program for nuclear power was halted after the disastrous reactor accident at Chernobyl, north of Kiev. It will still be some time before the full scale of contamination can be assessed, but there is already an alarming increase in the incidence of leukemia in children from the surrounding areas and from neighboring parts of Beloras.

The railroad network is concentrated in the Donets and Dnieper industrial areas, and there are good highways connecting all the main population centers. Busy ports such as Odessa on the Black Sea and Zhdanov by the Sea of Azov handle ocean freight, while major rivers and canals provide links to Central Europe. Airports at Kiev and elsewhere offer flights to national and international destinations.

Health, welfare and education services were well developed under the Soviet regime, although the promise of 10 years of free education for all children has not yet been fully realized. The Ukraine has many large and prestigious institutions of higher learning, including the Academy of Sciences in Kiev.

Russia

RUSSIA IS THE WORLD'S LARGEST COUNTRY, extending halfway round the globe. Much of this vast land is sparsely populated and has only limited natural resources. To the west Russia borders Finland, Norway, Estonia, Latvia and Belarus. The much longer southern frontier extends into central Asia and is shared with the Ukraine, Georgia, Azerbaijan, Kazakhstan, Mongolia, China and North Korea. To the north the country faces the Barents Sea and the East Siberian Sea, to the east the Sea of Okhotsk and to

NATIONAL DATA - RUSSIA

Land area	16,995,800 sq km (6,562,115 sq mi)			
Climate		Temperatures		Annual
	Altitude m (ft)	January °C(°F)	July °C(°F)	precipitation mm (in)
Moscow	156 (512)	-9 (45)	18 (63)	681 (26.8)
Sochi	31 (102)	7 (45)	23 (73)	1,351 (53.1)
Verkhoyansk	137 (449)	-51 (-60)	14 (57)	136 (5.3)

Major physical features highest point: Elbrus 5,642 m (18,510 ft); longest river: Yenisei 4,129 km (2,566 mi)

Population (2006 est.) 142,893,540

Form of government federal multiparty republic with two legislative houses

Armed forces army 395,000; navy 142,000; air force 160,000

Largest cities Moscow (capital - 10,406,578); St. Petersburg (4,601,000); Novosibirsk (1,405,569); Yekaterinburg = Sverdlovsk (1,304,251); Nizhniy Novgorod = Gorky (1,297,550); Samara = Kuybyshev (1,151,681); Omsk (1,142,773); Kazan (1,110,022)

Official language Russian

Ethnic composition Russian 79.8%; Tatar 3.8%; Ukrainian 2%; Bashkir 1.2%; Chuvash 1.1%; other or unspecified 12.1%

Religious affiliations Russian Orthodox 15-20%; Muslim 10-15%; other Christian 2%; non-affiliated 63%

Currency 1 Russian ruble (RUR) = 100 kopecks

Gross domestic product (2006) U.S. $1.723 trillion

Gross domestic product per capita (2006) U.S. $12,100

Life expectancy at birth male 60.45 yr; female 74.1 yr

Major resources oil, natural gas, coal, timber, antimony, asbestos, barley, basalt, bauxite, cobalt, copper, corn, diamonds, gas, gold, granite, iron ore, lead, livestock, marble, mercury, ,millet, molybdenum, nickel, oats, phlogopite, platinum, potassium salt, potatoes, rye, silver, sugar beet, sunflowers, tin, tobacco, tungsten, precious and semiprecious stones, vegetables, wheat, zinc

the southeast the Sea of Japan. There is also a short outlet to the Gulf of Finland in the west, as well as coastal borders with the Black and Caspian Seas in the southwest of Russia.

Until dissolution in 1991, Russia was the powerbase of the Soviet Union. The breakup of communist structures has left a legacy of political unrest, fueling ethnic and national rivalries among Russia's disparate population.

GEOGRAPHY

Russia can be divided into five main areas: the East European Plain, extending from Russia's western borders with Europe to the Ural Mountains; the Ural Mountains themselves; the West Siberian Plain between the Urals and the Yenisei river; the Central Siberian Plateau – the land between the Yenisei and Lena rivers – and the remote mountainous areas of southern and eastern Siberia.

The land

The East European (or Russian) Plain occupies most of European Russia. In the far south it meets the massive barrier of the Caucasus, which defines the country's southernmost borders with Georgia and Azerbaijan. From here the plain extends northward in a broad swathe as far as the Barents Sea (part of the Arctic Ocean) and the plateaus of Karelia and the Kola Peninsula. At the heart of the plain and to the west of Moscow are the Valdai Hills, where several major rivers, including the Volga, Dnieper, and Western Dvina, have their sources. The plain meets a natural

Winter to spring (*above*) Brown water formed by melting snow creates swamplike conditions over vast areas of the country, as in this nature reserve near the Oka river in Ryazan province, southeast of Moscow, where mixed forest merges into wooded steppe.

The snowy bulk of Koryakskaya Sopka (*left*), a 3,456 m (11,337 ft) high volcano on the Kamchatka Peninsula in Russia's far northeast, rises above Petropavlovsk-Kamchatskiy. This is a geologically volatile area with 22 active volcanoes, as well as geysers and hot springs.

eastern barrier at the Ural Mountains: the traditional boundary between Europe and Asia. East of the Urals, the West Siberian Plain is a vast and sometimes marshy lowland covering some 2.5 million sq km (1 million sq mi) and crossed by two of the world's longest rivers, the Ob and the Yenisei.

Farther east, central Siberia is dominated by an extensive, irregular plateau, that contains several highland areas. It is bounded by higher mountains to the south, and to the east by the central Yakut lowland, formed from the Lena and Vilyui river basins. In the north the plateau is interrupted by the swampy lowlands of the Taymyr Depression, beyond which the Byrranga Mountains on the Taymyr Peninsula rise above the Arctic shoreline. Offshore are the rugged,

desolate islands of Severnaya Zemlya.

Ranges of fold mountains rise to the south of the Central Siberian Plateau. The westernmost of these is the Altai range, covered by permanent snowfields. Farther east, the Mongolian border region is crossed by mountain ranges interlaced with troughs and depressions. Northeast from the Eastern Sayan Mountains is the ancient Lake Baikal, the deepest lake in the world at around 1,620 m (5,310 ft).

East of Lake Baikal, and south of the river Lena, the mountains sweep eastward in an arc through southeastern Siberia to the northwestern shore of the Sea of Okhotsk. Farther southeast, the highlands of Sikhote-Alin look out across the Sea of Japan.

Northeastern Siberia is dominated by the Verkhoyansk, Cherskogo and Kolyma ranges, reaching to more than 3,000 m (10,000 ft). In the extreme northeast the mountainous Chukot peninsula faces Alaska across the Bering Sea.

In the northeastern corner of Russia the Koryak highlands extend into the mountainous backbone of the great Kamchatka

Peninsula, which divides the Bering Sea to the east from the Sea of Okhotsk to the west. The mountains continue southwest through the Kuril Islands to northern Japan, and from there northward again along Sakhalin island (notorious as a penal colony in the late tsarist and soviet period) off the western shore of the Sea of Okhotsk. This area is prone to earthquakes and volcanic eruptions.

Climate

Russia's climate is less varied than its vast size might suggest. Generally a cold, continental climate prevails, except in the far southwest near the Black Sea (which enjoys almost Mediterranean warmth) and in the extreme southeast near Vladivostok. Winters are severe, with temperatures sometimes falling as low as –70° C (–94° F) in northeastern Siberia. Average summer temperatures range from 8° C (46° F) on the Arctic coast to 20° C (70° F) in the valleys of the Caucasus.

Rainfall is moderate to low, and is often unreliable, especially in the interior. Notable exceptions include the areas bordering the Black Sea and the Caucasus, and the far southeastern coastal strip, which receive heavy monsoon rains during a humid summer season.

Plants

In northern coastal areas few plants other than mosses, lichens and grasses can survive, creating a characteristic tundra landscape. Farther south the tundra gives

way to vast tracts of forest known as the taiga. In the zone between the tundra and the taiga, there are larch forests in the east, birch and fir woodlands in the west.

In central parts of the East European Plain, and in vast regions of southeastern Russia, the natural cover is mixed forest. Southern parts of the East European and West Siberian Plains are covered in forest steppe. Farther south there are three main areas of semidesert; these border the Caspian Sea and the Caucasus Mountains in the southwest, and the Altai mountains in the south.

Animals

The northern coast is home to Arctic species such as Polar bears, Bearded seals and walruses, as well as gulls and loons. On the tundra, Arctic foxes and owls prey on Arctic hares and lemmings, while reindeer herds roam freely. Wildlife in the forests is more diverse. Elks (moose), reindeer, wolves, brown bears, lynx, sables, squirrels, foxes and wolverines are found in the northern taiga.

The more southerly forests shelter Wild boar, mink and deer; Ussuri tigers inhabit the Primorski region in the far southeast. In the steppes, ground squirrels, jerboas and hamsters are hunted by Steppe polecats and Corsac foxes, while bird life includes falcons, cranes and eagles. In the Caucasus there are also many Mediterranean species.

SOCIETY

Throughout its long history Russia has been the crossroads of the Eurasian landmass. Under constant threat of invasion by nomadic horsemen from Asia for much of its early history, Russia built a vast transcontinental empire between the 16th and 19th centuries. Under the Soviet regime, built on the 1917 October Revolution, ethnic differences were forcibly submerged, to resurface in the 1990s.

History

From the 6th century AD, Finnic and Slavic peoples living in what is now western Russia became dominated by the Khazars, a group of Turkic and Iranian peoples who had migrated northward. The Khazars developed a prosperous empire including most of modern Ukraine and southern Russia, and in 737 established a new capital, Itil, near the mouth of the Volga on the Caspian Sea. Meanwhile, in the northwest, Viking explorers from Sweden pioneered a trade route from the Baltic to the Black Sea and Constantinople (modern Istanbul) by way of the Dnieper river. They founded first Novgorod in the north and then pressing south, established Kiev. Gradually the principal Slav peoples united around this city to form the powerful state of Kievan Rus. The Khazars were finally defeated in 965 by Grand Prince Svyatoslav of Kiev

Window on Europe (*above*) St Petersburg, seen here in an 18th-century engraving, was perhaps the greatest legacy left to Russia by Tsar Peter the Great. Founded in 1703 on a low and swampy site on the river Neva, it gave Russia an important outlet to the Baltic Sea.

Russia's imperial crest (*below*) – first adopted in Russia by Ivan the Great in the 15th century – is dominated by the double-headed eagle – a motif based on the Byzantine image of old and new Rome.

(d. 972). In 988 Svyatoslav's son Vladimir (c. 980–1015) was converted to Greek Orthodox Christianity. This event created close ties with the powerful Greek empire of Byzantium whose civilizing influence played a crucial role in shaping Russia's distinctive cultural identity.

The rule of Vladimir and his son Yaroslav the Wise (1019–54) saw the Kievan state's rise to prominence. But it was constantly plagued by attacks from its warlike neighbors in the south. In 1235 a new threat emerged from the east when Tartar Mongol horsemen advanced northward across the Caucasus. The Mongol progress across Russia was relentless, as they burned and sacked cities, slaughtered captives and exacted tribute. Novgorod in the northwest escaped the Mongol raids, but had to fend off attacks from the Swedes and the Germans in the west. The legendary Prince of Novgorod, Alexander Nevsky (c. 1220–63) eventually defeated his enemies in two great battles; one against the Swedes on the river Neva in 1240; and the other, against the Germans on the frozen Lake Peipus in 1242. By this time, however, the Mongols had reached Europe and established the Golden Horde empire, with its capital at Saray (near the Caspian Sea).

While eastern areas of the Kievan state remained under Mongol rule, the western parts were absorbed by Lithuania and Poland. However, the late 13th century witnessed the rise of a new and powerful state. When Alexander Nevsky's son, Daniel (1276–1304), became Prince of Moscow he founded a Muscovite dynasty, which ruled for the next three centuries until 1598. During this period Moscow rapidly expanded, in particular during the reign of Ivan III (Ivan the Great, 1440–1505) who finally ended his city's subjection to the Mongols in 1480. In 1472 he assumed the title of "Sovereign of all Russia" and adopted the emblem of the two-headed eagle. In 1547 his grandson Ivan IV (Ivan the Terrible, 1530–84) was crowned the first tsar of Russia.

The young tsar conquered the Tartar khanate of Khazan and began the great eastward expansion of the Russian empire, which took in Siberia and by 1649 reached the Pacific coast.

On the death of Ivan's son Fyodor I (1557–98), a Tartar noble, Boris Godunov (c. 1551–1605) who had been acting as regent, ruled as tsar. His reign was a period of intense upheaval in Russia known as the "Time of Troubles" (1595–1613); 18 years of civil war, famine, revolt and invasion by Poland, it ended when Michael Romanov (1596–1645) was elected tsar, founding the dynasty that went on to rule Russia until 1917.

Peter I (Peter the Great, 1672–1725), anxious to open up "a window on the west" and Europeanize his country, founded his new capital, St Petersburg,

The October Revolution: 1917

Russia's entry into World War I in 1914 at first stilled a rising tide of discontent with the tsar's reactionary government as all classes united against the German and Austro-Hungarian enemy. But the war went badly, casualties were heavy, defeats frequent and the economy crumbled under the strain.

On 10th March 1917 a shortage of bread led to strikes and riots in the capital. When the army joined the rioters, the tsar was forced to abdicate and the duma (parliament) was suspended. Power passed to two institutions: the Provisional Government, dominated by middle class "progressives", and the Petrograd Soviet of Workers' and Soldiers' Deputies, made up of a number of revolutionary workers' parties, including the Bolshevik Party. The two bodies shared power uneasily; democratic reforms were passed, but the war continued to go badly and the provisional government failed to address demands for land reform. Thus the provisional government alienated both right- and left-wing opinion. When, in July, the right-wing general Lavr Kornilov (1870–1918) attempted a military coup, the government was only saved by militant workers and its lack of real authority was exposed.

Only the Bolshevik Party, led by the formidable Marxist intellectual Vladimir Ilyich Ulyanov (1870–1924), codenamed Lenin, had consistently opposed the war and, campaigning under the emotive slogan "Peace, Bread and Land", it now began to gain support. By September they controlled the Petrograd and Moscow soviets and began to plan their coup. The moment came on the evening of 4 November (24 October under the old Russian calendar) when Red Guards began to take control of Petrograd. The provisional government took refuge in the Winter Palace, which fell without a fight the following night. On 7 November Lenin announced that the soviet had seized power in the name of the working class and immediately issued decrees to end the war and commence land reform.

However, the revolution was by no means secure: the Bolsheviks were still a minority in the country as a whole and final victory came only after four years of bloody civil war.

Revolutionary orator Lenin, founder of the Russian Communist Party, addresses a crowd in Petrograd (now St Petersburg) in March 1917. Such events won support for the Bolsheviks and the October Revolution.

on the river Neva next to the Gulf of Finland. Through his military successes, the introduction of Western technology and by draconian government reforms, Peter turned Russia into a major European force. The country's expansion and Europeanization continued under the rule of the German-born empress Catherine II (Catherine the Great, 1729–96).

Much of the reign of Catherine's grandson Alexander I (1777–1825) was preoccupied by war with Napoleonic France. Russia's defeat of Napoleon gave the country heroic status, but Russian society remained feudal and its industrial backwardness was exposed in the Crimean War (1853–56). In 1861 Alexander II (1818–81) abolished serfdom among the rural population, and attempted extensive social and administrative reforms. But dissatisfaction among the Russian intelligentsia grew, culminating in Alexander's assassination in 1881.

Under Alexander II Russia had conquered the Caucasus and Kazakhstan and penetrated deep into central Asia. By 1900 it controlled a vast empire of many nationalities extending to the borders of Persia, Afghanistan, India and China. After a disastrous war with Japan (1904–05), revolutionary unrest at home forced Tsar Nicholas II (1868–1918) to grant his people a *duma*, or parliament. The concessions proved to be too little, and too late.

The situation was exacerbated in 1914, when Russia plunged into World War I. Millions died in battle, the economy was overstretched and food supplies were soon seriously affected. In March 1917, after riots in St Petersburg (then called Petrograd), the tsar was forced to abdicate in favor of a provisional government. In November 1917 Vladimir Ilyich Lenin (1870–1924) at the head of the Bolshevik Party seized power in a remarkable coup that overthrew the provisional government. Lenin immediately withdrew Russia from World War I, and a devastating civil war (1918–22) ensued, which concluded in 1922 with the foundation of the Union of Soviet Socialist Republics.

After Lenin's death Joseph Stalin (1879–1953) won the power struggle for leadership, ruthlessly eliminating his rivals. In 1928 he inaugurated his first Five Year Plan: a period of intensive industrialization and the enforced collectivization of Russia's peasants. Agricultural production fell and a subsequent famine in the Ukraine in 1932 led to millions of deaths. Many of those who resisted Stalin's policies were either executed or deported to Siberian labor camps. In 1934 he embarked on the systematic elimination of all his political opponents leading to the Great Terror of 1936–38: estimates put the total deaths at some eight million.

In 1941, after the outbreak of World War II, an ill-prepared Soviet Union was invaded by Nazi Germany. The defeat of Germany was achieved by terrible sacrifice on the part of the Russian people and at a cost of more than 20 million lives. In the early postwar years a battle-scarred Soviet Union achieved considerable international power by establishing six satellite communist regimes in adjoining states of Eastern Europe. With the development of nuclear weapons and the race for military supremacy that accompanied it, the ensuing Cold War left the two great powers poised for mutual destruction. In 1962, under Soviet leader Nikita Khrushchev (1894–1971), war was narrowly averted after the United States and the Soviet Union contested the siting of Soviet nuclear missiles on the island of Cuba. Khrushchev's later efforts to improve relations with the West were cut short in 1964, when he was ousted by a collective Communist Party leadership that soon came to be dominated by Leonid Brezhnev (1906–82).

Unable to repair the growing rift between the Soviet Union and communist China, Brezhnev tried to establish a truce with the West. However, an increase

Church and state (*above*) Boris Yeltsin, the democratically elected president of the Russian Federation, greets the crowds after an Easter service celebration. Russian Orthodoxy was heavily suppressed by the communists, though many of the magnificent churches were preserved out of a sense of national pride. Under Yeltsin's liberal regime the traditionally strong religious feeling of the Russian people resurfaced and the church found new vigor.

Walled for safety (*left*), the magnificent Trinity Monastery of St Sergius with its large Baroque refectory, in Sergiyev Posad, 70 km (43 mi) northeast of Moscow, offers a dramatic assembly of religious architecture. Founded in 1337–40 by Sergius of Radonezh (c. 1321–91), a dispossessed Moscow aristocrat who became a hermit – and later one of Russia's patron saints – the monastery has been a spiritual beacon of the Russian Orthodox Church for more than 650 years.

in religious and political persecution during the Brezhnev years, particularly of the Jews, continued to alienate Western sympathies, and the Soviet invasion of Afghanistan in 1980 prompted widespread international condemnation.

This state of affairs continued until Mikhail Gorbachev (b. 1931) became General Secretary of the Communist Party in March 1985. He introduced a new and radical style of leadership, aiming at a restructuring of government and the economy. In 1988 he withdrew Soviet troops from Afghanistan, and initiated arms' reduction talks with the West. At home, however, he faced a collapsing economy, growing industrial unrest and the resurgence of long suppressed national identities and ethnic conflicts in many parts of the Union.

In the late 1980s, as communist regimes began to topple throughout Eastern Europe, Gorbachev requested greater presidential powers in an attempt to prevent the break-up of the Union. Boris Yeltsin (b. 1931) and more radical reformers warned against a "new dictatorship". However, early in 1990 the Communist party effectively renounced its monopoly on power, and as support for the party dwindled, Yeltsin became president of the Russian Federation, the largest of the Soviet Union's constituent republics, in May 1990.

Meanwhile Gorbachev's diplomatic successes in the United States and Great Britain signaled the end of the Cold War, and new reductions in armaments. Although popular abroad, Gorbachev faced mounting unpopularity at home. As the constituent republics of the Union declared their independence, Gorbachev fought to maintain a degree of central authority in a new Union treaty, but in August 1991 a group of hard-line conservatives attempted to seize power.

Following mass demonstrations in Moscow and St Petersburg (then Leningrad), the coup collapsed after three days. Gorbachev, who had been under house arrest in the Crimea, returned to the capital, but his power had been overthrown by Yeltsin who had opposed the coup in Moscow. The failure of the coup confirmed that the Communist Party's

The Second Russian Revolution

In June 1990, following a highly successful Soviet–United States summit in Washington, the Western world celebrated the end of the Cold War between the two great power blocs. At the time, few Western politicians foresaw the growing instability within the Soviet Union itself.

During his presidency, Mikail Gorbachev had fought hard to reform the Soviet system. The two Russian words *glasnost* (openness) and *perestroika* (restructuring) had become universally familiar as the expression of his vision of a more open society. But to achieve this goal Gorbachev had to walk a dangerous tightrope between the conservatives of the Communist Party and the new radical democrats. Despite all his efforts, ominous cracks appeared in the structure of the Union.

Just two months after the Washington summit, the Soviet republics of Armenia, Turkmenistan and Tajikistan declared their independence. Gorbachev struggled to preserve central authority by drafting a revised Union treaty to be voted on at a referendum.

However, Armenia, Georgia, Moldova and the Baltic republics refused to consider Gorbachev's treaty. Instead, Georgia, Latvia and Estonia voted for full independence from the Union. Meanwhile, a major food-supply crisis was fueling discontent in Russia and other regions.

On 19 August 1991, the day before the Union treaty was due to be signed, Radio Moscow announced that "ill health" prevented President Gorbachev from carrying out his duties. An eight-man "emergency committee" of communist hardliners took control, curbing demonstrations, strikes and the press. As tanks rumbled into Moscow, a crowd of protesters surrounded the Russian parliament building. At the forefront was the radical Boris Yeltsin, who condemned the coup, and called for a general strike.

Unwilling to turn their guns on the people, the army refused to obey the orders of the emergency committee, and without military backing the coup quickly collapsed. Gorbachev's policy of *glasnost* had borne unexpected fruit: an upsurge of support for the radicals broke the power of the Communist Party, and severed the remaining links of the Union.

authority was broken and the great monolith of the Soviet Union began to collapse. It was formally dissolved in December 1991, and 11 of its 15 constituent republics, including the Russian Federation, formed the Commonwealth of Independent States (CIS). Gorbachev had no option but to resign.

Yeltsin found the Russian parliament reluctant to embrace sweeping changes he proposed, and in 1993 armed conflict erupted when demonstrators seized the parliament building. Forces loyal to the president successfully aborted the attempted coup and a new parliament was elected later in the year under a new constitution, which gave the president greater powers. In the late 1990s, a state of continuing economic crisis was accompanied by a number of government corruption scandals. In 1999 Yeltsin, who had been suffering ill health, retired, naming Vladimir Putin as his successor. Putin was elected president in 2000.

Government

The Russian Federation has taken over many of the enterprises, resources and responsibilities of the Soviet Union. However, its political structure is still undergoing major adjustment. Perhaps the most dramatic change was the introduction of multiparty democracy at all levels of government, a process begun in 1989.

There are 10 autonomous areas, 49 provinces, 6 territories and 21 officially independent republics. The republics are highly autonomous and each elects its own president. Provinces and territories also elect their own governors.

People

The Russian Federation contains within its boundaries an enormous diversity of peoples. There are more than 60 different nationalities, who can be broadly divided into four main linguistic groups. The Indo-European group – those languages that developed in Europe and southern Asia – includes some 150 million Russian speakers, and accounts for the great majority of the population from the Baltic to the Pacific. The other three language groups are the Finno-Ugric group, widely dispersed across the forest and tundra of the north and in the Balkans; the Turkic languages spoken in southeastern Europe and Asia; and the Caucasian group encompassing the languages spoken in the west of the Caucasus Mountains, as well as Chechen, Ingush and Dagestanian languages spoken in the east of the Caucasus.

The majority of Russians differ marked-

Waiting in line for food (*above*) at a mobile stall in the St Petersburg suburbs, where housewives gather to buy potatoes trucked in from a collective farm. In the early 1990s a fall in agricultural output, plus huge price rises, made daily living extremely hard.

ly from many of the non-Slav groups, in culture and religion as well as in language. Since the early 1990s there is now a tremendous resurgence in religious practice – considered reactionary and actively suppressed during most of the Soviet period – as the Russian people rediscover their spirituality. A strong Orthodox Christian tradition has survived within the Slavic communities, Baptists are numerous, and many ethnic and national groups have retained their own traditional religions; Islam among some of the Turkic peoples, and Buddhism among many of the Kalmyks and Buryats.

In wintry Siberia (*above*) a herd of domesticated reindeer are rounded up. Under Soviet rule the distribution of reindeer meat was collectivized, but herders were allowed to keep to the ancient custom of following the seasonal migration routes of the animals.

Cooperation in space (*below*)The American and Russian shuttles *Atlantis* and *Mir* docked together in 1995; here, Russian cosmonauts Solovyev and Budarin smile at *Atlantis'* departing crew before the hatch closes it off again from *Mir*.

ECONOMY

Russia was the main driving force behind the economy of the former Soviet Union. Its rich deposits of minerals supplied heavy industry with raw materials, and its energy resources supplied power to the other republics and Eastern Europe.

In the 1990s Russia struggled to establish a modern market economy hampered by an entrenched bureaucracy and a dilapidated industrial base. Falling international oil prices and a failing currency led to a financial crisis in 1998 as a result of which the ruble was devalued and Russia decided to default on its international debt.

Agriculture

Wheat, barley, oats and rye form the major part of the agricultural sector, but the country cannot grow enough for its own needs. Only about a tenth of the land area is under cultivation, and most of this lies to the west of the Urals. The most suitable land for intensive livestock breeding, including beef and dairy cattle also lies in the west. Russia also has the world's largest reserves of timber.

The Russian fishing fleet, operating from the Arctic waters to the Pacific, is one of the biggest in the world.

Industry

Heavy industry is the backbone of the Russian economy and draws on the country's enormous resources of raw materials. There are rich deposits of iron ore on the Kola Peninsula, in the Kursk Magnetic Anomaly near the Ukrainian border, and in Siberia. There are also major deposits of nonferrous metals in the Urals and Siberia including gold, copper, zinc and various rare metals.

In addition, Russia possesses vast energy reserves. There are huge coal fields in eastern Siberia, the Kuznetsk basin north of the Altai, and the Urals. Northwestern Siberia has the world's greatest reserves of oil and natural gas. The Volga and Yenisei rivers supply hydroelectric power to large stations nearby, and more hydroelectric plants are being built. The nuclear power program suffered a setback in 1986, when a major accident at Chernobyl in the Ukraine (then part of the Soviet Union) exposed safety problems in many installations.

Under the former Soviet system, Russia met most of the country's demands for industrial, agricultural and transportation equipment; few of these items were ever imported. But Russian-built machinery is only just now beginning to benefit from the microprocessor revolution.

Transportation and communications

For the Russian economy to function at all, massive quantities of food and raw materials have to be carried over enormous distances. Since the abolition of central planning, distribution has been one of the greatest challenges facing economic reformers. Much needed foodstuffs are frequently left to rot in warehouses before they can be transported to the cities.

The heart of Russia's transportation system is the railroad network, which carries some three-quarters of all freight. Most track is in the west, especially around Moscow; the best-known link with the far east is the Trans-Siberian Railroad from Moscow to Vladivostok.

Surfaced roads cover six times the distance of the entire railroad system, but are less used. Most bulk freight is carried by rail or by water, and buses account for a high proportion of passenger traffic. Relatively few people own a car.

Foreign trade is chiefly conducted by a sophisticated merchant fleet. In the far north and east, shipping provides a lifeline between communities scattered along some 50,000 km (30,000 mi) of coastline. Although most harbors become icebound in winter, some, such as Murmansk, keep open all year. Airline services ensure year-round contact with the most isolated regions. An international airline flies to many major destinations.

Radio, television and the press are still adapting to the recent relaxation of censorship. Even before the breakup of the Soviet Union though, there was a more open approach to news reporting.

Welfare and education

Under Soviet rule, medical care was free throughout the Russian Federation. Although housing was inadequate and the standard of living relatively low, unemployment was practically nonexistent. The collapse of the Soviet system has thrown the old social welfare network into chaos. Russia now faces the same kind of social problems that have plagued the capitalist West: inflation, unemployment and a soaring crime rate.

Hyperinflation in the early 1990s brought a dramatic decline in living standards, and more and more Russians found themselves living below the poverty line. The welfare system is being reformed into an insurance-based system of benefits for sickness and unemployment and retirement pensions. Health and average life expectancy have declined and infant mortality has increased.

Georgia

REPUBLIC OF GEORGIA

RUSSIA

Sukhumi

Rioni

Kutaisi • • Tskhinvali

Black Sea

Poti •

GEORGIA • Gori

■ Tbilisi

Alazani

Batumi •

Kura

• Rustavi

TURKEY

ARMENIA

AZERBAIJAN

Black Sea

T HE REPUBLIC OF GEORGIA LIES AT THE eastern end of the Black Sea. To the north is the Russian Federation, to the southeast Azerbaijan and Armenia, and Turkey lies to the south. Its decision not to join the Commonwealth of Independent States after the breakup of the Soviet Union in 1991 was largely due to political and interethnic conflicts, but it also reflected the independent spirit of the nation. Membership was ratified in 1994.

GEOGRAPHY

Georgia is a country of mountains. The northern frontier with Russia runs along the massive mountain barrier of the Great Caucasus. These snowy peaks overlook the Black Sea in the northwest and have numerous spurs and secondary ranges cut by deep gorges. A central plateau called the Kartalinian Plain divides these mountains from a parallel range known as the Little Caucasus, whose lofty peaks and plateaus extend south into Turkey and Armenia. Farther east, beyond the capital, Tbilisi, the Kura river and its tributaries drain southeast into Azerbaijan through a series of upland valleys. The swampy plains of Kolkhida fringe the Black Sea coast.

The mountainous terrain produces a varied climate. Generally it is cold in the mountains in winter, with heavy snows, but the Great Caucasus also block still colder air coming from Russia to the north. Balmier winds from the Black Sea bring a damp, almost subtropical climate to the Kolkhida lowlands, whereas the eastern areas have much drier weather.

Much of Georgia is wooded, from sparse birch cover on the mountain slopes to dense broadleaf forests of beech, oak and chestnut at lower levels. In the drier east the forest gives way to grassland. Wildlife flourishes too, especially forest species such as roe deer, Caucasian deer, brown bears and wolves. Wild goats and their close relatives, tahrs, live in more mountainous areas.

SOCIETY

The Georgians pride themselves on a history that goes back to some of the earliest known records, and on a unique language with its own distinctive script. To the ancient Greeks, Georgia's western plain of Kolkhida was known as Colchis – the land whose fabled wealth inspired the legend of the Golden Fleece. Indeed, until recently sheepskins were still used to

Golden fields of corn lie within clear sight of the foothills and snow-capped peaks of the Caucasus Mountains in western Georgia. This region, with its almost subtropical climate, produces many agricultural products including tea, tobacco, flowers and wine.

catch the gold dust in its rivers. After coming under the domination first of the Greeks and later of the Romans the country converted to Christianity in 330 AD. The Georgians have since retained their distinctive Orthodox faith, along with a strong sense of national identity, despite many centuries of domination by their powerful neighbors – the Persians and the Ottoman Turks.

By the 18th century Georgia had been fragmented into several principalities. Its rulers sought Russian help in their struggle for independence, but the Russian response was simply to take over the country piecemeal during the 19th century. At the end of the century Georgia was reunited – but under Russian rule. Despite this, the Georgians remained resistant to Russian influences. Under later more liberal-minded governors a renewed sense of national identity emerged, including an artistic revival and movements for social reform. One such Marxist-inspired group was joined in 1896 by a young Georgian called Joseph Dzhugashvili (1879–1953); in later years he became known to the world by his adopted name "Stalin" – meaning "steel".

Following the 1917 Russian Revolution, Georgia was established as an independent state. But in 1921, after a short occupation by the British, the Red Army invaded and declared a Soviet republic. Many nationalists, and even some communists, were murdered, and several peasant uprisings were brutally put down. During Stalin's purges of the 1930s still more Georgians were killed.

Candles are lit during a service of commemoration. Christianity was adopted in Georgia in the early 4th century; the Georgian Orthodox Church (one of the world's oldest Christian communities) survives – with Christian Armenia – in a region dominated by Islam.

Glasnost (openness) in the 1980s gave Georgian nationalism a new voice that led in 1989 to a peaceful demonstration calling for independence. This was savagely crushed, but in April 1991, following a national referendum, the Georgian Supreme Soviet declared the country independent. Its chairman, the former dissident, Zviad Gamsakhurdia (1939–93) was elected president. A month later, however, his dictatorial methods provoked opposition, and soon afterward the country was plunged into armed conflict.

Early in 1992, after weeks of bloodshed, Gamsakhurdia was forced to flee the country. A new government was installed, under the former Soviet foreign minister Eduard Shevardnadze (b. 1928), himself a Georgian, who promised a return to stability. Gamsakhurdia returned to western Georgia in 1993 to ferment a rebellion which was only quelled with Russian help. Conflict with the self-declared Abkhazian Autonomous Republic in the northwest over its desire to secede from Georgia remains unresolved, Abkhazia again declaring independence in 1994. Shevardnaze was re-elected in November 1995.

ECONOMY

Georgia produces tea, citrus fruits, tobacco and fine wines; additional products include sugar beet and essential oils from roses and other flowers. Sheep are the main livestock. The collectivization of agriculture had less impact here than in other former Soviet republics, so privately owned plots have flourished.

Georgia's rich mineral and energy resources encourage local industrial activity. There are deposits of coal, natural gas and peat, as well as fast-flowing rivers, ideal for generating hydroelectric power. Large manganese-ore deposits support most of the mining industry, but many other metals and minerals are also extracted. Traditional industries have included steelworking, the construction of locomotives, engineering and chemicals. The richness of Georgian arts and culture can be seen in a long tradition of icon-painting and metalworking, embroidered textiles, and wood and stone carving.

Despite the rugged terrain, transportation is well developed, with links between the busy ports of Batumi and Poti and the densely populated hinterland. Healthcare and education in Georgia enjoy a high reputation among the former Soviet states. In particular, the widespread teaching of the ancient Georgian language helped the republic to retain a strong national identity and separate culture throughout the Soviet period. Tourism on the Black Sea coast has declined.

NATIONAL DATA – GEORGIA

Land area	69,700 sq km (26,911 sq mi)			
Climate		Temperatures		Annual
	Altitude m (ft)	January °C(°F)	July °C(°F)	precipitation mm (in)
Tbilisi	490 (1,608)	1 (34)	24 (75)	500 (19.7)

Major physical features highest point: Shkhara 5,201 m (17,063 ft); longest river: Kura (part) 1,514 km (941 mi)

Population (2006 est.) 4,661,473

Form of government multiparty republic with one legislative house

Armed forces army 7,042; navy 1,350; air force 1,350

Largest cities Tbilisi (capital - 1,026,874); Kutaisi (173,000); Batumi (115,806); Rustavi (105,393)

Official language Georgian

Ethnic composition Georgian 83.8%; Azeri 6.5%; Armenian 5.7%; Russian 1.5%; other 2.5%

Religious affiliations Orthodox Christian 83.9%; Muslim 9.9%; Armenian-Gregorian 3.9%; Catholic 0.8%; other 0.8%; none 0.7%

Currency 1 lari (GEL) = 100 tetri

Gross domestic product (2006) U.S. $17.79 billion

Gross domestic product per capita (2006) U.S. $3,800

Life expectancy at birth male 72.8 yr; female 79.87 yr

Major resources coal, natural gas, forests, hydropower, manganese, copper, citrus fruits, grapes/wine, sugar beet, livestock, vegetables

Armenia

REPUBLIC OF ARMENIA

M ODERN ARMENIA IS A TINY, LANDLOCKED republic in the mountains to the south of the Caucasus. Its people are survivors of an ancient civilization that was once fought over by Mongols, Turks and Persians and of a nation that was all but destroyed. Armenia shares borders with Georgia to the north, Azerbaijan to the east, Iran and the Azerbaijani region of Nakhichevan to the south, and Turkey to the west. Armenia also claims the Armenian populated enclave of Nagorno-Karabakh, which lies across its eastern frontier in Azerbaijan.

NATIONAL DATA – ARMENIA

Land area	28,400 sq km (10,965 sq mi)			

Climate		Temperatures		Annual
	Altitude m (ft)	January °C(°F)	July °C(°F)	precipitation mm (in)
Yerevan	990 (3,248)	-4 (25)	20 (68)	277 (10.9)

Major physical features highest point: Aragats 4,090 m (13,418 ft); longest river: Aras (part) 914 km (568 mi); largest lake: Lake Sevan 1,360 sq km (525 sq mi)

Population (2006 est.) 2,976,372

Form of government multiparty republic with one legislative house

Armed forces army 45,000; air force 3,160

Largest cities Yerevan (capital – 1,086,174); Gyumri (151,450); Vanadzor (104,242)

Official language Armenian

Ethnic composition Armenian 97.9%; Yezidi (Kurd) 1.3%; Russian 0.5%; other 0.3%

Religious affiliations Armenian Apostolic 94.7%; other Christian 4%; Yezidi (monotheist with elements of nature worship) 1.3%

Currency 1 dram (AMD) = 100 luma

Gross domestic product (2006) U.S. $15.99 billion

Gross domestic product per capita (2006) U.S. $5,400

Life expectancy at birth male 68.25 yr; female 76.02 yr

Major resources gold, copper, molybdenum, zinc, alumina, antimony, arsenic, barley, chromites, citrus fruits, cotton, figs, iron, limestone, livestock, magnesium, mercury, potatoes, pumice, silver, sugar beet, tobacco, wheat, wine/grapes

GEOGRAPHY

Armenia is occupied almost entirely by the mountains of the Little Caucasus, whose name belies their considerable height. The rugged landscape includes extinct volcanoes and high lava plateaus cut by deep ravines; frequent earthquakes show that mountain-building is still going on. The highest peak, Aragats at 4,090 m (13,418 ft), is in the northwest. Farther east is Lake Sevan, nearly 2,000 m (some 6,000 ft) above sea level. The Gegam mountain range to the west of the lake overlooks the capital, Yerevan, and the broad Ararat plain, which is crossed by the river Aras as it follows its southeasterly course along the borders with Turkey and, in the far south, Iran.

The country has a dry, continental climate; short, sharp winters and long, hot summers on the plains. The mountains are cooler, and winter conditions there are much harsher.

Steppe vegetation predominates, with drought-resistant grasses and sagebrush at lower altitudes, where jackals, wildcats and sometimes leopards are found. Oak woodlands in the southeastern forests merge with beechwoods in the more humid northeast. Here the animals include squirrels and brown bears. Alpine grassland on the higher slopes provides good summer grazing for wild goats and mouflons (wild sheep).

SOCIETY

The Armenians are descended from the various Indo-European peoples that inhabited the Ararat plain and surrounding areas from the 12th to the 6th century BC. Here, legend has it, Noah landed his ark after the Flood and founded the ancient city of Yerevan. The kingdom of Urartu (as Armenia was then known) flourished here between the 9th and 6th centuries BC, and became one of the most powerful states in the Middle East. In 520 BC the kingdom was conquered by the Persian empire, and then from 330 BC it was ruled by the Macedonian Alexander the Great (356–323 BC) and the Greek dynasty that succeeded him. Ultimately, Armenia fell under the dominion of Rome in 67 BC, becoming a buffer state between the Roman empire and the empire of Parthia – Rome's major rival for power in the east. When Armenia adopted Christianity in about 300 AD, it became a bulwark of Christianity in Asia.

In the centuries that followed, Armenia

struggled to retain its independence and cultural identity as it became a theater of war between the Persians, the Seljuk Turks and the Mongols. The population became scattered across the world, but nothing could destroy the strong sense of nationhood kept alive by the depth of its faith in the Armenian Church.

In the early 19th century, as the Russian empire advanced its frontiers across the Caucasus, the Armenians looked to the tsar for liberation from Persian and Turkish domination. By 1829 Russia had taken all the land now occupied by Armenia and Azerbaijan, except for western Armenia, which remained under Turkish rule. In subsequent decades there was a resurgence of Armenian culture, but following the Russo-Turkish War (1877–78) fear of Armenian nationalism led to repression in both Russian and Turkish Armenia. In 1915, during World War I, 1.75 million Armenians living in the western region ruled by Turkey were deported from their homes and c. 600,000 were massacred by the Turks. The establishment of the Soviet Republic of Armenia in 1920 prevented a Turkish takeover of the whole region and in 1921,

joined the Commonwealth of Independent States in 1991 and became a member of the UN in 1992. Conflict broke out again in Nagorno-Karabakh in 1992, with Armenian forces seizing much of the territory. A ceasefire agreement ended the conflict in 1994, but sporadic outbreaks of violence continued. These troubles exacerbated the already acute problems of supply and distribu-tion during the restructuring of the postcommunist economy and the introduc-tion of multiparty democracy.

The vast majority of the republic's population are Armenians, who despite long periods of Muslim rule have retained their strong Christian tradition. They belong either to the Armenian Orthodox Church or to the much smaller Armenian Catholic Church, which is in communion with Rome.

ECONOMY

Agriculture is the main source of income, especially in irrigated areas of the Aras valley and around Yerevan. Here the crops include grapes, almonds, figs and olives. Apples, pears and cereals are grown on higher ground, while sheep and cattle are raised in the mountains.

In the second half of the 20th century the country industrialized rapidly, exploiting its valuable deposits of copper, zinc and molybdenum, and developing its capacity to generate hydroelectricity. The chief industries are mining, metallurgy, mechanical engineering and chemicals. Armenia also exports precision instruments and textiles, as well as quality brandies and wines.

In December 1988 a devastating earthquake destroyed many towns in northern Armenia and seriously affected its economic infrastructure. Numerous factories and power stations were lost and much of the country's financial resources had to be diverted to meet the immediate demands of humanitarian aid to earthquake victims. The disruption of power services, exacerbated by the continuing conflict with neighboring Azerbaijan, and a subsequent blockade on fuel imports caused much suffering for Armenia's population.

Several years later, in the early 1990s, factories and hospitals were still unheated, power cuts were routine and for increasingly longer periods each day. The government reopened its antiquated nuclear power plant at Metzanor, which had been closed since the 1988 earthquake, in an attempt to solve the country's fuel crisis.

Spiritual home (*above*) The Armenian church of Khor Virap, in the shadow of Turkey's Mount Ararat, marks the site where the evangelist Gregory the Illuminator (257–332) was imprisoned by the pagan king Tiridates III, whom he later converted to Christianity.

Words in stone (*left*) Youngsters examine a statue of St Mesrob (c. 350–440), founder of the Armenian alphabet. It stands in front of the Matenaradan library, Yerevan, whose archives hold a fine collection of ancient Armenian manuscripts, such as the one shown.

after the Russian Revolution and civil war (1917), the Bolsheviks gave the regions of Nagorno-Karabakh (which was mostly populated by Armenians) and Nakhichevan to Azerbaijan. The following year, Armenia and its neighbors Azerbaijan and Georgia were united in the Transcaucasian Soviet Federative Republic.

Armenia achieved separate status as a Soviet republic in 1936, but its territory did not include either Nagorno-Karabakh or Nakhichevan. This continued separation of ethnic Armenians from their native state created tensions between Armenia and Azerbaijan that the Soviet regime contained for many years until they erupted violently in the late 1980s.

Both republics achieved independence on the break-up of the Soviet Union and

Azerbaijan

REPUBLIC OF AZERBAIJAN

L YING ON THE WEST COAST OF THE CASPIAN Sea, Azerbaijan occupies the southeast corner of Transcaucasia, with Russia to the north and Georgia to the northwest. Armenia is to the west and Iran to the south. Its territories include the autonomous oblast region of Nagorno-Karabakh in the southwest, which is also claimed by Armenia, and the autonomous republic of Nakhichevan, which lies farther southwest, beyond a strip of Armenian territory. It withdrew from the CIS in 1992 and rejoined in 1993, but is also making strong links with Turkey.

GEOGRAPHY

Northern Azerbaijan is dominated by the peaks of the Great Caucasus Mountains that extend southeastward, partly along the Russian border, as far as the capital and chief port, Baku. In the foothills of the Caucasus, the Kura river valley drops southeastward from the Georgian border, opening onto a broad floodplain that lies mainly below sea level. The mountains and plateaus of the Little Caucasus occupy Nakhichevan and western Azerbaijan, while in the far south the Talish Mountains on the Iranian border overlook the Caspian coastline.

In the lowlands the climate is dry and subtropical, with mild winters, long, hot summers and frequent droughts. Mountain areas that have a more continental climate experience cold winters with heavy snow; farther south it is wetter and more humid. The vegetation varies from steppe grassland in the drier lowlands to forests in the mountains and marshlands in the southeast. Wildlife includes leopards, gazelles and ground squirrels.

SOCIETY

The first people to inhabit Azerbaijan were Iranians, and for most of its long history Azerbaijan belonged to the Persian

Trapped innocent (*above*) A young girl from Nagorno-Karabakh – an Armenian Christian enclave in Azerbaijan – with straw brooms for sale. Like many thousands, she is caught by the fighting between Armenians and Azerbaijanis over the enclave's political status.

The waterfront of Baku (*right*) Azerbaijan's capital and major port on the west coast of the Caspian Sea. The city is heavily dependent on the nearby oil fields for its industry.

Olives from reclaimed land (*below*) Azerbaijani men and women clear weeds from a newly planted olive grove near Baku. The land they are tilling has been reclaimed partly from the desert and partly from an oil field that has ceased production.

NATIONAL DATA - AZERBAIJAN

Land area	86,100 sq km (33,243 sq mi)
Climate	humid subtropical to dry continental
Major physical features	highest point: Bazardyuze 4,480 m (14,698 ft); lowest point: Caspian Sea -28 m (-92 ft); longest river: Kura (part) 1,510 km (940 mi)
Population	(2006 est.) 7,961,619
Form of government	multiparty republic with one legislative house
Armed forces	army 56,840; navy 2,000; air force 7,900
Largest cities	Baku (capital – 1,118,904); Gäncä (304,472); Sumgayit (268,336)
Official language	Azerbaijani
Ethnic composition	Azeri 90.6%; Dagestani 2.2%; Russian 1.8%; Armenian 1.5%; other 3.9%
Religious affiliations	Muslim 93.4%; Russian Orthodox 2.5%; Armenian Orthodox 2.3%; other 1.8%
Currency	1 Azerbaijani manat (AZM) = 100 gopiks
Gross domestic product	(2006) U.S. $58.1 billion
Gross domestic product per capita	(2006) U.S. $7,300
Life expectancy at birth	male 59.78 yr; female 68.13 yr
Major resources	petroleum, natural gas, iron ore, nonferrous metals, alumina, cotton, tobacco, fruit, fisheries

empire. Azerbaijanis themselves, many of whom live farther south in the Iranian province of the same name, are a people of Turkic origin who arrived with the 11th-century Seljuk invaders who conquered much of Asia Minor. Persians and Turks fought for control of Azerbaijan throughout the ensuing centuries until finally the Russians gained a foothold during the early 19th century. The Treaty of Turkmenchai in 1828 divided the country between Persia (which gained the south) and Russia (the north).

During the chaos that followed World War I and the Russian Revolutions of 1917, the Azerbaijanis set up an independent anti-Bolshevik republic, but were invaded first by Turkey and then, in 1922, by the Soviet Red Army. In the subsequent resistance to Soviet rule, many religious and political leaders were killed and collectivization of farmland was fiercely resisted by Azerbaijan's peasantry. For a brief period from 1922 it became part of the Soviet Transcaucasian Federation, but it was eventually incorporated into the Soviet Union as a constituent republic in 1936.

Today most of the inhabitants are Muslim Azerbaijanis. However, there is a sizable minority of Armenians, who are mainly Christian, concentrated in the Nagorno-Karabakh enclave. In 1988 a long-standing dispute with Armenia over this area erupted into violence and resulted in the departure of thousands of Armenians from the country. When Azerbaijan and Armenia both achieved political independence on the dissolution of the Soviet Union in 1991, their conflict over Nagorno-Karabakh escalated into a war. A 1994 cease-fire brought temporary peace, but the status of the enclave remains unresolved. Azerbaijan has a 125-member parliament elected by majority vote and a directly elected president. There were several attempted coups in the 1990s and it remains politically unstable.

ECONOMY

Azerbaijan has little arable land, but nevertheless is a major producer of cotton, tobacco, grapes and other fruit. The Caspian Sea is a ready source of fish, especially sturgeon for its highly-prized caviar, but this important industry is now threatened by severe pollution.

Other natural resources include metal ores, natural gas and vast reserves of oil, which is refined mostly in the Baku area. Baku once provided half the world's supply of oil, and the petroleum and chemical industries expanded rapidly around this area until it became the Soviet Union's fifth largest city. The burgeoning industrial sector is now extremely diversified, from chemicals and gasoline to textiles, and electrical and other consumer goods. A market-based economic reform program and increased oil production was encouraging foreign investment and boosting GDP growth rates in the late 1990s.

Transportation includes good railroads and an extensive road network. There are few inland waterways, but shipping lines from the busy port of Baku serve Russian, central Asian and Iranian destinations. Free health, welfare and education services were well developed under the Soviet regime. Azerbaijan's strong intellectual tradition is shown in its prestigious scientific institutions.

Turkmenistan

REPUBLIC OF TURKMENISTAN

T HE DESERT REPUBLIC OF TURKMENISTAN IN south-central Asia extends southeastward from the eastern shore of the Caspian Sea to its frontier with Afghanistan. Kazakhstan and Uzbekistan lie to the north, and Iran to the south.

GEOGRAPHY

The land rises sharply in the west from below sea level on the Caspian shore to the Krasnovodsk plateau. Southeast of the plateau, the mountains of the Kopet Dag follow the Iranian border. The mountains are prone to earthquakes, but are flanked by a narrow belt of oases – the most popular area for settlement. The capital,

Ashkhabad, formerly a lonely outpost of the Russian empire, is in this belt. The rest of Turkmenistan is mostly a featureless wilderness; 80 percent of the country is covered by the bleak plains of the Kara Kum desert. Toward the Afghan frontier in the southeast, the land rises to meet other mountains in the far eastern corner.

The climate is generally arid with little rainfall, except in the southern oasis belt, and with hot summers and cold winters. Vegetation consists mainly of poplars and willows along the river banks, and sparse woodland in the Kopet Dag.

SOCIETY

Turkic nomads have roamed the deserts of central Asia since the 9th century AD. Today the Turkmen are mainly a Muslim people, probably descended from the Seljuk and Ottoman Turks.

The rivalry between individual Turkmen peoples was often intense. When Russian troops invaded the eastern shore of the Caspian Sea in 1869, they were able to turn these divisions to their advantage. Even so, it took 30 years and a major battle, in which 150,000 Turkmen were killed, before the region was finally absorbed into Russian Turkistan. After the 1917 Russian Revolutions, British forces helped in the attempt to create an independent Transcaspian state. When the British troops withdrew in 1920 the Soviet Red Army quickly seized control and in 1924 the Turkmen Soviet Socialist Republic was proclaimed.

Under Soviet rule the Turkmens were united for the first time, but anti-Soviet guerrilla warfare continued into the 1930s. Aspirations toward self-determination were not fulfilled until 1991, when the Soviet Union collapsed. That same year, Turkmenistan joined the Commonwealth of Independent States. Government, however, remains under the autocratic control of President Niyazov (b. 1940), who was re-elected in 1994 until 2002, and his former Communist Party.

ECONOMY

Turkmenistan's economy has been dependent on its cotton crop and also on sales of oil and gas. In the late 1990s it suffered from a reduced income as a result of the inability of its customers in the former Soviet Union to meet their oil bills. Foreign debt was increased by Russia's refusal to export Turkmen gas. Hope for economic growth in the future is centered on the construction of new pipelines

Advancing sand dunes in Turkmenistan's Kara Kum desert. Covering some 340,000 sq km (130,000 sq mi), the Kara Kum consists mainly of bleak plains and some farmland, with only a small area – less than 10 percent – containing the immense, crescent-shaped dunes.

through Iran and Turkey. Cotton and wheat account for two thirds of agricultural production. Maize, rice, silk and fruit are also important. Local sheep and camels provide wool and hides for making the region's famous Bokhara carpets and rugs.

Turkmenistan has rich mineral resources including lead, zinc, copper and gold. It possesses the world's fifth largest reserves of natural gas. In the late 1990s, Turkmenistan signed oil development agreements with several international oil companies. Other industries include engineering, cotton textiles and food processing. The growing road network is supported by a major railroad link from the port of Krasnovodsk through Ashkhabad to the oases of the interior. The 800 km (500 mi) long Karakumskiy Canal, the world's longest irrigation and shipping canal runs parallel to the railroad.

The Soviet regime aimed at providing a comprehensive welfare and education system, and considerable improvements were made to the literacy levels of the population. The Turkmen language has now been restored as the state language, and the country's mosques and religious institutions – long suppressed under Soviet rule – have been reopened.

NATIONAL DATA – TURKMENISTAN

Land area	488,100 sq km (188,456 sq mi)
Climate	Continental
Major physical features	highest point: Mount Ayrybaba 3,137 m (10,292 ft); longest river: Amu Darya (part) 2,539 km (1,578 mi)
Population	(2006 est.) 5,042,920
Form of government	multiparty republic with two legislative houses
Armed forces	army 21,000; navy 700; air force 4,300
Largest cities	Ashgabat (capital - 848,444); Türkmenabat (240,880)
Official language	Turkmenian
Ethnic composition	Turkmen 85%; Uzbek 5%; Russian 4%; other 6%
Religious affiliations	Muslim 89%; Eastern Orthodox 9%; unknown 2%
Currency	1 Turkmen manat (TMM) = 100 tenge
Gross domestic product	(2006) U.S. $45.11 billion
Gross domestic product per capita	(2006) U.S. $8,900
Life expectancy at birth	male 58.43 yr; female 65.41 yr
Major resources	petroleum, natural gas, sulfur, salt, alfalfa, clay, cotton, grapes, gypsum, limestone, melons, sheep, wheat

Uzbekistan

REPUBLIC OF UZBEKISTAN

THE FORMER SOVIET REPUBLIC OF UZBEKISTAN lying amid the deserts of south-central Asia, was once crossed by the fabled silk road to China. It is bounded by Kazakhstan to the north, and Turkmenistan and Afghanistan to the south. In the east a complicated frontier separates Uzbekistan from Tajikistan and Kyrgyzstan.

GEOGRAPHY

The northwestern frontier crosses the undulating Ustyurt Plateau, east of which is the salty Aral Sea, with the broad Amu Darya (Oxus) river delta at its southern end. From here the vast, sandy desert of the Kyzyl Kum stretches as far as the foothills of the eastern mountains. In the extreme east, beyond the capital Tashkent and the Chatkal mountains, is the fertile Fergana basin.

Uzbekistan has a mostly semi-arid climate, with rainfall largely confined to the east. Typically, a short winter with severe frosts is followed by a long, intensely hot summer. Vegetation ranges from desert shrubs to woodland in the east.

SOCIETY

Although the Uzbek people are of Turkic origin, they took their name from the Mongol khan Öz Beg who ruled 1313–40, and who may have converted them to Islam. They moved southward into their present territory in the 15th century, establishing khanates on lands that had once belonged to the powerful Tartar conqueror Tamerlane (1336–1405).

In the late 19th century the area was conquered by Russia, but Uzbekistan's present boundaries only took shape in 1924, when it became a constituent republic of the Soviet Union.

Although Uzbeks form the majority of today's population, there are some 60 other ethnic groups – a cause of much conflict. After the Soviet Union's disintegration in 1991, Uzbekistan joined the new Commonwealth of Independent States as a fully independent republic. Despite a multiparty constitution, the former Communist Party led by President Karimov (b. 1938), and re-elected in 1995 until 2000, tolerates little opposition.

ECONOMY

Intensive irrigation has made Uzbekistan one of the world's largest cotton producers, but it has also reduced the flow of the Amu Darya, causing the level of the Aral Sea to fall dramatically, devastating the local fishing communities. Fruit and rice are cultivated in oasis areas, and some cereals are grown in the east.

Mineral resources include huge deposits of uranium, gold and various metal ores, as well as some coal, oil and natural gas. Previously in short supply, oil and gas production was expanding in the late 1990s. Major manufactures include farm machinery, textiles, chemicals and food products.

The road and rail networks are extensive, and Tashkent's international airport is an important refueling point for flights between Europe and southern Asia. The beautiful architecture of Uzbekistan's medieval Islamic cities – Samarkand, Bukhara and Kiva – is now a considerable attraction to tourists.

NATIONAL DATA – UZBEKISTAN				
Land area 425,400 sq km (164,248 sq mi)				
Climate		Temperatures		Annual
	Altitude m (ft)	January °C(°F)	July °C(°F)	precipitation mm (in)
Tashkent	478 (1,569)	1 (34)	25 (77)	417 (16.4)
Major physical features highest point: Beshtor Peak 4,299 m (14,104 ft); largest lake: Aral Sea (part) 66,500 sq km (25,700 sq mi)				
Population (2006 est.) 27,307,134				
Form of government multiparty republic with one legislative house but under authoritarian presidential rule, with little power outside the executive branch				
Armed forces army 40,000; air force 15,000				
Largest cities Tashkent (capital – 1,959,190); Namangan (446,237); Andijon (321,622); Samarqand (312,863); Bukhara (249,037); Nukus (240,734)				
Official language Uzbek				
Ethnic composition Uzbek 80%; Russian 5.5%; Tajik 5%; Kazakh 3%; Karakalpak 2.5%; Tatar 1.5%; other 2.5%				
Religious affiliations Muslim 88% (mainly Sunnis); Eastern Orthodox 9%; other 3%				
Currency 1 Uzbekistani soum (UZS) = 100 tiyn				
Gross domestic product (2006) U.S. $54.81 billion				
Gross domestic product per capita (2006) U.S. $2,000				
Life expectancy at birth male 61.19 yr; female 68.14 yr				
Major resources natural gas, petroleum, coal, gold, uranium, silver, copper, lead and zinc, tungsten, molybdenum, cotton, livestock, fruit, grapes, iron ore, maize/corn, melons, rice, silkworms, sulfur, vegetables, wheat				

No shortage of bread A trio of Uzbek market stall-holders keenly proffer their freshly made flat, round loaves to passersby. Bread, a staple part of the diet, is the traditional accompaniment to the typical Uzbek meal of rice, mutton and vegetables.

Tajikistan

REPUBLIC OF TAJIKISTAN

I N THE HIGLANDS OF SOUTH-CENTRAL ASIA lies the small mountainous republic of Tajikistan, which shares borders with Afghanistan to the south and China to the east. To the north is Kyrgyzstan and to the northwest Uzbekistan.

GEOGRAPHY

Most of eastern Tajikistan, dominated by the icy wastes of the great Massif known as the Pamirs, is more than 3,000 m (10,000 ft) above sea level. The two highest mountains within the Commonwealth of Independent States are in this area. The highest, Qullai Garmo (previously Communism Peak), rises to 7,495 m

NATIONAL DATA - TAJIKISTAN				
Land area 142,700sq km (55,097 sq mi)				
Climate		Temperatures		Annual
	Altitude m (ft)	January °C(°F)	July °C(°F)	precipitation mm (in)
Dushanbe	796 (2,611)	3 (37)	27 (81)	653 (25.7)
Major physical features highest point: Communism 7,495 m (24,590 ft); longest rivers: Amu Darya (part) 2,539 km (1,578 mi), Vakhsh 800 km (497 mi)				
Population (2006 est.) 7,320,815				
Form of government multiparty republic with one legislative house				
Armed forces army 7,600				
Largest cities Dushanbe (capital - 533,495); Khudzhand = Khujand (142,513)				
Official language Tajik				
Ethnic composition Tajik 79.9%; Uzbek 15.3%; Russian 1.1%; Kyrgyz 1.1%; other 2.6%				
Religious affiliations Sunni Muslim 85%; Shi'a Muslim 5%; other 10%				
Currency 1 Tajikistan Somoni (TJ or TJK) = 100 dirams				
Gross domestic product (2006) U.S. $9.405 billion				
Gross domestic product per capita (2006) U.S. $1,300				
Life expectancy at birth male 62.03 yr; female 68 yr				
Major resources hydropower, some petroleum, uranium, mercury, brown coal, lead, zinc, tungsten, silver, gold, cotton, wool, cattle, antimony, apricots, fruits, livestock, niobium, phosphates, rice, silkworms, tantalum, vanadium				

(24,590 ft). In the north, other mountain ranges extend into Uzbekistan from Kyrgyzstan across Tajikistan, and north of these lies the fertile Fergana basin, famed for growing a variety of exotic fruits. In the southwest a number of rivers drain southward into the Amu Darya river.

The climate is continental, with cold winters, and long hot summers in the valleys; the higher mountains are markedly colder. Precipitation is largely confined to the high valleys bordering the Pamirs, from where water carries downstream to the much drier areas. Plant and animal life varies according to altitude.

SOCIETY

Most of the peoples of Central Asia have a language and culture strongly influenced by a history of Turkic invasion. The ethnic roots of the Tajiks, however, go back to ancient Persia, and the people speak a language that is Iranian rather than Turkic in origin. In the 16th century the area became part of the Bukhara emirate of the Mongol empire until it was overrun by Afghans in the mid 18th century. Although absorbed into the Russian empire in the 19th century, there was considerable resistance to Soviet domination in Tajikistan in the 1920s, and it did not become a constituent republic of the Soviet Union until 1929.

The population of Tajikistan is an ethnic mixture that includes Uzbeks, Russians and Tartars; the Tajiks themselves are a combination of several older ethnic groups. The years leading up to full independence in 1991 were overshadowed in Tajikistan, as in other parts of the

A party of Tajik men prepare to celebrate the circumcision of a young boy – part of his initiation into the Islamic faith – in traditional style. Soviet suppression of religion was in practice greatly relaxed in the Central Asian republics.

crumbling Soviet Union, by ethnic unrest and violence. The most serious clashes involved the Muslim Uzbeks, who form a majority in the Fergana basin territories. Heavily backed by Russian forces, the Tajik government has been in conflict with the rebels since 1992. Thousands died or were forced to flee before a UN-sponsored ceasefire was agreed in 1996. Further outbreaks of fighting occurred in 1997 and 1998.

ECONOMY

Agriculture in Tajikistan relies on extensive irrigation. Cotton is the chief crop, but cereals, rice and many kinds of fruit and nuts are also grown. There is also livestock farming mainly for the production of wool and meat. The country is rich in mineral resources, including iron ore, coal, oil and natural gas. Along with hydroelectricity (generated from the country's fast-flowing rivers), these fuels provide power for domestic consumers and for the factories that produce cotton textiles, clothing, carpets, food and machinery. Carpet making is still an important source of foreign revenue.

Transportation by road and railroad is adequate for local needs despite the difficult terrain. More remote districts are reached by air, and there are regular international flights to neighboring states. Health, welfare and education services are well developed only in the towns.

Kyrgyzstan

REPUBLIC OF KYRGYZSTAN

KYRGYZSTAN IS A MOUNTAINOUS REPUBLIC in central Asia. It has Kazakhstan on its northern frontier, Uzbekistan to the west, Tajikistan to the southwest and a long border with China to the southeast.

GEOGRAPHY

Kyrgyzstan is dominated by a massive mountain range, the Tien Shan, which extends the length of Kyrgyzstan's frontier with China from Pobedy Peak at 7,439 m (24,406 ft) in the east to the Alay mountains in the west, near the southwestern border with Tajikistan. The rest of Kyrgyzstan is largely made up of a series of parallel mountain ranges divided by deep

river valleys and basins. The deepest basin is in the northeast, where the clear, waters of Lake Issyk-Kul are enclosed by a ring of snowy mountains. In the west, the central Fergana valley opens out into a broad basin, most of which lies within Uzbekistan.

The mountains bring plentiful water to an otherwise desert area. Their higher slopes are wet and cold, with heavy winter snowfall. In contrast, summers in the valleys and basins are hot and dry, but runoff and meltwater from the mountains make irrigation possible on a wide scale. Vegetation ranges from dry scrub and woodland in the valleys to alpine meadow on the higher slopes. Wildlife is plentiful in the mountains.

SOCIETY

The origins of the mainly Muslim Kyrgyz people are uncertain. Their Turkic language resembles the language spoken in neighboring Kazakhstan, and many believe that both groups arrived from the northeast in the 13th century, seeking refuge from Mongol invaders. The pastoral nomadic life of the indigenous population changed little over the ensuing centuries, until Russians colonized Kyrgyzstan in the 19th century. The area was subsequently annexed region by region, until in 1876 Kyrgyzstan was made part of Russian Turkistan. Much of the best land went to Russians, who settled in the low-lying, fertile regions. The ruthless enforcement of collectivization that followed the 1917 Russian Revolution irrevocably altered the traditional way of life of the Kyrgyz people. In 1924 the area became an autonomous oblast (region), and in 1936 was incorporated into the Soviet Union as the Kirghiz Soviet Socialist Republic. But resistance to Soviet rule, carried out by local guerrilla groups or *basmachi*, continued into the 1930s.

Today the Kyrgyz account for little more than half the country's mixed population. The largest minority group, after the Russians, are the Uzbeks, who in 1990 were involved in violent clashes with the Kyrgyz as deepseated interracial tensions resurfaced with the relaxation of Soviet rule. The previous year the Kyrgyz had clashed with the Tajiks over a border dispute. In July 1991 the Kirghiz Supreme Soviet instituted its own presidential

A temple to the arts, the imposing opera house in Bishkek (formerly Frunze), capital of Kyrgyzstan. Founded as late as 1878, the city is built on a grid pattern, with wide tree-lined boulevards and parks. The large industrial quarter was developed in the 1960s.

government, and before the end of the year the newly independent republic of Kyrgyzstan joined the Commonwealth of Independent States. A revised constitution was adopted in 1994, with first elections in 1995 for the bicameral parliament.

ECONOMY

The main agricultural activity is livestock breeding, especially cattle, sheep and goats. The Kyrgyz, renowned for their horsemanship, use horses for rounding up and tending livestock. The country is self-sufficient in grain, and other crops include tobacco, cotton and poppy (the opium-producing species). Agriculture is highly mechanized, and makes use of extensive irrigation.

Industrial development has benefited from plentiful supplies of hydroelectricity and fossil fuels including coal, oil and natural gas. Industries are based on local resources and include food processing, weaving, textiles and manufacturing footwear. The mountainous terrain poses communication problems, mostly resolved through the development of the road network. Medical and educational facilities, made readily available under the old Soviet regime, include modern hospitals and welfare centers, schools and institutes of higher education.

NATIONAL DATA – KYRGYZSTAN

Land area	191,300 sq km (73,861 sq mi)

Climate	Altitude m (ft)	Temperatures January °C(°F)	July °C(°F)	Annual precipitation mm (in)
Bishkek	828 (2,716)	-3 (27)	25 (77)	442 (17.4)

Major physical features highest point: Pobeda Peak 7,439 m (24,406 ft); longest river: Syr Darya (part) 2,204 km (1,370 mi)

Population (2006 est.) 5,213,898

Form of government multiparty republic with two legislative houses

Armed forces army 8,500; air force 4,000

Capital city Bishkek (933,763)

Official languages Kyrghyz, Russian

Ethnic composition Kyrgyz 64.9%; Uzbek 13.8%; Russian 12.5%; Dungan 1.1%; Ukrainian 1%; Uygur 1%; other 5.7%

Religious affiliations Muslim 75%; Russian Orthodox 20%; other 5%

Currency 1 Kyrgyzstani Som (KGS) = 100 tyiyn

Gross domestic product (2006) U.S. $10.49 billion

Gross domestic product per capita (2006) U.S. $2,000

Life expectancy at birth male 64.48 yr; female 72.7 yr

Major resources hydropower, gold, uranium, coal, oil, natural gas, nepheline, mercury, bismuth, lead, zinc, antimony, cotton, fruit, vegetables, grapes, livestock, sugar beet, tobacco

Kazakhstan

REPUBLIC OF KAZAKHSTAN

THE REPUBLIC OF KAZAKHSTAN EXTENDS across central Asia from the Caspian Sea in the west to China in the east. It shares a long frontier with Russia to the north, and borders Turkmenistan, Uzbekistan and Kyrgyzstan to the south. After Russia, Kazakhstan is the second largest of the former Soviet republics, and it has proved to be one of the most influential members of the Commonwealth of Independent States, established in 1991.

NATIONAL DATA – KAZAKHSTAN

Land area   2,669,800 sq km (1,030,816 sq mi)

Climate	Altitude m (ft)	Temperatures January °C(°F)	July °C(°F)	Annual precipitation mm (in)
Almaty	775 (2,543)	-6 (21)	24 (75)	641 (25.2)

Major physical features   highest point: Khan-Tengri 7,199 m (23,620 ft); lowest point: Mangyshlak Depression -132 m (-433 ft); longest river: Irtysh (part) 4,400 km (2,760 mi)

Population   (2006 est.) 15,233,244

Form of government   multiparty republic with two legislative houses but with authoritarian presidential rule

Armed forces   army 46,800; air force 19,000

Largest cities   Almaty (1,227,059); Chimkent (420,435); Karaganda (404,451); Taraz (366,761); Astana (capital - 356,886); Pavlodar (330,165)

Official language   Kazakh

Ethnic composition   Kazakh (Qazaq) 53.4%; Russian 30%; Ukrainian 3.7%; Uzbek 2.5%; German 2.4%; Tatar 1.7%; Uygur 1.4%; other 4.9%

Religious affiliations   Muslim 47%; Russian Orthodox 44%; other 9%

Currency   1 tenge (KZT) = 100 tiyin

Gross domestic product   (2006) U.S. $138.7 billion

Gross domestic product per capita   (2006) U.S. $9,100

Life expectancy at birth   male 61.56 yr; female 72.52 yr

Major resources   coal, oil, natural gas, iron ore, bauxite, copper, nickel, lead, gold, uranium, cereals, cotton

GEOGRAPHY

Kazakhstan's predominantly flat landscape is broken by several mountain ranges in the east and southeast on the borders with Russia, China and Kyrgyzstan. Most of eastern Kazakhstan is an eroded tableland characterized by shallow uplands, depressions and lakes. The west is lower-lying, with the northern basins of the Aral Sea in the center, and farther west beyond the Mugodzhary Hills, the Caspian Sea.

The country's climate is characterized by intense winter cold and summer heat. Precipitation levels are generally low. More than two-thirds of Kazakhstan is desert or semidesert. The central lakes are salty, as are the lowland marshes in the west, and in the center there are few permanent rivers. The grass steppelands in the north are the country's most fertile regions. Wildlife ranges from wolves and bears to gerbils and mole-voles that inhabit the steppes and the desert.

SOCIETY

Kazakhstan is named after the Kazakh peoples, Turkic-speaking nomads who established a powerful khanate in the region during the 15th and 16th centuries. In the 18th and 19th centuries their empire broke up and was gradually overrun by the Russians. From 1861 onward, newly emancipated Russian and Ukrainian peasants settled in the fertile northern steppes. Kazakh nationalism culminated in a bitter civil war in Kazakhstan after the 1917 Russian Revolutions. In 1936 it was incorporated as a Soviet republic.

The Soviet regime became increasingly repressive in the late 1930s when Kazakhstan's traditionally nomadic peoples were collectivized on huge cattle farms. Many fled to China and Afghanistan. In the 1950s and 1960s vast tracts of uncultivated territory were transformed into arable land. This was known as the "Virgin Lands" project; its longterm effect on the ecological balance was disastrous. The Soviet regime also set up sites in the desert for the Soviet space launching program and for nuclear weapons testing. With the relaxation of Soviet rule in the late 1980s, nationalist sentiments resurfaced. Kazakhstan's refusal to recognize

A Kazakh shepherd Kazakhstan is home to some 200,000 nomadic shepherds and herdsmen, who, like the Tartars before them, pack their yurts (collapsible houses) on their horses and travel with the changing seasons to favorable pastures for their flocks.

the Soviet coup of 1991 was a major reason for the Soviet Union's breakup. Kazakhstan regained its independence in December 1991. Elections in 1994 were declared invalid in 1995, after which President Nazarbaev (b. 1940) took power to rule by decree. Today's population includes almost as many ethnic Russians as Kazakhs, which creates tensions.

ECONOMY

Kazakhstan is extremely rich in mineral resources including iron ore, bauxite, copper and nickel. It has proven oil reserves of 2 billion tonnes, and the successful exploitation of these will make the country one of the biggest oil producers in the world in the 2000s.

In the late 1990s Kazakhstan signed agreements with China and with a number of international consortia to explore and develop oil and natural gas fields. This should help Kazakhstan out of the economic collapse it suffered after the breakup of the Soviet Union. Chemical production, metal smelting and the manufacturing of heavy machinery are important industries as well as the more recent electronics and pharmaceuticals.

Agriculture employs around one quarter of the population. The production of grain and cotton has now replaced the traditional cattle, sheep and horse breeding as the major activity.

Mongolia

REPUBLIC OF MONGOLIA

A burgundy-robed Buddhist monk in the monastery at Erdeni Dzuu in central Mongolia, southwest of the capital, Ulan Bator. Buddhism was introduced into Mongolia by Tibetan teachers in the 13th century at the request of Genghis Khan's grandson, Godan Khan.

MONGOLIA IS A VERY LARGE LANDLOCKED republic with China on its southern frontier and Russia to the north. It is sometimes known as Outer Mongolia, to distinguish it from Inner Mongolia, an Autonomous Region of China.

GEOGRAPHY

Most of the country is over 1,000 m (3,300 ft) above sea level. The highest peaks are in the Altai mountains, in the lofty central massif of the Hangayn range, and in the Hentiyn range, which runs northeastward from the capital, Ulan Bator. Much of the country is prone to earthquakes.

Nearly four-fifths of Mongolia is dry steppe grassland, and much of the rest is cold, arid desert. Most of the south is occupied by the Gobi, an area of stony desert and low mountain ranges. Forest, mostly on northern mountain slopes, covers less than one-tenth of the land. The winters are long and extremely cold, with snow in the higher mountains. Wildlife is plentiful. Lynx, bears and deer inhabit the northern regions, while the steppes support marmots and gazelles. The snow leopards of the Altai are becoming endangered.

SOCIETY

For thousands of years Mongolia was inhabited by nomadic peoples from central Asia and southern Siberia. In the 13th century they were united under the warrior Genghis Khan (c. 1162–1227). He and his successors established a vast empire that extended across China, much of central Asia and Russia. In 1368, however, the empire collapsed and Mongolia came under Chinese domination.

After the Russian Civil War (1918–22), the Chinese were eventually driven out of Outer Mongolia, and the Mongolian People's Republic was established in 1924. From its foundation, the People's Republic was closely tied to the Soviet Union. Government was along Soviet lines, with only one political party. A new constitution, which introduced a multiparty 76 seat elected parliament and guaranteed freedom of speech, was approved in 1992 with first elections later that year. A second free election was held in 1996 with the Democratic Union party winning, finally ending communist control.

Most people in the country are Mongols. The Kazakhs are the largest non-Mongol minority. Religious practice has been suppressed for many years, but other cultural traditions have been kept alive.

ECONOMY

Mongolia's grasslands support a huge livestock population, including sheep, cattle, goats and horses. Crop farming is minimal. The forests provide some roundwood for fuel and for industry.

The country has large deposits of coal (the chief energy source) and fluorite, as well as copper, molybdenum and other metal ores. Most industries process raw materials. The principal exports are minerals and animal products; imports include machinery and a range of consumer goods. Most trade is with countries of the former Soviet Union.

The Trans-Mongolian Railway, linking Mongolia with Russia and China, is the country's major transportation route. There are few surfaced roads and camels are still used in the south. Ulan Bator has an international airport and offers internal flights to other cities.

The state provides free health, welfare and education services. Literacy levels are high, and most children receive schooling.

NATIONAL DATA – MONGOLIA				
Land area 1,564,116 sq km (603,909 sq mi)				
Climate		Temperatures		Annual
	Altitude m (ft)	January °C(°F)	July °C(°F)	precipitation mm (in)
Ulaanbaatar	1,337 (4,385)	-20 (4)	18 (64)	271 (10.6)
Major physical features highest point: Hüyten (Nayramdal) Peak 4,374 m (14,350 ft); longest river: Selenga 998 km (620 mi)				
Population (2006 est.) 2,832,224				
Form of government mixed parliamentary/presidential multiparty republic with one legislative house				
Armed forces army 7,500; air force 800				
Capital city Ulaanbaatar (881,218)				
Official language Khalkha Mongolian				
Ethnic composition Mongol (mainly Khalkha) 94.9%; Turkic (mainly Kazakh) 5%; other (including Chinese and Russian) 0.1%				
Religious affiliations Buddhist Lamaist 50%; none 40%; Shamanist and Christian 6%; Muslim 4%				
Currency 1 tögrög/tugrik (MNT) = 100 mongos				
Gross domestic product (2006) U.S. $5.781 billion				
Gross domestic product per capita (2006) U.S. $2,000				
Life expectancy at birth male 62.64 yr; female 67.25 yr				
Major resources oil, coal, copper, molybdenum, tungsten, phosphates, tin, nickel, zinc, fluorspar, gold, silver, iron, livestock, cereals, hay fodder, lignite, potatoes, vegetables, wolfram				

The Middle East

Black Sea

Istanbul
Izmit
Samsun
Bosporus
Ankara
TURKEY
Erzurum
Izmir
Lake Tuz
Kizil Irmak
Elazig
Ararat 5165
Aras
Anatolia
Taurus Mts
Adana
Lake Van
Tabriz
Antalya
Urmia
Lake Urmia
Rasht
Caspian Sea
Aleppo
Mosul
Elburz Mountains
Mashhad
Concord Peak 5407
Baghlan
Hindu Kush
Nowshak 7485
Mediterranean Sea
Tripoli
SYRIA
Homs
Kirkuk
Tehran
Damavand 5671
Shah Fuladi 5143
Herat
Hari Rud
Kabul
LEBANON
Beirut
Damascus
Dasht-e-Kavir
IRAQ
Baghdad
Tigris
IRAN
AFGHANISTAN
ISRAEL
Tel Aviv
Amman
Isfahan
Zagros Mountains
Karun
Dasht-e-Lut
Kandahar
Jerusalem
Dead Sea -400
An Najaf
Euphrates
Helmand
JORDAN
Syrian Desert
Basra
Kerman
Seistan
Aqaba
Shiraz
An Nafud
Kuwait
KUWAIT
Persian Gulf
Bandar Abbas
Hejaz
Nejd
Dammam
BAHRAIN
Al Manamah
QATAR
Strait of Hormuz
Tropic of Cancer
Medina
Doha
Dubai
Riyadh
Abu Dhabi
Muscat
UNITED ARAB EMIRATES
Sur
Cape Hadd
SAUDI ARABIA
Jedda
Mecca
Arabian Sea
Arabian Peninsula
Masirah
OMAN
Cape Madrakah
Red Sea
Rub al Khali
Kuria Muria Islands
Salalah
San'a
YEMEN
Hadhramaut
Al Mukalla
Aden
Socotra
Gulf of Aden

■ capital city
● major town

height of land (meters)

5000
3000
2000
1000
500
200
0 (sea level)

▲ mountain peak
▼ depression

Luxury in an arid land (*left*) Water is one of the most precious resources in the Middle East. Here a Bedouin nomad waters his herd of camels at a well in Qatar. But it was the discovery of another liquid – oil – that brought the Arabian Peninsula into the modern world and made it rich beyond measure.

Turkey

REPUBLIC OF TURKEY

TURKEY SPANS TWO CONTINENTS, LINKING mainland Europe with Asia across the narrow straits of the Bosporus and the Dardanelles. European Turkey (Thrace), to the north of the Sea of Marmara, shares borders with Bulgaria and Greece. Asian Turkey, much of which is known as Anatolia or Asia Minor, is larger in area and borders Syria and Iraq to the south and southeast, Iran and Armenia to the east, and Georgia to the northeast. The long coastlines are washed by the Black Sea to the north, the Aegean to the west and the Mediterranean to the south.

NATIONAL DATA – TURKEY

Land area	770,760 sq km (297,592 sq mi)			
Climate		Temperatures		Annual
	Altitude m (ft)	January °C(°F)	July °C(°F)	precipitation mm (in)
Ankara	861 (2,825)	1 (34)	23 (73)	384 (15.1)
Izmir	28 (92)	9 (48)	28 (82)	686 (27)

Major physical features highest point: Ararat 5,165 m (16,945 ft); largest lake: Lake Van 3,675 sq km (1,419 sq mi); longest river: Kizil Irmak 1,150 km (715 mi); (Euphrates and Tigris also rise in Turkey)

Population (2006 est.) 70,413,958

Form of government multiparty republic with one legislative house

Armed forces army 402,000; navy 52,750; air force 60,100

Largest cities Istanbul (10,291,102); Ankara (capital – 3,641,931); Izmir (2,615,568); Bursa (1,504,817); Adana (1,294,460)

Official language Turkish

Ethnic composition Turkish 80%; Kurdish 20%

Religious affiliations Muslim 99.8% (mainly Sunni); other 0.2% (mainly Christians and Jews)

Currency 1 Turkish lira (YTL) = 100 kurus

Gross domestic product (2006) U.S. $6.27.2 billion

Gross domestic product per capita (2006) U.S. $8,900

Life expectancy at birth male 70.18 yr; female 75.18 yr

Major resources coal, iron ore, copper, chromium, antimony, mercury, gold, barite, borate, celestite (strontium), emery, feldspar, limestone, magnesite, marble, perlite, pumice, pyrites (sulfur), clay, cotton, tea, tobacco, wheat, barley, rice, sugar beet, olives, grapes/sultanas, figs, hazelnuts, tourism, hydropower

GEOGRAPHY

Frequent earthquakes, together with geysers and other volcanic phenomena, are evidence that this is a tectonically active region.

The European part of Turkey is characterized by fertile rolling plains surrounded by low mountains. On the Asian side, western Anatolia is crossed by long mountain ridges separated by deep valleys. South of the central Anatolian plateau, the three main ranges of the Taurus Mountains flank the Mediterranean coast. Farther east a complex of even higher mountain ranges culminates in the massive cone of Mount Ararat. To the southwest is Lake Van.

Annual rainfall reaches about 2,500 mm (100 in) along the Black Sea coast, where temperatures are relatively high all year. The central plateaus of the interior are semiarid. In the mountainous east the winter snows may last for several months.

Thick, scrubby undergrowth characterizes the Mediterranean south and west of the country. The Black Sea coast supports dense forest, while the drier interior is covered mainly in steppe grassland. Bears and red deer are still found in the forests and colorful flocks of greater flamingoes thrive on some of the lakes.

SOCIETY

Asia Minor has seen the rise and fall of many of the world's great empires. However, by the early 20th century Turkey was a weak and impoverished country. It was the country's first president, Kemal Atatürk (1881–1938), who laid the foundations for a stronger, modern nation.

The arrival of Indo-European-speaking Hittites from central Asia in about 1900 BC marked the beginning of the first great Anatolian empire. Anatolia was absorbed into the Persian empire in the 6th century BC. Alexander the Great (356–323 BC) then brought the region under Greek control before it fell to the Roman empire in the 1st century BC.

From the 11th century AD Asia Minor was invaded by the Muslim Seljuks – Turkic rulers from the east. Osman I (1258–1326), the ruler in northwestern Anatolia, is traditionally regarded as the founder of the Ottoman state. By the later 15th century the Ottomans had taken the old Roman capital, Constantinople, which they renamed Istanbul, and gained control of the greater part of Asia Minor. Later rulers, notably Suleiman the Mag-

Straddling Europe and Asia (above), Istanbul is strategically positioned at the entrance to the Black Sea, controlling a major shipping route. Among its many attractions are the colorful bazaar, the Topkapi Palace, and the great Roman walls and underground cistern.

Colossal ruins atop Mount Nemrut (below) in the far east of Turkey. The huge statues of gods were commissioned by Antiochus I of Commagene (62–32 BC) – together with one of himself – "in commemoration of [his] own glory and that of the gods".

The terraces of Pamukkale (*right*) in west-central Turkey descend for some 90 m (300 ft). As the hot, mineral-rich spring water flows down the slope, it evaporates, depositing the minerals in fantastic formations.

nificent (1494/5–1566), pushed the empire's boundaries still farther into the Middle East, Northern Africa and Europe.

By the 19th century the Ottoman empire had begun to disintegrate, and defeat in the Balkan Wars of 1912–13 reduced Turkey's European territories to their present size. In 1918, following its defeat in World War I, the Ottoman empire was dismembered. Meanwhile a nationalist revolutionary movement led by Mustafa Kemal had set up a government in Ankara. As Turkey's first president, Kemal, who took the name Atatürk, or "Father of Turkey", transformed and Westernized Turkish society.

In 1951 Turkey became a member of the North Atlantic Treaty Organization (NATO). A long-running dispute with Greece over the Mediterranean island of Cyprus has been an obstacle to Turkey's aim of joining the European Union (EU). In 1980 there was a military coup but from 1983 political parties were allowed to operate again. During the late 1980s there was growing insurgency among the largely Kurdish population of the eastern provinces. This situation was complicated further by the 1991 Gulf War between Western Allies and Iraq over the invasion of Kuwait. Turkey imposed sanctions in accordance with NATO policy, at consid-

erable cost to its own economy, but was unhappy about having foreign troops stationed on Turkish soil. Moreover, when Kurdish refugees crossed the border from Iraq, Turkey was obliged to help those who had long been regarded as a dangerous minority.

Under the 1982 constitution the 550 members of the Grand National Assembly (GNA) are elected by proportional representation to serve a five-year term. Executive power rests with the president.

Most of the population are Turkish-speaking Sunni Muslims. Kurdish communities in the southeast constitute the only large minority. Since 1984 Turkey has been fighting Kurdish separatists.

ECONOMY

Turkey's economic development has been slowed by the need to import some of its fuel and food. Local coal is available but petroleum has to be imported. Up to half the nation's energy comes from hydro-electric power. Plans to build large dams

and hydroelectric plants will increase Turkey's energy production by 70 percent. Irrigation in the coastal lowlands has enabled the country to produce the largest crops of hazelnuts and sultanas in the world; these and other cash crops such as cotton, tea and tobacco are vital to its export trade.

Iron, aluminum and copper ores supply raw materials for industry. The textile industry, which produces yarn, fabrics, rugs (from locally grown cotton) and carpets (from wool), accounts for about 40 percent of the country's exports. Tourism is growing rapidly. Turkey has the best-developed road and railroad systems in the Middle East. Istanbul is the country's largest port. The state airline, Turkish Airlines, serves most major international and domestic airports.

The national welfare provision includes health insurance and retirement pensions. Healthcare is free for the poor. Housing is in short supply. Free education is provided at primary and secondary levels. There are some 30 universities.

Syria

SYRIAN ARAB REPUBLIC

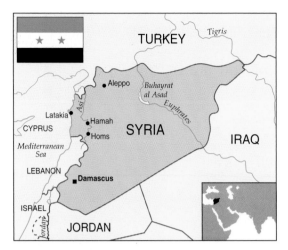

S YRIA LIES AT THE EASTERN END OF THE
Mediterranean Sea. To the north is
Turkey, to the east Iraq, to the south
Jordan, to the southwest Israel, and to the
west Lebanon and the Mediterranean
coast. Ever since its frontiers were drawn
up at the end of World War I by the Allies,
the country has played a leading role in
Middle Eastern politics; the government's
hardline Arab nationalism has led Syria
into frequent conflicts with its neighbors,
particularly Israel.

GEOGRAPHY

Syria's Mediterranean seaboard extends
for some 180 km (110 mi) between Turkey
and Lebanon and is one of the country's
most fertile and densely populated areas.
Inland are the mountains of the Jebel
Ansariye. Across the mountains, the Asi
(Orontes) river flows through the deep
trench formed by the Ghab Depression.
The mountainous terrain continues along
the Lebanese frontier as far as the Golan
Heights in the southwest, presently
under Israeli military occupation. Extending north and east from the mountains is
a broad, semiarid plateau – crossed by the
Euphrates river in the southeast – that
turns into the Syrian Desert.

The climate is generally dry with hot
summers and mild winters. Rain falls
mostly in the western highlands between
September and May, but varies considerably from year to year, with frequent
summer drought. Hot dry khamsin winds
from the desert blow in summer.

The vegetation varies from thin desert
scrub in the east and dry grassland in the
north to Mediterranean maquis in the
west, and forests on the highest mountains. Gazelles, jerboas and lizards live in
remote desert areas; the high mountain
forests are home to Wild boars, wolves,
badgers and porcupines, as well as
several species of warbler.

SOCIETY

Situated on ancient trade and military
routes between the Mediterranean and
Mesopotamia (present-day Iraq), Syria
from about 2000 BC was part of Akkadia –
one of the earliest urban civilizations.
Over time it became a province of neighboring empires such as those of the
Egyptians, the Hittites in the northwest,
and the Assyrians and Babylonians. The
Persians were driven out by Alexander
the Great (356–323 BC); his successors
introduced Greek culture. In 64 BC Syria
became a Roman province.

From 395 AD Syria was ruled from
Constantinople, the capital of the eastern
Roman empire, until it was overrun by
Muslim Arabs in 634. Most Syrians converted to Islam under the Umayyad dynasty, and Damascus became the political
and cultural center of the Islamic world.
The region fell to the Seljuk Turks in the
11th century. During this period Syria
was the scene of many battles between
Western Christian crusaders and Muslims. The crusaders seized much of Syria
in the 12th century, but were expelled in
1187 by the Kurdish general Saladin (c.
1137–93). Saladin's Ayyubid descendants
and their Mameluke successors ruled until the Ottoman Turkish conquest of 1516.

Syria came under a French mandate in
1920, after Turkey's defeat in World War
I, and its present boundaries were established. During World War II, Syria was
briefly controlled by the pro-German
Vichy French regime, before being occupied by a joint British and Free French
invasion force. The French did not with-

A peaceful mosque courtyard (*above*) in the heart of
Syria's capital, Damascus, a city that has been
continually inhabited since 2500 BC. Its ancient crafts
of weaving and metal working are major industries.

The traditional domed roofs of mud brick (*below*) –
derived from a scarcity of wood – in a rural settlement
near Aleppo. Although many agricultural workers have
moved to the cities in search of work, about half of
Syria's workforce still makes a living from agriculture.

NATIONAL DATA – SYRIA

Land area 184,050sq km (71,062 sq mi)

Climate	Altitude m (ft)	Temperatures		Annual precipitation mm (in)
		January °C(°F)	July °C(°F)	
Damascus	720 (2,362)	7 (45)	27 (81)	133 (5.2)

Major physical features highest point: Mount Hermon 2,814 m (9,232 ft); longest river: Euphrates (part) 3,596 km (2,235 mi)

Population (2006 est.) 18,881,361

Form of government multiparty republic with one legislative house

Armed forces army 200,000; navy 7,600; air force 40,000

Largest cities Aleppo (1,649,694); Damascus (capital - 1,592,206); Homs (822,440); Hama (495,390)

Official language Arabic

Ethnic composition Arab 90.3%; Kurds, Armenians, and other 9.7%

Religious affiliations Sunni Muslim 74%; Alawite, Druze, and other Muslim sects 16%; Christian (various sects) 10%; Jewish (communities in Damascus, Al Qamishli, and Aleppo)

Currency 1 Syrian pound (SYP) = 100 piastres

Gross domestic product (2006) U.S. $75.1 billion

Gross domestic product per capita (2006) U.S. $4,00

Life expectancy at birth male 69.01 yr; female 71.7 yr

Major resources petroleum, phosphates, chrome and manganese ores, asphalt, iron ore, rock salt, marble, gypsum, hydropower, barley, cattle, cotton, fruit, goats, natural gas, potatoes, sheep, sugar beet, vegetables, wheat

draw until 1946, when Syria became independent. Three years later Syria's new civilian government was overthrown, and in 1963, after a succession of military coups and a brief union with Egypt (1958–61), the Arab Socialist Baath (Renaissance) Party took power.

The Baathist regime has survived despite internal unrest and conflict with foreign powers. The country suffered major defeats by Israel in the Arab–Israeli War of 1967, when Israel captured the Golan Heights, and in the Yom Kippur War of 1973. Syria became deeply embroiled in the civil war in Lebanon after 1975, leading to more than a decade of further clashes with Israel until a fragile stability was achieved in Lebanon in 1992. During The Gulf War of 1991 Syria joined the coalition forces against its old rival Iraq, and so improved its erstwhile stormy relations with the West.

Hafez al-Assad (1928–2000) was president until he died in 2000. He was succeeded by his son, Bashar al-Assad, who polled 97 percent of a referendum vote. The president, elected for a seven-year term, appoints two vice-presidents and a cabinet. The 250-member People's Assembly is elected every four years; all parties within it are officially linked as the National Progressive Front, which in turn is dominated by the Baath Party. Although the 1973 constitution defines Syria as a "socialist popular democracy", Baathist domination effectively controls the elections.

Apart from Kurdish communitites in the north, the population consists predominantly of Arabic-speaking Arabs. Sunni Muslims (who include the Kurds) are by far the largest religious group, but political power rests mainly with the 'Alawites – a Shi'ite Muslim sect to which the Baath leaders belong. There are many other religious minorities, among them Druze Muslims and Christians.

ECONOMY

The Syrian economy, burdened by military spending that consumes some two-thirds of the budget, is still predominantly agricultural despite a program of industrialization since World War II. Modest petroleum reserves provide the leading source of export revenue.

Agriculture is concentrated along the coast and beside the Asi and Euphrates rivers, but dry farming is also important. Cotton is the chief cash crop, and also the most reliable; other crops include sugar beet, lentils, olives, cereals and high-quality tobacco. Sheep, goats and camels feed on pastures irrigated from artificial terraces that collect rainwater: a technology some 4,000 years old.

There are many mineral resources. Hydroelectric stations on the Euphrates, supplemented by thermal stations burning oil, supply electricity. Industries include textile production, food processing and engineering.

Road and rail networks are well developed. Damascus, the capital, and the northwestern city of Aleppo both have an international airport. Syrian Arab Airways serves a small domestic network. The Syrian press is subject to strict censorship, and broadcasting is directly controlled by the state.

Healthcare, concentrated in the cities, is free for the poor. Water quality and sanitation remain inadequate, and infant mortality is fairly high. A limited welfare system provides retirement, injury and invalidity pensions. Education, which is free of charge, is compulsory for six years; some secondary and higher education is also available.

Lebanon

REPUBLIC OF LEBANON

THE SMALL, MOUNTAINOUS COUNTRY OF Lebanon on the eastern Mediterranean coast is bounded to the north and east by Syria and to the south by Israel – the two most mutually antagonistic countries in the Middle East. The deeply divided Lebanese nation is struggling to rebuild its economy and preserve a fragile stability. The long and disastrous civil war which ended in 1991 devastated the economy and the capital, Beirut. Violence in southern Lebanon in the late 1990s continued to hamper recovery.

GEOGRAPHY

The Lebanon Mountains rise from a narrow coastal plain, culminating in Qurnet es Sauda at 3,087 m (10,128 ft) in the north, and falling away southward to the Galilee hills in Israel. East of these mountains, the fertile Bekáa valley – part of the Great Rift Valley system – separates the Lebanon Mountains from the Anti-Lebanon and Hermon ranges along the Syrian border. The Litani river flows southwest through the Bekáa valley to the Mediterranean; the Asi (Orontes) river rises at the valley's northern end and flows northward into Syria.

Lebanese summers are hot and dry while the winters are mild and wet. The climate is generally cooler and wetter in the mountainous areas.

Lebanon's famous cedar trees can be found in the northern mountains, where they are strictly protected; little is left elsewhere of the original forest cover. Today's vegetation consists largely of shrubs and small trees, particularly oak, cypress, fir, juniper and carob. There are still a few bears roaming the mountains, and much smaller mammals, including hedgehogs, squirrels and dormice, are quite common, along with their larger predators such as wildcats and martens.

SOCIETY

Lebanon was settled in about 3000 BC by the Phoenicians, whose writing system formed the basis of the Greek and Roman alphabets. The Phoenician city of Tyre fell to Alexander the Great (356–323 BC) in 332 BC. During the 1st century BC Phoenicia became fully absorbed into the Roman empire.

Early in the 7th century AD, a group of Syrian Christians founded a Maronite community in what is now northern Lebanon. Arabs settled in the south, adopting the Druze faith (derived from Shi'ite Islam); Sunni Muslims predominated in the coastal towns.

Lebanon was conquered by Western Christian crusaders in the 11th century. Later, it became part of the Muslim Mameluke empire of Egypt and Syria, before falling to the Ottoman Turks (also Muslims) in 1516–17. By the 19th century the Maronites had expanded into southern Lebanon. In 1861 the French compelled the ruling Ottomans to establish Mount Lebanon as an autonomous Christian district.

Turkey's defeat in World War I forced its withdrawal from Lebanon. In 1920 Lebanon's present boundaries were defined under a French mandate, creating a new balance of Christians and Muslims, but French control was decisive. Lebanon achieved independence in 1946.

Pan-Arabism in the 1950s made Lebanon's pro-Western government increasingly unpopular among Muslims. From the 1960s an influx of rural Muslims into the cities heightened tension, while Palestinian refugees entered Lebanon from Israeli-occupied lands farther south.

Civil war erupted in 1975, when Palestinians operating against Israel from inside Lebanon clashed with Phalangist Christian forces. Shi'ite Muslims joined the Palestinians. Syrian troops intervened in 1976 and were joined by United Nations' peacekeeping troops. In 1982 an Israeli invasion drove the Palestinians back into the capital, Beirut, which then became a battleground for rival militias. After 1990 the Lebanese army gradually resumed control.

Promised financial aid by Saudi Arabia, Lebanon joined the Arab–Western alliance against Iraq during the 1991 Gulf War. Constitutional government was restored under Syrian protection, though a "security zone" set up by Israel in 1985 remained under the control of its Christian allies, the South Lebanon Army.

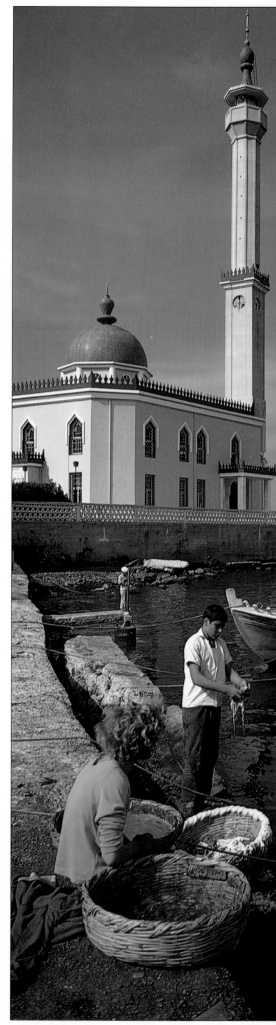

The harbor of Tripoli (*left*) Lebanon's second city after Beirut. Founded by the Phoenicians in the 7th century BC, it is still an important trading center, particularly for silk, fruit and tobacco. It is also the terminus for the oil pipeline from Iraq, and oil-refining in the city is an important industry.

Grapes, pomegranates, dates, plums, apples and citrus fruits (*right*) are just some of the cash-crops grown on the fertile coastal plain and in the Bekáa valley for sale in Beirut, Lebanon's capital, where traders have had to adapt to the unpredictable conditions of sudden violence on the streets.

Executive power rests with a president chosen by a two-thirds majority of the 128-member National Assembly. Its members are directly elected, with an equal division of seats between Christians and Muslims. By convention the president is a Maronite, the speaker a Shi'ite and the prime minister a Sunni.

Armenian immigrants from Turkey, immigrants from other Arab countries and successive waves of Palestinian refugees have made Lebanon the most densely populated country in the Middle East. However, Arabic-speaking Lebanese still form the great majority of the population. Religious allegiances among the Lebanese are highly complex, reflecting the histories of the different communities. Muslims outnumber the Christians.

ECONOMY

Decades of bitter conflict left the once prosperous Lebanese economy in ruins. Once known as the "Switzerland of the Middle East", Lebanon lost most of its international banking business and although this has largely recovered it is still heavily dependent on foreign aid.

Agriculture is mainly confined to the irrigated lowlands along the coast, the Bekáa valley and adjoining slopes. Food production suffered badly during the civil war. Wheat is the most important cereal; livestock includes sheep, goats, poultry and cattle. The main cashcrops are apples, grapes and citrus fruits.

The few exploitable minerals include brown coal (lignite), building stone and high-grade iron ore. A dam on the Litani river supplies water to a major irrigation scheme, and also provides hydroelectric power. Petroleum-fired power stations provide most of the energy for industry and a pipeline is under construction to import natural gas from Syria. Oil refining and food processing are the chief indus-

tries while telecommunications are growing in importance. The tourist industry is slowly reviving.

Most of the country's roads are surfaced but have been poorly maintained. The railroad system is state-run. The principal cargo and passenger port and the airport are both in Beirut.

There is a serious housing shortage. Healthcare facilities, which are concentrated in urban areas. suffered badly during the war. Lebanon has the highest literacy rate in the Arab world. Education is free, and in 1998 the World Bank loaned $60 million for the restructuring of technical and vocational education.

NATIONAL DATA – LEBANON

Land area 10,230 sq km (3,950 sq mi)

Climate		Temperatures		Annual
	Altitude m (ft)	January °C(°F)	July °C(°F)	precipitation mm (in)
Beirut	34 (112)	13 (55)	25 (77)	873 (34.4)

Major physical features highest point: Qurnat as Sawda 3,088 m (10,131 ft); longest river: Litani 145 km (90 mi)

Population (2006 est.) 3,874,050

Form of government republic with one legislative house and no party system

Armed forces army 70,000; navy 1,100; air force 1,000

Largest cities Beirut (capital – 1,295,078); Tripoli (237,909)

Official language Arabic

Ethnic composition Arab 95%; Armenian 4%; other 1%

Religious affiliations Muslim 59.7% (Shi'a, Sunni, Druze, Isma'ilite, Alawite or Nusayri); Christian 39% (Maronite Catholic, Greek Orthodox, Melkite Catholic, Armenian Orthodox, Syrian Catholic, Armenian Catholic, Syrian Orthodox, Roman Catholic, Chaldean, Assyrian, Copt, Protestant); other 1.3%

Currency 1 Lebanese pound (LBP) = 100 piastres

Gross domestic product (2006) U.S. $21.45 billion

Gross domestic product per capita (2006) U.S. $5,500

Life expectancy at birth male 70.41 yr; female 75.48 yr

Major resources limestone, iron ore, salt, water-surplus state in a water-deficit region, grapes, citrus fruits, olives, tobacco, apples, cotton, goats, potatoes, sheep, sugar beet, wheat, hydroelectric power

Israel

STATE OF ISRAEL

T HE SMALL, POPULOUS STATE OF ISRAEL WAS established as a Jewish homeland in Palestine in 1948. Its people, who for many centuries have been scattered all over the world, have historical ties with the region that date back more than 3,000 years. However, the Jews' claim to the land conflicts with that of the Palestinian Arabs, whose historical ties are no less ancient, and a state of hostility between Israel and neighboring Arab countries has existed ever since.

NATIONAL DATA – ISRAEL

Land area 20,330 sq km (7,849 sq mi)

Climate	Altitude m (ft)	Temperatures January °C(°F)	July °C(°F)	Annual precipitation mm (in)
Jerusalem	757 (2,484)	9 (48)	24 (75)	544 (21.4)

Major physical features highest point: Mount Meron 1,125 m (3,962 ft); lowest point: Dead Sea –418 m (–1,371 ft); longest river: Jordan (part) 320 km (200 mi)

Population (2006 est.) 6,352,117

Form of government multiparty republic with one legislative house

Armed forces army 125,000; navy 7,600; air force 35,000

Largest cities Jerusalem (capital – 743,512); Tel Aviv-Jaffa (388,288); Haifa (265,796)

Official languages Hebrew, Arabic

Ethnic composition Jewish 76.4% (Europe/America-born 22.6%; Israel-born 67.1%; Africa-born 5.9%; Asia-born 4.2%); non-Jewish 23.6% (mainly arab)

Religious affiliations Jewish 76.4%; Muslim 16%; Arab Christians 1.7%; other Christian 0.4%; Druze 1.6%; unspecified 3.9%

Currency 1 new Israeli shekel (ILS) = 100 agorot

Gross domestic product (2006) U.S. $166.3 billion

Gross domestic product per capita (2006) U.S. $26,200

Life expectancy at birth male 77.33 yr; female 81.7 yr

Major resources timber, potash, copper ore, natural gas, phosphate rock, magnesium bromide, clays, sand, tourism, citrus fruit, vegetables, cotton, groundnuts (peanuts), diamond processing, crude oil, figs, grapes, livestock, olives, sugar beet, vegetables, wheat

GEOGRAPHY

Israel is bounded by the Mediterranean Sea to the west, Lebanon to the north, Syria to the northeast, Jordan to the east and Egypt to the southwest.

Israel can be divided into four main areas: the fertile densely populated strip of the Mediterranean coastal plains; the central hills extending from Galilee in the north to Judea in the center; the Great Rift Valley running the length of the eastern borders to the Gulf of Aqaba; and the Negev desert in the south. The river Jordan flows south into the Dead Sea.

Winter brings heavy rainfall to the cooler northern highlands, but very little to the south. Summers are hot and dry. Although most of the original evergreen forests have been cleared over the centuries, there has been widespread reforestation, with conifers, eucalyptus and citrus trees. Scrub is the main vegetation, including Mediterranean maquis in the central hills. Several wildlife nature reserves have been established to preserve the enormous variety of wild life.

SOCIETY

The hinterland of Palestine is significant not only for Jews throughout the world, but also for Christians and Muslims; Jerusalem is a holy city for all three faiths.

Modern Israel occupies about three-quarters of the area known as Palestine. The Old Testament of the Bible (the Jewish Torah) describes how nomadic Hebrew peoples (the Jews) escaped Egyptian captivity to settle in the land of the Canaanites (from whom the modern Palestinian Arabs claim descent).

By 40 BC all of Palestine was under Roman domination. Palestine's Jewish population was greatly reduced and scattered. By the time Palestine was conquered

by Muslim Arabs in the 7th century, Jews formed only a small minority in a largely Christian population. The Ottoman Turks captured the region in 1516 and ruled for some 400 years.

During the 19th century Jewish nationalists called for a homeland in Palestine. Ever since the 6th century BC, millions of Jews had left Palestine to settle all over the Middle East, the Mediterranean and Europe. In 1917 the British, who occupied Palestine during World War I, declared support for a national Jewish homeland.

Persecution of European Jews by Nazi Germany between 1933 and 1945 accelerated Jewish migration to Palestine, though this was bitterly opposed by its Arab population. In 1947 the United Nations partitioned Palestine and in 1948 the Jewish state of Israel was declared. Since then there has been almost continuous strife between Israel and its Arab neighbors. Several wars (1948, 1967 and 1973) ended with Israel making territorial gains – the Gaza Strip, the West Bank of the river Jordan (with East Jerusalem) and the Golan Heights on the Syrian border. A peace accord in 1993 led to progressive withdrawal of Israeli troops, and expansion of Palestinian self-rule into occupied West Bank and Gaza. However, a "final

The barren Negev desert (*left*) extends across much of southern Israel, but some parts of it have been made to bloom; water from the huge underground aquifers is used to irrigate crops such as pears and tomatoes.

Hasidic Jews in Jerusalem (*below*) during Sukkot, a festival associated with the Jews' exodus from Egypt. Many holy days in the Jewish calendar are observed as religious and public holidays in Israel.

status" agreement has yet to be reached. Violence continues between Israelis and Palestinians with terrorist attacks occurring daily. Assassinations of Israeli ministers have also come from both sides opposing the peace process – in November 1995 Prime Minister Yishtak Rabin was assassinated by an extreme Jew and in 2001 Tourism Minister Rehavam Zeevi was killed by a Palestinian militant.

The Israeli parliament, or Knesset, is elected every four years by proportional representation. Members of parliament choose the president for a five-year term. Power rests with the cabinet, appointed by the president from the parties in government and chaired by the prime minister.

Jews constitute the majority of the population, and more than half of them are Israeli-born. However, a strong ethnic divide exists between Ashkenazi Jews, originally from Central and Eastern Europe, and Sefardic Jews from Asia and the Mediterranean.

ECONOMY

Israel's economic growth and stability has been greatly helped by financial aid from the United States and support from Jewish communities across the world.

Israel won international admiration for its pioneering collective and cooperative agricultural settlements – *kibbutzim* and *moshavim* – and for their achievements in "making the desert bloom". Productivity has benefited from improved efficiency and mechanization, while irrigation has increased the area of cultivable land. Crops include citrus fruits (a major export), cotton and peanuts. Local fishing is supplemented by Atlantic trawling.

The Dead Sea yields valuable bromides and magnesium salts. Some petroleum is also extracted locally. Food processing is the principal manufacturing activity, but textiles, chemicals and electronics are also important along with armaments. Tourism brings valuable foreign currency.

The road and railroad networks are excellent. There are major ports on the Mediterranean and at Elat on the Gulf of Aqaba and an international airport near Tel Aviv. Newspapers and magazines come in Hebrew, Arabic, Yiddish and many European languages.

The welfare system includes a full range of health and social security services. The state provides free schooling at Hebrew and Arab schools. Many students go on to higher education. Literacy rates and standards of health and housing tend to be higher among the Jewish population.

Jordan

HASHEMITE KINGDOM OF JORDAN

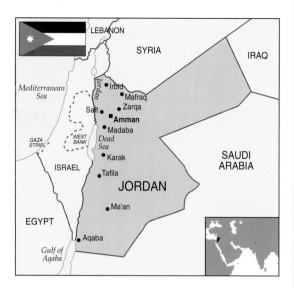

THE SMALL ARAB KINGDOM OF JORDAN shares a frontier with Israel to the west, Syria to the north, Iraq to the northeast and Saudi Arabia to the east and south; a short coastline in the far southwest borders the Gulf of Aqaba, giving access to the Red Sea and beyond. Until the Arab–Israeli war of 1967 Jordan held an area to the west of the river Jordan known as the West Bank, which included the Old City of Jerusalem. The king formally renounced his country's claim to this territory in 1988; while it is still under Israeli occupation, Palestinian self-rule was being progressively introduced in 1995/96.

GEOGRAPHY

Jordan's western frontier with Israel runs along the bottom of the Ghor depression, a deep rift valley that contains the Dead Sea. This large saline lake is the lowest point on the Earth's land surface at –400 m (–1,312 ft); it is fed from the north by the river Jordan and from the south by the Wadi Araba. The valley is overlooked by mountains, beyond which lie the barren plateaus and basins of the great North Arabian desert, which merges in the northeast with the rolling lands of the Syrian Desert.

Jordan is an arid country, though the western highlands receive some rainfall during the cooler winter season. In summer it is hot everywhere except on the highest slopes. Between April and October there is little rain anywhere, and water is often in short supply.

Vegetation ranges from Mediterranean plants in the mountains to sagebrush, grasses and scattered shrubs in the dry steppes. Some hardy plants grow in the eastern deserts, especially in basins watered by seasonal streams from the mountains.

SOCIETY

The part of the Middle East now called Jordan was originally claimed by the ancient Moabites and Edomites (told of in the Old Testament of the Bible) – traditional enemies of their western neighbors, the Hebrews. From 800 BC the region was ruled successively by the Assyrians, the Babylonians, the Persians and the Greek Seleucids. It was conquered by the Romans in 64–63 BC, and by Muslim Arabs in the 7th century AD. Muslims clashed with invading Christian

"The rose-red city" of Petra, cut from sandstone rock, was founded about 1000 BC by the Edomites. One of the best-preserved parts is the Street of Façades, which includes the treasury of Khazneh (shown here); in fact it was probably a temple.

crusaders during the 12th century. From the 16th century the region became an imperial province of the Ottoman Turks.

During World War I an Arab rebellion drove out the Ottoman Turks. In 1921 the British persuaded the Hashemite emir Abdullah ibn-Hussein (1882–1951) to establish an Emirate of Transjordan in part of Palestine. This was proclaimed an independent constitutional state in 1923, but it remained a British protectorate. Full independence was not achieved until 1946, when Abdullah became the first king of the Hashemite Kingdom of Transjordan. He joined forces with other Arab leaders to fight the creation of a Jewish state in neighboring Palestine in 1948. In the following year an armistice was signed with Israel, whereby Jordan (as it then became known) retained part of central Palestine (the West Bank), along with the quarter of Jerusalem known as the Old City.

King Hussein ibn-Talal (1935–1999) inherited the throne in 1952 at the age of 17, after his grandfather Abdullah had been assassinated and his mentally ill father deposed. Hussein's own position remained insecure. Raids on Israel by Palestinians living on Jordanian territory provoked sharp reprisals. In 1967 Jordan again joined Arab neighbors in a war against Israel, but this resulted only in the loss of all Palestinian territories west of the river Jordan. Hussein survived a short but bitter civil war with his Palestinian subjects (including refugees from the West Bank). His relations with other Arab countries were also frequently troubled. He supported Iraq during the Iran–Iraq War (1980–88), but was neutral in the 1991 Gulf War against Iraq, when a huge influx of Palestinian refugees from Kuwait strained the already beleaguered Jordanian economy. A peace treaty was signed between Jordan and Israel in 1994, and Israeli forces withdrew from occupied Jordanian land in the Arava Valley in 1995. Hussein died in 1999 and was succeeded by his son Abdullah ibn al-Hussein.

Jordan is a constitutional monarchy. The prime minister, the cabinet and the 30 members of the upper parliamentary chamber are all appointed by the reigning monarch; the 80 members of the lower house are directly elected by the people.

The vast majority of Jordan's inhabitants are Arabic-speaking Arabs, among whom are an ever-increasing number of Palestinian refugees cared for by United Nations' agencies. Although most Jordanians are Sunni Muslims, there is a significant Christian minority.

ECONOMY

The loss of agricultural land and tourist revenue from the West Bank in 1967 had a significant impact on the economy, and the country is now heavily dependent on foreign grants and loans. Natural resources are scarce and Jordan imports half of its food needs.

Irrigation in parts of the Jordan valley and adjoining mountain valleys has made arable farming possible, but dry farming is also important. Although the spread of agriculture has reduced the pastureland available to livestock, some sheep and goats are kept for meat and milk. There is some fishing in the Gulf of Aqaba. Reforestation in the eastern highlands has increased the area devoted to forestry.

The discovery and exploitation of phosphates and potash reserves have boosted the country's industrial development. Manufactures range from foodstuffs and textiles to phosphatic fertilizers, pharmaceuticals, cement and electrical goods. Electricity is generated by thermal power stations burning oil imported mostly from Iraq. Jordan's ancient archeological sites, notably the ruins of Petra in the southern mountains, attract tourists from all over the world.

There is a well-developed network of surfaced roads, and the Hejaz railroad runs north–south between the Syrian and Saudi borders through the capital, Amman, with a branch line to the port of Aqaba. There are international airports at Amman and Aqaba.

Although the welfare system is generally good, government-funded healthcare is available only in the cities. Housing, food and sanitation are often inadequate, with diseases such as typhoid, hepatitis and dysentery not yet fully under control. Education is compulsory up to the age of 14, whether at state, private or refugee schools. However, literacy rates among women have remained low. Further education is available to a limited number.

Sunlight on the hills of Amman Jordan's capital is a new town, built almost entirely since the 1880s over the ruins of the Roman city of Philadelphia. Its modern infrastructure is now severely strained by the presence of thousands of Palestinian refugees.

NATIONAL DATA - JORDAN

Land area	91,971 sq km (35,510 sq mi)			
Climate	Altitude m (ft)	Temperatures January °C(°F)	July °C(°F)	Annual precipitation mm (in)
Amman	771 (2,529)	8 (46)	25 (77)	273 (10.7)

Major physical features highest point: Jebel Ramm 1,754 m (5,755 ft); lowest point: Dead Sea -418 m (-1,371 ft); longest river: Jordan (part) 320 km (200 mi)

Population (2006 est.) 5,906,760

Form of government multiparty constitutional monarchy with two legislative houses

Armed forces army 85,000; navy 500; air force 15,000

Largest cities Amman (capital - 1,331,028); Zarqa (494,655); Irbid (327,543)

Official language Arabic

Ethnic composition Arab 98%; Circassian 1%; Armenian 1%

Official religion Islam

Religious affiliations Sunni Muslim 92%; Christian 6% (mainly Greek Orthodox, but some Greek and Roman Catholics, Syrian Orthodox, Coptic Orthodox, Armenian Orthodox, and Protestants); other 2% (small Shi'a Muslim and Druze populations)

Currency 1 Jordanian dinar (JOD) = 1,000 fils

Gross domestic product (2006) U.S. $28.89 billion

Gross domestic product per capita (2006) U.S. $4,900

Life expectancy at birth male 75.9 yr; female 81.05 yr

Major resources phosphates, potash, shale oil, olives, citrus fruits, grapes, tourism, olive oil, vegetables, wheat

Iraq

REPUBLIC OF IRAQ

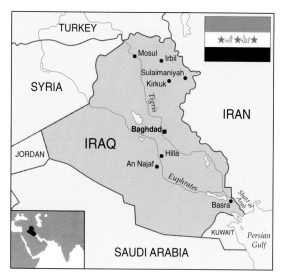

A blue and yellow glazed bull, characteristic decoration of the ruined city of Babylon. Dating from before the 18th century BC, the city was rebuilt by Nebuchadnezzar II in the 6th century BC, and is famous for having been the site of the Hanging Gardens, one of the Seven Wonders of the Ancient World.

I RAQ IS AN ARAB REPUBLIC AT THE NORTH-west end of the Persian Gulf, bordered by Kuwait and Saudi Arabia to the south, Jordan to the west, Syria to the northwest, Turkey to the north and Iran to the east.

GEOGRAPHY

Southwestern Iraq is desert, while the northeast is mountainous. In between is the fertile Tigris–Euphrates river basin; the rivers flow southeast in parallel, meeting to form a vast swamp on their way into the Persian Gulf.

Toward the center of the country the land rises gently through steppeland northwest of Baghdad. In the far north is higher ground, while northeast of the Tigris the land rises to the Hamrin up-lands and high mountains on the Turkish and Iranian frontiers. The western deserts slope gently upward toward Saudi Arabia and Jordan.

Summers are hot and humid in the Tigris–Euphrates basin. The winters – from October to May – are cooler but also humid. Northeastern Iraq receives more rainfall, and seasonal temperatures are generally lower, with snow in winter.

The steppe vegetation of the north and east gives way to thorny tamarisks and salt-resistant plants in the south and west. Willow and poplar trees line the river banks, while sedges and reeds flourish in the Tigris–Euphrates marsh-lands. During the winter migrant birds flock to the swamps and large lakes.

SOCIETY

Mesopotamia, the fertile land between the Tigris and Euphrates, was the cradle of ancient civilization. Agriculture cre-ated the wealth that led to the first city-states in about 3450 BC. The Akkadian ruler Sargon (reigned c. 2334–2279 BC) united these city-states in the first of many empires to dominate the region. Babylon, which defeated its thousand-year-old rival Assyria in 612 BC, was itself conquered by the Persians in 538 BC.

The Persians, in turn, were conquered by Alexander the Great (356–323 BC), and Mesopotamia became part of the Greek Seleucid empire. Roman, and later Per-sian, occupation followed, before the region fell to the Muslim Arabs in the mid 7th century AD. The city of Baghdad quickly became the greatest cultural and

The persecuted Marsh Arabs are seminomadic marsh dwellers who have lived for thousands of years in the marshes of southeastern Iraq, herding water buffalo and hunting wildfowl. Their elaborately structured houses are supported by arches woven from reeds.

commercial center of the Muslim world. Following the Mongol invasion of 1258, Baghdad was destroyed and Mesopot-amia lost much of its importance as a political and commercial center. In 1638 it was taken by the Ottoman Turks.

Following Turkey's defeat in World War I, Britain established an Iraqi mon-archy under British protection. Eleven years later Iraq achieved full indepen-dence under King Faisal I (1885–1933), but pro-German activities led to another short British occupation during World War II. In 1958 the monarchy was overthrown in the first of a series of army coups.

In 1967 Iraq fought against Israel in the Arab–Israeli war. The following year the Baath Party took power and, in 1979, its leader Saddam Hussein (b. 1937) became president. In 1980 Iraq invaded Iran fol-lowing a dispute over control of the Shatt-el-Arab waterway (where the Tigris and Euphrates meet along the Iraq–Iran border), hoping for a quick victory over a country paralyzed by revolution. Instead, the war dragged on until 1989, leaving Iraq nothing to show for heavy casualties and a huge war debt. In 1990 Saddam Hussein ordered the invasion of Kuwait to seize control of its rich oil fields. An international force led by the United States drove the Iraqis out of Kuwait in 1991, crippled Iraq's infrastructure and inflicted a massive defeat on its armed forces. Because of the war and Saddam's continued persecution of the rebellious Kurds in the northeast and Shi'ites in the southeast, the United Nations imposed stringent economic sanctions on Iraq and set up protected zones, including "no-fly-zones". Incursions of these in 1993 led to bombing of strategic targets by the US-led alliance. In 1994 a build-up of Iraqi forces near Kuwait was countered by the

increased presence of American and other forces, and the Iraqis moved away. US troops patrolled the area once more in 1996 following Iraq's installation of anti-aircraft systems in the "No-fly" Zones. In late 1998 American and British planes bombed Iraqi bases after Iraq blocked United Nations weapons inspectors.

Under Iraq's 1968 provisional constitution, power rests with the nine-member Revolutionary Command Council (RCC); one member is elected president. The RCC and a 220-member elected assembly legislate. Saddam Hussein was re-elected president in 1995.

Arab peoples form an overwhelming majority of the population. The largest single minority is the semi-nomadic Kurds. More than 95 percent of the country is Muslim. Shi'ite Muslims (most of them Arabs) now outnumber the Sunnis (including Arabs, and Kurds), but Sunni Muslims continue to wield political power. Some 3.5 percent of the population are Christian.

Baghdad, Iraq's hot, dusty capital, expanded rapidly as soon as the seasonal flooding of the Tigris river was brought under control. Historic mosques and bazaars now stand in the shadow of high-rise apartment blocks.

ECONOMY

Oil is the mainstay of the Iraqi economy although since 1991 sanctions have held back oil sales of around $100 billion. Agriculture has been severely affected by sanctions with output down by up to 50 percent. Cultivated land is concentrated in the Tigris–Euphrates valley, and approximately one-third of the farms are irrigated. Vegetables and cereals are the most important crops, but international sanctions, still in place in 1996, have damaged trade. About one-tenth of the land area provides grazing for livestock. Local waters supply fish for domestic needs.

Apart from petroleum, Iraq's chief mineral resources are phosphates, used in the production of fertilizer. Manufactured goods include building materials, textiles and petrochemical products. Most of the country's power is supplied by hydroelectric stations on the Tigris. A nuclear reactor was destroyed by Israel in 1981 and has not been replaced. Following the Gulf War, the UN placed a total ban on any further nuclear research in the country, although Iraq is suspected abroad of secretly developing nuclear weapons.

The war years inflicted severe damage on the transportation network. Only one-third of roads are surfaced, but the western deserts can support even heavy vehicles. The railroad runs from Baghdad to the southeastern port of Basra, and northwest to Turkey and Syria. Ships operate on the Tigris as far as Baghdad, and some goods are still carried downstream by raft on the Tigris and the Euphrates. Baghdad and Basra both have international airports. Iraq's media are strictly controlled.

War damage and sanctions have adversely affected health. Food and medicine shortages have led to increased mortality, especially among children, and health care facilities are limited. Welfare includes retirement, sickness and unemployment benefits. Children are supposed to receive six years' free compulsory education, but many rural areas lack facilities, and illiteracy is widespread.

Kuwait

STATE OF KUWAIT

KUWAIT IS AN INDEPENDENT ARAB EMIRATE at the head of the Persian Gulf, sandwiched between Saudi Arabia to the south and Iraq to the west and north, with a coast facing east across the Persian Gulf. In 1990 it was occupied by Iraq, regaining its sovereignty only after a war in which both countries suffered considerable economic and environmental damage.

GEOGRAPHY

The landscape consists largely of undulating sandy plains that rise gently to the southwest toward the Saudi frontier. To the northeast there are a few salt marshes. Most of the country is dry, barren desert, except for a few basins that occasionally

NATIONAL DATA - KUWAIT

Land area	17,820 sq km (6,880 sq mi)			

Climate	Altitude m (ft)	Temperatures January °C(°F)	July °C(°F)	Annual precipitation mm (in)
Kuwait City	5 (16)	13 (55)	37 (98)	99 (3.9)

Major physical features	highest point: ash-Shaqaya 290 m (951 ft)

Population	(2006 est.) 2,418,393

Form of government	nonparty constitutional monarchy with one legislative house

Armed forces	army 11,000; navy 2,000; air force 2,500

Capital city	Kuwait City (67,103)

Official language	Arabic

Ethnic composition	Kuwaiti 45%; other Arab 35%; South Asian 9%; Iranian 4%; other 7%

Religious affiliations	Muslim 85% (Sunni 70%; Shi'a 30%); Christian, Hindu, Parsi, and other 15%

Currency	1 Kuwaiti dinar (KD) = 1,000 fils

Gross domestic product	(2006) U.S. $52.17 billion

Gross domestic product per capita	(2006) U.S. $21,600

Life expectancy at birth	male 76.13 yr; female 78.31 yr

Major resources	petroleum, fisheries, shrimps, natural gas, chemical fertilizers, dates, fruits, goats, sheep, vegetables

catch rainwater. The little rain, which falls mainly in winter, brings a welcome coolness after the extreme summer heat.

Vegetation on the coast is limited to salt-tolerant plants, while in urban areas green oases have been created through the irrigation of imported soil. Wildlife was severely depleted as a result of oil pollution during the 1991 Gulf War.

SOCIETY

Kuwait takes its name from the Arabic word *kut*, meaning fort. Its capital city, also called Kuwait, is in a strategic location on the coast that may have been settled in ancient times. The present Arab inhabitants are probably the descendants of nomads who arrived from the nearby desert in the early 18th century seeking permanent settlement. In 1756 the Sabah family established the emirate that rules today. In the late 19th century the country was threatened by the colonial ambitions of both Germany and the Ottoman Turks, and thus formed an alliance with Britain, becoming a British protectorate in 1914. The discovery of oil in the 1930s brought with it unexpected prosperity.

Kuwait became independent in 1961 and was immediately claimed by Iraq. The claim was soon dropped, but was revived in 1990, when the Iraqi leader Saddam Hussein (b. 1937), following a dispute over oil prices, invaded on the pretext of supporting native insurgents. This invasion met with international condemnation. When Iraq refused to withdraw, a United Nations' coalition sent in a military task force consisting primarily of American, British, French and Saudi troops. The Iraqis were expelled in 1991 with heavy casualties, but the damage done to Kuwait was immense. Vast quantities of oil were spilled into the Persian Gulf and hundreds of wells were deliberately set on fire by the Iraqi troops or accidentally by Allied bombing. However, by late 1993 oil production was back to its quota level. Iraq formally recognized the state of Kuwait in November 1994.

After the Iraqi withdrawal, the emir, an absolute monarch under Islamic law, was pressed to initiate democratic reforms. Before the invasion more than half of the country's population consisted of foreign workers denied the same rights as Kuwaitis. Some were killed in the occupation, but many left at the time or were subsequently encouraged to leave by the Kuwaitis. Kuwait's Palestinian population, many of whom had supported Iraq, were distrusted and harshly treated.

ECONOMY

Kuwait's economy had largely recovered from the effects of the Gulf War by the late 1990s. Livestock farming is of some importance, but intensive farming using hothouses and hydroponics (growing plants without soil) provides large crops of melons, tomatoes and dates. Commercial fishing has been important, especially the export of prawns, but catches have been reduced by pollution in the Persian Gulf following the war. Oil accounts for 93 percent of export revenue; chemical fertilizers and natural gas supply most of the remainder. The modern road system was badly damaged in the war but has since been repaired.

Excellent health benefits and free education, which is compulsory to the age of 14 and extends to university level, are available to native Kuwaitis. Press and broadcasting are privately owned but subject to government controls.

The desert city of Kuwait Until the 1960s the city had only one main street and a surrounding mud wall; since then, the country's vast oil revenue has been used to build a modern city with luxurious air conditioned buildings such as the dramatic Kuwait Towers.

Iran

ISLAMIC REPUBLIC OF IRAN

The snowcapped Elburz Mountains of northern Iran run from the capital, Tehran, to the Caspian Sea in Northern Eurasia. The country's highest peak, at 5,671 m (18,606 ft) is the dormant volcano Damavand. Oak and beech trees cover the lower mountain slopes.

I RAN, A MOUNTAINOUS COUNTRY IN SOUTH-western Asia, is one of the largest of The Gulf states. It shares frontiers with 10 neighboring countries. A short coastline in the north borders the Caspian Sea, while the much longer southern coast borders the Persian Gulf. The Iranian plateau was the home of the ancient Persian civilizations, and its cultural heritage goes back 5,000 years.

GEOGRAPHY

Much of Iran consists of a massive semi-arid plateau ringed by mountains. It is subject to frequent earthquakes. Fertile lowland is extremely scarce; the only large area is the Karun river basin in the southwestern province of Khuzistan. The river basin is bounded to the northeast by the Zagros Mountains. East of the Zagros range lies a vast plateau at an average height of 1,200 m (4,000 ft) above sea level, characterized by extensive salt deserts and occasional oases. Its southeastern side is fringed by coastal mountain ranges, while to the east are the peaks of the Baluchistan and Khorasan mountains. The Elburz Mountains of the north and northeast include several active volcanoes. North of the Elburz range a narrow lowland fronts the marshy shore of the Caspian Sea, broadening to the east into the Turkoman steppe.

The climate varies from continental to subtropical. In the north the plains and mountains have rain all year round. By contrast, precipitation in the mountainous west and northwest is confined mainly to winter. The rest of the country is much drier. The southern lowlands have mild winters and hot, humid summers, whereas the interior is hot in summer and bitterly cold in winter.

Plant life thrives on the shores of the Caspian Sea, and the northern slopes of the nearby Elburz Mountains are covered in broadleaf forests; dense forest also covers the western mountains. Elsewhere the vegetation is sparse.

Iran has a rich variety of animal life and many wildlife sanctuaries. The shoreline of the Caspian Sea is a refuge for numerous water birds. Marine life in the Persian Gulf is still recovering from the effects of oil released during The Gulf War in 1991.

SOCIETY

The Iranian plateau has seen the rise and fall of many civilizations: archeological evidence of village life dates back as far as c. 4000 BC. The earliest-known civilization in the area, that of the Elamites in present-day Khuzistan, was founded in the 3rd millennium BC. There followed periods of control by the Medes, Persians, Greeks and Parthians. In 224 AD Ardashir Papakan (d. 241), ruler of the vassal kingdom of Persis (Fars), defeated the Parthians and founded the Sasanian dynasty of Persian rulers.

NATIONAL DATA – IRAN

Land area	1,636,000 sq km (631,663 sq mi)			
Climate	Altitude m (ft)	Temperatures January °C(°F)	July °C(°F)	Annual precipitation mm (in)
Tehran	1,220 (4,002)	3 (37)	30 (86)	230 (9)
Abadan	2 (7)	12 (54)	36 (97)	193 (7.5)

Major physical features highest point: Damavand 5,771 m (18,934 ft); longest river: Karun 850 km (528 mi)

Population (2006 est.) 68,688,433

Form of government theocratic republic

Armed forces army 350,000; Islamic revolutionary guard corps 125,000; navy 18,000; air force 52,000

Largest cities Tehran (capital – 7,185,831); Mashhad (2,463,393); Esfahan (1,600,554); Karaj (1,602,350); Tabriz (1,496,319); Shiraz (1,307,552)

Official language Farsi (Persian)

Ethnic composition Persian 51%; Azeri 24%; Gilaki and Mazandarani 8%; Kurd 7%; Arab 3%; Lur 2%; Baloch 2%; Turkmen 2%; other 1%

Official religion Islam

Religious affiliations Shi'a Muslim 89%; Sunni Muslim 9%; Zoroastrian, Jewish, Christian, and Baha'i 2%

Currency 10 Iranian rial (IRR) = toman

Gross domestic product (2006) U.S. $610.4 billion

Gross domestic product per capita (2006) U.S. $8,900

Life expectancy at birth male 68.86 yr; female 71.74 yr

Major resources petroleum, natural gas, coal, chromium, copper, iron ore, lead, manganese, zinc, sulfur, building materials, cement, fisheries, textiles, timber, cereals, sugar beet, fruits, nuts, cotton

The Sasanian emperors presided over a golden age of Persian culture, until in the mid 7th century they succumbed to an Arab invasion. In the 13th century the Arabs were overrun by the Seljuk Turks, and later by Mongol invaders from the north. The Iranian Safavids achieved supremacy in 1502, adopting the Shi'ite form of Islam, but fell to an Afghan usurper in 1722. During the reign of the Turkmen warrior Nader Shah (1688–1747) the empire expanded into India.

During the Qajar dynasty (1794–1925) Persia was threatened by Russia to the north, the Ottoman Turks to the west and the British in India to the east. In 1906, following a revolution, Persia's first constitution ended the absolute rule of the Qajars. The discovery of oil in 1908 by a British consortium intensified colonial interest, and from then on the country became a semicolonial state dominated by both Great Britain and Russia.

In the chaotic aftermath of World War I an army officer, Reza Khan (1878–1944), overthrew the ruling Qajar dynasty and had himself elected shah in 1925. He adopted the name Reza Shah Pahlavi, changed the country's name from Persia to Iran, and sought to create a modern state free of foreign domination. However, during World War II British and Soviet occupation forces forced him to

Weavers at work Two girls of the Turkmen minority from northeastern Iran weave a carpet, one of Iran's most ancient crafts. Much in demand by affluent foreigners, the rugs are a valuable export. According to tradition a flaw is deliberately woven into each design, because "only Allah is perfect".

abdicate in favor of his son and heir Mohammad Reza Shah Pahlavi (1919–80).

After the war Iran reasserted its independence under Mohammad Mossadeq (1880–1967). In 1951 he nationalized the petroleum industry, forcing the British Oil Company to withdraw. His attempt to depose the shah led to his downfall. In 1961 the shah initiated the so-called "White Revolution", a program of radical economic and social change that caused major unrest. In 1978 antigovernment demonstrations were put down by force, fueling opposition by Islamic fundamentalists who were followers of the exiled religious leader Ayatollah Ruhollah Khomeini (c. 1900–1989). At the start of 1979 the shah was forced into exile, and Khomeini returned in triumph.

Khomeini's call for an Islamic republic was overwhelmingly supported by referendum. That same year Islamic militants stormed the United States' embassy and seized 66 American hostages, most of whom were not released until 1981. The West, supported by conservative Arab states fearing the spread of the Islamic revolution, responded with diplomatic and economic sanctions. Meanwhile, Iraq had invaded the petroleum-producing southwestern province of Khuzistan, starting a bitter and bloody war that lasted until 1988. Islamic fundamentalism continued to be a powerful force in Iran after Khomeini's death in 1989.

Under the 1979 constitution, supreme authority in all matters rests with the Islamic spiritual leader, or *valiy-e faqih*. The president and the 270-member par-

liament are both freely elected for a four-year term. The president appoints a cabinet, subject to the approval of the Majles and the *valiy-e faqih*. Legislation is vetted by the 12-member Council of Guardians to ensure it conforms to the Islamic constitution.

Speakers of Farsi, the official language, account for less than half the population. A few Armenian Christians remain in the cities; they are the only non-Muslim group apart from some Jews and Zoroastrians.

ECONOMY

Iran's economy is almost totally dependent on petroleum, which for many years brought great prosperity. Economic development was disrupted by the Islamic revolution and the Iran–Iraq war and in the 1990s fluctuating oil prices have prevented stable economic progress.

Despite widespread irrigation, good arable land is in short supply. Yields are low, so that much of the country's food has to be imported. Forestry and fisheries are both well developed.

Iran's mineral resources include coal and various metal ores, though only its vast reserves of oil and natural gas have been fully exploited. Iran is self-sufficient in energy. Apart from traditional textiles – Persian carpets and rugs – most of the manufacturing sector has been developed since 1954. Products range from processed food and clothing to petrochemicals, electrical machinery and armaments.

Railroads cater largely for long-distance travel on major routes; most local passenger and freight traffic relies on poorly maintained roads. Regular shipping services operate on Lake Urmia in the northwest and from ports on the Caspian Sea, but the only navigable river is the Karun in the southwest. Bandar Beheshti in the Gulf of Oman is now the chief trading port.

Radio and television are strictly controlled. Broadcasts are in Farsi, several local languages and also some foreign languages. Press freedom is guaranteed under the constitution, but strict censorship is imposed to ensure adherence to Islamic principles. Healthcare is free to the poor, but many rural areas lack medical supplies and trained staff. Some welfare is available from the state.

Education is compulsory and free at elementary level, but illiteracy remains widespread. There are a number of universities, together with teacher-training colleges and technical schools.

Yemen

REPUBLIC OF YEMEN

Brightly painted fishing boats are unloaded at Al Hudaydah, a port on the Red Sea coast of western Yemen. Most of the fresh catch is taken directly to the local fish market and is consumed locally, but some is dried for sale to villages in the interior.

T HE REPUBLIC OF YEMEN WAS CREATED IN 1990 by the unification of the former Yemen Arab Republic (Yemen), whose coast faces west across the Red Sea, and the People's Democratic Republic of Yemen (Southern Yemen), which overlooks the Gulf of Aden to the south.

GEOGRAPHY

The Yemeni landscape is one of contrasts, from the fertile highlands in the west to the arid coast and rocky uplands of the east. The narrow sandy plain along the Gulf of Aden is intersected by wadis – seasonal watercourses that flood in summer, providing water for irrigation.

A hot, dry desert climate prevails except in the western highlands, which have high rainfall during the summer. The Red Sea coast sometimes experiences fierce dust storms. Drought-resistant plants are widespread. More fertile areas support fruit trees such as fig and carob.

SOCIETY

Ancient Yemen was ruled by a number of kingdoms – including the Kingdom of Saba (Sheba), which flourished from about 750 to 115 BC. Yemen submitted to the rule of Muhammad in 631 and the inhabitants began to adopt Islam.

In the 16th century Yemen became part of the Turkish Ottoman empire, but in 1635 the Turks were expelled. The British took the port of Aden in 1839 while the Ottomans gradually occupied northwestern parts of Yemen until 1914, when they were again overthrown, and northwestern Yemen fell to a number of despotic *imams* (Islamic leaders).

Aden became a British crown colony in 1937, and the protectorate later extended to neighboring states to form the Federation of South Arabia. Following independence in 1967, a Marxist republic was established, called the People's Democratic Republic of Yemen.

In northwestern Yemen the last of the despotic imams was deposed in 1962 and the Yemen Arab Republic was proclaimed. Disputes with Southern Yemen during the 1970s led to a war in 1979.

When the war ended in the same year, discussions began between the two states that resulted in their unification in May 1990. Civil war broke out between Northern and Southern forces in May 1994 after years of tension. Peace returned in July when Northern forces were victorious. The new Yemeni constitution provides for a multiparty republic with a president as head of state and a government led by a prime minister.

The vast majority of Yemenis are Muslim Arabs. Three-quarters of the population live in rural areas. The culture is based on traditional Islamic values.

ECONOMY

More than half the workforce is engaged in agriculture. Subsistence farming dominates, using traditional methods. Staples include millet, sorghum, fruit and nuts. Coffee and cotton are cash crops.

Petroleum was disovered in the region during the 1980s, but so far industry is confined largely to Aden, which has petroleum and food-processing plants. There are no railroads and few surfaced roads, most of which are in the west. Press and media are limited and few people have access to telephones.

Welfare and education are in need of much greater investment. Birthrates are high but so too is infant mortality, while life expectancy is low. Education is free, but is only accessible to a limited number of children in the towns and cities.

NATIONAL DATA – YEMEN				
Land area 527,970 sq km (203,850 sq mi)				
Climate		Temperatures		Annual
	Altitude m (ft)	January °C(°F)	July °C(°F)	precipitation mm (in)
Aden	7 (23)	25 (77)	31 (88)	47 (1.8)
Major physical features highest point: Jabal an-Nabi Shu'ayb 3,666 m (12,028 ft)				
Population (2006 est.) 21,456,188				
Form of government multiparty republic with one legislative house				
Armed forces army 60,000; navy 1,700; air force 5,000				
Largest cities San'a (capital – 2,015,515); Tai'zz (729,583); al-Hudaydah (713,458); Aden (646,180)				
Official language Arabic				
Ethnic composition Arab 92%; Afro-Arab 3%; Indian 3%; others 2%				
Religious affiliations Sunni Muslim 53.0%; Shi'ite Muslim 46.9%; others 0.1%				
Currency 1 Yemeni rial (YER) = 100 fils				
Gross domestic product (2006) U.S. $20.38 billion				
Gross domestic product per capita (2006) U.S. $900				
Life expectancy at birth male 60.23 yr; female 64.11 yr				
Major resources petroleum, natural gas, rock salt, marble, coal, gold, lead, nickel, copper, cotton, coffee, hides and skins, fisheries, fruit, millet, sesame, sorghum, sugarcane, wheat				

399

Afghanistan

ISLAMIC STATE OF AFGHANISTAN

A FGHANISTAN IS A LANDLOCKED STATE IN the mountains of south-central Asia, sharing borders with Pakistan to the southeast and Iran to the west. Turkmenistan, Uzbekistan and Tajikistan lie across the northern frontier, and China lies to the northeast at the end of the narrow Wakhan corridor. Afghanistan has been plagued by civil war for decades and has endured a harsh Islamic regime since 1996. In 2001, US-led air attacks in response to Afghanistan sheltering the terrorist Osama bin Laden led to the fall of the Islamic Taliban government. A UN-brokered deal led to an interim council and a peacekeeping force was sent to stabilize Kabul.

NATIONAL DATA - AFGHANISTAN

Land area	647,500 sq km (250,001 sq mi)			
Climate		Temperatures		Annual
	Altitude m (ft)	January °C(°F)	July °C(°F)	precipitation mm (in)
Kabul	1,815 (5,955)	–1 (30)	24 (76)	311 (12.2)

Major physical features highest point: Mount Nowshak 7,485 m (24,557 ft); longest river: Helmand (part) 1,400 km (870 mi)
Population (2006 est.) 31,056,997
Form of government Islamic republic
Armed forces army 27,000
Largest cities Kabul (capital – 3,199,091); Kandahar (411,727); Mazar-i-Sharif (326,737)
Official languages Pashto, Dari Persian
Ethnic composition Pashtun 42%; Tajik 27%; Hazara 9%; Uzbek 9%; Aimak 4%; Turkmen 3%; Baloch 2%; other 4%
Religious affiliations Sunni Muslim 80%; Shi'a Muslim 19%; other 1%
Currency 1 afghani (AFA) = 1 puls
Gross domestic product (2004) U.S. $21.5 billion
Gross domestic product per capita (2004) U.S. $800
Life expectancy at birth male 43.16 yr; female 43.53 yr
Major resources natural gas, petroleum, coal, copper, chromite, talc, barites, sulfur, lead, zinc, iron ore, salt, precious and semiprecious stones, barley, cotton, fruit, goats, maize/corn, nuts, rice, sheep, sugar beet, vegetables, wheat

GEOGRAPHY

Afghanistan is at the heart of southern Asia's great mountain belt. The Hindu Kush and neighboring ranges cover eastern and central Afghanistan, a rugged and inhospitable region prone to earthquakes. Most of the population lives in irrigated valleys on the fringes of the mountains, particularly in the fertile valley of the Kabul river in the southeast, after which the capital is named.

The northern plains adjoining the Amu Darya river are fertile and densely populated. South of the mountains, the deserts and saltflats of the broader southwestern plateau are sparsely watered by rivers flowing out of the Hindu Kush.

The extreme climate of cold winters and hot summers is typical of semiarid steppe country, though there is considerable local variation; the winter months are coldest in the northeastern mountains. The southwest is the driest area, and the eastern valleys are humid.

Vegetation is sparse in the semiarid southwest and in some mountain areas. Elsewhere the mountains are forested with pine, cedar and fir, while walnut and oak grow at lower levels. Many large mammals – including tigers – have all but disappeared, but bears, gazelles and snow leopards still survive. Bird life is plentiful, especially birds of prey.

SOCIETY

Afghanistan has long been disputed by foreign powers anxious to control its strategic position astride the great land route to India via the Khyber Pass. The present boundaries are relatively recent.

From the 6th century BC Afghanistan was part of the great Persian empire. After its conquest by Alexander the Great (356–323 BC) in 328 BC, the region was the center of the Greek kingdom of Bactria, which saw a remarkable fusion of Greek, Hindu and Buddhist cultural influences. After 135 BC the region fell to a succession of empires – Parthian, Kushan and Sasanian Persian – before being conquered by the Arabs in the 7th century AD. The Arabs introduced Islam, which by the 9th century had become the dominant cultural force in Afghanistan. The Mongols under Genghis Khan (c. 1162–1227) overran the country in the 13th century, and his descendant Babur (1483– 1530) later made Kabul the headquarters for the Mughal invasion and conquest of India.

When the Persian monarch was assas-

sinated in 1747, the chief of his Afghan bodyguard, Ahmad Khan Abdali (1722– 72), was elected shah by Afghan chieftains, and became known as Ahmad Shah Durrani. Afghanistan became an independent state, though it continued to be disturbed by factional rebellions and conflicting British and Russian territorial ambitions. In 1880, following the second Anglo-Afghan war, new boundaries made Afghanistan a buffer state between British India and Russia. After the third Anglo-Afghan war in 1919, Afghanistan gained full independence, and formed a special relationship with the new Soviet regime that was to last until 1979.

Later Afghan rulers attempted to bring stability to the fragile monarchy through economic and constitutional reforms, but with little success. Unrest increased after a severe famine in 1972; the shah was deposed the following year, and the coun-

A Kyrghyz seamstress (*above*) with her sewing machine. The Kyrghyz are a small group of Turkic speakers in the extreme northeast of Afghanistan. There is great ethnic diversity among the isolated, often feuding, communities each with their different culture.

Liberation of Kabul On November 13 2001, the Northern Alliance entered Kabul greeted by cheering crowds. Soon after, the strict, Islamic Taliban regime collapsed and Afghanis were free to sell music and men were able to visit barber shops to shave off their beards.

try was declared a republic. In a further coup in 1978, the president was killed and power passed to the Soviet-backed People's Democratic Party of Afghanistan (PDPA). The new regime was violently opposed by Afghan Islamic groups, and in December 1979 Soviet forces invaded Afghanistan to install a new PDPA government. Fighting between Soviet troops and Afghan rebels continued for nine years.

The last of the Soviet troops were withdrawn in February 1989, and a military council was established under President Mohammad Najibullah (b. 1936). Peace proved elusive as many different rebel factions continued to wage war. During late 1996 the fundamentalist Taliban militia emerged, bringing a measure of stability to the country. However, the Taliban were internationally criticised over their harsh implementation of Islamic law. In 2001, the Taliban were accused of harboring the al-Qaeda terrorist leader Osama bin Laden who was blamed for the horrific September 11 terrorist attacks on the United States. In October the US initiated air attacks on Afghanistan hoping that bin Laden would be flushed out. The attacks allowed opposition groups such as the Northern Alliance to drive the Taliban from power but bin Laden remained elusive. An interim power-sharing council was set up until an elected representative government and new constitution could be adopted. An international peacekeeping force was also set up to guarantee security for the Afghan capital, Kabul.

The ethnic diversity of the Afghan people reflects a history of conquest and migration. The Uzbeks and Turkmens of the far northwest retain their Turkic culture and language.

ECONOMY

Afghanistan's economic development has been severely retarded by war and political instability, and the country remains one of the poorest in the world, with agriculture the mainstay of the economy. Although irrigation projects increased the amount of cultivable land, much of this has been lost through war. The principal crop is wheat, but rice, barley, fruit and cotton are also grown.

The most vital mineral resource is natural gas. There are also sizable deposits of iron and other metal ores, but their potential, like that of hydroelectricity, has not yet been fully exploited. The main industrial activity is the manufacture of woolen and cotton textiles; traditional handicrafts and woven carpets are important exports.

With no railroads and few navigable rivers, transportation is normally by road. There are domestic air services to most provinces, and international airports at both Kabul and Kandahar.

Education, healthcare and welfare provision are all inadequate, especially in rural areas; nearly all healthcare provision depends on foreign aid. Literacy levels are low, particularly among women.

Saudi Arabia

KINGDOM OF SAUDI ARABIA

SAUDI ARABIA, NAMED AFTER THE RULING royal dynasty of Sa'ud, occupies most of the Arabian Peninsula. This desert kingdom, rich in petroleum, borders Jordan, Iraq and Kuwait to the north and Bahrain and Qatar along The Gulf coast. Southeast are the United Arab Emirates and Oman, and to the south is Yemen.

GEOGRAPHY

The Arabian Peninsula is largely made up of a vast plateau. To the west the plateau forms a steep escarpment running parallel to the Red Sea coast; in the northwest there is virtually no coastal plain. The highest mountain peaks are in the Asir highlands in the southwest.

East of the escarpment, the barren plateau drops gently to the shallow waters of The Gulf, whose shores are lined with swamps and salt flats. The plateau is mostly sandy desert, with some areas of bare volcanic rock. A vast swathe of desert known as the Empty Quarter occupies the whole of the south.

Saudi Arabia has no permanent rivers. The Empty Quarter can go without rain for 10 years at a time. Other regions experience the occasional downpour, especially near The Gulf coast in the spring, and in the summer in the Asir highlands. Inland the winters are cool, often with frost at night, but in summer temperatures in excess of 48°C (120°F) are not unusual.

In most parts of the country vegetation is limited to small herbs and shrubs, with date palms in scattered oases. The Arabian oryx has been hunted almost to extinction, and the Arabian tahr (a goat-like antelope) and some birds of prey are now endangered. But wild cats, desert-dwelling sandgrouse, burrowing rodents such as jirds and jerboas, and diverse reptiles and insects are common. The coastal waters of the Red Sea teem with life, particularly among the coral reefs.

SOCIETY

In 622 AD Muhammad (c. 570–632) – a merchant from the caravan city of Mecca who believed himself to be divinely inspired – traveled north to Medina and founded a new religion: Islam. He returned with an army of converts to capture Mecca, and by the time of his death all Arabia was under the sway of Islam. In 661, however, the center of Islamic power was transferred northward to newly conquered lands. Soon Arab unity was broken, and for many centuries Arabia became a battleground for rival rulers, until the Ottoman Turks added it to their growing empire in 1517.

In the 18th century a Muslim scholar, Muhammad ibn 'Abdal-Wahhab (1703–92), founded the Islamic fundamentalist Wahhabi movement. His ultrapuritanical teachings were spread by the Sa'ud princes of Najd as far as Yemen. During the 19th century the area they controlled was greatly reduced, but in 1904 Ibn Sa'ud (c. 1880–1953) restored Saudi rule in Najd, and in 1924 he invaded the British protectorate of Hejaz in the west, fulfilling his ambition to take Mecca and Medina. Three years later the British

The tall, narrow houses of Al Kana, a small oasis village in southern Saudi Arabia, show a strong Yemeni influence in both their design and painted decoration. On the flat rooftops vegetables are dried, bread is baked and handwoven rugs are cleaned.

The changing role of women (*above*) Formal education for women in Saudi Arabia was introduced as part of the progressive reform program of King Faisal (1906–75), who came to power in 1964. Veils are not worn in the state-run girls' schools.

Riyadh airport (*right*), built at great expense for the burgeoning capital – the seat of the royal family, and until recently a small oasis town. The airport is a symbol of Saudi Arabia's new-found wealth and its entry into the international arena.

acknowledged his kingship over the regions of Najd and Hejaz, which he formally united in 1932 to form the Kingdom of Saudi Arabia.

The country remained neutral in World War II, and played little part in the postwar affairs of other Arab states. But when Iraq invaded neighboring Kuwait in 1990, the Saudi king invited Western and Arab forces to support his army along the northern frontier. Victory in The Gulf War of 1991 strengthened relations between Saudi Arabia and its Arab allies. The war made little impact on the prosperous petroleum-based Saudi economy

despite the damage caused by pollution and Iraqi missiles.

The Saudi king, whose official title is Custodian of the Two Holy Mosques (at Mecca and Medina), rules as absolute monarch. His heir and deputy, the crown prince, is selected from among the Sa'ud family by its leading members in consultation with the *ulama*, or supreme religious council. The king governs according to Islamic law, choosing a council of ministers, many of whom are Sa'uds.

The population is predominantly Arab, including many immigrants from other Arab states, especially Yemen. There is a large transient population made up of pilgrims from other Muslim countries visiting Mecca, and also large numbers of foreign industrial workers. Most Saudis are Sunni Muslims – chiefly the Wahabi interpretation – apart from some Shi'ites in areas bordering the Persian Gulf. Islam is strictly enforced (though not, critics say, in some areas concerning human rights), affecting every aspect of Saudi life from its laws to its culture and its visual arts. It is forbidden to practice any non-Muslim religion openly.

ECONOMY

The Saudi economy is based on petroleum exploitation. Revenue from this finances heavy spending elsewhere.

High government subsidies have helped produce a surplus of wheat for export. But two-thirds of all food still has to be imported. Apart from a tiny area of fertile farmland in the mountains of Asir, agriculture is dependent on old and new irrigation techniques. Major crops include cereals, bananas, dates and coffee. Sheep, goats and camels are herded by the country's few remaining nomads.

Saudi Arabia possesses one-quarter of the world's oil reserves as well as significant quantities of natural gas and various metal ores. Manufacturing industry, powered by electricity from oil-fired stations, produces petrochemicals, fertilizers and rolled steel.

The Saudi road network connects all main centers of population, with further links to Yemen, Jordan, Kuwait and Qatar. A railroad runs northeast from the capital, Riyadh, to the major Gulf port of Dammam. Pipelines from the nearby port of Jubail carry oil and natural gas across the country to Yanbu on the Red Sea coast, west of Medina. Jeddah, another major port on the Red Sea, near Mecca, has an international airport, as do Riyadh and Dhahran (next to Dammam).

Wide-ranging welfare services include free medical care for all citizens and visitors. Healthcare is excellent in the cities, but in rural areas diseases such as malaria, trachoma and bilharzia are all common. Education is free but not compulsory, and literacy levels are generally very low. However, students graduating from secondary school have a wide choice of university education, or can opt for technical or vocational training.

NATIONAL DATA – SAUDI ARABIA

Land area	2,149,690 sq km (830,000 sq mi)			
Climate	Altitude m (ft)	Temperatures January °C(°F)	July °C(°F)	Annual precipitation mm (in)
Riyadh	609 (1,998)	14 (58)	35 (95)	113 (4.4)

Major physical features highest point: Jabal Sawda 3,133 m (10,279 ft)

Population (2006 est.) 27,019,731

Form of government absolute monarchy

Armed forces army 75,000; navy 15,500; air force 18,000

Largest cities Riyadh (capital – 4,453,447); Jedda (3,002,839); Mecca (1,383,707)

Official language Arabic

Ethnic composition Saudi 82.0%; Yemeni 9.6%; other Arabs 3.4%; others 5.0%

Religious affiliations Muslim 96.7%; Christian 2.9%; others 0.4%

Currency 1 Saudi riyal (SAR) = 100 hallalah

Gross domestic product (2006) U.S. $374 billion

Gross domestic product per capita (2006) U.S. $13,800

Life expectancy at birth male 73.66 yr; female 77.78 yr

Major resources petroleum, natural gas, iron ore, gold, copper, wheat, dates, bananas, coffee, alfalfa, cattle, goats, camels, grapes, poultry, sheep, sorghum, watermelons

Bahrain

STATE OF BAHRAIN

THE TINY INDEPENDENT MONARCHY OF Bahrain consists of a group of some 30 islands in the Gulf of Bahrain, a small branch of the Persian Gulf between Saudi Arabia and the peninsula of Qatar.

GEOGRAPHY

The main island of Bahrain is joined by causeways to the neighboring islands of Muharraq and Sitrah, and to Saudi Arabia. Its low-lying landscape ranges from barren rock to sandy plains and salt marshes. A little rain falls mainly during the warm winters; the hot summers are also humid. Underground springs and

The King Fahd causeway, 25 km (16 mi) long, links the tiny island nation of Bahrain to Saudi Arabia, with which it maintains close relations. To offset its dependence on oil revenue, it also has a major industrial complex and shipbuilding and repair yard.

desalination plants provide the water needed for irrigation. Tough desert and saltmarsh plants thrive on the island.

SOCIETY

Bahrain has been a flourishing trading center for over 4,000 years. Since the 7th century AD it has been an Arab Muslim country. From 1521 until 1783 it was ruled by the Portuguese and then the Persians. Bahrain became a British protectorate in the mid 19th century, and achieved full independence in 1971. The emir presides as head of state and prime minister over an appointed cabinet. Before 1975 there was an elected National Assembly, but this was dissolved following tension between Shi'ite and Sunni Muslims. There were Sh'ite protests in 1994/95 demanding its reinstatement.

There are large numbers of foreign workers, mainly from Iran and southern Asia. Shi'ite Muslims are in the majority, but the ruling family and many of the richest Bahrainis are Sunnis. The culture remains traditionally Arab, despite the predominantly urban lifestyle and widespread Western television and cinema.

ECONOMY

Bahrain's prosperity is based on natural gas and oil, but with reserves dwindling, aluminum processing and financial services have been growing in importance. Agricultural production meets domestic

demand for both fruit and vegetables.

Transportation and telecommunications are highly developed. The principal port is close to the capital, Al Manamah, and there is an international airport on Muharraq. Oil revenue has enabled Bahrain to offer free medical care and schooling up to secondary level. Tertiary-level education is also available.

NATIONAL DATA – BAHRAIN

Land area	665 sq km (257 sq mi)			

Climate		Temperatures		Annual
	Altitude m (ft)	January °C(°F)	July °C(°F)	precipitation mm (in)
Al Manamah	6 (18)	17 (63)	34 (93)	70 (2.7)

Major physical features	largest island: Bahrain 583 sq km (225 sq mi); highest point: Durkan Hill 134 m (440 ft)

Population	(2006 est.) 698,585

Form of government	constitutional hereditary monarchy with one appointed council

Armed forces	army 8,500; navy 1,200; air force 1,500

Capital city	Al Manamah (148,622)

Official language	Arabic

Ethnic composition	Bahraini 62.4%; non-Bahraini 37.6%

Official religion	Islam

Religious affiliations	Muslim (Shi'a and Sunni) 81.2%; Christian 9%; other 9.8%

Currency	1 Bahraini dinar (BHD) = 1,000 fils

Gross domestic product	(2006) U.S. $17.7 billion

Gross domestic product per capita	(2006) U.S. $25,300

Life expectancy at birth	male 71.97 yr; female 77 yr

Major resources	oil, natural gas, aluminum processing, fisheries, pearls, dates, eggs, livestock, lucerne, vegetables

Qatar

STATE OF QATAR

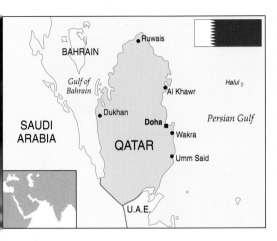

THE SMALL EMIRATE OF QATAR OCCUPIES THE peninsula of the same name on the southwestern coast of the Persian Gulf. It shares borders with Saudi Arabia to the southwest and the United Arab Emirates to the southeast. The island state of Bahrain lies immediately to the west.

GEOGRAPHY

The long Qatar peninsula is mainly flat except for a few hills in the west and some low cliffs in the far northeast. Rainfall is scarce, apart from rare showers during the winter. Summers are generally hot and humid, but winter nights are often chilly. Vegetation, restricted to plants that can survive prolonged drought, is most abundant in the north. Farther south salt-flats and sand dunes predominate. The wildlife includes desert creatures such as gerbils and other sand-burrowing rodents, reptiles and insects.

SOCIETY

Qatar, one of many emirates established in Arabia during the spread of Islam in the 7th century AD, was ruled for many centuries by the Khalifah dynasty. However, the transfer of the emir's headquarters to nearby Bahrain in 1783 led to a rift with his Qatari subjects, and eventually to war in 1867. The British intervened, establishing a separate emirate under a member of the leading Qatari family, al-Thani. In 1916 Qatar became a British protectorate under the terms of a treaty. The discovery of oil in 1939 changed the emirate's fortunes. Qatar chose not to join the United Arab Emirates after the withdrawal of British troops in 1971, and opted for complete independence. In 1995 the emir was deposed by his son, Sheikh Hamed ibn Kalifa ath-Thani, in a bloodless coup.

The emir of Qatar rules as absolute monarch, though his authority depends on the cooperation of the ruling family, whose members make up two-fifths of the indigenous Qatari Arab population. Islam is the official religion though religious law is enforced only within the indigenous community.

ECONOMY

Revenue from the limited oil reserves is increasingly being supplemented by exploitation of natural gas. The country's industries, which are highly dependent on these primary resources, include oil and gas refining and petrochemical, steel and fertilizer production. The use of irrigation and high-technology allow fruit and vegetables to be grown, and the local fishing industry is also being modernized. However, most foodstuffs still have to be imported. Qatar is also a major banking center.

The road network in Qatar is excellent, and there are many sea and air services. Welfare and education are provided for all residents, and university education is also available to a limited number.

Flames from Qatar's oil wells light up the sky The small peninsula has both onshore and – increasingly important – offshore reserves. Since the discovery of oil in 1939, Qatar, formerly one of the poorest countries in the world, has become one of the richest.

NATIONAL DATA – QATAR

Land area	11,437 sq km (4,416 sq mi)			
Climate		Temperatures	Annual	
	Altitude m (ft)	January °C(°F)	July °C(°F)	precipitation mm (in)
Doha	11 (36)	17 (63)	35 (94)	81 (3.2)

Population	(2006 est.) 885,359
Form of government	absolute monarchy
Armed forces	army 8,500; navy 1,800; air force 2,100
Capital city	Doha (358,098)
Official language	Arabic
Ethnic composition	Arab 40%; Indian 18%; Pakistani 18%; Iranian 10%; other 14%
Official religion	Islam
Religious affiliations	Muslim 92.4%; Christian 5.9%; Hindu 1.1%; others 0.6%
Currency	1 Qatari rial (QAR) = 100 dirhams
Gross domestic product	(2006) U.S. $26.05 billion
Gross domestic product per capita	(2006) U.S. $29,400
Life expectancy at birth	male 71.37 yr; female 76.57 yr
Major resources	petroleum, natural gas, fish, steel and cement, petrochemicals, fertilizer, fodder, fruit, livestock, vegetables

United Arab Emirates

THE UNITED ARAB EMIRATES (UAE), FORMERLY the Trucial States, are a federation of seven states ruled by emirs on the southern coast of the Persian Gulf. Qatar is to the northwest of the UAE, Saudi Arabia to the southwest and Oman to the east. From west to east the emirates are Abu Dhabi (by far the largest), Dubai, Sharjah, Ajman, Umm al-Qaiwain, Ras al-Khaimah and Fujairah. They all face north across the Persian Gulf except for Fujairah, which fronts the Gulf of Oman on the east coast of the Musandam peninsula.

NATIONAL DATA – UNITED ARAB EMIRATES

Land area	83,600 sq km (32,280 sq mi)			
Climate		Temperatures		Annual
	Altitude m (ft)	January °C(°F)	July °C(°F)	precipitation mm (in)
Abu Dhabi	5 (16)	18 (64)	35 (95)	798 (88.9)

Major physical features	highest point: Jabal Hafib 1,189 m (3,901 ft)

Population	(2006 est.) 2,602,713

Form of government	multiparty republic with one appointed council

Armed forces	army 44,000; navy 2,500; air force 4,800

Largest cities	Dubai (1,225,137); Abu Dhabi (capital - 633,136); Sharjah (584,286)

Official language	Arabic

Ethnic composition	Emirati 19%; other Arab and Iranian 23%; South Asian 50%; others 8%

Official religion	Islam

Religious affiliations	Sunni Muslim 80%; Shi'ite Muslim 16%; Christian 3.8%; others 0.2%

Currency	1 Emirati dirham (AED) = 100 fils

Gross domestic product	(2006) U.S. $129.4 billion

Gross domestic product per capita	(2006) U.S. $49,700

Life expectancy at birth	male 72.92 yr; female 78.08 yr

Major resources	petroleum, natural gas, camels, cattle, fish, fruit, vegetables, goats, sheep, tourism

GEOGRAPHY

The landscape is generally sandy and low-lying except for coastal salt flats and a range of steep mountains in the east – an extension of Oman's Hajar Mountains that runs northward along the Musandam peninsula. The climate is hot and dry all year round, though more humid near the coast; strong seasonal winds bring fierce dust storms that can cause considerable damage. Apart from date palms in oases in the northeastern emirates, vegetation is generally sparse and scrubby; the wildlife includes many sand-dwelling animals such as jerboas.

SOCIETY

The Muslim emirates that make up the federation were probably founded between the 7th and 8th centuries AD, reflecting local divisions among the different Arab peoples. From the early 17th century onward the British were the most influential Europeans in the area. They established a military base to suppress piracy in the early 19th century. A treaty in 1892 gave Britain control over foreign policy in what became known as the Trucial States, but each emirate maintained its traditional local rule.

In the 20th century the tapping of vast petroleum reserves made the emirates extremely rich. They began to cooperate more closely and in 1971, when British forces left the Persian Gulf area, most chose to join the UAE, though Bahrain and Qatar did not. The union maintained close links with both Britain and neighboring Arab states, with whom they sided in the 1991 Gulf War against Iraq.

The indigenous Arab population is today outnumbered by immigrant workers, mainly from southern Asia and neighboring Arab states. Islam, including both Sunni and Shi'ite sects, is the official religion, and forms the basis of law. Each emirate is an absolute monarchy, whose ruler is a member of the Supreme Council. The emirs elect a federal president and vice-president from among themselves, in addition to appointing a 40-member advisory council.

ECONOMY

The wealth from petroleum reserves has made this otherwise unproductive land one of the most prosperous countries in the world. Only a tiny proportion of the land area is suitable for agriculture: either

Abu Dhabi, largest of the 7 emirates and capital of the UAE, is a thriving modern city of wide boulevards, high-rise buildings and extensive public services. Its wealth is founded on oil, but other traditional industries include fishing and pearl harvesting.

around the oases or where irrigation has been possible. Crops include dates and alfalfa, along with other fruit varieties and cereals. Some crops are grown using high-technology methods such as hydroponics (without soil). Fish are plentiful in neighboring waters, but livestock has been limited mainly to goats and camels because of widespread overgrazing.

Most food has to be imported, but the cost is small in comparison to huge revenues from the oil and gas industries. Aluminum smelting, trade and financial services are also important. Economic development has been further assisted by good road and shipping facilities.

Social services, generously funded by petroleum revenues, are the main source of employment for local people. Healthcare and education to university level are both provided free of charge. However, these developments are relatively recent, and many foreign teachers and doctors are employed.

Oman

SULTANATE OF OMAN

OCCUPYING THE SOUTHEASTERN PORTION OF the Arabian Peninsula, Oman looks southeast across the Arabian Sea and northeast across the Gulf of Oman. It shares borders with Yemen, Saudi Arabia and the United Arab Emirates. The small Omani enclave of Ru'usal-Jibal, at the tip of the Musandam peninsula, guards the entrance to the Persian Gulf.

GEOGRAPHY

In ancient times this part of the Arabian Peninsula was green and fertile, but major climatic changes then turned the land to desert. The only highland areas are the limestone Hajar Mountains in the north, overlooking a fertile coastal plain and the Gulf of Oman, and a lower range in the southern province of Dhofar. In the mountains, especially in Dhofar, summer monsoons bring cooling mist, and in the desert rare but torrential rainstorms fill the dry riverbeds or wadis, causing occasional flash floods.

SOCIETY

Muscat, on the Gulf of Oman, was a flourishing trade center as early as the 5th century BC. From the 2nd century AD the land was overrun by warring Arab peoples. With the spread of Islam in the 7th century, power passed to the leaders of the Islamic faith, the imams. After a royal dynasty was established in 1154, Oman grew into a maritime power. In 1507 the Portuguese sacked the capital, Muscat, but in 1650 the imams regained power, spreading their maritime empire. A sultanate established in 1744 ruled from Zanzibar (an Omani conquest) until 1861, when the monarchy split.

From 1798 onwards Oman established close treaty ties with the British, and it retained strong links with Britain even after it ceased to be a British protectorate in 1951. In 1970 Sultan Qabus Ibn Sa'id (b. 1940) deposed his father. The country gained full independence in 1971, and Qabus began to introduce liberal reforms. In 1991 a new consultative assembly was formed consisting of a nominated president and 59 nominated representatives.

Omani Arabs – mostly Sunni Muslims – form the majority of the population. There are also large numbers of foreign workers from Africa and Asia.

ECONOMY

Oman's petroleum reserves have been very important in building up the economy. The many associated manufacturing industries burgeoned during the 1970s.

Fishing and subsistence agriculture (supported by irrigation wells) provide the main source of revenue for local people. Crops include wheat and alfalfa, along with fruits such as dates, limes and coconuts. Goats are the main livestock, with sheep and cattle in irrigated areas.

Since 1970 transportation has been improved and welfare and healthcare provision extended. Education has been expanded at all levels and remains strongly bound to Islamic scholarship.

NATIONAL DATA – OMAN

Land area	212,460 sq km (82,031 sq mi)			
Climate		Temperatures		Annual
	Altitude m (ft)	January °C(°F)	July °C(°F)	precipitation mm (in)
Muscat	5 (16)	22 (72)	34 (93)	98 (3.8)

Major physical features highest point: Jabal Akhdar 3,074 m (10,086 ft)

Population (2006 est.) 3,102,229

Form of government monarchy with one appointed council

Armed forces army 25,000; navy 4,200; air force 4,100

Capital city Muscat (252,864)

Ethnic composition Omani Arab 73%; Indian 13%; Pakistani 7%; Egyptian 2%; others 5%

Official religion Islam

Religious affiliations Muslim 88%; Hindu 7%; Christian 4%; others 1%

Currency 1 Omani rial (OMR) = 1,000 baiza

Gross domestic product (2006) U.S. $43.88 billion

Gross domestic product per capita (2006) U.S. $14,100

Life expectancy at birth male 71.14 yr; female 75.72 yr

Major resources petroleum, copper, asbestos, marble, limestone, chromium, gypsum, natural gas, fisheries, alfalfa, bananas, coconuts, dates, fruit, vegetables, wheat, tourism

Goats for sale in the livestock market at Nazwa in the northern interior of Oman. Monsoon rains in the summer create plentiful pasture for livestock, and most families in rural areas keep goats and cows. Half the population lives by subsistence agriculture.

The Indian Subcontinent

COUNTRIES IN THE REGION
INDIA · BHUTAN · PAKISTAN · BANGLADESH · NEPAL
MALDIVES · SRI LANKA

DEPENDENCIES IN THE REGION
BRITISH INDIAN OCEAN TERRITORY

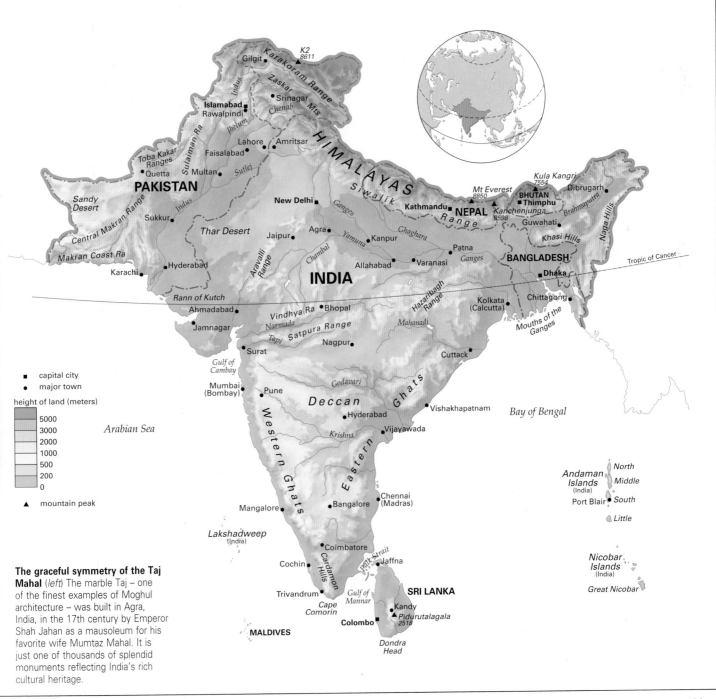

capital city
major town

height of land (meters)
5000
3000
2000
1000
500
200
0

mountain peak

The graceful symmetry of the Taj Mahal (*left*) The marble Taj – one of the finest examples of Moghul architecture – was built in Agra, India, in the 17th century by Emperor Shah Jahan as a mausoleum for his favorite wife Mumtaz Mahal. It is just one of thousands of splendid monuments reflecting India's rich cultural heritage.

India

REPUBLIC OF INDIA

THE INDIAN SUBCONTINENT, A LARGE PENIN-sula in southern Asia, covers an area about one-third the size of Europe. To the north the massive mountain ranges of the Himalayas form a barrier to the rest of Asia. To the west India is bounded by the Arabian Sea; the Indian Ocean is to the south and the Bay of Bengal to the east. India boasts one of the world's oldest civilizations (the Aryans and the Dravidians), whose history has been recorded as far back as the 15th century BC.

The modern republic of India occupies the greater part of the Indian subcontinent, bordering Pakistan to the west, Tibet (now occupied by China), Nepal and Bhutan to the north, Bangladesh and Burma to the east.

GEOGRAPHY

The varied landscapes of India are a prime example of plate tectonics (the movements of continental land masses). The Himalayas – the highest mountains in the world – and the great trench to the south of them were created by the collision of two ancient continental "plates".

The land
The Himalayas descend southward to a fertile alluvial plain that crosses the country in a broad band from east to west. The most northerly band of the Himalayas runs across northern Kashmir and deep into neighboring Tibet. It is the source of the great Indus and Brahmaputra rivers. Farther south lie the Greater Himalayas, whose mountains are so high that they are permanently covered in snow and ice.

Still farther south the Lesser Himalayas give way to the lower sandstone ranges of the Siwalik Hills, fringing India's northern plains. These were formed from a trench, originally up to 2,000 m (6,500 ft) deep, that is now largely filled by stones and alluvium washed downstream from the mountains by rivers. Of the main rivers, the Indus runs southwest into Pakistan, while the Ganges flows southeast along the plains toward the Bay of Bengal. India's other great river, the Brahmaputra, flows southwest to meet

Dwarfed by the mountains of Ladakh, part of the mighty Himalayan chain that extends across northern India, Thikse monastery perches on a barren hill. The high fertile plains and deep mountain valleys of Ladakh provide some of the most dramatic scenery in India.

the Ganges, creating a massive delta in Bangladesh and West Bengal. In the south lies the massive Deccan plateau, which contains some of the world's oldest rocks.

Climate
From December to March, cold winds blowing southward out of the Himalayas

produce cool, dry winter conditions in all regions except the far south and west, which remain hot and humid. Spring, in April and May, brings a change in the prevailing winds, drawing in warmer air from the south. Much of India has little or no rain during this, the hottest time of the year. The summer monsoon rains come as a welcome relief. Most of the country receives at least some rain, but if the monsoon (usually between June and September) fails, drought and famine are the result. The highest rainfall levels are recorded in the northeast and along the southwest coast. In October and November the winds turn once again, sometimes bringing with them violent tropical cyclones.

Plants and animals

Today less than one-fifth of India is forested. Most river deltas have mangrove forests along their seaward edges. The Sundarbans, at the mouth of the Ganges, is one of the world's largest mangrove swamps, covering 4,000 sq km (1,540 sq mi). Dry tropical forest is found on higher ground with low rainfall. Assam in the northeast and the Western Ghats along the southwest coast have moist tropical rainforest. In the wetter parts of the eastern Himalayas, evergreen oak and chestnut predominate, and on higher ground rhododendrons.

India has a remarkably rich variety of wildlife. But as forests are reduced, animal habitats are destroyed. Protected animals include the Asian elephant, the Indian rhinoceros, the gaur (a kind of wild ox), the Asiatic lion, tigers, snow leopards, musk deer, the great Indian bustard and the gharial, a large crocodile.

A caravan at rest in the Thar Desert, northwestern India, the eastern extension of the world's largest arid region that encompasses the Sahara and the Arabian desert. Covering some 259,000 sq km (100,000 sq mi), the Thar is home to nomadic cattle herders.

Lakshadweep – land of the giants

Some 300 km (200 mi) off the Malabar coast of southwestern India are the Lakshadweep, the "hundred thousand islands" (also known as the Laccadives). In fact there are only 14 islands, of which 10 are inhabited, the rest of the chain comprising coral reefs. The islands do have a mystery though.

A meter or so (a few feet) beneath the light coral sand there should be a layer of coral some 40 cm (16 in) thick, and yet on all the main islands this coral stratum, at some point in the remote past, was completely dug away. This was a massive task for an area measuring as much as 30 sq km (12 sq mi). As a result, the underlying sands were exposed, creating the ideal growing medium for crops. Local tradition says that the excavation work was done by giants; the truth may never be known.

The people of Lakshadweep are Moplahs, of mixed Indian and Arab descent who are Muslims but speak the Malayalam language of neighboring southwest India. From 1498 the islands were in Portuguese hands, but the Moplahs shook off their rule in 1545 and gave their allegiance to the raja of Cannanore on the Malabar coast. After 1792 the southern group of islands won self-government in return for paying tribute. However all the islands were sequestrated by the British in 1877, and came under Indian control following the withdrawal of the British in 1947. In 1956 they were designated an Indian "union territory".

The islanders are very skilled boat-builders and sailors. Only local craft make the dangerous journey through the encircling reefs to the Indian mainland. Fishing is the chief occupation, and grain, pulses, bananas, vegetables and coconut palms are farmed. The Moplah women make coir from the fibers of coconut husks.

NATIONAL DATA – INDIA

Land area 2,973,190 sq km (1,147,955 sq mi)

Climate	Altitude m (ft)	Temperatures January °C(°F)	July °C(°F)	Annual precipitation mm (in)
New Delhi	216 (708)	14 (57)	30 (86)	797 (31.3)
Cherrapunji (near Shillong)	1,313 (4,309)	12 (54)	20 (68)	10,799 (425.1)
Chennai (Madras)	16 (51)	25 (77)	30 (86)	1,266 (49.8)

Major physical features highest point: Kanchenjunga 8,586 m (28,169 ft); longest rivers: Brahmaputra (part) 2,900 km (1,800 mi), Ganges 2,505 km (1,557 mi)

Population (2006 est.) 1,095,351,995

Form of government multiparty republic with two legislative houses

Armed forces army 1,100,000; navy 55,000; air force 170,000

Largest cities Mumbai/Bombay (13,073,926); New Delhi (capital – 11,505,196); Bangalore (5,281,927); Kolkata/Calcutta (4,643,011); Chennai/Madras (4,376,400); Ahmadabad (3,819,497); Hyderabad (3,665,106)

Official languages National: Hindi; English State: Bengali; Telugu; Marathi; Tamil; Urdu; Gujarati; Malayalani; Kannada; Oriya; Punjabi; Assamese; Kashmiri; Sindhi; Sanskrit

Ethnic composition Indo-Aryan 72%; Dravidian 25%; Mongoloid and other 3%

Religious affiliations Hindu 80.5%; Muslim 13.4%; Christian 2.3%; Sikh 1.9%; other 1.8%; unspecified 0.1%

Currency 1 Indian rupee (INR) = 100 paise

Gross domestic product (2006) U.S. $4.042 trillion

Gross domestic product per capita (2006) U.S. $3,700

Life expectancy at birth male 63.9 yr; female 65.57 yr

Major resources coal, iron ore, manganese, mica, bauxite, titanium ore, chromite, natural gas, diamonds, petroleum, limestone, tea, coffee, rubber, pepper, coconuts, jute, cotton, barley, fruit, gemstones, ground nuts, millet, potatoes, pulses, rice, sorghum, sugarcane, tea, vegetables, wheat, livestock

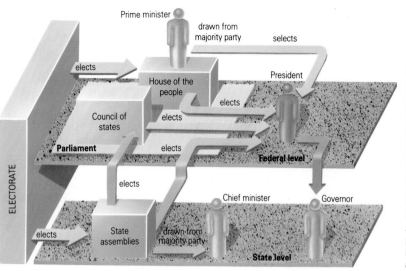

SOCIETY

The long and eventful history of the Indian subcontinent has left a rich legacy of art and culture. It has also left a very divided society; the gap between rich and poor is very marked.

History

The Indian subcontinent's first trace of a major civilization – in the Indus valley – dates back to 2500 BC. Many Hindu beliefs and customs derive from this Indus culture. The Indus civilization was probably ousted by an invasion of Aryan peoples from the Iranian plateau after about a thousand years. They built towns and cities along the Ganges plains, spreading their culture and their Sanskrit language far and wide. Their concept of social classes, or castes, were vital to the development of later Indian societies.

Two new religious movements arose in the 6th century BC: Jainism, which still claims many adherents among India's merchant community, and Buddhism, which later spread across much of southern and eastern Asia. During the following century the great Kingdom of Magadha, south of the lower Ganges, was established. King Chandragupta Maurya (reigned c. 321–297 BC) expanded its boundaries west of the Indus and south into the Deccan plateau. Under his grandson Ashoka (reigned c. 273–232 BC) the Mauryan empire grew till it dominated the subcontinent.

In the 4th century AD the Gupta dynasty, based at Magadha, ushered in a "Golden Age" of Hindu civilization but Hunnish invasions in the 5th century caused its collapse. In 712 the northwest was occupied by Arab Muslim invaders and in 1001 a Turkish sultan, Mahmud (971–1030) took Islam eastward into the Punjab and beyond. After the 12th century, the Ghurids (from Afghanistan) captured what remained of Mahmud's

Indian empire, and built their own empire. This became the powerful Sultanate of Delhi, until a Mongol invasion from the northwest in 1398 greatly reduced its powers. The 14th century also saw the foundation of the Sikh religion.

From 1519 Babur "the lion" (1483–1530), a Muslim of Mongol extraction, crushed what was left of the Delhi sultanate, and in 1526 founded the Mughal empire. His grandson Akbar (1542–1605) extended it throughout most of northern India and part of the Deccan.

In the 16th century the Portuguese established trading posts and bases in India, and retained a monopoly on trade with India until the 17th century, when the Dutch, English and French East India Companies set up trading stations of their own. By 1799 the British East India Company had become the dominant power in India, and by 1849 virtually all of the subcontinent was under British control.

In 1857 an uprising among Indian soldiers, known as the Indian Mutiny, gained the support of peasants, landlords and princes, but the Sikhs stayed loyal to Britain. After this, the British transferred power in India from the East India Company to the British Crown.

India was now part of the British empire, and in 1877 Queen Victoria was crowned Empress of India. Later, in 1885, the Indian National Congress was formed, the first political organization to claim members from all over the country. By the early 20th century its more radical members had begun to question Britain's right to rule India.

Indian hopes for independence were crushed by the Rowlatt Acts of 1919, which authorized imprisonment without trial. Mohandas Gandhi (1869–1948), a leading member of the Congress Party, resisted the new law by organizing massive displays of civil disobedience and noncooperation. Britain passed the Government of India Act in 1935, intro-

Colorful prayer flags (*above*) flutter above the walls of a Tibetan Buddhist monastery in Leh, capital of India's northern Ladakh district. Every flag has a prayer printed on it, and it is believed that each flap of the cloth is equivalent to the prayer being recited.

A high-caste Hindu (*below*), identified by the sacred yellow thread around his neck, performs ritual ablutions in the river Ganges at Varanasi. According to Hindu myth the river once flowed through heaven, and is worshiped as a goddess able to wash away sins.

ated by Sikh members of her own body-guard. Her elder son Rajiv Gandhi (1944–91) was appointed prime minister and leader of the Congress Party, but in 1991 when his Congress Party was returned with a fragile majority, Gandhi himself was assassinated during the elections.

Government and people
In 1950 India became the first republic to achieve independence within the British Commonwealth, and the world's longest written constitution came into force.

The lower house of the two houses of parliament is called the House of the People (Lok Sabha). Members are elected for a five-year term and there may be up to 550 members. The Council of States (Rajya Sabha), the upper house, can have up to 250 members. The head of state is the president, who is chosen by elected members of both houses, as is the vice-president. Real power, however, rests with the prime minister and council of ministers, who are both answerable to the House of the People.

India has a complex racial mix and a rich diversity of languages. Linguists list more than 700 different languages and dialects; 15 major languages have official state recognition. The largest group is the Indo-Aryan languages, which are all related to Sanskrit and are spoken throughout northern and western India. Among them Hindi is used as a first or second language by more than four-fifths of the population, and after independence it became the official language of India alongside English, which

Fishermen carry in the nets from their boat in southern India. The rich waters off India's coasts offer a bountiful harvest of fish, such as sardines, anchovies and bummalo (also known as Bombay duck). In 1995 the country's total fish catch was 4.9 million tons.

ducing a new federal constitution. But this satisfied no one. Congress won the elections of 1936 and in 1939, when the British viceroy declared India to be at war with Nazi Germany, without having consulted local leaders, all the Congress ministers resigned in protest. A long and troubled series of negotiations, punc-tuated by rioting and bloodshed, finally brought full independence in 1947. The price of Hindu–Muslim enmity was the partitioning of the country. India retained the states with a Hindu majority, and the mainly Muslim areas to the east and northwest became the separate state of Pakistan. An estimated 8.5 million people were uprooted by partition, of whom some 400,000 died.

India's first prime minister was Jawa-harlal Nehru. He tried to avoid align-ment with either the Soviet Union or the West, but a border dispute with China in 1962 forced him to appeal for help to Britain and the United States. In 1965, after Nehru's death, a dispute with Pakis-tan flared into open war, and after Neh-ru's daughter Indira Gandhi (1917–84) became prime minister in 1967, fighting broke out for a second time in 1971. The

war led to the birth of a new nation – Bangladesh – formerly East Pakistan.

Indira Gandhi became increasingly un-popular, and in 1977 was voted out of office, but in 1980 she was re-elected in triumph. In 1984 Indian troops attacked Sikh militants who had occupied the Golden Temple at Amritsar, the Sikhs' holiest shrine, killing 400 people. Shortly afterward, Indira Gandhi was assassin-

A steam train, en route for Agra, hurtles across a bridge in northern India while below women and men wash clothing in the shallow Yamuna river. India's well-developed railroad system, created by the British in the mid 19th century, is the world's fifth largest.

is preferred in the south. Other major Indo-Aryan languages include Bengali, Marathi, Gujarati and Assamese.

India's dominant religion is Hinduism, but more than one-tenth of the people are Muslim, and there are also significant Christian, Sikh, Buddhist and Jain minorities. The caste system divides Hindus into some 3,000 different groups according to their birthplace and occupation. Members of different castes cannot intermarry without one of them "losing caste". The lowest caste are the "untouchables". Untouchability has been unconstitutional since 1950. However, in 1997 Kocheril Raman Narayanan was elected the first "untouchable" president.

ECONOMY

In recent years India's economy has picked up with exports increasing and an improvement in the balance of payments situation. Investment by developed countries has rocketed and a significant decline has been achieved in foreign debt. There is less need to rely on foreign aid.

Agriculture and industry

More than two-thirds of the total labor force work on the land. Many farms still consist of smallholdings worked by peasants. Floods and droughts are a constant threat but the Green Revolution of the 1970s and 1980s produced better harvests, and irrigation is increasing. India is now self-sufficient in food production.

Rice is the main food crop; over 4,000 varieties are grown. Wheat is cultivated in drier areas. Major cash crops such as tea, coffee, rubber and pepper are grown on large plantations. Others include coconuts, jute and cotton.

India has the largest livestock population in the world, but milk yield per cow is among the lowest in the world because of a shortage of fodder. India's fishing industry has grown rapidly.

Massive deposits of iron ore are being exploited, but iron production has not yet reached its peak. India also has bauxite, manganese, copper, sulfur and salt, and gemstones remain a major export.

There are also huge reserves of coal and some oil. Petroleum production satisfies about one-half of the country's fairly modest requirements. Around three-quarters of India's electricity is supplied by thermal stations, some by hydroelectric power and a small amount by nuclear and

wind power. Manufacturing is the fastest-growing sector of the economy. Consumer products such as textiles, bicycles and electrical appliances are made mostly by private companies. State-owned ventures concentrate on heavy industries, but are increasingly open to private and foreign investors.

Transportation and communications

India's rail network is the country's largest state enterprise. Steam locomotives are still widely used, but new equipment is being introduced. Some areas still rely on river transportation. India's major seaports include Kolkata (Calcutta), Chennai (Madras), Mumbai (Bombay) and Cochin. The press functions with few restrictions, but because of the huge number of languages many newspapers have small circulations.

Health and welfare

Poverty and malnutrition are widespread, and medical resources are overstretched. Despite efforts to promote birth control, the population continues to rise. Education is generally free. Literacy has more than doubled since independence. India has a large reservoir of highly educated workers many of whom are employed overseas.

Bhutan

KINGDOM OF BHUTAN

I N THE EASTERN HIMALAYAS, SANDWICHED between Tibet (now occupied by China) to the north and India to the south, is the tiny kingdom of Bhutan. Like Tibet, Bhutan has long been isolated from international politics and commerce, and is regarded by the rest of the world as a secret and mysterious kingdom.

GEOGRAPHY

Bhutan is a country of high mountains and deep valleys that divides naturally into three main geographic regions. The lowlands along the southern border are known as the Duars Plain. From here the land rises sharply to the rugged mountains and broad valleys of the Lesser Himalayas. Above these tower the Great Himalayas, topped by massive peaks along the Tibetan border.

The climate ranges from subtropical in the south to Alpine in the north; the summer rains are heavy in places, falling as snow on the higher peaks. Most of the people live in the temperate central valleys, where there is seasonal pastureland. The southern lowlands and slopes are a mixture of dense bamboo forests and savanna, which support wildlife such as elephants and tigers. The mountains are home to hardy mountain animals such as yaks and tahrs (a species of wild goat).

SOCIETY

Bhutan's origins are shrouded in legend, but historians believe that the land was settled by migrants from Tibet in the 9th century AD. In about the 16th century a Tibetan monk is thought to have assumed both temporal and spiritual authority over a group of mountain and hill peoples. After him, power was divided between a spiritual leader, the *Dharma Raja*, who was chosen as the reincarnation of his predecessor, and a secular leader, the *Deb Raja*. The last *Dharma Raja* died in the early 20th century, and no new spiritual leader was identified. In theory the *Deb Raja* was elected by a council of governors, but from the 19th century military commanders fought each other for political power. In 1907 the British, the colonial power in the subcontinent, encouraged Bhutan to put an end to this situation by installing a hereditary monarchy – *Druk Gyalpo*, the Dragon King. Bhutan's external affairs were managed by the British until they withdrew from the subcontinent in 1947.

The first Dragon King's grandson, Jigme Dorje Wangchuk (1928–72), introduced limited social reform aimed at improving living standards without sacrificing traditional culture. Slavery was not abolished until the 1960s, when the first real towns were developed. The reign of his son, Jigme Sangye Wangchuk (b. 1954), has been troubled by unrest among refugees from Tibet, and later among the Nepalese minority, who wanted democratic reforms like those being implemented in Nepal. Few concessions were granted. The Nepalese-speaking Hindus of southern Bhutan won little support for change among the majority Bhutia, who are Buddhists and speak Tibetan dialects.

In the early 1990s hundreds of thousands of Nepali-speaking Hindus were expelled and hundreds killed in pro-democracy demonstrations. In 1998 the king gave the National Assembly the right to dismiss him. Tourism is tightly controlled by the government and limited to package tourists with a pre-planned itinerary. Nearly 5,500 foreign tourists visited Bhutan in 1997.

ECONOMY

Agriculture, at a little above subsistence level, is the mainstay of the economy. Most Bhutanese are farmers, for whom rice, potatoes and buckwheat are among the major crops. Livestock farming and forestry are also significant.

The chief mineral and energy reserve is coal, but dolomite, limestone and gypsum are also mined. Industrial output consists largely of food, timber and pulp products. Along with hydroelectricity these are the main exports. Almost all trade is with India, which supplies the country with manufactured goods.

There are few roads in Bhutan, and air travel is largely confined to government helicopters. Welfare, healthcare and education are free but unavailable to most people, and there is widespread illiteracy.

Hot peppers or chilies lie spread out to dry on a roof in a mountain valley in Bhutan. Only 9 percent of the country is cultivated, the rest being either forested or too mountainous. Yet the Bhutanese are largely self-sufficient in food, having to import mainly rice. Together with staple crops they grow oranges, apples and cardamom.

NATIONAL DATA – BHUTAN

Land area	47,000 sq km (18,147 sq mi)			
Climate	Altitude m (ft)	Temperatures January °C(°F)	July °C(°F)	Annual precipitation mm (in)
Thimphu	2,265 (7,431)	4 (39)	17 (63)	3,000 (118)

Major physical features highest point: Kula Kangri 7,554 m (24,784 ft)

Population (2006 est.) 2,279,723

Form of government constitutional monarchy

Armed forces army 6,000

Capital city Thimphu (79,334)

Official language Dzongkha (a form of Tibetan)

Ethnic composition Bhote 50%; ethnic Nepalese 35% (includes Lhotsampas–one of many Nepalese ethnic groups); indigenous or migrant tribes 15%

Religious affiliations Lamaistic Buddhist 75%; Indian- and Nepalese-influenced Hinduism 25%

Currency 1 ngultrum (BTN) = 100 chetrum

Gross domestic product (2003) U.S. $2.9 billion

Gross domestic product per capita (2003) U.S. $1,400

Life expectancy at birth male 55.02 yr; female 54.53 yr

Major resources timber, hydropower, gypsum, calcium carbonate, apples, barley, cardamom, coal, corn, dolomite, maize/corn, millet, oranges, potatoes, rice, buckwheat, yaks

Pakistan

ISLAMIC REPUBLIC OF PAKISTAN

PAKISTAN OCCUPIES THE NORTHWESTERN corner of the Indian subcontinent. To the west is Iran, to the northwest Afghanistan, to the northeast China and to the east India.

GEOGRAPHY

Rising in the Himalayas to the northeast, the Indus river drains the whole of Pakistan from the high northern plateaus to the southern plains. Along the frontier with Afghanistan, the Khyber Pass opens out into an irregular plateau along the foot of the Himalayas. Farther south are the plains of the Punjab – the fertile heartland of Pakistan – watered by the Indus and its tributaries. In the southwest is the arid plateau of Baluchistan.

Pakistan has a continental climate, characterized by very hot summers and cool winters with large daily variations in temperature. Most of Pakistan's rain falls between July and September, when monsoon winds blow from the southwest.

Over much of the country vegetation is sparse. In the northwest, there are forests of pine, deodar and holly, while deciduous trees such as plane and poplar grow in the far north. The mountains are rich in wildlife, including bears, leopards and wild sheep. But with forest cover reduced to 3.5 percent of the land area, animal habitats are threatened.

SOCIETY

Pakistan was created in 1947, when the Muslim-dominated northeastern and northwestern states of India were given autonomy at the end of British rule. Part of Bengal in the east, and part of Punjab in the west were partitioned off from India, creating the new divided nation of East Pakistan and West Pakistan. In 1956 an Islamic republic was proclaimed. Two years later, following a military coup, the country was placed under martial law. In 1965 disputes with India over the Kashmiri border erupted into war. During the elections of 1971 East Pakistan demanded its independence. A civil war followed, in which India supported East Pakistan, and in 1972 that territory became the independent state of Bangladesh.

Zulfikar Ali Bhutto (1928–79) became Pakistan's new head of state, but his election victory in 1977 roused a storm of protest that ended with the renewal of martial law under General Zia-ul-Haq (1924–88). Bhutto was arrested, convicted of attempted murder, and executed.

After Zia-ul-Haq's death in 1988 Bhutto's Pakistan People's Party (PPP) was returned to power under the leadership of his daughter Benazir Bhutto (b. 1953). During the 1990s Bhutto was twice elected and dismissed amid corruption allegations. In 1999 she and her husband were convicted and sentenced to prison. A coup in October 1999 returned Pakistan to military rule under General Musharraf.

Pakistan has two legislative bodies; the national assembly and the senate. The national assembly is headed by the prime minister, and its members are directly elected by popular vote. Members of the senate are chosen by the four provincial assemblies. The president, who is head of state and commander of the armed forces, is chosen by an electoral college.

The people of Pakistan are racially mixed, owing to centuries of invasions and migrations. The main linguistic division is between Iranian languages such as Baluchi and Pashto, and the Indo-Aryan group, which includes Punjabi, Sindhi and Urdu, the official language. The official religion is Islam, which is practiced by most people.

ECONOMY

More than half of Pakistan's population is employed in agriculture, though this now accounts for less than a quarter of national income. The new varieties, techniques and machinery introduced in the "green revolution" of the 1970s have helped to provide a surplus of wheat, rice and sugar, all available for export. Livestock are numerous, but the yields of milk and meat are relatively low because of the shortage of fodder. Fishing is a significant growth industry.

Pakistan has some coal, oil and high-grade iron ore as well as copper and bauxite. Limestone is mined in quantity and is the basis of a growing cement industry. Hydroelectric plants meet about three-fifths of the country's energy requirement, but large reserves of natural gas are now being exploited.

The cotton industry developed rapidly after independence, and is still very important today. Other products including leather, wool and sugar are now processed for the home market and for export. Heavy industry has been government-controlled since 1971.

Pakistan's main road and rail routes link the principal seaport, Karachi, to Lahore, Islamabad (the capital), and ultimately to Peshawar. The national carrier, Pakistan International Airlines, offers an internal and international service. Press censorship in one form or another is generally the rule. The government-controlled radio network is the main source of news and opinion.

Healthcare in Pakistan suffers from a shortage of trained staff. There is some basic welfare provision. Small pensions are paid to the old and disabled. Primary education is free and widely available, but only a minority receive secondary or higher education. Literacy rates remain comparatively low.

Verdant terraces (*left*) in the Hushe river valley near Skardu in northern Pakistan, where barley is grown in irrigated plots beneath the barren scree slopes of the mountains. Snowmelt feeds the Hushe river, a tributary of the Shigar, which itself feeds the mighty Indus.

Muslim faithful face toward Mecca (*below*) as they gather in the courtyard of the Mahabuk Khan mosque in Peshawar, northwestern Pakistan; some have already prostrated themselves in prayer. Islam is the dominant and official religion of Pakistan.

NATIONAL DATA – PAKISTAN				
Land area 778,720 sq km (300,665 sq mi)				
Climate		Temperatures		Annual
	Altitude m (ft)	January °C(°F)	July °C(°F)	precipitation mm (in)
Islamabad	511 (1,644)	10 (50)	29 (84)	1,142 (44.9)
Karachi	4 (13)	20 (68)	30 (86)	168 (6.6)
Major physical features highest point: K2 8,611 m (28,251 ft); longest river: Indus 2,896 km (1,800 mi)				
Population (2006 est.) 165,803,560				
Form of government federal multiparty republic with two legislative houses				
Armed forces army 550,000; navy 24,000; air force 45,000				
Largest cities Karachi (12,315,843); Lahore (6,659,543); Faisalabad (2,658,181); Rawalpindi (1,843,418); Multan (1,506,424); Gujranwala (1,459,139); Hyderabad (1,447,275); Islamabad (capital - 955,629)				
Official languages Urdu, English				
Ethnic composition Punjabi 48.2%; Pashto 13.1%; Sindhi 11.8%; Saraiki 9.8%; Urdu 7.6%; others 9.5%				
Official religion Islam				
Religious affiliations Muslim 97% (Sunni 77%, Shi'a 20%); Christian, Hindu, and other 3%				
Currency 1 Pakistani rupee (PKR) = 100 paisa				
Gross domestic product (2006) U.S. $427.3 billion				
Gross domestic product per capita (2006) U.S. $2,600				
Life expectancy at birth male 62.4 yr; female 64.44 yr				
Major resources coal, cement, cotton, fertilizers, food processing, livestock, fisheries, petroleum, natural gas, sugarcane, yarn, fabric, iron ore, copper, salt, limestone				

Bangladesh

PEOPLE'S REPUBLIC OF BANGLADESH

The small densely populated country of Bangladesh has India to the west, north and east. The southern coastal area fronts the Bay of Bengal.

GEOGRAPHY

Bangladesh occupies the eastern two-thirds of the Ganges–Brahmaputra delta. The land is extremely fertile but highly vulnerable to flooding from both the rivers and the sea especially in the monsoon season. In recent years soil erosion from deforestation upstream has made this flooding even more devastating. Bangladesh is mostly a country of flat plains criss-crossed by rivers. The Sundarbans, a huge mangrove forest in the south, is one of the largest swamplands in the world. The flatness of the land is broken only in the far northeast and in the southeast, where the higher Chittagong Hills run parallel to the Myanmanese border. The Chittagong region in the southeast includes offshore islands and the world's longest beach, at Cox's Bazaar.

Most of the annual rainfall occurs in the summer monsoon season, from June to October. In winter the northeasterly winds bring dry, warm days and cool nights. The winds change direction late in March, when there are torrential thunderstorms. In April and May severe cyclones create massive storm surges causing flooding in coastal areas.

Bamboo and rattan grow in the eastern forests, and central and southern areas are rich in plants adapted to swampy conditions. Fruit trees, including date and coconut palms and mangoes, are cultivated in the densely populated inland areas of Bangladesh. The country's wildlife includes several endangered species, notably the Bengal tiger, the clouded leopard and the Asiatic elephant.

SOCIETY

The name Bangladesh means "land of the Bengalis". Throughout their history the Bengali people have played an important part in the history and culture of the Indian subcontinent.

In the 4th century Bengal was part of the Mauryan Indian empire, ruled by the Buddhist and later Hindu kingdoms and empires of northern India. After the 13th century Islam became the majority religion of Bengal. A British colony from 1760, Bengal was still administered by the Hindu ruling classes, increasing tension between Hindus and Muslims. In 1905 the British partitioned Bengal into a Hindu-dominated western province and a Muslim-dominated eastern province. Western Bengal objected strongly, and in 1912 the two provinces were reunited.

When British rule in India ended in 1947, the country was partitioned along religious lines by creating the state of Pakistan. East Bengal became East Pakistan, despite being separated from West Pakistan by some 1,500 km (1,000 mi). In elections in East Pakistan in 1970, the Awami League, a political party campaigning for Bengali autonomy, won virtually every seat. But the Pakistan government, instead of convening the Assembly, sent troops into East Pakistan. A war broke out, in which India intervened. West Pakistan was defeated, and in December 1971 the new state of Bangladesh was created.

Shaikh Mujib ur-Rahman (1920–75), leader of the Awami League, became the first prime minister of Bangladesh, but he was assassinated in a coup in 1975. Several more coups followed and martial law continued until 1991, when Begum Khaleda Zia (b. 1945), leader of the Bangladesh National Party, was elected prime minister.

Although the original 1972 constitution established a parliamentary government under an elected prime minister and president, years of martial law disrupted this system. In the 1990s the country reverted to an elected parliamentary system where 30 of the 330 seats in the national assembly are now reserved for women.

The great majority of Bangladeshis are Muslim ethnic Bengalis. However, there is a significant Hindu minority, and Buddhists in the Chittagong region. The official language is Bengali. Classical Indian culture, particularly music and dance, exists alongside Bengali culture and Western cinema and sport.

Fishermen near Kurigram (*above*) in northern Bangladesh wait patiently for fish to enter their nets. The fish breed in the rivers, and when the waters flood the neighboring paddyfields – as the Brahmaputra river has done here – the protein-rich fish are caught.

Ox-power (*below*) A Bangladeshi farmer urges his two bullocks on as a youngster feeds sugar cane into a primitive, but effective, crushing machine that is being powered by the animals. Sugar cane – a source of sugar – is one of the country's principal crops.

ECONOMY

Bangladesh is one of the world's poorest countries. Its essentially agrarian economy is vulnerable to the vagaries of the climate; ruined harvests, drought and floods have brought frequent famines. The principal crop is rice, followed by jute, tea and sugar cane. Fishing is also important. Less than 10 percent of the land is forested.

As Bangladesh has few mineral resources, the steel, chemical and textile industries depend on importing raw materials (bought with aid). The industries using local raw materials include jute processing (making rope and other products) and paper-making from bamboo.

An extensive railroad network was built during the years of colonial rule, but today the inland waterways are more important for transporting cargo. Cycle rickshaws are a major form of transport in towns. The main seaport is at Chittagong on the southeast coast.

Comparatively few people own a television or radio; all broadcasting is controlled by the state. By contrast, the press is privately owned.

Healthcare, education and family planning programs are provided by government and private agencies; facilities are based in Dhaka and Chittagong in the east of the country. Even so, malnutrition is rife everywhere. Many Bangladeshis cannot read or write. The government provides five years of free, but not compulsory, primary education. Secondary and higher education are available only to a minority.

Nepal

KINGDOM OF NEPAL

NEPAL IS A SMALL LANDLOCKED KINGDOM in southern Asia lying between India to the south and Tibet (now occupied by China) to the north.

GEOGRAPHY

The Tarai lowlands along Nepal's southern border give way to the Churia foothills farther north, and then to the higher Mahabharat ranges. Still farther north, across the Kathmandu Valley, are the Himalayas, including Mount Everest, the world's highest peak at 8,850 m (29,035 ft). The climate ranges from subtropical in the south to Alpine in the north. The summer brings heavy mon-

A cornucopia of vegetables at a market in Kathmandu, Nepal's capital city. Most Nepalese live as subsistence farmers in scattered communities, often producing crops on heavily fertilized, thin-soiled terraces that have been carved out of the plains and lower hillslopes.

soon rains, and snow in the higher mountains. The western half of the country tends to be drier than other areas.

More than half of Nepal is still forested, ranging from acacia and sissoo forests in the Churia, to pines, oaks and rhododendrons in the Mahabharat and lower Himalayas. The Terai is rich in wildlife, with tigers, leopards and even a few rhinoceroses. Above the tree line are wild sheep and goats, snow leopards and yaks. The brilliantly colored Himalayan monal – a species of pheasant – is the national symbol of Nepal.

SOCIETY

Situated between the Indo-Aryan culture of northern India and the Mongoloid culture of Tibet, Nepal has been greatly influenced by both throughout its history. From the 10th to the 18th century, Nepal was ruled by the Malla dynasty. Then in 1769 it was conquered by the expanding Gurkha principality, which established the capital at Kathmandu. In 1860 Nepal faced the prospect of British rule, and agreed, as a price for continued self-government, to allow Gurkhas to be recruited into the British army. In 1959 King Mahendra (1920–72) introduced a constitution giving him greater power

but in 1990 his son Birendra (b. 1945) was forced by popular unrest to approve a new, more democratic constitution. In June 2001 King Birendra and most of the royal family were shot dead by Crown Prince Dipendra in a family quarrel. Birendra's brother Gyanendra assumed the throne when the prince himself died of gunshot wounds.

Most Nepalese are of Indian descent. Southern Nepal is the birthplace of the Buddha (563–438 BC). Today, the majority of the Nepalese population is Hindu.

ECONOMY

Most of the population live by subsistence farming. Crop yields are relatively low, though Nepal is self-sufficient in staple foods. Commercial forestry generates products for export, but deforestation and subsequent erosion is a problem.

Nepal's mineral resources include coal, iron ore and copper, though reserves are small. Manufacturing is mostly local and on a small scale. The largest factories are processing plants for agricultural produce such as jute, sugar, meat and rice. Tourism is expanding, especially around Kathmandu. Many new roads have been built, and the Royal Nepal Airline Corporation operates both domestic and international flights.

Outside the main cities healthcare is rudimentary. The government's school-building program and its drive to increase literacy are having some impact.

NATIONAL DATA - NEPAL

Land area	143,181 sq km (55,208 sq mi)			
Climate		Temperatures		Annual
	Altitude m (ft)	January °C(°F)	July °C(°F)	precipitation mm (in)
Kathmandu	1,334 (4,376)	26 (79)	25 (77)	1,461 (57.5)

Major physical features highest point: Mount Everest 8,850 m (29,035 ft); longest river: Ghaghara (part) 915 km (570 mi)

Population (2006 est.) 28,287,147

Form of government multiparty constitutional monarchy with two legislative houses

Armed forces army 69,000

Capital city Kathmandu (856,340)

Official language Nepali

Ethnic composition Chhettri 15.5%; Brahman-Hill 12.5%; Magar 7%; Tharu 6.6%; Tamang 5.5%; Newar 5.4%; Muslim 4.2%; Kami 3.9%; Yadav 3.9%; other 32.7%; unspecified 2.8%

Religious affiliations Hindu 80.6%; Buddhist 10.7%; Muslim 4.2%; Kirant 3.6%; other 0.9%

Currency 1 Nepalese rupee (NPR) = 100 paisa

Gross domestic product (2006) U.S. $41.92 billion

Gross domestic product per capita (2006) U.S. $1,500

Life expectancy at birth male 60.43 yr; female 59.91 yr

Major resources quartz, water, timber, hydropower, lignite, copper, cobalt, iron ore, barley, cattle, fruit, jute, medicinal herbs, pepper, potatoes, rice, sugarcane, tobacco, wheat

Maldives

REPUBLIC OF THE MALDIVES

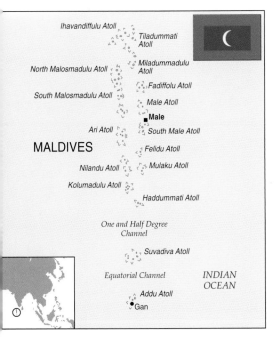

Parts of the Maldives resemble an idyllic island paradise. The islands are crowned with rich forests of breadfruit and palm, and are ringed with sandy beaches and clear lagoons.

The Maldives were settled some time between 500 BC and 500 AD, probably by fishing people from India and Sri Lanka. The Maldivian language, Divehi, resembles Sinhalese, the majority language of Sri Lanka. Later Arab settlers introduced Islam, which is now the state religion.

The islands fell to the Portuguese in the 16th century. They then became a sultanate under Dutch protection, and in 1796 they became a British protectorate which was formalized in 1887. In 1953 a republic was declared, but the president was soon deposed and the monarch restored. The Maldives attained full independence in 1965. President Maumoon Abdul Gayoom, who was first elected in 1978, survived an attempted coup in 1988 and secured a fifth term in 1998, polling over 80 percent of the vote.

Tourism accounts for 20 percent of GDP. Nearly half a million tourists visited the Maldives in 1998. Fishing provides most other national income. Agriculture is mainly at subsistence level. Crops include

The idyllic atolls of the Maldives lie scattered about the Indian Ocean. At the center of many of them lies a small island – sometimes inhabited, sometimes not – surrounded by a pellucid blue lagoon and enclosed by a reef. For many travelers they are paradise incarnate.

NATIONAL DATA - MALDIVES				
Land area 300 sq km (116 sq mi)				
Climate		Temperatures		Annual
	Altitude m (ft)	January °C(°F)	July °C(°F)	precipitation mm (in)
Male	1 (3)	23 (73)	23 (73)	1,911 (75.2)
Population (2006 est.) 359,008				
Form of government nonparty republic with one legislative house				
Armed forces army 900				
Capital city Male (89,287)				
Official language Divehi				
Ethnic composition Sinhalese, Dravidian, Arab, African				
Official religion Islam				
Religious affiliations Sunni Muslim 100%				
Currency 1 rufiyaa (MVR) = 100 laari				
Gross domestic product (2002) U.S. $1.25 billion				
Gross domestic product per capita (2002) U.S. $3,900				
Life expectancy at birth male 63.08 yr; female 65.8 yr				
Major resources fisheries, tourism, coconuts/copra, fruit, vegetables				

THE MALDIVES ARE A CHAIN OF SOME 19 island groups in the Indian Ocean to the southwest of India and Sri Lanka, extending over 764 km (475 mi) south to just beyond the Equator.

The islands are atolls – coral reefs surrounding the peaks of a submerged volcanic mountain range. In all there are almost 2,000 islands, but only about 200 are inhabited. The climate is mainly hot and humid. Nearby barrier reefs provide the only protection from occasional violent monsoon-related cyclones.

millet, cassava and yams, and coconuts are gathered for food and copra.

Vital imports, chiefly from India, include food (especially rice), manufactured goods and fuel for energy. Transportation is mainly by boat, but air travel is increasing. Religious and secular schools provide elementary education, and literacy levels are high. Healthcare still relies largely on traditional medicine.

Sri Lanka

DEMOCRATIC SOCIALIST REPUBLIC OF SRI LANKA

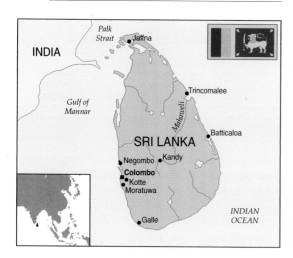

S RI LANKA IS AN ISLAND-STATE IN THE INDIAN Ocean off the southeast coast of India. Once a popular resort for tourists, in recent years it has been torn by ethnic violence.

GEOGRAPHY

A narrow strip of shallow water – the Palk Strait – separates Sri Lanka from the southeast coast of the Indian peninsula. Sri Lanka once formed part of an ancient landmass and its rock formations are some of the oldest in the world.

The landscape is dominated by the central mountainous region, rising to the highest point near the middle of the island at Pidurutalagala 2,518 m (8,261 ft) above sea level. Many small rivers radiate from the highlands, tumbling in waterfalls down onto the plains below. The rest of the country is mainly low-lying, interrupted by steep ridges and by *inselbergs*, ancient eroded blocks of stone. The coastline is partly fringed with coral islands and lagoons, especially in the far northwest around the Jaffna Peninsula and Mannar Island.

Sri Lanka is hot and humid throughout the year. The summer monsoon blows from the southwest between May and October, bringing rain to most of the island. In the autumn the winds gradually switch round to the northeast, with heavier rain in some northern and eastern districts. The monsoon is often unreliable, and severe droughts followed by floods are not uncommon.

The wettest areas support rich rainforests, while drier parts have hardier monsoon forests, including valuable trees such as ebony and satinwood. Much of Sri Lanka's forest has been cleared. The island's elephant population has now fallen to only a few hundred. Other typical wildlife includes leopards, bears, jackals, wild pigs and monkeys.

SOCIETY

Sri Lankan society throughout its history has retained its own religious faith and strong cultural identity. In about the 5th century BC the island was settled by the Sinhalese people from northern India who later established a Buddhist kingdom. From about 1200 AD Tamils from southern India founded a Hindu kingdom in the north.

In the early 16th century the island was invaded and colonized by the Portuguese. However, the Sinhalese managed to maintain an independent mountain kingdom around Kandy. The Portuguese were later ousted by the Dutch and in 1796 the island, which Europeans called Ceylon, became a British colony.

Ceylon achieved dominion status at the same time as India in 1947, and was governed by the United National Party (UNP). In the 1956 elections the UNP was defeated by the Sinhalese nationalist Sri Lanka Freedom Party (SLFP), whose leader, Solomon Bandaranaike (1899–1959) was later assassinated. His widow Sirimavo Bandaranaike (b. 1916) stood for election and in 1960 became the

world's first woman prime minister. When her daughter Chandrika was elected president in 1994 Chandrika gave up her own premiership in favor of her mother Sirimavo once again.

Ceylon became the Republic of Sri Lanka in 1972, but could not prevent the slide into economic decline. Throughout the 1980s Tamil opposition to Sinhalese domination increased, as Tamil groups began to fight for an independent homeland in the north and east of the island. Violence continued into the 1990s, and in 1993 President Ranasinghe Premadasa (1924–93) was assassinated by a suicide-bomber. The Tamil capital Jaffna was recaptured in 1995.

Sri Lanka is an independent republic within the British Commonwealth. The 1978 constitution established a presidential system in which the head of state is directly elected for a six-year term. The president has the power to appoint the prime minister and other ministers, and to dissolve the 168-member national assembly whose members are directly elected by proportional representation.

The majority of the population are ethnic Sinhalese, but Tamils form a sig-

Tropical palm trees rising from the long grass are a striking feature of the landscape around Anuradhapura in north-central Sri Lanka. This lowland region is cut by several rivers that rise from the central highlands where they are fed by heavy rains.

nificant minority. Both Sinhalese and Tamil communities have a caste system similar to that of India.

Generally, the Sinhalese are Buddhists, the Tamils are Hindus, and the Sri Lankan Moors practice Islam. However, there are also some Christians within most ethnic groups. The Tamil language now has some official recognition alongside Sinhala, and English is widely spoken. Traditional culture is based on Buddhist and Hindu art, dance and drama.

ECONOMY

In the late 1990s, plantation crops – principally tea, rubber and coconuts – contributed 20 percent of exports. The chief subsistence crop is rice, though harvests still fall short of the country's needs. Fruit, vegetables and spices are grown as staples and for export. Fishing is important, and reforestation has produced timber mainly for domestic use.

The island depends on hydroelectric power and imported oil to generate electricity. Graphite is a major export. Other mineral resources include iron ore, kaolin and gemstones. Cement manufacturing and the construction industry draw on the limestone from the north of the island. Telecommunications, banking and insurance have increased in importance. Tourism is another potential revenue-earner, but has suffered in the civil war.

Sri Lanka has extensive road and railroad networks. The principal port is Colombo, on the west coast; Trincomalee in the northeast handles much of the export trade in tea. Air Lanka, the national carrier, operates internal and international flights from the main airport near Colombo. There are radio broadcasts and daily newspapers in Sinhalese, Tamil and English.

A social welfare program covers benefits for unemployment, disability and old age but there is a shortage of urban housing. Although medical care is provided free of charge, malnutrition and disease are both widespread.

Education, compulsory up to the age of 14, is free and literacy rates are high.

Dependencies in the region

BRITISH INDIAN OCEAN TERRITORY
UNITED KINGDOM

The British Indian Ocean Territory is a tiny oceanic colony whose strategic military importance is out of all proportion to its size. It consists of a group of islets in the center of the Indian Ocean some 1,600 km (1,000 mi) from southern India and about 600 km (nearly 400 mi) due south of the Maldives.

Like the Maldives, the islands are coral atolls formed around the summits of submerged volcanic mountains. Most are uninhabited, and even the largest and best-known of them, Diego Garcia – first discovered by the Portuguese in the 16th century – occupies only 27 sq km (10 sq mi). The climate on the islands is both warm and humid, and the natural vegetation consists primarily of coconut palms, which until the 1970s provided the only economic resource.

All the people on Diego Garcia, most of whom were migrants working on the copra plantations, were relocated to Mauritius to clear the way for the construction of a major support and refueling station for British and United States' armed forces. Today there is no permanent civilian population on the islands, only military personnel.

NATIONAL DATA - SRI LANKA				
Land area 64,740 sq km (24,996 sq mi)				

Climate		Temperatures		Annual
	Altitude m (ft)	January °C(°F)	July °C(°F)	precipitation mm (in)
Colombo	7 (23)	27 (81)	27 (81)	2,524 (99.3)

Major physical features highest point: Pidurutalagala 2,524 m (8,281 ft); longest river: Mahaweli 332 km (206 mi)

Population (2006 est.) 20,222,240

Form of government multiparty republic with one legislative house

Armed forces army 78,000; navy 15,000; air force 18,000

Largest cities Colombo (capital – 650,875); Dehiwala (219,114); Moratuwa (189,155); Jaffna (172,812) (Kotte – legislative capital 119,364)

Official language Sinhala, Tamil

Ethnic composition Sinhalese 73.8%; Sri Lankan Moors 7.2%; Indian Tamil 4.6%; Sri Lankan Tamil 3.9%; other 0.5%; unspecified 10%

Religious affiliations Buddhist 69.1%; Muslim 7.6%; Hindu 7.1%; Christian 6.2%; unspecified 10%

Currency 1 Sri Lankan rupee (LKR) = 100 cents

Gross domestic product (2006) U.S. $93.33 billion

Gross domestic product per capita (2006) U.S. $4,600

Life expectancy at birth male 70.83 yr; female 76.12 yr

Major resources limestone, graphite, mineral sands, gems, phosphates, clay, hydropower, tea, rubber, coconuts, tourism, cassava, coconuts, fish, fruit, iron ore, rice, semi precious gemstones, spices, timber

China and its neighbors

COUNTRIES IN THE REGION

CHINA · TAIWAN

SPECIAL ADMINISTRATIVE REGIONS

HONG KONG · MACAO

Behind the mask (*left*) A Taiwanese opera singer in full makeup and elaborate costume. His character is immediately recognizable to a Chinese audience, while foreign observers are mystified. The traditional customs of the inward-looking Orient have been a source of fascination to other cultures for over 1,000 years.

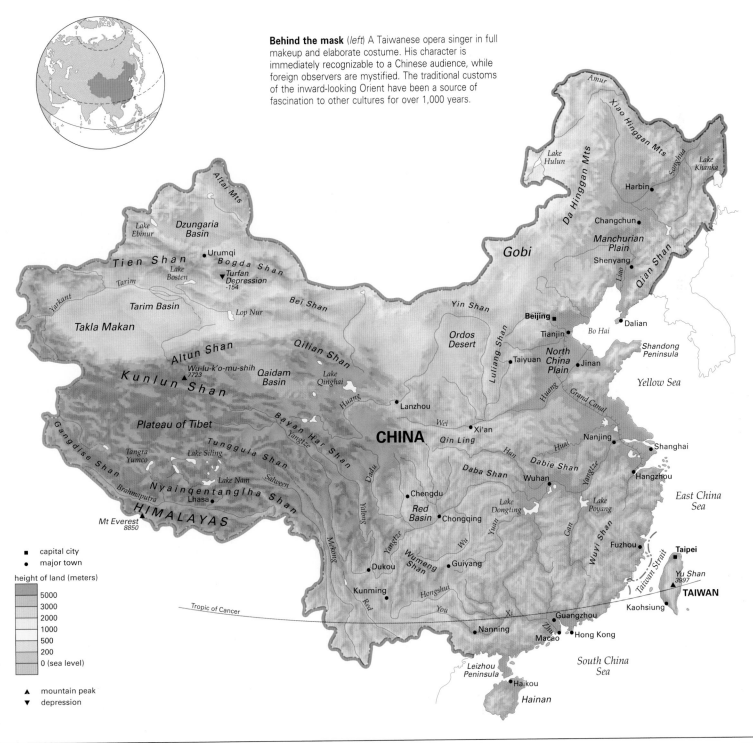

- ■ capital city
- ● major town

height of land (meters)

5000
3000
2000
1000
500
200
0 (sea level)

- ▲ mountain peak
- ▼ depression

China

PEOPLE'S REPUBLIC OF CHINA

CHINA IS THE THIRD LARGEST COUNTRY IN the world, occupying almost one-quarter of the Asian landmass and bordering on 14 countries. It is also the most populous nation on earth, with more than 1.2 billion people – almost one-fifth of the world total. The country boasts one of the oldest civilizations, dating back about 5,000 years, as well as many of the most important scientific and technical discoveries.

During the 20th century major social and political upheaval brought sweeping changes to Chinese society, but much of the country's past survived including its great corpus of art and literature.

GEOGRAPHY

China embraces a huge diversity of landscapes. Among its most distinctive regions are the Himalayas and the Plateau of Tibet in the southwest; the series of desert basins of the remote northwest, north and northeast; the densely populated lowlands of the north China plains; and the tropical southeast coast.

The land

The Plateau of Tibet in southwestern China – known as "the roof of the world" – is the highest plateau on earth. The greater part of the plateau stands between 4,000 m (13,000 ft) and 4,500 m (15,000 ft) above sea level, and is encircled by a series of higher mountain ranges. On its southern edge the Himalayas form a massive rampart along China's entire southwestern frontier. On the eastern side of the Plateau of Tibet are the sources of the Mekong river of Southeast Asia, and China's two major rivers, the Yangtze (Chang) and Huang (Yellow).

The lower-lying desert wastes to the north and northeast of the plateau comprise the two Autonomous Regions of Xinjiang and Inner Mongolia. In Xinjiang, the vast Takla Makan desert and Tarim Basin are almost completely ringed by high mountains. Immediately to the northeast of the Tarim Basin lies the lowest and hottest place in China, the –154 m (–505 ft) deep Turfan Depression. Farther northeast, on the Inner Mongolian plateau, is the southwestern fringe of the Gobi Desert.

Beyond the Gobi Desert, lies the Manchurian Plain – a rich agricultural area. Southwest of Manchuria, the alluvial flats of the north China plains surround the waters of Bo Hai – a vast inlet of the Yellow Sea. China's capital, Bei-jing, and Tianjin are both situated on this densely populated lowland.

Farther inland, an irregular plateau extends westward, cut by the valley of the great Huang river. This plateau is covered with a layer of fertile loess – a loamy soil – that in places is 250 m (800 ft) thick. It even colors the watercourses, giving the Huang or Yellow river its name.

Serene tranquility (*above*) on the Li river in Guangxi province, southern China. Many of the mist-shrouded pinnacles of limestone rise almost vertically to 100 m (330 ft). Their haunting beauty has long been an inspiration to Chinese landscape painters.

Watering the desert (*right*) in Gansu province, north-central China, a vast semiarid region lying between Inner Mongolia to the north and the Qilian Shan range to the south. Occasional oases can be found along the route of the centuries-old Silk Road.

To the south lies the Red or Sichuan Basin, a region of hills, ridges and valleys. The Yangtze river enters the basin from the Tibetan plateau and flows out eastward through a gap in the mountains, finally reaching the East China Sea just northwest of Shanghai.

The highlands to the southwest of the Red Basin occupy the central-southern provinces of Yunnan and Guizhou. In Guizhou, the limestone has eroded into a karst landscape of sinks, ravines and underground streams. Earthquake activity in China is most frequent in the high mountains of the southwest. China's southern coast is irregular and often steep; most of the rivers here drain into the great delta of the Zhu river.

Climate

The southerly winds of the summer monsoon bring high temperatures, humidity and heavy rainfall to the southeast; about four-fifths of the annual rainfall in this area falls between May and October. The rainy season grows progressively shorter to the northwest. The summer is shorter farther north, though in July both the temperature and humidity may approach levels similar to those in the south.

In winter bitter northerly winds from Siberia bring freezing temperatures to most of western and northern China. The southeast, by contrast, is pleasantly warm and dry during this season. Tropical cyclones and typhoons often ravage the southeast coast during the summer, and in the eastern lowlands the summer rains can cause severe flooding.

Plants and animals

Extensive forests remain in the northeast, and there are large areas of woodland in the south and on the eastern flanks of the Tibetan plateau. South of the Yangtze river, mixed forest gives way to evergreen broadleaf trees and then to tropical rainforest, with some mangroves on the south coast. North of the Yangtze river, deciduous broadleaf trees are replaced by mixed northern hardwoods and finally coniferous trees. In the desert basins of the northwest, a transitional belt of open woodland or grassland at the rim gives way to steppe and salt-tolerant desert scrub before reaching the "dead heart", where nothing will grow.

Animal life is equally diverse. Several species are unique to Tibet and its immediate environs, notably the yak (a mountain ox) and the takin, a relative of the musk ox. The forests of southern Tibet and Sichuan are the exclusive habitat of red pandas. Their bear-like relatives the giant pandas are found only in the bamboo forests of Sichuan, and are among the rarest species of all. Wildlife in the north and west includes wild asses, gazelles, wolves and Bactrian camels. In the southeast, typical mammals are leopards, tigers, buffaloes, civets and monkeys.

SOCIETY

For more than 3,000 years Chinese society had resisted serious change. The Chinese had built up traditions in thought and behavior that they believed were the best

NATIONAL DATA – CHINA

Land area	9,326,410 sq km (3,600,947 sq mi)			
Climate		Temperatures		Annual
	Altitude m (ft)	January °C(°F)	July °C(°F)	precipitation mm (in)
Hami (near Urumqi)	738 (2,421)	-11 (12)	27 (81)	35 (1.4)
Guangzhou	63 (201)	14 (56)	28 (83)	1,678 (66.1)
Beijing	51 (167)	-5 (23)	26 (79)	635 (25)
Harbin	172 (564)	-19 (-3)	23 (73)	521 (20.5)

Major physical features highest point: Mount Everest 8,850 m (29,035 ft); lowest point: Turfan depression -154 m (-505 ft); longest river: Yangtze (Chang) 5,525 km (3,434 mi)

Population (2006 est.) 1,313,973,713

Form of government communist state

Armed forces army 1,600,000; navy 255,000; air force 400,000

Largest cities Shanghai (15,434,642); Beijing (capital – 7,724,932); Hong Kong (6,940,432); Wuhan (4,287,693); Chengdu (3,972,509); Tianjin (3,755,249); Shenyang (3,564,751)

Official language Standard Mandarin

Ethnic composition Han Chinese 91.9%; Chuang 1.4%; Manchu 0.9%; Hui 0.8%; Miao 0.7%; Uighur 0.6%; Yi 0.6%; Tuchia 0.5%; Mongolian 0.4%; Tibetan 0.4%; others 1.7%

Religious affiliations Nonreligious 71.2%; Chinese folk religion 20.1%; Buddhist 6.0%; Muslim 2.4%; Christian 0.2%; others 0.1%

Currency 1 yuan (CNY) or Renminbi (RMB) = 10 jiao = 10 fen

Gross domestic product (2006) U.S. $10 trillion

Gross domestic product per capita (2006) U.S. $7,600

Life expectancy at birth male 70.89 yr; female 74.46 yr

Major resources coal, iron ore, petroleum, natural gas, mercury, tin, tungsten, antimony, manganese, molybdenum, vanadium, magnetite, aluminum, lead, zinc, uranium, hydropower, fisheries, tourism, asbestos, copper, cotton, jute and hemp, livestock, mercury, phosphate rock, rice, salt, soybeans, sugar beet, sulfur, tea, timber, tobacco, wheat

in the world. The fall of the last Chinese emperor in 1912 marked the first stage in a long process of social and political upheaval that was capped by the establishment of a communist republic in 1949.

History

Chinese civilization was born in the lower reaches of the Huang river valley about 1700 BC with the first historical dynasty, the Shang. During its 600 years, the Chinese perfected the wheel, developed writing and calendars and created a technology based on bronze working.

In the early 12th century BC the Shang were conquered by their southwestern neighbors, the Zhou, who ruled until the 3rd century BC, when the former vassal state of Qin (or Chi'n) achieved supreme power. The influential ideas of the great philosopher-teacher Confucius (551–479 BC) were to set the pattern of Chinese society for many centuries to come. Other developments during the Zhou dynasty

included the invention of iron casting, metal coinage and silk.

In 221 BC the Qin ruler Zhao Zheng (c. 259–210 BC) claimed the title of Shi Huangdi (the First Sovereign Emperor). He welded Chinese society into a unified nation and fortified his northern boundaries with a massive system of interlinked fortifications – the Great Wall of China. He also standardized the written Chinese language and laid the foundation for China's great administrative bureaucracy. He was buried with a symbolic army of some 6,000 terracotta figures. Today, these sculptures are one of China's foremost tourist attractions.

In 206 BC, following a rebellion, the peasant-born Liu Bang (256–195 BC), took the imperial title to become the founder of the Han dynasty, which lasted for more than four centuries. During this time, China's borders extended almost to their present-day limits. Paper (in 105 AD) and the seismograph were invented, and steel was manufactured for the first time. Under

the Han, people turned to the new religion of Buddhism from India.

After 220 AD China split into warring kingdoms. It was the short-lived Sui dynasty (581–618) that finally reunited China. The Sui emperors also built a major part of the Grand Canal system which linked the north with the fertile rice-growing Yangtze valley. Under their successors, the Tang, who held power for almost 300 years, China became the largest empire on earth. The arts – especially poetry and ceramics – thrived; the earliest known printing began, and paper money was first issued.

In the Song (Sung) dynasty, which began in 960 gunpowder was used for the first military rockets. China reached its golden age: cities grew, and trade with the

Provinces of China
- ■ national capital
- ● provincial or municipal capital
- ▢ province
- ▢ municipality
- ▢ autonomous region
- ▢ special administrative region

China's administrative divisions Administratively China is divided into 22 provinces, 4 municipalities, 5 "autonomous regions;"inhabited by China's national minorities, and 2 special administrative regions.

The Great Wall

For nearly three centuries of its early history (475–221 BC) China was a divided nation of feuding kingdoms, threatened by raids from fierce Central Asian nomads to the north, the Xiongnu (later known as the Huns, who were the scourge of Rome). Attacks became so frequent that the northern kingdoms of Yan and Zhao built earthwork walls to protect themselves. But the Xiongnu grew in strength and extended their territory, and by the 3rd century BC they had a 300,000-strong army of mounted archers.

Qin Shi Huangdi (c. 259–210 BC), the first emperor of unified China, responded to this increasing threat with a grand gesture. Recruiting millions of conscripted laborers, he linked the scattered defensive walls into a single 2,000 km (1,200 mi) line stretching from the eastern coast near modern Beijing deep into central Asia. Watchtowers 12 m (40 ft) high along the 7 m (23 ft) wide wall were built for defense and to allow smoke signals to be passed to the capital. Where possible the wall was built along the crest of mountain ranges, which provided a formidable natural barrier.

The wall failed to stop the Xiongnu attacks, but its presence was a reminder of Chinese military strength. By forcing raiders to dismount, the wall prevented their swift escape and limited the booty they could carry. On their return raids they had to scale its

Symbol of Chinese power The Great Wall, originally a scattered line of fortifications in the northeast, was extended by China's first emperor in the 2nd century BC. It is the only man-made structure in the world that is visible from space.

heights again. Although even a small force could take one of the forts by assault, the Chinese could mobilize along the wall and counterattack.

Later dynasties expanded and fortified Qin Shi Huangdi's work. Much of the massive structure that survives today was built from the 14th century onward by the Ming emperors (1368–1644) who rebuilt the sections closest to their new capital at Beijing. But the wall could offer no protection against domestic unrest, of which the Manchus took advantage in 1644, breaking the power of the Ming.

The last emperor

Emperor Puyi (1906–67), formally known as Xuan Tong, was the last of his line, the Manchu dynasty that had ruled China for 268 years; and the last emperor in a 2,000-year-old tradition. He was crowned emperor in 1908 at the age of two, with his father as regent. But his precarious position mocked his traditional title of "Son of Heaven."

China by the late 19th century was in serious decline. Its government was massive, cumbersome and corrupt; and its army was undisciplined and lacked modern weapons. Conservative and liberal factions at court were bitterly divided. The child emperor Puyi and his immediate family were subordinate to the powerful princes of the Manchu Clan Council, who increasingly depended on the protection of a viceroy and warlord, Li Hongzhang (1823–1901), the richest man in China; and later, on the ambitious general Yuan Shikai (1859–1916).

In 1912 a succession of uprisings threatened the dynasty. The Council called in General Yuan Shikai to quell the rebellion. Yuan instead opened up negotiations that rapidly whittled away the power of the imperial court and ended with his own installation as president of the new Chinese republic after Sun Yat-sen resigned. Puyi was forced to abdicate at the age of six. He continued to live in the Forbidden City – the emperor's traditional residence in a luxurious walled compound within

Royal but vulnerable China's last imperial rulers in a photograph taken in 1909. The emperor Puyi, age 3, stands holding the hand of his father, the regent Prince Chun. Three years later the dynasty collapsed and China became a republic.

Beijing – but despite his new title of Manchu Emperor, he was effectively powerless.

Puyi later became the puppet emperor of the Japanese-controlled state of Manzhouguo (Manchuria; 1931–45). Captured at the end of World War II, he served a 9-year sentence as a war criminal in a Chinese prison before being released to live in Beijing as a private citizen. In the last years of his life he worked in a botanical garden and wrote his autobiography, entitled *From Emperor to Citizen*.

length of 6,400 km (4,000 mi). The emperors built many palaces, including the Imperial Palace at Beijing, and dispatched great sea-going junks to explore as far as the Red Sea. In 1557 the Portuguese were able to establish a settlement in Macao, and they were followed by Jesuit missionaries. Beijing fell to the Manchu, descendants of the Ruzhen, in 1644 and Ming resistance was finally crushed in 1683.

The Manchu, or Qing, dynasty pursued an active policy of rapid expansion. They captured the southeastern island of Taiwan (1683), annexed large areas to the northwest, and posted a garrison in Tibet to put a firm check on Mongol power. The Qing won the loyalty of their subjects but they created growing hostility abroad because of their trading restrictions.

In 1779 the British East India Company acquired a monopoly on opium. The drug's popularity in China depleted imperial reserves and the trade was banned. In 1839 the Chinese authorities seized a consignment of more than 20,000 chests of opium and destroyed it. This triggered the first Opium War, which ended with the Treaty of Nanjing in 1842 granting new privileges to foreign merchants. A second Opium War (1856–60) led to temporary British and French occupation of Beijing in 1860.

The Qing had domestic problems, too. Rural poverty and an exploding population – from 100 million in 1650 to 420 million in 1850 – caused the Taiping Rebellion of 1851–64 in which at least 25

The Boxer Rebellion Between 1898 and 1900 groups of ardent Chinese nationalists known as the Boxers attacked European missionaries and killed thousands of Chinese converts in northern China. The port city of Tianjin, near Beijing, was captured briefly before the rebellion was crushed by an international force.

its golden age: cities grew, and trade with the outside world expanded as junks carried cargoes of tea, silk and porcelain to the East Indies, Africa and India.

In 1126 the Ruzhen (Juchen), a seminomadic people from Manchuria, conquered northern China and set up their own dynasty there. The Song retreated and established a new capital at the port of Hangzhou, south of Shanghai. Early in the 13th century the Ruzhen were attacked by the Mongol leader Genghis Khan (c. 1155–1227), and were finally routed in 1234. Hangzhou also fell to the Mongols in 1276.

Under the Ming (or "Brilliant") dynasty (1368–1644), which followed the overthrow of the Mongols, the Great Wall was repaired and extended to its present

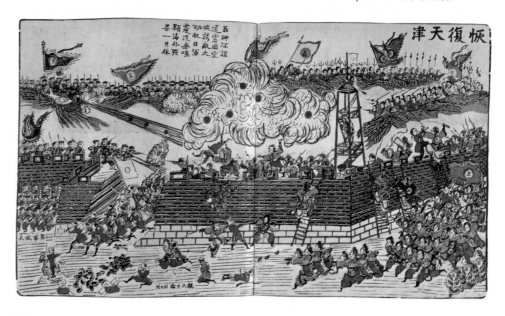

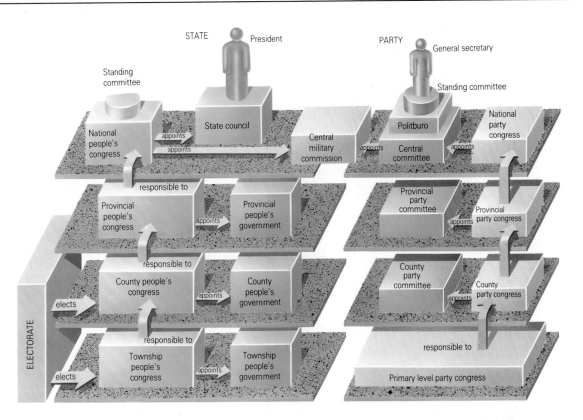

STATE President

PARTY General secretary

Standing committee

Standing committee

National people's congress

State council

Central military commission

Politburo

Central committee

National party congress

appoints

appoints

appoints

appoints

responsible to

Provincial people's congress

Provincial people's government

Provincial party committee

Provincial party congress

appoints

appoints

responsible to

County people's congress

County people's government

County party committee

County party congress

appoints

appoints

ELECTORATE

elects

responsible to

Township people's congress

Township people's government

responsible to

elects

appoints

Primary level party congress

China's government (*right*) State and Party organizations mirror each other at all levels. Legislative power is in the hands of the National People's Congress (NPC), which is elected via the provincial and county assemblies for a five-year term. However, as the NPC meets only two or three weeks a year, its day-to-day business is carried out by a 155-member standing committee which it appoints. The NPC also chooses the president, the prime minister and, on the prime minister's advice, the members of the State Council. Its function is to execute the laws enacted by the NPC. The State Council also has its own standing committee, made up of the prime minister, 12 or more vice-premiers and a secretary-general. A parallel organizational structure operates within the Communist Party. The 348 members of its central committee elect six of their number to a politburo, which lays down guidelines for every government action.

Yihe Quan (Righteous and Harmonious Fists – hence their Western name "Boxers") began murdering Chinese Christians. In 1900, they laid siege to foreign legations in Beijing. The affected governments responded by sending a massive international army.

Disillusionment with the Qing was growing. In 1911 a minor army mutiny in Wuchang (Hubei province) triggered widespread opposition and on 1 January 1912, Sun Yat-sen (1866–1925), a Western-educated leader of the budding Nationalist Party, was declared president of the new Republic of China, with its capital at Nanjing.

The republic did not endure, however, and China slid into civil war between regional warlords, whose armies caused immense damage and millions of casualties. In 1919, following World War I, there was a great upsurge of nationalist feeling against the presence of foreign powers within China. Marxism gained in popularity, and the Chinese Communist Party (CCP) was founded in 1921.

The Nationalist Party allied with the communists to confront the warlords in the north and reunite the country. But in 1927 Chiang Kai-shek (1887–1975), leader of the nationalists, turned against his communist allies and massacred many of them. In October 1934, some 100,000 communists began their epic 9,500 km (5,903 mi) retreat northward that has become known as the "Long March". The survivors, numbering fewer than 8,000, established their new Red Army headquarters in the north-central province of Shaanxi under the leadership of Mao

Zedong (1893–1976). Meanwhile the Japanese had invaded Manchuria in 1931, and in 1937 they overran northern and central China. The nationalists were compelled to form a united front with the communists against Japan.

With the end of World War II and the defeat of Japan in 1945, the struggle for power soon developed into an open civil war. The nationalists were completely defeated and the People's Republic of China was proclaimed on 1 October 1949 under Mao Zedong.

China was crippled by decades of conflict, and lacked modern industrial resources. Mao Zedong planned to develop the country's economy by increasing agricultural production and using the surplus for industrial growth. New political structures were modeled on those of the Soviet Union, and Soviet loans boosted initial growth. Tibet – which had expelled Manchu pretenders to Suzerainty in 1912 – was occupied in 1950 by the People's Liberation Army.

In 1958, Mao launched the "Great Leap Forward" – an attempt to speed up economic growth by mobilizing the rural workforce on a massive scale. But the program was a disastrous failure and had to be abandoned. Then in 1960 the Soviet Union cut off military and economic aid and openly criticized Mao.

To regain the ideological initiative, Mao, in 1966, launched the "Great Proletarian Cultural Revolution". It led to anarchy: intellectuals and artists became victims of inquisitions by youthful Red Guards; many were tortured and humiliated. A wave of terror led to the deaths of at least

400,000 Chinese. This, along with fears of a Soviet invasion in 1968, caused the leadership to call in the army to restore order.

China was admitted to the United Nations (UN) in 1971, and was visited by the United States' president Richard Nixon the following year. Mao's death, at the age of 82, on 9 September 1976 was followed by a bitter leadership struggle, which was won by the moderates. After 1982, a new constitution was drawn up under Deng Xiaoping (1904–97).

Deng Xiaoping's resignation from the Party's central committee in 1987 was followed by an upsurge of agitation for democracy. When, in May 1989, the army was called in to disperse a peaceful sit-in in Beijing's Tiananmen Square, 200 were killed according to official figures. Since then, China has made progress on economic liberalization while imposing further political repression, including a campaign against religious sects and the arrest and imprisonment of opposition leaders.

Government
Since 1949 China has been a single-party state, in which the Chinese Communist Party controls most aspects of daily life through mass-membership organizations such as trade unions, youth leagues and professional associations. The 1982 constitution vested legislative power in the 3,000-member National People's Congress (NPC), but in effect more power resides with the politburo of the Central Committee of the Communist Party.

China has a highly complex regional and local government structure. The primary divisions are the 22 provinces,

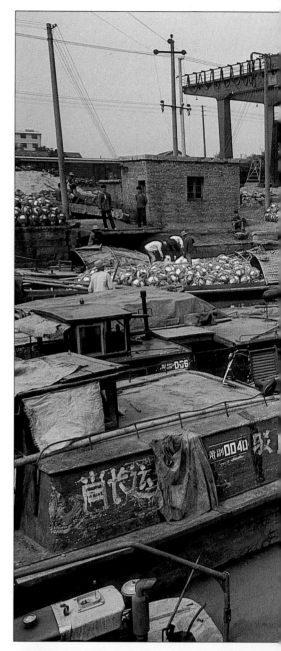

Geese by the barrowload (*left*) in Chengdu, the capital of Sichuan province. Customers buy the geese live and take them home either to cook or to keep for their eggs. Sichuan, with fertile soil and a mild climate, is almost self-sufficient in food production.

The world's longest canal (*right*) China's Grand Canal system extends from Beijing to Hangzhou, through five major river basins and the most populated of China's eastern cities. It was begun in the 5th century BC, but the main sections were built in the 7th century AD and it was completed in the late 13th century. It was built to connect the north with the rice-growing Yangtze valley. Renovated in the 1960s, it is now a major transportation route.

Elbow grease (*below right*) Since the economic reforms of 1978, light industry – including the manufacture of bicycles (shown here), radios, sewing machines, cotton yarn and cameras – has grown rapidly. But despite attempts at modernization, many factories such as this one at Changshu on the Grand Canal in Jiangsu province retain old-fashioned production methods.

and the four city municipalities of Beijing, Tianjin, Shanghai and Chongqing. These areas are divided in turn into prefectures, then into counties and various smaller units. There are also five nominally "autonomous regions" (Tibet and Xinjiang in the west, Inner Mongolia and Ningxia in the north, and Guangxi in the south).

People and culture

Although Han Chinese form some 92 percent of the population, 55 different minority groups and nationalities are recognized in the country. The largest minority is the 13-million strong Chuang (Zhuang) people of Guangxi (bordering Vietnam). Others include the Tibetans of Tibet, the Turkic Uighurs and Kazakhs of Xinjiang, the Mongols of Inner Mongolia, and the Hui people of Ningxia in the upper Huang valley.

Given the large number of minorities, there is a bewildering array of languages and dialects. Even Chinese is spoken in many "dialects", which constitute major languages in their own right. Mandarin is the dominant tongue, and is understood throughout China. All forms of Chinese share a single written language.

Since the creation of the People's Republic, religious practice has been discouraged by the Communist Party. However, Buddhism is still very strong in Tibet, and Islam in Xinjiang and Ningxia

has also resisted communist secularism.

Traditional Chinese religions or philosophies, notably Taoism and Confucianism, are still openly practiced by about one-fifth of the population, and the small Christian minority includes Protestants and some 3 million Chinese Catholics.

Chinese writing – an art in itself in the form of calligraphy – has provided the primary vehicle for a rich and diverse heritage of painting, poetry, drama and music. Chinese porcelain has for many centuries been the envy of the world, and Chinese cuisine is also much admired and imitated.

ECONOMY

Since the late 1970s the Chinese economy has moved steadily away from the Soviet command economy model toward a market economy, where privately owned produce can be sold for private profit (a percentage tax is paid to the state). Most medium to large industries, however, are still state-owned or cooperative.

Agriculture

Only one-tenth of China's land area is suitable for cultivation, and half of this is given over to paddy fields or irrigated for other crops. Rice, the main crop, is grown mostly in the southern provinces, and China is among the world's leading pro-

west and around Hainan Island, and plans to build a distribution grid by 2010.

Most of China's electricity is generated from coal. Most hydroelectric production is in the southwest; however, China, is building the world's largest dam on the Yangtze Jiang (Yangtze River). There are two nuclear power plants and three more are being built. There is also heavy investment – some foreign – in renewable sources such as wind and solar energy.

By 2000 China had moved out of heavy industry and become a leading producer of fertilizers, textiles and clothing as well as electronics and transport equipment.

China's most important trading partners are Japan and the United States. Manufactured goods account for around 70 percent of exports, while imports consist mainly of industrial supplies and capital goods.

Transportation and communications

Every province and "autonomous" region has railroad connections, and the total road system now amounts to about 1 million km (600,000 mi). However, most roads outside the cities are unsurfaced and most Chinese still rely on bicycles.

Waterways carry more than one-third of all internal freight. The nine principal ports and other depots around the coast serve an ocean-going merchant. fleet of more than 400 vessels. There are many international airports, the most important being Beijing, Shanghai, Guangzhou and Hong Kong.

The press and broadcasting media are entirely under state control. There are hundreds of radio and television stations. The 1990s saw a boom in telecommunications and by 2000 there was an estimated 10 million Internet users.

Health and education

China has made significant progress in improving healthcare. Medical teams are regularly sent to rural areas, where the so-called "barefoot doctors" are given short but intensive training for work in the country. As a result, serious infectious diseases such as cholera and typhoid have been largely eliminated. A variety of contributory welfare plans, including medical insurance, unemployment insurance and a national pension scheme, have been introduced by the government to support an ageing population.

All children are officially required to attend six years of primary schooling and three years of secondary schooling, though secondary education is sometimes unavailable outside the cities. All university fees are paid by the state.

ducers. Almost one-third of China is pasture, supporting a variety of livestock raised mainly for meat, though milk is taken from sheep, goats and water buffalo as well as from cattle.

Fish are an important part of the Chinese diet. About two-thirds of the catch is obtained from the neighboring seas. The rest comes from freshwater fisheries, mostly in the Yangtze and Zhu river deltas. There are still extensive forests in the Greater and Lesser Hinggan ranges, but nevertheless China has to import timber to meet its needs.

Industry

China's vast mineral wealth includes the largest reserves of antimony in the world. China is both the largest producer (1.2 billion tons in 1998) and consumer of coal in the world. Petroleum reserves are considerable, especially in the north and west, but demand outstrips supply. China has large reserves of natural gas in the north,

Taiwan

THE INDEPENDENT ISLAND STATE OF TAIWAN, formerly known as Formosa, faces the southeast coast of China across the 160 km (100 mi) wide Taiwan Strait, with the Philippines to the south and Japan's Sakishima Islands, part of the Ryukyu Islands to the northeast. Taiwan claims to be the sole legitimate Republic of China (its official name), a reflection of the fact that in 1949 it became the last refuge of China's nationalist government after its defeat by the communists on the mainland. The Taiwan government also has jurisdiction over 20 or so small islands that extend along the Chinese coast, and about 60 more in the Pescadores, just to the west of Taiwan itself.

NATIONAL DATA - TAIWAN

Land area	32,260 sq km (12,460 sq mi)			
Climate		Temperatures		Annual
	Altitude m (ft)	January °C(°F)	July °C(°F)	precipitation mm (in)
Taipei	9 (30)	16 (61)	29 (84)	2,129 (83.8)

Major physical features	highest point: Yü Shan 3,997 m (13,113 ft)
Population	(2006 est.) 23,036,087
Form of government	multiparty republic with two legislative houses
Armed forces	army 200,000; navy 45,000; air force 45,000
Largest cities	Taipei (capital – 2,468,705); Kaohsiung (1,519,748); Taichung (1,124,514); Tainan (737,271)
Official language	Mandarin Chinese
Ethnic composition	Taiwanese 84%; mainland Chinese 14%; Aboriginal 2%
Religious affiliations	Chinese folk religion 48.5%; Buddhist 43%; Christian 7.4%; Muslim 0.5%; others 0.6%
Currency	1 New Taiwan Dollar (NTD) = 100 cents
Gross domestic product	(2006) U.S. $668.3 billion
Gross domestic product per capita	(2006) U.S. $29,000
Life expectancy at birth	male 74.67 yr; female 80.47 yr
Major resources	coal, natural gas, dolomite/limestone, marble, asbestos, salt, fisheries, rice, bananas, citrus fruits, pigs, pineapples, poultry, sulfur, sugarcane, sweet potatoes, timber

The "Beautiful Island" Taiwan's outstanding natural beauty deserves the name *Ilha Formosa* bestowed on it by Portuguese explorers in the late 16th century. Lush vegetation covers much of the island, from the mountainous interior to the sunny coasts.

GEOGRAPHY

Although Taiwan is close to the Chinese mainland, its landscape is different; the high mountainous interior with its lush vegetation is fringed with flat coastal plains, sandy beaches and forests.

The mountains of the Central Range, the tops of a submerged mountain chain, rise steeply from the eastern coast. The highest peak is Yu Shan at 3,997 m (13,113 ft). On the western side of the range the land drops gently away to a coastline fringed with dunes and lagoons. The far north, near the capital Taipei, is a volcanic area. Yangmingshan National Park, north of the city, is notable for its volcanic peaks, sulfurous craters, lakes and rivers.

Taiwan has a subtropical monsoon climate. Fall and spring are warm, and summer is hot and humid. There is heavy seasonal rainfall between July and September, when typhoons strike on average three or four times a year. Winters tend to be mild and dry, though the north and east receive some rain.

More than half the island is covered in forest. The mountain vegetation changes with altitude, ranging from laurels and acacias on the lower slopes, to coniferous forests on the higher peaks. In the lowlands, tropical evergreen trees grow alongside bamboos and palms.

Animal species include the Asian black bear, mongooses, foxes, various monkey species and deer; the Taiwan macaque is unique to the island. Bird life, too, is plentiful, including 220 native species.

SOCIETY

Taiwan has been a refuge for those fleeing the Chinese mainland and at other times has been dominated by European and Japanese colonists. Today the island is discovering an identity of its own.

The Chinese knew about Taiwan in the 7th century AD, but at that time made no attempt to settle there. In 1590 Portuguese explorers called it *Ilha Formosa*, or "Beautiful Island". The first successful European settlers were Dutch traders, who arrived in 1624. In 1661 the Dutch settlements were expelled by Zheng Cheng-gong (1624–62), a refugee of China's defeated Ming dynasty. The Qing regime captured Taiwan in 1683, and soon a flood of Chinese immigrants arrived from the mainland. The island's economic importance grew, and in the 1880s it received provincial status.

Following the Sino-Japanese War of 1894–95, Taiwan was ceded to the Japanese, but after the defeat of Japan at the end of World War II, China's nationalist government took control of the island. In 1949 the Chinese communists routed the nationalists on the mainland, forcing them to flee to Taiwan, where they set up a government under the leadership of Chiang Kai-shek (1887–1975). The US provided economic and military aid, and in 1958 prevented a communist attack on the island. The nationalist government, who claimed to be the legitimate rulers of China, imposed martial law for almost 40 years.

In 1971 Taiwan lost the Chinese seat at the United Nations (UN) to China, and in 1979 the United States recognized the communist mainland government as the sole legitimate authority in China. At the same time the transfer of leadership from Chiang Kai-shek to his son Chiang Ching-kuo (1910–88) brought a more populist style of government to the island. He was succeeded in 1988 by his Taiwan-born vice-president Lee Teng-hui (b. 1923), who, in 1991, introduced a program of democratic reform and formally ended the 40-year state of war with the People's Republic. A series of threatening Chinese military exercises off the coast of Taiwan in spring 1996 ended without incident following a heavy US presence in the area.

The president serves a six-year term as head of state and is elected by popular vote. There are also five yans (yuans), or councils. The first democratic elections to the Legislative Council were held in 1989. The Executive Council acts as a cabinet, and includes the prime minister. The other councils manage the judiciary, civil service and government administration.

Taiwan has a population of some 22 million and is one of the world's most densely populated countries. Most of the people are ethnic Taiwanese (84 percent), descended from 17th-century Chinese settlers. A large Chinese minority (14 percent) – called mainlanders – is made up of postwar refugees and their descendants, who speak Mandarin Chinese, which is now the official language. The principal religions are Buddhism, Taoism and Confucianism.

ECONOMY

Cheap, plentiful labor and a vigorous promotion of Taiwanese products overseas has fueled dramatic growth of the economy since the 1950s, making Taiwan one of the most powerful economies in Asia.

Farming was once the lifeblood of the Taiwanese economy, but today some food has to be imported to meet demand. Rice, the main crop, is in surplus. Although over half the island supports forest, many areas are inaccessible to loggers, and timber quality is poor. Fish, notably tuna, are plentiful in the surrounding waters.

Mineral resources are limited. Marble and dolomite are quarried, and salt is produced in coastal evaporation pans. Hydroelectric and nuclear power have been developed. Manufacturing industries are the mainstay of the economy. Electrical and electronic goods, textiles and garments are important exports. Taiwan is a world leader in producing computer hardware and semi-conductors.

A well developed road network serves much of the island. The capital, Taipei, suffers from heavy traffic congestion, and many scooter and motorcycle riders wear face masks as protection against exhaust pollution. The main railroad encircles the greater part of the island; a final southern section will complete the ring. Of the five major ports, three can berth large ships. There are internal air services between all major cities, with international airports at Taoyuan in the north and Kaohsiung in the southwest.

Government control of the media has now been considerably relaxed. There are some 30 daily newspapers, 30 radio companies and three television stations.

Taiwan has a well-developed healthcare system, and government programs have ensured that there is adequate housing for most of the population. The welfare program provides a comprehensive range of benefits, which are available to a growing proportion of the population. Life expectancy for men is 73.8 years, and 79.5 for women.

Education is free and compulsory from the age of 6 to 18, with increasing opportunities for higher education at college, and also for adult and special education.

The rice harvest, carried out by farm workers from the indigenous Taiwanese Ami people from eastern Taiwan, produces a surplus to requirement, and many paddyfields are now used to grow vegetables and fruit for the local population.

Special Administrative Regions

HONG KONG

Hong Kong is a small but extremely prosperous former British colony on the south coast of China to the east of the Zhu (Pearl) river estuary. With one of the world's foremost centers for trade and finance this capitalist enclave was returned to communist China in July 1997.

Measuring just 1,069 sq km (413 sq mi) in area, it consists of Hong Kong Island and the Kowloon peninsula – one of the most densely populated places in the world. To the north of these lie the more rural New Territories, and there are more than 230 nearby islands. The whole former colony is part of a semisubmerged mountain chain. The Sham Chun river forms the border with mainland China.

Hong Kong's subtropical monsoon climate creates marked seasonal variations. Summers are hot and humid, with heavy rainfall and occasional typhoon storms between June and October. Winters are relatively cool and dry.

Most of the naturally occurring forest has been cleared, but some new woodland – including pines, palms and eucalyptus – has been planted and the "sacred groves" of trees surrounding many New Territories villages have been left undisturbed.

British merchants first arrived in the area in 1821, and were quick to realize the economic potential of the deep-water harbor. The Chinese ceded Hong Kong Island and Kowloon to the British in the treaties that ended the two Opium Wars (1839–40 and 1856–60). Then in 1898 the New Territories were leased to Britain for 99 years. Apart from the four-year Japanese occupation during World War II, Hong Kong remained under British rule, with a governor appointed by the British monarch. In 1949, after the communist victory in China, tens of thousands of Chinese poured into the colony, many of them industrialists from Shanghai.

In 1984 Britain and China signed a treaty outlining the colony's future and that in 1997 the whole of Hong Kong would revert to Chinese sovereignty, and become a "special administrative region". China agreed to leave the present capitalist economic and social systems intact for 50 years. However, in its first fully democratic election for a Legislative Council in 1995, pro-democracy parties clearly won putting a smooth transition in jeopardy.

The great majority of Hong Kong's huge population of 7.3 million people are ethnic Chinese, most of whom speak a Cantonese dialect. English is an official language alongside Chinese. Urban settlement is concentrated around Victoria Harbor, both along the north coast of Hong Kong Island and in Kowloon, with its suburbs in the New Territories (New Kowloon). The chief religions are Buddhism and Taoism, but there is also a sizable Christian minority. British influence has created a culture that combines traditional Chinese and modern Western elements.

Hong Kong is the world's third largest financial center and its twenty-seventh largest trading economy. With few natural resources, its manufacturing industry is heavily dependent on raw materials from mainland China; principal products are textiles and electronic equipment, much of it for export. Hong Kong is also Asia's number one tourist destination.

Hong Kong boasts one of the world's

East meets West (*above*) in Hong Kong, where junks still sail in Victoria Harbor, framed against the distinctive skyline of the central district. In the hills above the harbor is one of Hong Kong's affluent residential districts.

finest natural harbors and one of its largest container ports. A new international airport opened in 1998 at Chek Lap Kok replacing the congested airport at Kai Tak. The world's longest road and rail suspension bridge, Tsing Ma Bridge, opened in April 1997, linking the airport to Kowloon. Most people travel by bus, ferry or subway. With space for housing at a premium, there is also very little land for agriculture. Fishing is well developed.

The media produce newspapers, magazines and television programs with an international appeal. Education and health services are well maintained, but there is little welfare provision. Housing conditions in the numerous high-rise tenements are often extremely cramped.

MACAO

Macao is a tiny special administrative region on the southern coast of China, facing west toward Hong Kong across the mouth of the Zhu river. Smaller and less prosperous than Hong Kong, the territory reverted to China in 1999.

Macao has a total area of only 16 sq km (6 sq mi). It is linked to China's Guangdong province by a short isthmus, and comprises a narrow mainland peninsula, mostly occupied by the city of Macao, and the two small islands of Taipa and Colôane. The peninsula and islands are both hilly in the interior with outlying flatlands, some of which have been extended by land reclamation. There are no rivers, and water must either be imported from China or collected during the heavy rains that fall in the hot and extremely humid summer monsoon season. Winters are

pleasantly cool and dry. Macao's original monsoon forest has long since been cleared, but there has been some attempt at reforestation on Colôane.

The Portuguese began trading with China at the beginning of the 16th century, and in 1557 they leased Macao from China as a trade center and supply station between China, Japan and Europe. For many years it grew and flourished, but in the mid 19th century, as Hong Kong grew in importance, Macao declined, and most of the merchants left.

The Sino-Portuguese Treaty of 1887 established Macao's status as a Portuguese colony, which in 1961 became a Portuguese overseas province. In 1974 its status was again redefined as that of a Chinese territory under Portuguese administration, controlled by a Portuguese-appointed governor. In 1987, after talks between Portugal and China, it was agreed that Macao would be restored to Chinese rule on 20 December 1999, with an autonomous status similar to that of Hong Kong.

Macao has a population of about 450,000, the majority of whom live in the 6.5 sq km (2.5 sq mi) of the mainland peninsula. Although Portuguese is the official language, most people there are Chinese – including thousands of refugees – speaking a Cantonese dialect.

Most of the inhabitants are Buddhists, but some are Roman Catholics; Macao was once the seat of Roman Catholic missionary activities in Asia.

There is little farmland in Macao, and few natural resources apart from fish, so it depends for its prosperity upon its status as a free port and on its exports of clothing, toys and other consumer-based products. Tourism is also important; millions of visitors are attracted each year by Macao's seven casinos, its easy-going lifestyle, its historic churches and temples, and cheap consumer goods.

The 64 km (40 mi) long Macao–Hong Kong sea route is one of the busiest in the world: there are regular ferry, hydrofoil and jetfoil services between the two territories. There is also a frequent but expensive helicopter service from Hong Kong. A new international airport opened – with Taiwanese investment – in Macao in 1999. Local media are largely based in Hong Kong, apart from a few newspapers and a radio and television station. Healthcare and welfare are provided, along with education for students at all levels including university.

The quiet waters of Macao's inner harbor (below) With the arrival of the Portuguese in 1513, Macao became the principal port for international trade with China and Japan. It was eclipsed in the 19th century by the rise of the neighboring Hong Kong.

Southeast Asia

A trio of Buddhist monks (*left*) stroll back to their monastery in Myanmar (formerly Burma). The mainland peninsula and roughly 20,000 tropical islands of Southeast Asia present a vibrant mosaic of different faiths and cultures, the mystical often jostling with the secular.

■ capital city
● major town

height of land (meters)

3000
2000
1000
500
200
0

▲ mountain peak

439

Myanmar (Burma)

UNION OF MYANMAR

MYANMAR (BURMA) IS THE WESTERNMOST of the south-east Asian countries, with Bangladesh and India to the north-west, China to the northeast, and Laos and Thailand to the east. In 1989 the military government changed the official name of the country to Union of Myanmar, which is gradually obtaining international acceptance.

GEOGRAPHY

Myanmar consists, broadly, of a central lowland area enclosed by parallel ranges of forbidding mountains to the east and west. To the north a complex knot of mountains – part of the great Himalayas range – radiates out into branches.

The westernmost branch runs southwest along the Indian frontier, forming a series of densely forested ridges. The central ranges in the north dwindle rapidly, while the higher eastern branches run southeastward into China. From here, the Irrawaddy river flows south for some 2,100 km (1,300 mi), passing the ancient city of Mandalay and the modern capital, Rangoon, and ending in a broad flat delta where it enters the Andaman Sea.

To the east the land rises sharply, the broad uplands of the Shan Plateau adjoining further mountain ranges along the Laotian and Thai borders. Along the long neck of land in the southeast, which Myanmar shares with Thailand, are further mountainous ridges, notably the Tenasserim range.

Myanmar has a tropical monsoon climate moderated to a certain extent by altitude; only the coastal areas are consistently hot throughout the year. The mountains in the far north are subject to freezing winter

temperatures, but act as a barrier to cold air from the north. Between May and October the southwest monsoon brings high humidity and torrential rains; up to 95 percent of the average annual rainfall of 2.5 m (15 ft) occurs during this period. A cooler dry period from October to February gives way to a hot season from March to mid-May when temperatures can soar to over 46°C (115°F) in Pagan, in central Myanmar. Rainfall is heaviest on the coastal ranges, and lightest in the central lowlands around Mandalay. About half the land area is still forested. Forests range from evergreen hardwoods in the wettest areas to hardier monsoon forest where the rainfall is less torrential. Scrubland is characteristic of the driest areas, and mangroves flourish around the deltas. Above 900 m (3,000 ft), pine and evergreen oak predominate, giving way to rhododendrons in the northern mountains.

Myanmar is rich in wildlife. Large mammals include elephants, tigers, leopards and bears, as well as small populations of rhinoceroses, wild water buffaloes and gaurs (Indian bison). The forests teem with gibbons and several kinds of monkeys. Reptiles include endangered crocodiles and gharials in the rivers, turtles along the coasts, and many species of snake, some of them highly venomous.

SOCIETY

Myanmar's two earliest civilizations were the Mon culture, founded in the south in about the 3rd century BC by the Buddhist Mon people from Thailand and Cambodia; and the Pyus people from the north, who established themselves in the Irrawaddy lowlands near Prome in central Myanmar. In the 9th century AD the Pyu lands were conquered by their northern

Coils of brass (*above*) twisted around her neck from the age of five give this Karenni woman the elongated neck considered elegant by her people. The weighty loops do not stretch the neck, but make it appear long by forcing the collarbones downward.

Ruined by time and earthquake (*left*) the great Buddhist temples of Pagan (the capital of Burma 1044–1287) testify to its ancient splendor. The city, sprawling for 16 sq km (6 sq miles), was founded in 849 AD but declined after being sacked by the Mongols.

neighbors and former subjects, the Burmans. In the 11th century a Burman king, Anawrahta (reigned 1044–77), united the whole country under his rule, with the city of Pagan its capital and heart of a Buddhist kingdom that flourished for 250 years. Pagan was eventually destroyed by the Mongols in 1287.

The Chinese invaded in the middle of the 17th century, once more plunging the whole area into war. Subsequently, the Burman leader Alaungpaya (ruled 1752–60) and his successors emerged as rulers. They began to extend their territory along the coast and westward into the Indian state of Assam, where they confronted the British in the first Anglo-Burmese war of

Basic technology A worker in a Myanmar lacquer factory uses an elementary machine to apply a coat of lacquer to a drum. Although its economy is improving, Myanmar remains technologically backward due to a longterm self-sufficiency that rejected Western aid.

NATIONAL DATA – MYANMAR (ex-BURMA)

Land area 657,740 sq km (253,955 sq mi)

Climate	Altitude m (ft)	Temperatures January °C(°F)	July °C(°F)	Annual precipitation mm (in)
Rangoon	23 (75)	26 (78)	27 (79)	2,731 (107.5)

Major physical features highest point: Mount Hkakabo 5,881 m (19,296 ft); longest river: Irrawaddy 2,092 km (1,300 mi)

Population (2006 est.) 47,382,633

Form of government military republic

Armed forces army 350,000; navy 13,000; air force 12,000

Largest cities Rangoon/Yangon (capital - 4,668,755); Mandalay (1,266,453); Mawlamyine (463,401); Bago (253,489); Pathein (246,228); Monywa (189,617)

Official language Burmese

Ethnic composition Burman 68%; Shan 9%; Karen 7%; Rakhine 4%; Chinese 3%; Indian 2%; Mon 2%; other 5%

Religious affiliations Buddhist 89%; Christian 4% (Baptist 3%; Roman Catholic 1%); Muslim 4%; animist 1%; other 2%

Currency 1 kyat (MMK) = 100 pyas

Gross domestic product (2006) U.S. $83.84 billion

Gross domestic product per capita (2006) U.S. $1,800

Life expectancy at birth male 58.07 yr; female 64.03 yr

Major resources petroleum, timber, tin, antimony, zinc, copper, tungsten, lead, coal, marble, limestone, precious stones, natural gas, hydropower, beans, chromium, cotton, groundnuts/peanuts, gold, gypsum, jute, nickel, pulses, cereals, rice, rubber, sugarcane

1824–26. As a result, the British acquired Assam, together with the Burmese states of Arakan and Tenasserim along the coast. A second war (1852–53) led to further losses; the British took Rangoon, and annexed Lower Burma. After the third Anglo-Burmese war (1885–86), the British made the entire Kingdom of Burma a province of India, a decision that greatly angered the Burmese. In 1937 Britain separated Burma from India, making it a crown colony, but independence was not granted until 1948, three years after the end of Japanese occupation (1942–45) in World War II.

In 1962 the country's civilian government was overthrown in a military coup led by General Ne Win (b. 1911). Over the next 12 years Ne Win and his Revolutionary Council nationalized banks and businesses and isolated the country from foreign contacts. Following a further military coup, a multiparty election in May 1990 brought victory to the opposition parties. However, the military rulers refused to give up power, and imprisoned many supporters and the party leader Aung San Suy Kyi.

Opposition leader, Aung San Suu Kyi, was released from house arrest in 1995 but restrictions on political freedom remain. The name of the government was changed to the State Peace and Development Council (SPDC) in November 1997. The military regime has made no commitment to democracy.

Of the country's 41.9 million people the Burmans make up about 69 percent. They live mainly in the central lowlands and coastal districts. Among the many ethnic minority groups, the Karens of the far southeast, and the Chins and Kachins of the west and north, have close affinities with the Burmans. The ancient Mon people are now confined to central coastal districts. Altogether 100 or more different languages are spoken. The Karens have been fighting for over 50 years for their independence; in 1995 Burmese forces over-ran their last outpost in Kawmoora.

Myanmar has a long literary tradition, based on a writing system adapted from that of the Mon, who in turn inherited an even older Indian script. Buddhism is the dominant religion among most ethnic groups, and its beliefs have inspired Myanmar's artistic culture.

ECONOMY

Rice has been Myanmar's major crop and its principal export for well over a century. Other significant crops include cereals,

Traders at an inland floating market on Lake Inle, central Myanmar, sell to customers as they float by lakeside villages built on stilts. A relatively small percentage of the land in Myanmar is farmed for fruit and tree crops; bananas are sent to this area in bulk.

cotton, peanuts and various pulses. Forest hardwoods earn valuable foreign currency, as does the flourishing opium crop. Cattle are the main livestock, and fish, mostly caught at sea, are a vital part of the local diet.

Myanmar's rich mineral resources include sufficient natural gas and petroleum to meet domestic needs, as well as various metal ores, notably copper, tungsten and silver, and jade from the northern mountains. The country is also one of the world's chief sources of rubies and sapphires. Electricity is produced by both oil-fired and hydroelectric power plants.

Since independence there has been heavy investment in industry, especially petroleum and metal refining. The food and timber processing industries have also received new investment, and cot-

tage industries have been subsidized.

Myanmar's roads and railroads are largely in the central and southern coastal lowlands. Few roads are surfaced, and most goods are carried by water. The Irrawaddy forms the main artery of a system of waterways that can bring small boats to any paddy that has surplus rice to sell; Rangoon is the chief port and trading center. The national airline operates domestic flights from Rangoon to other major cities, as well as international routes to Southeast Asian and Indian destinations. All the news media are owned and operated by the government.

Healthcare is free, but there is a serious shortage of medical supplies and modern drugs. Most of the population are still at the mercy of diseases such as malaria, cholera, hepatitis, tuberculosis and polio, and malnutrition is common among children. Primary education is free and compulsory, and some secondary education is available. There are six universities as well as medical and technical education.

Laos

LAO PEOPLE'S DEMOCRATIC REPUBLIC

L AOS, BETWEEN THAILAND AND VIETNAM, IS the only landlocked state in Southeast Asia. The population, about 5.6 million, is one of the lowest in the region.

GEOGRAPHY

Northwestern Laos rises from the Mekong valley lowlands to the Xieng Khouang Plateau. North and east of the plateau the landscape becomes extremely mountainous. In the southwest the Mekong river forms the frontier with Thailand. The monsoon season between May and October is generally hot and humid with heavy rainfall. In the cooler dry season from November to late April

The thatch-roofed village of Nong Het, in northeastern Laos, suggests a rural idyll. In fact, life is basic for the majority of Laotian peasants who derive much of their income from rice – the country's principal crop – which they grow on small, irrigated holdings.

humidity and rainfall are low. More than half the land is forested.

SOCIETY

The first Lao peoples arrived from southwestern China in about the 8th century AD. In the following centuries Laos was invaded by the Burmese, Vietnamese and Thais before it was established as a French protectorate starting in 1893 and completed by treaties in 1904 and 1907.

Laos achieved independence in 1953 but the constitutional monarchy was opposed by the Communist Pathet Lao movement, which controlled the north. The conflict escalated in the 1960s, with the war in Vietnam. A ceasefire was signed in 1973, and in 1975 the Pathet Lao gained power. In 1991, a new constitution provided for a president to be elected five yearly by the national assembly but political parties are banned.

Four ethnic groups are officially recognized in Laos. The official Lao language is spoken primarily by the Lao-Lu, or valley Lao. The majority of Laotians are Theravada Buddhists.

ECONOMY

Agriculture accounts for more than three-quarters of the workforce. The largest single crop is rice. Opium, grown in northern Laos, has become an important (unofficial) cash crop; others are coffee, tobacco and cotton. The forests yield hardwood, spices and oils. There are reserves of coal, metal ores and precious

stones, but the manufacturing sector is limited to processing raw materials.

Most of the roads are unsurfaced, and there are no railroads in the interior. River traffic is the country's lifeline, linked to railroad terminals on the Thai bank. Health conditions are poor and life expectancy is short. Primary, secondary and some further education are available, but literacy levels are generally low.

NATIONAL DATA - LAOS				
Land area 230,800 sq km (89,112 sq mi)				
Climate		Temperatures		Annual
	Altitude m (ft)	January °C(°F)	July °C(°F)	precipitation mm (in)
Vientiane	162 (531)	25 (77)	28 (82)	895 (35.2)
Major physical features highest point: Mount Bia 2,818 m (9,245 ft); longest river: Mekong (part) 4,200 km (2,600 mi)				
Population (2006 est.) 6,368,481				
Form of government one-party (Communist) republic with one legislative house				
Armed forces army 25,600; air force 3,500				
Capital city Vientiane (202,908)				
Official language Lao				
Ethnic composition Lao Loum (lowland) 68%; Lao Theung (upland) 22%; Lao Soung (highland) including the Hmong and the Yao 9%; ethnic Vietnamese/Chinese 1%				
Religious affiliations Buddhist 60%; animist and other 40% (including various Christian denominations 1.5%)				
Currency 1 kip (LAK) = 100 at				
Gross domestic product (2006) U.S. $13.43 billion				
Gross domestic product per capita (2006) U.S. $2,100				
Life expectancy at birth male 53.45 yr; female 57.61 yr				
Major resources timber, hydropower, gypsum, tin, gold, gemstones, rice, coffee, cotton, opium, spices, tobacco				

Thailand

KINGDOM OF THAILAND

THAILAND, KNOWN AS SIAM UNTIL 1939, IS AT the heart of the Indochinese peninsula in Southeast Asia. An independent kingdom for many centuries, Thailand maintains strong ties with the West and with neighboring countries.

GEOGRAPHY

Thailand is a tropical country of high mountains and rainforests, broad floodplains and sandy beaches. The highest land is in the far north, where a series of fold-mountain ranges are the last bastion of the great Himalayan chain. A range of mountains continues southward along the Myanmar border and runs the full length of the southwestern isthmus. These mountains drop steeply on the western side toward the Andaman Sea. The east coast, on the Gulf of Thailand, is gentler but equally indented, with hundreds of sandy bays along its length. Both coasts are fringed with beautiful islands, including Phuket, one of the largest.

The fertile central plains of the Chao Phraya river delta are the most densely populated part of the country, where most of its agricultural and commercial activities are concentrated. This area in particular suffers regular and often severe seasonal flooding. Eastern Thailand is occupied by the Khorat Plateau, a tilted platform made up of rolling sandstone hills. Hills in the far southeast form the continuation of the Cardamom Mountains of southern Cambodia.

Thailand has a tropical monsoon climate, with most of the annual rainfall occurring between May and September. In general, rainfall is lightest in the northeast and heaviest in the south and over the mountains, but this can vary greatly because of exposure to winds.

With much of the natural forest cover cleared for agriculture, less than one-third of the land is now forested, and most of this is in the more inaccessible northern areas. There, hardwoods predominate, particularly teak and resin-producing trees. Game hunting has reduced the wildlife populations considerably. Leopards and tigers are now rare, and rhinoceroses and tapirs nearly extinct.

SOCIETY

Thailand, or "land of the free", though never colonized has not escaped foreign cultural and political influences, both from its neighbors and from the West. The Thai people are believed to have arrived from southwestern China about a thousand years ago. In the 13th century the Sukhothai empire – with its own highly developed culture – was established in the north of the central plain. This was later absorbed by the Buddhist kingdom of Ayutthaya, which emerged farther south in about 1350. In 1767 a Burmese victory and subsequent occupa-

tion ended Ayutthaya rule. The Burmese were soon ousted, however, and the Chakkri dynasty of monarchs was established in 1782 at the new capital of Bangkok; their succession has continued to the present day. Under the scholar-king Mongkut (reigned 1851–68) and his son Chulalongkorn (1868–1910) Thailand was ushered into the modern age: treaties were signed with the West, slavery was abolished and study abroad encouraged.

In 1932 a bloodless coup established a constitutional monarchy. During World War II, the country was forced into an alliance with Japan, but afterward, Thailand received massive economic and military support from the United States. Political instability was increased by wars in neighboring countries, guerrilla activity in border areas and a steady influx of refugees. In 1991 General Suchinda Kraprayoon led a military coup. He was appointed prime minister in April 1992, but was forced to resign the following month after demonstrations calling for democratic reform were brutally crushed. However, in 1996 a new constitution lim-

Sewing to survive (*right*) A woman belonging to the Meo hill tribe near Chiang-Mai, in northern Thailand, sits at her sewing machine as she makes clothes for the local townspeople. Such work, which brings in much-needed money, helps to complement the traditional (but illegal) earnings from the harvest of opium poppies.

White sands, clear water (*left*) The beach at Krabi Ao Pranang, on Thailand's rugged southwest coast beside the Andaman Sea, offers the kind of coastal beauty that lures tourists from around the world. Thailand receives several million foreign visitors every year, making tourism the country's largest source of foreign exchange

ited the power of the military and separated the executive, legislative and judicial branches of government.

The monarchy is hereditary and there is a bicameral National Assembly. The head of government is the prime minister, who is required by a 1992 constitutional amendment to be an elected member of the House of Representatives. Under the 1996 constitution the Senate will consist of a 200-member elected body with members serving six-year terms and the House of Representatives, with 392 members, will become a 500-member body elected for four-year terms after 2000.

Most of the population (about 61.8 million) are ethnic Thais, including some 27 percent of Laotian origin. The large Chinese population (about 12 percent) lives mainly in urban areas, whereas people of Malay ancestry – who practice Islam – are concentrated mainly in the south. The northern hills are inhabited by half a million non-Thai peoples, such as the Karen, Meo Lahu and Akha.

The vast majority of the population are Buddhist, and Thai culture reflects this in its ornamental temples and religious art and in the popular festivals that accompany religious and royal ceremonies.

ECONOMY

Around one-half of the labor force are in agriculture, most of them working in the country's extensive paddy fields. Rice is the nation's primary staple and principal export; other major crops include maize, tapioca, sugar cane and soybeans. The hill peoples of northern Thailand have traditionally grown poppies for opium, but there have been recent efforts to introduce alternative cash crops, such as tea and tobacco. Cattle and buffalo are used mainly as draft animals, while pigs and poultry are raised for meat. Fish from the sea and freshwater varieties form a major part of the Thai diet. Forestry contributes to the export trade.

Thailand is one of the largest producers of tin and a major tin exporter. It also has other mineral resources such as iron ore, gypsum and tantalite (from which the rare, acid-resistant metal tantalum is extracted). The main source of power is offshore natural gas, reserves of oil, and brown coal (lignite). Manufacturing is smallscale and industrial activity is concentrated on the production of clothing, electrical goods, beer, cement, chemicals and motorcyles. Since 1982 tourism has been Thailand's largest revenue earner.

Bangkok is the center from which all transportation systems radiate. Railroads are well developed, whereas roads are often inadequate. Rivers and canals, the traditional mode of transportation, are now less significant than the railroads. There are over 20 ports besides Bangkok. The capital is at the center of an extensive internal flight network.

Broadcasting is controlled by the govern-

ment, and several radio and television stations are owned by the military. There are over 30 daily newspapers, several of them in Chinese.

Social welfare systems are generally inadequate, but some help is given to those in most need. The government has tried to provide primary healthcare throughout the country, but there are still major health problems. Education is free and compulsory for children aged from 7 to 15, and literacy levels are high. Several universities and institutions offer higher education.

NATIONAL DATA – THAILAND

Land area 511,770 sq km (197,596 sq mi)

Climate	Altitude m (ft)	Temperatures January °C(°F)	July °C(°F)	Annual precipitation mm (in)
Bangkok	2 (7)	27 (81)	29 (84)	1,498 (58.9)

Major physical features highest point: Mount Inthanon 2,585 m (8,481 ft); longest river: Mekong (part) 4,200 km (2,600 mi)

Population (2006 est.) 64,631,595

Form of government multiparty constitutional monarchy

Armed forces army 190,000; navy 70,600; air force 46,000

Largest cities Bangkok (capital – 4,819,253); Nonthaburi (404,805); Samut Prakan (393,217); Udon Thani (255,754); Chon Buri (228,227); Nakhon Ratchasima (209,458)

Official language Thai

Ethnic composition Thai 75%, Chinese 14%, other 11%

Religious affiliations Buddhist 94.6%, Muslim 4.6%, Christian 0.7%, other 0.1%

Currency 1 baht (THB) = 100 satang

Gross domestic product (2006) U.S. $585.9 billion

Gross domestic product per capita (2006) U.S. $9.100

Life expectancy at birth male 69.95 yr; female 74.68 yr

Major resources tin, rubber, natural gas, tungsten, tantalum, timber, lead, gypsum, lignite, fluorite, maize/corn, rice, soybeans, sugarcane, cassava, sorghum, tea, tobacco, jute, livestock, fisheries, tourism, textiles

Cambodia

STATE OF CAMBODIA

Riverside bustle (*above*) The shores of the Mekong river in Phnom Penh, capital of Cambodia, throng with people as fresh produce is unloaded from wooden barges, and a passenger boat, carrying cargo and bicycles on the top deck, docks at the pier.

Cleared for agriculture (*below*) Well-watered paddyfields in central Cambodia glint in the sun. The lowland forests have been cut down and replaced by fields of rice, the staple crop, which rely on seasonal flooding by the Mekong and Tonle Sap (Great Lake).

C AMBODIA IS THE HOMELAND OF THE KHMER peoples of southern Indochina (the country is known by its own people as Kampuchea). One of the poorest states in the region, Cambodia suffered severely from involvement in the Vietnam War, and faces an uncertain future after three decades of internal and external conflict.

GEOGRAPHY

Cambodia is a mainly lowland country bordering Thailand to the northwest, Laos to the northeast, and Vietnam to the east and southeast; to the southwest it faces the Gulf of Thailand. Tonle Sap (the Great Lake) at the heart of Cambodia lies to the west of a wide alluvial plain fed by the waters of the Mekong river and its tributaries. Southwest of Tonle Sap, the Cardamom and Elephant Mountains overlook a narrow coastal plain.

The climate is generally hot, with high humidity and heavy rainfall between mid May and early October – the time of the southwest monsoon. From early November to mid March the dry season brings low humidity, sparse rainfall and gentle northeasterly winds.

Some 70 percent of Cambodia is covered by forest, ranging from broadleaf evergreens and deciduous woodlands in the north to dense tropical rainforests on the mountains overlooking the sea. Lowland areas support open forest and savanna grassland, with paddies, reeds and bamboo forest in the wetter floodplains of the rivers. Animal life includes elephants, tigers, leopards and bears, as well as many water birds along the rivers.

SOCIETY

About the 6th century AD the emerging Chenla empire took control of the area that is present-day Cambodia. Gradually, Khmer peoples arrived from the north,

and after the breakup of Chenla in the 8th century a powerful Khmer dynasty emerged. Known as the Angkorian dynasty, it lasted until the 15th century. In 1863 Cambodia, by then little more than a puppet of Siam (modern Thailand) and Vietnam, was obliged to become a French protectorate, retaining its monarchy.

· The Japanese occupied Cambodia in 1941, but French rule was reimposed after World War II. In 1955 the French granted independence, and in 1960 Prince Sihanouk (b. 1922), on the death of his father, became head of state. For the next 10 years, during the Vietnam War, Cambodia remained officially neutral, but Sihanouk allowed the communist North Vietnamese to operate against United States' forces from secret bases within the country. In 1970 Sihanouk was overthrown in a rightwing, pro-United States' coup led by General Lon Nol. The government tried to eradicate the Vietnamese presence but it was unsuccessful. The United States' Air Force took over the job, subjecting the country to an intense and often indiscriminate aerial bombardment that lasted until 1973. This drove many Cambodians to join the Khmer Rouge ("Red Cambodians") – a leftwing movement led by Pol Pot that soon took power in the countryside. In 1975 the capital Phnom Penh fell to the Khmer Rouge, and the government was ousted.

With Pol Pot as premier, the Khmer Rouge began to transform the entire society in the most brutal way. Whole

urban populations were moved onto the land and forced to work on vast irrigation and agricultural projects. Anyone regarded as in any way hostile to the regime was killed. By 1979, when Vietnamese troops and Cambodian rebels captured Phnom Penh, more than 2 million people had died through disease, starvation and mass execution. Although the Cambodian people were relieved to see the Khmer Rouge overthrown, the West and China did not welcome the establishment of a Vietnamese-backed regime at Phnom Penh. When, in 1982, the Khmer Rouge formed an alliance with other opposition groups, it won the overt diplomatic, and covert military, support of the West and China. Under pressure from the United States the Vietnamese withdrew in 1989. The long and bloody civil war came to an uncertain end in 1991, when Prince Sihanouk won an agreement to introduce power-sharing among all of the major factions, including the hated and feared Khmer Rouge. The first free elections were held in June 1993 under United Nations' supervision, with a coalition led by Sihanouk winning a majority. Under a new constitution in 1993 Sihanouk was elected king and appointed a new government. The Khmer Rouge was outlawed in 1994. Pol Pot was captured by Khmer Rouge dissidents in late 1997. He evaded being tried by an international court by committing suicide in 1998.

The great majority of Cambodia's people are Khmers, who speak their own language. Many are Buddhists, despite the suppression of all religion under the Pol Pot regime (1975– 79). The significant ethnic minorities are the Chinese and the Chams, said to be of Indonesian origin, and who are traditionally Muslims. Some 350,000 Cambodians live across the Thai border as refugees from the civil war.

ECONOMY

Cambodia lives by its rice crop, rice fields accounting for nearly 90 percent of arable land. Most people live and work as subsistence farmers. Besides rice, they also grow some fruit, and raise livestock; water buffalo are reared as draft animals. Tonle Sap is one of the world's richest sources of freshwater fish. Cambodia's vast forests are threatened by indiscriminate local tree-felling.

The country has limited mineral reserves, and electricity generation relies entirely on imported fuel. Manufacturing is confined to local processing of rubber and wood, and agricultural products.

Cambodia's road network dates from French colonial days. Few roads are surfaced, and many areas have no roads at all. Flooding is a problem in the rainy season, also affecting the country's vital waterway links. There are two railroads connecting Phnom Penh with Bangkok in Thailand and with Kompong Som, the country's only maritime port. Phnom Penh has an international airport.

Healthcare is gradually improving. However, diseases such as cholera and malaria cannot be controlled and malnutrition is common. Education was severely hit by the wars but children now receive six years of compulsory schooling and literacy is improving. Many primary, secondary and higher educational institutes have been reopened.

NATIONAL DATA – CAMBODIA				
Land area 176,520 sq km (68,155 sq mi)				
Climate		Temperatures		Annual
	Altitude m (ft)	January °C(°F)	July °C(°F)	precipitation mm (in)
Phnom Penh	12 (39)	27 (81)	29 (84)	1,536 (60.4)
Major physical features highest point: Mount Aôral 1,813 m (5,949 ft); longest river: Mekong (part) 4,200 km (2,600 mi); largest lake: Tonle Sap 2,590 to 24,605 sq km (1,000 to 9,500 sq mi) depending on the season				
Population (2006 est.) 13,881,427				
Form of government multiparty democracy under a constitutional monarchy				
Armed forces army 75,000; navy 2,800; air force 1,500				
Capital city Phnom Penh (1,781,113)				
Official language Khmer				
Ethnic composition Khmer 90%; Vietnamese 5%; Chinese 1%; other 4%				
Religious affiliations Theravada Buddhist 95%; other 5%				
Currency 1 riel (KHR) = 100 sen				
Gross domestic product (2006) U.S. $36.78 billion				
Gross domestic product per capita (2006) U.S. $2,600				
Life expectancy at birth male 57.35 yr; female 61.32 yr				
Major resources oil, natural gas, timber, gemstones, iron ore, manganese, phosphates, hydropower potential, bananas, bauxite, cotton, fisheries, jute, livestock, maize/corn, pepper, rice, rubber, sugar palms, tobacco				

Vietnam

SOCIALIST REPUBLIC OF VIETNAM

VIETNAM HAS EXPERIENCED CENTURIES OF foreign occupation, war and strife, and, more recently, partition. Today Vietnam is reunified and is striving to rebuild its economy.

GEOGRAPHY

Vietnam is a long, narrow and largely mountainous country sharing a frontier with China to the north and bordered by Laos and Cambodia to the west. Central Vietnam is dominated by the rugged mountains known as the Annam Highlands, whose crest mostly follows the western border. To the east lies a narrow coastal plain, and at either end sprawling floodplains, formed by the Red river delta in the northeast and the great Mekong river delta in the south. These fertile rice-growing plains are densely populated, even though the rivers often overflow their banks. The highest mountain ranges are in northwestern Vietnam extending westward and northward into Laos and China.

Vietnam has a tropical monsoon climate, but with considerable local variation. Summers are hot and humid everywhere; but in winter, only the north is relatively cool, while the south stays hot. Heavy rains fall between May and October.

Tropical forests cover two-fifths of the country, but deforestation has been remorseless – hardwoods are felled and exported for much-needed hard currency – and this has led to serious flooding and erosion. In addition the vegetation was severely damaged by the United States' military action (1965–73) in the Vietnam War, when defoliants were used.

Vietnam's wildlife is rich and varied.

The forests are home to a number of big cats – including tigers and leopards – as well as elephants, civets, bears, monkeys and deer. Sadly, uncontrolled and illegal hunting, as well as destruction of habitats has led to the extinction of many local animal populations.

SOCIETY

Throughout their history, the Vietnamese people have had to struggle for survival in the face of invasion and colonization by more powerful nations.

By 200 BC a Vietnamese nation had already been established on the delta plain of the Red river. However, for more than a thousand years the people were under the domination of China. It was not until 900 AD that the Vietnamese achieved independence. French missionaries began to arrive in the 17th century and by 1883 Vietnam was a French colony made up of three territories: Tonking in the north, Annam in the center, and Cochin China in the far south. French rule was deeply resented by the Vietnamese, giving rise to strong nationalist movements.

During World War II Vietnam was occupied by Japan. After Japan's defeat, Ho Chi Minh (1892–1969), leader of the communist-dominated League for Vietnamese Independence (Viet Minh), proclaimed an independent republic. But the French would not let go, and for eight years fought to regain their former possessions, until they were finally defeated at Dien Bien Phu in 1954. The country was partitioned, giving Ho Chi Minh's Viet Minh government control

north of the 17th parallel, while anticommunists backed by the French were granted all the land to the south of the dividing line, soon to become South Vietnam, an independent republic.

North Vietnam, backed by the Soviet Union, began a long military campaign to reunite the country. As guerrilla activity intensified, the United States intervened with economic and military aid for South Vietnam. American involvement in the war escalated. Eventually a combination of factors – opposition to the war at home, the loss of American lives and the persistence of the North Vietnamese – forced the United States to withdraw in 1973. The fighting continued, however, until the fall of Saigon in 1975 when the whole of Vietnam came under communist control. Following a series of single-party elections in 1976, the country was renamed the Socialist Republic of Vietnam. The war had devastated the land and left millions homeless. From the 1980s onward natural disasters, border disputes, and the flight of the "boat people" across the South China Sea to Hong Kong continued to impede economic recovery.

The 1992 constitution reaffirmed Communist Party rule, but also formalized free market economic reforms. The 395-member National Assembly, with candidates mainly proposed by the Communist Party, and sitting three times a year, appoints a prime minister and cabinet and elects a president as head of state. The State Council, headed by the president, issues decrees when the National Assembly is not in session. Local government at the provincial, district and village levels is conducted by people's councils.

Most of the country's people are ethnic Vietnamese, but there are as many as 60 minority groups. The Vietnamese language is related to Khmer rather than to Chinese. Vietnamese is the official language, but French and the various ethnic minority languages are all in current use.

Buddhism is the principal religion, but Chinese Taoism and Confucianism are also important, and have strongly influenced local Buddhist practices. The large Roman Catholic minority (7 percent of the population) is a consequence of the past work of French missionaries.

ECONOMY

Vietnam is one of the poorest countries in the world. It is struggling to develop a modern economy, but is heavily dependent on foreign aid.

Agriculture employs 67 percent of the labor force. Farming is concentrated in the two river delta areas, where rice, the principal staple and export crop, is grown; other crops include cassava and yams. Tea, coffee and rubber plantations have been established on the mountain slopes, and tropical fruit is cultivated throughout the country. Fish is the main staple apart from rice, with shellfish a substantial export commodity. Forestry is expanding.

Vietnam is rich in natural resources, having large amounts of coal, phosphates and chromites as well as smaller deposits of other minerals, including gold. Crude oil production was nearly 8 million tonnes in 1995 and there are natural gas reserves. Apart from manufacture of cement and steel, most industry is light, concentrating on chemicals, food processing and textiles. During the early 1990s Vietnam began to establish strong trading links with Japan, Taiwan, Thailand and Singapore, and also to attract substantial foreign investment. However, the economy suffered in the Asian financial crisis of 1997–98 and political adherence to central planning continued to slow economic growth.

Road and rail transportation is inadequately developed. The terrain presents difficulties and the war left a legacy of damage. Most roads are unsurfaced and heavy rains can turn these into quagmires. Rail links with China were reopened in 1996. There is an extensive waterways system and there are international air services from Hanoi and Ho Chi Minh City. Press and broadcasting are under direct government control.

Health and welfare provisions have been greatly improved in recent years, and now extend to previously neglected rural areas. All education is now free, and literacy rates are well over 90 percent for both sexes. There are seven universities, two distance-learning universities and nine for specialized study.

Paddyfields at twilight (*left*) reflect the dying rays of the sun. Paddy rice is overwhelmingly the most important crop throughout the region. Production exceeds 100 million tonnes across Southeast Asia, supplying domestic demand and export markets.

NATIONAL DATA – VIETNAM

Land area 325,360 sq km (125,622 sq mi)

Climate	Altitude m (ft)	Temperatures January °C(°F)	July °C(°F)	Annual precipitation mm (in)
Hanoi	16 (52)	17 (63)	29 (84)	1,667 (65.6)

Major physical features highest point: Fan-si-pan 3,141 m (10,306 ft); longest rivers: Mekong (part) 4,200 km (2,600 mi), Red (part) 805 km (500 mi)

Population (2006 est.) 84,402,966

Form of government one-party communist republic with one legislative house

Armed forces army 412,000; navy 13,000; air force 30,000

Largest cities Ho Chi Minh City (3,525,282); Hanoi (capital – 1,472,717); Haiphong (620,936); Da Nang (485,147)

Official language Vietnamese

Ethnic composition Kinh (Viet) 86.2%; Tay 1.9%; Thai 1.7%; Muong 1.5%; Khome 1.4%; Hoa 1.1%; Nun 1.1%; Hmong 1%; others 4.1%

Religious affiliations Buddhist 9.3%; Catholic 6.7%; Hoa Hao 1.5%; Cao Dai 1.1%; Protestant 0.5%; Muslim 0.1%; none 80.8%

Currency 1 dong (VND) = 10 hao = 100 xu

Gross domestic product (2006) U.S. $258.6 billion

Gross domestic product per capita (2006) U.S. $3,100

Life expectancy at birth male 68.05 yr; female 73.85 yr

Major resources phosphates, coal, manganese, bauxite, chromate, offshore oil and gas deposits, hydropower, anthracite, bananas, cassava, cattle, coffee, fisheries, gold, iron, lignite, limestone, maize/corn, pigs, pineapples, rice, rubber, salt, sweet potatoes, tea, timber, titanium, tobacco

Malaysia

MALAYSIA IS A VIBRANT AND ETHNICALLY complex nation. It consists of West Malaysia on a southern limb of the Southeast Asian mainland; and East Malaysia, comprising two states – Sabah and Sarawak – on the island of Borneo.

GEOGRAPHY

West Malaysia borders Thailand to the north. Sarawak occupies the northwest of Borneo facing the South China Sea, while neighboring Sabah in the far north of the island lies close to the Philippines. Northern and central West Malaysia are predominantly mountainous. The south-eastern corner of the peninsula is characterized by broad river valleys and extensive coastal plains. In East Malaysia the mountainous backbone of Borneo rises to 4,094 m (13,452 ft) at the summit of Mount Kinabalu in northern Sabah. A complex pattern of secondary mountain ranges descends to foothills and broad plains along the coast.

East and West Malaysia share a similar equatorial climate, with high temperatures and high humidity throughout the year. Annual rainfall is high. Between November and March the northeast monsoon blows, followed by the southwest monsoon from April to October.

The dense rainforests support thousands of plant species. There are mangroves on sheltered coasts. Animal life in West Malaysia includes tigers, tapirs and the now rare Javan and Sumatran rhinoceroses. East Malaysia has yet more species, notably sun bears, proboscis monkeys and the endangered orangutan in lowland rainforest.

SOCIETY

Malaysia has a long history of invasion and settlement by different indigenous peoples. The complex ethnic mix has created a rich culture, but also conflict. Today it is a federal constitutional monarchy made up of 13 states. The earliest ancestors of the Malays came from southern China in about 3000 BC. By about 100 AD the Malayans had established trade links with China and India, absorbing both Hinduism and Buddhism. Islam was brought by Persian traders about the 7th century.

In 1511 the Portuguese captured and occupied the city-state of Malacca, which soon became a major center for the lucrative spice trade. The Dutch took Malacca in 1641, but by then the spice trade was in decline. Malacca was captured by the British in 1795, and in 1819 they settled Singapore Island. North Borneo (later, Sabah) ceded to Britain in 1762, and in 1841 the British gained Sarawak.

The free port of Singapore prospered, and in 1867 the Strait Settlements (Pinang, Malacca and Singapore) became a British crown colony. Meanwhile the arrival of Chinese migrants on the peninsula in search of tin provoked violent conflicts with Malays. In the early 20th century British investment led to growth

Living on the edge In Mengkabong, a coastal village in Sabah, East Malaysia, the houses are built on stilts to prevent flooding. Although a tourist attraction, the village has no services, is polluted, and the fishing in nearby waters is poor.

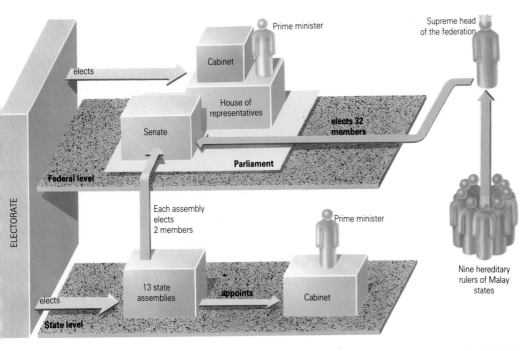

Labels in diagram: Prime minister, Cabinet, elects, House of representatives, Senate, Parliament, Federal level, ELECTORATE, Each assembly elects 2 members, Prime minister, 13 state assemblies, appoints, Cabinet, elects, State level, elects 32 members, Supreme head of the federation, Nine hereditary rulers of Malay states

An unusual federal arrangement Malaysia is a federation of 13 states, each with its own constitution, head of state and elected assembly. Nine retain their traditional rulers, the sultans. The constitution provides for these to elect one of their number to be supreme head (king) of the federation for a term of five years. The remaining four heads of state – of Malacca, Pinang, Sabah and Sarawak – are elected or appointed by their individual state legislatures. Effective power rests with the prime minister and cabinet drawn from the majority party or coalition in the house of representatives (Dewan Rakyat) of the two-chamber federal parliament. The lower house comprises 192 members elected by popular vote for five years, and the upper house 69 members appointed by the king and prime minister.

in rubber and tin production, encouraging a wave of settlers from India.

During World War II the Japanese occupied Malaya, Singapore and Borneo. After the war, British efforts to create a single Malay state brought strong opposition. In 1948, when a federation was formed between Pinang and the peninsular states, the Chinese-dominated Malayan Communist Party (MCP) rebelled. An anticommunist, anticolonial coalition of political parties was formed, which in 1955 won all but one of the seats in the federation parliament. Two years later, on 31 August 1957, Malaya achieved total independence. The constitution favored Malays but all ethnic groups were given freedom of worship and basic rights. In 1963 Sarawak and Sabah joined Malaya in the new independent Federation of Malaysia; Singapore left the federation in 1965. The constitution, amended in 1993, provides for one of the rulers of the Malay States to be elected Supreme Head of the Federation for a period of five years. The Deputy Supreme Head of State is similarly elected.

The largest group are the Malays, who form a slight majority in West Malaysia. Most are Muslim, and their language has been adopted as the national standard. In East Malaysia the Chinese, who make up nearly one-third of the country's total population, are the largest group; most are Taoist or Buddhist. The Indian populations of West Malaysia include Hindus, Muslims and Sikhs, as well as Christians. West Malaysia has a few smaller groups of native non-Malay peoples, most of them living in isolated mountain areas. In East Malaysia there are about 25 such ethnic groups, which account for more than half its population.

ECONOMY

Malaysia's economy is one of the strongest in Southeast Asia. Farming and forestry are important employing about one-third of the total labor force. Vast rainforests provide ample supplies of roundwood for export. The main crop is rubber, of which Malaysia is the world's leading supplier. Palm oil is the other major cash crop.

Malaysia is one of the world's leading tin producers. Bauxite, copper and iron ore are mined, and petroleum provides more than one-tenth of export earnings. Manufacturing includes electronic components for export alongside rubber goods and petroleum products. Most of Malaysia's electricity is generated by oil-fired power stations.

On the peninsula broad highways link

Flighty amphibian Wallace's flying frog can glide 12–15 m (40–50 ft) by using the expanded webbing between its fingers and toes. It is one of a multitude of animal species living in Mulu National Park, Sarawak; many more have yet to be discovered.

major population centers. Buses, taxis and trishaws – bicycle rickshaws – are principal means of transportation. Railroads are equally well developed. In East Malaysia, rivers provide the best means of communication. The ports along the Strait of Malacca are important to international shipping.

Healthcare is free for those living near government clinics or hospitals. In rural areas animal-borne diseases such as malaria are still common. Welfare programs assist the poor, the disabled and the old. About nine-tenths of the country's children attend primary school; after this they may go on to secondary school. Malaysia also has five universities.

NATIONAL DATA – MALAYSIA

Land area	328,550 sq km (126,854 sq mi)			
Climate		Temperatures		Annual
	Altitude m (ft)	January °C(°F)	July °C(°F)	precipitation mm (in)
Kuala Lumpur	39 (128)	27 (81)	28 (82)	2,431 (95.7)

Major physical features	highest point: Mount Kinabalu 4,101 m (13,455 ft); longest river: Rajang (Borneo) 565 km (350 mi)

Population	(2006 est.) 24,385,858

Form of government	federal multiparty constitutional monarchy with two legislative houses

Armed forces	army 80,000; navy 15,000; air force 15,000

Largest cities	Kuala Lumpur (capital - 1,509,699); Subang Jaya (1,091,600); Klang (995,926)

Official language	Bahasa Malaysian

Ethnic composition	Malay 50.4%; Chinese 23.7%; Indigenous 11%; Indian 7.1%; others 7.8%

Official religion	Islam

Religious affiliations	Muslim 52.9%; Buddhist 17.3%; Chinese folk religions 11.6%; Hindu 7%; Christian 6.4%; others 4.8%

Currency	1 ringgit (MYR) = 100 sen

Gross domestic product	(2006) U.S. $308.8 billion

Gross domestic product per capita	(2006) U.S. $12,700

Life expectancy at birth	male 69.8 yr; female 75.38 yr

Major resources	tin, petroleum, timber, copper, iron ore, natural gas, bauxite, cocoa, fish, palm oil, pepper, pineapples, rice, rubber

Singapore

REPUBLIC OF SINGAPORE

SINGAPORE IS ONE OF THE SMALLEST, MOST densely populated and most prosperous countries in the world. Most of the population of over 4 million people live on Singapore Island; the rest inhabit 54 neighboring islets.

Singapore Island has a low but undulating landscape with a hillier district in the center. The climate is uniformly hot and humid throughout the year, with high rainfall. Indigenous mammals include the crab-eating macaque (a type of monkey).

Modern Singapore was founded in 1819 by Sir Thomas Stamford Raffles (1781–1826) of the British East India Company, and later became a British crown colony.

In 1963 it was incorporated into the Federation of Malaysia, but two years later left the federation to become an independent republic.

The head of state is a directly-elected president, and most legislative power rests with the 87-member parliament, 81 elected every five years. However, the People's Action Party, is effectively the only political party. The three main ethnic groups – Chinese, Malay and Indian – speak a variety of languages, including English, and practice various religions.

Agriculture and fisheries are relatively unimportant; many staple foods, including fish, have to be imported. The prosperous and expanding manufacturing sector is based on imported raw materials. Major industries include petrochemicals, textiles and shipbuilding. Financial services, banking and tourism are also major money earners.

There are road and rail links to neighboring parts of Malaysia, and a busy international airport. Healthcare, welfare and education are all of a high standard.

Singapore's gracious colonial buildings are dwarfed by huge international hotels, tax-free shopping complexes and office blocks in the commercial heart of the city. Already a busy trading center in the 14th century, modern Singapore is an economic colossus.

NATIONAL DATA – SINGAPORE

Land area 683 sq km (264 sq mi)

Climate	Altitude m (ft)	Temperatures January °C(°F)	July °C(°F)	Annual precipitation mm (in)
Singapore	10 (33)	27 (81)	28 (82)	2,315 (91.1)

Major physical features largest island: Singapore 541 sq km (209 sq mi); highest point: Timah Hill 162 m (531 ft)

Population (2006 est.) 4,492,150

Form of government multiparty republic with one legislative house

Armed forces army 50,000; navy 4,000; air force 13,500

Capital city Singapore (3,654,103)

Official languages Chinese, Malay, Tamil, English

Ethnic composition Chinese 76.8%; Malay 13.9%; Indian 7.9%; other 1.4%

Religious affiliations Buddhist 42.5%; Muslim 14.9%; Taoist 8.5%; Hindu 4%; Catholic 4.8%; other Christian 9.8%; other 0.7%; none 14.8%

Currency 1 Singapore dollar (SGD) = 100 cents

Gross domestic product (2006) U.S. $138.6 billion

Gross domestic product per capita (2006) U.S. $30,900

Life expectancy at birth male 79.13 yr; female 84.49 yr

Major resources fisheries, deepwater ports, shipbuilding, financial services, tourism

Brunei

SULTANATE OF BRUNEI, ABODE OF PEACE

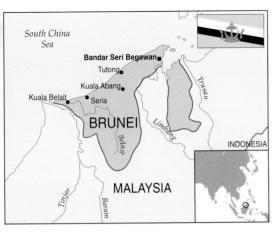

B RUNEI – AN "ABODE OF PEACE" ACCORDING to its unique official title – is an independent sultanate on the northwest coast of Borneo, surrounded and divided in two by the Malaysian state of Sarawak.

GEOGRAPHY

The two parts of Brunei are separated by a few kilometers of coastline where the Limbang river enters Brunei Bay. Both share a landscape of hills and valleys bordering a narrow and often swampy coastal plain. The smaller eastern enclave is more rugged. The humid tropical climate brings heavy monsoon rains between November and March.

Most of Brunei is covered in dense tropical forest, rich in hardwoods. Inaccessibility makes it a haven for wildlife such as monkeys, apes and birds.

SOCIETY

Brunei is known to have had trading links with China since before the 6th century AD. The Portuguese navigator Ferdinand Magellan (1480–1521) was the first European visitor in 1521. At that time the Sultanate of Brunei effectively controlled the whole of Borneo and part of the Sulu Islands and the Philippines to the northeast. In 1839 Sarawak was ceded to the "White Raja" James Brooke (1803–68) and his descendants, while other areas were lost to the British North Borneo Company; Brunei finally became a British protectorate in 1888. In 1929 the first petroleum reserves were discovered.

Following Japanese occupation during World War II Brunei reverted to British rule. In 1959 it adopted a constitutional government and in 1984 achieved full independence within the Commonwealth. Since then, all effective political power has been vested in the sultan.

With over 340,000 people occupying 5,765 sq km (2,226 sq mi), Brunei has the lowest and least dense population of any Southeast Asian country. Malay-descended Muslims make up the majority of the population, followed by Chinese and Indians. The small indigenous population includes Ibans, Dyaks and others.

ECONOMY

Petroleum is the mainstay of the Brunei economy, giving the country one of the highest per-capita incomes in Southeast Asia. There are also natural gas reserves. Other income derives chiefly from small-scale agriculture (pepper and rice), and forestry (cork and rubber).

Main transportation routes are along the coast and rivers, apart from a few roads along the coast. Excellent health-care is available free of charge, even in isolated locations by means of a "flying doctor" service. Free education also extends to remote areas.

NATIONAL DATA – BRUNEI				
Land area 5,270 sq km (2,035 sq mi)				
Climate		Temperatures		Annual
	Altitude m (ft)	January °C(°F)	July °C(°F)	precipitation mm (in)
Bandar Seri Begawan	27 (80)	27 (81)	27 (81)	2,913 (114.6)
Major physical features highest point: Pagon Peak 1,850 km (6,070 ft)				
Population (2006 est.) 379,444				
Form of government nonparty constitutional monarchy with one advisory body				
Armed forces army 4,900; navy 1,000; air force 1,100				
Capital city Bandar Seri Begawan (69,984)				
Official language Malay				
Ethnic composition Malay 67%; Chinese 15%; indigenous 6%; other 12%				
Official religion Islam				
Religious affiliations Muslim 67%; Buddhist 13%; Christian 10%; traditional beliefs and other 10%				
Currency 1 Bruneian dollar (BND) = 100 cents				
Gross domestic product (2003) U.S. $6.842 billion				
Gross domestic product per capita (2003) U.S. $23,600				
Life expectancy at birth male 72.57 yr; female 77.59 yr				
Major resources petroleum, natural gas, rubber, timber, bananas, cassava, coffee, cork, fisheries, rice, pepper				

Carried in splendor (*right*) Brunei's absolute monarch Hassanal Bolkiah waves to onlookers from his ornamental float during a procession through the streets of the capital. The sultan's wealth is derived from Brunei's petroleum and natural gas resources.

Indonesia

REPUBLIC OF INDONESIA

T HE REPUBLIC OF INDONESIA FORMS A LONG island bridge between the Asian and Australian continents. It comprises a vast archipelago of 13,677 islands extending 5,100 km (3,200 mi) from Sumatra in the west to West Papua (previously Irian Jaya) on New Guinea in the east, off the north coast of Australia. Two of the larger islands are shared with other countries: the northern coastal section of Borneo belongs to Malaysia, while the eastern half of New Guinea forms the main part of Papua New Guinea. The eastern part of the smaller southeastern island of Timor became independent of Indonesia in 2002.

NATIONAL DATA - INDONESIA

Land area 1,826,440 sq km (705,192 sq mi)

Climate	Altitude m (ft)	Temperatures January °C(°F)	July °C(°F)	Annual precipitation mm (in)
Jakarta	8 (26)	27 (81)	29 (84)	1,659 (65.3)

Major physical features highest point: Mount Jaya (New Guinea) 5,040 m (16,535 ft); longest river: Barito (Borneo) 885 km (550 mi); largest lake: Lake Toba (Sumatra) 1,300 sq km (502 sq mi)

Population (2006 est.) 245,452,739

Form of government multiparty republic with two legislative houses

Armed forces army 233,000; navy 45,000; air force 24,000

Largest cities Jakarta (capital - 8,568,838); Surabaya (2,358,111); Bandung (1,651,840); Medan (1,763,874); Bekasi (1,619,775); Palembang (1,257,449)

Official language Bahasa Indonesian

Ethnic composition Javanese 45%; Sundanese 14%; Madurese 7.5%; Coastal Malays 7.5%; other 26%

Religious affiliations Muslim 88%; Protestant 5%; Roman Catholic 3%; Hindu 2%; Buddhist 1%; other 1%

Currency 1 Indonesian rupiah (IDR) = 100 sen

Gross domestic product (2006) U.S. $935 billion

Gross domestic product per capita (2006) U.S. $3,800

Life expectancy at birth male 67.42 yr; female 72.45 yr

Major resources petroleum, tin, natural gas, nickel, timber, bauxite, copper, coal, gold, silver, bananas, cassava, copra, coffee, fish, ground nut, maize/corn, palm oil, rice, rubber, soybeans, spices, sugarcane, sweet potatoes, tea, tobacco

GEOGRAPHY

Indonesia lies at the junction of three sections of the Earth's crust, and its animal life is sharply differentiated between Asian and Australian species.

The land

Indonesia's islands are strewn across 8 million sq km (more than 3 million sq mi) of tropical seas. Geologically, the country divides into three sections. The western section includes Borneo in the north and the whole southern chain of islands, from Sumatra in the west to Timor-Timur in the southeast; these together form part of the Sunda shelf, a mostly submerged, southeastward extension of the Asian continent. In the center, Sulawesi (east of Borneo), and the southern Moluccas to the east of it, are part of a long, partially submerged mountain chain that extends northward to the Philippines. The Sahul shelf in the east, which includes New Guinea and the northern Moluccas immediately to the west, forms a northern extension of Australia.

The Indonesian part of Borneo, Kalimantan, is dominated by mountains in the north bordering the Malaysian states of Sarawak and Sabah. The thickly forested central highlands descend to broad, flat alluvial swamplands.

The island of Sumatra rises from swampy, forest-covered lowlands in the east to the Barisan Mountains running along the whole of the southwest coast; they include 10 active volcanoes and beautiful crater lakes, notably Lake Toba in the north.

Southeast of Sumatra, across the narrow Sunda Strait, lies the mountainous but heavily populated island of Java. A long volcanic range, forming the backbone of the island, includes 50 active volcanoes and a further 17 that have only recently become dormant. The volcanic arc stretches eastward across the Lesser Sunda Islands of Bali, Lombok, Sumbawa and Flores. Sumba and Timor-Timur, lying to the south of the main island chain, are rugged but not volcanic.

Sulawesi island is made up of mountainous peninsulas radiating from a highland core. The long, highly volcanic peninsula of Minahassa extends to the northeast. The islands of the Sahul shelf – New Guinea and the northern Moluccas – are structurally similar to those of the Sunda shelf. There are mountains on the northern side, including some active volcanoes, whereas southern New Guinea has large areas of low-lying swampland.

Climate

The entire archipelago lies close to or on the Equator. Temperatures are high throughout the year, and are influenced by altitude rather than latitude. Lowland areas can be extremely hot, but the highlands are cooler; only the central mountains of West Papua (previously Irian Jaya) rise high enough for snow to fall.

Most of Indonesia receives rainfall throughout the year, but the amount depends on exposure to the prevailing seasonal winds. The northwest monsoon blows from December through February, while southeasterly winds predominate from June through August. Winds are lighter and more variable during the two transitional periods.

have been devastated in many places through logging and settlement. They contain some 30,000 plant species.

This region is also home to some of the world's most varied wildlife, including 1,480 bird species (16 percent of the world's total), of which 370 are endemic. Of the 500 mammal species, 100 are unique to the archipelago. Also unique is the Komodo dragon, the world's largest lizard at 3 m (10 ft) long, found on Komodo and neighboring islands. The wildlife is divided between Australian and Asian species by an imaginary boundary, known as the Wallace line, that runs from north to south.

SOCIETY

Indonesia represents the union of many islands and peoples with their own diverse and often conflicting histories.

History

The Indonesian archipelago is one of the oldest inhabited areas on Earth. Early human beings called Java man (*Homo erectus*) lived there over 500,000 years ago. The ancestors of most Indonesians probably arrived from Asia from about 1000 BC. Later, Indian traders and exiles came to settle.

Between the 9th and late 12th centuries the region was ruled by the powerful Sailendra princes followed by the great Majapahit empire of Java, which reached its height in the mid 14th century, but broke up a century later. The decline of the Majapahits was partly due to the influence of the Islamic faith introduced by Arabic traders from the west.

In the early 16th century the Portuguese seized the Moluccas, or Spice Islands. Then, in 1608, the Dutch East India Company's fleet forced the Portuguese to capitulate, allowing the Netherlands to establish effective economic control of the Moluccas, Java and Sumatra. The Dutch colonies were taken by Britain in 1811, but were returned to the Netherlands by the Treaty of Vienna in 1814–15. As Dutch territorial expansion continued into the 20th century, it gave rise to Western-influenced nationalist movements. In 1927 Ahmed Sukarno (1901–70), an engineering graduate, founded the first of several Indonesian nationalist parties. In 1942, during World War II, the Japanese invaded and installed Sukarno as one of their administrators. When the Japanese surrendered in 1945, Sukarno declared the islands independent as the single country, Indonesia. However, the Dutch kept western New Guinea (was

The home of the Toradjans (*above*), one-time headhunters thought to be descendants of early Austronesian settlers, is in the interior of the lush Indonesian island of Sulawesi. They now farm the volcanic soils for rice, sugar cane, copra and rattan.

The Sulawesi cuscus (*left*), a marsupial with long claws and a prehensile tail for gripping the tree trunks, is well adapted to life in the rainforest. Sulawesi's unique animal life also includes the babirusa (a species of wild pig) and the anoa, or dwarf buffalo.

Plants and animals

The muddy coastlines bordering lowland areas are lined by mangroves that broaden out into vast mangrove swamps in southern Borneo and eastern Sumatra, where the adjoining seas are unusually shallow. Inland, much of the archipelago is covered in forest. These are among the richest habitats in the world but they

Irian Jaya now West Papua) until 1962.

Sukarno's long rule was marked by corruption, economic problems and conflict with Malaysia. Hostile to all Western countries, he espoused a communist-style ideology, assumed increasingly dictatorial powers, and eventually alienated his former allies, the military. In 1965 the army took control under General Suharto (b. 1921); in 1960 Sukarno was put under house arrest until his death in 1970.

Suharto, appointed president in 1968, remained in power until 1998 despite promising a swift return to democracy. He successfully opened up the economy to Western investment but ran an authoritarian regime with military aid. In 1975 Indonesia invaded East Timor and suppressed a revolt on Irian Jaya. In 1998 riots erupted following serious economic problems and Suharto resigned in favor of vice president Baccharuddin Jusuf Habibie. Habibie delivered liberal, political and economic reforms. In 1999, Habibie revoked laws banning political parties and announced East Timor could hold a referendum on independence. The first free, multiparty elections since 1955 were held in May with the majority seats won by the Democratic Party of Struggle (PDIP) but following a corruption scandal and international censure over East Timor, Habibie lost the presidential election in October to Abdurrahman Wahid. The East Timor referendum took place under UN auspices on August 30 and resulted in a majority vote for independence. However, paramilitary violence erupted requiring UN intervention. In May 2002 the UN finally handed over power to the new government of East Timor. Indonesia also face demands for independence from the provinces of Aceh in northern Sumatra and Irian Jaya, which became West Papua in 1999.

A colorful Balinese procession (*above*) Elaborate towers of fruit are carried by local Balinese villagers to a Hindu festival as offerings to the gods. These annual ceremonies often involve a banquet and may be accompanied by a traditional gamelan orchestra.

Symbol of mystery, a silhouetted Hindu temple on an island in a Balinese lake. In the 16th century, when Islam triumphed over Hinduism on neighboring Java, Bali became a haven for exiled Hindus. Today, it is the only bastion of Hinduism in the Indonesian archipelago.

Government

Executive power lies with the president and vice president who are elected every five years by the People's Consultative Assembly. The president appoints a cabinet and departmental heads. The legislative body, the House of People's Representatives (DPR) is made up of 400 elected and 100 appointed members who serve a term of five years. The People's Consultative Assembly, which meets every five years, is made up of the DPR plus 200 appointed members. Its functions are to elect the president and to determine national policy and the constitution.

People

The Indonesians comprise 200 to 300 ethnic groups, most of which probably have Austronesian (Malayo-Polynesian) ancestry. The exception are the Papuan peoples of New Guinea, who are of Australoid origin, like the Aborigines.

Although the Javanese are the dominant ethnic group, the official language, Bahasa Indonesia, is based on an east Sumatran form of Malay that spread to Malaya and coastal areas of Borneo; it is now spoken as a second language by most educated people. The only significant immigrant group are the ethnic Chinese.

Most of the population are Muslims but have been heavily influenced by Buddhism and Hinduism.

ECONOMY

Indonesia's economy suffered badly in the 1997 Asian financial crisis, but by 1999 had dramatically improved. Economic prosperity, however, relies on political stability.

Agriculture and industry

Agriculture employs about 40 percent of the labor force and produces about 20 percent of GDP. Rice is the major crop, other staples such as cassava and soybeans are grown. Cash crops include rubber, of which Indonesia is one of the world's largest suppliers, coffee, tea and spices.

Indonesia's vast forests have until recently supplied an ever-increasing demand for timber and its products.

However, there is international concern at the accelerating rate of Indonesia's deforestation – 1 million ha (2.5 million acres) per year between 1950 and 1981 and 5.4 million ha (13.3 million acres) between 1990 and 1995. Petroleum and natural gas from Sumatra, Borneo and the Java Sea provide vital export revenue; and, along with Sumatran coal, fuel for the electricity industry. The chief manufactures are chemicals, electronic components, rubber tires, and textiles.

Transportation and communications
Shipping is the most important means of transportation, providing vital links across Indonesia. Java and Sumatra both have state-operated railroads, but only Java has an adequate network of roads.

Indonesia has had a thriving free press and communications industry since 1999.

Welfare and education
Most communicable diseases in Indo-nesia are gradually being brought under control. Health problems relate to malnutrition and to severe housing shortages.

The majority of Indonesians are liter-ate, following a concerted government education program. Most children go to primary school for six years, and many spend a further six years in secondary education, which includes vocational, technical and agricultural training. There are public universities in every province as well as private higher education.

East Timor
DEMOCRATIC REPUBLIC OF EAST TIMOR

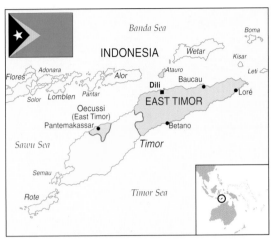

ON MAY 20 2002 EAST TIMOR BECAME AN independent republic after two years of United Nations Transitional Administration and 25 years of brutal occupation by Indonesia.

A former Portuguese colony, East Timor lost a quarter of its population to violence and famine during Indonesian military rule. In 1999 the new Indonesian government allowed East Timor to hold a referendum. Despite violence and intimidation at the polls 78.5 percent of over 450,000 voters chose independence. In response, pro-Jakarta militias began a campaign of arson and theft. By the time UN forces intervened most of East Timor's infrastructure had been destroyed.

In April 2002 East Timor elected its first president, independence hero Xanana Gusmao. He will lead in rebuilding his country with help from the international community. However, there is potential for revenues from offshore oil and gas fields due to be exploited from 2004.

NATIONAL DATA – EAST TIMOR

Land area	15,007 sq km (5,794 sq mi)
Climate	tropical with monsoon rain and pronounced dry season
Major physical features	highest point: Mount Tata Mailau or Ramelau 2,950 m (9,679 ft)
Population	(2006 est.) 1,062,777
Form of government	multiparty republic with one legislative house
Armed forces	army 1,250; naval element 36
Capital city	Dili (166,903)
Official languages	Tetum, Portuguese
Ethnic composition	Timorese 78% (including Maubere); Indonesian 20%; Chinese 2%
Religious affiliations	Roman Catholic 90%; Muslim 4%; Protestant 3%; Hindu 0.5%; Buddhist; Animist 2.5%
Currency	1 lira (Lit) = 100 centesimi
Life expectancy at birth	male 63.96 yr; female 68.67 yr
Major resources	fish, timber, maize/corn, rice, cassava, sweet potatoes, coffee, coconuts, vanilla, mangoes, livestock, gold, petroleum, natural gas, manganese, marble

Philippines

REPUBLIC OF THE PHILIPPINES

THE REPUBLIC OF THE PHILIPPINES, LYING some 800 km (500 mi) off the southeast coast of Asia, encompasses an archipelago of 7,107 islands.

GEOGRAPHY

The Philippines lie in an area of great tectonic instability – the "ring of fire" that surrounds the Pacific Ocean – and within the path of violent tropical storms.

The archipelago, which extends 1,770 km (1,100 mi) north to south, is roughly triangular, and bounded by the South China, Sulu and Celebes Seas. The Philippine Trench, 10,500 m (34,500 ft) deep, runs along the eastern coastline. The islands are generally mountainous, have narrow coastal plains and some are heavily populated. Just 11 islands constitute 94 percent of the total land area.

Luzon in the north is the largest island, much of it mountainous and rugged. The Zambales Mountains on the west coast partly enclose a broad central plain – the country's only extensive lowland – running from the Lingayen Gulf in the north to Manila Bay. Southeastern Luzon has a landscape of scattered volcanoes.

Between Luzon and the second largest island, Mindanao in the southeast, is a scattered range of islands including those of the Visayas group, such as Samar, Leyte, Cebu, Negros, Panay, Romblon and Mindoro. A chain of small volcanic islands, the Sulu Archipelago, extends southwest. North of Sabah (northern Borneo) is the island ridge of Palawan.

Like much of Southeast Asia, the Philippines are subject to seasonal monsoon winds blowing from the southwest between May and October, and from the northeast between November and Feb-

ruary. Temperatures do not vary greatly from north to south, but they are affected by altitude; the high ground is always cooler, and also generally wetter. Tropical typhoon storms, normally occurring between June and December, are often severe and accompanied by flooding.

Forests range from tropical rainforest near the coasts to subtropical evergreens on the slopes, and pinewoods around the peaks of northern Luzon. Many areas have been cleared. The islands are rich in wildlife, and isolation has encouraged the evolution of many unique animals.

SOCIETY

The islands were first settled by various groups of peoples about 30,000 years ago. These early Filipinos provided themselves with food by hunting, fishing and

Mount Mayon, overlooking paddy fields in southern Luzon (the largest of the Philippine islands) presents a quiet but simmering threat. Many areas face constant danger from active volcanoes, but ash from the eruptions makes soil fertile for growing crops.

slash-and-burn cultivation. The islanders lived in extended family groups called *barangays,* each led by a chief.

Between the 8th and 10th centuries the islands were trading with Chinese, Japanese and Malay merchants. However, the first major foreign influence was the arrival of Islam from Brunei (a sultanate on the island of Borneo) in the 15th century. In 1521 the seafarer Ferdinand Magellan (1480–1521) visited the archipelago, where he claimed the country for Spain. In 1565 the first permanent Spanish settlement was established on the central island of Cebu, and the capital, Manila, was founded six years later on Luzon. The Spanish also named the islands after their king, Philip II (1527–98). As Spanish rule became more firmly rooted, the majority of Filipinos were converted to Roman Catholicism.

In the 1830s Manila was opened to foreign trade, ending centuries of Spanish monopoly. Demand for coffee, sugar and hemp encouraged the growth of new estates, many of which were run by Chinese–Filipino mestizos. By the 1880s the seeds of nationalism were being sown, and in 1896 there was an abortive armed revolt against Spanish rule. During the Spanish–American War of 1898 the Americans encouraged the drive toward independence. But with the Spanish defeat, the islands were ceded to the United States for $20 million. After resisting for three years, the Filipinos were forced to accept United States' rule in 1901.

Filipinos took an increasingly active role in government. In 1935, an interim government was established for independence, but this was interrupted by the Japanese invasion in 1941. The United States recaptured the islands in October 1944 and on 4 July 1946, the Republic of the Philippines was finally declared. However, the United States obtained by negotiation a long lease on several army, navy and air bases.

In 1969 Ferdinand Edralin Marcos (1917–89) became the first president since independence to be elected to a second term of office. However, the election of the so-called Constitutional Convention in the next year was unpopular and students took to the streets in protest. Martial law was declared in 1972. The state of emergency officially ended in 1981, but Marcos retained enormous personal power. The assassination of the opposition leader, Benigno Simeon Aquino (1932–83), brought matters to a head. In the 1986 election Marcos was opposed by Benigno Aquino's widow, Corazon Aquino (b. 1933). Marcos was declared the winner, but public outrage at manifest electoral fraud developed into a popular revolution, which toppled him from power. He and his wife Imelda fled.

On assuming the presidency, Corazon Aquino faced a parlous economic situation, widespread poverty and threats from communist and Muslim insurgents. Her government drafted a new constitution, similar to that of 1935, which was ratified by referendum in 1987. Fidel Ramos succeeded to the presidency in 1992. In 1996 a peace agreement ended 25 years of civil unrest by Muslim insurgents. Joseph Estrada was elected president in 1998 but was toppled in a "people's revolution" in 2001 following a corruption scandal. Vice-president Gloria Arroyo took over the presidency.

The 1987 constitution provides for a two-chamber parliament: the Congress of the Philippines. The 250 members of the lower house are mostly elected every three years from local districts. The 24 senators are elected on a national basis, and may hold office for, at most, two six-year terms. The head of state is the president, who is elected for a single six-year term and is responsible for appointing the cabinet, though these appointments are also subject to parliamentary approval. There is a large degree of autonomy in local administration.

Few of the descendants of the first inhabitants – the Negritos – are left. Most of today's Filipinos are of Mongoloid origin whose ancestors came from Taiwan and the surrounding region. Over the centuries, however, there has been intermarriage with Chinese immigrants and with Spanish and American colonists.

Seventy or more languages are spoken in the Philippines. The Tagalog group of dialects has the majority of indigenous speakers. However, the national language, Pilipino, which is based on the Manila form of Tagalog, is fast gaining ground as both the first and second spoken language. The vast majority of Filipinos are Catholic. There is a vigorous Muslim community in the south.

ECONOMY

Development in farming and industry have both been encouraged by government investment and tax concessions. However, industrial output has tended to benefit foreign investors more than the local economy.

The country is still predominantly agricultural, with a large rural labor force. The mountain slopes are terraced to grow rice. Other major crops include tobacco and tropical fruits, such as pineapples; sugar, coconuts and copra from coconuts are significant export crops.

Fishing contributes very significantly to exports and to the Filipino diet. The output of timber, however, has declined with reduced forest cover.

The country is well endowed with mineral resources, especially nickel and copper; gold is also found in significant quantities. Nickel, copper, tin, zinc and lead are processed domestically in smelting and refining works. Manufacturing industry is still at a relatively early stage of development. Many factories established by foreign companies use local labor to finish goods manufactured abroad. The chief local manufactures include electronics, textiles, clothing, chemicals and machinery.

Only the islands of Luzon, Panay and Negros have railroads, but good road networks exist in all of the heavily populated areas. Manila is the main port, followed by Cebu city; each has an international airport nearby. All the main islands are linked to Manila by domestic airlines and shipping lines.

The Philippines' lively free press has enjoyed something of a renaissance since the overthrow of the Marcos regime, with newspapers published in English, Pilipino and main regional dialects. Local radio and television companies belong to a single national association, but are not subject to government control.

Healthcare is centered on Manila; elsewhere treatment is often hard to obtain. Malnutrition remains a problem, especially in outlying areas, and there is a serious shortage of housing and proper sanitation around Manila. Primary education is free and compulsory, and secondary schooling is also free, with a majority of children attending. There are about 500 state-run universities, and a few private higher education institutions.

NATIONAL DATA – PHILIPPINES

Land area	298,170 sq km (115,124 sq mi)			
Climate		Temperatures		Annual
	Altitude m (ft)	January °C(°F)	July °C(°F)	precipitation mm (in)
Manila	15 (49)	27 (75)	28 (82)	2,131 (83.8)

Major physical features largest island: Luzon 108,171 sq km (41,765 sq mi); highest point: Mount Apo (Mindanao) 2,954 m (9,692 ft)

Population (2006 est.) 89,468,677

Form of government multiparty republic with two legislative houses

Armed forces army 66,000; navy 24,000; air force 16,000

Largest cities Manila (capital - 10,665,584); Davao (1,278,148); Cebu (829,023); Antipolo (582,896)

Official languages Philipino, English

Ethnic composition Tagalog 28.1%; Cebuano 13.1%; Ilocano 9%; Bisaya/Binisaya 7.6%; Hiligaynon Ilonggo 7.5%; Bikol 6%; Waray 3.4%; other 25.3%

Religious affiliations Roman Catholic 80.9%; Evangelical 2.8%; Iglesia ni Kristo 2.3%; Aglipayan 2%; other Christian 4.5%; Muslim 5%; other 1.8%; unspecified 0.6%; none 0.1%

Currency 1 Philippine peso (PHP) = 100 centavos

Gross domestic product (2006) U.S. $443.1 billion

Gross domestic product per capita (2006) U.S. $5,000

Life expectancy at birth male 67.32 yr; female 73.24 yr

Major resources timber, petroleum, nickel, cobalt, silver, gold, salt, copper, coconuts, copra, fisheries, tobacco, rice, sugarcane, fruit

Japan and Korea

The gateway to a Shinto shrine (*left*) on the shore of Miyajima island in the Inland Sea, a symbol of the Shinto belief in the sacred power of nature. Colorful festivals celebrating the numberless spirits (*kami*) that reside in the sky, in rocks, trees, waterfalls and islands are held throughout the year in Japan.

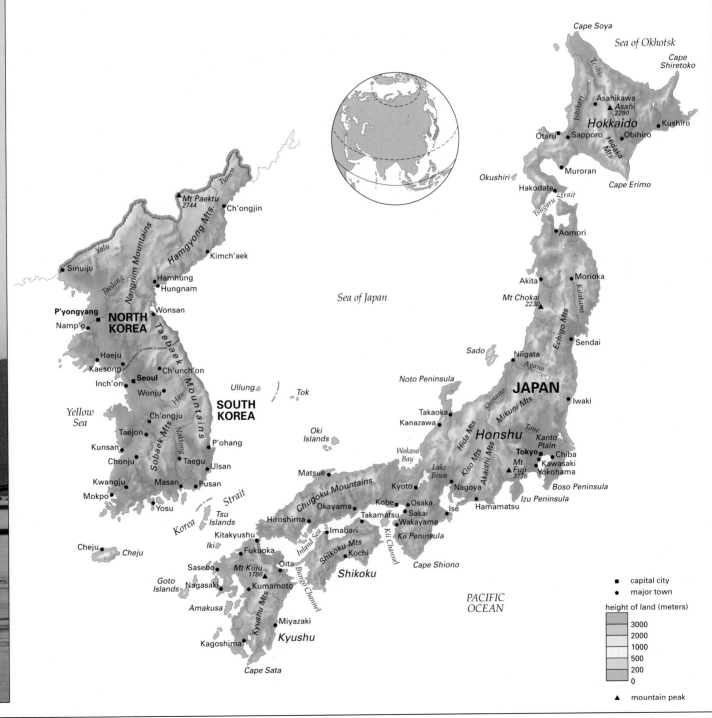

Cape Soya

Sea of Okhotsk

Cape Shiretoko

Teshio

Asahikawa
▲ *Asahi 2290*

Hokkaido

Kushiro

Otaru • Sapporo • Obihiro

Okushiri

Muroran

Hidaka Mts

Cape Erimo

Hakodate

Tsugaru Strait

Mt Paektu 2744

Ch'ongjin

Tumen

Hamgyong Mts.

Nangnim Mountains

Kimch'aek

Yalu

Sinuiju

Hamhung
Hungnam

Taedong

P'yongyang ■ Wonsan

NORTH KOREA

Namp'o •

Taebaek Mountains

Haeju

Kaesong
Seoul ■ Ch'unch'on

Inch'on • Wonju

Ullung *Tok*

SOUTH KOREA

Yellow Sea

Ch'ongju

Sobaek Mts

Naktong

Han

Taejon

P'ohang

Kunsan

Chonju Taegu Ulsan

Kwangju Masan Pusan

Mokpo

Yosu

Cheju • *Cheju*

Goto Islands Nagasaki

Tsu Islands

Iki

Kitakyushu

Fukuoka

Sasebo

Mt Kuju 1788

Amakusa

Kyushu Mts

Kumamoto

Miyazaki

Kagoshima *Kyushu*

Cape Sata

Korea Strait

Sea of Japan

Aomori

Akita Morioka

Mt Chokai 2230 ▲

Kitakami

Echigo Mts

Sendai

Sado

Niigata

Agano

Noto Peninsula

JAPAN

Iwaki

Shinano

Takaoka *Mikuni Mts*

Kanazawa

Hida Mts *Tone* *Kanto Plain*

Honshu

Kiso Mts *Akaishi Mts*

Tokyo ■ Chiba

Oki Islands

Wakasa Bay

Mt Fuji 3776 ▲ Kawasaki

Yokohama

Lake Biwa

Kyoto Nagoya Hamamatsu *Izu Peninsula*

Matsue

Boso Peninsula

Chugoku Mountains

Okayama Kobe • Osaka

Takamatsu Sakai

Hiroshima Wakayama Ise

Imabari

Inland Sea

Shikoku Mts Kochi

Kii Peninsula

Oita

Bungo Channel

Shikoku

Kii Channel

Cape Shiono

PACIFIC OCEAN

■ capital city

• major town

height of land (meters)

3000
2000
1000
500
200
0

▲ mountain peak

Japan

JAPAN IS AN ARCHIPELAGO OF SOME 4,000 islands that lie across a fault line off the eastern coast of mainland Asia. The Korea Strait between Japan and South Korea (the nearest mainland point) is about 180 km (110 mi) wide. Japan's relative geographical and cultural isolation – the country cut itself off from other cultures for 200 years until the mid 19th century – has contributed to the development of a remarkably uniform society.

Since earliest times Japanese civilization has been challenged by the destructive power of the elements: earthquakes, volcanoes, tidal waves and hurricanes. The country's mountains and thick forests cover two-thirds of the land area and have effectively confined settlement, agriculture and industry to the coastal plains. Despite these difficulties, Japan has transformed itself from a 20th-century feudal state into one of the world's most dynamic 21st-century industrial nations.

GEOGRAPHY

The islands of Japan form an arc some 2,500 km (1,500 mi) long. To the southwest are the East China Sea and the People's Republic of China; to the west is the Sea of Japan, beyond which lies South Korea, North Korea and Russia; to the north, across the Soya Strait, is the disputed Russian island of Sakhalin; while to the northeast, the Nemuro Strait separates Japan from the Russian Kuril Islands. The country's eastern and southeastern shores are washed by the Pacific Ocean.

The land

The most northerly of the four main islands of the archipelago is Hokkaido. Its chief mountain ranges are the Kitami in the north and the Hidaka in the south. A third range to the west is separated from the others by the floodplain of Japan's longest river, the Ishikari.

Southwest of Hokkaido, across the Tsugaru Strait, is Honshu, the largest and most populous of the islands. From the island's northern end the mountains of Hokkaido continue southward as the Kitakami range, which overlooks the Pacific coast of northern Honshu. At the heart of the island are the Japanese Alps, and southeast of here the tallest single peak – Mount Fuji, a volcano rising to 3,776 m (12,388 ft). Japan's largest lowland area, the Kanto Plain, lies east of Fuji covering some 13,000 sq km (5,000 sq mi). The capital, Tokyo, stands on its southern side, on Tokyo Bay.

West of the central mountain complex, the smaller Nobi Plain surrounds the city of Nagoya at the northern end of Ise Bay, and still farther west is Japan's largest lake, Lake Biwa. Osaka, Japan's second largest city, lies in Osaka bay. To the north and west of Osaka the Chugoku Mountains occupy the greater part of the

Dusk falls over the sleeping dragon (*above*) Sacred Mount Fuji, almost perfectly symmetrical, rises majestically from a flat plain. Last active in 1707, the volcano has long been celebrated in Japanese art and literature, and is a place of pilgrimage.

island's westernmost peninsula. A narrow coastal plain on its south side faces Shikoku, an island across the Inland Sea.

The most westerly island, Kyushu, is separated from Honshu to the north by the narrow Shimonoseki Strait, and from Shikoku to the east by the Bungo Channel. Kyushu's central mountains rise to several peaks above 1,500 m (5,000 ft). South of Kyushu the Ryukyu Islands form an arc measuring 1,000 km (600 mi) from end to end. Two other island groups also belong to Japan: the Bonin group, 1,000 km (620 mi) south of Tokyo, and, farther south the three Volcano Islands.

Japan has many rivers, most of them short, swift-flowing and draining comparatively small areas. In spring, they are swollen by melting snow and ice from the mountains, and in summer often by heavy rainfall. Flooding is a common problem in the low-lying coastal regions.

Japan also experiences volcanic eruptions and earthquakes. At least 60 volcanoes have erupted since records were kept. The major active zones are in south-

New year symbol of longevity and strength A pine tree clings to one of Japan's densely forested mountain slopes. Ranging from subtropical to alpine, the forests contain a great variety of species, including those of fir, spruce, beech, oak, maple, palm, bamboo and cherry.

ern Kyushu and in northern and central Honshu. A major earthquake occurs on average about every five years, the latest hit Kobe and Osaka in January 1995 killing almost 5,000 people.

Climate

Japan has a monsoon-type climate. In winter the prevailing northwesterly wind blowing from continental Asia gives bitterly cold weather with temperatures below freezing. The cold air, as it approaches from the milder waters of the Sea of Japan, gathers moisture, creating heavy clouds that release snow along the western coast and mountains. In the east, winters are drier and clearer. A narrow region of the southeast-facing coast is also warmed by the offshore current. Average winter temperatures are higher in the south; farmers on the island of Kyushu can grow a winter crop.

Toward the end of March, as the Asian continent warms up, the winds begin to switch round to the south and east, bringing several weeks of torrential rain (known as the *baiu* or "plum" rains) around mid-June or July. The summer is extremely hot and humid, with frequent sea fogs off Cape Inubo in the southeast, where the colder Oyashio Current meets the warm Kuroshio Current. In the north, the summers are cooler and the transitional season shorter. By late August or early September the winds are beginning to change yet again, bringing violent storms (typhoons) and heavy rain. These typhoons can be highly destructive, especially in the south, and have been known to devastate the main rice crop. By October the cold of winter has returned. Because of Japan's mountainous terrain and its maritime position, precipitation throughout the year is high, with permanent snow at high altitudes.

Plants and animals

Some two-thirds of the country is covered by forests, though few areas preserve the original forest cover. Bamboo forests, found as far north as Tokyo, provide useful wood and the root is a vegetable staple. Cherry trees, much prized for their blossom, are planted all over the country, and grow wild in the mountains. There are five zones of vegetation. Farthest south – and nearest the Equator – the Ryukyu, Bonin and Volcano Islands support semitropical rainforest, including camphor, tree ferns and mulberries. The second subtropical zone begins on the higher slopes of the southwestern islands with the 2,000-year-old Japanese cedars

on Yaku just to the south of Kyushu. This zone reaches as high as 1,000 m (over 3,000 ft) on Kyushu, but drops to sea level in northern Honshu.

Broadleaf deciduous and mixed forests make up the third zone, which reaches almost 2,000 m (about 6,000 ft) above sea level on Shikoku. Typical trees are beech, birch, katsura and oak. The fourth zone, consisting of mainly coniferous boreal forest, begins where the average annual temperature is below 6°C (43°F), notably in the highlands of central Honshu and Shikoku, and in southwestern Hokkaido. Boreal forest trees include Sakhalin spruce, Sakhalin fir and Blue fir. At high altitudes the forest gives way to a fifth zone of shrubs, creeping pine and alpines.

The forested mountains of the interior sustain a wide variety of animals. Wild boars root through the undergrowth, while bears and raccoon dogs hunt for hares, deer, antelope and monkeys (the Japanese macaque lives farther north than any other monkey). Reptiles and amphibians are also plentiful, among them lizards, turtles and snakes, including the venomous sea kraite and poisonous *habu* and *mamushi* species. The Japanese giant salamander, which lives on Honshu and Kyushu, can grow to more than 1.5 m

NATIONAL DATA – JAPAN

Land area	374,744 sq km (144,689 sq mi)			
Climate		**Temperatures**		**Annual**
	Altitude m (ft)	January °C(°F)	July °C(°F)	precipitation mm (in)
Sapporo	18 (59)	-4 (74)	21 (90)	1,128 (44.4)
Tokyo	6 (20)	6 (43)	26 (79)	1,467 (57.7)

Major physical features highest point: Mount Fuji 3,776 m (12,388 ft); largest lake: Biwa 673 sq km (260 sq mi)

Population (2006 est.) 127,463,611

Form of government multiparty constitutional monarchy with two legislative houses

Armed forces army 45,600; marine self-defense force 44,000; naval aviation 9,800; ground self-defense force 148,200

Largest cities Tokyo (capital - 8,403,512); Yokohama (3,632,023); Osaka (2,588,578); Nagoya (2,197,711); Sapporo (1,908,433); Kobe (1,541,229)

Official language Japanese

Ethnic composition Japanese 99%; others 1% (Korean, Chinese, Brazilian, Filipino, other)

Religious affiliations Shinto and Buddhist 84%; Christian 0.7%; other 15.3%

Currency 1 yen (JPY) = 100 sen

Gross domestic product (2006) U.S. $4.911 trillion

Gross domestic product per capita (2006) U.S. $33,100

Life expectancy at birth male 77.96 yr; female 84.7 yr

Major resources coal, edible seaweeds, fisheries, fruit, potatoes, poultry, rice, shellfish, silk, sulfur, sweet potatoes, tea, timber, vegetables

(5 ft) long. Japan is also rich in insect life – much of it a seasonal problem to the population – such as mosquitoes, cockroaches and fireflies.

Bird life too is abundant, with some 150 different species of songbird. The many sea birds include albatrosses, shearwaters and cormorants. The waters to the east of Cape Inubo in the southeast support some of the richest fishing grounds in the world, with cod, salmon and numerous crustaceans.

SOCIETY

Until the mid 19th century foreigners risked death if they tried to enter Japan, and any Japanese who left were forbidden to return. Following the enforced end to isolationism in 1853 the Japanese embarked on rapid social and economic change, freely adapting Western technologies. Since World War II Japanese society has achieved overwhelming industrial success within one generation.

History

The legendary history of Japan begins in 660 BC, with the accession of the first emperor, Jimmu. He was probably the ruler of the Yamato, a people that gave their name to a province in southwest-central Honshu.

By the 5th century AD, the Japanese had occupied the southern part of Korea and made their first contact with China. From this highly developed civilization came Japan's earliest forms of writing.

By the mid 6th century the Yamato court had lost its grip on Korea, and from there Buddhism, a new cultural influence, reached Japan. The traditional national religion was Shinto, or the Way of the

The first Shogun Minamoto Yoritomo (1147–99) founded the system of military dictatorship by which Japan was ruled for 700 years. He is remembered less as a general than as a politician and administrator.

Gods. Buddhism flourished toward the end of the 6th century under the government of the statesman and scholar Prince Shotoku (574–622).

In 794, Heian-kyo (modern Kyoto in west-central Honshu) became the imperial capital, and continued in that role until 1868. As Japan's isolation grew, the emperor became increasingly subordinate to powerful regent families, especially the Fujiwara, who in effect held power throughout the 11th century. This period was also significant for the rise of the *samurai* – a warrior class that became the only effective focus of local government. In 1192 Minamoto Yoritomo (1147–99) became Japan's first shogun, or military dictator, with his power base at Kamakura (south of present-day Tokyo).

The system of military feudalism in Japan was challenged by two attempted Mongol invasions from Korea, in 1274 and 1281. Both invasion fleets were scattered by typhoons – the so-called *Kamikaze*, or "Divine Wind" – but the shogunate's elaborate preparations for war led to financial collapse. In 1333 the shogunate, or *bakufu*, was overthrown in favor of a new emperor. A new *bakufu*

The code of the Samurai

In films such as "Throne of Blood" (1957), the internationally acclaimed director Akiro Kurasawa (b. 1910) gave a vivid picture of Japan's ancient warrior class – the *samurai*. His images combined Hollywood-style spectacle with the emotional intensity of traditional Japanese theater. But the role of the *samurai* in Japanese culture is not restricted to the theater of heroic warfare. Their code of conduct embodies a way of life admired by all walks of Japanese society.

Originally the term *samurai* referred to a small group of aristocratic provincial warriors; from the 12th century it included the entire warrior class then emerging within the Japanese feudal system. In the chaos of frequent power struggles among ruling aristocrats, the *samurai* quickly became a powerful political force. They developed a highly disciplined culture of unconditional loyalty based on an unwritten code of conduct, gradually enshrined as

The fearsome warrior class A 19th-century woodblock print shows a *samurai* in full costume, with the sword that *samurai* alone were privileged to carry. *Samurai* as a social class included women, though they rarely fought in battle.

The arrival of the Portuguese (*above*) in the mid 16th century. Jesuit missionaries from Iberia converted many powerful feudal lords interested in both Catholicism and making trade and cultural links with the West.

The Kinkakuji or Temple of the Golden Pavilion (*left*) at Kyoto, capital of Japan from 794 to 1868 and one of Japan's most beautiful cities. Built in the late 14th century as part of the shogun's villa, the temple was destroyed by fire in 1950 and rebuilt in 1955.

bushido – "the way of the warrior." Zen Buddhism, imported from China in the 15th and 16th centuries, influenced *bushido*, taming some of the more violent expressions of the code such as a warrior killing himself after his lord's death, or carrying out vendettas. In the 17th century, under the Tokugawa shogunate (1603–1867), *samurai* became the highest of four castes within a rigid class system.

The true *samurai* was most renowned for awe-inspiring courage that stemmed from deep stoicism and fierce loyalty to a leader. The *samurai* valued his honor above his life; *samurai* who neglected to commit ritual suicide (*seppuku*) after defeat or dishonor were cast out as *ronin* – masterless *samurai* who became mercenaries or bandits and were objects of contempt and loathing. Family devotion also played an important part in the *samurai*'s austere life, but as warriors their chief duty was to their overlord.

In the Tokugawa era, *samurai* still wore two swords to signify their rank, but their duties shifted from waging war to serving as civil administrators and setting a good moral example to the lower castes. With the growth of cities and the rise of the merchant class, the *samurai* became impoverished. In 1868 the feudal system was abolished and *samurai* lost their privileged status. *Bushido* was adopted as a code of behavior for the whole of Japanese society, and it formed the basis of the cult of emperor worship that was taught until 1945. In World War II *bushido* inspired Japanese soldiers to fight with often deliberately suicidal ferocity and to prefer death to the shame of surrender. However, these same values led the Japanese to show a callous disregard for the lives of Allied servicemen taken prisoner by them during the war.

Although ritual suicide is now rare (the writer Yukio Mishima committed *seppuku* in 1970 after urging the rearmament of Japan, banned by the post-World-War-II constitution), the old virtues of stoicism, loyalty and obedience have flourished. They can be observed daily in the classrooms, workplaces, and households of contemporary Japan.

was established in the Muromachi district of Kyoto.

In 1542 Portuguese traders from Macao became the first Europeans to make contact with Japan, and their arrival brought the feudal lords a powerful new weapon, the musket. Still more influential was the introduction of Christianity by the Spanish Jesuit Francis Xavier (1506–52), followed by a stream of missionaries. There was savage fighting as the struggle for supremacy intensified, until Oda Nobunaga (1534–82) entered the capital, Kyoto, in 1548 and set the country on the path toward unity. On Oda's death, his able general Toyotomi Hideyoshi (1536–98) took control, and in 1587 he made Nagasaki – a port on the far western side of Kyushu – an imperial city. Soon Nagasaki attracted regular Spanish, Dutch and Portuguese trading ships.

Hideyoshi's successor Tokugawa Ieyasu (1543–1616) continued to unite the country and went on to establish the Tokugawa shogunate, which lasted from 1603 until 1867. At first Ieyasu favored foreign trade and was open to Western influence, but later he became afraid of the spread of Christianity. After 1636 all contact with European traders and missionaries was forbidden, except at Nagasaki, and Japanese Christians were persecuted.

As the *bakufu* became more repressive and more corrupt, foreign pressure and internal discontent eventually combined to break Japan's isolation. Two major uprisings in 1836 and 1837 were put down, while European and American ships continued to press for a renewal of foreign trade. Finally, in July 1853, a squadron of United States' warships under Commodore Matthew Calbraith Perry (1794–1858) arrived and anchored in Tokyo Bay. There they stayed until the Japanese agreed to a trade and diplomatic treaty with the United States. Further

The Japanese tea ceremony (*above*) at its most elaborate, with participants wearing court dress. The highly ritualized ceremony, perfected in the 16th century, is generally associated with the austerity of three masters who were influenced by Zen Buddhism.

resistance to foreign interference was met by the shelling of Japanese ports. In 1867, the young crown prince Mutsuhito (1852–1912) defeated the forces of the last shogun, and in 1868 restored direct imperial rule as the Emperor Meiji. Guided by his powerful court, Meiji abolished feudalism, replaced the samurai military class with a conscript army and started industrialization. In 1889 he introduced a new constitution on Western lines, but retained the notion of the divine emperor as head of state.

War broke out with China after Japan seized the Ryukyu Islands in the East China Sea in 1879 and further refused to withdraw Japanese troops sent to Korea. The Chinese fleet was quickly overcome and much of it was destroyed. At the peace treaty of 1895 the defeated Chinese were forced to give up Taiwan. During China's Boxer Rebellion of 1900, Japanese troops helped to rescue foreign nationals. Japan won a spectacular victory in the Russo-Japanese war of 1904–5, enabling it to annex Korea in 1910.

During World War I Japan seized the German Pacific Islands (the Mariana, Caroline and Marshall Islands) and the German-held areas of the Shandong Peninsula of eastern China. When China pressed for their return, Japan made demands aimed at increasing its stranglehold on Chinese trade. Japan's allies, alarmed by this aggressive attitude, negotiated for Japan's withdrawal from Shandong, but Japan became increasingly militaristic and expansionist.

Dragon festival at the Cannon temple in Tokyo. The dragon is one of many motifs shared with the Chinese, who influenced early Japanese culture. On festival days local temples come alive with elaborately costumed dancers and crowds of camera-clicking onlookers.

In 1931 Japan occupied Manchuria and created the puppet state of Manzhouguo, and in 1937 renewed its war with China. The Japanese seized all China's large coastal cities and occupied many inland areas. In 1940 Japan signed an alliance with Nazi Germany – the Tokyo–Berlin Axis – and in 1941 entered World War II with a surprise air attack on the American fleet at Pearl Harbor, Hawaii.

Soon there were Japanese forces of occupation in the Philippines, Hong Kong, IndoChina, Siam (Thailand), the Malay Peninsula, Burma, Borneo, Sumatra and Java. But in May 1942 the tide began to turn as the Japanese started to suffer heavy losses. By 1945 many Japanese cities were suffering intense aerial bombardment. However, Japan refused Allied demands for unconditional surrender and on 6 August 1945 Hiroshima in western Honshu was destroyed by an atomic bomb; a second atomic bomb was dropped on Nagasaki three days later. On 14 August, six days after the Soviet Union had invaded Manchuria, Japan finally surrendered and was occupied by United States' troops.

In 1946 Emperor Hirohito (1901–89) renounced his divine status, and the government adopted a new and fully democratic constitution. Japan signed a mutual security treaty with the United States and most of its World War II opponents in 1951 and American forces of occupation were withdrawn the following year. In due course Japan regained many of the neighboring islands that it had lost at the end of the war, though the Soviet Union retained Sakhalin and the Kuril Islands.

By the 1960s Japan's rapid industrial recovery, backed by new technology, had

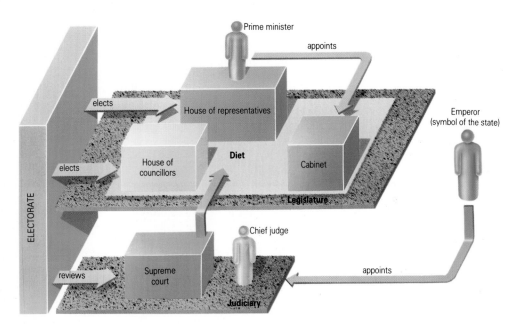

tional product in the world. Japan was badly affected by the oil crisis of the 1970s. Since then growth has continued, and Japan has maintained a large trade surplus in the international market. The 1990s featured a distinct slowdown with a financial crisis leading to severe recession in 1997–98.

Government
Under the 1947 constitution, the emperor remains titular head of state. The two-chamber Diet, or *Kokkai*, has a House of Representatives (*Shugiin*) with 500 members elected for a four-year term, and an upper House of Councillors (*Sangiin*) with 252 members serving a six-year term. The emperor appoints the prime minister recommended by the Diet. At local level there are 47 prefectures administered by governors and elected assemblies; each prefecture is subdivided into cities (*shi*), towns (*muchi* or *cho*) and villages (*mura* or *son*), and each of these has its own assembly. Most members of the Diet are elected by proportional representation from prefectural districts, but 300 seats in the House of Representatives are elected from single-seat constituencies and 200 by proportional representation within 11 regions. Everyone aged 20 or over has a vote. Japan's constitution forbids it from maintaining armed forces for aggressive purposes, and renounces war as a means of settling disputes.

People
Most Japanese are of Mongoloid origin. However, the Ainu people of Hokkaido are the survivors of an older Caucadian people that once lived all over Japan. The other significant minority groups are Koreans (still classified as resident aliens) and the *burakamin* – ethnic Japanese descended from a once outcast class.

The national language, Japanese, is

Japan's parliamentary system The Diet (parliament) consists of two chambers: the House of Representatives and the House of Councillors. Each bill passes through a committee system in each house before being approved by both houses together.

totally unlike any of the other Asian languages except perhaps Korean. The writing system was originally based on Chinese script, but has been greatly modified down the centuries. Since World War II a standard form of Japanese based on the Tokyo dialect has been established nationwide. However, throughout Japan many other dialects are still spoken.

Shinto is the indigenous polytheistic religion. It coexists with Christianity, various forms of Buddhism, and several *shinko shukyo* (new religions) that have emerged since the 19th century. Shinto was the state religion from the Meiji Restoration until the end of World War II.

No religion receives state support any longer and religious teaching in schools is forbidden, but there is full freedom of worship.

The Japanese have a highly developed and refined cultural tradition, but in recent years Western ideas have made considerable inroads. Conversely, many Japanese traditions have found a world-wide appeal, notably in recreation and sport: judo and karate, for example. Japanese cinema, too, has achieved international recognition as an art form. During the 20th century Japan has become increasingly urbanized and today more than three-quarters of the population live in towns and cities. Age-distribution among the population is also changing. Birth and death rates have fallen; the under-15s now account for less than one-fifth of the population, while the over-65 age group is expected to reach similar proportions by the year 2050.

ECONOMY

To build a modern industrial society, Japan has had to overcome many problems, including the survival of traditional practices in agriculture, forestry, fishing and the older industries.

Agriculture
Today the systematic use of fertilizers is practiced on a wide scale and agriculture employs only a very small proportion of the labor force. Exports of food account for only a tiny percentage of income.

The Ryukyu Islands

Until the 14th century the Ryukyu Islands, a long island chain that extends 650 km (450 mi) southwestward between Japan's Kyushu island and Taiwan, formed an independent kingdom. The islanders seem to have been of mixed Japanese and Southeast Asian stock; their language, though related to archaic forms of Japanese, cannot be understood by modern Japanese speakers, and their culture was originally quite distinct.

In 1372 the islands came under the political influence of China, and from the mid 15th century they were forced to pay tribute to both China and Japan.

They became a Japanese protectorate in 1609, and by 1879 had been incorporated into Japan. In World War II the largest island, Okinawa, became a bloody battleground as entrenched Japanese forces tried to beat off a determined amphibious assault by the United States; the Americans eventually won. After the war the islands were placed under United States' administration. They ran their own civil government from 1951, and by 1972 the last of the islands had been returned to Japan. However, United States' airbases were maintained on Okinawa.

Today the islands are a peaceful backwater, largely given over to farming, tuna fishing and traditional industries such as producing lacquer goods and pottery.

Modernization was hindered by the small size of most holdings and the scarcity of cultivable land (one-eighth of Japan's total land area). Most of the larger farms are on Hokkaido; elsewhere traditional terraces on lower mountain slopes are part of the modern landscape.

The warmth and humidity of summer is ideal for the cultivation of irrigated rice (wet rice or paddy). The development of frost-resistant strains has helped increase the potential yield, and rice is now grown all over most of Japan.

The retail price of rice has been kept artificially high – a policy that led to a massive rice surplus in the late 1960s and early 1970s. Since then, international pressure has forced a reduction in the rice support rate, and farmers have been encouraged to diversify. Tractors and mechanical cultivators have replaced traditional methods, and insecticides and artificial fertilizers are freely used. Before World War II, silk cocoons were the most important agricultural export after rice. Today livestock is in second place. Other important products include vegetables, barley, wheat, potatoes, fruit and tea.

Fishing and forestry

Although Japan's catch is among the largest in the world, its fishing enterprises are mostly smallscale ventures. Government plans to increase deep-sea fisheries at the expense of coastal fishing were frustrated when other nations declared a 370 km (200 nautical mi) economic zone around their coasts. Today imports have overtaken exports. Clams and oysters are popular Japanese food, as are crabs, prawns and shrimps.

The Japanese have strongly resisted worldwide controls on whaling. In 1982 Japan agreed to a universal whaling ban, while reserving the right to kill whales for "scientific" purposes, but on a number of occasions Japan, Iceland and Norway have argued for a resumption of whaling.

Japan has more hectares of forests than countries such as Finland, where forestry is a main source of revenue. However, Japan's forests are very much less accessible, and two-thirds of all forested land is in small private holdings, complicating efficient overall management; reforestation is largely confined to the public holdings. Timber is imported on a large scale to meet the country's needs.

Industry

Japan has few mineral resources, and mining accounts for only a small fraction of the national income. Despite govern-

Nagasaki rebuilt (*above*) A thriving port and ship-building center, Nagasaki was devastated by the second atomic bomb dropped by the United States in 1945. A peace park marks the site of the blast.

Vision robot technology (*right*) Some of Japan's most spectacular economic growth has occurred in high technology industries. Its top quality products now dominate the world market, and substantial investment in research and development ensures future success.

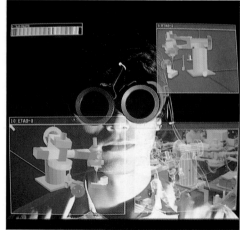

ment subsidies, the coal industry is losing out to foreign competition. Thin, badly faulted seams make Japanese coal difficult and expensive to mine, and the oilfields of northern Honshu meet only a tiny proportion of the country's fuel demands. Most fuels and raw materials have to be imported, including iron and copper ore, coking coal and bauxite.

Most of Japan's electricity is generated from oil and petroleum products, about one-fifth from coal and one-third from natural gas. Nuclear power supplies just over one-tenth of the electricity on the grid, and hydroelectricity is accountable for around one-tenth.

More than one in three of the workforce is in manufacturing or construction, which together provide nearly two-fifths of gross national product. Industrial growth began late in the 19th century, and the subsequent opening of new markets in Taiwan (Formosa), Korea and Southeast Asia led to rapid expansion. In the 1920s the first family corporations (*zaibatsu*) were formed; they are still a feature of Japanese industry today. The industrial centers were subjected to heavy bombing toward the end of World War II, but afterward Japan rebuilt them. During the Korean War (1950–53) Japan supplied United States' forces with arms.

Subsequently, Japan transformed its industrial strategy. The emphasis shifted from the manufacture of cheap, mass-produced goods to producing high-technology, high-quality goods, which have since captured a large share of the world market. Electrical and electronic goods – notably computers, audio and video equipment and cameras – are the country's leading products, closely followed by machinery, processed foods, motor vehicles and chemicals. Japan is also one of the world's major shipbuilding nations. Synthetic fibers and resins, paper, cement, iron and steel are other important manufactures. Unemployment has traditionally been low but it slowly rose during the recessions of the 1990s.

The main industrial belt extends along the south coast of Honshu from the Kanto lowlands in the east to northern Kyushu in the west. Industry is most heavily concentrated around Tokyo, Nagoya and Osaka (with Kyoto to the north).

Trade and commerce
Since 1969 Japan has maintained a balance-of-payments surplus. Its imaginative and aggressive marketing tactics and competitive prices have won an ever-larger share of the world market. An important factor in Japan's success has been the close cooperation between the manufacturers, the commercial banks who fund them, and the central Bank of Japan, from which the commercial banks borrow much of their venture capital. Visitors to

Japan's most visible export A selection of Nissan automobiles lined up awaiting export in the shadow of the Japanese flag. Cars from "the big three" – Honda, Toyota and Nissan – are in demand throughout the world, eclipsing all other global producers.

Japan provide useful income, but expenditure by Japanese tourists overseas far exceeds this.

Transportation and communications
Until the 1920s most Japanese roads were used almost exclusively by pedestrians. Today the construction of turnpikes, or toll highways, is proceeding rapidly in densely populated areas, where motor vehicle traffic is sharply on the increase. There are expressways linking the major urban centers.

Japan's first railroads were built in the 19th century. The building program accelerated to increase the network, and included subway trains in Tokyo and Osaka. After World War II the subway service was extended, so that today smaller cities have their own subways.

The first high-speed inter-city passenger train ran between Tokyo and Kyoto in 1964. That same year Japan's first monorail came into service in Tokyo; many more followed. There are plans to extend the high-speed rail network across the whole country. All four main islands have road and rail connections between them. Kyushu and western Honshu are linked by the world's first ever undersea rail tunnel (built in 1941), a two-decker road tunnel (1958) and a suspension bridge (1973). The Tokyo Bay Aqualine (1997) is a 4.4 km (2.7 mi) bridge and a 9.4 km (5.8 mi)

tunnel which spans the bay.

Japan is one of the world's principal seagoing nations. Each of the two main urban areas has its own port; Yokohama serves Tokyo, while Kobe serves Osaka and Kyoto. Air transportation centers on Tokyo, with other major airports at Osaka, Nagoya, Sapporo (Hokkaido) and Fukuoka (Kyushu).

The Japanese postal and telecommunications systems are among the finest in the world. Radio and television broadcasting is highly developed; the Japanese Broadcasting Corporation (Nippon Hoso Kyokai; NHK) – a public corporation financed by license fees – broadcasts in 20 different languages, alongside private corporations and satellite and cable companies. There is no government censorship of the press, but until the 1970s criticism of government was muted. Since then the Japanese press has been more ready to seek out and expose corruption.

Health and welfare
Most of Japan's population is covered by public or private health and welfare programs, with benefits available for injuries at work, sickness or disability, maternity, old age and unemployment. Health conditions are excellent: life expectancy ranks among the highest in the world.

After nine years of compulsory education, the great majority of students go on to upper secondary school for a further three years. On graduation they can choose a five-year course at technical college, a four-year university course or two to three years at junior colleges.

North Korea

THE COMMUNIST REPUBLIC OF NORTH KOREA occupies the northern half of the Korean peninsula on the northeastern coast of Asia. It forms half of a nation that was partitioned during the years following World War II.

GEOGRAPHY

The irregularly shaped Korean peninsula divides the Yellow Sea on its western shore from the Sea of Japan on its eastern shore. North Korea lies mainly to the north of the 38th parallel. Its northern border adjoins Jilin, a province of northeastern China.

North Korea is a country of rugged uplands and mountainous terrain. The highest point is on the border with China, where the extinct volcano Mount Paektu rises to 2,744 m (9,003 ft).

The two longest rivers – the Yalu and Tumen – rise at Mount Paektu and form the northern border. On the western side of the country are several other large rivers, whose plains form the only extensive lowland areas; the capital, P'yongyang, is on these plains. The west coast has many natural harbors.

Winters are long and cold, and the prevailing northwesterlies bring severe frosts but little rainfall. Winter temperatures vary considerably across the country, the lowest being in the mountainous north. Summers show less of a regional variation. The wind changes direction to the southeast, drawing in warm, humid air from the Pacific. More than two-thirds of the country's annual rainfall occurs between June and September, and there are occasional violent storms (typhoons) during the fall.

The Kaema plateau in the mountains of the northeast is densely forested with a range of coniferous trees such as spruce, larch and the Siberian fir. Largescale clearance in the lowland parts of North Korea has greatly reduced the variety of plant species, but the government now has introduced a reforestation program.

Antelopes, tigers, leopards and panthers are still found in the more remote forested areas. Bird life remains plentiful.

SOCIETY

Throughout its history Korea has been greatly influenced by its more powerful neighbors: China, Japan and Russia.

Korea was unified in the late 7th century AD and some two centuries later the kingdom of Koryo was established. It survived until 1392, when the Confucian Yi dynasty took power.

For hundreds of years control of Korea was disputed between China and Japan until, in 1910, Japan assumed supremacy by annexing Korea. Japanese occupation ended in 1945, when Japan was defeated at the end of World War II. The Allies agreed that Japanese forces north of the 38th parallel would surrender to the Soviet Union, while those to the south would surrender to the United States. The United Nations tried to secure an agreement for the unification of the country, but failed. In 1948 the Soviet Union backed the creation of a communist republic in the north, while a separate republic was established in the south.

Kim Il Sung (1912–94) became the first premier of the Democratic People's Republic of North Korea. In 1950 North Korean troops invaded South Korea in a bid to unify the country. A three-year war ensued, in which some four million people died. United Nations' troops, mainly from the United States' armed forces, supported the South, while the North was aided by China. Since then talks on reunification have regularly occurred. In June 2000 an historic south-north summit took place between North and South Korea. Kim Il Sung was informally succeeded in 1994 by his son Kim Jong Il (b. 1942) who is now the country's "chief of state".

The Democratic People's Republic is a single-party communist state. Officially, the highest authority is the single-chamber Supreme People's Assembly whose 687

Mount Paektu (*left*), Korea's highest peak, is an extinct volcano with the clear waters of a crater lake at its summit. The border with China runs through the lake; on the north side is the Chinese province of Jilin. Much of North Korea is mountainous, especially in the far north where there are a number of peaks over 2,440 m (8,000 ft).

The old guard (*right*) An assembly of North Korean army generals at a political rally. As in China and the former Soviet Union, with which North Korea has close ties, the military reinforces the political status quo. Since the 1960s North Korea has emphasized military preparedness, and both men and women perform national service.

members are elected. In practice, real authority lies with a 39-member politburo appointed by the Korean Workers' Party and under the direct command of the president.

The population is uniformly Korean, with no ethnic minorities apart from a few Chinese. The Korean language has been influenced by centuries of Chinese scholarship. After partition in the 1940s, the North Korean government banned the use of Chinese ideographs for the writing of Korean, favoring a more straightforward syllabic system.

The traditional religions are Confucianism and Buddhism. The communist regime has repressed Christian activity, but has encouraged Ch'ondogyo, a 19th-century religion combining elements of Confucianism, Buddhism and Christianity. Government institutions are responsible for supporting fine arts, music and literature, and the country's museums.

North Korea is less densely populated than its southern neighbor, but the pattern of migration from the country to the towns is similar to that of South Korea.

ECONOMY

Since 1948 the communist government has aimed to transform North Korea from an agricultural country into a self-sufficient industrialized state. This has met with only limited success, and the country was critically dependent on international food aid in the late 1990s.

The government has also modernized and reorganized the agricultural sector, with a consequent rise in production levels. Private farming has been virtually abolished, and nearly all farmers are organized into cooperatives, whose produce is distributed by the government. Rice is the principal food crop, followed by millet and other grains. Farmers also grow fruit and vegetables as well as oilseed rape, flax and cotton. The fishing industry has expanded considerably, and fish is now a major export.

Iron ore is the most important mineral deposit; iron and related products are major exports. Three-fifths of the country's power comes from hydroelectricity. There are also extensive coal deposits. Manufactured goods include both agricultural and industrial machinery, as well as chemicals and cement. However, industry still depends on imported machinery and fuel. The former Soviet Union was a major supplier of these before 1990. China, Japan and Russia have since become important.

Transportation is mainly by railroad. The network consists primarily of north–south routes in lowland and coastal areas; east–west travel is hampered by the mountains. Some freight is moved by river. There is an international airport near P'yongyang.

All media are controlled by the government. Telecommunications services are widely available. Almost everyone has a radio, and more people are acquiring television sets. Newspapers are published by the government, the Korean Workers' Party and the trade unions.

Welfare provision is a government priority; state insurance schemes cover all basic benefits. Housing has improved greatly since the 1950s, and free medical facilities are widely available. Education is free, compulsory between the ages of 5 and 16. Higher and adult education are also well provided for. Government aims and planning for education are closely linked to the country's labor needs.

NATIONAL DATA - KOREA, NORTH

Land area 120,410 sq km (46,490 sq mi)

Climate	Altitude m (ft)	Temperatures January °C(°F)	July °C(°F)	Annual precipitation mm (in)
P'yongyang	27 (89)	-8 (18)	25 (77)	824 (32.4)

Major physical features highest point: Mount Paektu 2,750 m (9,002 ft); longest river: Yalu 806 km (501 mi)

Population (2006 est.) 23,113,019

Form of government one-party communist republic with one legislative house

Armed forces army 950,000; navy 46,000; air force 110,000

Largest cities P'yongyang (capital – 3,059,678); Hamhung (569,994); Namp'o (458,903)

Official language Korean

Ethnic composition Korean 99.8%; Chinese 0.2%

Religious affiliations Nonreligious 67.9%; traditional beliefs 15.6%; Ch'ondogyo 13.9%; Buddhist 1.7%; Christian 0.9%

Currency 1 North Korean won (KPW) = 100 chon

Gross domestic product (2006) U.S. $40 billion

Gross domestic product per capita (2006) U.S. $1,800

Life expectancy at birth male 68.92 yr; female 74.51 yr

Major resources coal, lead, tungsten, zinc, graphite, magnesite, iron ore, copper, gold, pyrites, salt, fluorspar, hydropower, barley, beans, cattle, fisheries, grain, pigs, poultry, rice, tobacco, timber, wheat

South Korea

REPUBLIC OF KOREA

SOUTH KOREA OCCUPIES THE SOUTHERN half of the Korean peninsula in north-eastern Asia. The country also includes the island of Cheju to the southwest.

GEOGRAPHY

South Korea is characterized by rugged and often mountainous terrain. The country's population is crowded into the few lowland areas.

South Korea's east coast is relatively straight. Parallel to it and running from north to south are the Taebaek mountains, which contain the source of several westward-flowing rivers, notably the Han and the Naktong. The densely populated basin of the Han in the northwest includes the capital, Seoul and Inch'on, South Korea's main port of entry. From here the river Han flows north to enter the sea at the North Korean border.

The western and southern coasts are a maze of channels and islands. The many creeks and indentations contribute to some of the world's largest tidal ranges: up to 9 m (30 ft) at Inch'on. The highest point in South Korea is the extinct volcano Halla-san (1,950 m; 6,398 ft) on the island of Cheju, which has a crater lake.

Winter brings cold, dry winds from northern Asia, which keep temperatures very low inland. Coastal areas experience milder weather, especially in the south. In summer the wind switches round to the southeast, drawing in warm and very humid air from the Pacific. The wettest period is from June to September, known as the summer monsoon. Rainfall is heaviest in the south, and lightest in the

"Land of the morning calm" – a traditional name for Korea – typifies the tranquil rural landscape of southwestern South Korea. But such scenes belie the country's dynamic industrial expansion since the 1960s and a population density higher than that of Japan.

lee of the mountains. As summer turns to fall there is a risk of typhoons.

Some two-thirds of South Korea's land area is covered in forest. Subtropical broadleaf forests survive along the southern coast, but most other areas have now been planted with coniferous trees. Wildlife populations have suffered from the destruction of their habitats. Wild boars and deer remain, but bears, lynx, tigers and leopards are very rare.

SOCIETY

Since South Korea was founded in 1948 it has had numerous constitutions and protracted periods of military rule. However, the constitution of 1988 provided for greater democratic freedom, and reunion with the north is a longterm possibility.

Korea became a single nation as early as the 7th century AD. However, from the 14th century onward its history was dominated by its two more powerful neighbors – China and Japan.

The Japanese annexed Korea in 1910 but were expelled after their defeat in World War II. The peninsula was divided along the 38th parallel, with Soviet occupation forces in the northern zone and United States troops in the south. The south held elections in 1948, and declared a republic under the presidency of Syngman Rhee (1875–1965). Meanwhile a separate communist republic was set up in the north. In 1950 North Korea invaded the south and tried to reunite the peninsula by force. A three-year war followed, in which the United States supported the south with military aid, backed by a United Nations' mandate.

After the war, President Rhee resisted attempts to establish a parliament, and in 1954 made himself president for life. However, his fourth election in 1960 sparked off violent unrest, forcing him to flee the country. A second republic, based on a parliamentary system, was toppled only nine months later by a military junta led by Major-General Park Chung Hee (1917–79). Park became chairman of the Supreme Council of National Reconstruction (SCNR), and on his election as president in 1963 established a third republic and constitution giving himself sweeping powers. In 1971 he declared a state of emergency, and in 1972 replaced the constitution yet again. Park was assassinated in 1979, and in 1981 Chun Doo Hwan (b. 1931) was elected president. He was followed, in 1987, by Roh Tae Woo (b. 1932). He introduced more democratic changes. In 1997, Kim Dae Jung was elected president and in 2000 he was awarded the Nobel Peace Prize for his lifelong commitment to democracy and human rights in Asia.

The Republic of Korea's present constitution provides for a single-chamber national assembly with 299 members, each elected for a four-year term. The president is elected for a five-year term, and governs with an appointed state council, led by a prime minister.

Koreans form a single ethnic group speaking a single language; there are virtually no minority groups apart from a few Chinese. The Korean language probably originated in northern Asia. The writing system was based on Chinese, but has been substantially modified to suit the very different structure of Korean. The modern system, known as *han'gul*, is basically a phonetic system for writing; some educated South Koreans still combine it with Chinese ideographs.

There are a number of different religious groups, and members of the same family may differ in their faiths. The major religions are Buddhism, Confucianism and Christianity, together with *Ch'ondogyo*, a combination of the other three. Shamanism also survives in many rural areas, and *Ch'angga Hakhoe* (a derivative of the Japanese militant Buddhist Value Creation Learning Society) has spread in the cities since the 1960s.

South Korea is very densely populated and highly urbanized. More than two-thirds of the population live in the towns and cities, a quarter in Seoul.

ECONOMY

Despite government backing for heavy industry, economic success has been limited. Growth was not encouraged by the 1997–98 Asian financial crisis, but in 1999 an IMF loan helped improve prospects.

Farming is practiced in lowland areas and most farms are now family-owned. South Korea is almost self-sufficient in rice, the chief staple, whose price is subsidized by the government. Silk and tobacco are produced for the export market. Fishing provides a surplus catch for export.

In spite of its limited mineral resources, South Korea exports some iron ore and tungsten. Low-grade anthracite (coal) is mined, but petroleum and most metals have to be imported. Most power stations are oil-fired, but there is also hydroelectric power. The manufacturing sector forms the backbone of the South Korean industry and economy. Chemicals, metals and machinery are especially important. The electronics sector is expanding rapidly, and the country is now the world's second largest maker of silicon chips.

Until 1960 the government-owned railroads were the principal means of travel, but now the road network carries most of the passengers and freight. At the main ports of Pusan, Inch'on and Cheju freight-handling facilities have been expanded. Kimp'o airport near Seoul is one of four international airports. The state-controlled Korean Broadcasting Service is in charge of most radio and television output. Newspapers are privately owned.

Healthcare is expensive, and concentrated largely on Seoul and Pusan. Housing is in short supply. Tuberculosis and pneumonia are the major diseases, along with infant diseases. The limited welfare programs provide some care for war victims, the old, and homeless people. A national pension system is now in place. Unemployment insurance was introduced in 1998. Most children receive six years of free (and compulsory) primary education, and about three-quarters of them spend three years at middle school. Of the many that then go on to high schools, about a half continue into higher education.

NATIONAL DATA – KOREA, SOUTH				
Land area 98,190 sq km (37,911 sq mi)				

Climate	Altitude m (ft)	Temperatures January °C(°F)	July °C(°F)	Annual precipitation mm (in)
Seoul	295 (968)	-5 (23)	25 (77)	1,250 (49.2)

Major physical features highest point: Halla San (Cheju island) 1,950 m (6,398 ft); longest river: Han 470 km (292 mi)

Population (2006 est.) 48,846,823

Form of government multiparty republic with one legislative house

Armed forces army 560,000; navy 63,000; air force 64,000

Largest cities Seoul (capital - 10,451,281); Pusan (3,663,421); Inch'on (2,657,698); Taegu (2,591,996); Taejon (1,515,270)

Official language Korean

Ethnic composition Korean 99.9%; Chinese 0.1%

Religious affiliations no affiliation 46%; Christian 26%; Buddhist 26%; Confucianist 1%; other 1%

Currency 1 South Korean won (KRW) = 100 chon

Gross domestic product (2006) U.S. $1.18 trillion

Gross domestic product per capita (2006) U.S. $24,200

Life expectancy at birth male 73.61 yr; female 80.75 yr

Major resources coal, tungsten, graphite, molybdenum, lead, hydropower, barley, beans, fish, fruits, vegetables, gold, iron ore, lignite, maize/corn, millet, potatoes, rice, sorghum, sugar, timber, tobacco, wheat

Australasia, The Pacific Islands and Antarctica

COUNTRIES IN THE REGION

AUSTRALIA · NEW ZEALAND · PAPUA NEW GUINEA · SOLOMON
ISLANDS · TUVALU · NAURU · KIRIBATI · FIJI · VANUATU · TONGA
SAMOA · PALAU · MARSHALL ISLANDS · MICRONESIA

DEPENDENCIES IN THE REGION

GUAM · WAKE ISLAND · MIDWAY ISLANDS
JOHNSTON ATOLL · NORTHERN MARIANA ISLANDS ·
AMERICAN SAMOA · NEW CALEDONIA · WALLIS AND FUTUNA
FRENCH POLYNESIA · ANTARCTICA

A spectacular sunset (*left*) over the Marshall Islands in
the western Pacific Ocean. The vast Pacific region
encompasses a dramatic range of peoples and
landscapes, from the ice-world of Antarctica to the
parched outback of Australia and Micronesia's
palm-fringed atolls.

Australia

COMMONWEALTH OF AUSTRALIA

Australia is a federation of six states, two mainland territories and additional island territories. The states are: Queensland in the northeast; New South Wales along the central-eastern coast; Victoria in the southeast; Tasmania, an island state off southern Victoria; the central-southern state of South Australia; and Western Australia, occupying the western third of the continent. The internal territories are the Northern Territory (north-central Australia) and the Australian Capital Territory in southeastern New South Wales. Only a few of the island territories are inhabited.

Economically, Australia is a major world producer of raw materials. Japan, highly industrialized and poor in resources, is an important customer. Australia's own manufacturing sector has suffered through formidable competition with its Southeast Asian neighbors.

GEOGRAPHY

The island continent of Australia lies between the Indian Ocean, the Southern Ocean, and the Tasman and Coral Seas of the Pacific Ocean. Its land area almost equals that of the United States (excluding the two states of Alaska and Hawaii). The continent as a whole is the least mountainous in the world and also the driest; almost one-third of its land is desert and much of the rest dry.

The land and climate

Australia can be divided into three chief geographical areas: the Great Western Plateau, occupying more than half the country; the central-eastern lowlands; and the eastern highlands including the Great Dividing Range.

The Great Western Plateau is generally uniform apart from the Kimberley Plateau and Arnhem Land near the north coast, and the MacDonnell and Musgrave Ranges near the center of the continent. It rises near the west coast and then drops gently eastward into the "red center" of the continent, named after the distinctive red-brown landscapes. Between the edge

Ayers Rock (*above*), now called Uluru, is an outcrop of rock 600 million years old, rising from the red desert southwest of Alice Springs in Australia's Northern Territory. At over 300 m (1,000 ft) high and 10 km (6 mi) around, it is the world's most spectacular monolith.

Satellite view of fires near Sydney (*below*) Although common during summer, bush fires can become severe. The image here shows the extent of such fires during 2001/2002 with smoke drifting over Sydney from the Royal National Park (south) and Blue Mountains (west).

upland ridges. The Eyre Basin is by far the largest, draining toward Lake Eyre near its southwestern corner. The Murray Basin, the smallest of the three, contains Australia's two longest and economically most important rivers, the Murray and its tributary the Darling.

East of the central lowlands the land rises to the Great Dividing Range, then drops more steeply toward the coastal plains, where over half the population lives. From northern Queensland to the south coast of Victoria, the highlands run parallel to the shoreline, and then reappear on the island of Tasmania. They include the highest peaks of the Australian Alps in the southeast, at a little over 2,000 m (6,500 ft). In the south, the Great Dividing Range rises to the Blue Mountains, west of Sydney, and in the northeast touches the coast again near the town of Cairns. Parallel to Queensland's northeastern coast is the largest coral structure in the world, the Great Barrier Reef stretching some 2,300 km (1,400 mi) to Papua New Guinea. The mainland of Australia is relatively stable geologically, but its offshore territories include many islands that are of recent volcanic origin.

The climate ranges from tropical to temperate. It is coldest in the highlands of the southeast and Tasmania, the only places where snow falls regularly. January and February are generally the hottest months, with the summer season lengthening toward the north. The far north, nearest the Equator, experiences high temperatures throughout the year. Most western and central areas are arid or semiarid, with prolonged droughts. In the Lake Eyre Basin, annual rainfall can

The Great Barrier Reef extends 2,300 km (1,400 mi) along the coast of Queensland and supports a rich and varied marine life. Here, a colorful nudibranch – or sea slug – feeds on the coral. The reef is one of Australia's foremost tourist attractions.

be less than 150 mm (6 in). By contrast, northern coastal and northeastern Australia have a humid monsoon-type climate. The northwest and east coasts are liable to flooding and to occasional cyclones. Along the north and southeast

of the plateau and the Musgrave Range, is the world's best known single rock. Ayers Rock, now known by its original Aboriginal name Uluru, is 10 km (6 mi) around, more than 300 m (1,000 ft) high, and changes color in the sun.

The central eastern lowlands consist of three artesian basins – each a large reservoir of underground water – separated by

NATIONAL DATA – AUSTRALIA

Land area 7,617,930 sq km (2,941,299 sq mi)

Climate	Altitude m (ft)	Temperatures January °C(°F)	July °C(°F)	Annual precipitation mm (in)
Darwin	27 (89)	28 (81)	25 (77)	1,708 (67.2)
Alice Springs	548 (1,797)	29 (84)	12 (54)	289 (11.3)
Sydney	42 (138)	22 (72)	12 (54)	1,222 (48.1)

Major physical features highest point: Monte Kosciusko 2,228 m (7,310 ft); lowest point: Lake Eyre -16m (-52 ft); longest river: Murray-Darling 3,717 km (2,310 mi)

Population (2006 est.) 20,264,082

Form of government federal multiparty parliamentary state with two legislative houses

Armed forces army 26,035; navy 14,070; air force 15,670

Largest cities Sydney (4,490,662); Melbourne (3,829,400); Brisbane (1,938,851); Perth (1,497,487); Adelaide (1,079,558); Gold Coast (546,025); Canberra (capital – 325,888); Newcastle (502,338); Wollongong (262,894)

Official language English

Ethnic composition Caucasian 92%; Asian 7%; Aboriginal and other 1%

Religious affiliations Catholic 26.4%; Anglican 20.5%; other Christian 20.5%; Buddhist 1.9%; Muslim 1.5%; other 1.2%; unspecified 12.7%; none 15.3%

Currency 1 Australian dollar (AUD) = 100 cents

Gross domestic product (2006) U.S. $666.3 billion

Gross domestic product per capita (2006) U.S. $32,900

Life expectancy at birth male 77.64 yr; female 83.52 yr

Major resources bauxite, coal, iron ore, copper, tin, gold, silver, uranium, nickel, tungsten, mineral sands, lead, zinc, natural gas, petroleum, barley, cattle, diamonds, fish, fruit, maize/corn, manganese, oats, opals, pigs, rice, rutile, sheep, sorghum, sugarcane, timber, tobacco, tourism, vegetables, wheat, wine, wool

The Cocos Islands

In 1826 the Englishman Alexander Hare established a small colony at an isolated group of coral islands in the Indian Ocean, some 1,200 km (750 mi) southwest of Java. The following year the Scotsman John Clunies-Ross (d. 1854) brought a number of Malays to the islands and formed a second colony. The two atolls, also known as the Keeling Islands, were covered in coconut palms, and Clunies-Ross harvested them for copra. He stayed on, and the islands were controlled by his descendants until 1978 – Alexander Hare returned to Java in 1831. Al-

though they were annexed to the British crown in 1857, Queen Victoria (1819–1901) returned them to George Clunies-Ross and his heirs in perpetuity in 1886. Crown interests were administered from Singapore after 1903, and from Ceylon (now Sri Lanka) during World War II, but in 1955 authority passed to the Australian government. In 1978 Australia finally bought up the family interest – except the family home and grounds – and the islands became an external territory.

Today the copra harvest is gathered in by the descendants of the original laborers – known as Cocos Islanders or Cocos Malays.

coasts and in Tasmania annual rainfall averages over 2,000 mm (80 in).

Fed by this relatively high rainfall, the Murray River and its tributaries in the southeast serve the greater part of the country's irrigated land. Groundwater, found under much of the continent, is a vital resource for domestic and livestock requirements, especially in inland areas.

Plants and animals

Native forests and woodlands are found mainly on the northern and eastern coasts and along inland watercourses. The commonest trees are the gums (*Eucalyptus*) and wattles (*Acacia*). Inland, the drier landscape changes to scrubland and grassland, and eventually desert.

Vast areas of native vegetation have been cleared to introduce alien species, while woodlands have been cleared for cultivation or to provide building material. Over much of the continent, indigenous plants are losing out to millions of grazing livestock and non-native weeds introduced with European crops.

Most native mammals are marsupials – among them kangaroos, koalas, wombats and possums – whose young are born very early and suckled in protective pouches. Even more unusual are the monotremes – the duckbilled platypus and the echidna or spiny anteater – which lay eggs and suckle their young. Most bird species are endemic, including the flightless emu and the kookaburra.

Like the native plants, animal species, too, are suffering from competition with animals introduced by Europeans. The dingo, a wild dog, is the only animal introduced by the native Aboriginals. Australia has several species of venomous snakes and poisonous spiders.

Aboriginal rock engravings (*below*) depicting animals and hunting, at Mootwingee National Park in New South Wales. Ancient engravings such as these can be found throughout Australia, but the art form is no longer practiced, unlike other forms of native art.

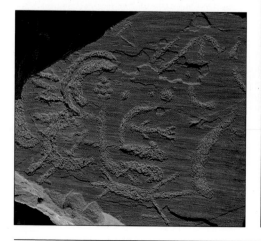

SOCIETY

Australia's first settlers, known as Aboriginals, probably came from Southeast Asia at least 50,000 years ago. Their hunter–gatherer society made only a modest impact on the ecology of the continent. Today, after two centuries of European settlement, the environment has undergone irreversible change.

History

It seems likely that the Aboriginals were the first of several peoples from Asia to discover Australia. In March 1606 the Dutch Captain Willem Jansz made the first positively known sighting by a European of the Australian mainland. A few weeks later Luis Vaez de Torres sailed the strait that now bears his name but almost certainly failed to sight Australia. A number of later Dutch explorers charted the region. Tasmania was first named Van

Captain James Cook (1728–79) (*above*) was a British explorer and navigator renowned for his charting of unexplored waters. During his first Pacific voyage (1768–71) he discovered and named New South Wales and claimed eastern Australia for the British Crown.

Sydney, New South Wales, (*below*) was a thriving town in the mid 19th century when Thomas Baines (1822–75) painted this view from the north shore. Founded as a convict settlement in 1788, today it is the largest Australian city.

Diemen's Land by Abel Tasman (c. 1603–59) who discovered the island in 1642. It was renamed Tasmania in 1855. Further exploration was undertaken by the British navigator James Cook (1728–79), who discovered and named Botany Bay in 1770.

At that time British criminals who had been sentenced to transportation were taken to America, but after the outbreak of the American War of Independence (1775–83) a new penal colony was needed. The British government first chose Botany Bay and then Port Jackson, just to the north, which eventually became Sydney, Australia's biggest city. Tasmania, Queensland, Victoria and Western Australia also began as penal colonies. The transportation of convicts ceased in 1868.

From the 1830s the boom in sheep farming caused Aboriginals to be driven from their ancestral lands. Then in 1851 and 1852 the discovery of gold in New South Wales and Victoria brought fresh

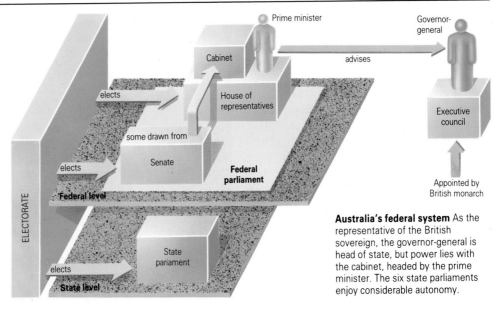

Australia's federal system As the representative of the British sovereign, the governor-general is head of state, but power lies with the cabinet, headed by the prime minister. The six state parliaments enjoy considerable autonomy.

waves of settlers. In 1854 miners clashed with troops near Ballarat, Victoria, after protests over license fees and voting rights. In 1901 the six self-governing colonies formed a federation known as the Commonwealth of Australia, with a parliament in Melbourne. In 1911 New South Wales ceded the Australian Capital Territory which became the site of Canberra, the new federal capital, and the parliament first met there in 1927.

During World War I Australian troops fought alongside the British. Like Britain, the country was badly hit by the Great Depression of the 1930s, but on the outbreak of World War II Australia rallied to the British cause. After the Japanese attack on Pearl Harbor, Hawaii, in 1941, the long and bitter battle for control of the Pacific brought Australia and the United States together.

After the war agricultural exports rose and many towns enjoyed new prosperity. As waves of immigrants arrived from Europe, Australia's ties with Britain weakened. However, racial prejudice against native Australians continued. The Aboriginals were not fully enfranchised until 1962, and were not included in the population census till 1971.

Norfolk Island

Captain James Cook (1728–79) discovered and named the uninhabited Norfolk Island in 1774. In particular he praised it for its pine trees, which he (wrongly) considered ideal for making ships' masts. Lying 1,600 km (1,000 mi) east of the New South Wales coast, the island was first settled in 1788 by a party of British soldiers and convicts. Between 1825 and 1856 it was a place of misery and brutality – a penitentiary for hardened and desperate criminals. After that, Norfolk Island became the stage for the final scene of a long drama that had started almost 70 years earlier with the notorious mutiny on the ship called the *Bounty*.

In April 1789, in the middle of the Pacific Ocean, a group of mutineers led by Fletcher Christian (dates unknown) cast their captain adrift on a longboat with several of the crew. Most of the mutineers stopped at the nearby island of Tahiti, but some sailed on eastward, along with several Tahitians, to settle on the isolated desert island of Pitcairn. There, after 18 years of violence and disease, a single European was left alive, and with him a small group of Tahitians. In 1856 the entire Pitcairn community, which by then consisted of 194 men, women and children, was transported to Norfolk Island. About one-third of today's islanders are descended from them, and today the former penal colony is a thriving tourist resort.

Every year more than 20,000 visitors come to the island, providing it with most of its income. Other sources of revenue include stamps, liquor sales and financial services. Seed of the Norfolk pine is exported and the island has a thriving forestry program.

Under the Norfolk Island Act of 1979 the island is an external territory of Australia, governed by its own nine-member assembly. Education is free and compulsory for children from 6 to 15, and teachers for the local school are provided from Australia. There is a small hospital providing healthcare.

Government

Australia has a three-tier system of government, with a federal parliament, six state parliaments and about 900 local authorities.

The federal parliament in Canberra is divided into an upper house (the Senate) and a lower houe (the House of Representatives). The cabinet, headed by the prime minister, includes members drawn from both chambers. There are 77 senators, who serve a six-year term and 148 seats in the House of Representatives which are elected for a three-year term.

Every adult aged 18 or over is required to vote in parliamentary elections. The Northern Territory has been self-governing since 1978; the Australian Capital Territory, which includes Canberra, became self-governing in 1989. Australia is a fully independent nation, but retains strong traditional links with the Commonwealth and the UK. In 1999, a referendum to change Australia's status, from a commonwealth headed by the British monarch to a republic, was defeated.

People

Most of the 19 million Australians are of European origin, and many are the descendants of British settlers. The Aboriginals account for 1 percent of the population; 7 percent are either Asian-born or of Asian descent.

After World War II migrants from mainland Europe – mainly whites – were encouraged to settle in Australia. In 1973 the Australian government formally declared an end to the discriminatory "White Australia" immigration policy.

Among Aboriginals continuing poverty, poor health and lack of education have reinforced their mistrust of a white-dominated society. Since the late 1970s, federal governments have given some Aborigines freehold to their land which is of great cultural importance to them. In 1997 a report into the removal of 100,000 Aboriginal children from their families was published, after which the government expressed its regret at the practice of forcing children into white foster homes (discontinued in the late 1960s) and held a "National Sorry Day". English is almost universal in Australia; all but a few of the 260 Aboriginal languages are dead. Christianity is the dominant religion.

ECONOMY

As a new nation Australia owed its early wealth to its extensive agricultural and mineral resources. Since World War II

The Camel train (*above*), nicknamed the "Ghan", is named after the Afghan camels and their drivers who originally explored the Australian desert. Today the federally administered Trans-Australian railway runs from Port Augusta, north of Adelaide, through the Simpson Desert to Alice Springs in the Northern Territory. In all, Australia has more than 40,000 km (24,000 mi) of railroad track.

Herefords at auction in Queensland (*right*) Australia is a major exporter of beef (principally to the United States), and the state of Queensland alone provides grazing for almost half of the country's 26 million cattle. Herefords are the most popular breed in the central and northern part of the state, but agricultural research continues to find new crossbreeds suited to Australian conditions.

Sheep shearing (*below*) is usually a contract job. A gang of about 18 men travel from station to station during the season, averaging 95 sheep per shearer per day. Among others, the gang includes a cook, a wool-classer and numerous "roustabouts" (all-purpose workers) to clean up.

Australia has tried to build up a manufacturing economy to process its minerals.

Agriculture

Much of the farmland is used as pasture and is unsuitable for crops. Water is often in short supply, and soils are generally low in fertility. Production has risen by more than 150 percent since 1950 because of the introduction of new farming methods and mechanization.

A relatively small area of Australia is under permanent cultivation. Sugar cane, mainly from coastal Queensland, and wheat from the southeast and southwest are the major crops. The wide variations in climate allow the cultivation of many kinds of fruit, and wine exports, mostly from vineyards in South Australia and New South Wales, are on the increase.

Almost one-third of the continent is given over to sheep farming, making Australia the world's leading wool producer. Cattle farming needs the taller pastures found in higher-rainfall areas. Beef cattle are mostly found in Queensland and New South Wales; dairy farming is concentrated in coastal New South Wales and southern Victoria.

Australia has extensive native commercial forests, mostly in the eastern highlands, Tasmania and Western Australia. Japan is the largest single customer for timber. Western Australia and Tasmania are the leading fishing states.

Industry

Australia has some of the world's biggest deposits of iron ore and bauxite and is also a major producer of uranium. New South Wales and Queensland have large reserves of coal, and petroleum and natural gas are found off the coast of Victoria and in northeastern South Australia. Iron ore and coal are the chief mineral exports worldwide. Australia is also a major supplier of energy to Japan.

Most of Australia's own energy is from coal or natural gas. Tasmania is almost totally supplied by hydroelectric power. Australia has one nuclear reactor plant, but no nuclear power stations.

Since the late 1970s manufacturing industries have contracted in the face of competition from, in particular, Southeast Asia. Today the most important industries include iron and steel; motor vehicle production; petroleum refining and petrochemical products; textiles and clothing manufacturing; electrical and electronic equipment; and agricultural machinery. Tourism is also a thriving industry, particularly on the Pacific coast.

Transportation and communications

Some four-fifths of domestic freight is carried on the roads, and there is heavy investment in road building. However, infrastructure in cities has not kept pace with the increase in traffic. Around 85 percent of people live in or near the country's six major cities (Sydney, Melbourne, Brisbane, Adelaide, Perth and Canberra).

The main function of Australia's railroads, apart from passenger services for city commuters, is the bulk haulage of minerals, grain and long-distance freight. There are about 70 commercially significant ports, most of them serving the state capitals and industrial and mining centers. Australia's scheduled domestic airlines operate from some 450 airfields; isolated areas rely on the use of small aircraft. The national carrier Qantas offers

Sailing in Sydney Harbor Yachting is a favored pastime in Australia, with over 150 different types of sailboats registered in local waters. Entrepreneur Alan Bond's victory in the 1983 Americas Cup with his yacht *Australia II* caused nationwide celebrations

a passenger service to 24 countries.

Australia has a vigorous and long-established free press, but the concept of a national newspaper is relatively new. The Australian Broadcasting Corporation is independent but financed by government grants; other radio and television stations are run commercially. An ethnic radio and television network is funded by government grants and advertising.

Health and welfare

Healthcare services are financed principally by the government. Australia's universal health insurance scheme, Medicare, entitles patients to reclaim 85 percent of the schedule fee of most medical expenses. There are separate community health programs for Aboriginals, funded by federal government but run and controlled by Aboriginals themselves. The famous Royal Flying Doctor Service, begun in 1928, serves isolated areas.

Australia, like New Zealand, was a pioneer of social security, introducing state retirement pensions as early as 1909. Today's benefits are much wider-ranging.

Education

Education is free up to and including secondary level. School attendance is compulsory up to the age of 15 (16 in Tasmania). About two-thirds of children are educated in state schools. Many younger children attend preschool centers organized at local level. Most secondary schools are coeducational high schools.

New Zealand

NEW
ZEALAND

Mt Cook and Mt Tasman, New Zealand's highest mountains, dominate the Southern Alps which were named by Captain Cook in 1770. This spectacular mix of snow-capped mountains and deep glassy lakes is a hallmark of South Island scenery.

NEW ZEALAND IS AN ISLAND STATE IN THE southwestern Pacific Ocean about 1,600 km (1,000 mi) southeast of Australia. It also has a number of small overseas territories in the Pacific, notably Tokelau, Niue and the Cook Islands.

GEOGRAPHY

The country consists mainly of two islands – the North Island and the South Island – divided by the narrow Cook Strait.

The land
The South Island is dominated by the great fold-mountain belt of the Southern Alps. The highest peaks are in the center, reaching 3,754 m (12,349 ft) on Mount Cook; the Tasman Glacier, the largest of New Zealand's many glaciers, flows down its eastern slopes. In the southwest, where the mountains meet the sea, the flooded valleys have formed steep fjords, to create Fiordland, New Zealand's largest national park. The only lowland areas are a few alluvial plains in the east.

The mountains of the North Island are generally much lower and much more fragmented. The central and eastern uplands are an area of considerable geothermal activity, with geysers, hot springs and several volcanoes, two of which are constantly active. The hills slopes, river terraces and scattered lowlands of the North Island are some of the country's best farmland. The capital, Wellington, and the northern city of Auckland are both built around large natural harbors.

Climate
New Zealand generally enjoys a temperate maritime climate with regular rainfall all year round. Summers are warm except in the southern mountains, with plenty of sunshine; winters are mild, especially in the North Island, and snow is common only in mountainous areas. In the South Island the western mountains, which face the prevailing winds, receive heavy rainfall while sheltered eastern areas such as the Canterbury Plains are sometimes unusually dry.

Plants and wildlife
Much of the country is still covered by evergreen rainforest. Nine-tenths of the indigenous plants are unique to the islands, including kauri pines, southern beeches, and certain varieties of ferns. Some other tree species have been introduced. At one time the only medium to large indigenous wildlife, apart from birds, comprised two species of bats and some 30 reptiles. Among the few native reptiles today is the tuatara, sole survivor of a group that died out elsewhere 60 million years ago.

In the absence of mammals and other threatening native predators, some remarkable flightless birds developed on the islands such as the kiwi and the kakapo – the world's largest parrot. Another large parrot is the kea which lives in the snow-covered Southern Alps. Today the islands are populated by a range of species brought by Maori and European settlers. They include cats, dogs, rats and rabbits. Introduced predators have threatened the survival of native species. Conservationists have ensured that the native saddleback bird is established on small predator-free islands.

SOCIETY

In the 19th and 20th centuries British dominion had a powerful influence on New Zealand society. Today, however, New Zealanders are keen to demonstrate their separate identity.

History
New Zealand was first discovered and settled between 800 and 1000 AD by the people now known as the Maori who came from the Polynesian islands. The Maori were farmers and settled mainly on the warmer North Island. The first European to arrive was the Dutch explorer Abel Janszoon Tasman (c. 1603–59),

Tokelau

At first sight the Tokelau island group, some 500 km (300 mi) north of Western Samoa, looks green and lush. Dense vegetation covers the three coral atolls – Fakaofo, Nukunonu and Atafu – and each atoll is made up of many islets, or *motu*. These curved strips of coral sand are never more than 5 km (3 mi) long, or higher than 5m (16 ft) above sea level. Yet, despite the growth of abundant coconut palms, pandanus and other trees and shrubs, there is no fresh water on the islands; the soil is too porous to retain it. However, that has not prevented people from living here since the first settlers arrived from Samoa, probably on drift voyages.

Like many Pacific islanders, the people of Tokelau have led a precarious existence. A raid by Peruvian slavers in 1863 was followed by an outbreak of disease that reduced the population to about 200. In 1889 a British protectorate was established and in 1948 the group became an island territory of New Zealand.

In the past, many islanders emigrated to Samoa, and some more recently to New Zealand, but the present population of Tokelau is relatively stable. Tokelau has a few exports – copra and traditional craft goods – and some revenue comes from stamps and coins. Most people practice subsistence agriculture and the islands receive aid from New Zealand. Even

Tokelau dancers The tiny island territory of Tokelau is made up of three coral atolls covering a mere 10 sq km (3.9 sq mi). Its 1,600 inhabitants are Polynesian with close linguistic and cultural ties to nearby Western Samoa.

today, the only water comes from heavy seasonal rains, which the people collect in tanks or in traditional rain catchments – hollowed-out trunks of coconut palms.

NATIONAL DATA – NEW ZEALAND

Land area	268,021 sq km (103,483 sq mi)			

Climate	Altitude m (ft)	Temperatures January °C(°F)	July °C(°F)	Annual precipitation mm (in)
Wellington	126 (413)	18 (64)	9 (48)	1,018 (40)

Major physical features	highest point: Mount Cook 3,764 m (12,349 ft); largest lake: Lake Taupo 606 sq km (234 sq mi)

Population	(2006 est.) 4,177,937

Form of government	multiparty constitutional monarchy with one legislative house

Armed forces	army 4,430; navy 1,980; air force 2,250

Largest cities	Auckland (1,442,101); Wellington (capital – 448,959); Christchurch (375,145); Hamilton (158,402); Napier (53,658); Dunedin (116,240); Tauranga (117,594)

Official languages	English, Maori

Ethnic composition	European 69.8%; Maori 7.9%; Asian 5.7%; Pacific islander 4.4%; other 0.5%; mixed 7.8%; unspecified 3.8%

Religious affiliations	Anglican 14.9%; Roman Catholic 12.4%; Presbyterian 10.9%; Methodist 2.9%; Pentecostal 1.7%; Baptist 1.3%; other Christian 9.4%; other 3.3%; unspecified 17.2%; none 26%

Currency	1 New Zealand dollar (NZD) = 100 cents

Gross domestic product	(2006) U.S. $106 billion

Gross domestic product per capita	(2006) U.S. $26,000

Life expectancy at birth	male 75.82 yr; female 81.93 yr

Major resources	natural gas, iron ore, sand, coal, timber, hydropower, gold, limestone, cattle, cereals, fish, fruit, vegetables, sheep, tourism

followed in 1769–70 by the British Captain James Cook (1728–79).

Later whalers and traders from Australia established settlements on the islands, and then came British and French missionaries. In 1840 the Maori chiefs ceded sovereignty to the British crown in return for protection and land rights. The following year New Zealand became a crown colony. The discovery of gold in the 1860s attracted many thousands more immigrants. In the North Island, farmer settlers came into increasing conflict with the Maori over land rights. Rebellion escalated into war, until Maori resistance was finally broken in 1870.

Europeans born in New Zealand were gaining a greater sense of nationhood, and in 1907 the colony was granted dominion status within the British Empire. New Zealand fought alongside Britain in World War I, and did not escape the worldwide depressions of the 1920s and 1930s. During World War II New Zealanders served in the Pacific, where their presence forged closer links with other Pacific nations.

Although a member of the Commonwealth in the postwar period, New Zealand sought new alliances with the United States and some Southeast Asian countries. During the 1970s and 1980s, an energy crisis and loss of trade with Britain caused economic recession. In 1984 a Labour government pursued a monetarist policy of market deregulation but lost the 1990 elections to the National Party.

Government

The members of New Zealand's single-chamber parliament, since a referendum in 1992, are elected under a mixed first-past-the-post and proportional representation system. Every voter has two votes, one for their local candidate and the other for their favored party. The system allows support for minor parties – the old first-past-the-post only system favored the two dominant parties, Labour and the National Party. The head of state is the British monarch, represented in New Zealand by the governor-general.

People

Most of the 3.86 million New Zealanders are of European origin – mainly British – but the Maori account for a significant minority (8.9 percent). In recent decades there has been some improvement in the legal and social status of the Maori, enabling them to recover some ancestral lands from state-owned enterprises. The Maori language has official status along-

Pitcairn Island

In the year 1790, nine British sailors, along with twelve women and six men from Tahiti, arrived on Pitcairn – a small volcanic island in the Pacific Ocean midway between New Zealand and Peru. The sailors were mutineers from the Bounty, and had left most of their companions on Tahiti, over 2,000 km (1,200 mi) to the northeast.

On arrival at Pitcairn, the settlers burned their ship – and for the next 18 years this tiny desert island, some 5 sq km (2 sq mi) in area, was the scene of violence, murder and disease. When a party of American whalers stumbled across the survivors in 1808, there was only one European along with several Tahitian women and a few children.

In 1856 the island's by then 194 inhabitants were moved to Norfolk Island off the Australian coast, but 43 of them later returned to Pitcairn. Today their descendants live in Adamstown, the island's only town with just over 50 inhabitants. They live on what they can grow, and what they can buy from passing ships in return for stamps and curios. Pitcairn is a British dependency. The governor is a British High Commissioner in Auckland, New Zealand, advised by the 10-member island council.

side English. There are also a considerable number of small ethnic groups of non-Maori Polynesians, as well as Chinese and Indians.

The majority of New Zealanders are nominally Christian. Among the small minority sects are two Maori churches that have incorporated aspects of traditional Maori beliefs. Cultural and leisure activities are European-influenced, and sport, particularly rugby football, has a

Sheep farming (*below left*), is ideally suited to New Zealand's mild climate. Refrigerated shipping revolutionized the industry in 1882, opening up the lucrative British market, and there are now some 70 million sheep – more than 20 per capita.

Maori tattooing (*below*) Today the facial tattoo, or *moko*, is applied with paint, but it was originally chiseled into the skin using a *uhi* knife made of human or albatross bone struck with a mallet. *Moko* were also used to adorn thighs, buttocks and waists.

strong following. However, Maori arts and crafts have been revived along with traditional music and dance.

ECONOMY

During the first half of the 20th century, New Zealanders enjoyed one of the highest standards of living in the world. However, in the 1970s the country suffered from the loss of export trade with European Union countries. In the late 1980s and 1990s New Zealand successfully expanded its industrial and service sectors.

Agriculture and fisheries

Agriculture employs 9 percent of the workforce and supplies 8 percent of GDP. Sheep farming is still the chief activity, and lamb and wool are the principal exports. Cattle provide exports of beef and butter, along with cereals, vegetables and fruit.

Plantation forestry and fishing are also important. Forestry, based principally on the introduced Californian Monterey pine, contributes to large paper, pulp, timber, and woodchip exports worldwide.

Industry and commerce

The emphasis has been on food processing and on the manufacture of agricultural machinery and consumer goods for domestic use. Other industries such as chemical manufacture and motor vehicle assembly

The thriving lumber industry (*below*) makes heavy use of the versatile Monterey pine, *Pinus radiata*. It matures in 26 years in New Zealand's favorable climate, making it a lucrative renewable resource. Introduced species account for nearly a million hectares of forest.

rely heavily on imports. Coal is mined for generating electricity and for exporting to Japan. Natural gas has also become an important fuel; converting it into gasoline has led to a reduction in petroleum imports. Hydroelectricity supplies most of the other energy needs.

Today Australia is New Zealand's principal trading partner, followed by Japan and the United States. Since the mid 1980s banking and financial services have become a more important economic sector

Transportation and communications

New Zealand has well maintained roads, an extensive railroad network and regular ferry services between the islands. Local air travel is also available, with international flights from Auckland, Wellington and Christchurch.

Radio and television broadcasting are run by the Broadcasting Corporation of New Zealand. Newspapers are published regionally rather than nationally, but give good news coverage.

Welfare and education

In 1938 the government brought in a system of health and welfare benefits that were then the most extensive in the world. State-provided healthcare remains at a high level, though private-sector medicine is on the increase.

Education is free and compulsory between the ages of 6 and 15, and standards of literacy are high. Subsidized preschool education is also widely available. Higher education establishments include seven universities.

Papua New Guinea

INDEPENDENT STATE OF PAPUA NEW GUINEA

P APUA NEW GUINEA, IN THE WESTERN Pacific, comprises the Eastern portion of the island of New Guinea and numerous smaller islands to the north and east.

GEOGRAPHY

More than four-fifths of the land area is on the New Guinea mainland. The island territories vary from the tiny Trobriand group in the southeast to the long ridges of New Britain and New Ireland.

Papua New Guinea is dominated by mountain ranges subject to earthquakes

NATIONAL DATA – PAPUA NEW GUINEA

Land area 452,860 sq km (174,850 sq mi)				
Climate		Temperatures		Annual
	Altitude m (ft)	January °C(°F)	July °C(°F)	precipitation mm (in)
Port Moresby	38 (125)	28 (82)	26 (79)	1,112 (43.7)

Major physical features highest point: Mount Wilhelm 4,508 m (14,790 ft); longest river: Sepik about 1,000 km (600 mi)

Population (2006 est.) 5,670,544

Form of government multiparty constitutional monarchy with one legislative house

Armed forces army 2,500; air force 200

Capital city Port Moresby (295,892)

Official language English

Ethnic composition Papuan 84%; Melanesian 15%; others 1%

Religious affiliations Roman Catholic 22%; Lutheran 16%; Presbyterian/Methodist/London Missionary Society 8%; Anglican 5%; Evangelical Alliance 4%; Seventh-Day Adventist 1%; other Protestant 10%; traditional beliefs 34%

Currency 1 kina (PGK) = 100 toea

Gross domestic product (2006) U.S. $15.13 billion

Gross domestic product per capita (2006) U.S. $2,700

Life expectancy at birth male 63.08 yr; female 67.58 yr

Major resources gold, copper, silver, natural gas, timber, oil, hydropower, fisheries, bananas, cassava, cocoa, coconuts, coffee, rubber, sago, sugarcane, sweet potatoes, tea, vegetables, yams

and volcanic activity. The only extensive lowlands are on New Guinea island, around the massive Central Range.

Papua New Guinea has a tropical monsoon climate – hot and humid, with heavy seasonal rainfall. Vegetation varies from coastal mangroves and lowland swamp forests to oak, beech and pine in the mountains. Wildlife includes many unique species and among the remarkable birds are the large, flightless cassowaries and colorful birds of paradise.

SOCIETY

Although New Guinea has probably been inhabited for about 50,000 years, it was unknown to Europeans until the 16th century. In the late 19th century, the British claimed the island's southeastern quadrant, known as Papua, while Germany took over the northeastern quadrant and the islands.

In 1906 the British gave Papua to Australia. Following World War I the German sector was mandated to Australia by the League of Nations in 1921. Papua New Guinea gained full independence in 1975 within the British Commonwealth.

The head of state is the governor-general, appointed by the British sovereign, but executive power lies with the prime minister and cabinet. Parliamentary elections are held every five years.

The people fall into two main categories – Papuan and Melanesian – but Micronesian, Polynesian and Negrito peoples also live on the island. They speak over 700 separate languages – more than in any other country in the world.

ECONOMY

Papua New Guinea is rich in mineral resources, including copper and gold. The chief economic activity is mining, and minerals account for most exports. Natural gas, oil and hydroelectric power supply the country's energy.

Agricultural activity tends to be either subsistence farming or intensive production of coffee, tea, cocoa and copra for export. Important forestry and fisheries exports include prawns and tuna, palm oil, logs and lumber.

Because of the islands' mountainous and swampy terrain, transportation is mainly by air. Shipping services operate between the coastal ports and Australia.

Telecommunications is a growing industry, but few people have telephones or television. Health, education and welfare facilities are not widely available.

Beautiful but endangered (*above*) This Raggiana bird of paradise is one of over 40 birds of paradise surviving in the tropical forests and mangroves of Australasia. Like most males of the species this one is brilliantly colored; the females are very plain by comparison.

Loading cargo at Port Moresby (*below*) Discovered in 1873 by the British captain John Moresby, this large and sheltered natural harbor has grown to be the commercial center and capital of Papua New Guinea housing over 125,000 people.

Solomon Islands

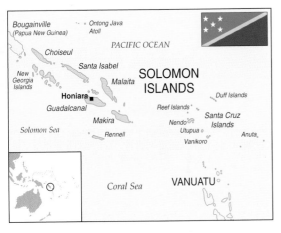

THE SOLOMON ISLANDS LIE IN THE WESTERN Pacific to the northeast of Australia. They were named in 1568 by a Spanish explorer who thought they might contain gold and other riches worthy of the biblical king, Solomon. Bougainville and Buka Island, part of the original group and sometimes known as the Northern Solomons, have been part of the territory of New Guinea for over a century.

GEOGRAPHY

The Solomons consist primarily of two parallel chains of volcanic islands running southeast of Bougainville. Farther east is the smaller Santa Cruz group including Nendo and Vanikoro. Several of the volcanic islands are still active.

The hot, humid climate is relieved by sea breezes, but temperature and rainfall are both very high in the cyclone season (November–April). Most of the land is rainforest. The wildlife includes large numbers of birds, and a unique lizard, the giant Solomon Island skink, with a prehensile tail.

SOCIETY

Melanesian peoples first settled the islands about 2000 BC. After Spanish settlements in the 16th and 17th centuries failed, British missionaries established a settlement in the 1870s. Soon after, Solomon Islanders were put to work on colonial plantations in Queensland, Australia, and on Fiji and other Pacific island colonies. In 1893 the British government declared the Solomons a protectorate.

During World War II the Solomons were the scene of bitter fighting between United States and Japanese forces, and a decisive battle was fought on the island of Guadalcanal. During the occupation nationalist feeling began to grow among the islanders, but full independence from Britain was not achieved until 1978.

Today the Solomon Islands is a member of the British Commonwealth. They have a single-chamber parliament whose 47 members are elected for a four-year term. Most Solomon Islanders are native Melanesians; over 60 indigenous languages are spoken, as well as English and an English-based pidgin.

ECONOMY

Agriculture, forestry and fisheries (the tuna trade in particular) are the largest export earners. Farmers produce copra, cacao, coconuts and palm oil as cash crops, and potatoes, yams, vegetables and rice for subsistence; pigs and cattle are raised for local consumption.

The islands have some gold and bauxite deposits, but no local energy resources. Light manufacturing industries (textiles, food-processing, batteries and fiberglass) flourish alongside traditional handicrafts, including shell-inlay and basketmaking. The islands have poorly developed roads,

Bringing in the catch Fishing – especially for tuna – is a mainstay of the Solomon Islands economy. Despite the introduction of boats with outboard motors, fishing techniques on the islands have hardly changed over the centuries.

but good shipping and air connections to Australia and other islands.

Healthcare is provided in clinics run by the government or by the church. The incidence of malaria is very high and increasing. Education is provided through church- and government-funded schools.

NATIONAL DATA – SOLOMON ISLANDS

Land area 27,540 sq km (10,633 sq mi)

Climate		Temperatures		Annual
	Altitude m (ft)	January °C(°F)	July °C(°F)	precipitation mm (in)
Honiara	2 (7)	27 (80)	26 (79)	2,094 (82.4)

Major physical features largest island: Guadalcanal 5,336 sq km (2,060 sq mi); highest point: Makarakomburu 2,447 m (8,028 ft)

Population (2006 est.) 552,438

Form of government multiparty constitutional monarchy with one legislative house

Armed forces no armed forces

Capital city Honiara (Guadalcanal – 58,492)

Official language English

Ethnic composition Melanesian 94.5%; Polynesian 3%; Micronesian 1.2%; other 1.1%; unspecified 0.2%

Religious affiliations Church of Melanesia 32.8%; Roman Catholic 19%; South Seas Evangelical 17%; Seventh-Day Adventist 11.2%; United Church 10.3%; Christian Fellowship Church 2.4%; other Christian 4.4%; other 2.4%; unspecified 0.3%; none 0.2%

Currency 1 Solomon Islands dollar (SBD) = 100 cents

Gross domestic product (2005) U.S. $9.413 billion

Gross domestic product per capita (2005) U.S. $2,500

Life expectancy at birth male 70.4 yr; female 75.55 yr

Major resources fish, gold, bauxite, phosphates, lead, zinc, nickel, coconuts, livestock, palm kernels, rice, sweet potatoes, taro, timber

Tuvalu

TUVALU

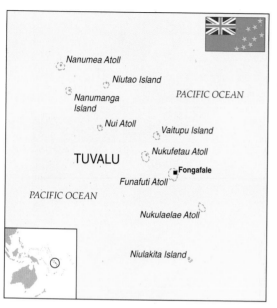

Dancing the night away Ritual and entertainment are exuberantly combined in the traditional dances of the tiny coral islands of Tuvalu. Pacific arts festivals and tourist interest have helped to keep many ancient customs alive.

Tuvalu is a scattered island chain in the southwestern Pacific between the Gilbert Islands of Kiribati to the north and Fiji to the south. Formerly the Ellice Islands, the group was part of the British colony of the Gilbert and Ellice Islands until 1978.

Tuvalu consists of nine islands, the tiny peaks of an undersea mountain range. Five of them are coral atolls; four are reef islands. They range in size from 4.9 sq km (1.89 sq mi) to 0.4 sq km (0.16 sq mi) and rise only a few meters (yards) above sea level. The islands' humid, tropical climate is cooled by the prevailing southeast trade winds. Annual rainfall is often high, but there are frequent severe droughts. The poor coralline soil supports only coconut palms, breadfruit and cassowary trees.

Wildlife includes mostly rats, lizards and sea birds, but the surrounding sea is rich in fish and other marine life.

It is likely that the islands were first settled by seafaring Polynesians, probably from Samoa, about the 14th century. In 1892 the British persuaded the islanders to join the Gilbert Islands protectorate, later the Gilbert and Ellice Islands colony. During World War II the United States set up bases on the Ellice Islands to combat the Japanese invasion of the Gilberts.

By the 1960s the mainly Polynesian Ellice islanders were becoming increasingly dissatisfied with the Gilbertese (Micronesian) domination of the colony. They eventually seceded in 1976, and the islands became independent as Tuvalu in 1978. Today Tuvalu remains in the British Commonwealth, with the British monarch as head of state.

Traditionally Tuvalu's economy rests on agriculture, mainly subsistence farming, and fishing. Copra, the only cash crop, is marketed by a cooperative body. Fishing is mostly for local consumption except for *beche de mer* (sea cucumbers) which are caught and processed for export. Many islanders go abroad to find work. However Tuvalu's fortunes altered in 2000 following the lease of its Internet domain name "tv." for $50 million.

NATIONAL DATA - TUVALU

Land area	26 sq km (10 sq mi)			
Climate	Altitude m (ft)	Temperatures January °C(°F)	July °C(°F)	Annual precipitation mm (in)
Funafuti atoll	2 (7)	28 (82)	28 (82)	3,544 (139.5)

Major physical features largest island: Vaitupu 5 sq km (2 sq mi)

Population (2006 est.) 11,810

Form of government nonparty constitutional monarchy with one legislative house

Armed forces no armed forces

Capital city Fongafale (Funafuti, 4,921)

Official languages Tuvaluan, English

Ethnic composition Polynesian 96%; Micronesian 4%

Religious affiliations Church of Tuvalu (Congregationalist) 97%; Seventh-Day Adventist 1.4%; Baha'i 1%; other 0.6%

Currency 1 Australian dollar (AUD) or 1 Tuvaluan dollar = 100 cents

Gross domestic product (2002) U.S. $14.94 million

Gross domestic product per capita (2002) U.S. $1,600

Life expectancy at birth male 66.08 yr; female 70.66 yr

Major resources fish, bananas, breadfruit, coconuts, copra, fish, papaya, root crops

Nauru

REPUBLIC OF NAURU

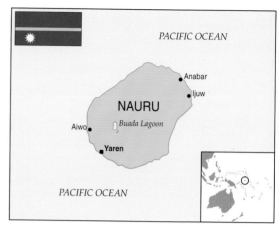

Nauru is a tiny single-island state in the western Pacific Ocean, immediately south of the Equator and about 700 km (450 mi) west of Kiribati. It is the world's smallest republic and in the mid-1980s had one of the highest per capita incomes in the world, thanks to a simple precious natural resource – phosphate.

The island is a raised coral reef with a central plateau, where rich phosphate beds have been created by bird droppings deposited over the centuries. The climate is hot and humid, though moderated by sea breezes. During the monsoon season (November to February) rainfall is heavy, but severe droughts can also occur. Vegetation consists mostly of coconut palms, breadfruit trees and scrub.

NATIONAL DATA - NAURU

Land area	21 sq km (8 sq mi)			
Climate	Altitude m (ft)	Temperatures January °C(°F)	July °C(°F)	Annual precipitation mm (in)
Nauru	27 (89)	28 (82)	28 (82)	2,051 (80.8)

Major physical features highest point: 69 m (225 ft)

Population (2006 est.) 13,287

Form of government nonparty republic with one legislative house

Armed forces no armed forces

Capital city No official capital but Yaren is the government seat (4,616)

Official languages Nauruan, English

Ethnic composition Nauruan 58%; other Pacific Islander 26%; Chinese 8%; European 8%

Religious affiliations Christian 100% (two-thirds Protestant, one-third Roman Catholic)

Currency 1 Australian dollar (AUD) = 100 cents

Gross domestic product (2005) U.S. $60 million

Gross domestic product per capita (2005) U.S. $5,000

Life expectancy at birth male 59.5 yr; female 66.84 yr

Major resources phosphates, fish, bananas, coconuts, pineapples

Kiribati

REPUBLIC OF KIRIBATI

Nauru was probably settled at various times by people from neighboring islands. In 1888 it was annexed by Germany. Some 10 years later phosphate was discovered and mining began in 1906, controlled by a British and a German company. After World War II, the island was made a United Nations' trust territory, and gained full independence in 1968. In 1970 the islanders took over control of all phosphate production.

Nauru's 18-member parliament is freely elected, and chooses a president as head of state. Native islanders are of mixed Pacific ancestry. Naurean and English are spoken.

Phosphate extraction remains the chief export earner – Japan, South Korea, New Zealand and Australia are major customers for phosphate rock which they use as a fertilizer. Although deposits will run out by 2008, government investment overseas should cushion the population from the sudden loss of revenue. In the 1990s, Australia, New Zealand and the UK agreed to compensate Nauru for environmental damage caused by phosphate mining. Much food is imported. Coconuts, bananas and pineapples are the main export crops. Welfare services are excellent on Nauru. Education and healthcare are free and the islanders pay no taxes toward funding these facilities.

KIRIBATI (PRONOUNCED KIREE-BASS) IS MADE up of three widely separated archipelagoes spread across a vast area of the central Pacific. The largest groups are the Gilbert Islands, the Phoenix group and the Line Islands.

The islands are coral atolls, founded on the summits of undersea volcanoes and encircled by reefs. They rise only a meter or so (a few feet) above sea level.

Kiribati's climate is hot and humid. Rainfall is frequent in the Gilberts, but is sparser elsewhere, and is generally unreliable. Only the Gilberts and three of the Line Islands are now inhabited. Vegetation is mostly confined to salt-resistant plants and palms, chiefly coconut and bananas. The reefs teem with marine life.

The Gilbert Islands were settled several thousand years ago by Micronesian peoples, followed later by Polynesian seafarers from the south. In 1900 the British annexed Banaba island for its phosphates and forcibly resettled its population. In 1916 the Gilbert and Ellice Islands colony was founded, and later extended to include the Phoenix and Line Islands.

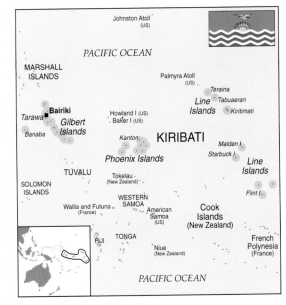

Earthly paradise A lagoon beach on Tabuaeran (one of the Line Islands). The turquoise waters and palm-fringed beaches have captivated visitors ever since the first Western mariners started to explore the South Pacific seas in the 17th century.

In 1975 the Polynesian-inhabited Ellice Islands seceded and became independent as Tuvalu in 1978. The remaining colony became independent as Kiribati in 1979. The vast majority of today's population are Gilbertese (I-Kiribati) speaking a distinct Micronesian language.

Phosphate deposits on Banaba were exhausted in the late 1970s, but the profits had created a substantial reserve fund. Otherwise, Kiribati's economy is based largely on subsistence agriculture and fishing, with copra the chief cash crop.

NATIONAL DATA – KIRIBATI

Land area	811 sq km (313 sq mi)			
Climate		**Temperatures**		**Annual**
	Altitude m (ft)	January °C(°F)	July °C(°F)	precipitation mm (in)
Tarawa	3 (10)	28 (82)	28 (82)	1,914 (75.4)

Major physical features largest island: Kiritmati (Line Is) 388 sq km (150 sq mi); highest point: Banaba 81 m (266 ft)

Population (2006 est.) 105,432

Form of government multiparty republic with one legislative house

Armed forces no armed forces

Capital city Tarawa (47,437)

Official language English

Ethnic composition Micronesian 98.8%; other 1.2%

Religious affiliations Roman Catholic 52%; Protestant (Congregational) 40%; Seventh-Day Adventist, Muslim, Baha'i, Latter-day Saints, Church of God 8%

Currency 1 Australian dollar (AUD) = 100 cents

Gross domestic product (2004) U.S. $206.4 million

Gross domestic product per capita (2004) U.S. $2,700

Life expectancy at birth male 59.06 yr; female 65.24 yr

Major resources bananas, breadfruit, coconuts, copra, fish, papaya

Fiji

REPUBLIC OF FIJI

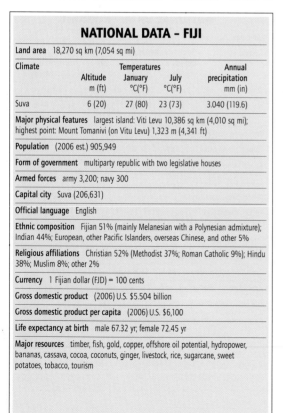

F IJI IS AN ISOLATED ARCHIPELAGO IN THE southwestern Pacific, some 2,100 km (1,300 mi) north of Auckland, New Zealand. Two of its islands – Viti Levu and Vanua Levu – account for nearly nine-tenths of the land area.

Like most of the islands of Fiji, Viti Levu and Vanua Levu are volcanic in origin with central mountain ranges rising steeply from narrow coastal plains and offshore coral reefs. From November to April the tropical climate brings hot, humid conditions and there is threatened danger from cyclones. The densely forested southeastern coastline receives heavy rainfall, brought by the trade winds. The western slopes, where there are open forests and grasslands, are much drier. More than two-thirds of the land area of Fiji is forested and only a small percentage of land is meadow or pasture.

When the Dutch navigator Abel Janszoon Tasman (c. 1603–59) reached Fiji in 1643, its inhabitants were a mixture of Melanesian and Polynesian peoples. In 1874 Fiji became a British crown colony, and from 1879 the British imported thousands of Indian laborers to work their sugar plantations.

After Fiji became independent in 1970, the political parties divided along racial lines creating considerable political instability. Indian parties won the election of 1987, but three subsequent military coups restored Melanesian dominance which was confirmed in the 1990 constitution.

Agriculture, tourism and light industry are striving to support an economy that suffered badly in the unrest and unfavorable conditions of the late 1980s – bad weather, low sugar prices, labor strikes and political unrest.

Major cash crops include sugar cane, copra, ginger and coconuts. Gold is mined on Viti Levu, and is another major export.

Gaily painted canoes moored against a jetty on the Wainibuki river on Fiji's main island, Viti Levu. Canoes are used for transport, fishing and provide endless photo-opportunities for the country's many tourists.

NATIONAL DATA – FIJI				
Land area 18,270 sq km (7,054 sq mi)				
Climate		Temperatures		Annual
	Altitude m (ft)	January °C(°F)	July °C(°F)	precipitation mm (in)
Suva	6 (20)	27 (80)	23 (73)	3.040 (119.6)
Major physical features largest island: Viti Levu 10,386 sq km (4,010 sq mi); highest point: Mount Tomanivi (on Vitu Levu) 1,323 m (4,341 ft)				
Population (2006 est.) 905,949				
Form of government multiparty republic with two legislative houses				
Armed forces army 3,200; navy 300				
Capital city Suva (206,631)				
Official language English				
Ethnic composition Fijian 51% (mainly Melanesian with a Polynesian admixture); Indian 44%; European, other Pacific Islanders, overseas Chinese, and other 5%				
Religious affiliations Christian 52% (Methodist 37%; Roman Catholic 9%); Hindu 38%; Muslim 8%; other 2%				
Currency 1 Fijian dollar (FJD) = 100 cents				
Gross domestic product (2006) U.S. $5.504 billion				
Gross domestic product per capita (2006) U.S. $6,100				
Life expectancy at birth male 67.32 yr; female 72.45 yr				
Major resources timber, fish, gold, copper, offshore oil potential, hydropower, bananas, cassava, cocoa, coconuts, ginger, livestock, rice, sugarcane, sweet potatoes, tobacco, tourism				

Vanuatu

REPUBLIC OF VANUATU

T HE SOUTHWESTERN PACIFIC REPUBLIC OF Vanuatu (formerly the New Hebrides) comprises 13 main islands and dozens of smaller ones, forming a Y-shaped group between the Solomon Islands to the north and New Caledonia to the south.

The main islands are Espiritu Santo and Efaté, site of the capital, Port-Vila. Like most of the country, these islands are mountainous and covered in rainforest, with raised coral terraces on some coastal slopes and offshore coral reefs. Earth tremors are frequent, and there are

490

Tonga

KINGDOM OF TONGA

several active volcanoes. The humid, tropical climate produces twice as much rainfall in the north as in the south, and seasonal storms can rise to hurricane force. In 1987 the worst cyclone in the country's history damaged or destroyed most of the houses on the islands.

Vanuatu's largely Melanesian inhabitants lived here for more than 2,500 years before Europeans discovered the group of islands in the 15th century. From the 1840s sandalwood traders and missionaries made their way to the islands. During the 20th century growing pressure from nationalist movements brought independence in 1980, a new constitution and a new name, Vanuatu, "Our Land Forever". Today, members of Vanuatu's single-chamber parliament are elected for a four-year term. Parliament and the heads of regional councils elect a president to serve a five-year term.

The economy depends largely on exports of copra, fish, beef and veal; most manufactured goods and fuel are imported. About three-quarters of the population work on the land, but most are subsistence farmers raising livestock and crops such as yams and bananas. Other exports include manganese, mined on Efaté, and pine lumber. Tourism is a growing source of hard currency, and Vanuatu's favorable tax laws have also made it a financial center.

Tonga, in the southwestern pacific, is a double chain of 169 islands to the southwest of Samoa and east of Fiji. In 1875 the king of Tonga created a constitutional monarchy, and his system of government continues virtually unchanged today.

The islands are in three main groups from north to south, Vava'u, Ha'apai and Tongatapu. The Vava'u group and one island in the Ha'apai group have mountainous landscapes of volcanic rock, and there are four intermittently active volcanoes. All the other islands are low-lying coral formations.

The semitropical climate supports a rich variety of mainly forest vegetation, including trees such as the paper mulberry. Bird life on the islands is plentiful, and some of the world's largest bats, flying foxes, are also found there.

The settlement of Tonga began in about 1000 BC, and since the 10th century AD kingship has been handed down from the original Tu'i Tonga royal line.

A British Methodist mission established in 1826 converted the majority of the islanders to Christianity. In 1900 the kingdom became a British protectorate;

Friendly faces The welcoming smiles that greet modern visitors reflect Tonga's longstanding tradition of hospitality – the British explorer Captain Cook named the islands "The Friendly Islands" when he landed there in the 1770s.

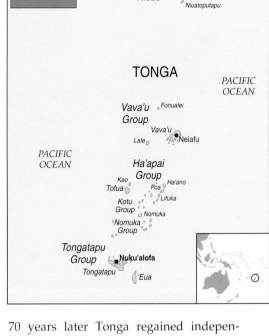

70 years later Tonga regained independence within the British Commonwealth.

Agriculture is central to the economy; cash crops include vanilla, coconuts and bananas, while yams, taro and cassava are among the chief subsistence crops. Traditional handicrafts cater for the growing tourist market, but many Tongans work abroad in New Zealand, Australia or the United States, and send remittances home.

NATIONAL DATA - VANUATU

Land area	12,200 sq km (4,710 sq mi)			
Climate		**Temperatures**		**Annual**
	Altitude m (ft)	January °C(°F)	July °C(°F)	precipitation mm (in)
Port-Vila	57 (187)	27 (81)	23 (73)	2,221 (87.4)

Major physical features highest point: Espiritu Santo 3,678 sq km (1,420 sq mi); highest point: Tabwémanasana on Espiritu Santo 1,880 m (6,167 ft)

Population (2006 est.) 208,869

Form of government multiparty republic with one legislative house

Armed forces no armed forces

Capital city Port-Vila (38,438)

Official languages French, English

Ethnic composition Ni-Vanuatu 98.5%; other 1.5%

Religious affiliations Presbyterian 31.4%; Anglican 13.4%; Roman Catholic 13.1%; Seventh-Day Adventist 10.8%; other Christian 13.8%; traditional beliefs 5.6% (including Jon Frum cargo cult); other 9.6%; none 1%; unspecified 1.3%

Currency vatu (VUV)

Gross domestic product (2003) U.S. $276.3 million

Gross domestic product per capita (2003) U.S. $2,900

Life expectancy at birth male 61.34 yr; female 64.44 yr

Major resources manganese, hardwood forests, fish, bananas, cassava, cattle, cocoa, coconuts, coffee, taro, yams, tourism

NATIONAL DATA - TONGA

Land area	718 sq km (277 sq mi)			
Climate		**Temperatures**		**Annual**
	Altitude m (ft)	January °C(°F)	July °C(°F)	precipitation mm (in)
Nuku'alofa	3 (10)	26 (78)	21 (70)	1,738 (68.4)

Major physical features highest point: Tongatapu 256 sq km (99 sq mi); highest point: on Kao Island 1,030 m (3,380 ft)

Population (2006 est.) 114,689

Form of government nonparty constitutional monarchy with one legislative house

Armed forces no armed forces

Capital city Nuku'alofa (23,854)

Official languages Tongan, English

Ethnic composition Tongan 96%; Part-Tongan 1.7%; European 0.6%; Fijian 0.2%; Samoan 0.2%; Indian, Chinese, Japanese 0.3%; Other 1%

Religious affiliations Christian 100% (Free Wesleyan Church claims over 30,000 adherents)

Currency 1 pa'anga (TOP) = 100 seniti

Gross domestic product (2004) U.S. $178.5 million

Gross domestic product per capita (2005) U.S. $2,200

Life expectancy at birth male 67.32 yr; female 72.45 yr

Major resources fish, bananas, cassava, citrus fruits, coconuts, sweet potatoes, taro, vanilla, tourism

Samoa

INDEPENDENT STATE OF SAMOA

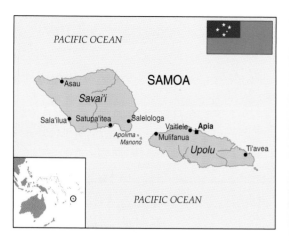

PACIFIC OCEAN

SAMOA

Asau

Savai'i

Sala'ilua • Satupa'itea • Salelologa

Apolima • Vaitele • Apia
Manono • Mulifanua

Upolu • Ti'avea

PACIFIC OCEAN

THE NINE ISLANDS OF SAMOA LIE IN the south-central Pacific Ocean, north of Tonga and northeast of Fiji. Only four islands are inhabited, Upolu, Savai'i, Apolima, and Manono. Today, Samoa's fragile economy depends on agriculture, tourism, and income from Samoans working abroad, especially in New Zealand.

GEOGRAPHY

On the largest islands – Upolu and Savai'i – the interior is mountainous, and coral reefs ring the narrow coastal plain. The islands are all volcanic, and Savai'i suffered some major eruptions in 1902 and 1911. The climate is generally warm and humid, with rainfall especially heavy on southeastern slopes exposed to the frequent trade winds. From November to March there is also a danger of cyclones. Almost half the land is covered with forest, and mangroves thrive in the lowland swamps. Wildlife on the islands is sparse, but includes a unique and rare tooth-billed pigeon.

Women feeding and tending pigs in a typical village in Samoa. Pigs have a special significance in many South Pacific cultures: they are highly valued not only as a source of food but also as a symbol of wealth. In some parts, pigs are used as dowries.

SOCIETY

The Samoan islands were probably settled by Tongans in about 1000 BC. The first European visitor in 1766 was the French explorer Bougainville (1729–1811) who named them the Navigator Islands. Local disputes on the islands in the 19th century were complicated by competing United States', German and British colonial interests. In 1899 the three colonial powers signed an agreement whereby Britain ceded its interests to Germany and the United States retained American Samoa to the east. After World War I (1914–18) the League of Nations mandated German Samoa to New Zealand as Western Samoa, which finally achieved full independence in 1962. The country's name was changed to Samoa in July 1997, but there have been strong objections to this from its neighbor American Samoa.

Samoa is a constitutional monarchy ruled by a paramount chief. He appoints a prime minister on the recommendation of the 47 members of the single-chamber legislative assembly. Two members of the assembly are elected by the communities of European and mixed-descent, while the rest are chosen by the Polynesian clan chieftains.

The great majority of the population are indigenous Samoans, whose language – Samoan – may be the oldest of all the Polynesian languages. English is the second official language.

ECONOMY

The economic mainstays are agriculture, fishing, lumber exports and tourism. Copra and cocoa are grown for export while yams, breadfruit and papaya are grown for local consumption. Fishing has also become an export industry. Timber supplies for local use and for export trade are maintained by reforestation programs. For power, the country relies mainly on hydroelectricity. The range of light industries is growing, while traditional handicrafts meet the demands of a brisk and growing tourist trade.

About one-fifth of Samoa's roads are surfaced, and the islands are linked by sea and air; they have two deepwater ports and one international airport. Healthcare is administered from the main hospital in the capital, Apia, on Upolu. Education, modeled on New Zealand's system, is free though not compulsory.

NATIONAL DATA - SAMOA

Land area	2,934 sq km (1,133 sq mi)			

Climate		Temperatures		Annual
	Altitude m (ft)	January °C(°F)	July °C(°F)	precipitation mm (in)
Apia	2 (7)	27 (72)	26 (79)	2,961 (116.5)

Major physical features largest island: Savai'i 1,813 sq km (700 sq mi); highest point: Silsili (Savai'i) 1,858 m (6,095 ft)

Population (2006 est.) 176,908

Form of government multiparty republic with two legislative houses

Armed forces no armed forces

Capital city Apia (41,204)

Official languages Samoan, English

Ethnic composition Samoan 92.6%; Euronesians 7% (persons of European and Polynesian blood); Europeans 0.4%

Religious affiliations Congregationalist 34.8%; Roman Catholic 19.6%; Methodist 15%; Latter-Day Saints 12.7%; Assembly of God 6.6%; Seventh-Day Adventist 3.5%; other Christian 4.5%; Worship Centre 1.3%; other 1.7%; unspecified 0.1%

Currency 1 tala (SAT) = 100 seniti

Gross domestic product (2002) U.S. $1 billion

Gross domestic product per capita (2005) U.S. $2,100

Life expectancy at birth male 68.2 yr; female 73.94 yr

Major resources hardwood forests, fish, hydropower, bananas, coffee, copra, tropical fruits, tourism

Palau

REPUBLIC OF PALAU

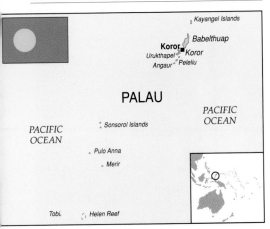

THE REPUBLIC OF PALAU, ALSO KNOWN BY its native name Belau, became independent in October 1994. Before then it was a UN trusteeship in free association with the United States. The territory comprises the westernmost portion of the Caroline Islands chain some 850 km (528 mi) southeast of the Philippines. The main islands are part of a coralline reef structure on a volcanic base, with over 300 islets surrounding a huge central lagoon.

The islands vary geologically from the mountainous main island of Babelthuap to low coral islands fringed by barrier reefs. The reefs are home to a variety of marine life, and the hot wet climate supports lush vegetation. Agriculture and fishing remain at subsistence level, though tuna is exported and copra is grown as a cash crop. Coconut oil is exported in small quantities. Transportation is by sea or air, and tourism has not been fully developed.

Marshall Islands

REPUBLIC OF THE MARSHALL ISLANDS

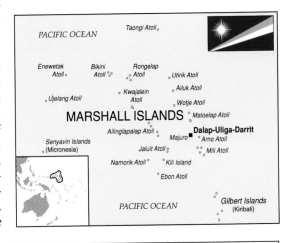

THE MARSHALL ISLANDS ARE A GROUP OF 5 single islands, 31 widely separated coral atolls (including the largest atoll in the world) and over 1,000 islets and reefs strung out in two chains in the western Pacific Ocean. During World War II the islands were occupied by the Allied Forces, and from 1949 they were part of the Trust Territory of the United States. In 1986 they became self-governing in free association with the US, an arrangement whereby the US retained responsibility for defense in return for assistance. In 1991 the Marshall Islands became politically autonomous.

Local politics are traditionally dominated by the island chiefs, and their high chief holds the post of president, but the economy of the Marshall Islands is still heavily dependent on aid from the United States, and the rent paid by its benefactor for the use of Kwajalein Atoll as a missile base. The tourist industry is an important source of foreign revenue and employment. High-grade phosphate is mined for export on Ailinglapalap Atoll but most other industry is limited to handicrafts, and processing fish or copra. Agriculture is mainly at subsistence level, but some cash crops (coconuts, tomatoes, melons and breadfruit) are grown in small quantities on local farms.

Living with nature (*below*) A typical house in the Marshall Islands is constructed of woven mats and roofed with coconut palm fronds. Coconut palms and breadfruit are the islands' main vegetation; the leaves and fruit serve a variety of purposes.

NATIONAL DATA – MARSHALL ISLANDS

Land area	181 sq km (70 sq mi)			
Climate		Temperatures		Annual
	Altitude m (ft)	January °C(°F)	July °C(°F)	precipitation mm (in)
Ujelang Atoll	10 (33)	27 (81)	28 (82)	1,976 (77.8)

Population	(2006 est.) 60,422

Form of government Constitutional government in free association with the United States of America

Armed forces no armed forces

Capital city Majuro (25,400)

Official language English

Ethnic composition Micronesian 100%

Religious affiliations Protestant 54.8%; Assembly of God 25.8%; Roman Catholic 8.4%; Bukot nan Jesus 2.8%; Mormon 2.1%; other Christian 3.6%; others 2.5%

Currency 1 U.S. dollar (USD) = 100 cents

Gross domestic product (2001) U.S. $115 million

Gross domestic product per capita (2005) U.S. $2,900

Life expectancy at birth male 68.33 yr; female 72.39 yr

Major resources coconuts, copra, fruits, deep-seabed minerals, phosphates, clams, fish, livestock, oysters, tourism

NATIONAL DATA – PALAU

Land area	458 sq km (177 sq mi)

Major physical features 26 islands and over 300 islets; largest island is Babelthuap

Population	(2006 est.) 20,579

Form of government constitutional republic with two legislative houses

Armed forces no armed forces

Capital city Melekeok (391) replaced Koror (11,559) in 2006

Official languages Palauan and English are official in all islands except Sonsoral (Sonsoralese and English are official), Tobi (Tobi and English are official), and Angaur (Angaur, Japanese, and English are official)

Ethnic composition Palauan (Micronesian with Malayan and Melanesian admixtures) 69.9%; Filipino 15.3%; Chinese 4.9%; other Asian 2.4%; white 1.9%; Carolinian 1.4%; other Micronesian 1.1%; other or unspecified 3.2%

Religious affiliations Roman Catholic 41.6%; Protestant 23.3%; Modekngei 8.8% (indigenous to Palau); Seventh-Day Adventist 5.3%; Jehovah's Witness 0.9%; Latter-Day Saints 0.6%; other religion 3.1%; unspecified or none 16.4%

Currency 1 U.S. dollar (USD) = 100 cents

Gross domestic product (2004) U.S. $124.5 million

Gross domestic product per capita (2005) U.S. $7,600

Life expectancy at birth male 67.26 yr; female 73.77 yr

Major resources forests, gold, marine products, deep-seabed minerals, fish, root crops, copra, coconuts, tourism

Micronesia

FEDERATED STATES OF MICRONESIA

THE FEDERATED STATES OF MICRONESIA are a group of scattered volcanic islands in the western Pacific Ocean, including the eastern and central Caroline Islands. They have a hot rainy climate encouraging lush tropical vegetation and extensive rainforest. In 1986 all the islands were released from a UN trusteeship administered by the United States, and became a federal republic in free association with the United States; in 1991 they became an independent republic. The main economic activities among the indigenous peoples are subsistence farming and fishing, but a small quantity of cash crops (copra, bananas, citrus fruits, peppers and taro) are produced for export. In the past, the islands depended on US aid, grants and military spending. In the future, however, they intend to become more self-sufficient, developing tourism and generating income from the sale of fishing rights and cash crops.

NATIONAL DATA - MICRONESIA

Land area	702 sq km (271 sq mi)			
Climate		**Temperatures**		**Annual precipitation**
	Altitude m (ft)	January °C(°F)	July °C(°F)	mm (in)
Palikir	0 (0)	27 (81)	27 (81)	4,907 (193.2)

Population	(2006 est.) 108,004
Form of government	federal republic
Capital city	Palikir (on the island of Pohnpei, 4,462)
Official language	English
Ethnic composition	Chuukese/Mortlockese 49%; Pohnpeian 24%; Kosraean 6%; Yapese 5%; Yap Outer Islands 5%; Polynesian 1%; Asian 2%; White 1%; Others 7%
Religious affiliations	Roman Catholic 50%; Protestant 47%; other 3%
Currency	1 U.S. dollar (USD) = 100 cents
Gross domestic product	(2002) U.S. $277 million
Gross domestic product per capita	(2005) U.S. $2,300
Life expectancy at birth	male 68.24 yr; female 71.95 yr
Major resources	forests, marine products, deep-seabed minerals, phosphates, tourism, bananas, citrus fruits, copra, peppers, taro, fish, livestock

Dependencies in the region

Riches of the sea (*above*) These fishermen on the island of Tahiti in French Polynesia are bringing in cages of oysters which have been cultivated for their pearls. One is wearing the traditional sarong, but western dress is becoming increasingly popular.

GUAM
UNITED STATES

Guam, the most southerly of the Mariana Islands, is a self-governing unincorporated territory of the United States. The landscape ranges from a thickly forested limestone plateau in the north to low, grass-covered volcanic hills in the southwest. The climate is warm, wet and equable except for the occasional violent typhoon. Guam's population is mixed, with the language of the indigenous Chamorro people having official status alongside English.

The economy is heavily dependent on the presence of the United States' Navy and Air Force which have used Guam as a major base since World War II. Most food and consumer goods have to be imported. Revenue from tourism is a growing sector of the economy.

WAKE ISLAND
UNITED STATES

Wake Island in the western Pacific Ocean is an unincorporated territory of the United States. It consists of three tiny coral islands linked by causeways around a lagoon; there is little rainfall and no natural water supply.

It is used as an emergency stopover point for commercial aircraft.

MIDWAY ISLANDS
UNITED STATES

The two Midway Islands are an unincorporated territory of the United States. Lying at the far northwestern end of the Hawaiian chain in the north-central Pacific Ocean, they are coral atolls formed around an undersea volcano. Their hot, humid equable climate makes them an ideal refuge for vast numbers of sea birds. The only inhabitants are United States' military personnel, and nature wardens in charge of the bird reserve.

JOHNSTON ATOLL
UNITED STATES

Johnston is a remote coral atoll consisting of two small islets (Johnston Island and Sand Island) some 1,126 km (700 mi) southwest of Hawaii. It is part of the United States' defense system and is controlled by the Defense Nuclear Agency and the United States' Fish and Wildlife Service. The Atoll is worked for its plentiful guano.

NORTHERN MARIANA ISLANDS
UNITED STATES

The Northern Marianas, in the western Pacific Ocean, consist of the whole of the Mariana island chain apart from the southernmost island, Guam. They are a self-governing commonwealth in free association with the United States. Under these terms the United States is responsible for security and defense, and agrees to provide financial assistance. The relationship can be ended by plebiscite. The islands form part of a long volcanic chain that runs due south from Japan. The climate is warm and humid, supporting lush

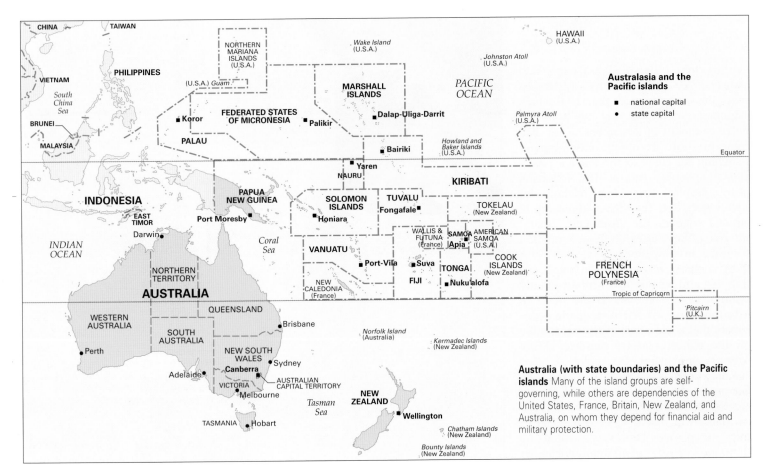

Australia (with state boundaries) and the Pacific islands Many of the island groups are self-governing, while others are dependencies of the United States, France, Britain, New Zealand, and Australia, on whom they depend for financial aid and military protection.

forest vegetation except in the drier north. The population, concentrated on Saipan island, is culturally diverse, but the majority of inhabitants are native Chamorros, a Micronesian people.

Tourism has taken over from agriculture as the mainstay of the economy. Farming and fishing are mainly for subsistence, though the islanders produce garments, copra, meat and fish for export.

AMERICAN SAMOA
UNITED STATES

American Samoa is an unincorporated territory of the United States consisting of seven islands along the east of the Samoan archipelago in the south-central Pacific Ocean. The largest, Tutuila, accounts for more than two-thirds of the total land area. The climate is hot and humid with occasional violent cyclones.

Every Samoan belongs to an extended family headed by a chief. Families are grouped in villages, each controlled by a council of chiefs according to ancient custom; the chiefs' councils elect the Senate within the national assembly.

Tuna canning and tourism are the islands' main industries, and canned tuna, accounts for almost all the country's export revenue. Standards of living and educational facilities are good.

NEW CALEDONIA
FRANCE

The islands of New Caledonia, the largest French Overseas Territory, lie in the southwestern Pacific about 1,770 km (1,100 mi) north of New Zealand. The warm, humid climate supports lush rainforests on the eastern slopes of the main islands; open forests and savanna grassland grow on the more sheltered western side. The rainy season, December to March, may also bring typhoons.

New Caledonia ranks with the world's leading nickel-mining nations; mining alone accounts for 12 percent of the national income. Tourism has also become an important growth industry. The largest ethnic group, Melanesians, still practice subsistence farming and fishing, but the territory remains dependent on French aid. Health and education facilities, which are concentrated in Noumea, the capital and main port, are generally good.

WALLIS AND FUTUNA
FRANCE

The French overseas territory of Wallis and Futuna comprises two small volcanic island groups in the southwestern Pacific Ocean. The humid climate and rich volcanic soils typical of both island groups

support abundant forest vegetation.

Local government is based on the three Polynesian kingdoms of Uvéa (on Wallis), Tua and Alo (both on Futuna). Subsistence farming is the chief economic activity but the islanders also export quantities of lumber and copra.

FRENCH POLYNESIA
FRANCE

French Polynesia consists of five scattered archipelagoes in the warm waters of the south-central Pacific Ocean. The islands include low coral atolls and reefs, as well as a number of larger volcanic cone-shaped islands covered in rich vegetation.

The islands were first inhabited by Polynesian peoples who settled from the Marquesas southward in the 5th century. In 1880 they became a French colony, and in 1946 an overseas territory. Today Tahiti, in the Society group of islands, is an expanding tourist center and home to more than two-thirds of the population. Light industry and commercial services are important, and copra, fish and cultured pearls are valuable exports.

Welfare and education resources are concentrated on Tahiti.

The resumption of nuclear tests by France in 1995/96 on Mururoa Atoll caused local and international outrage.

ANTARCTICA
INTERNATIONAL

During the International Geophysical Year (1957–58), scientists from 12 nations spent the winter in more than 50 research stations scattered across the southern polar continent of Antarctica. They had come to study its unique environment, searching for clues to the Earth's past – and its possible future. The following year, despite competing claims for sovereignty over the region, the same 12 nations signed the Antarctic Treaty, reserving the continent for nonpolitical scientific investigation. From the 1970s they imposed restrictions on local seal trapping and whaling, which had already reduced several species to the point of extinction. In 1981 scientists made a discovery that could affect every form of life on the planet: massive damage to the ozone layer above the South Pole.

Antarctica has a bigger land area than Europe or the United States and Mexico combined. About 95 percent is covered by an ice cap that is 4,000 m (13,000 ft) thick in places. This was created by thousands of years of accumulated snowfall, although annual snowfall is minimal, and accounts for some nine-tenths of all the world's glacial ice. The Antarctic landscape is an average of over 2,000 m (7,000 ft) above sea level, making it the world's highest continent, but in many areas as much as four-fifths of the overall height is due to the ice cap. During August and September, in the depths of winter, pack ice up to 3 m (10 ft) deep covers the surrounding Southern Ocean, doubling the area of the continent.

The long Transantarctic Mountain chain runs across the whole of Antarctica, passing close to the South Pole and dividing the continent into two unequal parts: Lesser Antarctica (mostly within western longitudes) and the massive semicircle of Greater Antarctica. From Lesser Antarctica the slender but mountainous Antarctic Peninsula snakes northeast toward the southern tip of South America to form the western boundary of the Weddell Sea. The Ross Sea forms a smaller gulf on the opposite side of Lesser Antarctica, almost directly south of New Zealand. The southern end of both these seas is covered by permanent ice shelves, constantly replenished by new ice forced through gaps in the Transantarctic Mountains by the ice cap beyond.

Several other mountain ranges emerge around the fringes of the ice cap, the highest being the Vinson Massif in Lesser

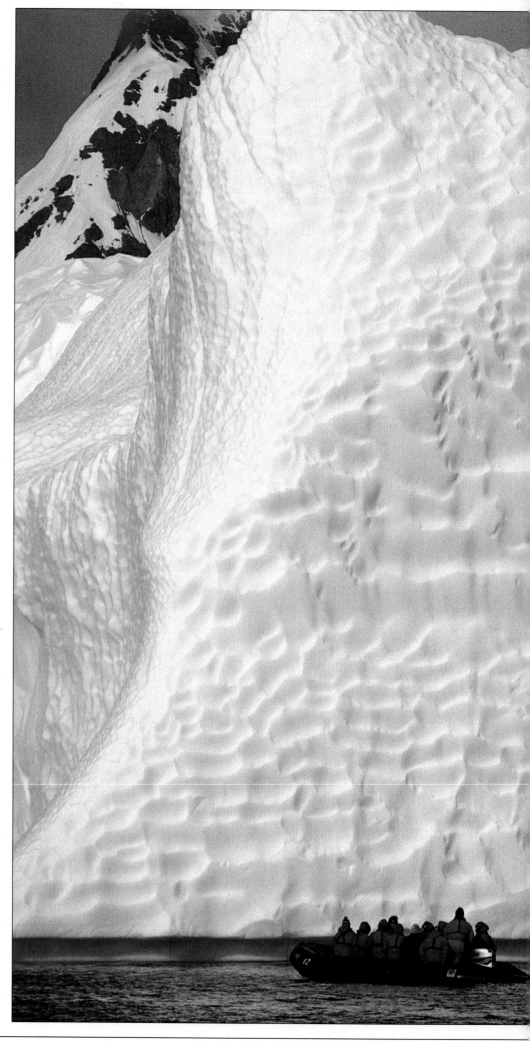

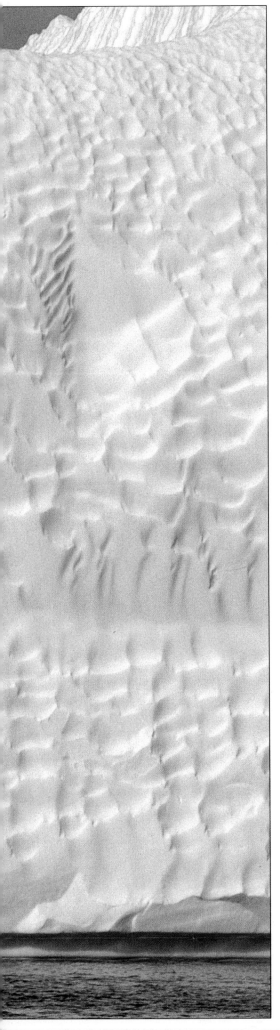

Antarctica – part of the Sentinel Range overlooking the Ronne Ice Shelf on the southern rim of the Weddell Sea. There are some volcanoes, which are largely in Lesser Antarctica, notably the islands north of the Weddell Sea; the active Mount Erebus, on Ross Island at the edge of the Ross Ice Shelf, is closely monitored.

The world's coldest continent suffers its lowest temperatures in August, dropping to -40° F to -95° F (-40° C to -70° C) on the inland ice sheet. On the Antarctic Peninsula, milder winds from neighboring oceans raise summer temperatures to 15° C (59° F), but the rest of Antarctica remains bitterly cold all year round. Rain is virtually unknown, and snowfall is rare except at the coast. Blizzards are frequent, when loose snow is whipped up by winds at the margins of the ice cap.

These hostile conditions, combined with long months of darkness, limit plant life to lichens, mosses, algae, and molds in the few ice-free areas. There are no land mammals in Antarctica, but whales and seals feed on masses of tiny, shrimp-like krill, while porpoises and dolphins are attracted by shoals of fish, predominantly Antarctic perches. With no land predators, the Antarctic coast is a haven for birds; Emperor penguins, Antarctic petrels and South polar skuas breed here and nowhere else, and more than 40 other species live in Antarctica.

A legendary first sighting of the icebound Southern Ocean is claimed by the Maori. The first European to cross the Antarctic Circle was the British Captain James Cook (1728–79) in 1773. In 1820 members of Russian, British and United States' expeditions to the Antarctic first set eyes on the mainland. Later explorers charted the Antarctic Peninsula and the seas around it; the French discovered and claimed Adélie Land in Greater Antarctica, while a British expedition led by James Clark Ross (1800–62) in 1839–43, discovered the sea and the ice shelf that bear his name.

Early this century several sledge expeditions set out to reach the South Pole at the heart of the interior. The Norwegian, Roald Amundsen (1872–1928), was the first to arrive on 14 December 1911. (The southern magnetic pole had been conquered in 1909). He was followed a month later by British explorers led by Robert Falcon Scott (1868–1912) who

Wall of ice (*left*) A tourist dinghy is dwarfed by a massive ice-covered cliff. The awesome scenery of the Antarctic attracts small but growing numbers of tourists, though tourism is strictly controlled to prevent environmental damage.

The leopard seal is perfectly adapted to the harsh conditions of the Antarctic – the coldest continent in the world. It shares its icy habitat with several species of seal, whales, dolphins, porpoises and a wide variety of seabirds.

perished in a blizzard on the return journey. In 1929 the American Richard Evelyn Byrd (1888–1957) became the first man to fly over the South Pole.

By 1942 seven nations – Argentina, Australia, Chile, France, New Zealand, Norway and Britain – had claimed sovereignty over sections of the Antarctic. But since the 1959 Antarctic Treaty those claims are largely ignored, and scientists from many countries work together. Whereas earlier adventurers pitted themselves against the bitter cold of the interior with sledges and dog teams, today's explorers travel in specially adapted vehicles known as snowcats, and use sophisticated seismic and remote-sensing equipment to map the contours of the hidden continent buried deep beneath the ice.

Other research includes planetary geology, hydrology, climatology, and meteoritics. In 1988 teams of scientists in the Antarctic traced the cause of the depletion of the world's ozone layer to the use of chlorofluorocarbons (CFCs) in aerosols, packaging materials, and refrigerants.

Detailed geological study has revealed the presence of many mineral resources. The Transantarctic Mountains include part of what could be the world's largest coalfield, and there are sizable reserves of iron ore in the Prince Charles Mountains near the coast of Greater Antarctica. Antimony, chromium, gold, molybdenum, and uranium have also been discovered, but so far it has been uneconomic to exploit any of these minerals, and potentially harmful to the fragile environment. In 1990 it was agreed to ban mining and mineral exploitation for 50 years.

A small but significant tourist industry has existed since the 1950s. Visits by cruise ships are strictly controlled to minimize environmental damage. Sealing and whaling activity is minimal, but fishing in Antarctic waters has increased greatly, including catching krill, which is the basis of the Antarctic food chain.

GLOSSARY

Terms in SMALL CAPITALS indicate a cross-reference to a main entry in another part of the glossary. Latin names and other foreign words are in *italic* type.

Acid rain Rain or any other form of PRECIPITATION that has become more acid by absorbing waste gases (for example, sulfur dioxide and nitrogen oxides) discharged into the atmosphere.

African-American A person of African ancestry born in the United States.

Agricultural economy An economy where most people work as cultivators or PASTORALISTS.

AIDS Acquired Immune Deficiency Syndrome, a disease that damages the body's natural immune system and therefore makes people more susceptible to disease. The Human Immunodeficiency Virus (HIV) is the name given to one of the viruses that can cause AIDS.

Alpine (1) A treeless environment found on a mountain above the tree line but beneath the limit of permanent snow. (2) A plant that is adapted to grow in the TUNDRA-like environment of mountain areas.

Amerindian A member of one of the many INDIGENOUS PEOPLES of Central and South America.

Ancestor worship The religious practice of worshiping the spirits of dead ancestors, who are believed to be able to intervene in human affairs.

Anglican A member of the PROTESTANT CHURCH – founded in England in the 16th century – including the Church of England and other churches throughout the world in communion with it.

Animism A religious belief that spirits inhabit natural phenomena such as the sun, sky, water and animals.

Apartheid A way of organizing society that keeps different racial groups apart. It was introduced in South Africa by the National Party after 1948 as a means of ensuring continued white political dominance, but has now been dismantled.

Aquifer An underground layer of permeable rock, sand or gravel that absorbs and holds groundwater.

Arctic (1) That part of the Earth lying north of latitude 66°32′N where for a time in summer the sun never sets and in winter never rises. It is generally cold and ice covered. (2) In biological terms it refers to the cold Polar regions where trees will not grow.

Arid (of the climate) Dry and usually hot. Arid areas generally have less than 250 mm (10 in) of rain a year. Rainfall is intermittent and quickly evaporates or sinks into the ground. Little moisture remains in the soil, so plant life is sparse.

Assembly As a LEGISLATURE of PARLIAMENT, a meeting of representatives convened to discuss GOVERNMENT proposals and enact laws.

Atoll A CORAL REEF enclosing a lagoon. The reef originally forms around a volcanic island, but as sea levels rise or the island sinks a lagoon is left in its place, surrounded by the reef.

Autocracy A form of rule in which one person, the "autocrat", possesses unlimited power.

Autonomy The condition of being self-governing, usually granted to a subdivision of, or territory belonging to, a larger STATE.

Balance of payments A statement of a country's transactions with all other countries over a given period.

Balance of power A theory of political stability based on an even distribution of power among the leading STATES.

Bantu Name given to a number of INDIGENOUS PEOPLES of Central and Southern Africa whose languages belong to a common sub-family.

Berber An INDIGENOUS PEOPLE of North Africa, particularly Algeria and Morocco.

Bible The book of scriptures of CHRISTIANITY and JUDAISM. The Jewish Bible contains many books in common with the Christian version describing historical events and prophetic teachings, but the latter also includes accounts of the life and teachings of Jesus Christ.

Bloc A group of countries closely bound by economic and/or political ties.

Boreal Typical of the northern climates lying between the ARCTIC and latitude 50°N, characterized by long cold winters and short summers. Vegetation in these regions is dominated by BOREAL FOREST.

Boreal forest The name given to the CONIFEROUS FORESTS or TAIGA of the northern hemisphere.

Brown coal A peat-like material, also known as lignite, which is an immature form of coal. It has a lower energy value than more mature forms of coal.

Buddhism A RELIGION founded in India in the 6th and 5th centuries BC and based on the teachings of Gautama Siddhartha (c. 563–483 BC), the Buddha or "Awakened One". It is widely observed in northern, southern and Southeast Asia.

Bureaucracy The body of STATE officials that carry out the day-to-day running of GOVERNMENT. It may refer to a system of administration marked by the application of rules, "red tape" and the proliferation of decisions.

Byzantine empire The Christian empire that succeeded the ROMAN EMPIRE in the eastern Mediterranean, centered on Byzantium, now Istanbul. It lasted from 395 to 1453.

Cabinet In a GOVERNMENT, the committee of leading ministers who oversee all major policy.

Capitalism The political and economic system based on the production of goods and services for profitable exchange in which labor itself is bought and sold for wages. Capitalist economies can be more or less regulated by GOVERNMENTS. In a capitalist mixed economy the government will own some of the country's utilities and industries as nationalized companies. It will also act as a major employer of labor.

Carnival A period of festivity usually accompanied by colorful and crowded processions and music, especially in Caribbean and many ROMAN CATHOLIC countries, where it precedes Lent (traditionally a period of fasting in the CHRISTIAN CHURCH).

Cash crop A crop grown for sale rather than for SUBSISTENCE.

Caste A system of rigid hereditary social divisions based on occupational groups. It is normally associated with the HINDU caste system in India, where an individual is born into the caste of his or her parents, must marry within it, and cannot leave it.

Caucasian (1) A racial classification based on white or light skin color. (2) An inhabitant of the Caucasus region or the Indo-European language of this people.

Celt (1) A member of an Indo-European iron-working group of peoples originating in southwest Germany and dominating large parts of Europe in the 6th and 5th centuries BC. (2) Speakers of Celtic languages such as Breton, Gaelic and Welsh are sometimes described as Celts today.

Cereal A cultivated grass that has been selectively bred to produce high yields of edible grain for consumption by humans and livestock. The most important are wheat (*Triticum*), rice (*Oryza sativa*) and maize (*Zea mays*).

Chaparral Scrub vegetation growing in the parts of California in the United States that have a mediterranean climate.

Christianity A RELIGION based on the teachings of Jesus Christ, and originating in the 1st century AD from JUDAISM. Its main beliefs are found in the BIBLE and it is now the world's most widespread RELIGION, divided into a number of CHURCHES and sects, including ROMAN CATHOLICISM, PROTESTANTISM and ORTHODOX CHURCHES.

City-state An independent STATE consisting of a single city and the surrounding countryside needed to support it. Singapore is an example of a modern city-state.

Cloud forest Forest growing in cool, damp TROPICAL highlands.

Colonialism The control of a foreign territory by a STATE or people for purposes of settlement and economic expansion.

Colony A territory under the sovereignty of a foreign power.

Commonwealth A loose association of STATES that are former members of the British EMPIRE, with the British MONARCH at its head.

Communism A social and economic system based on the communal ownership of property. It usually refers to the STATE-controlled social and economic systems in the former Soviet Union and Soviet BLOC countries and in the People's Republic of China.

Confucianism A RELIGION or moral code based on the teachings of the Chinese philosopher Confucius (c. 551–479 BC) that formed the foundations of Chinese imperial administration and ethical behavior; also followed in Korea and other east Asian countries.

Congress A political meeting; a LEGISLATURE, particularly of the United States.

Coniferous forest A forest of mainly coniferous, or cone-bearing, trees, frequently with evergreen needle-shaped leaves and found principally in the TEMPERATE ZONES and BOREAL regions. The timber they produce is known as softwood.

Conservation The use, management and protection of NATURAL RESOURCES so as to maintain their sustainable use and ecological diversity, without causing degradation, depletion or waste.

Conservative A political ideology against social change; now a general term for RIGHT-WING politics.

Constitution The fundamental statement of laws that defines the way in which a country is governed.

Constitutional monarchy A form of GOVERNMENT with a hereditary HEAD OF STATE or MONARCH and a CONSTITUTION.

Consumer goods Goods that are acquired for immediate use, such as foodstuffs, radios, televisions and washing machines.

Continental climate The type of climate associated with the interior of continents. It is characterized by wide daily and seasonal ranges of temperature, especially outside the TROPICS, and by low rainfall.

Coral reef A colony of vast numbers of coral polyps that are interconnected laterally and overlie the rigid calcium carbonate (limestone) skeletons of dead corals.

Counter revolution A movement to achieve a return to the political situation that existed before a REVOLUTION.

Coup d'etat A change of GOVERNMENT by unconstitutional means, usually involving a rebellion by the armed forces.

Creole (1) A sophisticated PIDGIN language that serves as a mother tongue and is spoken in the Caribbean, based on French, English or Dutch. (2) People of mixed descent living in the Caribbean or Latin America.

Cultural Revolution A radical attempt to continue the REVOLUTION in China from 1966 to 1976, involving the purging of the Communist Party and instigated by its chairman, Mao Zedong (1893–1976).

Decentralization The dispersion of power away from central GOVERNMENT to other levels of government.

Deciduous (of plants, trees, a forest etc) dropping their leaves in the winter or in the dry season.

Decolonization The transfer of GOVERNMENT from a COLONIAL power to the people of the COLONY at the time of political independence.

Deforestation Felling and clearing forested land for its conversion to other uses.

Delta A large accumulation of sediment, often fan-shaped, deposited where a river enters the sea or a lake. As the river enters the calmer waters of the sea, the water slows down and so is unable to transport the sediment it carries any farther.

Democracy A form of decision-making, or a form of GOVERNMENT, in which policy is made by the people (direct democracy) or on their behalf (indirect democracy). Indirect democracy usually takes the form of competition among political parties at ELECTIONS.

Denomination A group within, or branch of, a major RELIGION that sets up its own organization. For example, the ROMAN CATHOLIC CHURCH is a denomination of the CHRISTIAN RELIGION.

Dependency (1) A territorial unit under the jurisdiction of another STATE but not formally annexed to it. (2) An unequal economic or political relationship between two states or groups of states in which one side is dependent on and supports the other, often at the expense of its own interests.

Desert A very ARID area with less than 250 mm (10 in) of rainfall a year. Such areas may be hot or cold. In hot deserts the rate of evaporation is greater than the rate of PRECIPITATION and there is little vegetation as a result.

Desertification The creation of DESERT-like conditions usually caused by a combination of overgrazing, SOIL EROSION, prolonged drought and climate change.

Developed country Any country characterized by high standards of living and a sophisticated economy, particularly in comparison with DEVELOPING COUNTRIES. A number of indicators can be used to measure a country's wealth and material well-being; for example, the GROSS NATIONAL PRODUCT, the PER CAPITA consumption of energy, the number of doctors per head of population and the average life expectancy.

Developing country Any country that is characterized by low standards of living and a SUBSISTENCE economy. Sometimes called THIRD WORLD countries, they include most of Africa, Asia and Central and South America.

Dictator A leader who concentrates the power of the STATE in his or her own hands.

Druze A sect of SUNNI ISLAM founded in 1021 by the followers of Caliph al-Hakim bi-Amr Allah and practiced in parts of Lebanon and Syria.

El Niño The warming of the normally cold seawater off the coast of Peru in South America that occurs naturally every three to seven years for reasons that are not fully understood. Fish stocks are greatly reduced as a result.

Election The method of selecting persons for office or as representatives by voting.

Empire (1) A political organization of STATES and territories in which one dominates the rest. (2) The territory that constitutes such a group of states.

Endangered species A species whose population has dropped to such low levels that its continued survival is uncertain.

Endemic species A species that is native to one specific area, and is therefore often said to be characteristic of that area.

Erosion The process by which exposed land surfaces are broken down into smaller particles or worn away by water, wind or ice.

Ethnic group A group of people sharing a social or cultural identity based on language, RELIGION,

customs and/or common descent or kinship.

EU See EUROPEAN UNION.

Euro The common currency of the EUROPEAN UNION. Introduced in 1999, the euro operated at a fixed rate alongside the national currencies of member states until 2002 when the euro superseded them. To enter the euro trading zone countries need to meet financial convergence criteria set out in the Maastricht Treaty of 1992. Twelve countries have now joined the euro: Austria, Belgium, Finland, France, Germany, Ireland, Italy, Luxembourg, the Netherlands, Portugal, Spain and Greece. Membership in the euro zone involves ceding control of monetary policy to the European Central Bank, which some countries – notably Britain – remain reluctant to do.

European Community (EC) See EUROPEAN UNION.

European Union (EU) An alliance of western European NATIONS formed to agree common policies in the areas of trade, aid, agriculture and economics. The six founder members in 1957 were France, West Germany, Belgium, Holland, Luxembourg and Italy, when the organization was called the European Economic Community (EEC), renamed the European Community (EC) in 1967. A further three – Britain, Ireland and Denmark – joined in 1973, Greece in 1981 and Spain and Portugal in 1986. East Germany became a member when it was reunited with West Germany in 1990. The European Union (EU) came into being 1 November 1993 and Austria, Finland and Sweden joined January 1995.

Exotic (of an animal or plant) Not native to an area but established after being introduced from elsewhere, often for commercial or decorative purposes.

Exports Goods and services sold to other countries, bringing in foreign exchange.

Extinction The loss of a local population of a particular species or even the entire species. It may be natural or be caused by human activity.

Faction A distinctive group with its own policy within a larger political movement or PARTY.

Famine An acute shortage of food leading to widespread malnutrition and starvation.

Fault A fracture or crack in the Earth's crust along which there has been movement of the bedrock.

Federalism A form of CONSTITUTIONAL GOVERNMENT in which power is shared between two levels – a central or federal GOVERNMENT and a tier of provincial or STATE GOVERNMENTS.

Fetish An inanimate object thought to possess spiritual powers and used to influence the behavior of others, particularly in ANIMIST ceremonies.

Fjord A steep-sided inlet formed when a glaciated U-shaped valley is drowned by the sea. See GLACIATION.

Fossil fuel Any fuel, such as coal, oil and natural gas, formed beneath the Earth's surface under conditions of heat and pressure from organisms that died millions of years ago.

Franks A group of Germanic peoples inhabiting the Rhine region in the 3rd century who overthrew the ROMAN EMPIRE and in 800, under Charlemagne (742–814), founded the Holy Roman Empire.

Free trade A system of international trade in which goods and services are exchanged without tariffs, QUOTAS or other restrictions.

Fundamentalism The belief that religious scripture is divinely inspired and therefore infallible. It originally referred to those CHRISTIAN groups that believed in a literal interpretation of the BIBLE, but is now also applied to MUSLIM extremists.

GDP See GROSS DOMESTIC PRODUCT.

Ghetto A slum area in a city that is occupied by an ETHNIC minority. The word originally referred to the part of a city in medieval Europe to which JEWS were restricted by law.

Glaciation (1) The process of GLACIER and ice sheet growth. (2) The effect of these on the landscape.

Glacier A mass of ice formed by the compaction and freezing of snow and which shows evidence of past or present movement.

Global warming The increase in the average temperature of the Earth that is believed to be caused by the "greenhouse effect".

GNP See GROSS NATIONAL PRODUCT.

God A superhuman being or deity worshiped and believed to have divine powers over some aspect of the universe, or the creator of all things who is worshiped as such.

Government The primary political institution in a STATE, responsible for making and implementing laws and policies.

Green Revolution The introduction of high-yielding varieties of seeds (especially rice and wheat) and modern agricultural techniques to increase agricultural production in DEVELOPING COUNTRIES. It began in the early 1960s.

Gross Domestic Product (GDP) The total value of a country's annual output of goods and services, with allowances being made for depreciation. Growth in GDP is usually expressed in constant prices, to offset the effects of inflation. GDP is a very useful guide to the level of economic activity in a country.

Gross National Product (GNP) A country's GROSS DOMESTIC PRODUCT plus income from abroad.

Guerrilla A member of an irregular military force used to carry out harassment and sabotage.

Habitat The locality or environment to which an animal or plant is adapted and where it prefers to live, usually defined in terms of vegetation, climate or altitude eg grassland habitat.

Head of state The highest office of STATE, in which the holder personally represents the STATE in key functions such as declaring war or dissolving the LEGISLATURE. The role may be largely symbolic, as with the British MONARCH, or the holder may be the chief executive, as with the president of the United States.

Heterogenous Being composed of two or more very different ingredients or constituents.

Hinduism A body of religious practices, originating in India in the 2nd millennium BC. It emphasizes mystical contemplation and ascetic practices that are closely interwoven with much of Indian culture.

Homogeneous Being composed of parts that are very similar to each other, giving a uniform character.

Hunter–gatherers People who obtain their food by hunting wild animals and gathering wild plants and their fruits.

Ice age A long period of geological time in which cold climatic conditions prevail and snow and ice sheets are present throughout the year. There have been many ice ages in the Earth's history.

Imam In SUNNI ISLAM the leader of prayers in a MOSQUE. In SHI'A ISLAM the hereditary successors of Muhammad's son-in-law, Ali, and thus the legitimate leaders of ISLAM; sects differ as to whether there are 7 or 12 imams.

Imperialism The process whereby one country forces its rule on another country, frequently in order to establish an EMPIRE.

Imports Goods and services purchased from other countries.

Import substitution industry Any industry that has been set up (mainly in THIRD WORLD countries) to manufacture products that used to be imported. Import substitution industries are normally simple ones with an immediate local market such as the manufacture of cigarettes, soap and textiles. They are protected during their start-up phase by high tariffs on foreign rivals.

Inca Medieval ruling class of South American INDIGENOUS PEOPLE whose vast EMPIRE, centered on Peru, was overrun by the Spanish in the early 1530s.

Independence movement An organization that aims to achieve political independence for a COLONY.

Indigenous peoples The original inhabitants of a region, usually referring to pre-CAPITALIST or traditional peoples.

Inselberg A steep isolated hill found in ARID or semiarid landscapes.

Inuit An INDIGENOUS PEOPLE inhabiting the ARCTIC and TUNDRA regions of North America, Greenland and Siberia.

Islam A RELIGION based on the revelations of GOD to the Prophet Muhammad in the 7th century AD, as recorded in the QU'RAN. It teaches submission to the will of GOD, and is practiced throughout the Middle East, North Africa and parts of Southeast Asia.

Jew (1) A member of the Jewish people, descendants of the ancient Hebrews. (2) An adherent of the Jewish religion, JUDAISM.

Judaism A RELIGION that developed in ancient Israel based on GOD's law and revelations declared to Moses on Mount Sinai and contained within the Jewish books of the BIBLE.

Junta A self-appointed body that takes over the GOVERNMENT after a COUP D'ETAT.

Karst A bare LIMESTONE landscape formed by the chemical weathering of the rock and named after a typical region in Slovenia. Features include limestone pavements, caverns and gorges.

Lava The molten rock, or its later solidified form, produced from a volcano.

Left-wing A general term to denote antiestablishment political views, specifically used as a label for SOCIALIST or COMMUNIST parties.

Legislature The branch of GOVERNMENT responsible for enacting laws.

Limestone A sedimentary rock formed under the sea and consisting mainly of calcium carbonate. It is used as a building stone and in the manufacture of cement.

Lingua franca Any of various languages used, often for trading, as a means of communication between peoples of diverse speech.

Llanos The treeless grasslands of Venezuela and northern Colombia in South America (similar to the African SAVANNA).

Local government The lowest tier of GOVERNMENT dealing with the particular affairs of individual cities, towns and communities.

Lower house In a two-chamber LEGISLATURE, the part that is directly elected.

Lutheranism A PROTESTANT CHURCH founded on the principles of the German priest Martin Luther (1483–1546), with adherents mainly in Germany, Scandinavia and the United States.

Mammal A vertebrate animal belonging to the class Mammalia, having a four-chambered heart, fur or hair, and feeding its young on milk secreted by the mammae (nipples). With the exception of monotremes, mammals do not lay eggs, but give birth to live young.

Mangrove A dense forest of shrubs and trees growing on tidal coastal mudflats and estuaries throughout the TROPICS. Many mangrove species have aerial roots.

Maori The INDIGENOUS PEOPLES of New Zealand, of Polynesian descent.

Maquis The typical vegetation of the Mediterranean coast, consisting of aromatic shrubs, laurel, myrtle, rock rose, broom and small trees such as olive, fig and holm oak.

Maritime climate A generally moist climate, determined mainly by proximity to the sea. The sea heats up and cools down more slowly than the land, reducing variations in temperature so that the local climate is more equable than farther inland.

Market economy An economy in which most economic activities are transacted by private individuals and firms in largely unregulated markets.

Martial law Suspension of ordinary GOVERNMENT and the execution and implementation of laws by the armed forces.

Marxism The system of thought derived from the 19th-century political theorist Karl Marx (1818–83), in which politics is interpreted as a struggle between economic classes. It promotes the communal ownership of property, and so is commonly termed COMMUNISM.

Mestizo A person of mixed AMERINDIAN and Spanish or Portuguese descent in South America.

Methodism A PROTESTANT DENOMINATION of the CHRISTIAN CHURCH based on the evangelical teachings of the English theologian John Wesley (1703–91). Its adherents are found mainly in Britain and the United States.

Metis Peoples of mixed NATIVE-AMERICAN and French Canadian descent in Canada.

Migration (1) The permanent shift of people from one country, region or place to another for economic, political, religious or other reasons. (2) In animals, their seasonal movement to ensure secure food and water supplies.

Military regime A GOVERNMENT staffed and controlled by the armed forces.

Minister (1) In GOVERNMENT, an office usually with responsibility for a particular area of policy. (2) In the CHURCH it is a term often given to a member of the clergy.

Missionary A person who tries to convert others to his or her own religious beliefs.

Mogul A dynastic EMPIRE in India, 1526–1857, founded by invaders from Afghanistan with MONGOL ancestry; during the Mogul EMPIRE Indo-Islamic art and architecture flourished.

Monarchy A form of rule where there is a hereditary HEAD OF STATE.

Mongol A NOMADIC people from central Asia who conquered huge areas of the continent in the 13th century, from China to Europe; today they number 3 million and inhabit parts of China, the former Soviet Republics and Mongolia.

Monsoon (1) The wind systems in the TROPICS that reverse their direction according to the seasons; when they blow onshore they bring heavy rainfall. (2) The rain caused by these winds.

Montane The zone at middle altitudes on the slopes of mountains, below the ALPINE zone.

Moor (1) An inhabitant of North Africa of mixed Arab and BERBER descent. (2) Often refers to the 8th-century MUSLIM conquerors of Spain, who created an ISLAMIC civilization based in Cordoba before their defeat and expulsion in 1492.

Mosque The building in which MUSLIMS worship. It generally has one or more minarets, from which a summons to prayer is cried.

Multiparty system A system in which several PARTIES compete for votes in an ELECTION.

Muslim An adherent of ISLAM; literally "one who submits" to the will of GOD.

Nation A community that believes it consists of a single people, based on historical and cultural criteria. Sometimes used interchangeably with STATE.

Nationalism An ideology that assumes all NATIONS should have their own STATE, a NATION-STATE, in their own territory, the national homeland.

Nation-state A STATE in which the inhabitants all belong to one NATION. Most states claim to be nation-states; in practice almost all of them include minority groups.

Native-American The INDIGENOUS PEOPLES of North America.

Natural resources Raw materials created by the Earth's natural processes including mineral deposits, soil, water, timber and plants and animals. Most natural resources are exploited for use in agriculture, industry and economic activities.

Nazi A member of the National Socialist German Workers' Party founded in 1919 and used by Adolf Hitler (1889–1945) as his power base for political control of Germany in 1933.

Nomad A member of a usually PASTORALL people who move seasonally from one place to another in search of food, water or pasture for their animals.

Non-aligned movement A grouping of THIRD WORLD countries that refused to align themselves during the Cold War with either the United States or the former Soviet Union.

Official language The language used by GOVERNMENTS, schools, courts and other official institutions in countries where the population has no single common mother tongue.

One-party system A political system in which there is no competition to the GOVERNMENT PARTY at ELECTIONS, as in COMMUNIST and MILITARY REGIMES, and all but the government party is banned.

Orthodox church A group of CHRISTIAN CHURCHES that took the Patriarch of Constantinople as its head and split with the ROMAN CATHOLIC CHURCH in 1054. They include the national CHURCHES of Greece, Russia, Armenia, Bulgaria, Romania and Ethiopia, among others.

Ottoman empire An EMPIRE founded in the 13th century by the ISLAMIC Ottoman Turks. At its height it controlled an area extending from Hungary, southern Russia and the eastern Mediterranean to North Africa and lasted from 1453 to 1918. Turkey is its modern remnant.

Pampas The level, treeless and grassy regions of southern South America dominated by many species of feather grass. Much of the area is now cultivated with wheat or grasses for grazing.

Parliament An assembly of representatives, a LEGISLATURE.

Parliamentary democracy A political system in which the LEGISLATURE (PARLIAMENT) is elected by all the adult members of the population and the GOVERNMENT is formed by the PARTY that commands a majority in the PARLIAMENT.

Party An organized group seeking political power to implement an agreed set of policies.

Pastoralist A person following a way of life based on tending herds of animals such as sheep, cattle, goats or camels; often NOMADIC, it involves moving the herds according to the natural availability of pasture and water.

Per capita Per head.

Permafrost Soil and rock that remains permanently frozen, typically in the Polar regions. A layer of soil at the surface may melt in summer, but the water that is released is unable to drain away through the frozen subsoil and refreezes in colder conditions.

Phoenician empire Trading EMPIRE of the 2nd millennium BC on the east Mediterranean coast with colonies all around the Mediterranean Sea.

Pidgin A simplified language, often a LINGUA FRANCA, allowing for basic communication between persons whose mother tongues are mutually unintelligible; usually based on a European language and developed in former colonized countries.

Plateau A large area of level, elevated land. When bordered by steep slopes it is called a tableland.

Polder An area of level land at or below sea level obtained by land reclamation. It is normally used for agriculture.

Prairie The flat grassland in the interior of North America between 30°N and 55°N, much of which has been plowed and is used to grow cereal crops. It is characterized by low and sporadic rainfall and a wide annual range of temperature.

Precipitation Moisture that reaches the Earth from the atmosphere, including mist, dew, rain, sleet, snow and hail.

President A HEAD OF STATE, elected in some countries directly by voters and in others by members of the LEGISLATURE. In some political systems the president is chief executive, in others the office is largely ceremonial.

Prime minister The chief executive in a PARLIAMENTARY system of GOVERNMENT.

Productivity A measure of economic output in relation to the quantity of economic inputs (labor, machines, land, etc) needed for production.

Protestant Term describing a number of CHRISTIAN DENOMINATIONS that share a common rejection of the authority of the pope as head of the CHURCH, along with many ROMAN CATHOLIC practices. It originated during the 16th-century European Reformation.

Province Administrative subdivision of a STATE.

Puna scrub The sparse grass and bushes found in the high cold PLATEAUS of the Andes in South America.

Quota A limit imposed on the amount of a product that can be IMPORTED in a given time.

Qu'ran The sacred text of ISLAM that contains the revelations made by GOD to the prophet Muhammad; it contains the legislative, theological and ethical principles of ISLAM.

Radioactivity The emission of alpha-, beta- and gamma particles from atomic nuclei. This is greatest when the atom is split, as in a nuclear reactor. Prolonged exposure to radioactive material can cause damage to living tissue, leading to cancers and ultimately death.

Rainforest Forest where there is abundant rainfall all year round. The term is often associated with tropical rainforests, which are rich in species and where growth is lush and very rapid.

Rastafarianism A religious and political movement, originating in the Caribbean, which claims that Black Africans are the lost tribe of Israel and that Ethiopia is their spiritual home. They looked to Haile Selassie (1930–74), former emperor of Ethiopia, as their spiritual leader. Followers wear their hair in "dreadlocks" and have inspired *reggae* music.

Refugee A person who is forced to leave his or her country for either political or economic reasons. The chaos of war generally creates may refugees.

Religion A system of beliefs, practices and related institutions founded on the belief in, worship of and obedience to one or more GODS or spirits; it may also include prescriptions and practices addressing ethics, government and personal conduct, as well as accounts of the nature and origin of the world itself.

Republic A form of GOVERNMENT with a non-hereditary HEAD OF STATE.

Reservation/reserve An area of land set aside by a COLONIAL power for the sole use of INDIGENOUS PEOPLES, especially the NATIVE-AMERICANS of North America, as a means of removing them from their traditional lands and subjugating them. In such areas the inhabitants may retain their cultural identity and traditional ways of life.

Revolution A sudden and often violent change of regime.

Rift valley A long valley formed by the subsidence of a block of land between two parallel FAULTS.

Right-wing A general term to indicate pro-establishment political views. As a label it is usually given to the main PARTY in a country, either Christian Democrat or CONSERVATIVE, that opposes the SOCIALIST party.

Ritual A ceremony, usually of religious or cultural significance, that follows a set pattern of symbolic actions and words.

Roman Catholic The largest of the CHRISTIAN CHURCHES, headed by the pope in Rome, that traces its origin and authority to St Peter, one of

the disciples of Jesus Christ and the first Bishop of Rome. There are believers on all continents.

Romance language A sub-family of related languages, part of the Indo-European family, all of which are derived from Latin; includes Italian, French, Portuguese and Spanish.

Roman empire An EMPIRE founded in the year 27 BC from the Roman Republic, which began about 500 BC in present-day Italy. At its height it controlled the Mediterranean, large parts of western Europe and the Middle East. In the 5th century it divided, the eastern half becoming the BYZANTINE EMPIRE.

Savanna A HABITAT of open grassland with scattered trees in TROPICAL and SUBTROPICAL areas. Also known as TROPICAL grassland, it covers the area between TROPICAL RAINFOREST and hot DESERTS. There is a marked dry season and too little rain to support large areas of forest.

Secession The act of separating a territory from a STATE.

Self-determination The doctrine that people within a territorial unit or culture group with a distinctive political identity, usually a NATION, have a right to determine their own future status.

Senate The term given to the UPPER HOUSE in some LEGISLATURES.

Separatism A political movement in a STATE that supports the SECESSION of a particular minority group, within a defined territory, from that STATE.

Sephardic A JEW of Spanish or Portuguese descent or origin, whose religious traditions and prayer rites differ from the JEWS of Eastern and Central Europe.

Shifting cultivation A method of farming prevalent in TROPICAL areas in which a piece of land is cleared and cultivated until its fertility is diminished. The farmer then abandons the land, which restores itself naturally.

Shi'ite Muslim A member of the smaller of the two main divisions of ISLAM, prominent in Iran and Iraq. Followers recognize Muhammad's son-in-law, Ali, and his descendants, the IMMAMS, as his true successors and legitimate leaders of ISLAM. There are a number of sects, including the Isma'ilis, DRUZE and so-called Twelver Shi'ites.

Sikhism A RELIGION established by Guru Nanak in the 15th century in the Punjab, in the northwest of the Indian subcontinent, that opposed HINDUISM; it is now observed by the worldwide community of Sikhs, many of whom are fighting for their own autonomous homeland in the Punjab.

Slash-and-burn farming A method of farming in TROPICAL areas where the vegetation cover is cut and burned to fertilize the land before crops are planted. Often a feature of SHIFTING CULTIVATION.

Socialism A political ideology and economic system based on the principle of equality between people, the redistribution of wealth and property and equal access to benefits such as healthcare and education.

Soil erosion The removal of the topsoil mainly by the action of wind and/or rain. Sheet erosion is when a fine layer of soil is removed from the land. Gully erosion occurs when rain carves deep channels in the ground.

State The primary political unit of the modern world, usually defined by its possession of sovereignty over a territory and its people.

Steppe An open grassy plain with few trees or shrubs. Steppe is characterized by low and sporadic rainfall, and experiences wide variations in temperature during the year.

Subsistence A term applied to systems in which producers can supply their own needs for food, shelter, etc but have little or no surplus to trade.

Subtropical The climatic zone between the TROPICS and TEMPERATE ZONES. There are marked seasonal changes of temperature but it is never very cold.

Suffrage The right to vote.

Sunni Muslim A member of the larger of the two

main divisions of ISLAM. Its members recognize the Caliphs as the successors to Muhammad and follow the *sunna* or way of the prophet as recorded in the *hadith*, the teachings of Muhammad.

Tableland See PLATEAU.

Taiga The Russian name given to the CONIFEROUS FOREST and peatland belt that extends around the world in the northern hemisphere, south of the TUNDRA and north of the DECIDUOUS forests and grasslands.

Taoism A Chinese philosophy and RELIGION usually ascribed to the teachings of Lao Zi (Lao Tze) (7th century BC) and later mixed with Buddhism. It was banned by the Chinese COMMUNIST GOVERNMENT but widely practiced in Taiwan.

Temperate zone Any one of the climatic zones in mid latitudes, with a mild climate. They cover areas between the warm TROPICS and cold Polar regions.

Third World A term first used to refer to ex-COLONIES that were neither fully capitalist (First World) nor fully socialist (Second World). Now used to refer to the poorer, less industrialized countries of the DEVELOPING world.

Treaty An agreement between two or more STATES to pursue specific common policies, which may be military, political or economic.

Tribe A group of people united by a common language, RELIGION, customs and/or descent and kinship; often used to describe the social groups of peoples who have no developed STATE or GOVERNMENT and whose social organization is based on ancestry and extended family systems.

Tropics (tropical) The area between the Tropic of Cancer (23°30′N) and the Tropic of Capricorn (23°30′S). They mark the lines of latitude farthest from the Equator where the Sun is still found directly overhead at midday in midsummer.

Two-party system A political system in which two PARTIES alternate in GOVERNMENT.

Tundra The level, treeless land lying in the very cold northern parts of Europe, Asia and North America where winters are long and cold and the ground beneath the surface is permanently frozen. See also PERMAFROST.

Upper house That part of a two-chamber LEGISLATURE that is not necessarily directly elected. It is the second chamber in rank.

Vikings The sea warriors from Scandinavia who raided and then settled in much of northern and western Europe between the 9th and 11th centuries.

Voodoo A RELIGION based on the worship of West African GODS, mixed with ROMAN CATHOLIC beliefs, practiced mainly in Haiti. It is an ANIMISTIC cult, and trances induced by spirit possession are central to its RITUAL.

Welfare state A social and economic system based on STATE provision of, and responsibility for, such things as healthcare, pensions and unemployment benefit. These services are financed by contributions from the working population, and access is intended to be equally available to all, free of charge. It originated in Britain at the start of the 20th century and became widespread in Europe after WORLD WAR II.

Wetlands A HABITAT that is waterlogged all or enough of the time to support vegetation adapted to these conditions.

World War I A major war (1914–18) in which Germany, Austria–Hungary and the OTTOMAN EMPIRE (Turkey) fought Britain, France, tsarist Russia and later the United States.

World War II A major war (1939–45) in which Britain, France, the former Soviet Union and the United States were allied against the armed forces of Germany, Italy and Japan.

Further Reading

NORTH, CENTRAL AND SOUTH AMERICA
Bakvis, H. and Skogstad, G. (eds.) *Canadian Federalism: Performance, Effectiveness and Legitimacy* (Oxford University Press, Oxford, 2001)
Bone, R.M. *The Geography of the Canadian North* (Oxford University Press, Toronto, 1992)
Clark, D. *Post-Industrial America* (Methuen, New York, 1985)
Cummings, Jr., M.C. and Wise, D. *Democracy Under Pressure: an Introduction to the American Political System* 9th edn. (International Thomson Publishing, Cambridge, Massachusetts, 2001)
Faber, D.J. *Environment Under Fire: imperialism and the ecological crisis in Central America* (Monthly Review Press, New York, 1993)
Kotkin, J. and Moyers, S. (eds.) *The New Geography: How the Digital Revolution Is Reshaping the American Landscape* (Random House, New York, 2001)
Lopez, B. *Arctic Dreams: Imagination and Desire in a Northern Landscape* (Vintage Books, London, 2001)
Nash, G.B. et al. (eds.) *The American People* volumes I and II (Harper and Row, New York, 1990)
Peck, R.M. *Land of the Eagle: a Natural History of North America* (BBC, London, 1990)
Place, S.E. (ed.) *Tropical Rainforest: Latin American Nature and Society in Transition* Jaguar Books on Latin America, No.2 (Scholarly Resources, Wilmington, Delaware, 2001)
South America, Central America and the Caribbean 1993 4th edn. (Europa Publications, London, 1993)
Wiarda, H.J. *The Soul of Latin America: The Cultural and Political Tradition* (Yale University Press, New Haven, 2001)
Wilson, A. *The Culture of Nature: North American Landscape From Disney to Exxon Valdez* (Blackwell, Cambridge, Massachusetts, 1992)
Yeates, M. *The North American City* 5th edn. (Harper and Row, New York, 1997)

AFRICA
Adams, W.A. , Goudie, A. and Orme, A. (eds.) *The Physical Geography of Africa* (Oxford University Press, New York 1999)
Africa South of the Sahara (Europa Publications, London, 1990)
Cokc, J. and Koch, E. (eds.) *Going Green: People, Politics and the Environment in South Africa* (Oxford University Press, Cape Town, 1991)
Deegan, H. *The Politics of the New South Africa: Apartheid and After* (Pearson Education, New Jersey, 2000)
Middleton, J. (ed.) *Africa: An Encyclopedia for Students* (Charles Scribner's Sons, New York, 2001)
Mohamed Abdul Salih, M.A. and Tedla, S. *Environmental Planning, Policies and Politics in Eastern and Southern Africa* (St Martin's Press/Palgrave, New York, 1999)
Preston-Mafham, K. *Madagascar: a Natural History* (Facts On File, New York, 1991)
Tarver, J.D. *Urbanization in Africa* (Greenwood Publishing Inc., Westport, CT, 1994)
Weatherby, J.N. *The Middle East and North Africa: A Political Primer* (Longman, USA, 2001)

EUROPE
Benevolo, L., Le Goff, J. (ed.) and Ipsen, C. (translator) *The European City* Making of Europe Series (Blackwell Publishers, New York, 1995)
Carr, R. *Spain: A History* (Oxford University Press, Oxford and New York, 2000)
Cole, F. and Cole, J. *A Geography of the European Union* 2nd edn. (Routledge, London 1997)
Graham, B. *In Search of Ireland: A Cultural Geography* (Routledge, London, 1997)
Grant, N. *Eric the Red: The Viking Adventurer* (Oxford University Press, Oxford 1998)
Kaplan, M. *The Portuguese: The Land and its People* (Penguin, London, 1991)
Levy, J.D. *Tocqueville's Revenge: State, Society, and Economy in Contemporary France* (Harvard University Press, Cambridge, Massachusetts, 1999)
Morley, D. and Robins, K. (eds.) *British Cultural Studies: Geography, Nationality and Identity* (Oxford University Press, Oxford, 2001)
Sassoon, D. *Contemporary Italy: Politics, Economy and Society Since 1945* 2nd edn. (Addison Welsey Longman, London, 1997)
Strasser, H. and Kerbo, H.R. *Modern Germany* (McGraw-Hill, New York, 1999)
Wejnert, B. *Transition to Democracy in Eastern Europe and Russia: Impact on Politics, Economy, and Culture* (Praeger Publishers, Westport, CT, 2002)
Witte, E. Craeybeckx, J. and Meynen, A. *Political History of Belgium: From 1830 Onwards* (Standaard Uitgeverij/VUB University Press, Brussels, 2001)
Woodhouse, C.M. *Modern Greece: a Short History* 5th edn. (Faber and Faber, London, 2000)

ASIA
Aitken, S.R. and Leigh, C.H. *Vanishing Rain Forests: the Ecological Transition in Malaysia* (Clarendon Press, Oxford, 1992)
Amery, H.A. and Wolf, A.T. *Water in the Middle East: A Geography of Peace* (University of Texas Press, 1999)
Cumings, B. *Korea's Place in the Sun: A Modern History* (W.W. Norton and Company, New York, 1998)
Gilsenan, M. *Recognizing Islam: Religion and Society in the Modern Middle East* rev. edn. (Tauris, New York, 2000)
Madsen, S.T. (ed.) *State, Society and the Environment in South Asia, Volume 3* (Curzon Press Ltd., Richmond, UK, 1999)
McFaul, M. *Russia's Unfinished Revolution: Political Change from Gorbechev to Putin* (Cornell University Press, Ithaca, USA, 2001)
Smith, C.J. *China: People and Places in the Land of One Billion* (Westview, Boulder, Colorado, 1991)
Songqiao, Z. *The Physical Geography of China* (Wiley, New York, 1986)
Sparkes, J. *Realms of the Russian Bear* (BBC, London, 1992)
Tully, M. *No Full Stops in India* (Viking, London, 1991)
Varley, P. *Japanese Culture* 4th edn. (University of Hawaii Press, Honolulu, 2000)
Weightman, B. *Dragons and Tigers: A Geography of South, East and Southeast Asia* (Wiley, John & Sons, Toronto, 2001)

AUSTRALASIA AND ANTARCTICA
Drewry, D.J., Pyle, J.A. and Laws, R.M. (eds.) *Antarctica and Environmental Change* (Oxford University Press, Oxford, 1994)
Howe, K.R. *Nature, Culture and History: The Knowing of Oceania* (University of Hawaii Press, Honolulu, 2001)
Jeans, D.N. (ed.) *Australia – a Geography I: The Natural Environment* 2nd edn. (Sydney University Press, Sydney, 1986)
Jeans, D.N. (ed.) *Australia – a Geography II: Space and Society* 2nd edn. (Sydney University Press, 1987)
Robinson, G.M., Loughran, R.J. and Tranter, P. *Australia and New Zealand: Economy, Society and Environment* (Arnold, London, 2000)
Soons, J.M. and Selby, M.J. (eds.) *Landforms of New Zealand* (Longman Paul, Auckland, 1982)
Stonehouse, B. *The Last Continent: Discovering Antarctica* (Odyssey, Hong Kong, 2000)
Troy, P.N. *Australian Cities: Issues, Strategies and Policies for Urban Australia in the 1990s* (Cambridge University Press, Cambridge, 1999)

Acknowledgments

PICTURE CREDITS
Key to abbreviations
AA&A Ancient Art and Architecture Collection, Middlesex, UK; **AD** Art Directors Photo Library, London, UK; **BAL** Bridgeman Art Library, London, UK; **BCL** Bruce Coleman Limited, Middlesex, UK; **C** Colorific!, London, UK; **COP** Christine Osborne Pictures, London, UK; **E** Explorer, Paris, France; **EU** Eye Ubiquitous, Sussex, UK; **FSP** Frank Spooner Pictures, London, UK; **HDC** Hulton Deutsch Collection, London, UK; **HL** Hutchison Library, London, UK; **IP** Impact Photos, London, UK; **KP** Katz Pictures, London, UK; **M** Magnum Photos Limited, London, UK; **MEPL** Mary Evans Picture Library, London, UK; **MH** Michael Holford, Essex, UK; **NHPA** Natural History Photographic Agency, Sussex, UK; **OSF** Oxford Scientific Films Limited, Oxfordshire, UK; **PP** Panos Pictures, London, UK; **PEP** Planet Earth Pictures, London, UK; **RHPL** Robert Harding Picture Library, London, UK; **TIB** The Image Bank, London, UK; **TSW** Tony Stone Worldwide, London, UK; **WFA** Werner Forman Archive, London, UK; Z Zefa Picture Library, London, UK

t = top, b = bottom, l = left, r = right

1 A. Gin/Powerstock **2–3** M/Hiroji Kubota **10** Harald Mante **12–13** IP **13** RHPL **14–15** Karl-Heinz Raach **15t** Fotomas Index **15c** MEPL **16–17** MEPL **18–19** TIB **19c** TIB **19b** TIB **20** Z **21** E/P Gleizes **22–3** Z **22** RHPL **24** Z/R Beerenholtz **26–7** Z/Armstrong **27** RHPL/Walter Rawlings **28** TIB/John R Ramey **29t** TIB/Lynn M Stone **29b** Z/Armstrong **30** HDC **31** OSF/Stan Osolinksi **32tl** OSF/Stan Osolinksi **32r** TIB/John Lewis Stage **32b** OSF/W Gregory Brown **33l** Peter Newarks American Pictures **33r** Peter Newarks Historical Pictures **34t** BAL/Bristol Museum **34bl** Peter Newarks Military Pictures **34br** Brady Collection/Library of Congress **35** Peter Newarks Western Americana **36** Peter Newarks Western Americana **36–7** Z/Kurt Scholz **37t** Peter Newarks Western Americana **37b** HDC **38** M/Marlow **39tl** RHPL/Adrian Neville **39tr** RHPL/Walter Rawlings **39b** TRIP/Helene Rogers **40t** M/Steve McCurry **40b** Action Plus **41** TIB/Harald Sund **42t** M/Burt Glinn **42b** RHPL/Robert Francis **43** TIB/L D Gordon **44** TIB/Benn Mitchell **44–5** AGE Fotostock **45** Mark Wagner/aviation-images.com **46** TIB/Alvis Uptis **47l** HDC **47r** Z **48** RHPL/Michael J Howell **50–1** RHPL/John Gardey **51t** TIB/Grant V Faint **51b** Z **52t** TIB/David W Hamilton **52bl** Z/Kurt Goebel **52br** RHPL **53** RHPL/Robert Francis **54t** Z/Damm **54–5** RHPL **56–7** Z **57t** RHPL/Walter Rawlings **57b** Z/Damm **58–9t** BCL/Robert P Carr **58–9b** Z **60–1** Z **61** TIB/Guido Alberto Rossi **62** Z **63** Z/Damm **64–5** Joe Demaria/Corbis Sygma **64** Z **65** Z **66** Z **66–7** E/Richard Nowitz **68** TIB/Robert Kristofik **68–9** M/Richard Kalvar **69t** TIB/Janeart **69b** HL **70** M/Fred Mayer **70–1** M/Fred Mayer **71** M/Erich Hartmann

72l TIB/Janeart 72r Z 73 SP/Sander/Liaison 75 TIB 76–7 Z/Damm 77 RHPL 78–9 Z/ Klaus Benser 79 Z/H Grathwohl 80bl RHPL/Regior 80br RHPL/Paul Van Riel 81 RHPL 82 FSP/Prensa Grafica/Gamma 83bl TIB/Larry Dale Gordon 83br Z/Karl Kummels 84–5 RHPL/David Lomax 85 RHPL 86 RHPL 87l Z/Damm 87r RHPL 88 RHPL/David Lomax 90 TIB/Guido Alberton Rossi 91 TIB/Mel Di Giamoco 92 Carol Kane 93 RHPL 94 Z/Damm 94–5 RHPL 96 TIB/Kay Chernush 97 TIB/Jeff Smith 98 TIB/Back from Abroad 98–9 RHPL 100t Z/K Kerth 100c Z/S Chirol 100–1 Z/K Kerth 102 BCL/Gunter Zeisler 104–5 Z 105t TIB/Robert Phillips 105b TIB/Obremski 106 RHPL/Yoram Lehmann 106–7 PP/Adrian Evans 108 RHPL 109 Z/R Halin 110l TIB/Fred Ihrt 110r TIB/Larry Dale Gordon 111 TIB/Larry Dale Gordon 112–13 Z/ R Bond 113 Z/Gunter Heil 114 Z/K Murray 114–15 HL/Bernard Regent 116 TIB/ Fernanda Bergamaschi 116–17 M/Bruno Barbey 118–19 M/Michael K Nichols 120 M/ James Nachtwey 121 RHPL/Francois Gohier 122–3 RHPL/Robert Francis 123 TIB/Peter Miller 124–5 TIB/Walter Bibikow 125t TIB/Cesar Lucas 125b TIB/Walter Bibikow 126tl HL/K Rodgers 126tr M/Bruce Davidson 127 BCL/L C Marigo 128t Z/Kurt Goebel 128c M/Sebastian Salgado 128–9 RHPL 130 HL/Sarah Errington 132–3 Z 133 Z 134 HL 135t HL 135b HL 136–7 Z 137 Z 138–9 HL 139 HL 140 Z 141l Z 141r Z 142 RHPL 143 E 144 HL 145 Z 146 HL 147 PP/Jeremy Hartley 148 E/Philippe Maille 149 TIB/Eric Wheater 150 HL 151 HL/Leslie Woodhead 152 E 153 HL 154 Z/R Bond 156t RHPL 156b RHPL 157 RHPL 158 E/J Brun 159 E/Maja Koene 160bl HL/Alan Hutchison 160br FSP/Figaro/G De Laubier 161 PP/Jeremy Hartley 162 RHPL 162–3 RHPL/Hervy 163 RHPL/Hervy 164–5 RHPL 166 E/J Brun 166–7 RHPL/C Bryan 168–9 M/Abbas 169 RHPL/A Perrier 170–1 RHPL/Errath 171 RHPL/Ian Griffiths 172–3 E/J C Thoret 173 HL/Bernard Regent 174–5 RHPL 176 HL/Dr Nigel Smith 176–7 HL 178 RHPL 179 E/Y Arthus Bertrand 180 OSF/Richard Packwood 181 HL/P Goycolea 182–3 E/Y Gladu 184 RHPL 186–7 HL 187 HL 188–9 RHPL 189 RHPL/ Sassoon 190–1 RHPL/Sassoon 191 RHPL 193l PP/Chris Johnson 193r TIB/Pete Turner 194 RHPL 195 OSF/Patridge Films Limited/Carol Parneti 196 HL/Robert Aberman 196–7 HL/Lesley McIntyre 198 RHPL/G M Wilkins 199 HL/Robert Aberman 200–1 RHPL/Rolf Richardson 201 HL/Nancy Durrell 202 RHPL 203bl HL 203br HL/ Simon McBride 204–5 RHPL/Sassoon 206–7 HL 207 E/Geopress 208–9 TIB/Guido Alberto Rossi 209 E/Krafft 210–11 RHPL 212–13 HL 213 M/David Hurn 214 Knudsens Fotosenter 214–15 Knudsens Fotosenter 215 RHPL/Julia Thorne 216–17 OSF/Philippe Henry 217 OSF/Konrad Wothe 218–19 Tiofoto/Nils Johan Norenlind 219 Tiofoto/Hans Andersson 220 WFA/Statens Historiska Museum, Stockholm 220–1 Tiofoto/Hans Wretling 222–3 OSF/Konrad Wothe 224 HL/Bernard Regent 225 RHPL 226–7 RHPL/ Kim Hart 227 RHPL 228–9 Corbis Sygma/D F Joseph 229 RHPL 230 RHPL/Kim Hart 231 RHPL 232–3 Z 234 Viewfinder 234–5 Z 235 Viewfinder 236 Matthew Kneale 236–7 RHPL 237 Z 238l Z/W F Davidson 238r BAL 239t BAL 239b MEPL 240–1 Z 242l Colin McPherson/Corbis Sygma 242r Z 243b Eye Ubiquitous/Corbis 243t Z 244–5 RHPL 245 HL/Nigel Sitwell 246 HDC 247l Z 247r Z 248–9 TIB/Romilly Lockyer 250–1 RHPL/David Hughes 250 Daniel Philippe 252 E/Jean Michel Labat 252–3 OSF/ Martyn Colbeck 253 OSF/Ronald Toms 254 E/Jean Michel Labat 255 BAL/Chateau de Malmaison 256–7t Roger Viollet/L L Viollet 256–7b BAL 258 RF/Steve Wood 258–9 E/ Sylvain Grandadam 259 E/Franck Chazot 260 Comstock/Tor Eigeland 261 Action Plus 262–3 RHPL 264 Silvestris/Klaus Brauner 265 RHPL/Geoff Renner 266 Picture Box/ JWE 267 RHPL/RijksMuseum, Amsterdam 268 RHPL/Derrick Furlong 268–9 RHPL /Paul Van Riel 269 HL/Liba Taylor 270–1 Z/Starfoto 272–3 RHPL 273t RHPL/E/Loirat 273b RHPL 274–5 RHPL 276–7 RHPL 278 NHPA/V G Canselo 278–9 Z 279 Z/Rossenbach 280 RHPL/Vulcain 281 Z 282 AGE Fotostock 282–3 Z 284–5 Z 285 AGE Fotostock 286–7 HL/Robert Francis 287 Z/Sunak 289 AGE Fotostock 290 RHPL/Robert Francis 292–3 Matthew Kneale 293 RHPL 294 RHPL/Christopher Rennie 295 S 296–7 RHPL/ Duncan Maxwell 297 RHPL 298 RHPL 299 TRIP/Sefton Photo Library 300 Matthew Kneale 301 TSW/Shaun Egan 302–3 Z/Starfoto 303 RHPL/James Green 304–5 HL/ Juliet Hughet Briman 305t TRIP/Dave Saunders 305b RHPL 306 RHPL/Michael Jenner 307 RHPL/Michael Short 308 RHPL/Bildagentur Schuster/Rangnow 310–11 RHPL 311 Silvestris/Dr Christian Nowak 312t BAL/Prado, Madrid 312b BAL/Christie's, London 313 RHPL 314 RHPL/Richard G Elliott 314–15 RHPL/Rolf Richardson 315 RHPL 316–17 Z/Streichan 317 RHPL 318–19 RHPL 319 RHPL 320l FSP/Figaro 320r RHPL 321 Furstentum Liechtenstein 322–3 OSF/R J C Cannon 324–5 RHPL/Philip Craven 325 RHPL 326–7 Z/Tom 328–9t RHPL/Claye 328–9b IP/Peter Arkell 330 FSP/ Francolon/Mingam/Hires 330–1 Z 331 Z 332–3 Select/Liba Taylor 333 Z 334 HDC 335 Z 336–7 RHPL/Michael Short 339 HL/John G Egan 338–9 RHPL 339 HL/Liba Taylor 340 HL/Melanie Friend 341 RHPL/Rolf Richardson 342–3 Z/K Goebel 343 FSP/ Gamma 344–5 FSP/Keith Bernstein 346 Life File/F Ralston 347 Z/H Lutticke 348 Z 349 TRIP 350 FSP/Art Zamur 351 RHPL/Christopher Rennie 352 Matthew Kneale 354 TRIP/T Noorits 355 E/J P Courau 356 C/Lehtikuva 357 M/Bruno Barbey 358–9 Russia & Republics Photolibrary/Mark Wadlow 359 TRIP/Victor Kolpakov 360 TRIP/O'Brien-Thumm 360–1 E/Rosi Baumgartner 362–3 E/Trapman 363 Vadim Gippenreiter 364t MH 364b British Library 365 HDC 366–7 Vadim Gippenreiter 367 FSP/Gamma/F Lengaigne 368t E/Sylvain Grandadam 368b Corbis 368–9 M/Fred Mayer 370 M/Burt Glinn 371 C/Black Star/Peter Turnley 372–3 M/Dacat 373 M/Dacat 374t KP/REA/Dorigny 374b KP/Jeremy Nicholl 374–5 Russia & Republics Photolibrary 376 Vadim Gippenreiter 377 M/Abbas 379 M/Abbas 379 E/Maja Koene 380–1 TRIP 381 Dr Nick Middleton 382–3 HL/Bernard Gerard 384 EU/J Sergeant 384–5 COP 385 EU/Derek Redfern 386 COP 386–7 M/Abbas 388–9 COP 389 COP 390–1 EU/ Norman 391 COP 392 Matthew Kneale 393 HL/Jenny Pate 394t E/G Thouvenin 394b TRIP/Chris Kutschera 395 M/Steve McCurry 396 KP/REA/Saba/Ron Haviv 397 HL/Liba Taylor 398 HL/Christina Dowdell 399 HL/Melanie Friend 400–1 Luc Delahaye/Magnum Photos 401 HL 402t C/Randa Bishop 402b HL/Bernard Gerard 403 HL 404t COP 404–5 C/Piergiorgio Sclarandis 406 Life File/Alan Marchant 407 TRIP/Thelma Sanders 408–9 Z/W & D McIntyre 410 Matthew Kneale 411 Z/Eich.

Zinger 412b Z/Starfoto 412–13 Matthew Kneale 413 HL/Sarah Murray 414 M/Steve McCurry 415 RHPL 416 RHPL 417 RHPL 418–19 PP/Trygve Bolstad 419 PP/Cooper & Hammond 420 PP/Jeremy Hartley 421 PP/Dominic Sansoni 422–3 Z/Damm 424 AD 426–7 TRIP/K Sandwell 427 EU/Julia Waterlow 429 TIB/Harald Sund 430t TPS 430b ET Archive 432 TRIP/B Ashe 432–3t IP/Alain le Garsmeur 432–3b M/Michael K Nichols 434 EU/L Fordyce 435 PP/Chris Stowers 436–7 Telegraph Colour Library/VCL 437 M/Fred Mayer 438 HL/Jeremy Horner 440–1 HL/Liba Taylor 441t HL/Jeremy Horner 441b HL 442 COP 443 PP/D Hulcher 444–5 AD 445 AD 446 KP/David Stewart-Smith 446–7 HL/Rene Nicolas Giudicelli 448–9 Rob Judges 450 COP 451 PEP/ Philip Chapman 452 Z 453 M/Abbas 454–5 Matthew Kneale 455 PEP/Keith Scholey 456t COP 456c Matthew Kneale 456–7 IP/Dominic Sansoni 458 BCL/Brian Coates 460–1 Z 462–3 Matthew Kneale 463 AD 464l British Museum 464r ET Archive 464–5 Z/ Orion Press 465 AA&A 466r Matthew Kneale 466l EU/Frank Leather 468t TIB/ Toyofumi Mori 468b KP/REA/Gordon 469 M/Stuart Franklin 470–1 M/Hiroji Kubota 471 M/Hiroji Kubota 472 C/Heino Aga 473–4 Footprints 476–7 COP 476 Gamma/Katz 477 TRIP/Dave Saunders 478b NHPA/Otto Rogge 478b BAL 478–9 BAL/Royal Geographical Society, London 480t COP 480b TRIP 480–1 TRIP/Joan Wakelin 481 COP 482–3 M/Ian Berry 483 TRIP/EU/Matthew McKee 484–5 TRIP/Dave Saunders 485c TRIP 485b TRIP/Joan Wakelin 486 NHPA/Bruce Beehler 486–7 TRIP/CC 487 James Davis Travel Photography 488 TRIP/EU/Matthew McKee 489 PEP/Pete Atkinson 490 COP 491 COP 492 HL 493 Footprints 494 M/P Zachmann 496–7 Matthew Kneale 497 NHPA/Peter Johnson

SOURCES OF DATA

The main sources for the Fact Panels are: *The CIA World Factbook 2001* (www.cia.gov/cia/publications/factbook), *Britannica World Data* (Encyclopedia Britannica Inc., Chicago, annual); *Statesman's Yearbook* (Palgrave, London, 2001); *Whitakers Almanack* (J. Whitaker & Sons Ltd); and *Keesings Record of World Events* (Keesings, Redhill, updated throughout the year). Also used are *Webster's Geographical Dictionary* (Merriam-Webster, Springfield, Massachusetts, 1997); *The Times Atlas of the World* (Times Books, London, 1999); *The Economist Pocket World in Figures* (Profile Books/The Economist, 1999); *World Development Report* (World Bank, Oxford University Press, New York, annual); and *The World Gazetteer* (www.world-gazetteer.com).

Physical Geography Figures

Land area: *Britannica World Data* and *Statesman's Yearbook*.
Climate: *Statesman's Yearbook* (with altitude figures from *Webster's Geographical Dictionary* and *The Times Atlas of the World*.
Major Physical Features: *Webster's Geographical Dictionary*; *The Times Atlas of the World*; *Encyclopedia Britannica* (Encyclopedia Britannica Inc., Chicago).
Population: *The CIA World Factbook 2001* and *Statesman's Yearbook*.

Social and Economic Figures

Form of Government: *Britannica World Data*; *Statesman's Yearbook*; *Keesings Record of World Events*; *Whitakers Almanack*.
Armed Forces: *Britannica World Data*; *Statesman's Yearbook*.
Capital City: *The World Gazetteer*; *Britannica World Data*; *Statesman's Yearbook*
Official Languages: *Britannica World Data*; *Keesings Record of World Events*.
Ethnic Composition: *Britannica World Data*; *Statesman's Yearbook*.
Religious Affiliations: *Britannica World Data*; *Statesman's Yearbook*.
Currency: *Britannica World Data*; *Statesman's Yearbook*.
Gross Domestic Product: *World Development Report*; *The CIA World Factbook 2001*.
Per-capita income: *World Development Report*; *The CIA World Factbook 2001*.
Life Expectancy at Birth: *The CIA World Factbook 2001*.
Major Resources: *Britannica World Data*; *The Economist Pocket World in Figures*

Artists

Julian Baker (The Maltings Partnership); Janos Marffy

Cartography

Maps drafted by Euromap, Pangbourne

Flags

Produced by Lovell Johns, Oxford, and authenticated by The Flag Research Center, Wincester, Mass 01890, USA

INDEX

Numbers in *italics* refer to captions or maps; numbers in **bold** refer to major sections